Edited by Ferdi Schüth,
Kenneth S. W. Sing, and
Jens Weitkamp
**Handbook of Porous
Solids**

Volume 2

Further Titles of Interest

H.-P. Degischer, K. Kriszt (Eds.)

Handbook of Cellular Metals

Production, Processing, Applications

2002
ISBN 3-527-30339-1

U. Schubert, N. Hüsing

Synthesis of Inorganic Materials

2001
ISBN 3-527-29550-X

G. Hodes (Ed.)

Electrochemistry of Nanomaterials

2001
ISBN 3-527-29836-3

P. Braunstein, L. A. Oro, P. R. Raithby (Eds.)

Metal Clusters in Chemistry

1999
ISBN 3-527-29549-6

G. Meyer, D. Naumann, L. Wesemann (Eds.)

Inorganic Chemistry Highlights

2002
ISBN 3-527-30265-4

Handbook of Porous Solids

Edited by Ferdi Schüth, Kenneth S. W. Sing, and Jens Weitkamp

Volume 2

WILEY-VCH

Editors

Prof. Dr. Ferdi Schüth
Max-Planck-Institut für Kohlenforschung
Kaiser-Wilhelm-Platz 1
45470 Mülheim
Germany

Prof. Dr. Kenneth S. W. Sing
Fairfield
Whitelackington
Ilminster, Somerset TA 19 9EF
Great Britain

Prof. Dr. Jens Weitkamp
Institut für Technische Chemie
der Universität Stuttgart
Pfaffenwaldring 55
70569 Stuttgart
Germany

Library of Congress Card No.: applied for
A catalogue record for this book is available
from the British Library.
Die Deutsche Bibliothek – CIP
Cataloguing-in-Publication-Data
A catalogue record for this publication is
available from Die Deutsche Bibliothek

© WILEY-VCH Verlag GmbH, 69469
Weinheim (Federal Republic of Germany).
2002

Printed in the Federal Republic of
Germany.
Printed on acid-free paper.

Typesetting   Asco Typesetters, Hong Kong
Printing   betz-druch gmbh, Darmstadt.
Bookbinding   Litges & Dopf Buchbinderei
GmbH, Heppenheim.

ISBN   3-527-30246-8

Contents

Volume 3

Volume 4

Volume 5

4
Classes of Materials

4
Classes of Materials

4.1
Clathrates and Inclusion Compounds

Hermann Gies

4.1.1
The General Principle of Clathrate Formation

Clathrate formation dates back to 1811 when Davy [1] discovered the formation of the chlorine hydrate by bubbling chlorine gas into cool water. Other discoveries of the 19th century concerning clathrates are hydroquinone inclusion compounds [2], Hofmann's benzene compound [3], and Dianin's [4] compound. The principles of the formation of clathrates and the analysis of the chemical bonding involved in the crystalline solid became accessible through the application of crystallographic techniques most notably illustrated by the early work of Powell [5] who also coined the term "clathrate" compounds. Since Powell's time the number of clathrate compounds, better described as host–guest compounds, has increased steadily. Not only were more examples of the then-known families discovered, but also new host–guest systems were invented. In a number of highly recommended review articles and text books the historical development and the latest research results have been presented. The five-volume book series devoted to inclusion compounds [6] reviews all fields of inclusion phenomena and host–guest interaction from the very beginning to 1990. The series on *Comprehensive Supramolecular Chemistry* that appeared in 1996 [7] covers also almost all aspects of clathrate and inclusion chemistry and physics, starting from synthesis to structure and physical properties, organic, inorganic and biological systems. For more detail and in-depth information on clathrates and organic and inorganic inclusion compounds volumes 2, 3, 4 and 6 are recommended reading.

The overview given here will focus on those host–guest systems with 2- or 3D host frameworks. They may be built from molecules interacting with directional forces. The guest molecules then occupy the host voids to complete the structure. Hydrogen bonds are the most important directional interactions for the host

References see page 696

framework architecture. The following sections (4.1.2, 4.1.3.1 and 4.1.3.2) introduce the principles with examples of organic and inorganic inclusion compounds. Its abundance is demonstrated by the fact that well-known host–guest systems such as hydroquinone, urea, trimesic acid, water, diamondoid acids, dianin compounds, clathrate hydrates and acyclic diols belong to this group. The bonding interaction ranges from 40–20 kJ mol^{-1} in, for example, (O, N–H)–(O, N)-bonds for strong to conventional hydrogen bonds to 20–2 kJ mol^{-1} in e.g. (C–H–O)- and (O–H–C)-bonds for weak directional forces typical for biological systems. Weakly interacting directional forces also include halogen atom interaction as well as chalcogen atom interactions (S, Se)–(S, Se). A vast amount of work has been carried out on the various aspects of the nature of the hydrogen bond, which is best summarized in the monographs on the hydrogen bond by Jeffrey and Saenger [8] and Jeffrey [9].

In cyanometallates the directional bond is created by using the cyanide anion as a bridging ligand between metal centers. The linkage of square planar, tetrahedral or octahedral units to multidimensional host frameworks leads to open frameworks and produces voids that are occupied by guest molecules. The dominant metal centers are Ni, Cd, Hg and Fe. Other bidentate ligands are SCN$^-$ and NO$_3^-$ used together with the same metal centers.

Finally, porous frameworks will be discussed where the directional bond is established between tetrahedral units forming covalent bonds. The guest species are ions, ion pairs, atoms or molecules filling the void space in the open host framework. There are crystalline porous Si and Ge, and also SiO_2, GeO_2, $AlPO_4$, and similar compounds that are topologically related to various clathrate hydrates. The 3D host frameworks are thermally and chemically the most stable porous compounds. The family of porous SiO_2 clathrates, the clathrasils, are related to the technically important zeolites, where Si is partially substituted with, for example, Al introducing negative charges on the host framework.

4.1.2
Host Structures Built from Organic Molecules

The molecular organization in organic clathrate compounds to 2- or 3D host frameworks is achieved through directional, intramolecular bonds. The formation of the porous supramolecular array during crystallization requires the presence of suitable guest molecules. In general, there is only weak interaction between the host framework and the guest species so that the size and shape of the guest molecule determines the void volume and, therefore, the crystal structure to be formed. In the following sections an overview on the intensively investigated systems is given with references to recent review literature. An overview of highlights of the latest developments is included in this section. For more detailed accounts and a quantitative discussion of organic inclusion compounds including systems not covered here the reader is referred to *Comprehensive Supramolecular Chemistry*, Vols 1–11 [7].

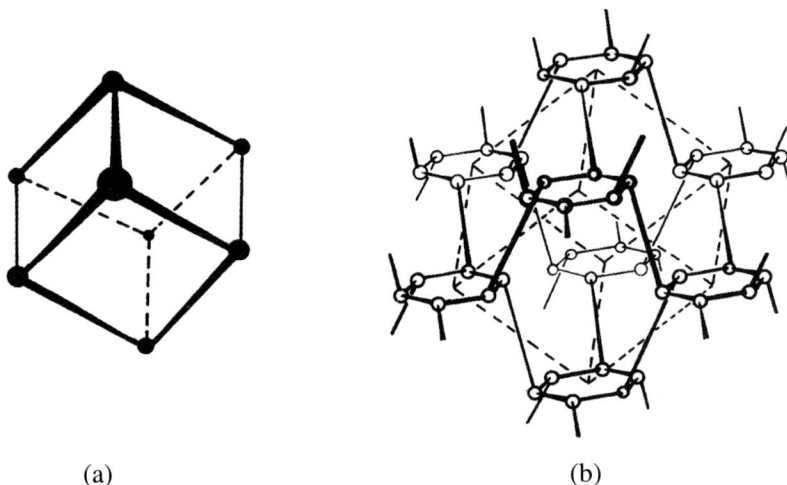

(a)   (b)

Fig. 1. Schematic drawing of the hydroquinone clathrate structure: (a) outline of the rhombohedral unit cell of the β-polymorph of hydrochinone; (b) each corner of the unit cell is occupied by a six-rings of $(OH)_6$ with the phenylene group represented as a connecting slanted line.

4.1.2.1
Hydroquinone Clathrates and Related Compounds

4.1.2.1.1 β-Hydroquinone Clathrates

Hydroquinone, p-$C_6H_4(OH)_2$, crystallizes in three polymorphic forms. The stable form at room temperature is the α-form. The γ-form can be obtained by sublimation or rapid evaporation of the solvent ether. In the presence of common solvents, e.g. methanol, inclusion compounds form. The β-form crystallizes with the solvent trapped inside cavities of the hydroquinone 3D host framework. Each hydroxyl group of the hydroquinone molecule is part of a hydrogen-bonded $(OH)_6$-ring connecting six neighboring hydroquinone molecules (Fig. 1). The center of mass of the $(OH)_6$-ring lies in the corner of a rhombohedron with the phenylene groups of the hydroquinone molecules as rhombohedron vertices pointing up and down alternately. This creates a large void inside the rhombohedron. In principle, there is an analogy with the open framework of the β-modification of elementary polonium, the only metal known to crystallize in a primitive cubic-packed structure. However, the open hydroquinone structure is unstable. In fact, a second hydroquinone network shifted by approximately 0.5, 0.5, 0.5 along the body-diagonal interpenetrates without directional bonding interaction. The void space in the interpenetrating structure is still large enough to accommodate small guest species such as Xe, H_2S, SO_2 or MeOH (Fig. 2). These are sandwiched between two $(OH)_6$-

References see page 696

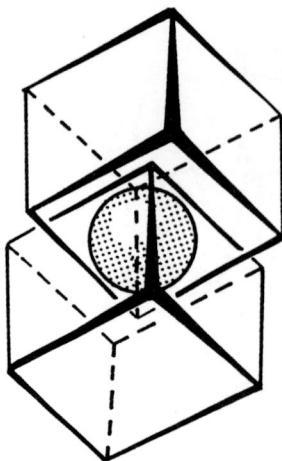

Fig. 2. Schematic representation of the interpenetration of the hydrochinone unit cells and the enclosure of the sphere-like guest species in the host framework.

rings building up van der Waals contacts. Depending on the symmetry of the crystal structure there are three hydroquinone type clathrates. Sphere-like guest species lead to space group R $\bar{3}$ for type I clathrates, more elongated molecules lower the symmetry to R3 for type II and, finally, guest molecules like MeCN lead to space group symmetry P3 for type III hydroquinone clathrates. Two recent review articles summarize the structural details of the host framework and give an overview of the variety of guest species for hydroquinone clathrate compounds [10, 11].

There is no report on new hydrochinone inclusion compounds in the recent literature. However, there are a number of studies investigating structural details of the dynamics of the host framework and the rotational tunneling of the guest species [12]. A survey may be obtained with the keywords "hydroquinone and clathrate" in the Web of Science [13] database.

4.1.2.1.2   *β*-Hydroquinone Clathrate Compounds with Fullerene Guest Molecules

The large void in the single *β*-hydroquinone inspired clathrate chemists to trap large molecules inside the 3D host framework. The void diameter of approximately 10 Å closely matches the dimension of buckminsterfullerene, C_{60}, which was predicted to be an ideal guest molecule. The C_{60}-clathrate was crystallized from hot benzene as stoichiometric compound. The crystal structure analysis in the ideal space group symmetry R $\bar{3}$m showed that the guest molecule is highly disordered with a sphere-like electron cloud of 3.5 Å radius (Fig. 3) [14]. C_{70}, the slightly larger buckminsterfullerene also forms an inclusion compound with hydroquinone [15]. The guest species is too large for the *β*-hydroquinone cavity, however, builds a new 3D host framework based on the same structural principles. The hydroquinone molecules connect via $(OH)_6$-ring hydrogen bonds, this time with three types of void. The large cage houses two C_{70} molecules, the medium cage one, and the small cage contains two solvent benzene molecules as guest species. As with the

Fig. 3. Projection along the 3-fold axis of the rhombohedral unit cell of the hydroquinone-fullerene inclusion compound. The highly disordered molecule is shown in the center of the β-hydroquinone cage.

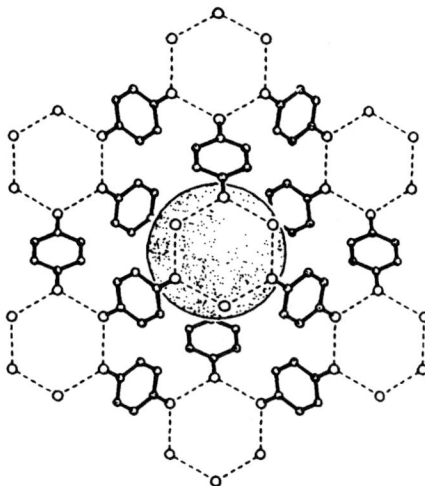

other hydroquinone inclusion compounds the material decomposes upon calcination. It remains to be seen whether other hydroquinone clathrates with other buckminsterfullerenes can be obtained.

4.1.2.1.3 Gossypol

A large variety of phenolic compounds has been isolated from cotton plant which form "coordination-assisted clathrates". The crystalline substance isolated from cottonseed oil, gossypol, Gp (**1**), has been investigated intensively (Fig. 4). Its triterpene structure is obviously capable of forming clathrate-type compounds where cavities between hydrogen-bonded neighboring molecules house the guest species. A review article covering the extensive chemistry of gossypol and related compounds and examples of the their crystal chemistry is contained in *Comprehensive Supramolecular Chemistry* [16].

Depending on the solvent Gp exists in three tautomeric forms, an aldehyde form, a quinoid form and a hemiacetal form. As is obvious from the number of

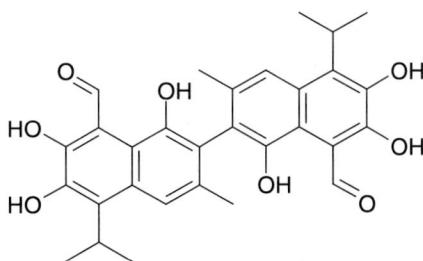

Fig. 4. Skeletal formula of the aldehyde form of gossypol.

References see page 696

hydroxyl groups in the molecule various intermolecular hydrogen bonds might form in the solid. So far three polymorphic guest free structures have been analyzed. In the presence of suitable, small molecules Gp builds up intramolecular hydrogen bond networks with voids containing guest species. Slight changes in temperature and concentration of the guest molecules or using different solvents also change the structure of the host framework. In many cases the guest species are involved in hydrogen bonds with the host structure. Therefore, many guest species are alcohols, carboxylic acids or other functional groups containing molecules. Until now more than 100 host–guest systems have been investigated showing the richness and versatility of the gossypol host framework structure. Upon calcination the host structure collapses and no guest-free framework can be obtained.

Using the Web of Science with the keywords "gossypol and inclusion" starting from 1995 27 hits were found. Most of the literature reports studies of structure and synthesis of gossypol inclusion complexes. This is, however, only a fraction of the "gossypol" literature, which accounted for more than 400 hits using the keyword "gossypol" alone.

4.1.2.1.4 Diamondoid Host Frameworks

Substitution of sp^3-carbon atoms in the diamond lattice by 4-connected molecular building blocks, thus retaining the 3D connectivity, creates diamondoid host frameworks with spacious voids. Most investigated are the host frameworks from adamantane-derived carboxylic acids such as adamantane-1,3,5,7-tetracarboxylic acid. Other derivatives contain functional groups on various carbon sites of the adamantane molecule or are alkali-salts of the tetracarboxylic acid (Fig. 5). The host framework is formed via intramolecular hydrogen bonds between two carboxylic acid groups building a diamondoid framework, also called superadamantane. The large open space in the structure of the organic host in its guest-free molecular structure is filled in a most fascinating way. Several diamondoid frameworks mutually interpenetrate giving a stack of translationally equivalent frameworks (Fig. 6). The adamantane-1,3,5,7-tetracarboxylic acid as parent molecule crystallizes as a 5-fold self-interpenetrating diamondoid network. The multiplicity of interpenetration decreases to 4, 3, or even 2 with cations or substituents connected to the ada-

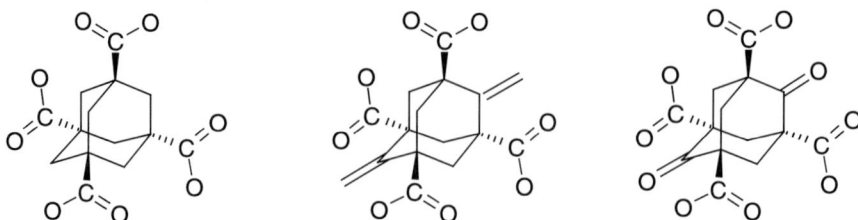

Fig. 5. Examples of host molecules for diamondoid frameworks. Molecules are shown without hydrogen atoms. Modification of the substitutends will yield more host molecules.

Fig. 6. Section of a doubly interpenetrating net of diamondoid structures.

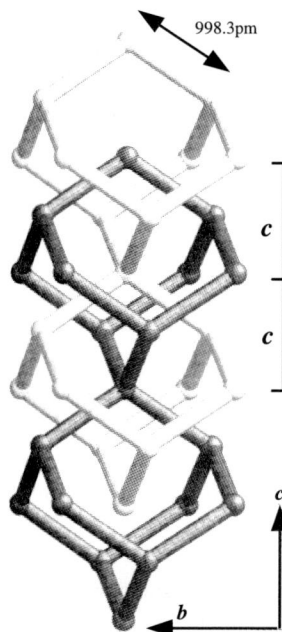

mantane building block and in the presence of suitable guest species. These now occupy the void space created by the removal of diamondoid nets. So far, the single host framework structure has not been obtained. The very open framework would need a large and rigid guest species fitting the void space in order to prevent the framework from collapse.

There are other molecules with tetrahedral symmetry and functionality that form diamondoid framework structures and self-penetrating lattices. One example is methane tetraacetic acid (Fig. 7). Replacing acetic acid as substituent with a pyridone derivative (tetrakis-[4-{(6-oxopyridone-2-yl)ethynyl}phenyl]methan) also yields a superadamantane structure with 7-fold interpenetration yet still incorporating the solvent butyric acid in the crystal structure. Here the intermolecular hydrogen bonds are formed between the pyridone entities giving rise to an extraordinarily large cavity. However, so far the open framework has not yet been obtained

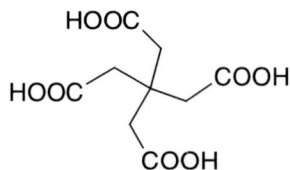

Fig. 7. Skeletal formula of the host molecule methane tetraacetic acid.

References see page 696

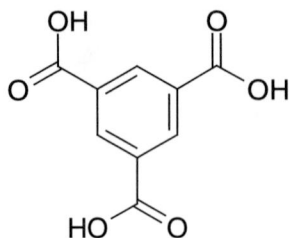

Fig. 8. Skeletal formula of the host molecule trimesic acid.

and also would require guest species with exceptional properties fitting the size and shape of the superadamantane void.

An excellent review by Mak et al. [17] on the chemistry and structural properties of these compounds has been published in Supraamolecular Chemistry. More recent literature is scattered and some is given in the list of references.

4.1.2.2
1,3,5-Benzenetricarboxylic Acid

1,3,5-Benzenetricaboxylic acid, trimesic acid (TMA) has a rich chemistry for the formation of binary or more complex products. The planar molecule (see Fig. 8) is capable of connection with other molecules via hydrogen bridges to 2D bonding networks. Crystallizing TMA from aqueous alkaline solutions leads to salt formation with many inorganic and organic cations in their hydrated form connecting to the carboxylic acid. In neutral, organic solvents and in the presence of molecules capable of forming hydrogen bonds TMA crystallizes as binary product with 2D slabs of hydrogen-bonded arrays. A recent review article on trimesic acid and trimesic acid based inclusion compounds by Herbstein [18] is available in *Comprehensive Supramolecular Chemistry*.

4.1.2.2.1 **Inclusion Compounds with 1,3,5-Benzenetricarboxylic Acid**
When TMA is crystallized from water the anhydrous α-form is obtained. The complicated crystal structure consists of the basic 2D "chicken wire"-motif (Fig. 9). In the 3D crystal structure the chicken-wire ring is interpenetrated by three equivalent units perpendicular to the 2D net (catenation). Those in their turn accommodate two more rings leading to a triply interpenetrated structure [19]. Looking at the packing of the individual molecules an essentially space-filling arrangement is achieved.

In the presence of guest molecules, catenation is suppressed and inclusion compounds form. There are two types of host structure, one with hydrate water participating in the hydrogen bonds between the carboxylate groups. Here, the TMA-molecule itself and, if added, picric acid serve as templates in the windows of the hydrogen-bonded 2D host structure [20]. The host–guest array stacks with the guest molecule in partial disorder leading to channel-like pores. It is interesting to

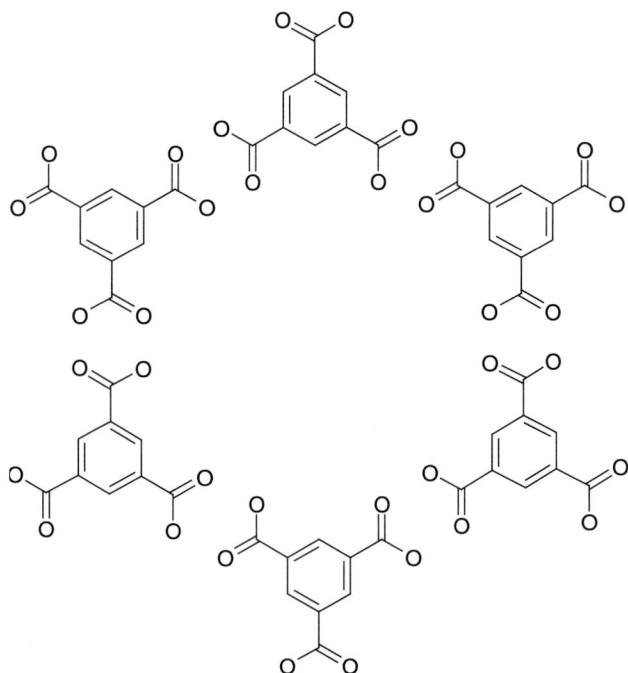

Fig. 9. Section of the chicken-wire arrangement of trimesic acid. Hydrogen atoms are omitted for clarity.

analyze the TMA molecule as guest species which is hydrogen bonded along the channel thus stabilizing the stacking arrangement of the inclusion compound.

The alternative type of host structure is the water-free chicken-wire hexagonal TMA network [21]. Representative guest molecules are squalene, octane, tetradecane etc., which crystallize from acetone in the presence of TMA. The layer-like TMA net stacks in a 3-layer repeat forming hexagonal channels with the disordered templates inside. In the presence of camphor an interrupted TMA-net forms with water molecules bridging two opposite hydrogen bonds of the 6-ring arrangement. The geometry of the channel changes to elliptical. The structure of the host framework was determined by X-ray diffraction yielding a two-layer repeat, however, the guest species camphor was not located. So the role of the template in the formation could not be analyzed.

A detailed account on the synthesis, structure, and crystal chemistry of TMA salts and TMA inclusion compounds covering the literature until 1994 is given in [18]. More recent results were obtained in a literature search (63 hits in Web of Science [13], search keywords: trimesic acid and inclusion compounds, since 1995). The survey shows that the analysis of the hydrogen-bond system in the host–guest system is of primary interest. Based on the principles described here,

References see page 696

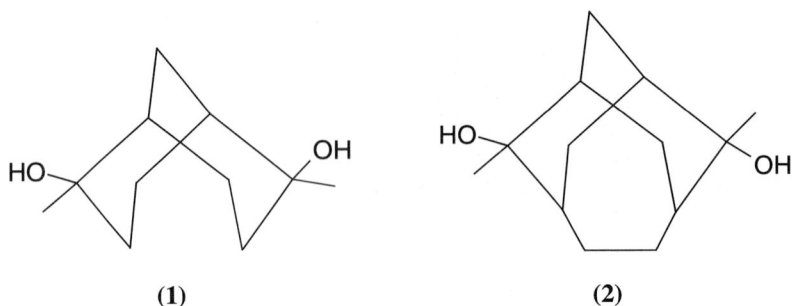

(1)   (2)

Fig. 10. Examples for host molecules of helical host structures formed by alicyclic diols: (1) 2,6-dimethylbicyclo[3.3.1]nonane-*exo*-2,*exo*-6-diol and (2) 2,7-dimethyltricyclo[4.3.1.1^{3,8}]undecane-*syn*-2,*syn*-7-diol. Only protons of hydroxyl groups are shown.

the synthesis and structural characterization of self-organized host–guest systems using derivatives of trimesic acid and other, new guest molecules also play an important role [22].

4.1.2.3
Inclusion Compounds with Alicyclic Diols

Helical host structures are formed with alicyclic diols in the presence of a wide range of guest molecules. The host structures are built from hydrogen-bonded bicyclic or tricyclic molecules such as 2,6-dimethylbicyclo[3.3.1]nonane-*exo*-2, *exo*-6-diol (1) or 2,7-dimethyltricyclo[4.3.1.1^{3,8}]un-decane-*syn*-2, *syn*-7-diol (2) (Fig. 10) and related compounds. A set of rules summarizes common sterical and chemical features of the host-building diol molecules allowing for the synthesis of new derivatives. More than 10 host molecules are known to crystallize with guest species as helical tubulate inclusion compounds. A summary covering the literature up to 1995 on synthesis aspects and the crystallographic analysis of the host–guest interaction by Bishop [23] who first discovered the unique property of these molecules is presented in a review article in *Comprehensive Supramolecular Chemistry.*

As examples for the host–guest systems a short summary of the structure and chemistry of the host molecules (1) and (2) will be provided. The tubular canals are arranged in hexagonal symmetry with the canal axis running parallel to the screw axis (Fig. 11). For (1) a series of more than 40 guest molecules such as acetonitrile, ethyl acetate, or toluene have formed inclusion complexes [24, 25]. All information on host–guest interaction comes from detailed X-ray single-crystal structure determination [26–28]. The X-ray analyses clearly showed that size and shape factors of the guest molecule predominantly determine the function of the guest. Hydrogen bond host–guest interaction has not been observed as a decisive feature. However, in several cases guest-guest and host–guest interaction has been observed and seemed to be important for the formation of the particular clathrate. In general, the flexibility of the host system with a maximum size of unobstructed channel

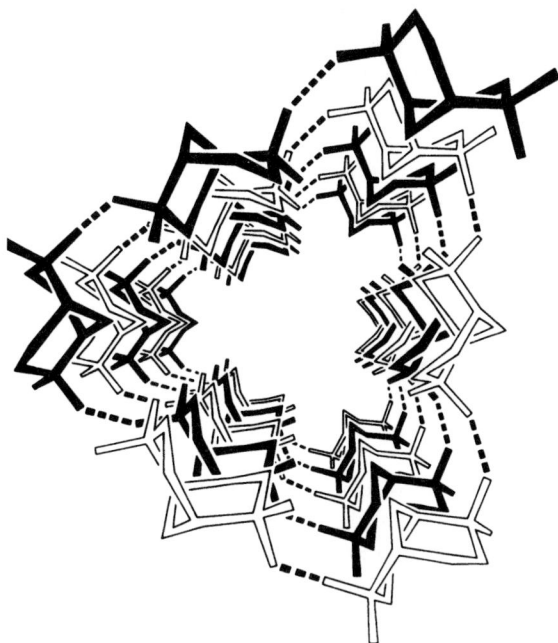

Fig. 11. Schematic drawing of the helix of (**1**) in the host framework of the tubular channel system. The black and the open line indicate two individual helices of a double-helical arrangement winding along the crystallographic screw axis in the channel center.

cross-sectional area (UCA) of 25 Å^2 for (**1**) as well as the nonspecific bonding interaction, i.e. the guest species is aligned in the center of the channel, are responsible for the versatility of the host–guest system.

For molecule (**2**) the free space in the channel varies between 19.8 and 29.2 Å^2. Hence the diol host framework can take up larger species and more than 20 have been discovered and investigated (e.g. [29, 30] and references therein). So far host frameworks with up to 34 Å^2 UCA have been found providing space for rather spacious guest molecules. Surprisingly, in the presence of small molecules, diol (**2**) forms an interpenetrating network with voids still large enough to accommodate the guest species. The channel-space is ellipsoidal now with tetragonal symmetry. Sets of four diols are arranged such that they can connect to four others building a distorted diamondoid framework. The second, independent framework is then shifted along the 4-fold screw axis such that they are related by an inversion center. The channel space is still large enough to take up molecules of the size of dichlorobenzene.

References see page 696

As for most organic inclusion compounds the host frameworks formed by diols are not stable upon calcination. As soon as the guest species evaporates the host structure disintegrates. These compounds might act as efficient substances for the separation of molecules capable of forming clathrates, however, because of their instability they can not be used as molecular sieves.

The major trend in the recent literature (13 hits in Web of Science [13], search keywords: diols and inclusion compounds, since 1996) is the modification of host molecules for new helical open host structures and the identification of new host–guest systems with various guest molecules.

4.1.2.4
Urea, Thiourea, and Selenourea Inclusion Compounds

4.1.2.4.1 Host Framework Structure
Urea inclusion compounds were discovered in 1940 [31, 32]. Since then they have attracted continuous interest until now, most convincingly demonstrated by the number of review articles that have been published on this subject (e.g. [33–35] and references therein). The most recent account also including thiourea and selenourea appeared in *Comprehensive Supramolecular Chemistry*, also citing the most relevant literature up to 1996 [35]. Urea inclusion compounds (UIC) precipitate in the presence of mainly linear hydrocarbons. Urea crystallizes in helical ribbons forming intermolecular hydrogen bonds (Fig. 12). In the center of the helix is a channel of hexagonal symmetry. Like the well-known double-helix in DNA each

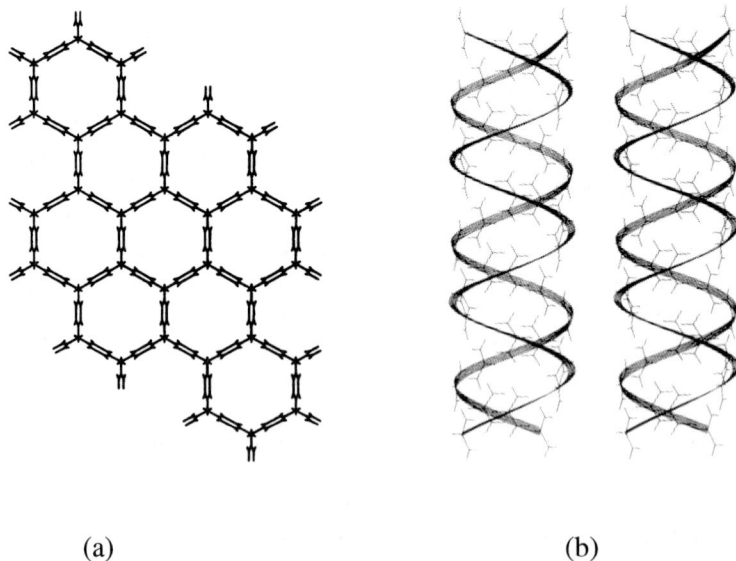

(a)   (b)

Fig. 12. The structure of the urea host framework: (a) seen as projection along *c*; (b) perpendicular to the double helix built by hydrogen-bonded urea molecules.

channel in UIC contains two ribbons of urea running in opposite directions. The two helices are defined by N–H–O hydrogen bonds and build a channel wall of hexagonal symmetry. Each crystal grows either as the right-handed (space group $P6_122$) or left-handed (space group $P6_522$) enantiomorph containing the right-handed or left-handed helical ribbons. The channel system seen in a projection along the hexagonal c-axis has a honeycomb geometry with a minimum tunnel diameter of approximately 5.5 Å (Fig. 12a). Thiourea and selenourea host frameworks are isostructural with slightly larger minimum tunnel diameters (5.8 Å) extending the variability of guest molecules.

4.1.2.4.2 Guest Molecules for Urea Inclusion Compounds and its Homologues

It is difficult to count the precise number of guest molecules used for the synthesis of urea inclusion compounds. Urea forms inclusion compounds with linear hydrocarbons, their alcohol, ester and ether derivatives. Furthermore, aldehydes, ketones, carboxylic acids, amines, nitriles, thioalcohols and thioethers can be used provided that the hydrocarbon chain exceeds more than six carbon atoms. As an example, the homologous series of linear carboxylic acids covers the range from C_4 to C_{18} in hydrocarbon chain length. There are even more exotic guest molecules acting as templates such as, for example, octadecylbenzene, although benzene as an isolated molecule never formed inclusion compounds. Because of its increased channel diameter, thiourea inclusion compounds also take up slightly larger guest molecules such as cyclohexane and derivatives, adamantane, organometallics such as ferrocene and compounds containing aromatic rings. Very little specific information is known about selenourea inclusion compounds perhaps because of the high toxicity. The similarity with the thiourea-adamantane system suggests close structural and chemical relationship between the two host–guest systems.

4.1.2.4.3 The Host–Guest Relation in Urea and Thiourea Inclusion Compounds

The van der Waals diameter in the urea channel varies only slightly between 5.5 and 5.8 Å. In addition, the hydrogen-bond network is established between neighboring urea molecules in the helical ribbon, leaving only weak van der Waals forces for the host–guest interaction. Hence, the packing requirements of the guest species inside the urea void determine the stacking periodicity of the molecules along the channel axis. Therefore, incommensurate crystal structures along the hexagonal c-axis are commonly observed in urea inclusion compounds. In addition, the guest molecules in neighboring channels might also order independently leading to a three-dimensional ordering of the guest molecule in the incommensurate urea inclusion compound (Fig. 13). In most cases additional dynamic disorder of the guest species inside the channel is observed. The superposition of these phenomena makes a detailed structural analysis very time-consuming and difficult. However, conventional X-ray structure analysis using the Bragg reflections maxima of the host structure unit cell yields the helical structure of the urea host for all systems investigated so far. The guest molecules are not resolved in their atomic

References see page 696

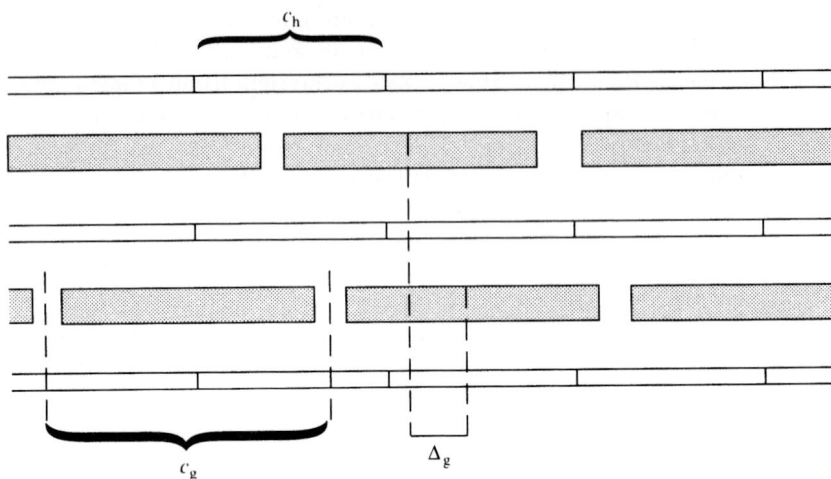

Fig. 13. Incommensurately ordered guest molecules in the urea host channel. The periodicity of the host framework along the channel axis is marked with c_h, the periodicity of the guest with c_g, the mismatch of guest arrangement with Δg. These incommensurations lead to diffuse scattering intensity in the diffraction experiment that can be analyzed in order to determine the host–guest arrangement.

structure but represented as smeared-out electron density averaging the positional and dynamical disorder of the molecule. Including superstructure reflections and diffuse intensities temperature dependent X-ray diffraction studies provide information for the local structure and dynamics of the guest molecule in its sublattice. So far only very few guest systems have been analyzed in detail giving insight into the host–guest interaction [36–38]. A detailed account of static and dynamic structural properties of a number of urea host-organoguest systems is given in the review article by Hollingsworth and Harris [35].

Unlike urea inclusion compounds thiourea inclusion compounds are centrosymmetric. The substitution by sulfur also changes the pore structure of the host framework considerably. There are prominent constrictions (approximately 5.8 Å) and bulges (approximately 7.1 Å) along the channel axis transforming the tunnel to a cage-like pore. In general, the ordering of the guest species is commensurate with the thiourea host structure. However, dynamic disorder of the guest molecules within the cavities is observed showing that the size and shape of the guest molecules determine the host–guest interaction. The structures of selenurea inclusion compounds are similar to their thiourea counterparts.

Like all other hydrogen-bonded organic clathrates urea, thiourea and selenourea inclusion compounds are not stable in their calcined form. Whereas the small molecules slowly evaporate from the inclusion compound and, as a consequence, disintegrate the host framework, the inclusion compounds with larger guest species are stable under ambient conditions. Calcination, however, leads to the decomposition of the clathrate.

In a literature search in the Web of Science [13] using the keywords urea and (inclusion compound or clathrate) 71 entries were reported since 1994. Most of the recent papers dealt with the detailed analysis of the interaction between host and guest and their 3D ordering with the well-known hexagonal urea structure as host framework. Particular attention was paid to the dynamic nature of the guest species inside the urea channel.

4.1.3
Inorganic Clathrates and Inclusion Compounds

4.1.3.1
Werner Clathrates

Werner clathrates in their initial composition have the general formula $MX_2A_4 \cdot G$, where M is a divalent transition metal, typically manganese, iron, cobalt, nickel, and copper, less often zinc or cadmium, X is an anionic ligand such as SCN^-, NO_3^-, halide or other, A stands for a pyridine derivative or an arylalkylamine and G is the guest species. The crystal structures of typical Werner clathrates are composed of the octahedrally coordinated transition metal complex with the X-ligands axially and the A-ligands equatorially bound to the metal center. In a 2D arrangement the A-ligands organize in a layer-like array with the anionic X-ligands as spacers between the layers. The void space created with the aid of the X-ligands is filled with guest molecules (Fig. 14).

The guest molecules inside the voids created by the Werner host complex interact weakly with the host and order in a space-filling fashion. The size and shape requirements clearly dominate the inclusion of the guest species. Since the experimental enthalpy of sorption is always twice or more that of the enthalpy of evaporation, a typical, clathrate-like nonspecific host–guest interaction is assumed. The desorption of guest molecules might lead to the decomposition of the host structure in cases where the interaction energy is high. On the other hand, those molecules that might be removed from the inclusion compound lead to a severe rearrangement of the molecular structure of the host. Upon desorption the crystal lattice relaxes, leading to a considerable contraction in unit cell volume. Sorption of guest species recreates the host structure again, however, by transforming the collapsed structure into the expanded host structure. This zeolite-like property makes Werner clathrates interesting compounds for separation processes.

Detailed review articles on the synthesis, structure and physical properties of Werner clathrates have been published recently [39, 40]. For quantitative and exhaustive information on the synthesis and structure of Werner clathrates including kinetic studies of sorption and desorption of guest molecules the reader is referred to those. In a literature search in the Web of Science [13] starting from 1995 using the keywords Werner and inclusion compound or clathrate there are 16 hits. An

References see page 696

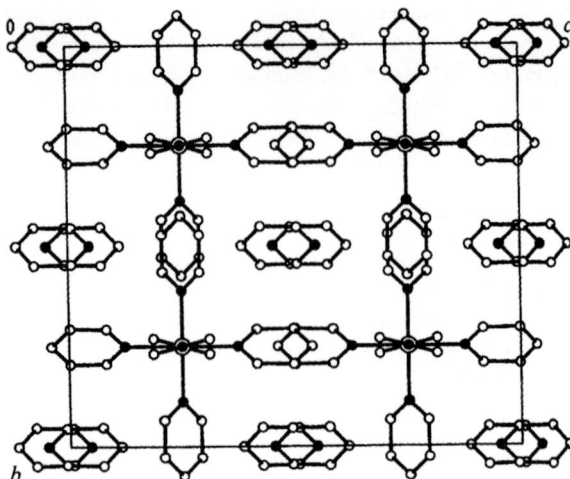

Fig. 14. Projection along *a* of a layer-like slab of the crystal structure of the Werner clathrate [Ni(NO$_3$)$_2$(py)$_4$]·2py along the *a*-axis of the unit cell. The central Ni cation is surrounded by four pyridine molecules in the *b,c*-plane. The two nitrate groups are seen superimposed together with the central Ni cation and keep the distance to the neighboring layers below and above. The two pyridine guest molecules occupy the void space in the rectangular channel.

article of Karunakaran [41] on the synthesis of new Werner-type clathrates and two articles by Manakov et al. [42, 43] on the structure-property relation of Werner clathrate cover most of the latest developments on this family of inclusion compounds.

4.1.3.2
Hydrate Inclusion Compounds

Hydrate inclusion compounds, as clathrate hydrates or gas hydrates were named in appreciation of the increasingly important role of inclusion phenomena, have a long history and have been the subject of review articles since 1927. The development in their crystal chemistry and host–guest interaction has been recently summarized by one of the pioneers in the field, Jeffrey [44, 45]. Later, hydrate inclusion compounds became a focus of interest again because of their natural abundance on the sea floor [46, 47]. There, they form in the presence of small hydrocarbons, which they store. It is estimated that these natural hydrocarbon resources account for over 50 % of the world reserves of organic carbon [48, 49]. The economic potential of the deposits and ecologic relevance on the change of the climate upon the release of the greenhouse gas methane is obvious. An excellent introduction into the geological aspects of hydrate inclusion compounds is given in references [46] and [49].

The host frameworks of the hydrate inclusion compounds are built from water molecules establishing a 3D 4-connected net of hydrogen bonds. The characteristic

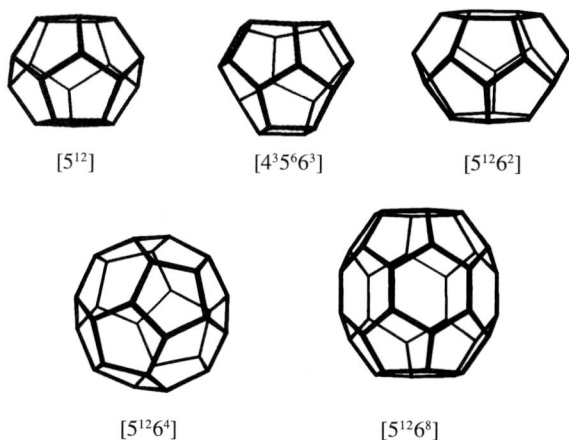

[5^{12}]   [$4^3 5^6 6^3$]   [$5^{12} 6^2$]

[$5^{12} 6^4$]   [$5^{12} 6^8$]

Fig. 15. A selection of polyhedral cavities as found in the host framework of clathrate inclusion compounds. The corner represents the site of an oxygen atom, the hydrogen atoms are located close to the connecting line and are omitted here for clarity. Nomenclature for cages gives the number of polygon as superscript for an *n*-hedron, e.g. pentagonal dodecahedron: [5^{12}].

feature of hydrate inclusion compounds is polyhedral cavities (Fig. 15) that are occupied by suitable guest species. The flexibility of the hydrogen bond allows for a number of different framework structure types to be formed. Depending on the size and shape of the structure directing guest molecule many different polyhedra might be obtained. The pentagonal dodecahedron is the most prominent polyhedron that has been observed in several hydrate inclusion compounds. Since the polyhedron is not space filling it occurs in a combination with others in order to obtain a 3D periodic crystal structure (Fig. 16). In the ideal hydrate inclusion compound the host framework is entirely built from water molecules. The tetrahedral network is complete and does not contain defects. These clathrates are called true clathrates. When, for example, alkylammonium salts are used as templates ionic clathrates form with defects in the tetrahedral host framework in order to compensate for the charge of the guest molecule. The host–guest interaction here is determined by the ionic bond between the template cation and the negatively charged water host framework. In addition, there are semiclathrates, layer and channel structures formed in the presence of, mainly, amines. Here, an open framework structure forms where the guest species are involved in hydrogen-bond interaction with the host constituent.

4.1.3.2.1 **True Clathrate Hydrates**

Rare gases, common gases and a wide range of neutral organic molecules with van der Waals diameter of less than 9 Å lead to the formation of true clathrates. There

References see page 696

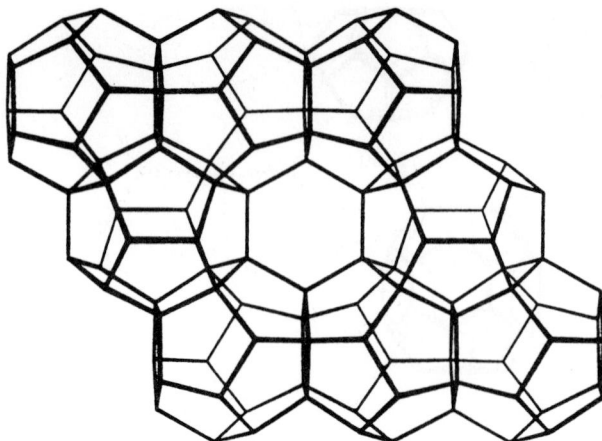

Fig. 16. Section of a polyhedral layer of pentagonal dodecahedra, [5^{12}]. The center shows a 6-ring window that belongs to a second type of cage arising from the arrangement of pentagonal dodecahedra. This layer is found in the clathrate hydrate type II structure and in the 1H type structure.

is no ionic or hydrogen-bond interaction between host structure and guest molecule. Depending on the size and shape of the guest molecule 10 different framework structure types have been obtained with almost 100 different guest molecules. This demonstrates the remarkable flexibility of the hydrogen-bond network of the water molecules. Nevertheless, the hydrogen bond in the different water host frameworks is similar to what is known from the ice structures with O–O-distances of approximately 2.75 Å. The polyhedral cages contain the guest species such as small atoms or molecules like, for example, Kr in the pentagonal dodecahedron. The maximum cavity size has been reached with a 20-hedron (cf. Fig. 15) housing, e.g. adamatanone, a molecule of 8.5 Å diameter. Most of the research on the true clathrates was performed between 1960 and 1980. The excellent review of Jeffrey [45] summarizes the historical literature and in particular the studies on the structural properties of the true clathrates.

The true clathrate hydrates decompose upon heating. The decomposition temperature might be regarded as a measure of the strength of the host–guest interaction. Whereas the ethyleneoxide clathrate decomposes at temperatures above 10 °C, the least stable clathrates with O_2, N_2, CO and other small gases as guest molecules only crystallize at temperatures below freezing point and at elevated pressure. An overview on the stability and other thermodynamic properties of hydrate inclusion compounds is given in the review article by Dyadin and Belosludov [50].

4.1.3.2.2 Ionic Clathrates

In the crystal structures of ionic clathrates the water host framework includes ions, generally anions such as OH^-, F^-, carboxylate anions, etc., the charge of which is

Fig. 17. Large polyhedral void of an interrupted tetrahedral water host framework created from 4 $[5^{12}6^2]$-cages (cf. Fig. 15). The tetraalkylammonium cation resides in the center of the cavity with the alkyl chains reaching into the side pockets. The charge of the cation is compensated by substitution of oxygen by fluorine in one tetrahedral center.

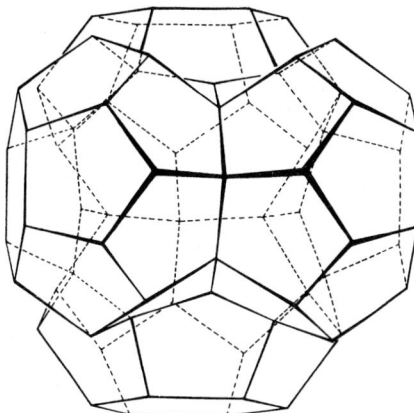

compensated by cationic guest species. The cations are spacious alkyl-onium salts occupying the polyhedral void space. The structural chemistry is much more diverse than for true clathrates. More than 10 framework structure types are known, some of which are unique for a particular guest molecule.

The substitution of H_2O by OH^- or other anions leads to an interruption of the tetrahedral hydrogen-bonded water network creating large cage-like voids. As an example, a 44-hedral cavity with tetrahedral symmetry is created by joining four 14-hedral cages and removing a common vertex (Fig. 17). The hetero-atom of the alkyl-onium salt now resides in the center of the cavity with the four alkyl-chains reaching into the four crevices of the former cages.

Another interesting group of ionic clathrates is formed by strong, monobasic acids such as HPF_6. Here the water framework is protonated and positively charged whereas the globular anion resides in the cages. It is interesting to note that two structures have host frameworks similar to the true clathrates, i.e. they have a complete tetrahedral water network with the excess proton disordered over the hydrogen bonds of the cage housing the counter anion. A complete account of the structural chemistry of ionic clathrates is given by Jeffrey [44, 45].

Because of the stronger host–guest interaction the decomposition temperature of ionic clathrates is higher, e.g. 35 °C for the (i-amyl)$_4$NF-hydrate. However, the high concentration of ionic species also leads to a lowering of the freezing point of water. Therefore, the decomposition temperature is not a valid unique indicator for the strength of the host–guest interaction. Extensive investigations of the thermodynamics of ionic clathrates is contained in the review by Dyadin and Belosludov [50] on thermodynamic properties of clathrate inclusion compounds.

4.1.3.2.3 Semiclathrates

More than 100 alkylamine hydrates are reported in the literature, however, only a few have been studied structurally in detail. There are six crystal structure analyses

References see page 696

(see in Jeffrey [45], and [51]) and all have been classified as semiclathrates. Most of the solid hydrates crystallize in structures unknown so far. Although the general structural features as pore space for guest species resemble those of gas hydrates, so far no single host lattice has been found that satisfies the requirements of different guest molecules. In the semiclathrates the amine guest molecule participates via hydrogen bonds in the hydrogen-bonded water framework with the hydrophobic tail occupying the void space. This leads to broken bonds in the water host framework yielding large and complicated polyhedral cavities. Therefore the structural characteristics are significantly determined by the individual structural characteristics of the guest species. An insight into the few structures analyzed so far and the preparative literature is given in Jeffrey's review article [45].

In a literature search using the Web of Science starting from 1995 over 250 hits were listed using the keyword "clathrate hydrate". This indicates that the field of research is very active and one of the most attractive in inclusion chemistry. The majority of the papers published recently deals with clathrate hydrates of small molecules and their physical properties. In many papers reference is made to the naturally occurring clathrate hydrates and their crystal chemistry as a function of the locality.

4.1.3.3
Cyanometallates

The building blocks in cyanometallate inclusion compounds are cyanometallate polyhedra with multidentate cyanide ligands coordinating metal centers such as Ag in linear $[Ag(CN_2)]^-$, Cu in trigonal $[Cu(CN)_3]^{2-}$, Cd in tetrahedral $[Cd(CN)_4]^{2-}$, Ni in square planar $[Ni(CN)_4]^{2-}$, and Fe in octahedral $[Fe(CN)_6]^{4-}$. The chemical bond in cyanometallate host frameworks is the linear Me–CN–Me bridge with a metal-to-metal distance of approximately 5 to 6 Å. The host framework structures are formed as multidimensional bonding networks of the above building blocks creating void space for guest molecules. Whereas the linear $[Ag(CN)_2]^-$ acts as a bridging ligand between two centers of higher coordination, the others connect to 2- or 3D host frameworks. The linkage of building blocks of the same metal center and also of different metal centers leads to various structural patterns demonstrating the variability of the crystal structures of the host framework.

Additional ligands such as NH_3, H_2O, unidentate aliphatic and aromatic amines, ambidentate α,ω-diaminoalkanes, piperazine, etc. play an important role in the stabilization of the 3D structure and in the volume of the void space that is occupied by guest species. As an example, the coordination of the square planar cyanometallate complexes is extended to 6-fold quasi-octahedral coordination introducing pillar-like ammonia spacers between 2D cyanometallate units. The guest molecule benzene occupies the void space between the cyanometallate layers (Fig. 18). The figure conveys clearly the tolerance of the crystal structure for tailor-made host systems using various components as building blocks. Recent reviews by Iwamoto [52, 53] summarize the history, the synthesis concepts, the crystal chemistry and the structural hierarchy of cyanometallate inclusion compounds.

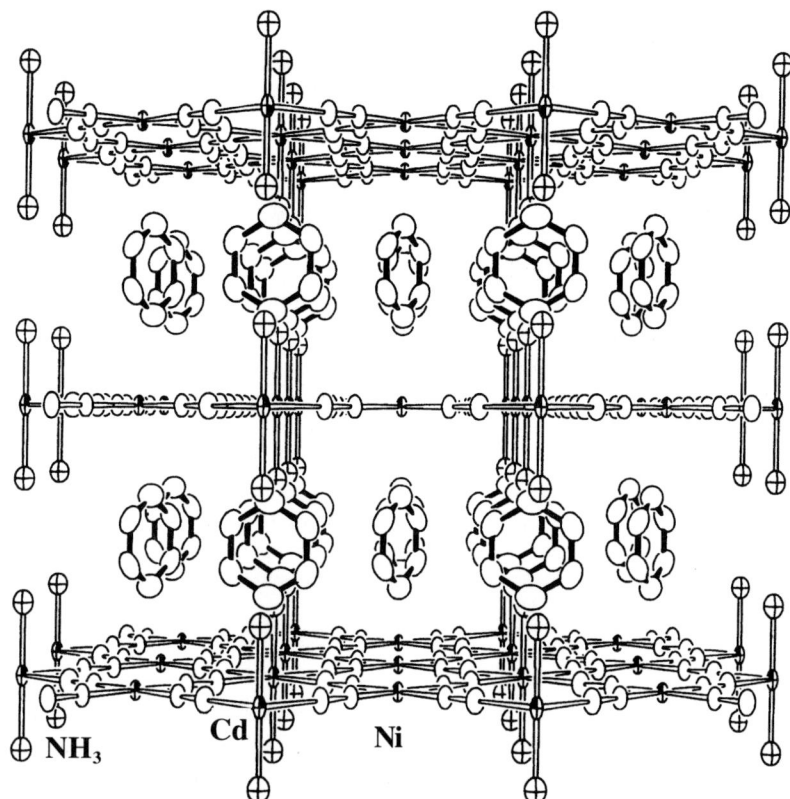

Fig. 18. Perspective view of structure of the Hofmann-type inclusion compound, $[Cd(NH_3)_2Ni(CN)_4]\cdot2C_6H_6$.

4.1.3.3.1 Host Frameworks with Square Planar $[Ni(CN)_4]^{2-}$

The parent structure of cyanometallate inclusion compounds with square planar $[Ni(CN)_4]^{2-}$ is $[Ni(NH_3)_2Ni(CN)_4]\cdot2C_6H_6$, Hofmann's benzene clathrate. Here a 2D network of corner-sharing $[Ni(CN)_4]^{2-}$ units are separated by NH_3 ligands bound to every second metal center increasing their coordination to 6. The interlayer space is filled by benzene guest molecules centered in the 4-ring pore of the cyanometallate layer (Fig. 18). Isostructural inclusion compounds have been obtained with M = Mn, Fe, Co, Cu, Zn, Cd coordinating to ammonia and M′ = Pd or Pt as cyanocomplexes and a variety of organic guest species. Attempts to replace ammonia as complementary ligand were also successful and increased the number of structural analogs.

Replacing ammonia with bifunctional α,ω-diaminoalkanes with alkylchain length from 2 to 9, or related diamino compounds such as 4,4′-bipyridine, led to a 3D bonding scheme with the amine as pillar between the cyanometallate layer. The

References see page 696

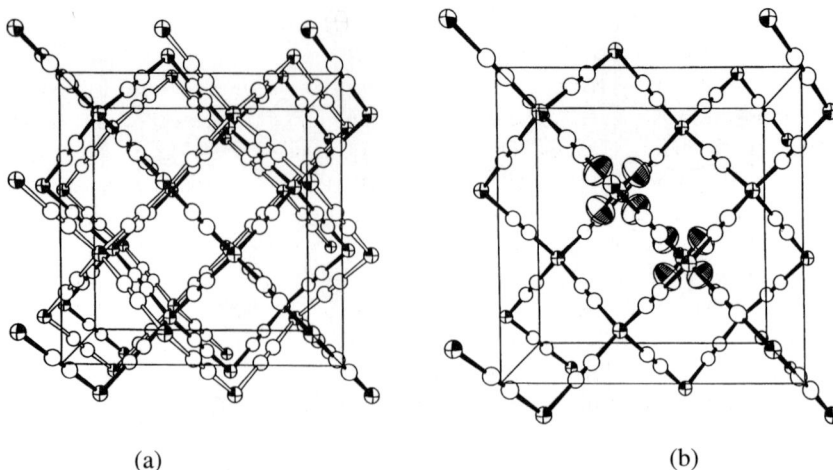

(a)   (b)

Fig. 19. (a) The tetrahedral framework of $Cd(CN)_2$ shown as a self-interpenetrating structure isostructural with cristobalite. The two bonding networks are indicated with open and filled bonds. (b) In the presence of tetramethylammonium cations interpenetration is avoided and an inclusion compound forms. The cristobalite topology of the host framework in the inclusion compound is maintained.

rational design of new cyanometallate inclusion compounds varying the different functionalities has been very fruitful as the number of successful syntheses and structure analyses shows. The very rich field of crystal chemistry reviewed by Iwamoto [52, 53] documents the efforts.

4.1.3.3.2 Host Frameworks with Tetrahedral $[Cd(CN)_4]^{2-}$

Pure $Cd(CN)_2$ crystallizes as an interpenetrating framework. Each 3D $Cd(CN)_2$ network has the same topology as high-cristobalite, a low-density polymorph of SiO_2. The relationship between these structures indicates that $Cd(CN)_2$ might show similar diversity of crystal structure as the crystalline polymorphs of silica in the silicates. In the presence of neopentane, the interpenetration is avoided and a clathrate-type analog with cristobalite topology crystallizes (Fig. 19). Similar to the Hofmann-type compounds, a number of different guest molecules have been incorporated and also the cyanometallate building blocks have been substituted. In addition, the combination with the square planar building blocks of cyanometallate inclusion compounds offers another way of systematic structural variation.

Replacing the square planar $Ni(CN)_4$ by tetrahedral $Cd(CN)_4$ in Hofmann-type inclusion compounds transforms the planar host structure into a 3D host framework. Varying the metal centers and the complementary ligands in the same fashion as in the Hofmann-type cyanometallate inclusion compounds yields another series of host–guest structures with a wide variety of guest molecules. As a common feature the different host frameworks have cage-like voids occupied by guest species. There is no direct bonding interaction between guest and host, so size

and shape of the guest molecule determine the pore size and geometry and, subsequently, the host framework type.

Cd(CN)$_2$ also form layered and framework structures with the tetrahedral building block only. Similar to silicates, layered cyanometallates intercalate guest molecules between layers. In analogy with zeolites, guest molecules are trapped in cage-
like voids in 3D porous cyanometallate framework structures. Of particular interest
are Ag-Cd-cyanometallate host framework structures. The linear [Ag(CN)$_2$]$^-$
building block acts as spacer bridging two metal centers. The widened pore space
is then filled with larger guest molecules. Up to three linear units in a row have
been successfully introduced into Cd(CN)$_2$ framework structures [54]. [Ag(CN)$_2$]$^-$
might also act as unidentate ligand or even as discrete anion. Gold complexes are
also linear, but have been investigated much less intensively. So far the concept has
been applied very successfully mainly to Cd-cyanometallates. For more detailed information the reader is referred to review articles on cyanometallates. The two
articles by Iwamoto [52, 53] provide a rich source of information on the crystal
chemistry and of the original literature.

The stability of the cyanometallates, in general, is only limited. They all decompose when the guest molecule is set free. Some are not stable in dry air. The guest
molecule evaporates leading to the subsequent disintegration of the cyanometallate
host structure. A brief account on thermodynamic data is given in [55].

A survey of the recent literature yields almost 100 hits in the Web of Science [13]
from 1993 with the keywords metallocyanide or cyanometallate or Hofmann type
showing that the researchers in the field are still productive. The publications are
dominated by preparative and structural chemistry, but there are also many papers
on vibrational spectroscopy, showing that the variation of building blocks continues to produce new cyanometallate inclusion compounds.

4.1.3.4
Silica-based Inclusion Compounds

Inclusion compounds with host framework composition SiO$_2$ are crystalline, porous silicas built from [SiO]$_4$–tetrahedra as building units. They link via oxygen
bridges and form 3D tetrahedral networks. The strongly covalent bond is responsible for the high thermal and mechanical stability of the silica host framework.
The host is hydrophobic and forms under hydrothermal conditions in the presence
of suitable guest molecules, so-called structure-directing agent, SDA. New silica
host frameworks have been discovered over the years mainly through innovative
synthesis strategies. The number of host framework types is now close to 40 [56].
In the same period more and more guest molecules for the known and the new
silica inclusion compounds were found. Theoretically, an infinite number of tetrahedral nets exists leaving a challenge for the synthetic chemists. A number of review articles updating and summarizing the synthesis and structure of crystalline
porous silicas, their physical properties and potential applications has been pub-

References see page 696

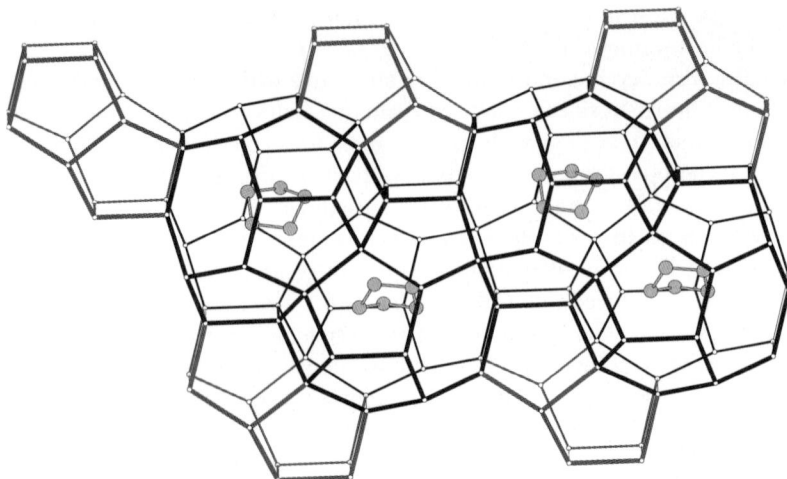

Fig. 20. Projection of a slab of the crystal structure of the clathrasil silica-RUB-10. The tetrahedral host framework with composition SiO$_2$ is shown as a skeletal drawing with the Si atoms in the corner of the polygonal windows. The oxygens are close to the connecting lines and are omitted here for clarity. The templating guest molecule pyrrolidine is shown as a skeletal model inside the large 16-hedral cage, [4^{4}5^{6}6^{5}8^1].

lished over the years. For detailed discussions on these aspects the reader is referred to the review articles and the references cited therein [57–59].

4.1.3.4.1 Silica Host Frameworks

Two classes of silica host frameworks are known, clathrasils with cage-like voids (Fig. 20) and zeosils (Fig. 21) with channel-like voids. Up to now almost 40 framework structure types have been synthesized using more than 100 different SDA. For details on the framework structure type and its IUPAC 3-letter code see [56]. The latest review on the crystal chemistry of porous silicas highlights the recent advances in the field [58]. Whereas the guest species of zeosils can easily be deliberated by calcination at temperatures of 550 °C, the making of guest molecule-free clathrasils requires calcination procedures at higher temperature and of sometimes several month exposure time. For example, dodecasil 3C can be obtained as a guest-free silica framework after 3 months at 1150 K emphasizing the extraordinary thermal stability of the host framework. An interesting property of all pure silica host frameworks investigated so far is their thermal volume expansivity, which is negative over a wide temperature range typical for every framework structure type [60, 61].

The pore diameter in zeosils ranges from 8-membered rings (8MR) made of [SiO$_4$]-tetrahedra, 8MR, with a free diameter of approximately 4.0 Å to 14MR with approximately 7.5 Å. reflecting the size of the structure-directing guest molecules. They range from extended aliphatic amines to complex heterocyclic molecules for the large pore channels. Because of the rigidity of the silica host framework the

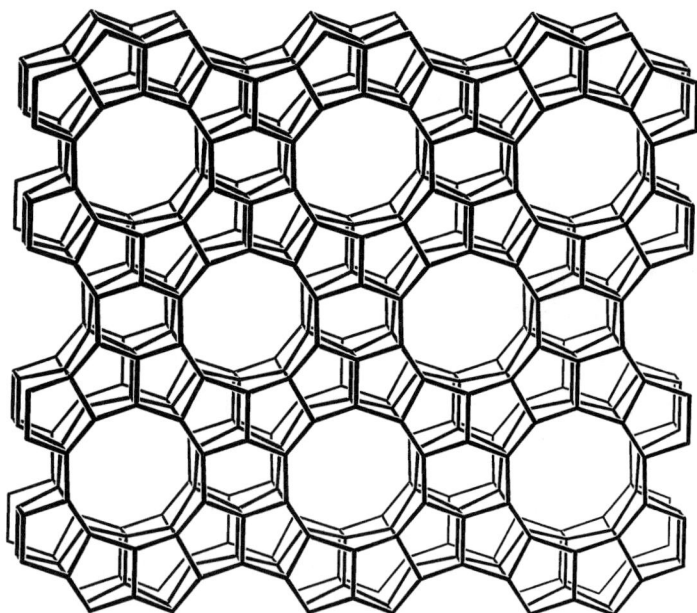

Fig. 21. Projection along the channel-like pore of a zeosil silica inclusion compound. The 10MR pore opening allows for the removal of the templating guest molecule without decomposition of the silica framework and the uptake of suitable sorbate molecules.

pore volume of the calcined host framework can be loaded with guest molecules of appropriate size. The hydrophobic nature of the electroneutral silica host restricts the uptake to hydrophobic sorbates making the zeosils suitable for sieving and separation purposes. The clathrasils have cage-like voids of free diameters ranging from 5.3 Å for atomic and diatomic guest molecules to approximately 10 Å for large multicyclic, rigid and spherical guest species. The guest-free clathrasil silica host framework can only be obtained after thermal degradation and evaporation of the degradation products. Because of the narrow pore opening of clathrasil silica frameworks, which is approximately 2 Å in diameter, no sorption properties are observed.

4.1.3.4.2 Guest Molecules for the Synthesis of Silica Inclusion Compounds

A systematic evaluation of the chemistry and the relevant properties of guest molecules for silica inclusion compounds revealed close similarities to clathrate hydrates. There is no specific host–guest interaction once the silica framework is formed. Size and shape of the guest molecules determine the pore void and dimension. Spherical SDA lead to the cage-like void in clathrasils (Fig. 22), extended, chain-like SDA lead to channel-like void in zeosils (Fig. 23). However, the synthesis

References see page 696

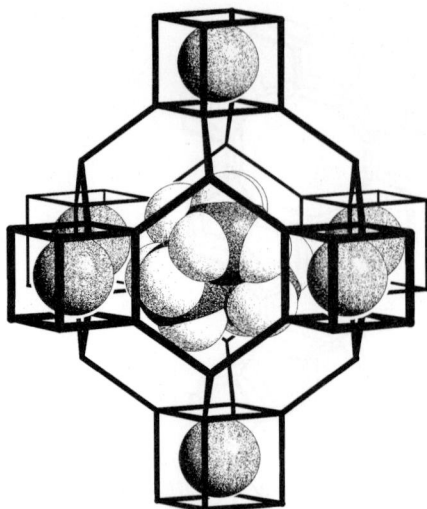

Fig. 22. Van der Waals model of the tetramethylammonium
cation as SDA in the 14-hedral cage, $[4^6 6^8]$ and F^- in the cube
of octadecasil (AST).

procedure restricts the number of guest species. Hydrothermal synthesis condition
in the temperature range of 140–250 °C require SDA, which are stable upon hy-
drolytic degradation. The templating guest molecules should also be amphiphilic
in order to establish interaction with the hydrophobic silica host framework and, at
the same time, allow for the continuous growth of the inclusion compound from

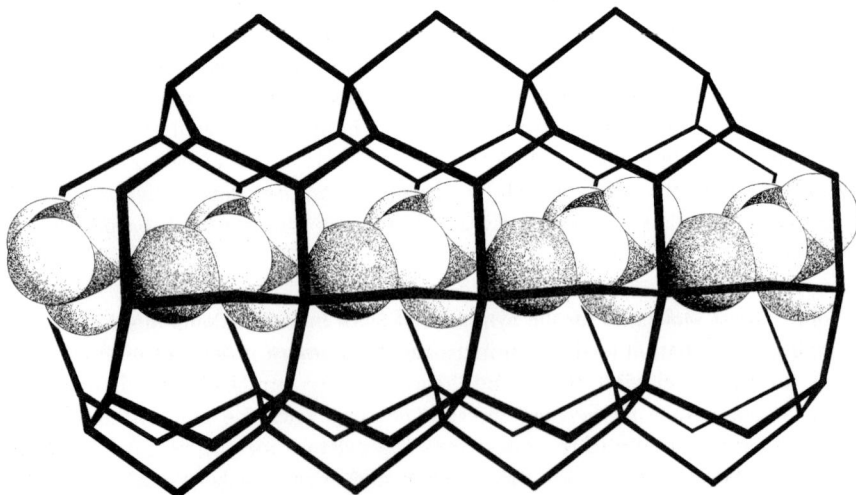

Fig. 23. Van der Waals model of the diethylenetriamine
template as SDA in the 10MR channel of the zeosil
silica-ZSM-22.

aqueous solution. The most suitable family of SDA are amines with their basic and hydrophilic amine-head group and the hydrophobic hydrocarbon tail. More than 100 examples have been used successfully so far.

The versatility of SDA might serve as an indicator for the nonspecific host–guest interaction. Many guest molecules serve as template at specific synthesis conditions for several silica host framework structure types. As an example, at synthesis temperatures lower than 180 °C piperidine acts as SDA for dodecasil 3C, above 180 °C dodecasil 1H is formed with a much larger cavity for the SDA. At the same time, one framework structure type is formed in the presence of a series of SDA. A summary of the specific conditions for the synthesis of silica inclusion compounds is given in the review by Gies et al. [58].

A survey of the more recent literature shows that research on silica inclusion compounds is very active. The natural proximity to the technically important zeolites that are used as sorbents, molecular sieves and catalysts also provides stimulation for continuing efforts in research on silica inclusion compounds. With the keywords clathrasils and zeosils and porosils and silica zeolites 174 hits since 1993 were compiled in a search in the Web of Science. Many of the publications are synthesis and application oriented and report on new host–guest systems and their potential use.

4.1.3.5
Relation between Different Inclusion Compounds with Tetrahedral Host Frameworks

There is an interesting relation between silica inclusion compounds and hydrate inclusion compounds [62]. Both form electroneutral tetrahedral host frameworks where the guest species is trapped without specific host–guest interaction. Not only are there nine framework structure types found in both systems, but there are also close similarities in the family of guest species, the synthesis conditions, and the type of host–guest interaction. The fact that other neutral tetrahedral frameworks are known, such as the above-mentioned $Cd(CN)_2$ and related cyanometallates, and other inclusion compounds not mentioned here, such as the crystalline porous elemental structures of Si and Ge with alkali metal atoms as guest species [63, 64] and their ionic clathrate counterparts [65, 66], $(Na, K)_x (Al, Ga, In)_x (Si, Ge, Sn)_{46-x}$ or $(Alk)_x (Al, Ga, In)_x (Si, Ge, Sn)_{132-x}$, the family of porous neutral phosphates [67, 56], e.g. $AlPO_4$, and arsenates [68, 56], e.g. $GaAsO_4$, demonstrates that there is a general principle of formation behind. In the presence of suitable guest molecules open tetrahedral framework structures with strong directional bonds form. The structure directing of the template is achieved by the shape and space requirements of the guest species. The host–guest interaction in this family of inclusion compounds is of van der Waals type and at least an order of magnitude weaker than the bond energy between the framework constituents. The inclusion compound in its as-synthesized form is a space-filling arrangement of guest and host, and thus, a product of an optimization process.

References see page 696

4.1.4
Final Remarks

This survey only summarizes key information on the authors' selection of host–guest systems. Space restrictions required a rigorous selection of topics. A more complete compilation is provided in the two book series mentioned earlier [7, 8]. With the efficient use of databases such as the Web of Science [13] there is also easy access to original literature once the keywords are selected specifically. The closing remark of every chapter should point out the progress in the field and provide an entry with a suitable set of keywords.

References

1 H. Davy, *Philos. Trans. R. Soc. Lond.* **1811**, *101*, 155.

2 F. Wöhler, *Liebigs Ann. Chem.* **1849**, *69*, 297.

3 K. A. Hoffmann, F. Küspert, *Z. Anorg. Allg. Chem.* **1897**, *15*, 204.

4 A. P. Dianin, *J. Soc. Phys. Chem. Russe.* **1914**, *46*, 1310.

5 H. M. Powell, *J. Chem. Soc.* **1948**, *1948*, 61.

6 a) Inclusion Compounds, J. L. Atwood, J. E. D. Davies, D. D. MacNicol (Eds), vol. 1–3, Academic Press, London, 1984; b) Inclusion Compounds, J. L. Atwood, J. E. D. Davies, D. D. MacNicol (Eds), vols 4–5, Oxford University Press, Oxford, 1991.

7 Comprehensive Supremolecular Chemistry, J. L. Atwood, J. E. D. Davies, D. D. McNicol, F. Vögtle (Eds), vols 1–11, Pergamon, 1996.

8 G. A. Jeffrey, W. Saenger, Hydrogen Bonding in Biological Structures, Springer, Berlin, 1991.

9 G. A. Jeffrey, An Introduction to Hydrogen Bonding, Oxford University Press, New York, 1997.

10 D. D. McNicol, in: Inclusion Compounds, J. L. Atwood, J. E. D. Davies, D. D. McNicol (Eds), Academic Press, 1984, vol. 2, Chapter 1, pp. 1–45.

11 T. C. W. Mak, B. R. F. Bracke, in: Comprehensive Supramolecular Chemistry, J. L. Atwood, J. E. D. Davies, D. D. McNicol, F. Vögtle (Eds), Pergamon, 1996, vol. 6, D. D. MacNicol, F. Toda, R. Bishop (Eds), chapter 2, pp. 23–35.

12 a) A. Detken, P. Schiebel, M. R. Johnson, H. Zimmermann, U. Haeberlen, *Chem. Phys.* **1998**, *238*, 301–314; b) A. Detken, H. Zimmermann, U. Haeberlen, *Mol. Phys.* **1999**, *96*, 927–940; c) K. Hermansson, *J. Chem. Phys.* **2000**, *112*, 835–840.

13 Web of Science, Institute for Scientific Information, http://www.isinet.com/isi/

14 O. Ermer, *Helv. Chim. Acta.* **1991**, *74*, 1339.

15 O. Ermer, C. Röbke, *J. Am. Chem. Soc.* **1993**, *115*, 10077.

16 M. Gdaniec, B. T. Ibragimov, S. A. Talipov, in: Comprehensive Supramolecular Chemistry, J. L. Atwood, J. E. D. Davies, D. D. McNicol, F. Vögtle (Eds), Pergamon, 1996, vol. 6, D. D. MacNicol, F. Toda, R. Bishop (Eds), chapter 5, 117–144.

17 T. C. W. Mak, B. R. F. Bracke, in: Comprehensive Supramolecular Chemistry, J. L. Atwood, J. E. D. Davies, D. D. McNicol, F. Vögtle (Eds), Pergamon, 1996, vol. 6, D. D. MacNicol, F. Toda, R. Bishop (Eds), chapter 2, pp. 37–60.

18 F. H. Herbstein, in: Comprehensive Supramolecular Chemistry, J. L. Atwood, J. E. D. Davies, D. D. McNicol, F. Vögtle (Eds), Pergamon, 1996, vol. 6, D. D. MacNicol, F. Toda, R. Bishop (Eds), chapter 3, pp. 61–83.

19 D. J. Duchamp, R. E. Marsh, *Acta Crystallogr. Sect. B* **1969**, *25*, 5.

20 F. H. Herbstein, R. E. Marsh, *Acta Crystallogr. Sect. B* **1977**, *33*, 2358.

21 F. H. Herbstein, M. Kapon, G. M. Reisner, *J. Inclusion Phenom.* **1987**, *5*, 211.

22 M. J. Zaworotko, *Chem. Commun.* **2001**, 1–9.

23 R. Bishop, in: Comprehensive Supramolecular Chemistry, J. L. Atwood, J. E. D. Davies, D. D. McNicol, F. Vögtle (Eds), Pergamon, 1996, vol. 6, D. D. MacNicol, F. Toda, R. Bishop (Eds), chapter 4, pp. 86–115.

24 A. T. Ung, D. Gizachew, R. Bishop, M. L. Scudder, I. G. Dance, D. C. Craig, *J. Am. Chem. Soc.* **1995**, *117*, 8745.

25 A. T. Ung, R. Bishop, D. C. Craig, I. G. Dance, M. L. Scudder, *J. Chem. Soc., Perkin Trans.* **1992**, *2*, 861.

26 I. G. Dance, R. Bishop, S. C. Hawkins, T. Lipari, M. L. Scudder, D. C. Craig, *J. Chem. Soc., Perkin Trans.* **1986**, *2*, 1299.

27 I. G. Dance, R. Bishop, M. L. Scudder, *J. Chem. Soc., Perkin Trans.* **1986**, *2*, 1309.

28 R. Bishop, I. G. Dance, *Top. Curr. Chem.* **1988**, *149*, 137.

29 A. T. Ung, R. Bishop, D. C. Craig, I. G. Dance, M. L. Scudder, *Struc. Chem.* **1992**, *3*, 59.

30 R. Bishop, D. C. Craig, A. Marougkas, M. L. Scudder, *Tetrahedron.* **1994**, *49*, 639.

31 M. F. Bengen, German Patent Appl. 1940, OZ 123438.

32 M. F. Bengen, W. Schlenk, Jr. *Experientia.* **1949**, *5*, 200.

33 F. Cramer, Einschlussverbindungen, Springer, Berlin, 1954.

34 K. Takemoto, N. Sonoda, in: Inclusion Compounds, J. L. Atwood, J. E. D. Davies, D. D. MacNicol (Eds), Academic Press, London, 1984, vol. 2, 47–67.

35 M. D. Hollingsworth, K. D. M. Harris, in: Comprehensive Supramolecular Chemistry, J. L. Atwood, J. E. D. Davies, D. D. McNicol, F. Vögtle (Eds), Pergamon, 1996, vol. 6, D. D. MacNicol, F. Toda, R. Bishop (Eds), chapter 7, 177–237.

36 S. van Smaalen, *Crystallogr. Rev.* **1995**, *4*, 79.

37 K. D. M. Harris, P. Jonson, *Chem. Phys. Lett.* **1989**, *154*, 593.

38 F. Guillaume, C. Sourisseau, A. J. Dianoux, *J. Chim. Phys. Phys. Chim. Biol.* **1991**, *88*, 1721.

39 J. Lipkowski, in Inclusion Compounds, J. L. Atwood, J. E. D. Davies, D. D. MacNicol (Eds), Academic Press, London, 1984, vol. 1, 59–103.

40 J. Lipkowski, in: Comprehensive Supramolecular Chemistry, J. L. Atwood, J. E. D. Davies, D. D. McNicol, F. Vögtle (Eds), Pergamon, 1996, vol. 6, D. D. MacNicol, F. Toda, R. Bishop (Eds), chapter 20, 691–714.

41 C. Karunakaran, K. J. R. Thomas, A. Shunmugasundaram, R. Murugesan, *J. Incl. Phenom. Macrocycl. Chem.* **2000**, *38*, 233–249.

42 A. Y. Manakov, J. Lipkowski, *J. Incl. Phenom. Mol. Recogn.* **1997**, *29*, 41–55.

43 A. Y. Manakov, J. Lipkowski, J. Pielaszek, *J. Incl. Phenom. Macrocycl. Chem.* **1999**, *35*, 531–548.

44 G. A. Jeffrey, in: Inclusion Compounds, J. L. Atwood, J. E. D. Davies, D. D. MacNicol (Eds), Academic Press, London, 1984, vol. 1, 135–190.

45 G. A. Jeffrey, in: Comprehensive Supramolecular Chemistry, J. L. Atwood, J. E. D. Davies, D. D. McNicol, F. Vögtle (Eds), Pergamon, 1996, vol. 6, D. D. MacNicol, F. Toda, R. Bishop (Eds), chapter 23, 757–788.

46 Gas Hydrates: Relevance to World Margin Stability and Climate Change, J.-P. Henriet, J. Mienert (Eds), Geological Society London, London, Special Publications, vol. 137, 1998.

47 Gas hydrates: Challenges for The Future, G. D. Holder (Ed.), Annals of the New York Academy of Sciences, New York, vol. 912, 2000.

48 K. A. Kenvolden, in: Gas Hydrates: Relevance to World Margin Stability and Climate Change, J.-P. Henriet, J. Mienert (Eds), Geological Society London, London, 1998, Special Publications, vol. 137, 9–30.

49 B. B. Buffett, *Annu. Rev. Earth Planet. Sci.* **2000**, *28*, 477–507.

50 Y. A. Dyadin, V. R. Belosludov, in: Comprehensive Supramolecular Chemistry, J. L. Atwood, J. E. D. Davies, D. D. McNicol, F. Vögtle (Eds), Pergamon, 1996, vol. 6, D. D. MacNicol, F. Toda, R. Bishop (Eds), chapter 24, 789–824.

51 D. Stäben, D. Mootz, *J. Incl. Phenom. Mol. Recogn. Chem.* **1995**, 22, 145–154.

52 T. Iwamoto, in: Comprehensive Supramolecular Chemistry, J. L. Atwood, J. E. D. Davies, D. D. McNicol, F. Vögtle (Eds), Pergamon, 1996, vol. 6, D. D. MacNicol, F. Toda, R. Bishop (Eds), chapter 19, 644–690.

53 T. Iwamoto, *J. Incl. Phenom. Mol. Recogn. Chem.* **1996**, *24*, 61–132.

54 T. Soma, T. Iwamoto, *Inorg. Chem.* **1996**, *35(7)*, 1849–1856.

55 T. Iwamoto, in: Inclusion Compounds, J. L. Atwood, J. E. D. Davies, D. D. MacNicol (Eds), Academic Press, London, 1984, vol. 1, 29–57.

56 Database of zeolite structures: http://www.iza-structure.org/databases/

57 H. Gies, B. Marler, in: Comprehensive Supramolecular Chemistry, J. L. Atwood, J. E. D. Davies, D. D. McNicol, F. Vögtle (Eds), Pergamon, 1996, vol. 6, D. D.

MacNicol, F. Toda, R. Bishop (Eds), chapter 26, 851–883.

58 H. Gies, B. Marler, U. Werthmann, in: Molecular Sieves, Science and Technology, vol. 1, Synthesis, H. G. Karge, J. Weitkamp (Eds), Springer, Berlin, 1998, 35–64.

59 M. A. Camblor, L. A. Villaescusa, M. J. Diaz-Cabanas, *Top. Catal.* **1999**, 9, 59–76.

60 S-H. Park, R. W. Große-Kunstleve, H. Graetsch, H. Gies, *Surface Science and Catalysis.* **1997**, *105 Part A–C*, 1989–1994.

61 P. Lightfoot, D. A. Woodcock, M. J. Maple, L. A. Villaescusa, P. A. Wright, *J. Mater. Chem.* **2001**, *11*, 212–216.

62 F. Liebau, in: The Physics and Technology of Amorphous SiO_2, R. A. B. Devine (Ed.), Plenum Press, New York, 1988, 15–35.

63 C. Cros, M. Pouchard, P. Hagenmueller, *Science.* **1965**, *150*, 1713–1714.

64 J. Gallmeier, H. Schaefer, A. Weiss, Z. *Naturforschung.* **1969**, *24*, 665–667.

65 H. G. von Schnering, H. Menke, *Angew. Chem.* **1972**, *84*, 30–31.

66 H. Menke, W. Cabrillo-Cabrere, K. Peters, E.-M. Peters, H. G. von Schnering, *Z. Kristallogr.*, *NCS* **1999**, *214*, 14, and related references.

67 S. T. Wilson, B. M. Lok, C. A. Messina, T. R. Cannan, E. M. Flanigen, *J. Am. Chem. Soc.* **1982**, *106*, 6092.

68 T. Loiseau, F. Taulelle, G. Ferey, *Microporous Mater.* **1997**, *9*, 83–97.

4.2
Crystalline Microporous Solids

4.2.1
Introduction and Structure

Wulf Depmeier

4.2.1.1
Introduction

In this contribution we will discuss basic structural and chemical properties of crystalline microporous solids. Since abundant structural information in the form of databases is readily available in the literature, we will refrain ourselves from going into too much detail, but will rather try to give some critical, and we hope, helpful commentaries. The same holds for the chapter on crystal chemistry, where we express some perhaps provocative opinions.

4.2.1.2
Structural Basis, Definitions and Databases

4.2.1.2.1 Basic Features of Crystalline Microporous Solids
An important subset of all solid materials is characterized by voids in their respective structures, with diameters exceeding typical atomic dimensions. The voids are called pores if they are spacious enough to accommodate at least a small molecule, such as water. Hence, by definition, their diameter $\varnothing$ must exceed 2.5 Å. The IUPAC has proposed a classification of pores, based on their size, which distinguishes between micropores, having widths $\varnothing < 20$ Å, mesopores with $20 \leq \varnothing \leq 500$ Å, and macropores with $\varnothing > 500$ Å [1]. We will maintain this classification.

Logically, the mere existence of pores, as well as their sizes and shapes, is defined by negation of the surrounding matter. In the case of microporous structures the latter is often called the "host structure", or simply "host". Since the frequently used atomic models of porous solids often resemble the skeleton of certain architecture, the descriptive terms "framework", or "framework structure", are also common. Note that these terms do not necessarily imply that the correspond-

References see page 734

ing structures extend over three dimensions, as in the case of framework or tecto-silicates.

The pores may be empty, or filled with various "guests". Possible guest species are atoms, simple or complex ions, neutral or charged molecules, or even some kind of supramolecular arrangement. In any case, the guests have to fit the geometrical, charge, and perhaps additional, constraints set by the host, and vice versa. Hence, for a given microporous material a differentiation between possible guest species can take place, namely between those that fit the requirements, and others that do not.

In a crystalline host the pores are periodically arranged. If, in addition, the apertures of the pores are smaller than 20 Å, then one deals, by definition, with "crystalline microporous solids", the title objects of this contribution. We note in passing that the pores of some mesoporous solids are, indeed, periodically arranged, although the internal structure of the host framework is amorphous.

Most commonly, the chemical bonds between the guests and the host are so weak that they are easily broken, and the guests are released. An important criterion for the definitions given further below is whether the released guests can migrate freely through the pore system, or are encapsulated in the pores. The latter is the case of so-called inclusion compounds.

Crystalline microporous solids in various matter, restriction to inorganic compounds Crystalline microporous solids occur in inorganic, as well as in organic matter. Most organic ones are inclusion compounds; the following examples are well known: 1) the urea complexes with n-alkanes, where hydrogen-bonded urea molecules are arranged into a supramolecular framework assemblage, with the alkane chains residing as guests in the channels of the urea host structure [2], and 2) the various cyclo-dextrine inclusion compounds, for example [3]. Many organic inclusion compounds are of high scientific or technical importance.

A highly interesting, and actually intensively studied, family of inclusion compounds of great interdisciplinary importance is that of gas hydrates [4]. It is also worth mentioning that host–guest interactions, however on the molecular level, are essential for understanding the physiological activity of many bio-molecules.

All these fascinating items cannot be touched on in this short contribution. We have rather chosen to focus our interest on particular inorganic hosts, the structures of which are more or less akin to a group of naturally occurring minerals, the zeolites. Our choice is justified by 1) an alleged preference of the readership for this class of materials 2) the fact that recently a large amount of structured data became available, 3) the author's personal interests and experiences, 4) space and time limitations. Terms like "zeolites", or "crystalline microporous solids" are commonly used rather loosely in the literature. Sharper definitions are available and will be given below.

Following Liebau [5], "crystalline microporous solids" will be called "poroates", henceforth.

Preliminary information on chemistry The strong chemical bonds of the poroate host structure are of covalent to ionic character. Typically, the host is composed of so-called TO_4-tetrahedra, where T represents Si or Al. Bold letters, for example T, are used as structure-site symbols throughout this article, in contrast to chemical element symbols, which are given as normal letters, for example O [6]. Normally the TO_4-tetrahedra are all-corner-connected, which implies for the host an average chemical composition TO_2. If all T atoms have formal charge 4+, as in the case of $T = Si$, then a neutral host structure results. If, on the contrary, Si^{4+} is partly substituted by Al^{3+}, then the resulting negative charge of the host has to be compensated by positively charged species, occupying as guests the pores of the host framework. Typical guest cations are Na^+, Ca^{2+} or K^+. The pores may also contain neutral guest molecules, water being of paramount importance for natural and also for synthetic poroates. Charged or neutral organic guest molecules are often used as structure-directing agents, so-called templates, in the synthesis of poroates.

Poroates, a suitable case for interdisciplinary research Poroates are intensively studied by a large, interdisciplinary, scientific community. The latter comprises mineralogists, crystallographers, chemists, mathematicians, information scientists, physicists, materials scientists, process engineers, geologists, and probably scientists from several other disciplines as well. Some of them study these materials out of sheer scientific curiosity, stimulated by their particular structures and ensuing properties. Others are primarily attracted by the various practical applications of poroates, many of them being already known and employed at present, or by their potential as high-technology materials, the practical importance of which will show up only in the future.

A long scientific history The long-standing scientific history of poroates began in the mid-18th century with the work of Cronstedt [7], who also coined the word zeolite. In an excellent review, Flanigen [8] describes clearly how the early observation of highly uncommon properties of natural zeolites raised the interest of researchers in these materials and stimulated the synthesis of new zeolites and similar compounds [9, 10]. These special properties were, for example, water loss upon heating, ion exchange, and selective adsorption of gases. However, it was the reversibility of these processes, and the fact that the host structure was apparently unaffected, which raised the highest interest.

Natural zeolites and synthetic poroates Natural zeolite minerals normally grow under mild hydrothermal conditions. Sometimes they occur as splendid crystals, or crystal aggregates, and are then very sought-after showpieces for mineral collections and museums. In particular cases certain zeolites occur in large quantities as massive deposits.

Because of the intensive efforts to prepare new, possibly useful poroates, the number of synthetic framework types known today exceeds by far that of natural ones. Interestingly, some of today's most popular poroates have never been found in nature. This is, for instance, true for the famous zeolite A. Drawings of its framework can be found ubiquitously in text books and review articles, where it is

References see page 734

often used to represent prototypically a zeolite framework structure. The equally important, and also often depicted, zeolites X and Y find their natural counterpart only in a rare mineral, faujasite.

Poroates as industrial products   The intensive research and development activities on poroates all over the world have made them important technical products of high economic value. The vast majority of all technical applications is taken by synthetic poroates, and it is probably fair to say that the higher the technological level, the higher the share of synthetic poroates. Only recently have some deposits of certain natural zeolites, notably clinoptilolite, begun to become exploited. As the material is cheap and available in large quantities, it is often used in low-level technologies, for example for the removal of bad odors, such as ammonia or fish smell from tanks or containers. A remarkable property of this zeolite is its strong affinity to certain radioactive isotopes making it attractive for the treatment of radioactive waste.

4.2.1.2.2 Various Definitions

The above-mentioned, quite diverse, community seems to have difficulties in agreeing over the question which materials should be included in the class of poroates, and which should not. Several definitions, partly mutually exclusive, are in use, and it seems that the time has not yet come for a unique, commonly accepted, definition. In the following we will present some different definitions, and comment on them, where appropriate. In particular, we will describe the different frontiers created by them.

A. The IMA definition   The Subcommittee on Zeolites of the International Mineralogical Association, IMA, defines [11]: "A zeolite mineral is a crystalline substance with a structure characterized by a framework of linked tetrahedra, each consisting of four O atoms surrounding a cation. This framework contains open cavities in the form of channels and cages. These are usually occupied by H_2O molecules and extra-framework cations that are commonly exchangeable. The channels are large enough to allow the passage of guest species. In the hydrated phases, dehydration occurs at temperatures mostly below about 400 °C and is largely reversible. The framework may be interrupted by (OH, F) groups; these occupy a tetrahedron apex that is not shared with adjacent tetrahedra."

Limits   It is quite natural that mineralogists will accept only such species as minerals that have been found to occur in nature. The restriction of the IMA definition to tetrahedral, oxygen-containing primary building units is caused by the historical circumstance that these are contained in all the classical zeolite minerals. This means, for example that the microporous mineral cetineite has to be excluded, as it contains not only nontetrahedral building units, but also sulfur as host framework anions. Cetineite is the prototype of a whole structural family that has been studied intensively recently. These materials are interesting not least because of their unusual photosemiconducting properties [12–15].

Note that the IMA definition tacitly assumes the existence of 3D-framework structures.

Even very common and important minerals having an oxidic tetrahedral microporous framework structure are excluded by the IMA definition, because they do not agree with the restriction that the channels should be large enough to allow the passage of guest species. Unfortunately, this part of the definition is not very precise, as it is difficult to define "large enough" and "guest species". In order to illustrate the problem we raise the question whether a microporous material which, for example, allows the passage of Li^+, but rejects K^+, should be called a zeolite in the sense of the IMA definition, or not. Obviously, it is difficult to justify the acceptance of the one, but the rejection of the other.

In addition, the use of the term "passage" is only justified if the flux of a given guest species through the host is significantly higher than a well-defined threshold value. However, the definition does not define such a value. Besides, even with a defined threshold value, a scientifically sound decision would require a very carefully planned and performed diffusion experiment. Such experiments are notoriously difficult, time consuming, and prone to error. The above definition is therefore only of limited practical value.

Obviously in order to overcome these difficulties, the "passage" criterion has been replaced – as it seems rather tacitly – by another, which, at a first glance, is no longer phenomenological, but rather relies on information which allows final answers. The new criterion uses a crystal chemical parameter that is automatically obtained as a by-product of the structure determination. The parameter in question is the size of rings of TO_4 tetrahedra in the host framework. It is expressed by the term n-ring, an n-ring containing n TO_4 tetrahedra, or T cations, respectively. Experimental evidence suggests that at ambient conditions the migration of water or cations in significant amounts is blocked up by any 6-ring in the intersections along the diffusion pathways. Because this is even more the case with still smaller rings, $n = 6$ has been chosen as the lower limit.

While this criterion allows clear-cut decisions, it is still not fully satisfactory, as it continues to employ basically the transport criterion, albeit by devious means. The borderline has been chosen rather arbitrarily, it excludes by definition all those important microporous mineral structures that contain only n-ring windows with $n \leq 6$.

This concerns, notably, the well-known sodalite family. The following observation demonstrates that the exclusion of sodalites is, indeed, somewhat arbitrary. Admittedly, at room temperature and atmospheric pressure the 6-ring windows of the sodalite framework do not allow significant transport of normal guest species. However, if the temperature is increased sufficiently then cations can migrate quite easily and the sodalite is ion exchanged. This has recently been demonstrated, once again, by the observation that a sodalite of composition $Na_8[Al_6Si_6O_{24}]Cl_2$, pressed into KBr pellets for the registration of high-temperature infrared spectra, showed rapid exchange of Na^+ against K^+ already around 650 K [16].

References see page 734

The exclusion of the sodalite family from the set of zeolite-type minerals is regretted by some people for the following reason. Its framework can be considered as the most simple of all microporous 3D frameworks. Its topological symmetry is cubic, it contains only one type of pore, and one tilt angle is sufficient to describe the conformational state of the framework. The tilt angle describes the orientation of the quasi-rigid TO_4 tetrahedra with respect to a certain zero-tilt state. For a given sodalite structure this corresponds to the case of largest expansion of the framework. Because of their simplicity, the members of the sodalite family exhibit many of the typical structural features, and also physical properties, more clearly than other large-pore poroates, where structures are more complex, and properties are often smeared out beyond recognition. Besides, the sodalite framework is depicted in the literature probably as often as that of zeolite A or faujasite. Sometimes, the term "zeolite-like" has been used for small-aperture framework structures such as sodalite, in order to distinguish them from "proper" zeolites.

B. The IZA-SC definition   Another definition was proposed by the Structure Commission of the International Zeolite Association, IZA-SC, acting under rules set up by an IUPAC Commission on Zeolite Nomenclature [17]. The materials of interest, that is zeolites in a broader sense, are generally defined as open, 4-connected, 3D-nets which have the general (approximate) composition AX_2, where **A** is a 4-connected atom and **X** is any 2-connected atom, which may, or may not, be shared between two neighboring **A** atoms. The latter statement allows for interrupted frameworks, where bridging O atoms are replaced by terminal OH^- groups.

The IZA Code and the Atlas   If a new framework topology (that is a 4-connected net) is accepted by the IZA-SC, it is given a unique code consisting of three capital letters. Examples are **SOD**, **LTA** and **MFI**, for the sodalite, zeolite A and the framework type to which belongs the important synthetic zeolite ZSM-5, respectively.

These codes serve as identifiers in a database maintained by the IZA-SC, and published in the so-called *"Atlas"* [18]. As an additional, invaluable, service a frequently updated website is maintained, with Ch. Baerlocher and L. B. McCusker as responsible members of the commission. This "Database of Zeolite Structures" [19] contains at present 133 framework types. (As of this writing. Meanwhile the number of framework types has increased to 135. The new codes **BCT** and **BEC** have not been considered throughout this contribution.) References [18] and [19] together will be cited as *"Atlas"*, henceforth. Some of its features will be described and discussed further below.

Limits   Note that synthetic materials are definitely admitted by the IZA-SC, and nonoxygen framework anions not explicitly excluded, by this definition. However, nets other than such 4-2 nets are not allowed. Hence, cetineite, albeit highly porous, is excluded.

According to the rules under which it acts, the final decision about acceptance or rejection of a given material is always left to the discretion of the IZA-SC. The commission acts in the interest of the zeolite scientific community at large. However, as such a common interest is difficult to define, it might happen in extreme cases that the interests of individual scientists, or of a whole subcommunity, do not match those of the commission.

C. A note on nets and graphs   "Net" in the IZA-SC definition means that the corresponding framework type can be mapped on an infinite, periodic graph, the nodes and the edges of which represent the T and O atoms, and the bonds between the atoms, respectively. Graphical representations facilitate the theoretical analysis of topologies [20–22]. The nets can be made simpler, if the nodes represent the T atoms at the centers of the TO_4-tetrahedra and the edges correspond to the connectivity of the TO_4-tetrahedra, that is to their linkage via T–O–T bonds.

Not all four-connected nets can exist as real matter, that is materialize as framework structures. In fact there exists an infinite number of such nets, but only a small number is relevant for poroates. The latter allow for "reasonable" interatomic distances and angles, and fulfil the constraint that fairly rigid TO_4-tetrahedra rather than mathematical points are the four-connected objects. Making use of these constraints, and employing the empirical argument that in real structures the number of topologically different TO_4-tetrahedra tends to be small, Klein [23, 24] was able to obtain a complete enumeration of already existing and hypothetical poroate frameworks for given numbers of independent tetrahedra.

In a recent study graph theory was combined with quantum-mechanical calculations, to derive theoretically the structures, properties and relative stabilities of 3D-connected, sp^2-bonded carbon polymorphs [25].

D. Differences between the IMA and IZA-SC concepts   The IMA and the IZA-SC definitions are conceptually different. Both use the term "open". Strictly speaking, "open" is not unambiguous, as it can mean "easily accessible", as well as "spacious". These two meanings express the different concepts of the IMA and the IZA-SC. Basically, both concepts use relationships between the structure of a poroate and its potential functionality. The functionalities to which the definitions refer are different. The IMA definition demands that the material can operate as a molecular sieve under ambient conditions. The IZA-SC definition requires that a microporous structure provides sufficient pore space, for example for the encapsulation of guests.

In order to illustrate the difference we can resort to images borrowed from our macroscopic world. Microporous structures in the sense of the IMA point of view can be compared with a mall, where it is essential for the operating company that as many people as possible have free access to do their shopping. The IZA-SC policy, on the other hand, can be compared with the requirement to provide as much space as possible, for instance for the installation of a large machine, such as a generator or a turbine, in a factory. Usually, such large items pass through the gates only once, and are probably taken to pieces, thus the aperture of the gates is less important. Besides, shopping malls have to be controlled sometimes, or even

References see page 734

closed for some reason. On the length scale of microporous structures this can be realized by cations that partly or totally block the access to the internal space.

E. The framework density The IZA-SC has chosen to identify the openness of a structure with a measure of its porosity, that is its internal space, regardless of whether this is accessible to guests, or not. The porosity is expressed by the ratio between the total pore volume and the crystal volume. For convenience, an easily calculable parameter, the so-called framework density F_d, has been introduced. It is defined as the number of tetrahedrally coordinated atoms per 1000 Å^3, **T** atoms/ 1000 Å^3, and is closely related, but not identical, to the porosity. Tetrahedral framework types with F_d values less than about 20–21 **T** atoms/1000 Å^3 are considered "open" by the IZA-SC definition. This range was chosen because it was observed that a plot of the F_d data of the known tetrahedral frameworks exhibits a conspicuous gap at about this value [26], see Fig. 1. The gap is considered to divide a field of open framework structures with $F_d < \sim 21$ **T** atoms/1000 Å^3, from a field of dense frameworks with $F_d > \sim 21$ **T** atoms/1000 Å^3. To the dense framework types belong, among others, the silica modifications quartz, tridymite, cristobalite, and the feldspars. The actual location of the gap depends slightly on the size of the smallest rings present in the framework; it increases from a value of about $F_d = 20$ **T** atoms/1000 Å^3 for the smallest ring size (3, plus some **T** atoms in larger rings), to around 21 **T** atoms/1000 Å^3, for the smallest ring size equal to six.

The smallest F_d values that have been observed for zeolite-like microporous structures are around 12 **T** atoms/1000 Å^3. Faujasite, **FAU**, and zeolite A (**LTA**) have $F_d = 13.3$ and 14.4 **T** atoms/1000 Å^3, respectively. Sodalites are accepted as poroate as their F_d values (16.7 **T** atoms/1000 Å^3 for sodalite proper) place them well into the stability field of open framework structures in the IZA-SC sense. In fact, their F_d value is considerably smaller than that of the well-known, and for practical reasons exceedingly important, ZSM-5 (**MFI**, 18.4 **T** atoms/1000 Å^3).

The essential difference between both frameworks types, with respect to the passage of cations, is the width of the diffusion pathways. **MFI** contains wide 10-ring channels, whereas sodalites contain only 6-ring apertures, as mentioned earlier.

It is not quite clear whether the gap has any physical meaning, or is just feigned by the relative scarcity of data points. It seems that no convincing arguments have been given as to why this gap should exist at all (see below).

A structure field of open framework structures? In addition to the alleged upper limit of the existence of open frameworks, the distribution of the experimental F_d data, as known until now, was interpreted to suggest also the existence of a lower one. Both borderlines taken together define an area in the diagram F_d versus smallest ring size that could be interpreted as a structure field of microporous structures in the sense of the IZA-SC definition, see Fig. 1. If the area were accepted as a structure field, its shape would suggest that the range where open framework structures may possibly be found decreases as the size of the smallest ring increases. To a first approximation, the probability of finding actual structures can be assumed to

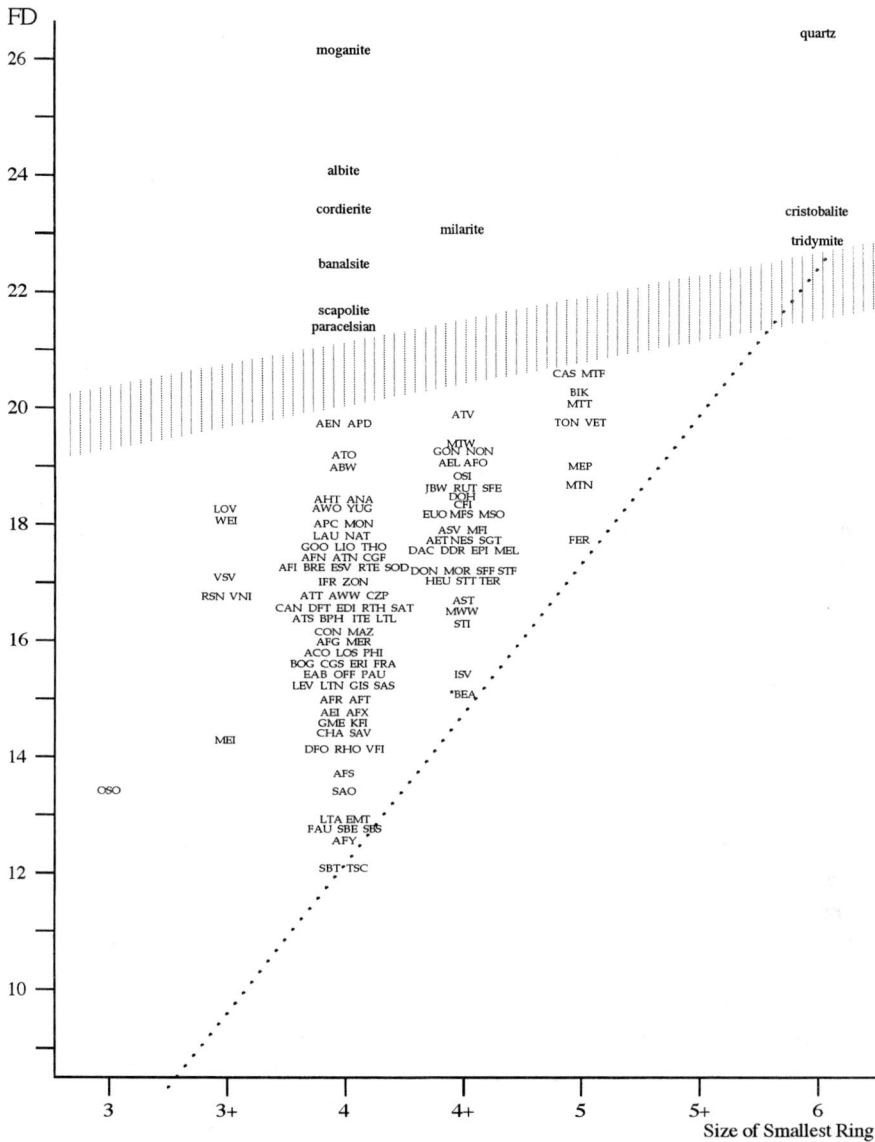

Fig. 1. A plot of the framework density F_d, versus the smallest ring size present, for various microporous structures as indicated by their respective IZA code. For comparison, some dense tetrahedral framework structures are also shown. Reproduced with kind permission of the authors of [18, 19].

References see page 734

depend on the width of the field. This is perhaps the reason why open framework structures with a smallest ring size of six or higher have not been found yet.

The variability of F_d   From the definition of F_d it is clear that its actual value for a given real structure not only depends on its topology, but also on its geometry. The geometry is determined by the chemical composition of the host, for example, larger **T** cations necessarily result in longer **T**–O bonds, and concurrently in larger unit cell volumes, and therefore smaller F_d values. This is strikingly demonstrated by the following example. Two compounds, $Cs_4Zn_4P_4O_{16}$ and $Cs_4Be_4P_4O_{16}$, were reported to have strongly different F_d values of 16.9 and 20.2 **T** atoms/1000 Å^3, respectively, despite the fact that both belong to the **ABW** framework type [27]. The values reflect clearly the differences in the lattice parameters, which in turn are determined by the different cation sizes, Zn^{2+} having about eleven times the volume of Be^{2+} (ionic radii of Zn and Be are 0.60 Å and 0.27 Å, respectively [28]).

A given open framework structure may change its F_d value considerably, even without any change of the chemical composition. For instance, certain flexible open frameworks are able to adjust their unit cell volume by cooperative rotations of the TO_4 tetrahedra, so-called tilts, under the influence of changing temperature or pressure conditions. The effect can be much more pronounced than that which results from the "normal" thermal expansion or compression of nonflexible structures. As an example of such an anomalously strong pressure effect, we report here recent experimental results on tetramethylammonium sodalite [29]. At ambient pressure the structure possesses a unit cell volume of about 720 Å^3, corresponding to a framework density of 16.7 **T** atoms/1000 Å^3. The unit cell volume is reduced to 620 Å^3, that is a change of 14 %, by applying hydrostatic pressure of about 5 GPa. No phase transition, nor pressure-induced amorphization, occurs up to this point; thus the topology remains the same all the way. The compression of the unit cell corresponds to an increase of the framework density up to a value of 19.4 **T** atoms/1000 Å^3.

Changes of the composition and/or concentration of the guest molecules may also strongly influence the framework density. This happens, for instance, if a microporous structure loses water by dehydration, or is ion-exchanged. An example that demonstrates the effect of replacing K^+ by large Cs^+ ions is provided by two compounds of the **ANA** framework type, namely leucite (19.4 **T** atoms/1000 Å^3, $K_{16}Al_{16}Si_{32}O_{96}$, [27]) and pollucite (18.4 **T** atoms/1000 Å^3, $Cs_{16}Al_{16}Si_{32}O_{96}$, [27]). The minimum and maximum values of F_d in the **ANA** framework type are 17.6 and 24.0 **T** atoms/1000 Å^3, respectively, for all kinds of framework composition, but this range is much smaller, 18.4 to 20.6 **T** atoms/1000 Å^3, for aluminosilicates [27].

F. The new IUPAC definition of ordered microporous and mesoporous materials   In addition to diverging definitions of poroate, and different concepts of "openness" as discussed before, there is also some disagreement in the literature concerning semantics. In particular, terms like cage, cavity, pore, void, channel or tunnel are used in a virtually uncontrolled way, thereby considerably adding to the confusion.

Recently, a commission was set up under the auspices of the IUPAC with the task to lay down binding definitions [30].

The commission worked under the following general conditions:

1. The definitions should cover all inorganic, ordered, microporous and meso-porous host structures. The host structures can extend not only in three, but also in two, one, or even zero dimensions.
2. The subset of classical zeolite framework types should be given special emphasis.
3. The minimum diameter of the pores should be set to 2.5 Å.
4. Nonzeolitic chemical composition of the framework should be allowed.
5. Polyhedral building units other than tetrahedra should be allowed.

Statements 1) and 2) express the determination of the commission to extent the applicability of the definition far beyond the classical zeolites. However, the latter are conceded their traditional special status, without doubt in recognition of their long-standing scientific history and actual technical importance. Note that zeolites constitute the only subset having been given an extra status. Condition 4) is less stringent than the mineralogical definition, but conforms to the IZA-SC rules. S, or other nonoxygen host anions, are formally allowed.

Point 5) is interesting, as it extends the class of possibly acceptable materials considerably. It would allow, for example, to include octahedral or mixed octahe-dral/tetrahedral framework structures, at least in principle. This means a liberal-ization with respect to both, the older IZA-SC and the IMA definition. One of the commission's premises was its pragmatic, rather than systematic, determination to put special emphasis on those structural features of microporous materials that are relevant for their use as molecular sieves, ion exchangers, or catalysts. In other words, preferential attention was paid to those properties that control the diffusion of guest species, or set space restrictions to reaction intermediates, at the expense of other structural features or properties.

In this respect, 14 of the 133 poroates framework types contained in the *"Atlas"* take a special position, as their host structures contain no rings larger than 6-rings. This applies to the framework types coded as **AFG, AST, DOH, FRA, LIO, LOS, LTN, MEP, MSO, MTN, NON, RUT, SGT** and **SOD**.

Nomenclature   In its proposal the Commission very successfully clears up several important aspects of the nomenclature. For the convenience of the reader some of the most important definitions will now be cited, but commented on where appropriate.

Topology, connectivity   The topology of a host structure describes the connectivity of its host atoms without reference to chemical composition or observed symmetry (including crystallographic translations).

References see page 734

Comment on connectivity   Connectivity is a less clear criterion than might appear at a first glance. The connection between the atoms in a host framework is normally provided by strong chemical bonds. In some cases the structure may contain additional atoms with the property that they are bonded specifically to some host atoms, but with weaker than normal bond strength. For example, it is known from the as-synthesized form of certain aluminophosphates that 4-fold coordinated Al can make additional, rather weak, bonds with water (or OH) in the pores, thus making it 5-fold, if not 6-fold coordinated. Upon heating the water is expelled from the pores, and Al loses its extra coordination partners, thereby being transformed back to the common tetrahedrally coordinated Al. Strictly speaking, the Al atoms in the pristine and in the calcined form of these compounds have different topologies.

The problem becomes less stringent when the Al cations are considered only in their role as central atoms of the polyhedral building units of the host structure. The connectivity of the polyhedra, and the associated topology, normally do not change upon the described dehydration.

Comment on translational symmetry, commensurate and incommensurate modulations   It is not quite clear why crystallographic translations have been given extra emphasis in this definition. These symmetry operations belong to the space group of the crystal in their own right, in the same way as, for example rotations or reflections. Possibly, it was the authors' intention to admit the formation of commensurate superstructures, which are, indeed, integer multiples of the basic unit cell.

However, this statement cannot be accepted without caveat, because it would exclude by definition any noncrystallographic translation and, concomitantly, any incommensurately modulated framework structure. The topology of the latter would have to be considered distinct from that of the corresponding nonmodulated basic structure. This is clearly untenable because displacively modulated structures are always reducible to the underlying basic structure by setting the amplitude of the modulation wave to zero, and substitutional modulations are reducible to the corresponding basic structures by allowing for appropriate disorder.

Incommensurate modulations occur in many dense tetrahedral frameworks, and have been found in at least one family of microporous structures, namely aluminosilicate and aluminate sodalites [31–34]. We cannot see any reason why the formation and occurrence of incommensurately modulated structures in other microporous framework structures should be forbidden, in principle. The question under which conditions such modulated phases may occur in large pore poroates, has been discussed elsewhere [35].

Symmetry   The highest possible symmetry for a host structure is the symmetry of its topology (topological symmetry). Although the symmetry of a particular material can be as high as the topological symmetry, it is often a subgroup thereof.

Whatever the observed symmetry, however, the number of framework atoms in the unit cell will be an integer multiple of the number in the topological unit cell.

Comment on symmetry   As stated above, the possibility of the formation of in-commensurately modulated frameworks must be accepted. The symmetry of modulated structures, incommensurate as well as commensurate, is best described in terms of so-called superspace groups [36]. Special group–subgroup relationships exist in superspace between the symmetries of the parent phase and the derived commensurately modulated daughter phase, see for example [33].

The situation would become more complex if composite, or misfit, structures were admitted. The formation of these structures is imaginable as the result of an interpenetration of two partial structures. Imagine a situation where each of two partial structures possesses a three-dimensional channel system, with the second partial structure occupying the channels of the first one, and vice versa. Furthermore, let both partial structures have identical *a*- and *b*-, but slightly different *c*-parameters, then the formation of a "misfit poroate" would be conceivable. In contrast to modulated phases misfit compounds lack a basic structure with zero amplitude. This means that their symmetry is not easily reducible to a normal 3D case. In order to cover cases of incommensurately modulated or misfit structures the last statement of the definition ("... integer multiple ...") has to be extended to real numbers.

Distortions   Distortions of the host structure due to the chemical composition of the host and/or to the presence of guest species in the pores are common.

Comment on the causes of distortions   Other possible causes for distortions are changes of temperature, pressure or stress. The distortions may change continuously as a function of such variables, or discontinuously, for example, as the result of a structural phase transition. In particular cases changes driven by an electrical or a magnetic field, or by still other parameters, may also be important.

Zeolite framework type and IZA codes   Microporous materials with an inorganic, 3-dimensional host structure composed of fully linked, corner-sharing tetrahedra and the same host topology constitute a zeolite framework type.

Each confirmed zeolite framework type is assigned a three-letter code ..., and details of these framework types are published in the Atlas of Zeolite Framework Types .... All framework types, including updates between editions of the Atlas are also published on the Internet at http://www.iza-structure.org/databases/. A three-letter code preceded by a hyphen has been assigned in a few cases to zeolite-like framework types that are not fully 4-connected ...

Comment on zeolite framework types   It is worth emphasizing that the codes must not be mistaken for actual materials. Framework types do not depend on composition, distribution of the T-atoms, cell dimensions or symmetry.

Because the condition of the accessibility is crucial here, sodalites and other small aperture frameworks are excluded by this definition. Ultimately these structures would have to be eradicated from the *Atlas*, see above.

References see page 734

It is interesting to note how easily the definition of microporous materials was adapted to a new situation when the IZA-SC decided to include a new, yet unregistered, structure in the *Atlas*, because it was considered sufficiently important for the zeolite community. This happened with the so-called interrupted frameworks that, strictly speaking, are not in keeping with the original definition. These host frameworks contain host anions that belong to only one basic building unit, normally a tetrahedron, instead of two. Currently this is the case for five out of the framework types in the *Atlas*.

Recently, even polytypic materials have found access to the *Atlas*. In these cases only the hypothetical fully ordered end members are reported; the corresponding code is preceded by an asterisk. Actually, there are eight examples given in the catalogue of disordered zeolite structures.

Fully linked, overlinked, and underlinked tetrahedron frameworks   In view of this development one might wonder, how the following discovery will be processed by the IZA-SC. Recently, two new synthetic microporous phases were reported, the tetrahedral framework structures of which do not only contain the familiar 2-fold coordinated oxygen atoms, $O^{[2]}$, but also so-called overbonded atoms, $O^{[3]}$ or $O^{[4]}$, where the numbers in square brackets indicate the coordination number. One of these structures contains additional terminal $O^{[1]}$ atoms. This means this structure is simultaneously over-bonded and interrupted (under-bonded). Both structures contain Zn in tetrahedral coordination, together with P or As.

This remarkable fact is of fundamental importance for a rigorous systematic classification. It was first realised by Liebau when he recast the composition of these compounds in a crystal chemical formulation that takes account of the coordination numbers: $(N_3C_6H_{17})[HZn_4^{[4]}P_3^{[4]}\ O^{[1]}\ O_9^{[2]}\ O_3^{[3]}]$ and $Cs_3[Zn_4^{[4]}As_3^{[4]}\ O_{12}^{[2]}\ O^{[4]}]\cdot 4H_2O$ [37–39]. In view of his discovery he suggested a new general subdivision of framework types built of linked TX_4 tetrahedra. The subdivision is based on the atomic ratio $z = \Sigma X:\Sigma T$, between the host anions **X** and the host cations **T**. Liebau distinguishes three cases, namely 1) $z = 2$, 2) $z < 2$ 3) $z > 2$, and emphasizes that from the global chemical composition, and without further structural information, only the *average* coordination number can be deduced, because the possibility of mixed coordination numbers cannot be excluded. Taking this into account he called 1) on average fully linked, 2) on average overlinked, and 3) on average underlinked tetrahedron frameworks.

By considering possible host cations and anions, and their respective crystal chemical character, he argues that tetrahedron frameworks consisting of small, highly charged **T** atoms, such as Al, Si and P, are unlikely to contain overbonded $O^{[3]}$ and $O^{[4]}$ atoms, as these connections are energetically unfavorable due to the strong repulsion between the **T** atoms. However, the larger the **T** atoms and the lower their valence (for example, Zn, Fe, Mn), and the larger the **X** atoms and the lower their (negative) valence (for example, S), the more likely becomes the existence of $X^{[3]}$ and $X^{[4]}$ species. This is in agreement with the compositions and structures of the quoted zincophosphate and -arsenate. More examples can be expected in the future, not least from sulfur- or nitrogen-containing frameworks.

Building units   The host structure can be constructed by linking basic building units (usually coordination polyhedra sharing corners, edges, or faces, but sometimes single atoms). In the case of zeolite structures, these basic building units (BBU) are tetrahedra, where the central atom ($_{ce}$H) is typically Si or Al, and the peripheral atoms ($_{pe}$H) are O. In the vast majority of microporous and mesoporous materials with inorganic hosts, the central atoms are cations and the peripheral atoms anions (for example, $[BO_3]$, $[SiO_4]$, $[AlO_4]$, $[PO_4]$, $[SnS_5]$, $[MnO_6]$). In a few cases, however, an anion is surrounded by cations.

It is sometimes useful to combine BBUs to construct a larger composite building unit (CBU) that is characteristic of the topology. For example, single rings, single chains or polyhedral building units, composed of a finite or infinite number of BBUs can be chosen. In some cases, more complex CBUs ... might be appropriate ...

Windows, cages, cavities and channels   The n-rings defining the faces of a polyhedral pore are called windows.

A polyhedral pore whose windows are too narrow to be penetrated by guest species larger than H_2O is called a cage. For oxides ($_{pe}$H = O), the limiting ring size is considered to be $n = 6$. For example, the $[4^6 6^8]$ polyhedron (sodalite cage) in zeolite A is a cage.

A polyhedral pore, which has at least one face defined by a ring large enough to be penetrated by guest species, but which is not infinitely extended (that is, not a channel), is called a cavity. For example, the $[4^{12} 6^8 8^6]$ polyhedron in zeolite A is a cavity.

A pore that is infinitely extended in one dimension and is large enough to allow guest species to diffuse along its length is called a channel. Channels can intersect to form 2- or 3-dimensional channel systems. For example, the channels in zeolite A intersect to form a 3-dimensional channel system.

The effective width of a channel is a fundamental characteristic of a microporous or mesoporous material that describes the accessibility of the pore system to guest species (that is, the "bottleneck"). It is generally defined in terms of either the smallest *n*-ring (topological description) or the smallest free aperture (metrical description) along the dimension of infinite extension. For example, the 3-dimensional channel system in zeolite A has 8-ring pore openings (topological description) in all three directions. The free diameter takes into account the size of the $_{pe}$H atoms (...). The pore width along the channel, topologically as well as metrically, can be constant or it may undulate (for example, if the channel consists of a series of cavities). In zeolite A, for example, the section of the channel in the center of the cavity is wider than it is at the 8-ring entrance to the cavity. However, it is the 8-ring that defines the size of guest species that can diffuse through the material (that is, the effective channel width). For channels defined by helices, by rings not perpendicular to the channel or by unusually distorted rings, the effective width must be given metrically.

References see page 734

Comment on building units   Note that CBUs are not identical with the so-called secondary building units, SBUs, of the *Atlas*. In any case, BBUs, CBUs and SBUs are only working models that have been invented because such models facilitate the description of the architecture of framework structures. The electronic version of the *Atlas* offers the opportunity to follow step-by-step the assembly of a number of framework types by joining such CBUs. An impressive example is that of the very complex zeolite framework of tschörtnerite, **TSC**:

http://www.iza-sc.ethz.ch/IZA-SC/Atlas/data/models/TSC_assembling.html

It must be stressed that BBUs, CBUs and SBUs should not be mistaken as real species that necessarily occur in solution, or during the crystal growth of a microporous material. Similarly, the described assembly of these units must not correspond to reality, but is a very useful, though artificial, tool to support the human imagination.

Note that the designation of the $[4^6 6^8]$ polyhedron, found in several frameworks, as sodalite cage or β-cage, is in agreement with the above definition. On the contrary, the central $[4^{12} 6^8 8^6]$ polyhedron in zeolite A should not be called "α-cage", as is often done. "α-cavity" would be the proper name.

In addition to these clear and useful definitions of terms that were often confused in the previous literature, the authors propose a versatile crystal chemical formula that does not only allow to distinguish between the various structural and chemical components of a microporous or mesoporous structure, but which is also highly flexible with respect to what information is relevant or irrelevant for the particular case under consideration. Thus, the chemical composition of the host and guests, the structures of both, host and pores, the symmetry, or the IZA code can be emphasized if needed, or left out if of no interest for the actual discussion, see below.

G. Liebau's definition   The drawing up of the new IUPAC recommendations was controlled by the commitment of the commission to satisfy the strong interest of the community in application purposes, the recommendations were therefore heavily biased towards zeolites. The IUPAC recommendations are largely based on a more general and more systematic classification scheme proposed by Liebau [5]. This scheme accounts for the chemical composition, classical crystal chemical parameters, as well as for information specific for microporous structures, for example on CBUs and on the characteristics of the pores, in very much the same way as the IUPAC rules.

Equivalences and nonequivalences between the IUPAC recommendations and Liebau's scheme   While basic structural terms and also the accessibility are defined in practically the same way, there is an important difference, as Liebau definitely does not discriminate against porous materials with nonaccessible pores. In his classification materials that have cages as the only pores are included as so-called clathrates. He suggests a pragmatic argument for his decision by noticing that clathrates are useful for practical purposes in their own right. For instance, some inorganic clathrates are capable of encapsulating, and thus rendering harmless, dangerous chemical species, and organic clathrates are used for the controlled sustained release of encapsulated drugs.

Information on accessibility: poroates, zeoates and clathrates As distinct from clathrates, microporous structures with accessible pores are called zeoates. Both, clathrates and zeoates together, form the set of poroates. Liebau's scheme allows essentially any coordination number of the central host atom, that is, it allows polyhedral BBUs other than tetrahedra.

The following examples illustrate some aspects of the nomenclature. Poroates-[6] comprise all clathrates and zeoates having coordination number six of the central host atoms, thus including all octahedral frameworks. Zeoates-[4] designate all poroates with accessible pores and coordination number four of the central host atoms. One nice feature, the full power of which may become operational only in the future, is that the nomenclature also allows mixed coordination numbers of the central host atoms. Thus, the designation poroates-[4, 6] indicates mixed 4- and 6-coordinated host cations, including mixed tetrahedral and octahedral framework structures.

An important feature for possible applications is the dimensionality of the pore system. This is used as a classifier, with values of 0 for clathrates and 1, 2 or 3 for zeoates. Less important, at least for the moment, is the possibility to specify the dimensionality of the host structure.

Information on the host composition: porosils, zeolites, clathrals, etc. For the actually most frequently encountered oxidic poroates-[4] with three-dimensional host frameworks Liebau proposes a nomenclature that allows the identification of the chemical composition of the host from the name's ending. Thus, porosils and poroals contain exclusively Si or Al, respectively, as central host atoms, whereas porolites contain both, Al and Si. With the prefixes "clathra" and "zeo" distinguishing between clathrates and zeoates, one arrives at the familiar zeolites as zeoates-[4] with three-dimensional host frameworks and Si and Al as central host atoms. It seems that the Al:Si ratio is not considered in this formulation. The well-known clathrasils contain only Si as host cations, and their pores are unaccessible. The structural family that has become known as aluminate sodalites, belongs to the clathrals. Zeolites with Si and Al are easily distinguished from other zeoates with different framework composition, for example zeo-APOs contain Al and P as host cations, zeo-SAPOs have additional Si, and so on.

Note that Liebau's classification is designed to allow also for host anions other than O, for example S.

4.2.1.2.3 **Databases**
After many decades of intensive research and development performed in the field of microporous structures by thousands of scientists such a wealth of structural information has become available that probably no single person will be able to

References see page 734

keep track of the already existing data and the ever-increasing flow of new information. Fortunately, recently databases have become available that facilitate the task considerably. The basic structural data, such as unit cell dimensions, space groups, atomic positions, derived parameters, as for example bond lengths, bond angles, etc., references to the literature, and much more information, are transformed into structured data. Special search tools allow strategic use of the data, thus allowing "metadata" or "data on data" to be determined. These allow, at least in principle, an understanding on a higher, or metalevel.

The following three databases will be described and given brief commentaries:

1. Landolt–Börnstein, [27, 40], henceforth abbreviated LBA and LBB, respectively.
2. The *Atlas* [18, 19]
3. The compilation of extra framework sites [41].

Landolt–Börnstein LBA [40] and LBB [27] are the products of a long-standing cooperation of the three authors, Baur, Fischer and Smith. LBA contains basic information on the topology of zeolites and related frameworks, based on the analysis of topological features of four-connected, 3D nets. Thus, the contents of this issue go far beyond the frameworks that are accepted under the IMA or IZA-SC definitions of open, tetrahedral framework structures as given before, but other than tetrahedral frameworks are not considered. The sometimes rather succinct introductions into basic mathematical and other concepts related to the topic, and the numerous useful references to the literature make this volume a highly valuable reference work. The largest part of the volume is devoted to the enumeration and presentation of the many different building units that result from disassembling the nets or frameworks in various ways. The building units are classified as polyhedra, 2D nets, or one-dimensional units, such as chains or columns. They are very useful for the description, discussion, prediction and understanding of the static architecture, and of distortion modes, of existing and hypothetical framework structures. Topologies conforming with an IZA-SC code are thoroughly analyzed with respect to their topology, in particular concerning the occurrence of rings and of the various one-, two-, and 3D building blocks. An aesthetically pleasing feature of this volume is the graphical representation of the various building units as line drawings, especially if they are given in perspective, as in the case of polyhedral and of one-dimensional units.

LBB is planned to be published in a total of four volumes. Only those framework types that have been accepted by the IZA-SC, and given a three-letter code, will be included. Only the first volume has been published at the time of writing, it covers the IZA-SC codes from **ABW** to **CZP**.

The authors strive to present for each framework type, and its representatives, a very extended and detailed set of information. This comprises the topology of the framework type, including perspective line drawings and illustrative tetrahedra plots of the framework, and skeleton models to emphasize the connectivity of the **T** atoms. An example is shown in Fig. 2.

A unique, highly welcome, aspect of LBB is the great care and importance that it

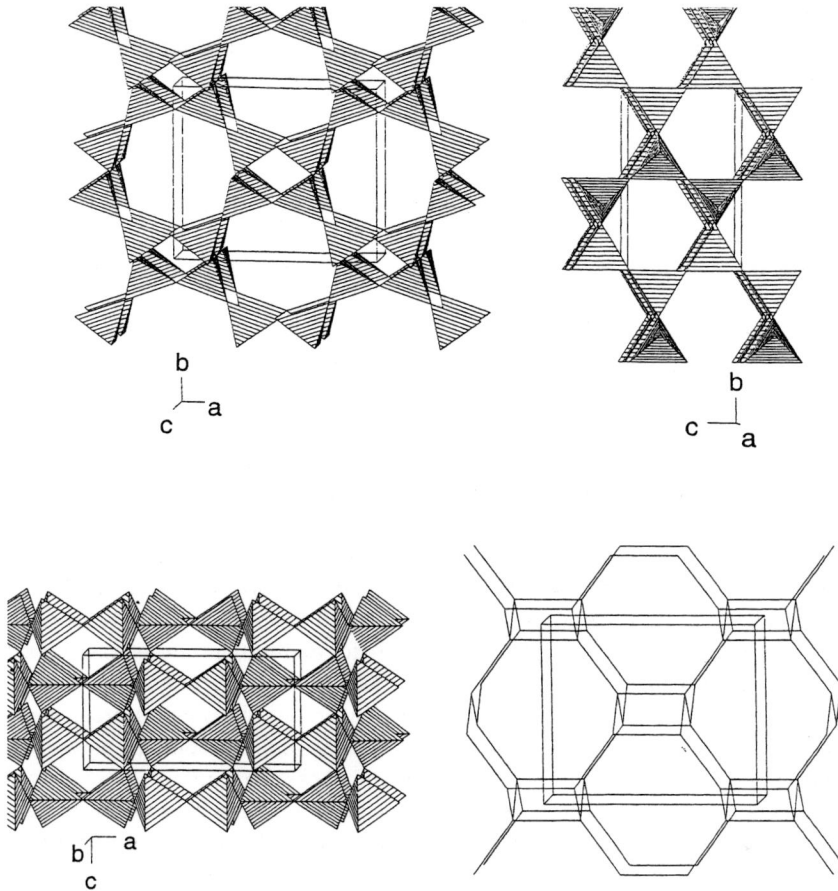

Fig. 2. Projections of the idealized **ABW** structure and a
corresponding skeleton model; from [27].

attaches to symmetry information. As stated earlier, the spacegroup of an actual
framework structure is often a subgroup of the topological spacegroup. This is so
because the real structure can be thought of to be derived from the ideal one by
only small atomic displacements without breaking bonds. The topological and the
real symmetry can be related by sequences of group–subgroup relationships, and
shown diagrammatically in a so-called Bärnighausen tree. The atomic site symmetries change correspondingly (Wyckoff splitting). For the practitioner it is a very
valuable tool that LBB gives these symmetry relationships explicitly.

Being a predominantly structural database, the experimental chemical and crystallographic data of structures belonging to a given zeolite framework type take a

References see page 734

prominent place in LBB. They are accompanied by additional important information, such as framework densities, etc., and, of course, complemented by the references to the original literature.

Already with a minimum of effort interesting "metadata" can be obtained from a search in LBB. For example, the tolerance of certain framework types towards the chemical composition is certainly somehow related with the sheer number of compounds belonging to this type. For example, for the framework type **ABW** not less than 86 entries are listed, whereas there is only one for **AHT**. It is perhaps typical of the dynamics of the actual activity in research on poroates that seven more compounds of the **ABW** framework type have become known since the first volume went to press.

For each zeolite framework type standardized structural data, namely atomic positions and selected interatomic distances, are given for one carefully chosen prototype in each subgroup of the topological spacegroup. Atomic coordinates and selected interatomic distances complete the information, as do perspective views and skeletal drawings of the framework.

A highly useful feature is the analysis of the chemical elements having been found to occur in a particular framework type, and histograms showing the distribution functions of framework densities, T–O distances, individual and mean T–O–T angles, see Fig. 3. An information block on the flexibility of the framework and on the apertures is very important, not only for application purposes.

Since the appearance of this first volume of LBB, two new framework types, **BEC** and **BCT**, have been accepted by the IZA-SC. In proper alphabetical order these entries would have had their place in this first volume of LBB. However, *a posteriori* adding of new items is a difficult matter in bound books. The authors have decided that these, and future, entries, as well as any possible errata, will probably be published in the last volume of LBB. The publication date of this volume is not known yet.

The LBB is a highly valuable work, full of interesting details and close to complete. Its completeness allows the determination of metadata with the potential of exciting new insights. The restricted flexibility of printed media, in general, and of bound books in particular, poses a certain problem. This point can partly be overcome by updates on CD-ROMs. A major problem with LBA and LBB is their rather prohibitive price that makes it difficult for many libraries in academia to purchase these otherwise extremely valuable works.

The *Atlas* A rather different concept is that of the *Atlas* [18, 19]. Originally being a printed version it grew steadily over the years until its present 5th edition. In the last few years this product was complemented by an electronic version offering all the virtues of electronically maintained databases: It is easily updated and maintained, it is open, and it is searchable with a suitable software. Links allow easy access to other databases, thus allowing data mining, for example by combining structural parameters with certain physical properties. Hyperlinks connect with definitions, help functions, more detailed or additional information, or with other databases, or electronic reference works. A wonderful feature are the graphical

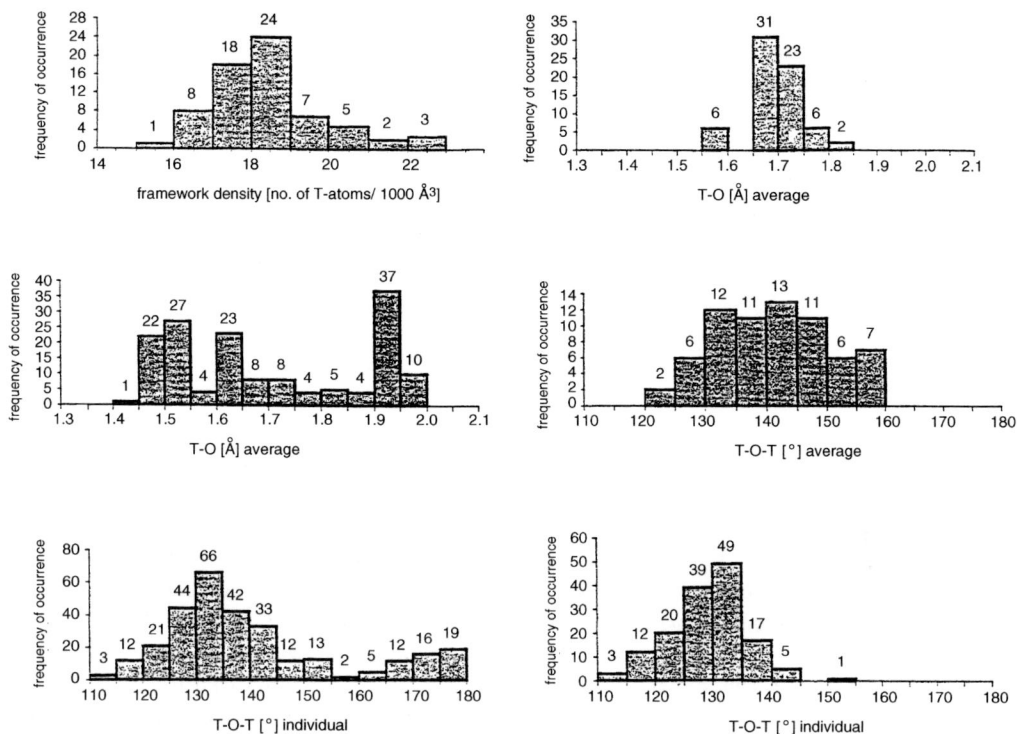

Fig. 3. Frequency distributions for the ABW framework type. From top left to bottom right: (i) framework density; (ii) average T–O and T–F distances; (iii) mean individual T–O and T–F distances; (iv) average T–O–T and T–F–T angles; (v) individual T–O–T and T–F–T angles; (vi) individual T–O3–T and T–F3–T angles, emphasizing the special role of the O3 (F3) atoms; from [27].

representations of the framework types in various perspective views or projections, emphasizing the most important channels and their apertures, sometimes with space-filling models, the assembly of a framework from smaller building blocks helps in grasping the sometimes very complex framework architectures.

A powerful tool for students, as well as for experienced researchers, is the Collection of Simulated XRD Powder Patterns for Zeolites. Obviously, the searchability of the *Atlas* is still in its infancy, but already at the present state it is a very useful tool, and was also used in the preparation of this contribution.

Compared with the LBB, the database of the *Atlas* is much smaller and less detailed. Therefore, it is, for example, not possible to derive distribution functions with the same degree of precision as with LBB. Furthermore, the symmetry aspects receive much less attention than in LBB.

There is an irrefutable argument in favor of the *Atlas*. It is available free of charge in the Internet, which makes it available to virtually everybody.

References see page 734

The Atlas of extra-framework sites The Atlas of extra-framework sites [41] is a compilation of published cation sites. Some of the data should be regarded and used with due caution. This is particularly true for those cations that are reported to reside on high symmetry sites with low occupancy. The problem of such dubious cation positions has been discussed elsewhere [42].

Future trends of databases Future trends are probably towards electronically maintained databases. It would be highly welcome if the information content contained in all three mentioned databases on poroates and other microporous structures could be extended. In particular, it would be desirable if, in the future, information on the distortion state of the various frameworks under various conditions could also be listed.

4.2.1.3
Crystal Chemistry and some Remarks on the Mineral Physics of Poroates

4.2.1.3.1 Basic Properties of Tetrahedral Frameworks
Probably the first materials that were found to possess microporous structures were the zeolite minerals [7]. Once X-ray diffraction was discovered, and structure determination became possible, it could be established that zeolites belong to the family of tecto-silicates, with 3D frameworks built from TO_4 tetrahedra. The chemical pseudo-formula TO_4 indicates that small, highly charged cations T are tetrahedrally coordinated by oxygen. In noninterrupted, 3D framework structures the TO_4 tetrahedra share common oxygens at all their corners.

In nature the most common T cations are Si^{4+} and Al^{3+}. If all the T-sites are occupied by Si^{4+}, the chemical formula of the corresponding structure writes $[SiO_2]$, with the understanding that the square brackets enclose all atoms that belong to the tetrahedron framework. $[SiO_2]$ represents therefore a modification of silica having a framework structure. We will see below that many different, more or less open, frameworks of silica composition exist.

Clearly, the formula $[SiO_2]$ does not tell us anything about how the architecture of the framework looks like. We understand "architecture" to mean both the idealized framework type, that is the general way in which the TO_4 tetrahedra are linked to each other, that is its topology, and its actual, measurable, geometry. The topology can be expressed by the specification of, for example, topological ring sizes, coordination sequences, or loop configurations (see *Atlas*, [18, 19]). The description of the actual geometry requires data on, for example, interatomic distances, intratetrahedral O–T–O angles, or intertetrahedral angles T–O–T (see LBB, [27]).

While the interatomic distances and intratetrahedral angles are more or less constant for all tetrahedra frameworks of a given composition, say $[SiO_2]$, the corresponding intertetrahedral T–O–T angles are very flexible, spanning the whole range from about 115° to 180°. Note that T–O–T angles of 180° are still a matter of debate. In many cases the high values of thermal parameters can be understood when the calculated value of 180° is only a time-, or space-averaged value, the true value at any instance being smaller. These relationships have been confirmed by a compilation of the corresponding data for a number of zeolite type structures, see, for example, Fig. 3. Indeed, the narrow distribution functions of $T-O$ bond lengths and $O-T-O$ bond angles contrast sharply with the broad distribution function of the angles $T-O-T$ [43–47]. The flexibility of the angles $T-O-T$ is a very important property, because it not only enables the tetrahedra to be linked in virtually infinitely many topologically distinct patterns, but also allows the frameworks to adapt their geometry in response to given conditions. The adaptation is managed by conformational changes, that is changes of the mutual orientation of the TO_4 tetrahedra in response to variable chemical composition, temperature, pressure, or additional chemical species. To a much lesser extent the shape of the tetrahedra may also be affected [48].

In general, the chemical bonds between the atoms in the host framework are strong, and of ionic-covalent nature. The covalent component accounts for the marked rigidity of the TO_4 tetrahedra.

Sometimes it is useful to identify the TO_4 tetrahedra with quasi-rigid bodies, and to visualize the function of the common oxygens as that of constrained hinge joints. This allows us to define specific distortion parameters, for example the so-called tilt angles, which describe the conformational state of a framework. In particularly simple cases the specification of just one tilt angle is sufficient to describe the conformation of the whole framework. An overview on structural distortions and modulations occurring in microporous solids has been published recently [35].

4.2.1.3.2 Basic Properties of Guests and of Host–Guest Interactions

The host frameworks of crystalline microporous solids contain periodic patterns of micropores having various sizes and shapes. These may, or may not, be occupied by additional chemical species, namely cations A^{m+}, water or other neutral molecules M, or, less common, anions X^{n-}. $m+$ and $n-$ express the respective charge. These additional species do not belong to the host framework, hence they are sometimes referred to as "extra-framework species".

In general, these extra species interact with the framework atoms only via (weak) Coulomb forces, hydrogen bonds, or van-der-Waals interactions. Normally, covalent bonds do not occur. Therefore, the interactions between the extra-framework species and the framework are weak, at least when compared with the strong bonds in the framework, and it costs only little energy for them to be broken. Often this happens already under ambient conditions. Once the bonds are broken, and suitable geometry provided, the extra species can easily migrate in the pore system of the framework, or even leave and re-enter the crystal. Most remarkably, this happens virtually with only minor effects on the structure of the latter. Therefore it has become customary to name the extra species "guests", the framework "host", and the whole composite a "host–guest" compound.

In contrast to the weak attractive forces, there are very strong, short-ranged, repulsive, interactions between guests and the host that put severe geometrical con-

References see page 734

straints onto the guests. This means their sizes and shapes have to fit into the pores of the host. A severe selectivity between different guests is the result and forms the basis for a plethora of unique properties found in host–guest compounds, and for ensuing applications.

4.2.1.3.3 A Functional Chemical Formula

Given the distinct functionality of the various chemical species in such a host–guest compound, it is reasonable to express this fact by introducing a functionalized chemical formula. Such a formula was recently introduced. It expresses the structural function of the different species in the following way [5, 30]. First, guest and host species are clearly distinguished by writing the corresponding chemical symbols with their stoichiometric coefficients inside different typographical symbols. Vertical bars and square brackets enclose the guests and the host, respectively. The guest species, namely cations A, anions X and neutral molecules M, are listed together with their stoichiometric coefficients a, x, m, in the order: $|A_a X_x M_m|_n$. The enumeration of the host species starts with optional information on interstitial species $(_iA, _iX, _iM)$, where A, X, M have the same meaning as before, and the subindex i indicates "interstitial". Interstitial species are by definition restricted to voids having a diameter of less than 2.5 Å, that is they are not able to migrate. They are not essential for the host function of the framework, and are often absent.

The central and peripheral host atoms, $_{ce}H$ and $_{pe}H$, traditionally the T and O atoms, are, of course, essential for the framework. The complete framework part of the formula then reads $[_iA_a \, _iX_x \, _iM_m \, _{ce}H_c \, _{pe}H_p]_n$. Information on oxidation states, coordination numbers, or other information, can be added, if desired. Very important is the information on the IZA code of the zeolite framework type. If assigned, it defines the topology of the framework. In order to avoid information overflow, it is strongly recommended to restrict the formula to a strict minimum of relevant data.

Following these rules the classical example of zeolite A reads $|Na_{12}(H_2O)_{27}|[Al_{12}Si_{12}O_{48}]$-**LTA**, and the formula of template-free silica sodalite is $[SiO_2]$-**SOD**.

4.2.1.3.4 Silica Frameworks

The last example is a representative of a whole family of open silica frameworks, all having the same composition, $[SiO_2]$, but different topologies. In its as-synthesized form silica sodalite accommodates molecules M, for example trioxane, in the pores of the framework. The formula then reads $|(C_3H_6O_3)_2|[Si_{12}O_{24}]$-**SOD**. The framework of silica sodalite contains only one pore type. In other open framework topologies usually more than one pore type of different geometries exists. For only one type of guest molecule M, the question will then arise which of the pores will accommodate this molecule. Clearly, the answer depends on the sizes and shapes of the different pores, and of M. On the other hand, if more than one kind of molecule M is present, then a selection between these molecules may become possible.

The organic guests in the pores of an open silica framework can be destroyed by careful heating of the material in an oxidizing atmosphere. After this so-called calcination, the empty silica framework $[SiO_2]$ is left behind. The calcination proceeds fairly easily in accessible silica frameworks, but in $[SiO_2]$-**SOD** the process is more difficult to perform, because of the narrow aperture of the sodalite windows that impede the removal of the products of the burning process [49].

Often microporous silica frameworks are thermally very stable, and resistant to harsh chemical environments. These properties make them attractive for technical applications, for example as catalysts, or as hydrophobic selective absorbents.

From a chemist's point of view, the composition of silica is simple. Structurally, however, this is less so, as this primitive compound crystallizes in many different structures, $[SiO_2]$. A search in the *Atlas* established that of the 133 framework types listed, at least 28 form with the silica composition, namely **ASV, CFI, DDR, DOH, DON, FER, GON, IFR, ISV, ITE, MEL, MEP, MFI, MTF, MTN, MWW, NON, RTE, RUT, SFE, SFF, SGT, STF, STT, TON, VET, AFI** and, obviously, **SOD**. Most of them can exist also with other, nonsilica, framework compositions. In addition, $[SiO_2]$ crystallizes with the structures of the denser polymorphs quartz, cristobalite, tridymite, mogenite, keatite, coesite and stishovite.

A. De-alumination This already surprisingly high number would be still higher if several highly siliceous frameworks were included. In their frameworks often less than a few per cent of Si atoms are replaced by Al, or sometimes other cations. As an example we mention the highly siliceous zeolite Y of **FAU** topology. It cannot be prepared directly, as the synthesis yields only products with a relatively low Si/Al ratio, that is high Al content. In order to reduce the amount of Al, it must be removed from the framework. This is achieved by so-called de-alumination. This after-synthesis treatment employs processes like steaming, acid leaching, or high-temperature treatment with gaseous, Si-containing compounds, for example $SiCl_4$.

An analytical problem concerns both scientists and process engineers who want to understand or control these processes. The Al that is removed from the framework in the process of de-alumination is often deposited as 6-fold coordinated Al, either on the external surface of the crystals, or in the micropores of the framework. This may result in problems if only bulk chemical analysis is available to determine the degree of de-alumination. Therefore, these methods have to be complemented by more sophisticated methods that are able to distinguish between 4-fold and 6-fold coordinated Al, that is between intra- and extra-framework Al. Very powerful tools in this respect are the various methods of solid-state nuclear magnetic resonance spectroscopy.

B. Al and catalytic processes The small amount of Al replacing Si in the TO_4 tetrahedra of highly siliceous zeolites is indispensable for their catalytic properties. Each Al^{3+} replacing Si^{4+} represents a missing positive charge. The necessary

References see page 734

charge compensation is often provided by protons. The latter being readily dissociated off, the protonated forms of high-silica zeolites act as solid acids. This is the chemical basis for their use in many technical applications, the micropores providing the structural basis. An example is the synthetic zeolite ZSM-5 that is used for the cracking of crude oil.

C. Substituting Al^{3+} for Si^{4+}   The ability of the framework of ZSM-5, and of other tetrahedral frameworks, to accept Al^{3+} on Si^{4+} sites is worth a moment of reflection, given that the charges and the ionic radii of these two cations are quite different. Taking the Shannon ionic radii 0.26 Å and 0.39 Å of tetrahedrally coordinated Si^{4+} and Al^{3+} [28], respectively, the volume of the Al^{3+} cation is more than three times that of Si^{4+}. Therefore, an AlO_4 tetrahedron exerts local stress onto the adjacent framework that responds by elastic deformation.

It is interesting to get an idea how far such a local disturbance will extend out into the framework. Experimentally, this is not easy to determine with only a few, furthermore randomly distributed, Al atoms, as in the case of ZSM-5. Fortunately, as a byproduct of a quantum-mechanical computational investigation of the reorientation potentials of tetramethylammonium cations as guest-molecules in the pores of tetramethylammonium sodalite, $|(N(CH_3)_4)_2|[Al_2Si_{10}O_{24}]$-**SOD** [29], data could be obtained that give a clue to the order of magnitude of the effect. The work was facilitated very much by the fact that the arrangement of the AlO_4 tetrahedra in the otherwise siliceous framework is periodic, in contrast to the case of ZSM-5.

The structure contains three symmetrically independent and topologically different **T** atoms, Si_0, Si_I and Al. The AlO_4 tetrahedron is linked at all four corners to Si_IO_4 tetrahedra, which are aditionally linked to one Si_0O_4 tetrahedron and two more Si_IO_4 tetrahedra. The Si_0O_4 tetrahedra have no direct contact with AlO_4, and are therefore considered to represent the silica matrix. Si_IO_4, as an intermediate, links Si_0O_4 on all four corners to AlO_4 tetrahedra. Similarly to Si_0O_4, the AlO_4 share corners with Si_IO_4 tetrahedra only.

The analysis of the geometrically optimized structure revealed that the framework responds to the stress mainly by two different mechanisms, namely local twisting of the AlO_4 tetrahedra, and distortion of the Si_IO_4 tetrahedra. The twisting of the AlO_4 tetrahedra reduces their size effect on the framework. The Si_IO_4 tetrahedra are most heavily affected by strong bond length and angular distortion. The Si_0O_4 tetrahedra are unaffected. As an underlying feature, all three kinds of TO_4 tetrahedra undergo an expected tetragonal tetrahedron distortion, induced by the topology of the **SOD** framework type [50]. From these findings it can be concluded that the size-induced strain drops to zero over a length scale of only three TO_4 tetrahedra, or half the edge length of the cubic unit cell, that is about 4.5 Å. The phenomenon is cooperative because of the 3D linking of the quasi-rigid TO_4 tetrahedra.

The relaxation mechanism in the framework of ZSM-5, or in other zeolite framework types, is certainly different from that in **SOD**, because of the different way of linking the tetrahedra. In addition, $|(N(CH_3)_4)_2|[Al_2Si_{10}O_{24}]$-**SOD** represents a special case because the AlO_4 tetrahedra are periodically distributed, in contrast to most other highly siliceous poroates. However, despite these differences, we cannot see any reason why the situation in other tetrahedral framework types should be fundamentally different.

Aluminosilicate frameworks   Most of the naturally occurring microporous structures belong to the class of 1:1 aluminosilicates. Their frameworks are electrically charged, for example $[SiAlO_4]^{1-}$, because for each atom Si^{4+} replaced by Al^{3+} there is obviously one positive charge missing. The necessary compensation is achieved by incorporation of guest cations into the pores, or smaller interstitial voids, of the host. The guest cations are usually of low charge, a very common guest being Na^+. One can imagine a hypothetical, coupled substitution of the kind $Si^{4+} + \blacksquare \leftrightarrow Na^+ + Al^{3+}$, where the symbol $\blacksquare$ represents an empty position in the pores. This hypothetical reaction does not change the topology of the host framework during the reaction. It should be noted, though, that the geometry might be affected, as mentioned before. In a vivid description of this kind of hypothetical substitution, such aluminosilicates are sometimes called "stuffed derivatives" of the corresponding "empty" silica modifications with the same framework topology. Stuffed derivatives are usually dense structures.

Clearly, the charge compensating cations have to fit into the pores. The substitution is therefore constrained by at least two conditions, one set by the sizes, the other one by the charges of the guest species. Because of its short-ranged nature the size constraint prevails over that of the charge. This is demonstrated by the following example: In many cases Na^+ can be quite easily replaced by the almost equally sized, but bivalent Ca^{2+}. On the other side, the replacement of Na^+ by monovalent, but significantly bigger cations, such as K^+, or even Cs^+, is possible only in very open and/or adaptable frameworks. For instance, [AlSi]-**SOD** frameworks easily accommodate Na^+ or Ca^{2+} equally well, whereas K^+ is accepted only with difficulties at higher temperatures; Cs^+ is virtually never accepted ($r(Na^+) = 1.02\,\text{Å}$, $r(Ca^{2+}) = 1.00\,\text{Å}$, $r(K^+) = 1.38\,\text{Å}$, $r(Cs^+) = 1.67\,\text{Å}$, all atoms 6-fold coordinated [28]).

By way of contrast, the highly adaptable zeolite rho accepts simultaneously guests of very different size, as shown by the following formula: $|(Na^+, Cs^+)_{12} (H_2O)_{44}|$ $[Al_{12}Si_{36} O_{96}]$-**RHO** [51, 52].

Note, that in the last example only one out of four T sites is occupied by Al, the other three by Si, that is Al:Si $= 0.333$. It is often, but wrongly, believed that the Al:Si ratio in silica or aluminosilicate frameworks is restricted to values ≤ 1, see below. Of course, the number of charge compensating guest cations must match the negative charges of the framework, provided that no guest anions require an extra contribution.

Loewenstein's rule   Replacement of Na^+ by Ca^{2+} deserves special attention. Each coupled substitution of the type $Na^+ + Si^{4+} + Al^{3+} \leftrightarrow Ca^{2+} + Al^{3+} + Al^{3+}$ results in two AlO_4 tetrahedra in the framework. Loewenstein [53] first reported that this

References see page 734

has some consequences. By evaluating the structural data on aluminosilicates available at that time, he was able to formulate what later became known, and famous, as the Al avoidance, or Loewenstein's, rule. He noticed that any two AlO_4 tetrahedra in aluminosilicate frameworks have the tendency to avoid direct contact with each other. As a corollary this means alternating AlO_4 and SiO_4 tetrahedra for 1:1 aluminosilicates.

By X-ray diffraction this rule is not very easy to prove, because Si^{4+} and Al^{3+} scatter X-rays with virtually the same power, making them difficult to distinguish. Bond length considerations, neutron diffraction or solid-state nuclear magnetic resonance are suitable methods to overcome this problem.

A theoretical study on the **ANA** framework of leucite lent theoretical support to Loewenstein's rule. By empirical computational methods it could be established that the number of Al–O–Al linkages and the relative free energy of the system are linearly correlated in such a way that the latter is the higher, the more such linkages are present [54].

The computer experiments thus confirmed Loewenstein's claim that it is favorable for a 1:1 aluminosilicate framework to avoid neighboring AlO_4 tetrahedra. In the zeolite community Loewenstein's rule has become a highly popular guiding principle. Unfortunately, some uncritical adepts have uncautiously interpreted the rule as an irrefutable natural law, despite clear evidences that Al–O–Al linkages in aluminosilicate frameworks, albeit slightly unfavorable, are not at all forbidden. For instance, such linkages can be stabilized entropically at high temperatures, or exist metastably at ambient or lower temperatures. Several 1:1 aluminosilicates show substitutional disorder of the **T** atoms in their respective high-temperature phase. Clearly, the disorder requires that Al–O–Al linkages have a certain probability > 0 to exist. Upon cooling the disorder can be frozen in and Al–O–Al linkages can thus persist metastably at room temperature for long, even geological, times. Al–O–Al linkages can even be thermodynamically stable, as shown further below.

Overinterpretation of Loewenstein's rule can be rather misguided, and may possibly lead to erroneous conclusions. It may even hamper scientific progress. For instance, large parts of the zeolite community obviously take it for granted that the existence of aluminosilicates with an excess of tetrahedrally coordinated Al over Si, that is Al:Si > 1 is forbidden, and argue that this is so because of Loewenstein's rule. In fact, the statement is not absolutely true, as shown by counterexamples. For instance, in the mineral bicchulite, $|Ca_8(OH)_8|[Al_8Si_4O_{24}]$-**SOD**, the ratio Al:Si $= 2$ [55], and in a number of synthetic hauynites Al:Si ratios of the order 3–5 have been reported [56]. In both cases the **T** atoms are reported to be disordered.

It is remarkable that until now the occurrence of counterexamples seems to be restricted to **SOD** framework types. At present, we are not able to exclude or to confirm that aluminosilicates with Al:Si > 1 may occur in other framework types as well. However, the next paragraph will show that the **SOD** framework type seems to be particularly favorable for Al–O–Al linkages.

Aluminate sodalites   The special case of the family of aluminate sodalites has been intensively discussed in the literature [50, 57], and will only briefly be commented on. A typical representative of this family is $|Ca_8 (WO_4)_2|[Al_{12}O_{24}]$-**SOD** [58]. At a first glance this formula could be misinterpreted as representing another massive violation of Loewenstein's rule. However, one moment of reflection will tell us that this is not the case. A given AlO_4 tetrahedron in a framework consisting entirely of AlO_4 tetrahedra does not exert stress in the same way as it does in a matrix of smaller SiO_4 tetrahedra. Therefore, there is no reason for applying Loewenstein's rule to aluminate sodalites.

We note in passing that aluminate sodalites experience other stresses more strongly than aluminosilicate sodalites. These stresses are responsible for marked tetragonal tetrahedron distortions with intratetrahedral angles O–Al–O of the order of $120°$, instead of typically $112°$ for TO_4 tetrahedra in [AlSi]-**SOD** [50].

Aluminate sodalites are interesting materials because of their physical properties, the occurrence of ferroic and nonferroic phase transitions, and of modulations [34].

Their sheer existence provokes the question of whether other microporous framework types with all-alumina framework composition may exist. To the best of our knowledge such all-aluminate microporous solids, other than aluminate sodalites, have not been reported yet.

D. AlPOs, templates   It was argued before that Loewenstein's rule is based on the fact that the Al^{3+} and Si^{4+} cations in aluminosilicate frameworks differ significantly with respect to their charges and sizes. With ionic radii for tetrahedrally coordinated Al^{3+} and P^{5+} of $0.39\,Å$ and $0.17\,Å$, respectively [28], the difference is even more pronounced in the case of a substitution $2\ SiO_4 \rightarrow AlO_4 + PO_4$. Framework structures of this type exist. In view of the previous discussion, it comes as no surprise that Loewenstein's rule seems to be strictly obeyed in such compounds.

With a P:Al ratio of 1:1, neutral framework structures $[AlPO_4]$ are formed. A dense structure of this type occurs naturally as the mineral berlinite, which is isotypic with quartz. For the present purpose we are more interested in microporous structures. Open framework structures $[AlPO_4]$ are often called AlPOs. The members of the meanwhile large family are usually synthesized in the presence of spacious, usually organic, and often neutral molecules, such as amines, furanes or thiophenes. The function of these molecules obviously controls the kind of framework formed. They act as structure-directing agents, and are often called templates. The mechanisms of how these molecules control the topology of the forming framework are not fully understood. In some cases steric effects have been shown to be probably responsible for the kind of framework formed [49], but from other cases it is clear that these effects, though important, cannot act alone. "Lone-pairs" of the frequently used amines are likely to contribute significantly to the function, as might molecular dipoles or polarizabilities.

References see page 734

4.2.1.3.5 **Zeolite Framework Types and Framework Composition**

The use of many different molecules as possible templates not only enabled the synthesis of the new members of the AlPO family, but also proved suitable for synthesizing new microporous frameworks of the aluminosilicate family.

Once a sufficiently high number of microporous framework structures of various compositions was known, the available data could be investigated for inter-relationships, for example between the composition of a framework and its topology. It was realised that some framework types are very compliant with respect to the composition, whereas others seem to be much more stubborn. According to the actual knowledge, the sodalite topology belongs to the most versatile frameworks. It is found in silica, aluminosilicate, and aluminate frameworks, thereby spanning the full range of the ratio Al:Si from 0 to ∞. Already this fact makes **SOD** possibly unique amongst the microporous framework types. Furthermore, **SOD** occurs also with AlPO composition (AlPO-20, [59]), and with many more framework compositions (see the *Atlas*). A **SOD** framework that contains only boron as T atoms has been known for a long time [60].

From the framework types listed in the *Atlas*, an interesting relationship between chemistry and topology made its appearance. The whole set of 133 framework types could be subdivided into two different subsets, one comprising silicates in a rather broad sense (silicas, aluminosilicates, germanates, and so forth), the other one phosphates, also in a broad sense (aluminum phosphates, aluminum arsenates, and so forth). The former subset counts 98 members, the second 60. 73 framework types of the silicates, and 35 framework types of the phosphates, belong exclusively to their respective subset. The intersection of both subsets, that is 25 different framework types, encloses those types that crystallize with silicate as well as with phosphate composition. It is interesting to note that the intersection contains precisely those framework types that belong to the best-known types at all, for example **ANA, CAN, CHA, ERI, FAU, GIS, LTA, RHO, SOD** and **THO**. If this possible subdivision can be maintained after future volumes of the *Atlas* or LBB have enlarged the basis of this observation, this will necessarily raise the question as to which special features make the framework types of the intersection so much versatile with respect to the chemical composition. Until now it seems that simple answers to this question are not available.

A. Possible T cations   Many microporous frameworks are able to form with more than two different types of **T** atoms, resulting in sometimes quite complex compositions. This is demonstrated by the arbritrarily chosen example of so-called MAPSO-46, $|(C_6H_{16}N)_8(H_2O)_{14}|[Mg_6Al_{22}P_{26}Si_2O_{112}]$-**AFS** [61].

Today an amazingly high number of elements are known to occupy **T**-sites in tetrahedral frameworks. According to LBB [27], the following elements occur in at least one of the 42 structure types listed in the first volume, (ABW through **CZP**), namely Li, Be, B, Mg, Al, Si, P, S, Ti, Cr, Mn, Fe, Co, Cu, Ni, Cu, Zn, Ga, Ge, As, and Cd. The occurrence of transition metals is noteworthy and might be interesting for certain physical properties, or some useful chemical reactions.

The various possible ordered and disordered distribution patterns of different numbers and types of T cations, including possible extensions of Loewenstein's rule, are appealing from an experimental and theoretical point of view.

Possible guest cations   In the first volume of LBB the following *non*framework cations are listed: Li, Na, K, Rb, Cs, Mg, Ca, Sr, Ba, Tl, Si, Mn, Fe, Co, Cu, Ag, Cd, NH_4. The occurrence of alkali and alkaline earth cations should not be surprising. Interestingly, some elements are in some respect ambivalent, as they do not only occur as framework cations, but may also act as guests. This is the case of Li, Mg, Si, Mn, Fe, Co, Cu and Cd. With the exception of Li and Mg, the listed guest species seem to be always integrated into molecular complexes, rather than being present as simple isolated or hydrated cations.

There are a few puzzling cases where a given element is reported to be found simultaneously in the host framework and in a guest species. This is, for instance, the case of Co in an aluminum phosphate of the **CHA** structure type [62].

B. Guest anions   Guest anions occur notably in the **SOD** family. A great variety of anions have been shown to be able to reside in the pores, for example the halogenides, pseudohalogenides, carbonate or simple oxygen complexes, like sulfate, chromate and so forth. Note that the interaction of the latter guest anions with the negatively charged **SOD** frameworks is determined by quite strong repulsive Coulomb forces, in addition to their size effect.

In this context it is worthwhile to mention a series of experiments for which the size of guest anions in **SOD** frameworks was decisive. One of the characteristics frequently attributed to the sodalite framework is its ability to adapt itself to the sizes of the guest ions via the so-called tilt mechanism. This can be expressed rather loosely by saying that large guest ions stretch the framework open, while smaller ones allow the framework to fold around them and reduce its volume. Given that the tilt mechanism is a cooperative effect with, in principle, infinite correlation length, one may wonder what would happen if a certain number of cage ions is missing. Stated another way, to what extent may guest cations or anions be extracted from the structure before the framework collapses?

A partial answer to this question was given by Buhl [63], who found that the solid solution $|Na_8(IO_3)_{2-x}(OH^-,H_2O)_x|[Al_6Si_6O_{24}]$, with x up to 1.0, exhibits virtually the same thermal expansion behavior as the anhydrous end member $|Na_8(IO_3)_2|[Al_6Si_6O_{24}]$. A discontinuity in the thermal expansion curves of either composition demonstrates their strong similarity. In either case all sodalite cages are filled with anions, but the compound with x = 1.0 has only 50 % of the cages occupied by the large IO_3^- anions. From this observation Buhl concluded that this portion is sufficient to prevent the framework from collapsing.

The just-mentioned sodalite exhibits another interesting behavior that is related to the framework distortion. When heated under carefully controlled conditions

References see page 734

the compound releases H_2O and O_2, until a composition $|Na_8I(O)_{0.5}|[Al_6Si_6O_{24}]$ is obtained. One half of the cages accommodate large I^- anions, whereas, respectively, one quarter of the cages are occupied by the much smaller O^{2-}, or are even empty. Similar to the above case the state of the structural collapse is determined by the large anions.

If the heating is carried out in an atmosphere containing common industrial waste gases, such as CO_2, NO or SO_2, compounds like $|Na_8I(CO_3)_{0.5}|[Al_6Si_6O_{24}]$ are formed. The author proposes that the gas molecules participate in the transformation of the original sodalite via intracage chemical reactions. In order to do so the gases have to diffuse into the sodalite cages via the 6-ring windows in the $\langle 111 \rangle$ directions. This is only possible, because at the high temperatures used for the reaction (973 K), the iodide anions (as decomposition products of the original IO_3^--anions) stabilize the framework in an expanded state, such that the windows are open for diffusion. After the intracage reaction is completed, and the sodalite cooled to room temperature, the waste gas molecules are trapped in the sodalite cages, because they can no longer escape anymore through the windows. Clearly, this interesting behavior bears some potential for future applications.

C. Host anions   From the previous discussion it is clear that microporous framework structures show considerable tolerance not only with respect to the various guest species, but also concerning the chemical composition of their host cations. On the other hand, surprisingly little is known about the variability of the host *anions*.

As discussed earlier, only O (and OH^-, or sometimes F^-, in interrupted frameworks) has traditionally been accepted as possible host anion. However, in the new IUPAC formula the peripheral host atoms are represented by $_{pe}H$, rather than simply writing O. This can be interpreted as the Commission's intention to admit the possibility of replacing oxygen by other anions. A few possible cases will now be discussed.

Nitridophosphate and oxonitridophosphate sodalites   A possible way to produce new microporous structures is replacing oxygen by nitrogen. This is easy to say, but more difficult to do. Since both elements differ significantly in their chemistry, new synthesis routes and preparation techniques have to be sought and worked out. Furthermore, it can be expected that the stabilities of such structures will be rather different and the structures may show new particularities. Schnick and coworkers [64] started their successful work by observing that the structure of the binary compound P_3N_5 consists of a framework of PN_4-tetrahedra. However, the tetrahedra shared not only corners, but also edges. The challenge was therefore to bring the PN_4-tetrahedra to join only by all their corners. This was successful with the composition $Zn_7[P_{12}N_{24}]Cl_2$, the structure of which turned out to be of the sodalite-type. Applying the new IUPAC rules to this compound, its formula reads $|Zn_7Cl_2|[P_{12}N_{24}]$-**SOD**. Considerable variability was found for the extra-framework cations and anions [65], thus lending support to the idea that these species can be regarded, indeed, as guests in a nitridophosphate host. Further evidence for host–guest behavior was obtained when a halogen-free sodalite of composition $|Zn_6|[P_{12}N_{24}]$ could be prepared which exhibits typical host–guest reactions, for example reversible hydrogen intake and release.

By replacing one quarter of all bridging N atoms in nitridophosphate sodalites by oxygen, oxonitridophosphate sodalites $|A_{8-m}H_mX_2|[P_{12}N_{18}O_6]$ with $X = Cl^-$, Br^- or I^- could also be prepared. The occurrence of the monovalent guest cations $A = Li^+$ or Cu^+ makes these materials potential ionic conductors. MAS/NMR-investigations and neutron powder diffraction indicate that the N and O atoms in the PON_3 tetrahedra are disordered.

Sulfide-based microporous solids   Some people believe that another promising route has been badly neglected in the past. This is the route to sulfide-based poroates.

In the past few decades oxides have received much more attention, and hence were much more intensively studied than their heavier homologs, namely sulfides, selenides, and tellurides. As a consequence, our present knowledge of the higher chalcogenides lags considerably behind that of oxides. Two major reasons for the different intensity with which the different classes were studied may be 1) the much more widespread use of oxides in traditional techniques such as catalysis, ceramics or glasses, and 2) probably the proven potential of many oxides for uses in modern high-technology applications, exploiting physical phenomena such as high-temperature superconductivity, ionic conductivity, ferroelectricity, ferroelasticity, or giant magnetoresistance. The great demand has required the employment of the full range of synthetic and analytical methods of solid-state sciences; some methods were even invented or specifically developed for this purpose.

However, presently we witness growing evidence for increasingly higher barriers, or even possible dead ends, on the oxide route, despite the still many successful applications, and continuing high level of research activity. In order to cope with future needs of mankind there can be no reasonable doubt that it is important to study other classes of materials with an effort that is comparable to that spent for the oxides.

From the chemistry point of view, sulfur and selenium are the higher homologues of the lightest chalcogen, namely oxygen. Despite their location in the same group of the periodic table a barrier divides the oxides from the sulfides and selenides. This can be exemplified by the fact that the structure of most sulfides, while being isotypic with that of the corresponding selenides, normally differs fundamentally from that of the corresponding oxide. In the language of classical chemistry this difference is attributed to the much more covalent character of the chemical bonds in sulfides/selenides, compared to the more ionic nature of the corresponding oxides. Correspondingly, the two classes differ also with respect to their electronic band structures. The higher chalcogenides have the tendency to be in, or to transform more easily into, the electronically semiconducting or even conducting state, whereas the oxides are chiefly insulators, or ionic conductors. Of course, many other physical properties differ as well. Taking into account both,

References see page 734

similarities and differences, it is clear that sulfides represent a natural choice for the search of possible alternatives to oxides.

Some activity in the field of metal sulfide-based microporous solids has already been reported [66].

Crystalline microporous metal sulfides were discovered which are based on Ge^{4+} and Sn^{4+} sulfide frameworks. They were synthesized hydrothermally in the presence of alkylammonium templates. Thiogermanate frameworks contained one or more of the following incorporated framework elements: Mn, Fe, Co, Ni, Cu, Zn, Cd, and Ga. X-ray powder diffraction patterns of these materials indicate that the obtained framework structure types are novel and have no analogs in microporous oxide chemistry, in accordance with the before-mentioned differences in the crystal chemistry of oxides and sulfides.

Some of the sulfide-based microporous structures seem to be very open, for example for one of the prepared compounds, MeGS-2, the framework density F_d has been calculated to be 7.27 $T/1000\,\text{Å}^3$. This is much less than the corresponding values of oxides, for example 11.1 $T/1000\,\text{Å}^3$ for **-CLO**, or 12.1 $T/1000\,\text{Å}^3$ for **SBT** and **TSC**.

It seems that these sulfide-based microporous structure types have not been given consideration by the IZA-SC and, hence, are not included in the *Atlas*.

New preparation strategies, for example, solvothermal synthesis, have been developed over the past few years, possibly clearing the way to the synthesis of potential new microporous structures on the basis of sulfides [67, 68].

Some of the newly prepared framework structures contain polyhedral building units other than tetrahedra. We recall that the new IUPAC recommendations allow for such nontetrahedral building units, and are sure that in future we will see many exciting new microporous structures also with nontetrahedral frameworks. An example, however, again from the oxide world is given in the next paragraph.

4.2.1.3.6 **New Families of Oxidic Nontetrahedral Microporous Solids**

Very interesting examples are found in the family of vanadates. In a particular system it could be demonstrated that the obtained microporous structure is able to undergo anion, rather than cation, exchange. The compound, with a framework composition $[H(VO)_{18}O_{26}(NO_3)]^{10-}$, has ellipsoidal pores and is composed of pyramidal OVO_4 building units. The enclosed anion controls the geometry of the framework [69]. Very recently, open-framework vanadium silicates have been reported that promise, amongst else, interesting catalytic properties [70].

4.2.1.3.7 **The Importance of Weak Forces**

We have mentioned before that guest species can have a marked influence on the framework, despite the general weakness of the attractive host–guest interactions. However, this should not come as a big surprise, because it is known from many diverse examples that, in general, it is the weak and short-ranged interactions that determine the fine details of a structure, rather than the strong, often covalent or ionic, bonds. The latter determine the basic features of a structure, for example the internal structure of building units. The weak forces control the actual symmetries, phase transitions, useful tensorial properties, modulations, the packing of molecules and the functioning of biological macromolecules.

In microporous structures, especially those with flexible frameworks, it can be expected that the guests control more or less the actual conformational state of the surrounding host structure. The effect can be local for a few guest molecules randomly distributed over the pores, or it can be cooperative, if the correlation length is high. In the latter case phase transitions may become possible.

Of particular importance are weak forces in the course of the synthesis of a poroate. Although the so-called template, or structure-directing, effect is not fully understood, it seems to be clear that by arranging monomeric and/or oligomeric building units on and around its surface, the templating species helps in the building of the host framework. It is thus these intermediate (weak) interactions that control the formation of the future poroate.

Phase transitions Numerous examples for phase transitions are found in dense framework structures, for example in the silica modifications quartz, tridymite and cristobalite, or in the feldspar structural family. In the field of microporous structures aluminate sodalites and their phase transitions have been studied in some detail [34]. ZSM-5 was one of the first examples of a medium pore zeolite where it could be demonstrated that calcination of its as-synthesized form results not simply in an isosymmetrical distortion of the framework, but that this is associated with a symmetry change, that is the occurrence of a phase transition [71].

The following is a short selection of zeolite framework types that are known to undergo structural phase transitions without topological change **ABW, AEL, AFI, ANA, APD, ATV, EDI, GIS, LTA, MFI, NAT, RHO, SOD**.

There can be little doubt that structural phase transitions can occur also in many other microporous structures. Possibly, they have not been detected yet, because they were hidden, notably because of 1) resolution problems due to the smallness of the effects, 2) poor crystal quality, 3) incomplete filling of the pores by guests. The latter point is equivalent to the presence of a high number of randomly distributed defects. These have been known for a long time to have the tendency to average out the corresponding effects.

RUMs Recently a model was proposed which is able to explain many properties of framework structures [72]. The model is based on the assumption that nearly rigid units build up the frameworks, and that conformational changes of the framework are brought about by cooperative movements of these building units. In the case of zeolite-type structures the building units are obviously the TO_4 tetrahedra. A new ingredient in this so-called "rigid-unit-mode", or RUM model, is the use of basic arguments and tools of lattice dynamics. With the help of the RUM model many static and dynamic properties of various framework structures – not only tetrahedral ones – could be explained.

One interesting outcome of applying the model was an answer to the question,

References see page 734

how zeolitic frameworks manage to fold locally around cations or molecules without distorting the rest of the structure. It could be established that this is accomplished by linear combinations of static RUMs. For example in the idealized high-symmetry sodalite structure there is a whole band of RUMs throughout the Brillouin zone. By forming localized wave packets, clusters of oxygens can be formed around a cation without cost of elastic energy. For zeolite A and faujasites an even greater number of localized RUMs can be expected and may be related with the known binding sites in these zeolites [73].

4.2.1.4
Conclusion

The science and technology of zeolites and other poroates have reached a high degree of maturity. However, as it seems that the investigations have been restricted to unnecessarily narrow limits it is proposed that the possibility to extend these limits should be seriously tested in future work. In particular, nonoxidic and non-tetrahedral framework structures, including mixed ones, should be synthesized and analyzed.

A great help for future work will be the availability of extended databases allowing metadata to be determined, and data mining to be performed.

Acknowledgement

Thanks to F. Liebau for critically reading the manuscript and many helpful discussions.

References

1 J. Rouquérol, D. Avnir, C. W. Fairbridge, D. H. Everett, J. H. Haynes, N. Pericone, J. D. F. Ramsay, K. S. W. Sing, K. K. Unger, *Pure Appl. Chem.* **1994**, 66, 1739.

2 K. D. M. Harris, *J. Mol. Structure* **1996**, 374, 241.

3 M. Añibarro, K. Gessler, I. Usón, G. M. Sheldrick, W. Saenger, *Carbohydr Res.* **2001**, 333, 251.

4 http://www.gashydrate.de/.

5 F. Liebau, *Microporous Mater.*, submitted for publication.

6 D. K. Smith, A. C. Roberts, P. Bayliss, F. Liebau, *Amer. Mineral.* **1998**, 83, 126.

7 A. F. Cronstedt, *Kongl. Svenska Vetenskaps Akademines Handlingar* **1756**, 17, 120.

8 E. M. Flanigen, in: Introduction to Zeolite Science and Practice. H. van Bekkum, E. M. Flanigen, P. A. Jacobs, J. C. Jansen (Eds), 2nd edn, *Studies in Surface Science and Catalysis*, Vol. 137, Elsevier, Amsterdam, 2001, p. 13–34.

9 R. M. Barrer, *Proc. Roy. Soc.* **1938**, A 167, 392.

10 R. M. Barrer, *J. Chem. Soc.* **1948**, 127, 2158.

11 D. S. Coombs, A. Alberti, T. Armbruster, G. Artioli, C. Colella, E. Galli, J. D. Grice, F. Liebau, J. A. Mandarino, H. Minato, E. H. Nickel, E. Passaglia, D. R. Peacor, S. Quartieri, R. Rinaldi, M. Ross, R. A. Sheppard, E. Tillmanns, G. Vezzalini, *Eur. J. Mineral.* **1998**, 10, 1037.

12 F. Liebau, X. Wang, *Beih. z. Eur. J. Mineral.* **1995**, *7*, 152.

13 X. Wang, Z. *Kristallog.* **1995**, *210*, 693.

14 U. Simon, F. Schüth, S. Schunk, X. Wang, F. Liebau, *Angew. Chem. Intern. Ed. Engl.* **1997**, *36*, 1121.

15 F. Starrost, E. E. Krasovskii, W. Schattke, J. Jockel, U. Simon, X. Wang, F. Liebau, *Phys. Rev. Lett.* **1998**, *80*, 3316.

16 J. Homeyer, C. H. Rüscher, O. Bode, J.-Ch. Buhl, Z. *Kristallogr. Suppl.* **2001**, *18*, 110.

17 G. O. Brunner, W. M. Meier, *Nature* **1989**, *337*, 146.

18 Ch. Baerlocher, W. M. Meier, D. H. Olson, Atlas of Zeolite Framework Types, 5th Edn., Elsevier, Amsterdam, 2001, 297 pp.

19 http://www.iza-structure.org/databases/.

20 A. F. Wells, Three-dimensional Nets and Polyhedra, Wiley, New York, 1977, 268 pp.

21 A. F. Wells, Structural Inorganic Chemistry, 4th edn, Clarendon, Oxford, 1975, 1095 pp.

22 M. O'Keeffe, B. G. Hyde, Crystal Structures. I. Patterns and Symmetry. Mineralogical Society of America, Washington, 1996, 453 pp.

23 H.-J. Klein, *Acta Crystallogr. Suppl.* **1996**, *A52*, C-551.

24 K. Goetzke, H.-J. Klein, *J. Non-Crystalline Solids* **1991**, *127*, 215.

25 B. Winkler, C. P. Pickard, V. Milman, G. Thimm, *Chem. Phys. Lett.* **2001**, *337*, 36.

26 http://www.zeolites.ethz.ch/Zeolites/Explanations.htm.

27 W. H. Baur, R. X. Fischer, Zeolite Structure Codes ABW to CZP, in: Microporous and other framework materials with zeolite-type structures, W. H. Baur, R. X. Fischer (Eds), Landolt–Börnstein/New Series IV/14, Vol. B (1) Springer, Berlin, 2000, 459 pp.

28 R. D. Shannon, *Acta Cryst.* **1976**, *A32*, 751; R. D. Shannon, C. T. Prewitt, *Acta Cryst.* **1969**, *B25*, 925; R. D. Shannon, C. T. Prewitt, *Acta Cryst.* **1970**, *B26*, 1046; F. Liebau, Structural Chemistry of Silicates – Structure,

Bonding, and Classification, Springer, Berlin, 1985, p. 308–318.

29 C. Griewatsch, Ph.D. Thesis, Universität Kiel, 1988, 161 pp.

30 L. B. McCusker, F. Liebau, G. Engelhardt, *Pure Appl. Chem.*, **2001**, *73*, 381.

31 H. Saalfeld, Z. *Kristallogr.* **1961**, *115*, 132.

32 H. Schulz, Z. *Kristallogr.* **1970**, *131*, 114.

33 D. Többens, W. Depmeier, Z. *Kristallogr.* **1998**, *213*, 522.

34 W. Depmeier, *Phys. Chem. Minerals* **1988**, *15*, 419.

35 W. Depmeier, in Structures and Structure Determination. H. G. Karge, J. Weitkamp (Eds), *Molecular Sieves, Science and Technology*, Vol. 2, Springer, Berlin, 1999, p. 113–140.

36 T. Janssen, A. Janner, A. Looijenga-Vos, P. M. de Wolff, in: International Tables for Crystallography: Mathematical, Physical and Chemical Tables, Vol. C, A. J. C. Wilson (Ed.), Kluwer, Dordrecht, 1992, p. 797–836.

37 K. O. Kongshaug, H. Fjellvåg, K. P. Lillerud, *Micropor. Mesopor. Mater.* **2000**, *39*, 341.

38 W. T. A. Harrison, M. L. F. Phillips, X. Bu, *Micropor. Mesopor. Mater.* **2000**, *39*, 359.

39 F. Liebau, Z. *Kristallogr. Suppl.* **2001**, *18*, 110.

40 J. V. Smith, Tetrahedral frameworks of zeolites, clathrates and related materials in: Microporous and Other Framework Materials with Zeolite-type Structures. W. H. Baur, R. X. Fischer (Eds), Landolt–Börnstein/New Series IV/14, Vol. A, Springer, Berlin, 2000, 266 pp.

41 W. J. Mortier, Compilation of Extra Framework Sites in Zeolites, Butterworths, London, 1982, 67 pp.

42 X. Hu, W. Depmeier, Z. *Kristallogr.* **1992**, *201*, 99.

43 W. H. Baur, *Acta Crystallogr.* **1978**, *B34*, 1751.

44 R. J. Hill, G. V. Gibbs, *Acta Crystallogr.* **1979**, *B35*, 25.

45 D. T. Griffen, P. H. Ribbe, *N. Jahrb. Miner. Abh.* **1979**, *137*, 54.

46 W. H. Baur, in: Proceedings of the 2nd Polish–German Zeolite Colloquium, M. Rozwadowski (Ed.), Nicholas Copernicus University Press, Torun, 1995, p. 171–185.

47 G. V. Gibbs, *Am. Mineral.* 1982, *67*, 421.

48 W. Depmeier, *Acta Crystallogr.* 1985, *B41*, 101.

49 C. M. Braunbarth, Ph.D. Thesis, Universität Konstanz, 1997, 322 pp.

50 W. Depmeier, *Acta Crystallogr.* 1984, *B40*, 185.

51 H. E. Robson, D. P. Shoemaker, R. A. Ogilvie, P. C. Manor, *Adv. Chem. Ser.* 1973, *121*, 106.

52 L. B. McCusker, Ch. Baerlocher, Proc. 6th Int. Zeolite Conf. 1984, 812.

53 W. Loewenstein, *Amer. Miner.* 1954, *39*, 92.

54 M. T. Dove, T. Cool, D. C. Palmer, A. Putnis, E. K. H. Salje, B. Winkler, *Amer. Miner.* 1993, *78*, 486.

55 K. Sahl, N. D. Chatterjee, *Z. Kristallogr.* 1977, *146*, 35.

56 J. Löns, Ph.D. Thesis, Universität Hamburg, 1969, 101 pp.

57 W. Depmeier, *Z. Kristallogr.* 1992, *199*, 75.

58 W. Depmeier, *Acta Crystallogr.* 1984, *C40*, 226.

59 E. M. Flanigen, B. M. Lok, R. L. Patton, S. T. Wilson, Proc. 7th Int. Zeolite Conf., 1986, 103.

60 P. Smith, S. Garcia-Blanco, L. Rivoir, *Z. Kristallogr.* 1964, *119*, 375.

61 J. M. Bennett, B. K. Marcus, *Stud. Surf. Sci. Catal.* 1988, *37*, 269.

62 Y.-H. Xu, Z. Yu, X.-F. Chen, S.-H. Liu, X.-Z. You, *J. Solid State Chem.* 1999, *146*, 157.

63 J.-Ch. Buhl, *Thermochimica Acta* 1996, *286*, 251.

64 W. Schnick, *Angew. Chem. Int. Ed. Engl.* 1992, *31*, 213.

65 W. Schnick, J. Lücke, *Z. Anorg. Allg. Chem.* 1994, *620*, 2014.

66 R. L. Bedard, S. T. Wilson, L. D. Vail, J. M. Bennett, E. M. Flanigen, in: Zeolites: Facts, Figures, Future, P. A. Jacobs, R. A. van Santen (Eds), *Stud. Surf. Sci. Catal.* Vol. 49, Elsevier, Amsterdam, 1989, p. 375–388.

67 M. G. Kanatzidis, *Current Opinion in Solid State & Materials Science* 1997, *2*, 139.

68 P. Stoll, P. Dürichen, C. Näther, W. Bensch, *Z. Anorg. Allg. Chem.* 1998, *624*, 1807.

69 A. Müller, *Nature* 1991, *352*, 115.

70 X. Wang, L. Liu, A. J. Jacobsen, *Ang. Chem. Int. Ed. Engl.* 2001, *40*, 2174.

71 H. van Koningsveld, J. C. Jansen, H. van Bekkum, *Zeolites* 1990, *10*, 235.

72 K. D. Hammonds, M. T. Dove, A. P. Giddy, V. Heine, B. Winkler, *Amer. Miner.* 1996, *81*, 1057.

73 K. D. Hammonds, H. Deng, V. Heine, M. T. Dove, *Phys. Rev. Lett.* 1997, *78*, 3701.

4.2.2
Synthesis of Classical Zeolites

Koji Nishi and Robert W. Thompson

4.2.2.1
Introduction

Since the pioneering work of Barrer et al. [1–10] and the fascinating achievements of Milton, Breck, Flanigen, and others in the Union Carbide laboratory [11–18], a wealth of zeolites and related microporous materials have been synthesized, and novel materials of this class will continue to be discovered. In almost all instances, hydrothermal synthesis is the method of choice for preparing zeolites. The techniques for hydrothermal synthesis of zeolites have reached a high level of sophistication, yet the scientific understanding of the very complex series of chemical events on route from the low-molecular weight reagents to the inorganic macromolecule remained somewhat obscure.

The objective of this section is to review the synthesis of classical molecular sieve zeolites, highlighting information regarding fundamental mechanisms and typical techniques for the classical zeolite crystallization.

A number of factors such as reaction temperature, composition of the reaction mixture, the nature of the reactants, and pretreatment of the amorphous precursors can influence not only the reaction kinetics but also the type of zeolite that forms. We focus on the zeolites A, X, Y, mordenite, ZSM-5, and zeolite Beta as the classical molecular sieve zeolites, and describe the features and the influences of factors in the crystallization process. However, at times we will include observations made using other zeolite systems for comparison or to illustrate key concepts. General aspects of zeolite synthesis are presented in Sect. 4.2.2.2. In Sect. 4.2.2.3 the typical syntheses of zeolites A, X, and Y are presented. In Sect. 4.2.2.4 the kinetics and mechanism of zeolite synthesis are presented. The text focuses especially, on the two important steps in the crystallization process, namely, nucleation and crystal growth. In Sects 4.2.2.5 and 4.2.2.6 the effect of seeding and aging of amorphous gel precursors on zeolite crystallization is discussed. The addition of seed crystals and aging of amorphous gel precursors can increase the rate of crystallization, and in some cases can direct the desired crystalline phases. The mechanisms of secondary nucleation caused by addition of seed crystals are also presented in this section. The nature and quality of the starting materials are factors that are found to be dominant in determining the kinetics and the size of final products. Section 4.2.2.7 focuses on the effect of the nature of reactants, particularly the silica source, on zeolite crystallization. The fact that the kinds and the amount of alkali cations influence the crystallization process is presented in Sect. 4.2.2.8. The addition of organic compounds such as triethanolamine (TEA) and tetraethylammonium (TMA) cation in the synthesis mixture influences the size of final crystals and is presented in Sect. 4.2.2.9. The roles of these organic additives are presented as well. Further, the observation of surface structure and crystal growth mechanism of single-crystal zeolite synthesized using organic additives are described in this section. The synthesis of zeolites from clay minerals, the synthesis of cubic and hexagonal analogs of zeolite Y using crown-ethers, and the synthesis using microwave heating are presented as somewhat specific techniques for crystallization of zeolites A, X, and Y in Sects 4.2.2.10 to 4.2.2.12, respectively. The synthesis of mordenite is presented in Sect. 4.2.2.13. The final section (4.2.2.14) focuses on the synthesis of high-silica zeolites such as ZSM-5 and zeolite Beta.

4.2.2.2
General Aspects of Synthesis of Zeolites

Currently, there are more than a hundred known zeolite structure types. These materials occur naturally in some cases, but most are prepared synthetically. Al-

References see page 804

though the synthesis literature of the past 50 years is voluminous, replicating a synthesis from the literature is often unsuccessful, even when done by researchers skilled in the art. The IZA (International Zeolite Association) Synthesis Commission reviewed the verified recipes for the synthesis of conventional zeolites including a series of articles on techniques of zeolite synthesis and characterization [19, 20].

The synthesis of classical molecular sieve zeolites is typically carried out in batch systems, in which a caustic aluminate solution and a caustic silicate solution are mixed together, and the temperature held at some level above ambient at autogeneous pressure for some period of time. Once a solution has been mixed, it is common for the original mixture to become somewhat viscous immediately, due to the formation of an amorphous gel phase. The viscous amorphous gel phase usually becomes less viscous as the temperature is raised.

As the synthesis proceeds at elevated temperature, zeolite crystals are formed by a nucleation step, and these zeolite nuclei then grow larger by assimilation of aluminosilicate material from the solution phase. The amorphous gel phase may be regarded as a reservoir of nutrients, dissolving to replenish the solution with aluminosilicate species as crystal growth occurs simultaneously. An induction period may exist during the early stage of the synthesis, in which no apparent nucleation or growth occur. It is assumed that reorganization of the nutrients into small entities having some of the characteristics of a zeolite structure occurs during that time. Such small entities, that is, crystal nuclei, appear to remain unstable with respect to dissolution until they have reached a critical size. Once the nucleus exceeds the critical size, the probability that it will grow into the macroscopic crystal size range is relatively high. The subsequent assimilation of mass from the solution and its reorientation into ordered crystalline material proceeds.

In the early reports [9, 17, 21, 22] of alkali aluminosilicate synthesis of low silica zeolites, it was proposed that hydrated alkali cations templated, or stabilized, the formation of the zeolite structure subunits. The unique structural characteristics of zeolite frameworks containing polyhedral cages had led to the postulate that the cation stabilizes the formation of structural subunits that were the precursors or nucleating species in crystallization. Syntheses were carried out as follows: alkali hydroxide, reactive forms of alumina and silica, and H_2O were mixed to form a gel, and the temperature held near 100 °C at autogeneous pressures for some period of time as shown in Fig. 1 [23]. Subsequently, the addition of quaternary ammonium cations to alkali aluminosilicate gels was reported to produce the high silica zeolites and "all-silica" molecular sieves. The synthesis of siliceous zeolites involves synthesis chemistry similar to that of early low-silica zeolites with two important differences: the addition of the quaternary ammonium cation, and crystallization temperatures higher than 100 °C, typically in the range 120–200 °C. Both syntheses of low-silica and siliceous zeolites were carried out at relatively high pH of 10–14.

Crystallization is generally agreed to proceed through two primary steps: nucleation of discrete particles of the new phase and subsequent growth of those entities. The former step can be subdivided further in the following way [24]:

a) Early Zeolite Synthesis

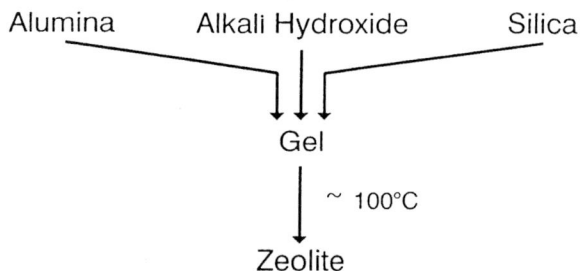

Alumina   Alkali Hydroxide   Silica

Gel

~ 100°C

Zeolite

Hydrothermal Crystallization of Reactive Alkali Aluminosilicate
Gels at Low Temperature and Pressure

b) Siliceous Zeolite Synthesis

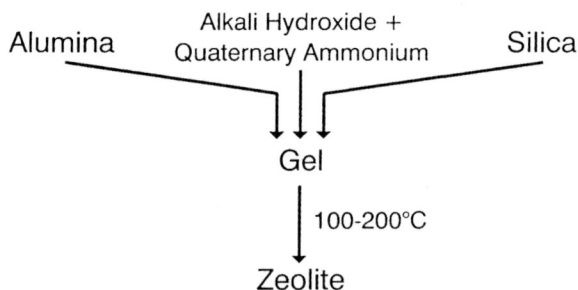

Alumina   Alkali Hydroxide +
               Quaternary Ammonium   Silica

Gel

100-200°C

Zeolite

Fig. 1. Schematic representation of synthesis method for (a)
early zeolites; (b) siliceous zeolites. Reprinted from [23] with
permission from Elsevier Science.

1. Primary nucleation
   a) Homogeneous nucleation
   b) Heterogeneous nucleation
2. Secondary nucleation
   a) Initial breeding
   b) Microattrition
   c) Fluid shear-induced nucleation

The term "primary nucleation" is used to describe the nucleation mechanisms
that produce nuclei whether or not suspended crystals are present. Nucleation
from a single-phase system is an example of primary nucleation. In "homogeneous nucleation", the mechanism is purely solution-driven, while "heterogeneous

References see page 804

nucleation" relies on the presence of an extraneous surface to facilitate a solution-driven nucleation mechanism. The extraneous surface is thought to reduce the energy barrier required for the formation of the crystalline phase [24].

The term "secondary nucleation" is used to describe any nucleation mechanism that requires the presence of the desired crystalline phase to catalyze a nucleation step. "Initial breeding" results from crystalline "dust" adhering to larger seeds introduced into the reaction medium. Nucleation by "attrition" is merely fracture of a lesser degree and results from crystal–crystal interaction at high suspension densities as well as from crystal-apparatus contact. Nucleation by "fluid shear" results when the fluid velocity relative to the crystal velocity is large and some of the adsorbed layer is removed. If this adsorbed layer is swept away into a sufficiently supersaturated environment, structured assemblies may become viable crystals and grow. Further details of crystal nucleation mechanisms, with numerous primary references, are described by Randolph and Larson [24].

Crystal growth from solution occurs by transfer of material from the solution phase, in which the solute has three-dimensional mobility, to the surface of the crystal lattice being formed, and incorporation thereon in a regularly ordered framework. Thus, individual species must diffuse to the crystal surface, and be incorporated into that crystalline structure for growth to occur.

Many studies on the gel and clear solution processes were undertaken to elucidate the mechanism of zeolite synthesis. Recent scattering studies of the evolution of colloidal particles, in particular of Silicalite-1, demonstrated the presence of nanoscale subcolloidal particles that were associated with nucleation and conversion into the final zeolite phase [25–31]. Two possible growth mechanisms were presented, including assembly of the lattice through (1) soluble small species from solution [32–34], and (2) aggregation and realignment of preassembled building blocks containing template molecule/aluminosilicate clusters [35–44]. The role of the subcolloidal particles might thus be as a source of nutrient for the solution phase that feeds the growing crystals with soluble species or as building units that participate in an aggregation/densification process. However, these issues are still in controversy.

Temperature, alkalinity, composition of reaction mixtures, the nature of reactants, and pretreatment of the amorphous gel can all affect crystallization kinetics and even the type of zeolite that forms. The degree of supersaturation of the synthesis solution is the principal driving force for nucleation. While much is known about the state of the solute in dilute solutions, the state of supersaturated solutions is not well understood, unfortunately.

The degree of supersaturation is determined by two main factors: nutrient concentration and temperature. As the nutrient concentration increases, the degree of supersaturation increases until the formation of an amorphous gel phase occurs. As the temperature of the system increases, the degree of supersaturation decreases due to the increase in solubility of the nutrients. However, increasing the temperature also increases reaction kinetics, thus nucleation and crystal growth rates may be accelerated in spite of the reduction of supersaturation. Theoretical modeling of zeolite syntheses has shown that the rate of nucleation can be very sensitive to changes in the temperature of the system [45]. Therefore, the rate at which the batch vessel is heated to its reaction temperature will certainly influence the nucleation step. This may be especially important in large vessels.

The nature of the reagents may be critical in determining the nucleation process. A nutrient having a monomeric form could more readily interact with other nutrients to form nuclei than one having a polymeric form. The presence of chemical impurities, for example, Fe^{3+} cation, alkali cations, alkali earth cations, etc. [24, 46, 47], and physical impurities, for example, dust, insoluble particles, etc. [24], can increase the nucleation potential for a given system.

The presence of seed crystals in the systems also affects the nucleation step. The addition of seed crystals can promote the nucleation by some secondary nucleation mechanism: microattrition, fluid shear, needle breeding, and initial breeding [24]. The added surface area of the seed crystals influences the more rapid consumption of reagents by growth. The aging of some amorphous gel solutions appears to provide a time during which the solution can form nuclei, though even at dramatically reduced rates, which can then become activated at elevated temperatures [e.g., 48, 49].

4.2.2.3
Synthesis of Zeolites A, X, and Y

Between 1949 and 1954 Milton et al. [11–16] discovered a number of commercially significant zeolites, specifically types A, X, and Y. Zeolite A is represented by the formula: $Na_{12}[(AlO_2)_{12}(SiO_2)_{12}] \cdot 27H_2O$. Its cubic structure consists of sodalite cages joined with double-four rings (D4Rs), and is characterized by a 3D network consisting of cavities 11.4 Å in diameter separated by circular openings 4.2 Å in diameter [12]. Typical zeolite X and Y compositions are represented by the formulae: $Na_{86}[(AlO_2)_{86}(SiO_2)_{106}] \cdot 264H_2O$ and $Na_{56}[(AlO_2)_{56}(SiO_2)_{136}] \cdot 250H_2O$, respectively. The number of aluminum atoms per unit cell of zeolite X varies from 96 to about 77, while in zeolite Y, the number of aluminum atoms per unit cell varies from about 76 to 48. The corresponding Si/Al ratio varies from 1 to 1.5 for zeolite X and from greater than 1.5 to 3 for zeolite Y. Based on X-ray powder diffraction data the framework structures of zeolites X and Y are similar to that of the mineral faujasite (FAU). The cubic zeolites X and Y and their recently discovered hexagonal polymorph EMT, have the most open zeolite framework known. The structure consists of sodalite cages joined with double-six rings (D6Rs). So-called supercages with a diameter of 13 Å are created within the framework. Along the [110] directions, channels with free diameter of 7.4 Å are formed by interconnection of the supercages [14].

Typically, zeolites A, X, and Y are crystallized from sodium aluminosilicate gels prepared by mixing aqueous sodium aluminate, sodium hydroxide, and sodium

References see page 804

Tab. 1. Synthetic zeolites Na_2O–Al_2O_3–SiO_2–H_2O system [18]. Reprinted by permission of John Wiley & Sons, Inc.

Zeolite type	Typical reactant comp. (moles/Al_2O_3)			Typical conditions reactants	Temp. (°C)	Zeolite comp. (moles/Al_2O_3)			Properties	Ref.
	Na_2O	SiO_2	H_2O			Na_2O	SiO_2	H_2O		
A	2	2	35	$NaAlO_2$ sodium silicate NaOH colloidal SiO_2	20–175	1	2	4.5	cubic, 1–2 μm, $n = 1.46$, $d = 1.99$	11,12
X	3.6	3	144	$NaAlO_2$ sodium silicate NaOH colloidal SiO_2	20–120	1	2.0–3.0	6	octahedra, $n = 1.45$–1.46, $d = 1.94$, faujasite-type	13,14
Y	8	20	320	$NaAlO_2$ sodium silicate NaOH colloidal SiO_2	20–175	1	>3.0–6.0	9	octahedra, $n = 1.45$, $d = 1.92$, faujasite-type	14,15

silicate solutions. As the amorphous gel forms, it is agitated to produce a homogeneous mixture. This is heated at an appropriate temperature until crystallization is completed as evidenced visibly by the separation of an extensive supernatant solution and the settling of the solids into a compact layer of crystals at the bottom of the vessel. By the relatively simple method, these zeolites have been synthesized as pure phases in the Na_2O–Al_2O_3–SiO_2–H_2O system as shown in Table 1 [18]. Zeolite A has been crystallized at temperatures ranging from 25 to 175 °C with the crystallization time varying from 14 days to 2.5 h. Recently, it was demonstrated that sufficiently aged synthesis mixtures could yield zeolite NaA, with crystal sizes ranging from 0.1 to 0.3 μm, even after 1 min in the microwave heating [50]. Zeolite X crystallizes at temperatures ranging from 25 to 120 °C with corresponding variations in the time required. Zeolite Y has been crystallized from aluminosilicate gels prepared from sodium hydroxide, sodium aluminate, sodium silicate, and colloidal silicas, either in the form of aqueous sols or amorphous reactive solids. It is well known that a preliminary aging of the reaction mixture at room temperature after gel formation followed by subsequent crystallization at higher temperature improves the crystallization process. Aging of the gel at room temperature yields a zeolite Y of greater purity [18].

From a study of many aluminosilicate gel compositions, relationships between the synthetic zeolite product and the composition of the starting reactant mixture have been established as shown in Fig. 2 [14]. Since the crystallization fields for zeolite A, X, and Y (faujasite) lie next to each other on the crystallization phase field diagram, they often grow under the same or similar conditions. Hence, these two types of zeolites will be discussed together.

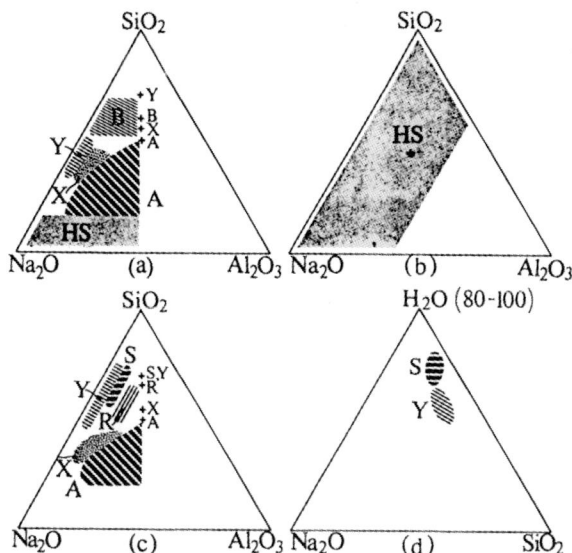

Fig. 2. Reaction composition diagrams [18]. Reprinted by permission of John Wiley & Sons, Inc. Original data were reported in [14]: (a) Projection of the $Na_2O-Al_2O_3-SiO_2-H_2O$ systems at 100 °C. H_2O content of gels is 90–98 mole %. Areas identified by letters refer to compositions that yield the designated zeolite. The points marked with (+) show typical compositions of zeolite phase; compositions in mole %. Sodium silicate used as a source of SiO_2. (b) Same as (a) with 60–85 mole % H_2O in the gel. (c) Same as (a). Colloidal silica used as a source of SiO_2. (d) Effect of water content in gel on synthesis of zeolites Y and S. Colloidal silica employed at 100 °C. Al_2O_3 content is 2–10 mole % of anhydrous gel composition. A = zeolite A; X = zeolite X; Y = zeolite Y; B = zeolite P; R = chabazite; S = gmelinite; and HS = hydroxysodalite.

4.2.2.4
Kinetics and Mechanisms

Zeolite crystallization experiments are quite commonly analyzed by means of a "crystallization curve". These curves, collected for batch zeolite crystallizer operations, represent the evolution of zeolite mass in the crystallizer in the course of an experiment. The data are frequently presented as the zeolite mass, the zeolite yield, or the percentage of zeolite in the solid phase, the remaining fraction of the solid phase is unreacted aluminosilicates, as a function of the crystallization time. These curves are characteristically sigmoid in shape and have an inflection point, which separates the primary autocatalytic (self-accelerating) stage of crystal mass growth from the final stage of a delayed growth.

In several early papers, it was reported that the crystallization of zeolites from aluminosilicate gels is a solution-mediated transformation process in which the

References see page 804

amorphous phase is a precursor for silicate, aluminate, and aluminosilicate species needed for the growth of the crystal phase [5, 51–53]. Zhdanov argued in a detailed review [51] in favor of a solution-mediated transport mechanism based on the results of investigation of the chemical structure of aluminosilicate gels, which showed that the nuclei of zeolite crystals begin to form in the liquid phase of gels or at the interface of gel phase. The growth of crystal nuclei proceeds at the expense of aluminosilicate hydrated anions present in the solution. These anions represent different combinations of Si–O and Al–O tetrahedra, as it was first put forward by Barrer et al. [5]. They proposed the nucleation to be the result of the polymerization of aluminate, silicate, and possibly more complex ions in the liquid phase, the ions being continuously supplied by the dissolution of the solid gel materials. This is analogous to the solution mechanism described by Kerr [52].

Kerr [52] reported on a study of the rate of crystallization of zeolite A in a reaction mixture prepared from an amorphous sodium aluminosilicate of composition Na $(AlO_2 \cdot 0.82\ SiO_2)$ and sodium hydroxide at 100 °C. The crystallization occurred rapidly after an induction period, which was concluded to be due to the formation of nuclei. It was found that the crystal growth occurred by deposition of some dissolved sodium aluminosilicate species on the crystal surface. He postulated that the rate of crystallization followed approximately first-order kinetics and was proportional to the quantity of crystalline zeolite present in the system. He also carried out an experiment that a solution of sodium hydroxide was circulated through an amorphous sodium aluminosilicate and then over zeolite NaA seed crystals, both suspended on filters. It was found that approximately 90 % of the amorphous substrate was dissolved and converted into zeolite NaA on the seeds. This result strongly indicated that dissolution of the amorphous substrate took place to sustain the crystallization of zeolite NaA.

In a further study, Ciric [53] showed that the solution phase concentration was essentially constant up to 80–90 % of conversion, and then suddenly dropped to a lower value. This indicated that the crystallization proceeded through the mass transfer of dissolved species from the bulk of the gel solution medium to the crystal surface and the driving force measured in terms of concentration really did not change very much until near the end of the synthesis. He also proposed a mathematical description of zeolite crystallization kinetics curves that showed autocatalytic features.

This solution-mediated transport mechanism was supported by several studies using Raman [54–56] and NMR [57] spectroscopy, and which suggested the existence of soluble aluminosilicate species. Angell and Flank [54], and Roozeboom et al. [55] observed changes in band intensities of aluminate and silicate species in the liquid phase during the crystallization period by Raman spectroscopy, whereas no spectral changes were observed by McNicol et al. [58] who proposed a solid phase transformation. Dutta et al. [56] showed the presence of aluminosilicate species in the liquid and solid phases during formation of zeolite Y by Raman spectroscopic study.

Recently, Antonic and Subotic [59] reported on the influence of the concentrations of aluminum and silicon in the liquid phase on the kinetics of crystal growth of zeolite A. The changes of the concentrations of aluminum (C_{Al}) and silicon (C_{Si}) in the liquid phase and dimension (L_m) of the largest crystals of zeolite A were measured during the crystallization of zeolite A at 80 °C. It was shown that the rate of crystal growth of zeolite A (dL/dt_c) could be expressed as:

$$dL/dt_c = k_g[C_{Al} - C_{Al}^*][C_{Si} - C_{Si}^*]$$

where C_{Al}^* and C_{Si}^* are the concentrations of aluminum and silicon in the liquid phase that correspond to the solubility of zeolite A at the given crystallization conditions. This revealed that the crystal growth of zeolite took place in accordance with the model of growth and dissolution proposed by Davies and Jones [60], and that the rate of crystal growth depends on the concentrations of both aluminum and silicon in the liquid phase.

In some early studies, the early part of the experimental sigmoid crystallization curves were described by the simple kinetic equation [51, 53, 61–63]:

$$Z = kt^n$$

where Z is the ratio of the mass of crystals formed in the gel at time t to their mass in the final crystallization product; k and n are constants for given experimental conditions. The constant n is related with the peculiarities of nucleation kinetics. At a constant rate of nucleation n should be equal to 4. The value $n > 4$ indicates an increasing rate of nucleation, and $n < 4$ a decreasing one. In many cases, the crystallization kinetic curves of zeolites were well described by this equation with values of $n > 4$ [49]. This equation describes the accelerating portion of the experimental sigmoid crystallization curves well, but does not contain quantitative information about the main factors responsible for crystallization kinetics: the rate of linear crystal growth and the rate of nucleation.

A method of analyzing the nucleation and crystal growth parts of such sigmoid curves was developed by Zhdanov and Samulevich [49] for the case of zeolite NaX. In their study, the nucleation history of the synthesis was determined by monitoring the growth of several of the largest crystals in the system over time, determining the crystal size distribution of the final crystalline zeolite product, and using both sets of data to estimate when each class of particles had been nucleated during the synthesis, as shown in Fig. 3 (from ref. [10]). In order to determine the times at which crystals in that mode started to grow, it was assumed that during the entire synthesis crystals of all sizes grew at the same linear rate. As one can see, the calculated crystallization curve is comparable to the experimental curve determined in the usual way from X-ray diffraction data. This similarity suggested that the assumption involved in the analysis about size-independent linear crystal growth was valid.

The nucleation kinetics curve in Fig. 3 indicated that nucleation began after some time had passed, most likely due to a transient heat-up time and some time

References see page 804

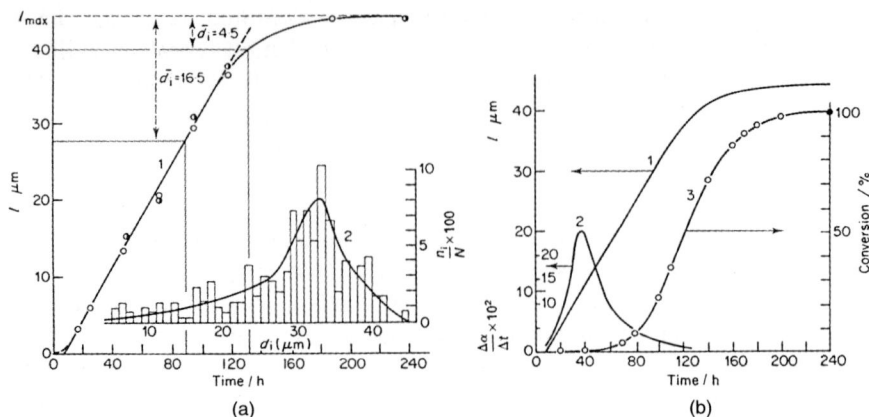

Fig. 3. (a) Curve 1 gives linear growth rate of faujasite NaX crystals. Curve 2 gives the histogram of crystal size distribution in the final product. (b) Curve 1 is curve 1 of (a). Curve 2 is the curve of nucleation rate against time derived from curve 1 and 2 of (a) and curve 3 gives the yield of faujasite as a function of time derived from curves 1 and 2. Reprinted from [10] with the permission of Academic Press Ltd. Original data were reported in [49].

required for dissolution of the amorphous gel to achieve some threshold concentration. However, it is most noteworthy that the rate of nucleation in zeolite crystallization systems was increasing only during its first period and ended when only about 50 % of aluminosilicate materials were consumed. The maximum rate of nucleation was reached at the crystallization time when only about 1–2 % of amorphous aluminosilicate precursor has been transformed into zeolite X. It is interesting that with in excess of 90 % of the unreacted aluminosilicate precursors left in the system, the nucleation process was diminished, while crystal growth proceeded for the duration of the synthesis. This observation led to further investigations of the autocatalytic (self-accelerating) stage of zeolite crystal mass growth.

Subotic et al. [61, 62, 64–67] proposed the "autocatalytic nucleation" mechanism and reported several studies supporting this mechanism. This mechanism is based on the early experimental results reported by Zhdanov [51]. The basis of the mechanism was the observation that even at early times there are very small (micro-) crystalline domains, which appeared to be contained in the amorphous gel phase. It was assumed that these autocatalytic nuclei lie dormant in the gel phase and begin growing after their release from the gel as it dissolved. As the cumulative zeolite crystal surface area increases due to crystal growth, the rate of solute consumption increases, which, in turn, increases the rate of gel dissolution, and results in increased rate of dormant nuclei activation.

It has been demonstrated that the original autocatalytic nucleation mechanism, assuming uniform distribution of the autocatalytic nuclei in the amorphous gel, cannot explain why the real nucleation occurred earlier in the synthesis. An empirical modification was made to the original autocatalytic nucleation hypothesis, by assuming that nuclei were located preferentially near the outer surface of the amorphous gel particles [68]. This empirical approach to modeling nucleation essentially allowed the nucleation phase to occur earlier in the process than for the earlier models of autocatalytic nucleation.

A kinetic analysis of nucleation and crystal growth processes during crystallization of zeolite ZSM-5 with 1,6-hexanediol ($R(OH)_2$) as a structure-directing agent showed the strong influence of the crystallization temperature on the crystal growth rate, on the nucleation rate, and on the fraction of zeolite crystallized [69]. On the other hand, the crystal size distribution of the final products crystallized from the same hydrogel ($30.6 \ Na_2O : 44.5 \ R(OH)_2 : Al_2O_3 : 106.4 \ SiO_2 : 4759.2 \ H_2O$) were almost the same. An increase in the temperature caused an increase in the rate of crystallization and of crystal growth. (0.257, 0.450, and $0.725 \ \mu m \ h^{-1}$ at 152, 160, and 170 °C, respectively). The maximum rate of nucleation was reached earlier and the initial density of nuclei decreased by increasing the temperature. The authors suggested that the temperature of reaction could directly affect the activation probability of the existing dormant nuclei. A similar effect was previously observed during crystallization of zeolite A at different temperatures [70]. The nucleation profiles exhibited an autocatalytic behavior, in which the initial increase in nucleation rate occurred before crystallinity had reached a significant level and then gradually dropped to zero. It was concluded that these results could be reasonably described by the nucleation model that assumed an empirical narrow distribution of dormant nuclei near the outer rim of the gel particles.

Recently, Subotic et al. [71] reported the analyses of the distribution of nuclei in amorphous gel matrices during crystallization of zeolites A, X. The system prepared by dispersion of X-ray amorphous aluminosilicate ($1.03 \ Na_2O : Al_2O_3 : 2.38 \ SiO_2 : 1.66 \ H_2O$) in the solutions containing $1.4 \ mol \ dm^{-3}$ NaOH (system I), $1.4 \ mol \ dm^{-3}$ NaOH $+ 0.045 \ mol \ dm^{-3} \ Al_2O_3$ (system I_a) and $1.4 \ mol \ dm^{-3}$ NaOH $+ 0.1 \ mol \ dm^{-3} \ SiO_2$ (system I_b) were heated at 80 °C until the amorphous aluminosilicate precursor completely transformed into zeolite A (system I and I_a) or zeolite X (system I_b). It was shown that the presence of aluminate anions in the liquid phase of the system (system I_a) caused an increase in the rate of crystal growth and of crystallization relative to the system in which aluminate is not present (system I). On the other hand, addition of silicate to the liquid phase of the system (system I_b) considerably decreased the rate of crystal growth and of crystallization. The maximum rate of nucleation was reached in all systems at the crystallization time when 50 % or more of the amorphous aluminosilicate precursor had been transformed into zeolite A (system I, and I_a) or zeolite X (system I_b). This indicated that the distribution of nuclei in the gel matrix was more or less homogeneous. The homogeneous distribution of nuclei in the gel matrix is possibly caused by the relatively high alkalinity under which the gel was precipitated [72]. Further, they carried out an analysis of the critical processes during crystallization of ZSM-5 reported by Falamaki et al. [69]. It has been shown that the concentration of nuclei was highest at the thin subsurface layer of gel particles and decreased towards the center of gel particles, as previously reported [68]. Thus, it

References see page 804

has been concluded that the number and distribution of nuclei in the gel matrix depend strongly on the physical and chemical conditions under which hydrogel is prepared. On the basis of their analysis [71, 72] it was concluded that the distribution of nuclei in most gel precursors is inhomogeneous, that is, the number of nuclei decreases from the surface of the gel particles to their interior. However, in some gels prepared at high alkalinity, the distribution may be more or less homogeneous.

Mintova et al. [73] showed only one single zeolite A crystal was nucleated in one amorphous gel particle at room temperature. The entire process of gel formation, nucleation, and growth of zeolite A in a clear solution model system that started at room temperature was followed by using high-resolution transmission electron microscopy (HRTEM) and dynamic light scattering (DLS). The nanosized single zeolite A crystals, which were only 10 to 30 nm in diameter, were nucleated and embedded in amorphous gel particles (40 to 80 nm) within 3 days at room temperature. During the continuation of the room temperature synthesis, the embedded tiny zeolite A crystallites grew at the expense of the surrounding amorphous gel agglomerates until the latter were completely consumed. The resulting nanoscale zeolite A crystallites obtained after 7 days at room temperature were fully crystalline particles with a diameter of 40 to 80 nm. These particles were single crystals without intergrowths of different lattice orientations. This indicated that each zeolite crystal was generated from one single nucleus in one isolated amorphous precursor gel particle. Much larger well-developed crystals of zeolite A (200 to 400 nm) were formed after subsequent hydrothermal treatment at 80 °C for 1 day. Although no amorphous gel phase was present in the suspension, very effective crystal growth was achieved. This indicated that zeolite A could grow through solution transport over large distances in which the nutrient pool must be the nanosized crystallites obtained at room temperature.

Zeolite nucleation is thought to occur via some primary mechanism, either homogeneous or heterogeneous nucleation. However, mathematical simulation of experimental results using population balance models [68, 69] suggested that the classical homogeneous nucleation mechanism probably does not apply in these syntheses [61, 62, 64–67]. Many studies have suggested that some precursor species form in the solution, and that these species involve more than just the aluminate ions, the silicate ions, or aluminosilicate oligomers. Recently, nanometer-sized particles have been observed in zeolite synthesis systems using very dilute solutions. Schoeman [28] reported that particles of 3 nm in size persisted throughout the synthesis of Silicalite-1. He also indicated that the zeolite crystal growth curve could be extrapolated back to about the same size, and that therefore, one might speculate that these particles were at least associated with nucleation, if not the nuclei themselves. Similarly, Gora et al. [74] reported that the nanometer-sized particles persisted throughout the synthesis of zeolite NaA. However, they noted that the same sized particles were observed to exist in the silicate solution prior to mixing with the aluminate solution, which itself did not contain any such particles. Both these reports give an indication that colloidal particles may participate in a form of heterogeneous nucleation.

Zhdanov [51] reported that the crystal growth rate was constant for some rather long period of time, and eventually slowed down as the reagent supply became depleted in the synthesis of zeolite A. That observation was made at several temperatures, and further demonstrated that the growth rate of zeolite crystals in these systems was independent of crystal size, at least from sizes as small as could be measured by optical microscopy. Several other experimental studies of zeolite crystallization have shown that linear, size-independent crystal growth during the early part of the crystallization process, as shown in Fig. 3, is typical for most zeolite synthesis [63, 66]. The size-independent crystal growth in the nanometer size range has been observed by laser light scattering techniques for several different zeolite systems [26, 75–78].

The general consensus from most studies of zeolite synthesis is that the crystal growth mechanism is a solution-mediated process. Consequently, it is possible that the crystal growth rate is governed by either diffusion of solutes to the crystal surface or surface reaction of soluble species.

Kacirek and Lechert [79, 80] noted the strict separation of the process of nucleation and growth for the crystallization of zeolite NaY over a wide range of Si/Al ratio of the final product. They proposed a cube root analysis of XRD crystallinities to determine crystal growth rate constants in the crystallization of zeolite NaY. It was shown that the apparent activation energies of growth were between 49 and 65 kJ mol^{-1} and increased with increasing Si/Al ratio. They concluded that since the rate of crystallization is dependent on the surface area of the crystals in the synthesis mixture the possibility of a diffusion-controlled process is ruled out. Barrer [10] stated that it is difficult to visualize the growth of porous structures such as zeolite Y via the condensation-polymerization of monomeric species. Further, he stated that a growth mechanism governed by diffusion control could be ruled out because of the high activation energies, for example, 65 kJ mol^{-1} obtained by measuring the linear growth rates in Kacirek and Lechert's study [80], whereas activation energies of about 12–17 kJ mol^{-1} would be expected for a diffusion controlled mechanism. Table 2 shows the values of the activation energy for linear crystal growth for several zeolite synthesis systems. It is found that the activation energies reported are of the same order of magnitude, that is, the values shown in Table 2 are all in the range of 45–90 kJ mol^{-1}, and clearly indicates surface kinetics as the rate-determining step rather than diffusion.

Schoeman et al. [26, 78] demonstrated this conclusion more convincingly by use of a chronomal analysis (dimensionless time parameter) of the conversion with respect to time, according to a technique suggested previously by Nielsen [87]. The method of Nielsen's chronomal analysis distinguishes between different growth mechanisms such as diffusion-controlled, surface reaction-controlled, dislocation-controlled, and compound growth mechanisms. The underlying principle in this approach is to rewrite the classical growth equations, for example, surface reaction and diffusion-controlled growth, as a linear relationship between synthesis time and conversion of nutrient to crystalline material. The crystallization of TPA-

References see page 804

Tab. 2. Comparison of activation energies for crystal growth of several zeolites.

Zeolite system	Activation energy/(kJ mol^{-1})	Ref.
NaA	46	14
NaX	59	
NaY	63	
NaA	44	49
NaX	63	51
NaA	71 (fresh preparation)	81
	75 (aged preparation)	
Mordenite	46	82
Silicalite	96	25
Silicalite	45	26
Silicalite	83	40
Silicalite	70	43
Silicalite	90	44
Silicalite	62.5 (length), 43.7 (width)	83
Silicalite	79 (length), 62 (width)	84
Silicalite	61 (length), 36 (width)	85
Analcime	75	86

Silicalite-1 was carried out in the temperature range of 80–98 °C, and the particle size in the range 30–95 nm and mass growth was measured by dynamic light scattering and ultracentrifugation, respectively. The chronomal for diffusion-controlled growth (I_D) and for surface reaction-controlled growth (I_p) of different kinetic order ($p = 1, 2, 3$) were calculated by the value of the degree of reaction obtained from experimental data according to Nielsen's approach. Figure 4 shows the results of their analyses using the four hypothesized model for crystallization as a function of synthesis time at a temperature of 98 °C [78]. If the plot of I_D or I_p against time were linear, the results would suggest that the hypothesized model is the limiting resistance to crystal growth. It is noted that the chronomal for a first-order surface reaction is linear during the period for which linear growth was observed. On the other hand, a diffusional mechanism can be ruled out as the rate-limiting growth mechanism, since no linear relationship between I_D and synthesis time existed. It was concluded that the crystallization kinetics recorded in the temperature interval 80–98 °C correlated with a first-order surface reaction controlled growth mechanism and the calculated apparent activation energy of 42 kJ mol^{-1} is reasonable under the conditions studied.

Using atomic force microscopy (AFM), Anderson et al. [88] investigated the fine surface structure of zeolite Y (FAU, EMT). A number of important features were observed in the AFM image of a (111) surface of the FAU structure (Fig. 5). This image revealed several previously unknown details of the mechanisms of crystal growth. On the crystal surface a host of triangular terraces was observed with the edges of the terraces rotated by 60° to the crystal edges. The height of the triangular terraces was quite uniform and was approximately 1.5 nm. The thickness of the

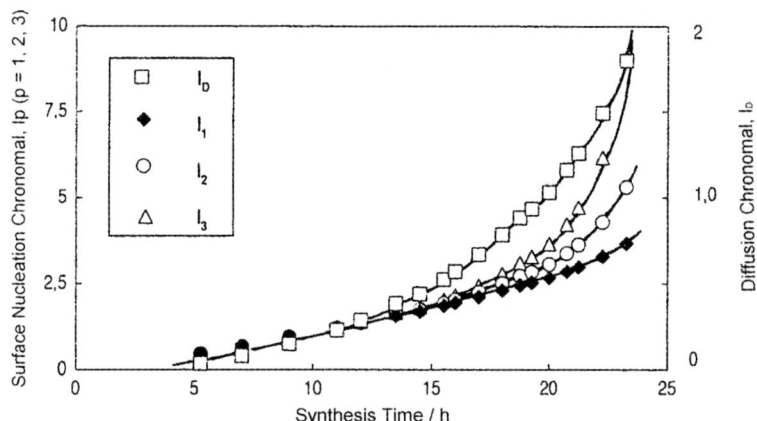

Fig. 4. The curves of the diffusion and surface nucleation chronomals as a function of synthesis time based on the data obtained from run at 98 °C. Reprinted from [78] with the permission of Butterworth-Heinemann.

triangular terrace was in very good agreement with the thickness of the sodalite cage plus a D6R of 1.43 nm, calculated crystallographically. The orientation of all triangular terraces was the same and the spacing between the terrace edges decreased towards the edges of the crystal. This was due to the fact that the area of the terraces grew at a constant rate and consequently the linear growth rate reduced as a function of growth time. Therefore, zeolite Y appears to grow by a layer growth mechanism.

Alfredsson et al. [89] showed the direct experimental evidence for a layer growth mechanism in zeolite synthesis. They showed 14.3-Å high steps on a (001) surface of FAU by using high-resolution transmission electron microscopy (HRTEM).

Ohsuna et al. [90] found, using HRTEM, that the external surface of zeolite Y contained a structure that corresponds to D6R. Furthermore, they also found that D6R units were present in the gel phase during the synthesis of zeolite Y by using ^{29}Si solid-state NMR. The D6R was, therefore, considered a key unit for crystal growth of zeolite Y [90].

The AFM study of crystal growth in zeolite A revealed striking similarities with zeolite Y. The AFM image of a (100) surface of a zeolite A crystal showed terraces, which in this case had ostensibly a square habit and terrace edges lying parallel to the edge of the cubic single crystals. Agger et al. [91] showed the height difference between successive terraces was uniformly equal to 1.20 ± 0.15 nm. This terrace height is consistent with one sodalite cage layer connected through D4R. Similarly, Sugiyama et al. [92] revealed that the surface of zeolite A crystals involved small steps of three different heights; 0.5, 0.7 and 1.2 nm. The 0.5 and 0.7 nm heights coincide with the shorter side (0.5072 nm) and diagonal (0.7201 nm) length of a

References see page 804

a)

b)

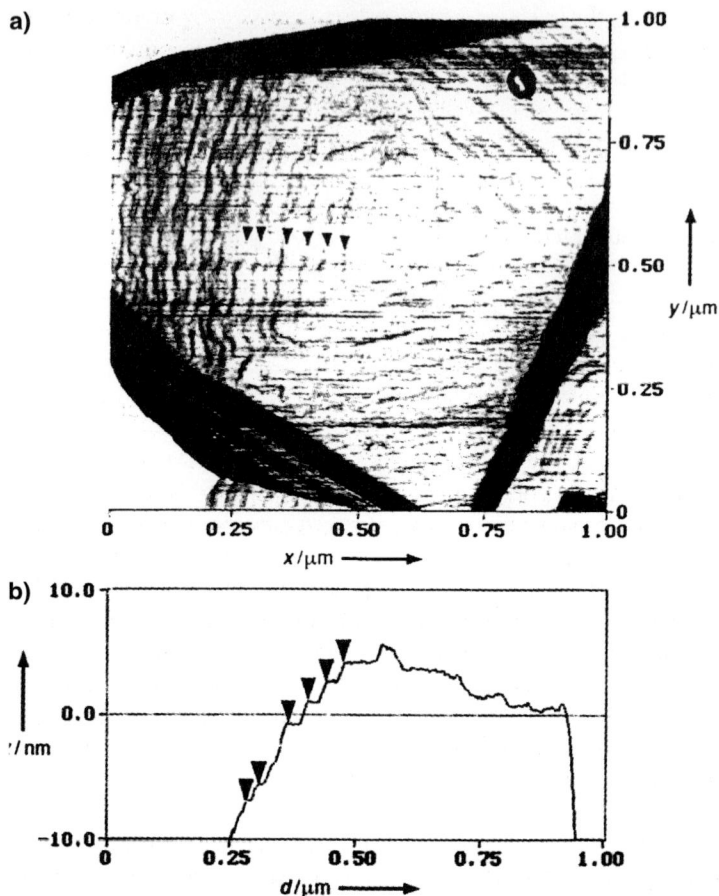

Fig. 5. (a) AFM image (tapping mode) of a faujasite structure showing the $(111)_c$ face. The triangular terraces are growth steps each about 1.5 nm in height. This value was derived from section analysis; (b), which shows height and separation of crystal steps (d: lateral shift in the xy plane). Reprinted from [88] with the permission of Wiley-VCH Verlag GmbH.

D4R, respectively. The highest step corresponds to the length of one side of a unit cell (1.2273 nm) of zeolite A. These results suggested that D4R is the key building unit for crystal growth of zeolite A as shown in the case of zeolite Y synthesis for the D6R.

In spite of the obvious differences in the crystal morphologies of zeolite A and Y, close parallels may be drawn between their crystal growth mechanisms. Both zeolites grow by means of a layer mechanism, in each case the height of layer being on the order of angstroms, that is, unit cell size. In the zeolite A, each layer comprises the equivalent of a sheet of sodalite cages and D4Rs. Similarly, each layer is composed of a sheet of sodalite cage and D6Rs in the case of zeolite Y. Both zeolites also show a linear relationship between terrace height and terrace area, indicative of a terrace-ledge-kink (TLK) mechanism.

Thus, it appears from the evidence revealed that transport by diffusion in the liquid layer is not the rate-limiting step in the zeolite crystal growth, but that the incorporation of solute by surface integration kinetics may well be. It also appears that the assembly units may be unit cells, parts of unit cells (D4Rs or D6Rs), or species that add these steps on the surface.

4.2.2.5
Effect of Seeding on Crystallization

It has been known since at least the late 1960s [52, 93] that adding seed crystals of desired zeolite phases to synthesis batches increases the rate of crystallization, defined simply as the slope of the crystallization curve, and shortens the time required for the crystallization to be completed. However, the mechanism of the rate enhancement has not been thoroughly understood. Kerr [52] noted that the induction period, during which nuclei form and grow to an observable size, could be eliminated by the addition of zeolite NaA to a batch designed to produce that zeolite. He also noted that the crystallization was completed in about 80 % of the time required for the nonseeded preparations, when the sufficient seed materials were placed in the sodium silicate solution before preparing the aluminosilicate gel [93]. Still more effective seeding could be achieved by saturating boiling solutions of reagents with seed material prior to formation of the amorphous substrate and by adding seed to the reaction mixture. It was noted that the initial aqueous phase of the reaction mixture had a deleterious effect on zeolite X; perhaps silica and/or alumina were preferentially dissolved from the crystals to alter at least a portion of the crystal surfaces that served as nucleation or growth sites [93].

Subsequently, several studies have reported on the crystallization rate enhancement for other zeolite synthesis systems [79, 94, 95]. The mechanism of the rate enhancement was not clearly delineated, but two explanations have been offered: (1) the added surface area of the seed crystals results in the more rapid consumption of reagents, reducing the supersaturation, even to the extent that nucleation of new crystals is prohibited [94], or (2) seeds promote nucleation by some secondary nucleation mechanism [24]. The additional area resulting from this enhanced nucleation, then, results in the faster consumption of reagents.

It was found that, on occasions, adding seed crystals of a particular zeolite phase forces the solution to produce that particular phase, rather than the phase the solution would otherwise form [93, 94, 96, 97]. Mirskii and Pirozhkov [96] reported on experiments in which seed crystals were added to normal batch zeolite synthesis mixtures. They added zeolite NaA to hydroxysodalite synthesis solution, producing zeolite NaA, and added hydroxysodalite to a zeolite NaA synthesis solution, producing hydroxysodalite. They also added mordenite crystals to a zeolite NaX synthesis solution, producing mordenite. Warzywoda and Thompson [97] studied the influence of adding seed crystals on crystallization of zeolite A in the presence of K^+ ions in nonagitated solutions. Na^+ cations were increasingly replaced by K^+

References see page 804

cations in a system that precipitated zeolite NaA when only Na^+ ions were contained. At levels beyond about 20 % replacement of the Na^+ ions with K^+ ions, mixtures of zeolites A, F, and G were formed, until 80 % replacement only zeolites F and G were synthesized. The nucleation of zeolite A decreased by replacing Na^+ cations by K^+ cations as shown by Meise and Schwochow [70]. Relatively low seeding levels, approximately 10–15 % by weight, added to the 50 % of Na/K mixture resulted in formation of a new population of zeolite A crystals which did not form in the unseeded system. The same low levels of seeding with zeolite A crystals in the 100 % K system resulted in the precipitation of zeolites F and G, but in remarkably shorter time than without seed crystals. That is, the presence of zeolite A seed crystals catalyzed the formation of zeolites F and G. However, at a seeding level of 72.5 % by weight in the pure K system, zeolite A was the only observable phase, regardless of whether seeds in the Na-form or the K-form were used. It was also obvious that the seed crystals had grown and that a new population of zeolite A crystals had been formed due to the presence of the seed crystals.

These results suggested that the presence of the seed crystals facilitated the precipitation of the new population, since it did not form with a smaller amount of seed crystals or no seed crystals present in the same batch composition.

The fundamental influence of the addition of zeolite seed crystals to ZSM-5 and/or Silicalite crystallization system was reviewed [98]. A population balance model was developed and solved numerically to generate predictions of the behavior of seed crystals added to a hydrothermal zeolite synthesis. Theoretical results were consistent with expectations, that is, added seed crystals provided more cumulative surface area for the assimilation of nutrient material from solution. The crystallizations, therefore, were predicted to proceed faster with increasing amounts of seed crystal surface area added, either in the form of more seeds or the same amounts of smaller seed crystals. The size evolution of seed crystals and newly formed crystals was shown to proceed along similar paths with time, and the final average crystal sizes were shown to be reduced with increased amounts of seed crystal surface area.

The two principal categories of nucleation from solution are denoted as "primary" and "secondary" nucleation. Within the category of primary nucleation fall homogeneous nucleation from solution and heterogeneous nucleation on foreign surfaces, also driven by a supersaturation in solution. Secondary nucleation mechanisms result from the presence of crystalline material in the medium and are typically of four types: initial breeding, needle breeding, fluid shear, and contact nucleation [24, 99, 100]. Needle breeding, which usually results from breakage of dendritic pieces from relatively larger crystals, could be ignored, since the dendrites rarely form on zeolite crystals. Thompson and Dyer [101] reported that contact nucleation, or microattrition, did not occur in their zeolite NaA syntheses.

Initial breeding results from microcrystalline dust being washed off seed surfaces into the crystallization solution. This microcrystalline dust is typically formed when the seed crystals are dried or during handling of the dry powder. Any residual droplet on the crystal after filtration might have dissolved reagents in it. These ingredients become more concentrated as the solvent evaporates and can eventually precipitate as crystallites. These microcrystalline fragments, initially held on seed surfaces by weak electrostatic forces, can become viable growing entities in a fresh synthesis solution. These fragments will appear as a new population of crystals once they have grown to observable sizes.

In the series of studies on the effect of adding Silicalite-1 seeds to an Al-free NH$_4$-ZSM-5 crystallization system, it was shown that initial breeding was responsible for the burst of nucleation [98, 102–105]. In particular, it was reported that seeds added to these unstirred systems resulted in populations of significantly smaller new crystals compared to the unseeded systems [104]. It was also shown that these entities could be separated from the seed crystal surfaces by slurrying the seed crystals in either water or mild NH$_4$OH solution followed by filtration [105]. The closely related phenomenon of polycrystalline formation was shown to occur when relatively large hydroxysodalite seeds were put into a clear solution containing dissolved aluminosilicates [103]. The product was predominantly hydroxysodalite; however, the well-defined habit was lost and a polycrystalline mass was formed over the seed crystals. Surface nucleation was also shown to occur on zeolite NaA seed surfaces in a clear solution environment [103]. In this situation, it appeared that the surface of the seed crystals covered with randomly oriented Losod crystals. In both cases, these surface crystals apparently adhered to the surface and could be dislodged from the surface with sufficient energy.

It was demonstrated that seed crystal growth alone is not sufficient to rationalize the rate enhancement caused by the addition of seed crystals, but that seed crystals carry with them much smaller entities, which also grow to become macroscopic crystals. These have been labeled "initial-bred nuclei", have been shown to reside on the surfaces of seed crystals [102], have been shown to be easily washed off the seed crystal surface [105], and have been shown to grow quite separately of the seed crystals in the synthesis solution [98]. One of the hypotheses regarding the origin of the initial-bred nuclei that were found on the seeds and in the washing solution from these seeds was that these small particles were unreacted aluminosilicate gel left from synthesis of the seeds. It was shown in a study using a zeolite NaA system that these small spheres form from residual aluminosilicate material, that they can grow, and finally that they form crystals of zeolite NaA [106].

Gora and Thompson [106] investigated the effect of adding seed crystals to clear synthesis solution using the same mass of the three different sized seed crystals. When the small seed crystals were added to a clear solution synthesis batch, the seed crystals simply grew with time and the new population nucleated from the solution and grew as if the seeds were not present. It was concluded that small seed crystals (about 1–3 µm) did not promote nucleation of a new population of zeolite crystals. However, much larger zeolite NaA crystals (about 40 µm) were demonstrated to promote zeolite crystal nucleation by an initial breeding mechanism. It was found that a new population of nuclei was created quite early in the synthesis and continued to grow as shown in Fig. 6. Small particles that were on

References see page 804

a. 0.75 h   b. 1.25 h

c. 1.75 h   d. 2.25 h

e. 3 h

Fig. 6. SEM micrographs of product obtained by using 0.02 g of 40-µm zeolite A seeds. Reprinted from [106] with permission from Elsevier Science.

the order of 0.1–0.2 µm in size were observed only on the larger crystal surfaces by SEM analysis. These small particles could be washed off from the large crystal surfaces and were observed in a filtrate solution by quasi-elastic light scattering spectroscopy (QELSS). The synthesis with washed large seed crystals resulted in less enhancement of the crystallization rate. It appeared that the washing procedure had removed most of the initial-bred nuclei from the seed crystal surfaces, and the nucleation enhancement caused by their presence had been eliminated almost completely. It was also shown that stopping the syntheses of seed crystals prematurely or adding the filtrate solution from washing the seed crystals instead of seed crystals were sufficient to promote new crystal nucleation. The authors concluded that the initial-bred nuclei might be residual aluminosilicate oligomer structures created in solution prior to incorporation into the solid crystalline phase, or they might arise during the cooling step prior to filtration.

Gora et al. [107] also studied the behavior of initial-bred nuclei in the reaction mixture by quasi-elastic light scattering spectroscopy and electron microscopy. Syntheses of zeolite NaA were carried out with clear aluminosilicate solutions adding prepared gel particles. The seed materials were prepared separately for 1 or 2 days using another batch composition. These seed materials were X-ray amorphous after one day, but contained about 11 % zeolite NaA crystalline in the gel particles after 2 days at 90 °C. The seed materials could exist and grow only in the complete synthesis solution, whereas dissolution occurred within 15 min in a NaOH solution, a silicate solution, and an aluminate solution, each having the same composition as that part of the complete synthesis solution. Despite the fact that the reaction solutions were filtered through a 0.2-μm membrane after the gel particles were added, particulates on the order of 130 nm were observed in the filtrate. These particulates were initially observed to shrink in size, and then to grow with reaction time. It was concluded that these particulates contributed to promotion of the crystallization rate and enhancement of the crystalline mass produced.

While adding seed crystals to a crystallization system can increase the rate of crystallization and improve the purity of the crystal product, the seeds need not be large, but may be quite small, even undetectable to the naked eye [25, 98, 108, 109]. These crystallization catalysts are sometimes called "directing agents" [109], and they serve the purpose of promoting the rate of formation of the desired phase. Several reports on zeolite synthesis have indicated that it was possible to synthesize a pure zeolite product in shorter crystallization times by adding a "clear aqueous nuclei solution" to the reaction mixture. It was found that a degree of control over the number of crystallites of zeolite Y [108] and Silicalite-1 [25] could be achieved by the addition of specific amounts of germ nuclei to a fresh synthesis solution. Gora and Thompson [110] conducted a systematic study, in which the strategy of controlling the nucleation process of zeolite NaA by adding a known amount of germ nuclei was demonstrated. The syntheses of zeolite NaA were carried out in batch systems that had been seeded with various amounts of an aged solution (30 min or 38.5 h at 25 °C) having the same batch composition as the standard fresh aluminosilicate solution. It was found that the crystallization yield increased with the amount and the aging time of added solution. When a solution aged for 30 min was added to a reaction mixture, a bimodal crystal population formed. As more of the solution aged for 30 min was added to the fresh solution, the relative amount of the smaller crystals, that is, the population nucleated by the fresh solution, seemed to become smaller. However, the zeolite crystallized from the reaction

References see page 804

mixture with the addition of a solution aged for 38.5 h produced a single-crystal size distribution. These results suggested that adding increasing amounts of secondary nuclei, presumed to be present as a result of the aging process, had an increasing tendency to suppress primary nucleation from the fresh solution. There seemed to be competition for active solute between the growth of the particles formed from the added aged solution and the nucleation and growth of the crystal formed by the fresh solution. If many secondary nuclei were added to the system with an aged solution, it is possible that most of the active solute would be used to grow these crystals at the expense of nucleating and growing a new population. On the other hand, too few secondary nuclei added to the reaction system would cause almost no interruption in the normal process of nucleation and growth of a new population resulting from conditions in the fresh solution.

The same solution aged for 38.5 h that precipitated zeolite NaA in the standard solution facilitated the nucleation of zeolite NaX in a different synthesis solution, which was richer in silica. Although that unseeded aluminosilicate solution by itself did not yield any zeolite after several hours at synthesis temperature, the addition of the standard aged solution precipitated zeolite NaX in relatively short periods. This suggested that nuclei existing in the aged solution did not have a precisely defined crystalline structure, because these nuclei on one occasion facilitated formation of zeolite NaA, while in another solution they facilitated formation of zeolite NaX. These results suggested that the nuclei could serve as a heterogeneous nucleation site, and that their structure might not be specific to the synthesis solution, as suggested in other studies [27, 28, 74, 111].

Therefore, the mechanism of the crystallization rate enhancement resulting from the use of seed crystals could be interpreted as follows; (1) nanometer-sized particles are quite probably present in most seed crystal samples, which can become viable growing zeolite crystals, (2) these nanometer-sized particles appear to be present in the mother liquor in which the seed crystals were formed, (3) they are physically separable from the seed crystals at the conclusion of the seed crystal synthesis, (4) they may have been present in the seed crystal batch prior to the conclusion of the seed crystal synthesis, and (5) they have been used separately from the seed crystals to increase the crystallization rate of a subsequent zeolite crystallization. Finally, these results illustrate some common features among various, seemingly different, zeolite systems. This observation reinforces the notion of developing fundamental concepts from specific systems that can be applied more broadly.

4.2.2.6
Effect of Aging of Amorphous Gel on Crystallization

Aging some aluminosilicate gels at room temperature can markedly influence the course of zeolite crystallization at the appropriate elevated temperature [48, 49, 63, 108, 112, 113]. The primary effects of the gel aging are the shortening of the induction period and the acceleration of the crystallization process [48, 49, 63, 112], but in some cases the gel aging also influences the types of zeolites formed [48, 108, 113].

In the study of crystallization of zeolite NaA from the gel aged at ambient temperature for 0–2 days, it was found that the time of aging did not influence the rate of linear crystal growth, whereas the duration of crystallization at 90 °C and the size of crystals in final products decreased with the time of aging [49]. It was concluded that the onset of nucleation at ambient temperature is the only reason for the decrease in duration of the crystallization after aging of the gel. From the kinetic analysis of the experimental data, Bronic et al. [63] showed that aging of the gel influenced only the nucleation process and not the growth rate of zeolite NaA particles. They calculated the number of both nuclei formed by heterogeneous nucleation in the liquid phase during the precipitation of gel (nuclei-I) and released from the gel during the crystallization, that is, autocatalytic nucleation (nuclei-II). The number of both nuclei increased with gel aging time. It was shown that their kinetic equations could be used to explain these increases in the number of nuclei and the calculated fractions of zeolite A, and that the analysis agreed well with the experimental results during the main part of the crystallization process. Therefore, it was concluded that the increase in the number of the quasi-crystalline nuclei inside the gel matrix and/or in the liquid phase during the gel aging resulted in the increase in the crystallization rate [63].

Gora et al. [74] investigated the influence of gel aging on crystallization of zeolite NaA from clear aluminosilicate solution using QELSS and electron microscopy. It was found that the induction time decreased with increasing aging times of the solution at room temperature, and for syntheses from a solution aged 6 days or more it was close to 0 min as shown in Fig. 7. Aging the reaction mixture for even as little as 1 min before elevating the temperature was shown to have the effect of reducing the induction time. Aging the synthesis solution produced a significant increase in the number density of growing crystallites. The number of particles, which was calculated based on the size of the particles and the mass of product,

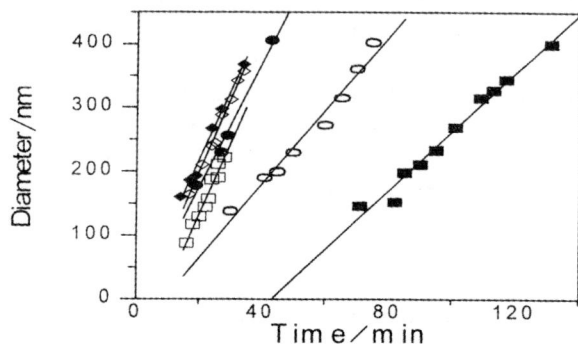

Fig. 7. Effect of aging the standard solution at 25 °C on subsequent crystallization kinetics at 60 °C. (■), nonaged; (○), aged for 7 h; (□), aged for 10 h; (●) aged for 1 day; (◇), aged for 2 days; (◆), aged for 6 days. Reprinted from [74] with permission from Elsevier Science.

References see page 804

increased 10-fold during the first 10 min aging (from on the order of 10^6 without aging to on the order of 10^7 particles per gram of solution). In a solution aged for 3 h or more the concentration of particles increased by another factor of 10.

Twomey et al. [25] proposed that during the induction time for Silicalite-1 the initial germ, or nonviable nuclei, were being generated from aluminosilicate species in solution and had not reached the critical size necessary for further growth to occur spontaneously. Thus, the induction time was reduced for aging times due to increased number of activated nuclei that began growing sooner than without aging as shown by Bronic et al. [63].

Unlike the induction time, the crystal growth rate of the crystallites measured by QELSS was influenced by the aging procedure, for aging times beyond approximately 7 h, increasing from 4.6 nm min^{-1} for unaged samples to about 11 nm min^{-1} after 10 h of aging [74]. Aging longer than 10 h did not change the growth rate of the crystals. It was demonstrated that prolonged aging of the synthesis solution resulted in increased levels of polycrystalline particle formation, probably due to agglomeration of the nuclei during the aging period.

The recent use of several novel analysis techniques such as, for example, small angle X-ray (SAXS) [35–43] and neutron (SANS) scattering techniques [43], and light scattering techniques [25–28, 44] have supplied further details on the mechanisms of zeolite growth. A SAXS and SANS study of an aging Silicalite-1 synthesis mixture identified small particles in the size range 1–10 nm. The evolution of particle size in the hydrothermal process was monitored by these scattering techniques. Many authors agreed on the formation of subcolloidal particles with a more or less internally ordered structure resembling the final product. These subcolloidal particles have been shown throughout the course of the crystallization of Silicalite-1. Several authors also observed the formation of larger intermediates in the crystallization process. Even though small particles and aggregates of such particles have been observed with many experimental techniques, their role in nucleation and growth is still poorly understood. The possible growth mechanisms discussed in the literature were that an existing crystal might grow by incorporation of small subcolloidal particles or aggregates of such particles or by monomer/oligomer addition from the solution.

In a series studies on TPA-Silicalite-1 Kirschhock et al. [114–118] have proposed the crystal growth mechanism by aggregation of intermediate nanoblocks in the synthesis solution (this subject will be described later in Sect. 4.2.2.14). The increases in the growth rates with aging time can be attributed to the agglomeration of nuclei, because polycrystalline particles have a higher surface area per particle, and, therefore, would be expected to assimilate material faster than single crystals with regular planar surfaces. However, this effect would be expected to become less important with increased aging due to limits on the number of nuclei activated and the limited extent to which particles can become agglomerated, since particle mobility decreases as agglomerates increase in size.

Zeolite NaP (gismondine-type) can co-crystallize with zeolite NaX and NaY (faujasite) if the solutions are not aged [93, 119]. However, when the crystallization was carried out using gels aged at ambient temperature for 1–10 days, zeolite NaX appeared as the first crystalline phase, thereafter zeolite NaP co-crystallized with zeolite NaX. After the maximum yield of zeolite NaX crystallized had been attained, the fraction of zeolite NaX slowly decreased as a consequence of the transformation of zeolite NaX into the more stable zeolite NaP. These results are consistent with the typical reaction sequence under the appropriate synthesis condition; amorphous – faujasite – gismondine [10]. However, in some cases, zeolite NaP appears as the first crystalline phase when freshly prepared gel has been heated at the appropriate temperature [79, 93]. On the other hand, faujasite can be crystallized either by adding seed crystals into the freshly prepared gel [79, 108], or by aging the gel at ambient temperature prior to the crystallization at the appropriate temperature [108, 120, 121].

Subotic et al. [62, 63, 65, 121, 122] showed that the formation of primary particles (nuclei) of zeolite takes place by two processes: the first is fast heterogeneous nucleation in the liquid phase (nuclei-I), and the second is the release of particles of a quasi-crystalline phase from inside the gel matrix, which act as potential nuclei (nuclei-II). Since the nuclei-I are assumed to be in full contact with the liquid phase they could start to grow immediately after the system has been heated, whereas the nuclei-II could become active only after their release from the gel dissolved during the crystallization, that is, when they come into full contact with the liquid phase. X-ray diffractograms and IR spectra revealed that the solid phase of freshly prepared gel contained a number of very small particles of quasi-crystalline phase having a structure close to the structure of the cubic modification of zeolite P [120]. It was also shown that the position of the amorphous maximum in the X-ray diffractograms of the solid phase of variously aged gels shifted toward lower X-ray diffraction spacing. This result indicated that structural changes took place in the solid phase of the gel during the aging period. As noted in the Raman spectroscopic study of the aging of the gel prepared for the crystallization of zeolite Y [56], it is reasonable to assume that the structural changes observed are the consequence of the slow formation of six-membered aluminosilicate rings and the possible formation of particles of quasi-crystalline phase (nuclei-II) with the structure similar to that of the faujasite inside the gel matrix.

The induction period of both zeolites NaX and NaP shortened and the maximum yield of zeolite NaX increased with the increased time of gel aging [63]. The average growth rate of zeolite NaX and of zeolite NaP was constant during the crystallization process and independent of the gel-aging period. The constancy of the crystal growth rate of both zeolites indicated that the shortening of the induction periods of the crystallization of both zeolites was the consequence of the increase in the number of nuclei during the gel aging. Similar increases in the number of nuclei during the aging were observed in their zeolite A crystallizing system [63].

The synthesis of zeolite Y from gels derived from colloidal silica requires room-temperature aging of the synthesis gel prior to heating to induce crystallization of the desired phase. Aging of such gels suppresses the formation of zeolite phases

References see page 804

other than faujasite, for example, zeolite P, R (chabazite-type), and S (gmelinite-type) [18, 119]. Aging results in higher crystallization rates and yields of zeolite NaY, and enables zeolite NaY to be made from batches containing smaller amounts of excess silica and base [18]. Based on studies of crystallization from aged colloidal gels, it has been proposed that structural rearrangements occur during aging that lead to the formation of zeolite NaY nuclei [48, 56, 120]. Fahlke et al. [120] proposed a "dissolution-precipitation" process as follows; (1) immediately after mixing the reactants the solid phase contains predominantly SiO_2 gel; (2) during aging time the SiO_2-rich primary gel goes into solution as low molecular weight silicate anions; (3) these silicate anions react with aluminate anions present in the solution and form aluminosilicate anions; (4) these aluminosilicate species precipitate from solution as a gel containing low molecular building units.

Ginter et al. [123, 124] investigated the physical and chemical transformations occurring during aging of gels produced from colloidal silica. The composition of gel was $4\ Na_2O : 1\ Al_2O_3 : 10\ SiO_2 : 180\ H_2O$. It was found that a gel was formed immediately on initial mixing of the colloidal silica sol with a sodium aluminate due to the flocculation of silica particles in the sol. Shearing of the initially formed gel created fine granules that were on the order of 1 mm in diameter, but did not alter the microstructure of the gel. There was no chemical transformation of silica on mixing, but during aging, the silica particles slowly dissolved, releasing monomeric silicate anions into the surrounding solution. These species reacted rapidly with aluminate anions to produce amorphous aluminosilicate precipitates having a Si/Al ratio of 1 and a Na/Al ratio of 2. Further aging resulted in the dissolution of any remaining silica and the restructuring of the aluminosilicate solid via the combined process of dissolution and precipitation. Prolonged aging brought about the incorporation of additional Si into the initially Al-rich aluminosilicate solid, gradually converting it to a less-hydrated solid. This gave rise to a larger number of smaller nuclei and resulted in a higher final yield of zeolite NaY. It is also significant to note that zeolite NaR and NaS were observed as the dominant phases when crystallization was carried out without prior aging of the gel using the batch composition noted above, as previously reported [18, 119].

4.2.2.7
Effect of Nature of Reactants

Freund [119] investigated the mechanism of zeolite X crystallization and, in particular, noted the effect of various silica sources. A series of syntheses was carried out with the molar ratio of reaction mixture $12.70\ Na_2O : Al_2O_3 : 8.60\ SiO_2$, 85.5 wt % H_2O at 80 °C. The crystalline product obtained had the composition $1.00\ Na_2O : 1.00\ Al_2O_3 : n\ SiO_2 : x\ H_2O$ ($n = 2.95$ for zeolite X; $n = 3.05$ for zeolite P1 (pseudo-cubic gismondine-type)). The silica sources used in his study were classified into two groups: "active" and "inactive". The former (particularly metasilicate hydrates) yielded zeolite X. The latter (mainly solid silicas and colloidal suspensions) yielded only zeolite P1. His results also varied if the aluminosilicate gels were aged at room temperature. When using an "active" silica source, zeolite X was crystallized without an aging period. Even though using an "inactive" silica source, the yield of zeolite X in an unstirred synthesis increased with the length of the aging period. Hence the effect of an "active" silica source could be to promote the rapid formation of a number of nuclei at the beginning of crystallization, or even before the system was brought to the crystallization temperature. He also found that the cause of the activity of some hydrated sodium silicates was related to their aluminum content. However, the activity of an "inactive" silicate was not promoted when aluminum oxide was added to pure silica. Hence aluminum must be present, at least partly, in tetrahedral coordination in the silica sources to promote activity. At that time, he suggested that the activity of the silica sources could be correlated with the immediate formation of a large number of nuclei, which he attributed to the presence of Al^{3+} impurities in the starting silica sources.

Lowe et al. [125] investigated the activity of various silicates in batch mixtures for zeolite A, X, and Y. The activity of sodium metasilicate hydrates was determined by examining the products obtained from a standard reaction in which a mixture of overall stoichiometry $5.15 \, Na_2O : Al_2O_3 : 4 \, SiO_2 : 242 \, H_2O$ was crystallized at 95 °C for 3 h with stirring. It was shown that the quantities of zeolite X and zeolite P1 in their final products depended on the amount of aluminum in the metasilicate and the age of silica source. The pure nonahydrate when 2 days old gave a completely amorphous product and the pentahydrate when 5 days old gave a very slight trace of zeolite X. However, a major amount of zeolite X was produced when the pure alumina trihydrate was added during the preparation of both hydrates. Consequently, they also concluded that the activity of the silica sources could be correlated with the level of Al compounds in the starting silicas. Both studies [119, 125] had determined that "active silicates" were those that had relatively high levels of aluminate impurities.

A systematic study of the effect of varying the starting silica source on the crystallization of zeolite X was reported by Hamilton et al. [126]. The batch composition of the synthesis mixture was the same throughout the study, and given by $4.76 \, Na_2O : Al_2O_3 : 3.5 \, SiO_2 : 454 \, H_2O : 2.0 \, TEA$, where TEA represents triethanolamine, used to stabilize the sodium aluminate solution and to produce slightly larger particles than otherwise would form. All of the sodium silicate solutions prepared had the same composition, regardless of the silica source. Separate sodium aluminate and sodium silicate solutions were filtered through 0.20-μm membrane filters prior to mixing the solutions together to make the amorphous gel phase. All the individual filtered solutions were clear, as the reagents were completely dissolved prior to the synthesis experiments, and residual particulate matter larger than 0.20 μm was filtered out. The syntheses were carried out in Teflon-lined autoclaves at 115 °C and autogeneous pressure. There were large differences in the results from the various batch synthesis mixtures, even though all parameters were identical, except the source of the silica. The synthesis times for each experiment (Fig. 8) and the ultimate particle sizes (Fig. 9) from each solution were quite different. The crystallization curves for four different silica sources,

References see page 804

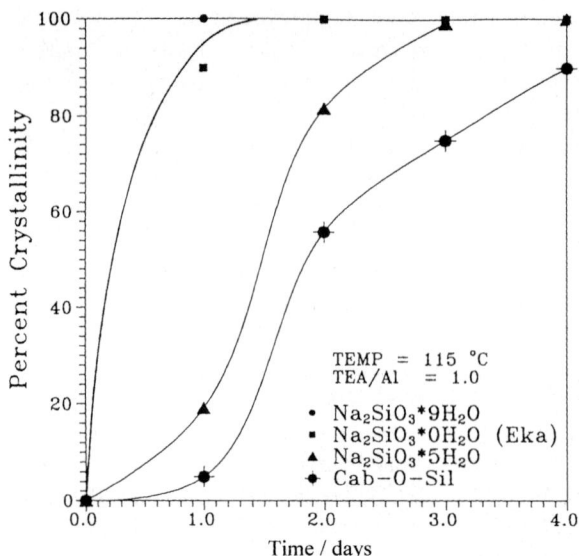

Fig. 8. XRD crystallinity versus synthesis time for four of the silica sources in synthesis of zeolite NaX. Reprinted from [126] with the permission of Elsevier Science.

sodium metasilicate nonahydrate, anhydrous sodium metasilicate, sodium meta-silicate pentahydrate, and Cab–O–Sil, showed some differences in the progress of the syntheses. The first two were essentially complete in about 1 day at the reaction temperature, the third was complete after about 3 days, and the fourth was still incomplete after 4 days as shown in Fig. 8 [126]. The fastest synthesis, with the sodium metasilicate nonahydrate, yielded the smallest particles (7.5 μm), whereas the slowest synthesis, with the Cab–O–Sil, yielded the largest particles (90 μm). Each system produced different numbers of nuclei, which consumed material from the solution at different rates, due to the different cumulative surface areas. The amorphous gel was, therefore, converted to crystalline zeolite X in different time periods. The results were interpreted in terms of inherent differences in the silicate solutions formed from the various silica sources, because all the silica sources were completely dissolved and filtered prior to combining with the alumi-nate solutions. It was speculated that the impurity levels contained in the silica sources might be responsible for promoting nucleation of zeolite X crystals. It was noted that the correlation of the number of nuclei formed with impurity levels was equally good with Al^{3+}, Fe^{3+}, Mg^{2+}, or Ca^{2+}, however, no one specific impurity was shown to correlate better than others. Similar impurities added deliberately to the silicate solutions did not have any observable effect on the outcome. Therefore, at that time, no convincing argument could be found persuasive to identify any particular impurity as the key ingredient in promoting nucleation in the system.

In the context of the studies by Freund [119] and Lowe et al. [125], an active sil-ica source for production of zeolite X would be one that promotes the formation of

Na$_2$SiO$_3$*9H$_2$O (Fisher)

Na$_2$SiO$_3$*5H$_2$O (Fluka, 490)

Na$_2$SiO$_3$*0H$_2$O (Eka)

Cab-O-Sil (Cabot)

Fig. 9. SEM micrographs of the final products from the syntheses shown in Fig. 8. All micrographs shown at the same magnification for comparison: (a) Na$_2$SiO$_3$·9H$_2$O (Fisher); (b) Na$_2$SiO$_3$·5H$_2$O (Eka); (c) Na$_2$SiO$_3$·0H$_2$O (Fluka, 490); (d) Cab-O-Sil (Cabot). Reprinted from [126] with the permission of Elsevier Science.

many zeolite X nuclei, that is, having a relatively large cumulative surface area available to assimilate material from solution. It was concluded that the most important factor affecting zeolite nucleation was most likely related to the levels of soluble impurities in the silica source, or insoluble particulate matter smaller than 0.2 μm in size.

Wiersema and Thompson [86] reported similar results on the influence of silica sources on nucleation and crystal growth of analcime. Hydrothermal syntheses of the zeolite mineral analcime were carried out in clear aluminosilicate solutions with batch composition 87 Na$_2$O : Al$_2$O$_3$: 84 SiO$_2$: 2560 H$_2$O at 160 °C. Four different silica sources such as Cab–O–Sil, puratronic silica, sodium silicate nonahydrate, and sodium silicate pentahydrate were used. The number concentrations of crystals nucleated using the different silica sources were slightly different in all four syntheses. These results suggested that there was something inherent in the silica sources that affected zeolite nucleation in these clear aluminosilicate sol-

References see page 804

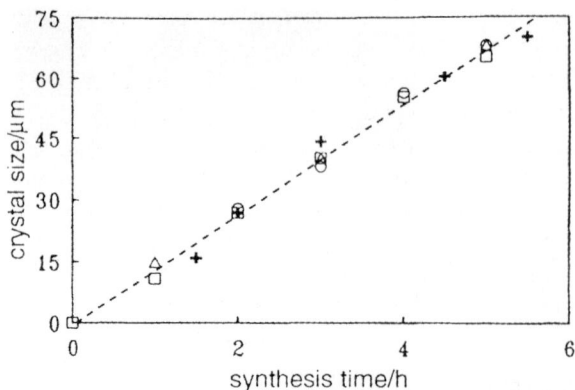

Fig. 10. Change in the linear dimension of analcime crystals at 160 °C using four different silica sources in synthesis using the standard batch composition. Symbols represent the silica sources: (□) Cab–O–Sil; (△) puratronic silica, (○) sodium silicate nonahydrate and (+) sodium silicate pentahydrate [86]. Reproduced by permission of The Royal Society of Chemistry.

utions, as in the previous case of zeolite NaX. Additionally, the differences noted in the analcime study correlated to metal impurities in the silica source as shown by Hamilton et al. [126]. However, Fig. 10 shows that the linear crystal growth rate was the same for the crystals in all four synthesis batches. This result suggested that the driving force for zeolite crystal growth was the same in all four experiments, and was a function of the material in the clear aluminosilicate solutions. That is, crystal growth rates did not vary with impurities in the silica sources, as the nucleation rates did.

Further, it was found that a second population of analcime crystals was nucleated once the first population had grown to such a large size that the crystals settled to the bottom of the autoclave, thereby removing crystal mass from the solution and reducing the consumption of solute. Therefore, it was suggested that impurities were present in sufficient concentration to catalyze the nucleation of the second population of analcime crystals. Twomey et al. [25] physically removed the first population of Silicalite crystals from the synthesis solution by filtration, and also observed that a second generation of nuclei formed in the remaining solution.

Antonic et al. [127] investigated the inherent differences in basic silicate solutions produced from various silica sources by ^{29}Si NMR and the molybdate method. They used four different silica sources, sodium metasilicate nonahydrate, sodium metasilicate pentahydrate, anhydrous sodium metasilicate, and silicic acid, from the same lots that had previously been used by Hamilton et al. [126]. All synthesis solutions were prepared in the same manner as shown by Hamilton et al. [126]. It appeared from the ^{29}Si NMR spectra that there were no significant differences in the silicate oligomer distributions in the freshly made silicate solution. Similarly, there were no significant differences in the ^{29}Si NMR spectra of the solution aged for 44 days, and all the spectra were almost identical to the corresponding spectra of the freshly prepared solutions. These observations suggested that there were no discernible changes in the silicate solutions over the aging period, at least none that could be determined by ^{29}Si NMR. They also analyzed the distribution of silicate species in silicate solutions by the molybdate method. In contrast to the insensitivity of the distribution of silicate anions to a short aging time, an aging time of 44 days considerably influenced the distribution of silicate anions in silicate solutions. The degree of polycondensation in the aged silicate solutions increased in the sequence: (silicic acid) < (anhydrous sodium metasilicate) < (sodium metasilicate pentahydrate) < (sodium metasilicate nonahydrate). The polycondensation increased with increasing content of crystalline water in the silica source. However, the distribution of silicate anions showed that approximately 50 % of the dimers and the remaining cyclic trimers and higher chain polymers in silicate solution, were not markedly influenced by the silica source. These results might suggest that there were minor differences, perhaps sufficient to cause differences in subsequent synthesis, but this could not be argued with much confidence from these observations. Therefore, questions about the influence of the silica source on the crystallization pathway and particulate properties of the crystalline are still not elucidated completely.

4.2.2.8
Effect of Alkali Cations

The behavior of alkali cations in solutions containing dissolved silicate and aluminate is well described in some reviews [10, 18, 22]. In zeolite synthesis alkali cations influence the crystallization process in two main ways: depending on their tendency to become hydrated, they modify the supersaturation of the solution (salting-out effect); depending on their structure-making or structure-breaking effects on water, they may orient and stabilize the aluminosilicate anions into particular configurations and direct the nucleation process (structure-directing effect).

Breck and Flanigen [14] reported that from gels of equivalent composition the substitution of K^+ for Na^+ resulted in the formation of zeolite L in place of zeolite Y that was formed in the sodium aluminosilicate system. They also showed that the substitution of K^+ for Na^+ in gels of equivalent compositions resulted in the formation of zeolite K–F in place of the zeolite A. They concluded that these results might be due to the size of the hydrated cation and its influence on the arrangement of the structural units of aluminosilicate polyhedra in the final zeolite framework structure. Meise and Schwochow [70] also reported that by replacing Na^+ ion by K^+ ion in a standard zeolite A batch composition, somewhat larger crystals of zeolite A were formed, even though the synthesis process took slightly longer. Similar results were reported by Warzywoda and Thompson [97], in which the Na^+ cation was increasingly replaced by K^+ ion in a system that precipitated zeolite NaA when only Na^+ ions were used. It was found that the crystallization process took longer with the increased fraction of K^+ ions in the system and that

References see page 804

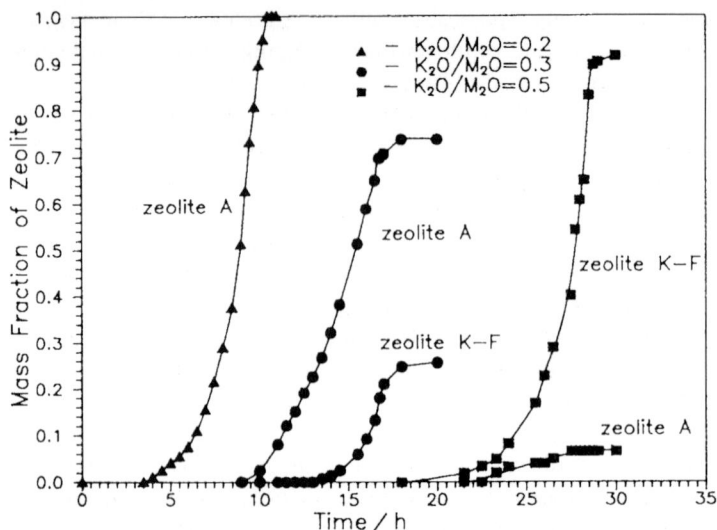

Fig. 11. Influence of potassium ion on zeolite A crystallization (zeolite A appears in the mixture with zeolite K–F). Reprinted from [97] with the permission of Elsevier Science.

the larger zeolite A crystals were formed with the increased K^+ ions level. It is thought that the observations were the result of Na^+ being required for nucleation of zeolite A; thus, at reduced Na^+ levels, fewer nuclei formed, resulting in larger crystals at complete conversion. Figure 11 shows that zeolite K–F co-crystallized with zeolite A and represented an increasingly larger fraction of the mass of the final product as the K^+ ion content was increased above 20 %. Zeolite A effectively stopped forming at replacement levels beyond 80 %, and zeolite K-G appeared as another coprecipitating phase. Therefore, it seems that the Na^+ ions facilitated, and the K^+ ions did not participate in the nucleation of zeolite A. The size of the hydrated K^+ ion and its influence on the structure of aluminosilicate structures in solution are probably responsible for these results [14].

Some interesting features of the zeolite synthesis in a mixed alkali ion system were shown in [128]. Starting with a sodium aluminosilicate composition that results in zeolite Y synthesis, the influence of K^+ and Li^+ ions on nucleation of this system was examined. On increasing the fraction of K^+ ions substituted for Na^+ ions, zeolite Y was replaced by zeolite D and finally chabazite was formed. The process appears to consist of competitive nucleation, in which Na^+ ions favor zeolite Y nucleation and K^+ ions favor chabazite. In the Na–Li system, it was found that at high Li^+ ion level, zeolite Z, which was similar to (Li, Na)-E as reported by Borer and Meier [129], is formed. However, at intermediate Li^+ ion level, the system remained amorphous for extended periods of time. The Raman spectra of the Na^+ and Li^+ exchanged forms of both zeolites X and Z revealed the structural changes in framework. It was proposed that Na^+ and Li^+ ions disrupt the nucleation of the competing zeolite framework, while at the same time nucleating zeolites on their own. Thus, it was demonstrated that in the presence of co-cations, not only can nucleation to a different zeolite occur, but that cations are able to prohibit the nuclei of co-crystallizing zeolites.

Schoeman et al. [76] reported the effect of sodium concentration on crystallization kinetics, product distribution, and particle size in the system of SiO_2–Al_2O_3–$(TMA)_2O$–Na_2O–H_2O, in which the alkali sources were tetramethylammonium hydroxide, TMAOH, and NaOH. A series of syntheses was performed with the molar ratio of reaction mixture $(2.5 - x)$ $(TMA)_2O : x$ $Na_2O:Al_2O_3 : 3.4$ $SiO_2 : 370$ H_2O $(x = 0.04–0.50)$. The sodium concentration in the synthesis mixture was adjusted by addition of NaOH solution. High sodium contents favored the formation of zeolite A, syntheses where the molar ratio of Na_2O/Al_2O_3 was more than 0.20 yielded zeolite A as the single phase. The reduction of sodium content caused the coexistence of zeolite Y and zeolite A in the products. A reduced sodium concentration, $Na_2O/Al_2O_3 = 0.04$, resulted in the formation of zeolite Y as the only crystal phase. The product distribution was essentially unchanged, a mixture of zeolite Y and A, even if sodium was added after the so-called induction period and the final sodium content was the same as in syntheses yielding a zeolite A. The product distribution, however, changed to zeolite A as the only crystal phase, if the addition of sodium took place at too early a stage in the crystallization process, during the induction time. It was shown that the time at which the sodium is introduced to the crystallizing system is crucial in determining which zeolite phase was obtained.

Relatively high Na_2O/Al_2O_3 ratios favored the formation of large crystals of zeolite A. Low Na_2O/Al_2O_3 ratios resulted in colloidal crystals of zeolite Y, whereas intermediate ratios yielded zeolite Y and/or zeolite A with a particle size of approximately 100 nm. The final particle size in the synthesis with high sodium content ($Na_2O/Al_2O_3 = 0.40$) was about 240 nm, whereas the size in the synthesis of lower sodium content ($Na_2O/Al_2O_3 = 0.04$) was about 100 nm. Higher sodium contents in the synthesis mixture resulted in shorter crystallization times and in a higher ultimate crystallinity. It was demonstrated that a carefully controlled sodium content in the TMA-aluminosilicate solutions enabled the synthesis of colloidal suspensions of zeolite Y and A or mixtures thereof.

On the other hand, another effect was also observed that the addition of sodium during the crystallization process permitted particle growth upon existing crystals without a secondary nucleation step, thereby resulting in increased zeolite yield.

4.2.2.9
Addition of Organic Compound

The synthesis of zeolites at around 100 °C usually produces crystals in the size range 0.1–10 μm. Charnell [130] showed that the synthesis of larger zeolite A and X crystals could be achieved with 2,2′,2″-nitrilotriethanol (triethanolamine, or TEA). TEA was added as a stabilizing, buffering, or complexing agent to the synthesis mixtures. The sodium metasilicate solution, the sodium aluminate solution,

References see page 804

and TEA solution were filtered through 0.20-μm filter membranes prior to mixing together to make the synthesis mixtures. The mixed gel solutions were hydrothermally treated in 1.9 l polypropylene jars, placed in a water bath at 75 to 85 °C. Crystallization was usually complete in 2 to 3 weeks for zeolite A and in 3 to 5 weeks for zeolite X. The crystals were obtained in sizes approaching 100 and 140 μm in width for zeolites A and X, respectively.

The role of triethanolamine (TEA) in zeolite syntheses has been the subject of a number of studies. Scott et al. [131] investigated the dominant role of TEA in zeolite A synthesis systems. Crystallization curves revealed an increase in the length of the induction time and in overall crystallization time in the presence of the amine. Particle size distributions of the products became steadily broader, furthermore, the average crystal size of the products steadily increased with increasing TEA. These results suggested that the presence of TEA suppressed the nucleation rate. Solution-phase ^{13}C NMR spectroscopy revealed that TEA complexed with aluminum in basic solution. Two sets of peaks were observed, one for free TEA (which was present in excess with respect to aluminum), and one for the proposed aluminum-TEA complex. A systematic investigation was made of the ^{13}C NMR spectra of solutions with a TEA/aluminum ratio of between 0.5 and 1.8. A plot of relative ^{13}C NMR peak intensity for the bound-TEA complex to the content of TEA in solution showed that no free-TEA in solution was observed at TEA/Al < 0.9, and the amount of unbound TEA increased at higher TEA/Al ratios. This result suggested that the TEA-aluminum complex had 1:1 stoichiometry. ^{27}Al NMR spectra showed the peak having comparable chemical shift to that of the aluminum-TEA complex, at around 63 ppm, suggesting that TEA chelated aluminum with tetrahedral symmetry (i.e., the hydroxyl groups of the molecule were directly involved in bonding to the metal ion) [132]. Since the peak due to $Al(OH)_4^-$ was hardly confirmed when the synthesis solutions containing TEA were heated to a typical reaction temperature of 358 K, it was demonstrated that this aluminum-TEA complex was stable to hydrolysis under these conditions. TEA was, therefore, thought to create a slow-release mechanism for aluminum in zeolite A synthesis via complexation of the metal ion [133]. At large TEA additions, $TEA/Al_2O_3 > 10$, zeolite X was produced in increasing amounts until almost pure zeolite X, which generally forms from more siliceous gels, was observed at $TEA/Al_2O_3 = 30$ as shown in Fig. 12. These results supported the aluminum complexing role of TEA.

The binding of aluminum reduced the effective concentration of free aluminum in solution, and thus reduced the degree of supersaturation of the synthesis mixture. It is the supersaturation of the solution phase that provides the driving force for nucleation and crystal growth. It is to be expected then, that the reduction of the level of supersaturation through removal of free aluminum species will reduce the tendency for nucleation and crystal growth. It can be assumed that the complexation of aluminum with TEA was equilibrated in the synthesis mixture. Thus, as the concentration of aluminum in solution was depleted due to crystal growth, the TEA-aluminum complex reaction reversed to replenish the aluminum in solution and allow crystal growth to continue.

Fig. 12. Effect of TEA on the crystallization of zeolite NaA (A) and NaX (X). Reprinted from [135] with the permission of Elsevier Science.

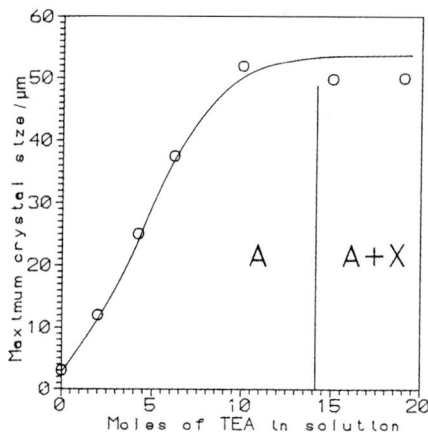

Charnell [130] first reported the growth of large crystals of zeolite A of approximately 60 µm in edge-length in the gel systems Na_2O–Al_2O_3–SiO_2–H_2O–TEA with the addition of TEA. Thompson et al. [131–133] have carried out extensive studies on the crystallization of zeolite A using TEA and some other tertiary alkanolamines as additives. They reported that the crystal sizes increased smoothly with increasing the molar ratio of TEA to alumina (TEA/Al_2O_3) to 10, at which point crystals up to 50 µm were obtained, while the crystals of 2 µm were obtained without addition of TEA. By optimization of factors affecting the crystal size, such as the TEA amount, SiO_2/Al_2O_3 ratio, Na_2O/Al_2O_3 ratio, silicon source, aging time and crystallization, large single crystals of zeolite A up to 80 µm with high quality were successfully crystallized in the gel mixture 1.12 SiO_2 : 1.0 Al_2O_3 : 2.55 Na_2O : 6.0 TEA : 280 H_2O using TEOS as the silica source [134].

The relative efficiency of four different tertiary alkanolamines, 1,1′,1″-nitrilotri-2-propanol (TPA), bis-(2-hydroxyethyl)-amino-2-propanol (BEP), 3-[N,N-bis-(2-hydroxyethyl)-amino]-2-hydroxypropane sulfonic acid, sodium salt (DIPSO), and bis-(hydroxymethyl)-2,2′,2″-nitriloethanol (BIS), as nucleation suppressors in the synthesis of zeolite A has been also investigated [135]. The four additives contained three hydroxyalkyl groups and possessed a hydroxyl group on the second carbon atom, preserving somewhat the geometry of the TEA molecule, while altering the length of the alkyl chains. The actual effect of each was found to be different with BEP offering no significant advantage over TEA, while TPA was less efficient than TEA as an aluminum chelator in zeolite synthesis. The other two compounds, DIPSO and BIS, were found to be better in this role than was TEA. The crystals prepared with 10 parts of BIS (i.e., 10 mol BIS per mole of Al_2O_3 in the synthesis solution) showed the co-precipitation of cubic zeolite A (about 10 µm) and octahedron-shaped zeolite X (about 30 µm). By contrast, the crystals prepared with 15 parts of BIS contained large (up to 80 µm) zeolite X, together with a small

References see page 804

amount of spherical particles. Since 30 parts of TEA were required in order to obtain pure zeolite X, BIS could be much more efficient as an aluminum chelator in zeolite synthesis solutions.

Zhu et al. [134] reported the synthesis of nanocrystals of LTA by using of tetramethylammonium hydroxide (TMAOH) as the hydroxide source and NaCl salt. It was found that, by using TMAOH as the hydroxide source, the zeolite A crystal size decreased significantly. By controlling the amount of NaCl in the gels with molar composition of 1.12 SiO_2 : 1.0 Al_2O_3 : 2.75 $(TMA)_2O$: x NaCl : 280 H_2O (x = 0.04–0.5), fine particles of LTA with the size ranging from 1 μm to less than 0.2 μm were crystallized. It was found that the crystal size of LTA decreased with the decrease of NaCl concentration. The existence of a trace amount of NaCl in the reaction mixture is essential for the formation of LTA, since the Na^+ ion is believed to play a role of stabilization of the D4R in the LTA structure. Higher temperature (100 °C) and longer aging time (2 days) favored the formation of nanocrystals.

The synthesis of zeolites in systems related to those that produce FAU (zeolite X and Y) was studied both in the presence and absence of TMA^+ cations by Hopkins [136, 137]. The effect of TMA^+ on the synthesis of zeolite X was investigated with the molar composition of 3.0 SiO_2 : 1.0 Al_2O_3 : 3.6 M_2O : 144 H_2O (M = TMA^+ + Na^+). TMA^+ was substituted for sodium, while maintaining the sum of the two constant. Substitution of more than one-half of the TMA^+ by sodium caused the product to change from FAU to LTA. A similar change, but at higher TMA^+ content, was observed in syntheses of zeolite Y. When sodium was the predominant cation the product was zeolite Y, but when there was more TMA^+ than sodium the product was ZK-4 (high-silica LTA). They also found the zeolite product changed first from gmelinite to omega and finally to HS as the TMA^+/Na^+ ratio increased in another reaction series. At low TMA^+ contents zeolites composed predominantly of D6R were synthesized, zeolite Y and gmelinite. The framework of gmelinite could be considered as being made up of parallel layers of D6Rs. As the TMA^+ content of reaction mixtures was increased structures that do not include D6R layers were formed. Sodalite units of zeolite Y were filled statistically by one TMA^+ or about two Na^+ ions based on the relative concentrations of the two ions in the reaction medium. This result suggested that sodalite units in zeolite Y could be formed without templating but cations were required for charge balancing during some step in the synthesis procedure. The presence of TMA^+ maintained higher crystallinity of the LTA structure even in high silica reaction environments. This indicated that the stabilization of the LTA structure at high reactant Si/Al ratios is one important role of TMA^+ in the synthesis. It was shown that essentially all sodalite cages in ZK-4 contain one TMA^+ cation by ^{13}C NMR [138]. These results, therefore, suggested that the role was as a template for sodalite cage formation. Templating of sodalite cages apparently, however, is not required for synthesis of zeolite A (Si/Al ratio of one) because the reaction is facile in the absence of TMA^+.

The mechanism of synthesis of zeolite A involving D4Rs joined to form the sodalite units was proposed [92, 139, 140]. This mechanism appears to be satisfactory because it provides the sodalite unit with six D4Rs to direct further reaction to the

LTA structure, but not to FAU or SOD. Analogously the synthesis of zeolite Y probably proceeds by formation of sodalite units by joining of four D6Rs [101, 102, 109]. Joining of D6Rs to form sodalite units appears to be facile and affected by the presence of TMA$^+$.

Molecular modeling shows that the spherical TMA$^+$ ion fills the sodalite cage well; TMA$^+$ also fits in the almost spherical gmelinite cage but somewhat more loosely. These fits suggest that TMA$^+$ may function as a template during the synthesis of these zeolites as well as acting as the counterion to the negatively charged framework [136].

Mabilia et al. [141] reported a series of complete structure optimizations of sodalite cages, using free-valence geometry molecular mechanics, in which both Si/Al composition and bonding topology were varied. Their major finding was that incorporation of Al atoms into the sodalite cage had little effect on the optimized molecular geometry, but played a major role on structural stability. As the amount of Al increased, the stability of the sodalite cage also increased. Consequently, TMA$^+$ is required to stabilize high-silica sodalite units, which are less stable than low-silica sodalite units, by templating.

4.2.2.10
Synthesis of Zeolite from Clay Minerals

It has been well known since the early 1960s that zeolites can be hydrothermally synthesized from clay minerals by reaction bases [6, 142, 143]. There are increasing numbers of reports related to successful synthesis of zeolites from clay minerals [10, 18]. Kaolinite, a clay mineral with an oxide formula of $Al_2O_3 : 2 SiO_2 : 2 H_2O$ is known to give hydroxysodalite on treatment with aqueous alkali below 100 °C. Metakaolinite, usually obtained by dehydroxylating kaolinite at temperatures from 500 to 700 °C, has been used as a convenient starting material for the synthesis of low-silica zeolites, for example, NaA, NaX, and NaY. Thus, the type of zeolite formed in hydrothermal reactions depends on the nature of the raw materials used even when the same molar compositions are maintained in the $Na_2O-Al_2O_3-SiO_2-H_2O$ quaternary system. Although much effort has been expended to optimize the reaction parameters of detergent-grade zeolite NaA, the mechanism of this reaction and the structural consequences of alkali and thermal treatments have not been sufficiently investigated.

Rocha et al. [144] investigated the changes of the metakaolinite structure in alkaline media. It was found that the aluminous matrix transformed very rapidly, and penta- and hexa-coordinated Al were converted into tetra-coordinated Al when metakaolinite was treated with caustic solution. However, the siliceous matrix was much more stable. They suggested the step that determines the kinetics of the synthesis of zeolite NaA from metakaolinite was the collapse of the siliceous matrix of metakaolinite. Indeed, Costa et al. [145] showed that the induction period of the synthesis of zeolite NaA from metakaolinite in alkaline media decreased when

References see page 804

the synthesis mixture was aged. This result suggested that the aging caused an increase in the nucleation from amorphous gel that enhances its reactivity, without any significant influence on crystallinity of zeolitic product.

Rees and Chandrasekhar [146] investigated the mechanism of zeolite crystallization from metakaolinite by comparing its structure and the composition of intermediates with those of the conventional gel systems. It was shown that the metakaolinite slowly dissolved in the alkaline medium to form a gel that was the direct precursor of zeolite NaA as observed in the conventional gel system. However, the crystallization seemed to start even before the metakaolinite gel dissolution was completed. The microscopic study revealed the difference in the intermediates and products of the two systems as follows. The crystals in the conventional gel system were well separated and more perfect in shape. The crystals in the metakaolinite system consisted of sharp-edged cubes, 1–6 μm, and larger crystals of layered growths, and exhibited a wide crystal size distribution. This suggested that a secondary nucleation occurred in the metakaolinite system.

It has been demonstrated that kaolinite treated by impact grinding becomes a very reactive product for aluminosilicate zeolite synthesis, as it behaves similarly to clay previously dehydroxylated by heating. Basaldella et al. [147] investigated the effect of milling on kaolinite structure and on its reactivity in zeolite synthesis, in comparison with kaolinite itself and the metakaolinite. Actually milling caused a gradual breakdown of the crystalline structure, and transformed the clay to an amorphous solid. Infrared and ^{27}Al NMR analyses evidenced modifications in the Al coordination, which changed from hexa- to penta- and tetra-coordinated. As the grinding treatment continued, the fractions of these latter species increased. The ground samples gave rise to the formation of zeolite NaA for short reaction times independent of the length of the grinding times, 750 to 9000 s. It should be noted that the nature and number of Si–O–Si, Al–O–Al, and Al–O–Si bonds in the starting materials, which was remarkably affected by heating or grinding treatment, was the key factor in the zeolite synthesis from clay minerals.

4.2.2.11
Synthesis of Cubic and Hexagonal Analogs of Zeolite Y using Crown-Ethers

The development of new zeolite structures and the optimization of the properties of known zeolites undoubtedly constitute the two main axes along which research in zeolite synthesis is being directed. The type of zeolite obtained and its properties depend on the selection of the so-called "structure-directing" agent. Most often, this agent consists of a preferred organic molecule that is incorporated in the voids of the zeolite during crystallization. The structure-directing role involves electrostatic and/or van der Waals interactions between the zeolite framework and the organic molecules, giving rise to stabilization by pore-filling and a real structure-directing effect [148, 149]. In general the analogs of high-silica zeolites have mainly been obtained using N-containing compounds, including both quaternary ammonium ions and neutral amines [150, 151].

Synthetic faujasite type zeolite Y has been known for several decades and is one of the most studied and commercially important zeolites. However, it was observed that a well-crystallized pure product with a high framework Si/Al ratio could hardly be expected unless a post-synthesis dealumination is performed [152]. Lechert's recent report [153] showed that the highest Si/Al ratio of the faujasite structure is close to 2.43. In addition, faujasite type zeolites with framework Si/Al ratio approximately equal to 3 often crystallize with impurities like gismondine or gmelinite-type zeolites [154].

On the other hand, a hexagonal variant has also been postulated even though it was not well defined as a zeolite with a cubic structure [18]. This hexagonal variant was reported in a number of zeolites such as ZSM-3 [155] and ZSM-20 [156, 157], which are intergrowths of FAU and the hexagonal polytype. The hexagonal structure was found to be generated locally at FAU twin planes [158]. The pure form of the cubic polymorph (Fig. 13a), FAU, was first synthesized by Delprato et al. [159] using the crown ether 15-crown-5 (1,4,7,10,13-pentaoxacyclopentadecane) as a structure-directing agent. It was also found that the pure form of the hexagonal polymorph (Fig. 13b), EMT, could be synthesized using 18-crown-6 (1,4,7,10,13,16-hexaoxacyclooctadecane). They tested a large number of organic species of the crown-ether family, which could serve as a substitute for a part of the Na^+ cations fitting the supercages and the sodalite-cages of zeolite Y. Only two pure and well-crystallized faujasite-type zeolites were obtained: a cubic faujasite with 15-crown-5 and a hexagonal faujasite with 18-crown-6. It was shown that eight 15-crown-5 molecules fit in the cubic unit cell (192 tetrahedra), and four 18-crown-6 molecules were accommodated in the hexagonal unit cell (96 tetrahedra), which corresponded to one crown-ether molecule per faujasite supercage. The 18-crown-6 is thought to have a structure-directing role; it is adsorbed on the growing surface and complexes with Na^+ cations creating a Na^+/18-crown-6 complex. ^{13}C CP/NMR and ^{23}Na MAS/NMR revealed that complexes [(crown-ether, Na)$^+$, OH$^-$] were present in the micropores of these zeolites. Such complexes were well known and 18-crown-6 or 15-crown-5 molecules were specific for complexing alkali-metal ions [159]. The ^{23}Na MAS/NMR chemical shift of as-synthesized products was approximately equal to that of the sample prepared with 18-crown-6 and $NaClO_4$. After removing the organic template by calcination, the chemical shift of the product was approximately equal to that of zeolite Y. This suggested that such complexes [(crown-ether, Na)$^+$, OH$^-$] were integrated into the micropores of growing crystals during the crystallization process. Subsequently, Dougnier et al. [160] showed that the as-synthesized samples were turned into microporous solids by removal of the crown-ether by calcinations at 450 °C for 4 h under air with a step at 200 °C for 2 h, and the H-form samples prepared by NH_4^+ ion exchange have the acidic properties which are similar to those of a classical USY zeolite obtained by a dealumination treatment. It is believed that the presence of Na^+/18-crown-6 complexes is essential for the synthesis of EMT [161–163]. The highest framework Si/Al ratio

References see page 804

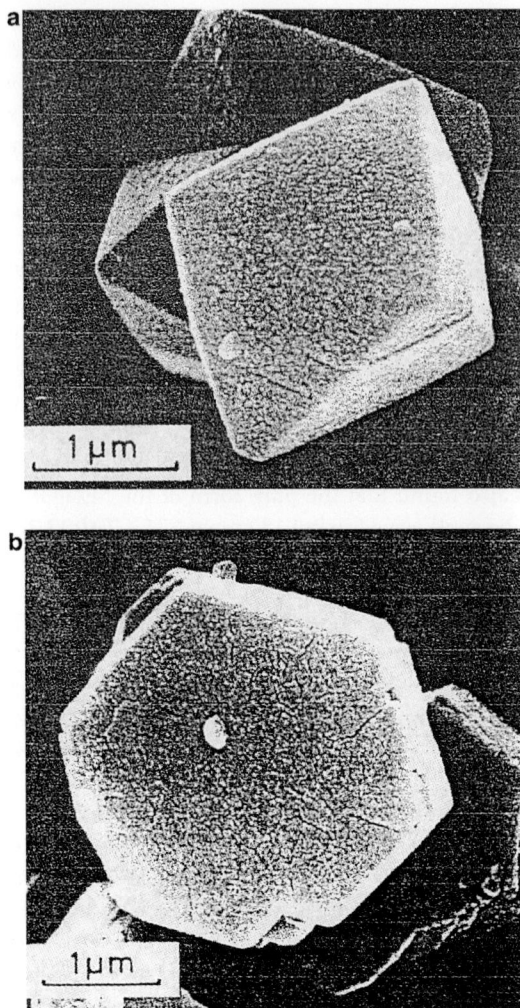

Fig. 13. Electron micrograph of the (a) cubic phase (FAU); (b) hexagonal phase (EMT). Reprinted from [159] with the permission of Elsevier Science.

that could be obtained was close to the value of 3.80, which seemed to be a limit, just as 2.43 was a limit for zeolite Y [153].

Intergrowths of the two polymorphs have been synthesized using mixtures of the crown-ethers, which were found to contain extended blocks of the cubic and hexagonal structures and not random stacking of the layers or overgrowths [164, 165]. An intergrowth of FAU and EMT can be considered as stacking of the sheets with a mixture of inversion and mirror relationships. In studies of competitive growth with mixed crown-ethers a mechanism for the crystallization of the two structures was proposed where growth is related to an oscillatory growth mechanism [161]. If 15-crown-5 is concentrated at the surface, the FAU structure is favored, but as it is depleted and its concentration drops, the growth switches to EMT when the 18-crown-6 occupies the surface. When the concentration of 18-crown-6 falls, the growth switches back to the FAU structure. It was suggested that concentration gradients of crown-ethers at crystal-growing surfaces are responsible for an oscillation of the structure between the two phases rather than a completely random intergrowth.

In recent work, Hanif et al. [166] showed that the final structure of a zeolite Y polymorph synthesized using different concentrations of two crown-ethers was affected by stirring the synthesis mixture during crystallization. With stirring the hexagonal structure became more favorable even when the concentration of 18-crown-6 was as low as 33 %. However, without stirring the preference was for intergrowth structures up to quite a high concentration of 18-crown-6. They suggested that the 18-crown-6 is a much stronger director than the 15-crown-5, and that the local concentration at the growth surface is an important factor to control the final product.

4.2.2.12
Synthesis using Microwave Heating

Microwave techniques have been extended to serve various areas in chemical research including drying of chemicals, dehydration of solid materials, acid hydrolysis of proteins, cleaning of metal surfaces, sintering of ceramics, and promotion of many inorganic and organic reactions. The use of microwave heating in the unseeded preparation of zeolite NaA and seeded preparation of ZSM-5 was first reported in the patent literature [167]. It was noted that microwave heating could be used for the rapid synthesis of zeolites. The authors reported the synthesis of zeolite NaA in 12 min. However, the products were contaminated with hydroxysodalite.

Jansen et al. [168] reported on the advantages of applying microwave heating in zeolite synthesis; zeolite NaA and NaX, hydroxysodalite, and TMA-hydroxysodalite. The synthesis mixtures were prepared according to the molar oxide ratios described by Breck [18]. The synthesis mixtures were first heated using a relatively high power of the magnetron, 800 W, to reach 120 °C in 40 s. Subsequently, the power was switched off, and after some initial cooling, the temperature was kept at 95 °C with a power adjustment to 100 W by partial shielding of the vessel using a perforated stainless steel cage. The complete crystallization of pure zeolite NaA was obtained after 10 min, which is fast compared to a conventional heating method, which normally requires several hours. Since the crystal size distribution was rather narrow, it was assumed that the main part of the nucleation took place in the high-temperature region at 120 °C, followed by crystal growth at 95 °C. However, the dissolution and crystallization of zeolite NaA took about a factor 10 longer when the syntheses were carried out using initial microwave heating to

References see page 804

120 °C followed by conventional heating at 95 °C. Moreover, the crystal size distribution showed substantially larger crystals than in the microwave synthesis. It was concluded that the short crystallization time in a microwave-heated system is mainly due to the fast dissolution of the gel. Smaller crystals in the microwave environment coupled with faster gel dissolution rates might be consistent with the facilitated release of the "autocatalytic nuclei" discussed by Subotic et al. [61, 62, 64–67].

Further, they demonstrated the effect of aging on the microwave synthesis of zeolite NaA [50]. It was found that aging is a prerequisite for the successful rapid synthesis of NaA, in contrast to the conventional synthesis. When the synthesis mixture had been aged sufficiently, the synthesis of NaA could be completed in 1 min at 120 °C with crystal sizes ranging from 0.1 to 0.3 μm. During the aging, mixing on a molecular scale was envisaged to allow the formation of nuclei necessary for the crystallization of NaA. In conventional synthesis, the mixing and formation of nuclei can take place during the heat-up, so that aging is not necessary for a successful synthesis.

They also demonstrated the synthesis of rather uniformly sized zeolite NaY with Si/Al ratio up to 5 and ZSM-5 crystals using microwave heating [169]. They proposed that microwave heating was a useful method for preparation of zeolite Y, allowing a wide range of Si/Al ratios compared to conventional heating method. In the study of zeolite NaY synthesis, small uniformly sized crystal aggregates with a maximum crystal size of 0.5 μm were obtained in 10 min, whereas 10–50 h were required by conventional heating techniques. Other crystalline phases such as zeolite P, gismondine, or gmelinite, which are often formed in conventional heating, were not found. Therefore, crystallization of zeolite NaY under microwave heating was faster and more selective than conventional heating.

It was found that zeolite NaA and sodalite could be synthesized in 10 min, NaX in 60 min, and NaY in 4 h using induction heating [170], a technique that uses an alternating magnetic field to generate heat. In the case of a zeolite synthesis, the induction heating was generated through the conducting properties of the synthesis mixture, that is, the ionic strength. The mobile ions formed the current, and, in collisions with neighboring molecules, heat was generated. Therefore, pure water cannot be heated by this method. Because the rapid synthesis of zeolite NaA and sodalite occurred with both microwave and induction heating, the acceleration of the synthesis was thought to be due not to an intrinsic microwave effect, but to a more efficient method of energy transfer to the sample.

Recently, the continuous synthesis of zeolite NaA and NaY has been achieved using a tubular reactor positioned in an oil bath [171]. The synthesis mixture was pumped with a membrane pump through a coil that was submerged in an oil bath. Syntheses were carried out with various residence times by varying the pumping rate (2–4 ml min^{-1}) and the tube length (6, 12, 18 m). NaA could be fully crystallized in 10 min at 110 °C, and NaY with a Si/Al ratio of 4 could be fully crystallized in 12 min at 140 °C. These results indicated that these zeolites can be made in conventional heating systems in synthesis times, which are similar to those found when using microwave heating. It was, therefore, concluded that the rapid crystallization obtained with microwave heating is not due to a microwave effect, but to a temperature effect. Rapid heat-up to high temperatures results in an acceleration of the synthesis.

4.2.2.13
Synthesis of Mordenite

Mordenite is a high-silica zeolite that can be synthesized hydrothermally from alkaline aluminosilicate reaction mixtures. Its typical unit cell composition is $Na_8[(AlO_2)_8(SiO_2)_{40}] \cdot 24H_2O$ [18]. Mordenite has parallel elliptical channels with free diameter of 6.95×5.81 Å, and these main channels are interconnected by small side channels of 2.9 Å free diameter. Mordenite is used in adsorption separations and in catalysis such as cracking or hydrocracking under severe environments. As a catalyst for the reactions at higher temperatures, and in particular where acidic components are involved, mordenite-type zeolites with high-silica contents are preferred [172].

Leonard [173] was the first to claim synthesis of a mordenite type zeolite in 1927 from feldspars and alkali carbonates. The synthesis was carried out at 200 °C and 15 atm pressure for a period of 7 days. However, his characterization of the product appeared to be questionable. Barrer [1] first reported a more reliable synthesis of mordenite in 1948. A sodium aluminate solution was poured into an aqueous suspension of silicic acid gel containing a trace of entrained alkali. The mixture was evaporated to dryness below 110 °C. Mordenite was crystallized in good yield from aluminosilicate gels prepared of the compositions: $Na_2O : Al_2O_3 : x\ SiO_2 : n\ H_2O$ ($x = 9.3–10.9$). Hydrothermal crystallization was carried out up to 300 °C, and the best yields were obtained at 265–295 °C for 2 or 3 days. It was noted that pH was as important as composition or temperature in ensuring a high yield, and the best results were for a range of pH between from 8 to 10 in the cold motherliquor after crystallization.

It was demonstrated that mordenite could be synthesized at relatively low temperatures, between 150 to 200 °C under certain conditions [82, 174]. Domine and Quobex [82] reported that a practically pure mordenite could be synthesized between 150 and 300 °C with the temperature affecting only the rate of crystallization. When the crystallization was carried out using mixtures of gels with the composition $SiO_2/Al_2O_3 = 12$ and $Na_2O/Al_2O_3 = 2$, pH > 13.5, mordenite was obtained after 1 h at 300 °C, between 12 and 24 h at 200 °C, and between 4 and 8 days at 150 °C. They also reported the evolution of the reaction as a function of time at different temperatures between 100 and 340 °C. The SiO_2/Al_2O_3 ratio of the reaction mixture was 10.9 and the pH was adjusted to obtain the value of 12.6 which was favorable to a fast crystallization. The induction period varied from about 1 h at 340 °C to 2 days at 200 °C and 4 weeks at 100 °C. Bajpai et al. [174] reported a similar result in the $3.5\ Na_2O : Al_2O_3 : 10\ SiO_2 : 219\ H_2O$ system at 135–165 °C. The induction period decreased from about 14 h at 135 °C to 10 h at 150 °C and

References see page 804

7 h at 165 °C, and the rate of crystallization also increased as the temperature was increased. It was observed that at higher temperatures mordenite crystallized from initial mixtures containing more silica or less alkali (Na_2O). It was explained that an increase in the solubilities of the silicate and aluminate ions resulted in the shift in the concentration of the liquid phase as reaction temperature increased. Zhdanov [51] noted that the concentration of the components, that is, silicate and aluminate species in the liquid phase of the gel was the main controlling factor in the synthesis of zeolites.

Most of the early studies resulted in zeolites with opening of about 4 Å diameter, approximate openings of the natural mordenites. Sand [175] first reported the synthesis of mordenite type zeolite with opening 7–8 Å, as the structure determined by Meier [176]. It was proposed that the smaller pore openings in natural and earlier prepared mordenites were due to stacking faults in the structure or to amorphous materials clogging the channels [176]. With the variation of the conditions, Sand [175] obtained two types of synthetic mordenites that were termed "large-port" and "small-port" mordenites. Small-port mordenites were crystallized as a single crystalline phase at higher temperatures, especially between 275 and 300 °C, with starting compositions consisting of gels of the mordenite composition ($Na_2O : Al_2O_3 : 9$–$10\ SiO_2$) and a large excess of water. The conditions for synthesis of large-port mordenites were considerably different. Large-port mordenites were synthesized at temperatures from 75 to 260 °C in which the starting compositions were on a compositional boundary between the composition of sodium silicate (0.3 $Na_2O : SiO_2$) and mordenite ($Na_2O : Al_2O_3 : 9$–$10\ SiO_2$). Crystallization of large-port mordenite at 75 °C required 168 h and a large ratio of sodium silicate in the mordenite reaction mixture. However, at 260 °C, complete crystallization required as little as 4 h with a starting composition lower in sodium silicate content.

It was known that the SiO_2/Al_2O_3 ratio of mordenite could be increased by dealumination with acid leaching of the framework aluminum or by steaming treatment [177]. However, a direct synthesis of high-silica mordenite with a SiO_2/Al_2O_3 ratio of 12–19.5 was first reported by Wittemore [178]. Furthermore, Ueda et al. [179] obtained siliceous mordenite with a SiO_2/Al_2O_3 ratio of 12.8 by aging a clear sodium aluminosilicate solution, and they also obtained mordenite with much higher SiO_2/Al_2O_3 ratio up to 25.8 by adding benzyltrimethylammonium ion into such a solution system [180]. Itabashi et al. [181] showed also that the SiO_2/Al_2O_3 ratio of mordenite crystallized from a mineral gel could be varied from 10.2 to 19.1 when the Na_2O/SiO_2 ratio of the gel was decreased from 0.145 to 0.108.

When Bajpai [182] reviewed the synthesis of mordenite extensively, he found that the synthesis conditions for mordenite were, in general, confined to narrow limits of the SiO_2/Al_2O_3 ratios of the gel; most of the studies on mordenite synthesis were reported with SiO_2/Al_2O_3 ratios ranging from 9 to 12. This SiO_2/Al_2O_3 ratio corresponds to that encountered in the natural mineral. This typical value was accounted for in terms of the preferential location of aluminum atoms in the four-membered rings of the mordenite framework, requiring a double occupancy of these structural elements. Self-consistent field molecular orbital calculations (SCF-MO) confirmed that aluminum was indeed energetically favored in the four-membered rings, and that diagonally paired aluminum sites across four-membered rings were more stable than isolated ions in the same rings [183]. Bodart et al. [184] showed that the aluminum content per unit cell decreased from eight to four atoms with increasing $[OH^-/H_2O]\cdot[SiO_2/Al_2O_3]$ ratios from almost zero to 0.5. No further decrease in the aluminum content per unit cell for higher $[OH^-/H_2O]\cdot[SiO_2/Al_2O_3]$ ratios was observed. This range of aluminum content corresponds to between two and one aluminum atoms per four-membered ring. A mordenite with two to one aluminum atoms per four-membered ring corresponds to a SiO_2/Al_2O_3 ratio varying from 10 to 22. It was concluded that alkaline or silica-rich media favor the formation of siliceous mordenite. Many reports of the siliceous mordenite synthesis in the presence of organic compounds were presented in the form of patents. Some important patented information was reviewed by Jacobs and Martens [185]. They showed that there was a direct proportionality between SiO_2/Al_2O_3 ratios in the synthesis gel and the zeolite phase for mordenite synthesized with these ratios in the range 10–30. This indicated that mordenite with high SiO_2/Al_2O_3 ratio greater than 22 up to 30, could be obtained at a high efficiency. It was suggested that a SiO_2/Al_2O_3 ratio of 30 then corresponds to an occupation of about one aluminum atom per every second four-membered ring and a probable upper limit for the SiO_2/Al_2O_3 ratio in mordenite.

A direct synthesis of high-silica mordenite without organic compounds was reported by Kim and Ahn [186]. Crystallization was carried out with the molar ratio of reaction mixtures (2–20) $Na_2O : Al_2O_3 : (10–100)$ $SiO_2 : (240–6000)$ H_2O at 150–170 °C. It was shown that mordenite with SiO_2/Al_2O_3 ratio of 31.5 was obtained from the substrate composition of 14 $Na_2O : Al_2O_3 : 80$ $SiO_2 : 1680$ H_2O. This directly synthesized high-silica mordenite showed a higher thermal stability compared to the dealuminated mordenite with almost same SiO_2/Al_2O_3 ratio. They also investigated the effect of aging and crystal seeding on the crystallization rates of mordenite synthesis at 170 °C. Aging treatment at room temperature had an adverse effect on crystallization rates, and, in addition, longer aging time led to larger mordenite crystals. The pH of the substrate decreased significantly during the aging step, as a consequence of the reaction of OH^- with silica particles producing silanol or soluble silicate anions. The crystallization rates and crystallinity of mordenite were found to be enhanced when amorphous silica powder (Zeosil) increasingly replaced the sodium silicate solution as silica sources in the synthesis mixture. They suggested that the fine silica particles promote the nucleation of mordenite better than do the dissolved silicate or aluminosilicate ions, whereas it was suggested that the dissolution of polymeric silica particles was the rate-determining step in the crystallization of mordenite using silicic acid [187] and ZSM-5 [188] as the silica sources.

The initial composition of any mixture is an important factor in governing the type of zeolite crystallized. In the $Na_2O–Al_2O_3–SiO_2–H_2O$ system, the SiO_2/Al_2O_3 ratio can decide the zeolite species obtained and their yield. Barrer [1] showed

References see page 804

that quartz often appeared in some quantity at the silica-rich end of the range ($Na_2O : Al_2O_3 : 12.3 \ SiO_2$) and analcime was found in increasing quantities at the silica-deficient end ($Na_2O : Al_2O_3 : 8.2 \ SiO_2$). Sand [175] reported the relationship between the composition of the starting reaction mixtures and some of the zeolite phases produced as shown in Fig. 14. The diagram shows that the compositions of aluminosilicate gels that yield large-port mordenite as a single crystalline phase were limited. In the higher silica content range, quartz, opaline silica, and crystalline sodium silicate hydrates were found as coexisting phases. Sodalite, albite, analcime, and phillipsite were formed from gels of lower silica content. Somewhat similar results on the synthesis of large-port mordenite were reported by Bajpai et al. [174]. They noted that the sequence of product formation was in the order, analcime to phillipsite/mordenite to amorphous material, as the SiO_2/Al_2O_3 ratio was increased. However, they did not observe the formation of quartz at higher SiO_2/Al_2O_3 ratio, as the temperature range was not high enough to crystallize quartz.

The alkalinity of the starting mixture is another important factor in the synthesis of zeolites. Within the stability field of a given zeolite, increasing alkalinity at constant temperature influences the kinetics like increasing temperature at constant alkalinity. Domine and Quobex [82] reported on the influence of the alkalinity on the rate of reaction at a temperature of 300 °C. Figure 15 shows the evolution of crystallization as a function of time for pH values increasing from 10.2 to 13.3. It was found that the induction times decreased strongly and the crystallization rates were accelerated owing to the increased pH up to a certain limit. At pH = 13.3 the mordenite first formed began to transform into analcime with increased reaction time.

The type of starting material was found to be critical for the kinetics of crystallization [82, 175, 187, 189, 190]. Sand [175] reported the influence of silica source on the rate of crystallization at 175 °C of mordenite from an anhydrous batch composition $2.6 \ Na_2O : Al_2O_3 : 15.6 \ SiO_2$ with water contents of 65, 75, and 85 mol %. When diatomite was used as a source of silica, the rate of crystallization was very fast, and the crystallization was essentially completed in about 16 h with 65 mole % water. An increase in water content resulted in the decrease of crystallization rates and the increase of overall crystallization time. However, when silicic acid was used, crystallization could be observed following an induction period of about 10 h and was still incomplete after 24 h in the case of 65 mole % water. An increase in water content caused the process to proceed more quickly, finishing in about 20 h with 85 mole % water.

Domine and Quobex [82] reported that the crystallization from oxides in a state of gel was faster than the crystallization of the amorphous compound at approximately identical pH value. Mordenite was obtained from gel within 1 h at 300 °C, while the crystallization of the amorphous product began only after 4 h.

Bajpai et al. [189] reported that under identical conditions of temperature and overall reaction composition, reactive silica from rice husk ash in the form of a silica solution resulted in shorter induction periods and overall mordenite crystallization times than did silica present partially as silica gel. They also reported that

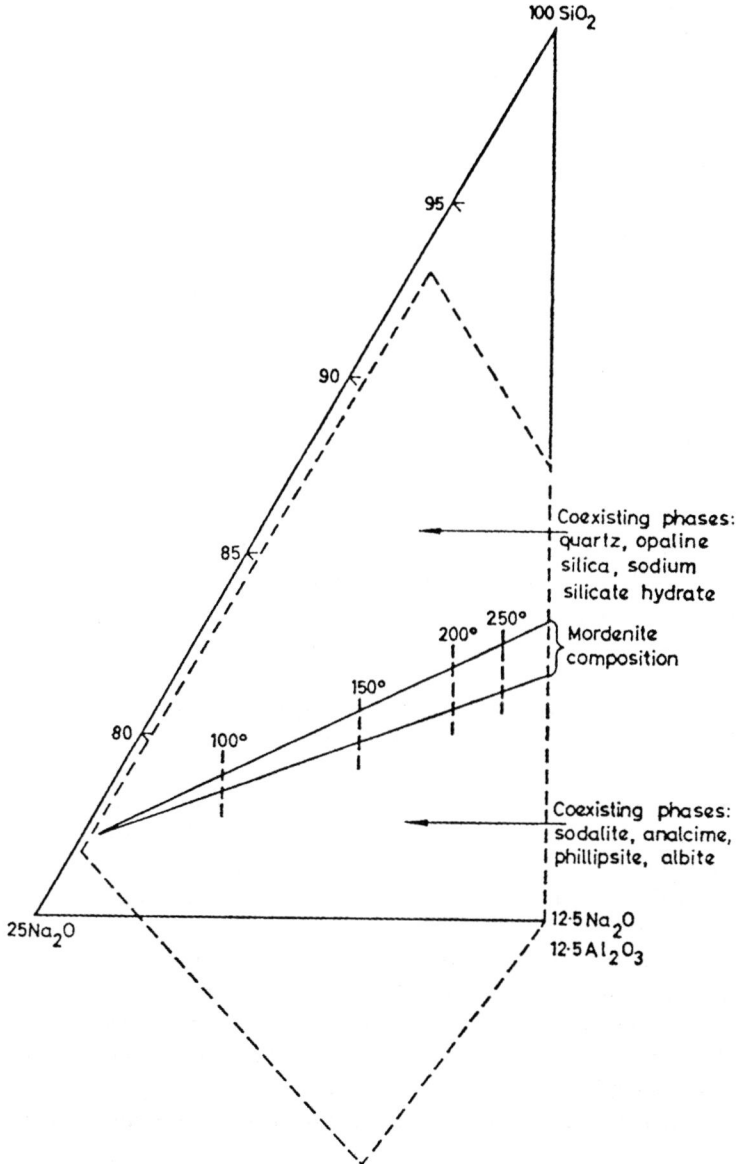

Fig. 14. Conditions for producing large-port mordenite as a single phase as function of temperature and anhydrous batch composition are shown as a wedge between the mordenite composition and a sodium silicate of the composition 0.3 Na_2O, SiO_2. Reprinted from [175] with permission from the Society of Chemical Industry.

References see page 804

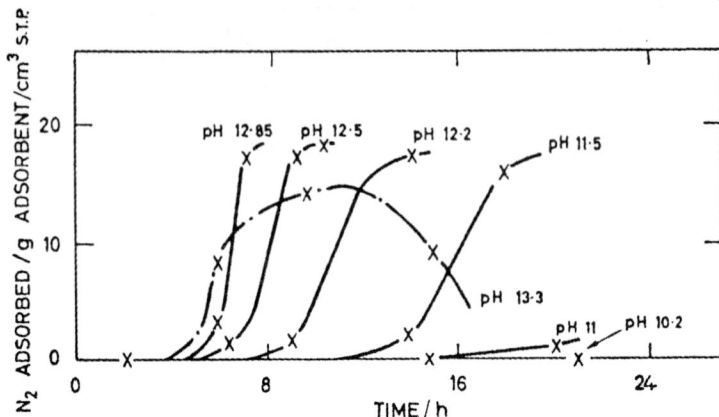

Fig. 15. Influence of pH on the crystallization rate of mordenite at 300 °C from amorphous compound. Reprinted from [82] with permission from the Society of Chemical Industry.

for synthesis of mordenite using reactive and amorphous silica from rice husk ash, relatively lesser Na_2O or greater SiO_2 content in the starting mixtures was required compared to that using silica from chemical sources. It was also observed that analcime appeared as a product while using silica from rice husk ash at temperatures of 150 °C and above, and SiO_2/Na_2O ratio less than 3. However, mordenite could be formed using these conditions when silica from a chemical source was used. It was suggested that the different results observed for these crystallizations might be caused by different solubilities of the silica sources in the final mixtures.

Sun et al. [190] reported that large single crystals of mordenite could be synthesized hydrothermally from clear homogeneous solutions by using both Aerosil and sodium silicate as the silica source. For instance, large crystals with a size of 185×125 μm were crystallized at 150 °C for 15 days by using the mixture with composition of 15 Na_2O : Al_2O_3 : 60 SiO_2 (Aerosil) : 15 SiO_2 (sodium silicate) : 4 NaCl : 550 H_2O. A difference in the reactivities of the two silica sources was claimed to facilitate crystallization of large single crystals.

Warzywoda et al. [187] reported that the use of different lots of silicic acid powders in synthesis mixtures resulted in different conversion rates and product particle sizes for mordenite from mixtures with the same initial batch composition (4.32 Na_2O : Al_2O_3 : 19 SiO_2 : 293.6 H_2O). Different overall crystallization times (lot A = 10 h, lots B and C = 28 h), conversion rates at the 50 % point (lot A = 44 % h^{-1}, lots B and C = 12 % h^{-1}), and maximum crystal sizes (lot A = 10–15 μm, lot B = 35–40 μm, lot C = 30 $\pm$ 35 μm) were observed (Fig. 16). It was shown that mordenite crystallization rates could not be correlated with the amount of impurities present in the investigated lots of silicic acid, since adding significant levels of impurities did not make the system active or promote mordenite nucleation nor retard the crystallization by silica stabilization. The different amounts of

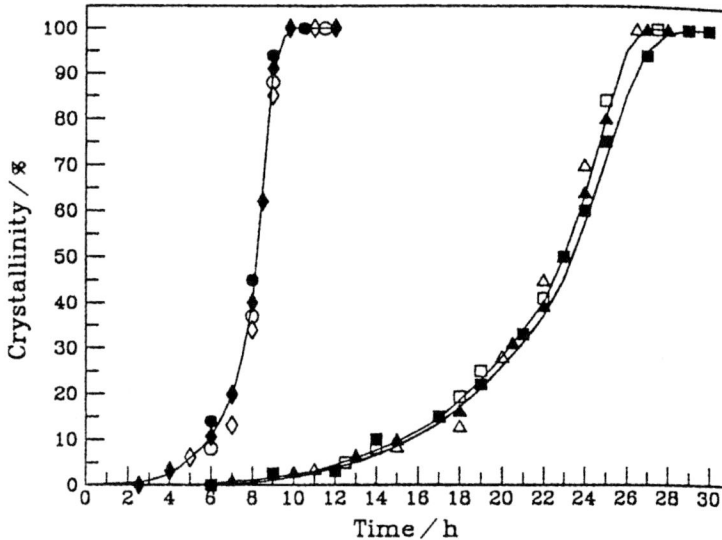

Fig. 16. Mordenite crystallization curves obtained using untreated silicic acid ($\blacklozenge$, $\diamondsuit$) lot A; ($\blacksquare$, $\square$) lot B; ($\blacktriangle$, $\triangle$) lot C; ($\bullet$) lot A fortified with oxides of Al, Ca, Mg, Na, Fe, and Ti; ($\bigcirc$) lot A fortified with salts of Al, Ca, Mg, Na, and Fe. ($\blacklozenge$, $\blacksquare$, $\blacktriangle$, $\bullet$, $\bigcirc$) 20-min aging; ($\diamondsuit$, $\square$, $\triangle$) 6-days aging. Reprinted from [187] with permission from Elsevier Science.

physically adsorbed and chemically bonded water in silicas and different particle sizes of aggregate silica substrates, that is, Carman permeability surface area, did not affect the kinetics of crystallization and the crystal sizes of final products. They suggested that the dissolution of amorphous silica particles was the rate-limiting step in the mordenite syntheses in their experiments. The different crystallization rates of mordenite were found to be the results of the different specific surface areas and the structure, for example, average pore diameter and the strength of Si–O–Si bonds, of the silicic acid powders. The dehydration of silicic acid powders by heat treatment caused a decrease in their specific surface areas and resulted in the synthesis of larger crystals of mordenite with all three lots of silicic acid. Thus, lot C heated at 850 °C prior to crystallization produced large crystals up to 250 μm in length.

Only a few studies have been reported in the literature dealing with the crystallization mechanism and kinetics of mordenite synthesis. In general, the crystallization curves show a sigmoid shape indicating an induction period followed by a rapid crystallization as shown in other general zeolite synthesis systems. The crystallization of mordenite from hydrogels is thought to occur in the liquid phase of the reaction mixture, and the gel dissolves progressively during the crystallization, supplying the nutrients directly to the growing crystals [191]. Culfaz and Sand [94] showed the mechanism of nucleation and crystal growth in the seeded systems.

References see page 804

Microscopic evaluation of the crystals grown in seeded systems indicated that the single crystals of seed grew only partially, and new acicular crystals were grown separately from the seed crystals. It was demonstrated that nucleation in the seeded systems apparently took place on the surface of seed crystals since the seed crystals caused the independent growth of new crystals in addition to their own growth and since the induction period was totally eliminated by seeding. The conversion rate to mordenite was progressively increased by using larger amounts of seed crystals and/or by using smaller size seed crystals since they provided larger total external surface area for nucleation. These results suggested that crystal growth of mordenite occurs from the solution phase rather than in the gel phase at the crystal/liquid interface. The presence of silicate anions in various solutions yielding mordenite was confirmed by ^{29}Si NMR [191, 192]. The various values of activation energies for nucleation and growth of mordenite reported in the literature ranged from 40 to 100 kJ mol^{-1} for nucleation and from 29 to 63 kJ mol^{-1} for crystal growth [82, 94, 174, 189]. It is suggested that the different rates of dissolution of silica or aluminosilicate gels in the mordenite reaction mixtures are very likely responsible for different values of activation energies for nucleation and growth of mordenite. These results would suggest that determining zeolite crystallization kinetics from mordenite crystallization curves may be misleading in view of the fact that the process can be limited by silica dissolution. In many cases, some portion of the silica necessary to prepare a mordenite reaction mixture is present, at least initially, in a polymerized form, either as a colloidal sol, amorphous gel, or powder. Therefore, the presence of undissolved silica during mordenite crystallization and the variety of forms of amorphous silica used in mordenite syntheses may result in different products being formed and in differences in apparent crystallization kinetics from the same starting composition as mentioned above.

4.2.2.14
Synthesis of High-silica Zeolites

Zeolite syntheses have been studied extensively throughout the last five decades. In the early work, low-silica zeolites (Si/Al = 1–5) were synthesized by crystallizing reactive aluminosilicate gels with alkali and alkaline earth metal hydroxides. The synthesis gels typically exhibited very high pH and were usually crystallized at 100 °C or less. Well-known examples of this group were zeolites A, X, Y, and synthetic mordenite (see preceding sections).

The major advance in the synthesis of new zeolite materials was the introduction of a new chemistry, that is, the addition of alkylammonium cations to synthesis gels. In the early 1960s, tetramethylammonium cation (TMA$^+$) was introduced as the first organic cation to be used in zeolite synthesis [6]. When large cations such as TMA$^+$ were incorporated, the zeolite A and faujasite produced would be more silica-rich than usual, because there was room for only a limited number of these large ions in the pore space available in the framework, and, therefore, the anionic framework charge necessarily had to be low [22]. Thus the first effect of the addition of the alkylammonium cations was to generate more siliceous framework compositions of previously known structure types.

The following stages in the research for more siliceous zeolites were achieved in the late 1960s and the early 1970s. In 1967, one of the high-silica zeolite (Si/Al > 10) family, zeolite Beta, was synthesized using the tetraethylammonium cation (TEA$^+$) [193]. Shortly after this discovery, the synthesis of other members of the high-silica zeolites, commonly referred to as the ZSM (Zeolite Socony Mobil) family, was reported. Some examples were ZSM-5, 8, 11, 12, 21, 34, 39, and 48 [194]. ZSM-5, 8, and 11 were synthesized with the addition of some number of alkylammonium cations, such as tetrapropylammonium cation (TPA$^+$) [195], TEA$^+$ [196], and tetrabutylammonium cation (TBA$^+$) [197] to highly siliceous gels, respectively. ZSM-5, 8, 11, 12, and 39 have structures not found among previously known low-silica zeolites. ZSM-21 and 34 have ferrierite-type and offretite-erionite-type structures, respectively, but with higher Si/Al ratios than for previously known materials. ZSM-48 has an X-ray diffraction pattern that resembles that of ZSM-12. These zeolites had Si/Al ratios from 10 to 100 or higher, and showed unexpected different surface characteristics. In contrast to the "low-silica" and "intermediate-silica" zeolites, the internal surface of high-silica zeolites provided an organophilic-hydrophobic environment. In 1978, the ultimate in siliceous molecular sieve zeolites was also achieved with the report of the synthesis of the first pure silica zeolite having the MFI structure, Silicalite-1, containing essentially no aluminum or cation sites [198]. Because of the unique and fascinating activity and selectivity of these materials for a variety of catalytic reactions currently processed in chemical industries, increasing attention has been devoted to a better understanding of the various mechanisms that govern the synthesis of ZSM-5.

Distinct synthesis mechanisms were suggested for low-silica and high-silica zeolites, respectively. In the case of low-silica zeolites, it was suggested that four- and six-membered rings and cages of aluminosilicate tetrahedra, stabilized by alkali metal cations, dominate the synthesis chemistry and appeared [10, 22] in the structures as mentioned before (in Sect. 4.2.2.4). On the other hand, a true templating or clathration mechanism was suggested in the synthesis of high-silica zeolites wherein the organic molecules, such as alkylammonium ions and amines, interact with silicate anions and stabilize a certain clathrasil and zeolite structure [199].

ZSM-5 has been one of the most widely studied and commercially important zeolites. The unit cell contents of the Na$^+$-form are Na$_n$[Al$_n$Si$_{96-n}$O$_{192}$]·~16H$_2$O, where $n < 27$ and typically about 3 [195]. Kokotailo et al. [200] first described the crystal structure of ZSM-5 (MFI) in 1978, and discussed further the principal features of the structure and important structure-dependent properties later [201]. The framework of ZSM-5 contained a novel configuration of linked tetrahedra shown in Fig. 17a and consisting of eight five-membered rings (D5Rs). These ZSM-5 units join through edges to form chains as in Fig. 17b. The chains could be connected to form sheets and linking the sheets leads to a 3D framework structure.

References see page 804

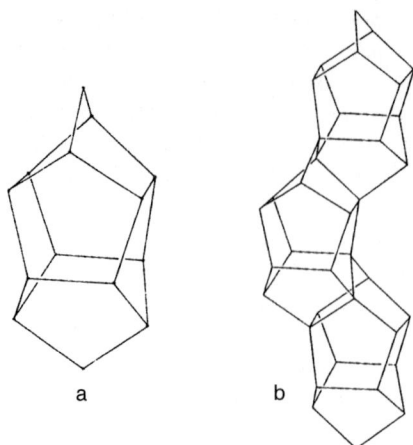

Fig. 17. (a) Characteristic configuration; (b) its linkage within chains in ZSM-5. These chains run parallel to [001]. Only T-atoms (Si, Al) are shown. Reprinted by permission from [200] Macmillan Magazines Ltd.

a   b

The 3D channel system consists of straight channels running parallel to [010] having 10-ring openings of about 5.4×5.6 Å free diameter and sinusoidal channels running parallel to [100] having 10-ring openings of about 5.1×5.4 Å.

The influence of the concentration of the different components in a typical gel for the synthesis of ZSM-5 was reviewed previously [202, 203]. The following factors might influence the rates of nucleation and crystal growth as in the case of low-silica zeolites: (1) the content of aluminum (the Si/Al ratio), (2) the degree of dilution (the H_2O/SiO_2 ratio), (3) the alkalinity of the gel (the Na^+/SiO_2, TPA^+/SiO_2, or OH^-/SiO_2 ratio), (4) the nature of the silica source or its degree of polymerization.

ZSM-5 has been synthesized with Si/Al ratios from about 10 to greater than 1000, and the aluminum content could be changed by several orders of magnitude with silica contents approaching and including that of essentially pure silica. Thus, this zeolite in its least siliceous form had about twice the Si/Al ratio of the most siliceous common zeolite, such as mordenite. Although it was known that isomorphous zeolite structures, which contain different concentrations of aluminum, such as in the series of zeolites X and Y, could be synthesized, the relative percentage change of aluminum content was relatively limited [14]. It was shown that the rate of crystallization of ZSM-5 became faster when, all other factors remaining unchanged, the aluminum content of the gel was lower [202]. This was interpreted to mean that the incorporation of aluminum into a ZSM-5 type of structure was of a disruptive nature and became increasingly difficult when the system contained more aluminum.

The rate of crystallization increased with increasing TPA^+/SiO_2 ratio, at least up to certain values. It was reported that a saturation value was about 0.05, which corresponded to approximately 3–4 TPA^+ molecules per unit cell, if all organics were retained by the structure and all silica was transformed into zeolite [202]. This corresponded to a situation in which every intersection of the ZSM-5 pores was filled by a TPA^+ cation and represented a situation of perfect pore filling.

TPA$^+$ cations had been recognized to be able to form complexes with either silicate or aluminosilicate species and to compete with Na$^+$ ions for charge compensation of these silicate or aluminosilicate species. These cations might stabilize the formation of certain subunits and then further cause replication of these primarily formed building units via a stereospecific hydrogen bond between the TPA$^+$ cation and oxygen anions. Hence, the presence of TPA$^+$ might be necessary for the formation of a particular structure or might be structure-directing. This effect was generally known as the "templating effect" in zeolite synthesis.

Although the addition of organic molecules such as amines and alkylammonium ions to zeolite synthesis gels had an enormous impact on the formation of high-silica forms of already known structures and on the formation of novel materials [194, 204], the exact role of the organic species and mechanism by which it affected the formation of the product structure remained to be elucidated. Only their templating or structure-directing role was emphasized frequently. The templating or structure-directing role was the phenomenon occurring during either the gelation or the nucleation process whereby the organic molecule organized metal-oxide tetrahedra into a particular geometric topology around itself and thus provided the initial building blocks for a particular structure type. However, it was indicated that other possible roles of the organic also have to be considered, because the clear absence of a 1:1 correspondence between the geometries of the organics used and the structure obtained [194]. It was found that frequently one template, for example, TMA$^+$ and TEA$^+$, directed various structures and that one structure, ZSM-5, could be formed in the presence of any one of more than 20 organic molecules, such as tetraalkylammonium ions, amines, alcohols, ketones, and glycerol. These results indicated that even though stereospecificity was probably one of the strongest arguments for the structure-directing theory, template fit was not the only factor in structure determination.

The effects of organic molecules on silica in aqueous solution were well known [205]. These effects were classified into at least three areas. First, some of the organics such as amines and quaternary ammonium hydroxides raised the pH of the solution and hence increased the solubility of silica. Some of the organics, such as catechol, for example, could complex silica species and increase the solubility in water. Second, some of the organics, such as polymers, adsorbed on the surface of colloidal silica particles and retarded dissolution to $Si(OH)_4$. Third, the organics formed various organosilicate or organoaluminosilicate species.

ZSM-5 and MFI-type zeolites have been most widely studied and can be synthesized with a wide variety of framework compositions via numerous synthetic routes, and with, or without, many different organic species. However, there is a marked specificity between the MFI framework and TPA$^+$ cation [206–210]. This specificity is pronounced in the case of pure-silica ZSM-5 or Silicalite-1, since pure-silica ZSM-5 has not yet been obtained without organic structure-directing agent and TPA$^+$ does not direct any other structure type in pure systems. In the resulting materials, the TPA$^+$ cations are located at the channel intersection with the

References see page 804

propyl chains extending into both the linear and sinusoidal channels [211]. The molecules are held tightly at these sites and can be removed only by calcination. Tight enclathration of the TPA^+ cations suggests that they must be incorporated into the silicate or aluminosilicate structure during the process of crystal growth. Based on this observation and on the close geometric correspondence between the TPA^+ cation and the channel intersections of ZSM-5 structure, a structure-directing role has been proposed for TPA^+ in the synthesis of ZSM-5 [206–210].

In the original report of the synthesis of Silicalite-1 [198], a crystallization mechanism of structure direction by TPA^+ cations was postulated based on geometric considerations. The proposed crystallization mechanism involved silica clathration of the hydrophobic organic cation analogous to the formation of crystalline water clathrates of alkylammonium salts [212]. The silica tetrahedra assembled into a framework in place of the hydrogen-bonded water lattice of the water clathrate, and surrounded the hydrophobic organic guest molecules. Since the initial geometric considerations of structure directing by TPA^+ cations involving reorganization of silicate species around the organic molecule to form the ZSM-5 structure, numerous studies have been carried out to understand the mechanisms that govern the nucleation and crystal growth. Raman spectroscopy [213], ^{29}Si NMR spectroscopy [214, 215], and thermal analysis [216] were used to demonstrate the existence of the organized organic-inorganic species. Burkett and Davis [34, 217, 218] reported the first direct evidence of specific intermolecular interactions that occurred within organized inorganic-organic composite structures during the synthesis of pure-silica ZSM-5 and prior to the development of long-range ordering. The role of TPA^+ as a structure-directing agent in the synthesis of pure-silica ZSM-5 was investigated by $^1H-^{29}Si$ CP/MAS/NMR. The NMR results indicated that short-range intermolecular interactions, that is, on the order of van der Waals interaction (approximately 3.3 Å for H to Si), are established during the heating of the zeolite synthesis gel prior to the development of long-range order indicative of the ZSM-5 structure. They proposed a mechanism of structure direction in zeolite synthesis for which the formation of inorganic-organic composite structures was initiated by overlap of the hydrophobic hydration spheres of the inorganic and organic components, with subsequent release of ordered water to establish favorable intermolecular interactions. Recently, the dynamics of the TPA^+ cations in the early stages of the synthesis of pure-silica ZSM-5 zeolite gel were investigated by *in situ* ^1H, ^{15}N, and ^{29}Si NMR spectroscopy [219]. It was shown that hydrogen bonds between the organic and water molecules, initially composing the structured clathrate around TPA^+, were progressively replaced by hydrophobic interactions between the organic and silicate species. This was in agreement with the mechanism proposed by Burkett and Davis [34, 217, 218] and the van der Waal's interactions between the alkyl groups of the organic molecule and the hydrophobic silica species, proposed for the structure-directing effect of organic species in the synthesis of silica-rich zeolites by Gies and Marler [199].

By using a combination of wide-, small-, and ultra-small-angle X-ray scattering techniques, de Moor et al. [38–42] showed a correlation between the formation and consumption of nanometer scale precursor particles during the hydrothermal synthesis of pure-silica ZSM-5 from a clear synthesis solution. The formation of nanometer-scale primary units (2.8 nm) was found on dissolution of the silica source. Aggregation of these particles to ∼10 nm sized particles was found to depend on the alkalinity of the synthesis mixture (in terms of OH^-/SiO_2 ratio). The synthesis mixture, which had only the 2.8-nm sized primary units and no aggregates, was not able to form viable nuclei, but normal growth on added pure-silica ZSM-5 seed crystals was found to occur [40]. These results illustrated that the aggregation of the primary units was an essential step in the nucleation process, but that these aggregated particles did not play a crucial role in the growth process. On the basis of these results, they provided a general scheme for organic-mediated zeolite crystallization as shown in Fig. 18 [42].

In ZSM-5 forming solutions, both at room temperature and at 100 °C, double four-ring (D4R) and double five-ring (D5R) silicate anions were present. It was shown that a D4R/D5R ratio was 2.6 at room temperature, but the relative amount of D5Rs increased at higher temperatures despite the fact that partial hydrolysis of D4R and D5R silicates had occurred [220]. Therefore, the formation of ZSM-5 frameworks could easily be envisaged starting from D5R and monomer silicate and/or aluminate anions. There were several observations compatible with the idea of a D5R acting as a precursor for ZSM-5 formation [220, 221]. At low OH^-/SiO_2 ratios, especially lower than 0.5, significant amounts of D5R were formed [220]. Furthermore, Mostwicz and Sand [188] reported that the induction period became shorter and crystallization became faster with decreasing OH^-/SiO_2 ratio, at constant TPA^+/SiO_2 ratio. This result was consistent with the observation of increased amounts of D5R silicates shown by Groenen et al. [220]. However, Keijsper and Post [222] concluded that the D5R silicate condensation mechanism was not generally operative in the nucleation and/or crystallization stage of the synthesis of ZSM-5 structure. They showed that the nucleation rate increased in order of TEA^+ < organic-free < TPA^+, whereas the order organic-free < $TEA^+ = TPA^+$ was expected, on the basis of the D5R concentrations in analogous silicate solutions. The rate of formation of Silicalite in the presence of dimethyl sulfoxide (DMSO) was slower, although the amount of D5R silicates in solution during the Silicalite synthesis in the presence of DMSO was much higher than without DMSO. DMSO was added as water-miscible organic solvent to promote the amount of DnR silicates in solution [223]. These results could refute the possibility of the precursor role of D5Rs during nucleation. Further, the result that defects due to missing T sites were present only in a few samples and that the defects appeared to be randomly distributed did not directly support a precursor role for D5R species during the crystallization step.

The recent use of several novel analysis techniques such as small angle X-ray scattering (SAXS) [31, 35–37, 43], small angle neutron scattering (SANS) [31, 43], and light scattering techniques [27–29] have supplied further details on the mechanisms of zeolite growth.

Regev et al. [35] studied the small-angle X-ray scattering (SAXS) from the

References see page 804

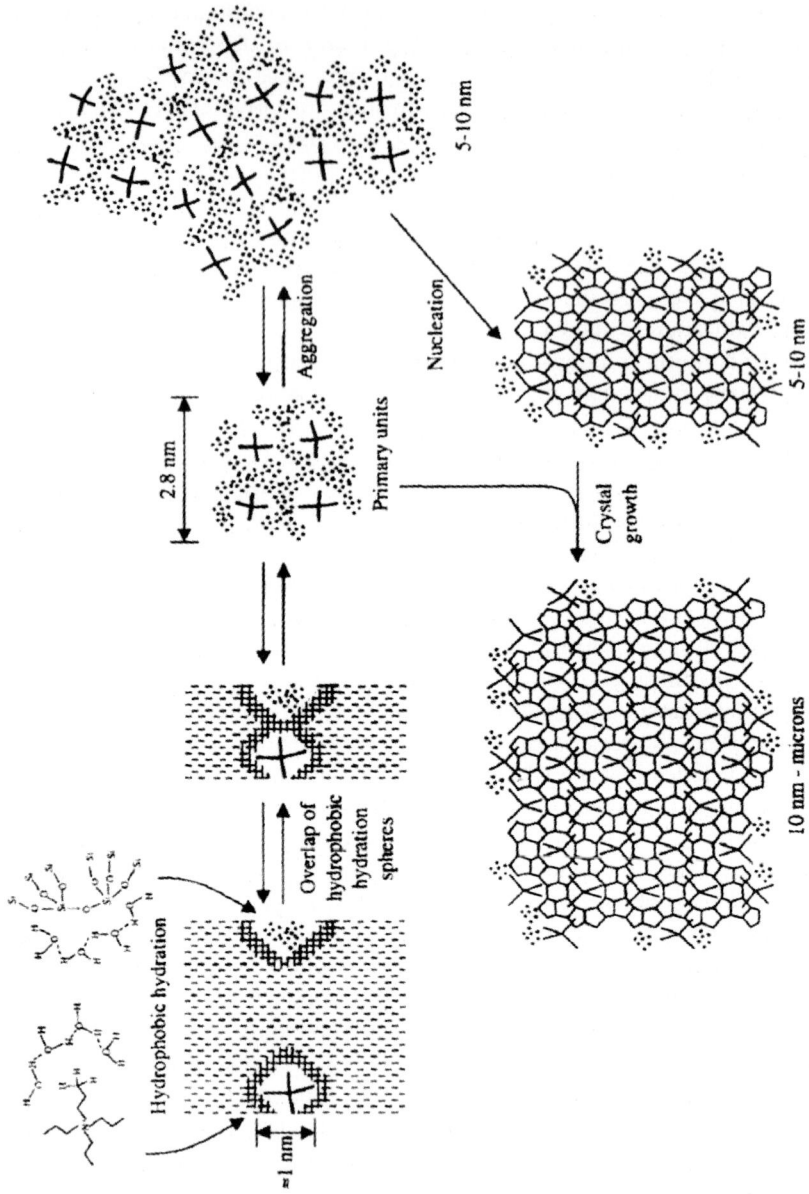

Fig. 18. Scheme for the crystallization mechanism of Si-TPA-MFI. Reprinted from [42] with the permission of Wiley-VCH Verlag GmbH.

mother liquor above a gel-based preparation of ZSM-5. The unheated mother liquor from this gel contained 5-nm spherical particles, and these particles were present throughout the hydrothermal reaction. They speculated that the "globular structural units" were comprised of several "tetrapods" though they did not confirm directly the inclusion of TPA^+ in the particles. They also found that neither the shape nor the size of the particles in the mother liquor changed after the first 2 h of heating, thereby defining their nucleation period. During the nucleation period, aggregation of the 5 nm globular structural units occurred with the formation of cylindrical particles of dimensions 8×22 and 16×44 nm. It was proposed that these aggregates settled into a disordered precipitate, but subsequently ordered into ZSM-5 with further heating.

Dokter et al. [37] monitored the nucleation and crystallization processes of Silicalite-1 synthesis in the homogeneous system by SAXS and WAXS studies. They proposed a crystallization model, termed "growth by aggregation". The main steps involved in this model were as follows; (1) tetrapropylammonium (TPA)-silicate clusters with a diameter less than 3.2 nm formed in solution, (2) primary fractal particles aggregated according to a reaction-limited cluster-cluster aggregation model to form larger secondary aggregate structures with a size of about 6.4 nm, (3) densification of these primary fractal aggregates to slightly larger and denser particles with a size of 7.2 nm (corresponding to about 20 Silicalite unit cells), (4) combination of the densified aggregates into a secondary fractal structure larger than 52 nm, at which time crystallinity was observed with WAXS, and (5) densification of the secondary aggregates and crystal growth.

By using SAXS and SANS studies, Watson et al. [43] showed that a cylindrical form factor gave the best fit to the measured scattering functions for the particles that developed and persisted from nucleation to the end of the induction period. They conducted two experiments that showed the same trend in growth with heating time, where a small increase in the radius of the cylinder (R) was recorded together with a large increase in the length of the cylinder (L) until reaching a maximum detectable size of $R = 4.4$ nm and $L = 34$ nm. They proposed a model for the nucleation and crystallization processes as follows; (1) a short initial period of formation of cylindrical primary nuclei, with radius of gyration $R_g < 4.3$ nm, that incorporate TPA^+ and silicate in a structure that had the MFI framework geometry; (2) an induction period during which primary nuclei fused coaxially and the resulting fused nuclei, termed primary crystallites, assembled end-to-end along the crystal c-axis into an average length of 33 nm while maintaining an average diameter of 8.3 nm; (3) a rapid crystal growth period during which primary crystallites fused, maintaining a strong degree of long-range orientational order, to form ellipsoidal polycrystalline particles that ultimately attain near-micrometer size.

Schoeman monitored the early stages of TPA-Silicalite-1 formation by dynamic light scattering (DLS) [27–29]. It was shown that there were two distinct particle populations present in the synthesis suspension. The average size of the large-size

References see page 804

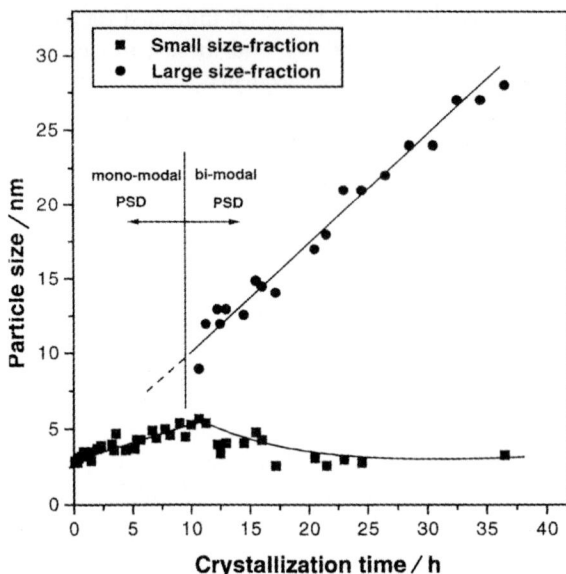

Fig. 19. The increase in the average particle size as a function of crystallization time. The particle size distributions measured before 9.75 h are monomodal and at later times, the light scattering technique is able to resolve two distinct particle populations. Reprinted from [28] with permission from Elsevier Science.

fraction continued to increase with crystallization time. The important observation was, however, that the average particle size of the small-size fraction remained essentially constant with an average size of approximately 3.3 nm throughout the synthesis as shown in Fig. 19 [28]. Deconvolution of the scattered light intensity data, which was recorded in the time interval during which it was not possible to resolve the particle size distributions, indicated that the growing Silicalite-1 particles were present in suspension at a very early stage in the crystallization process. There were indications that certain subcolloidal particles might possess a short-range structure such that they increased in size on hydrothermal treatment and might thus be termed zeolitic nuclei. Further, he investigated the possibility that Silicalite-1 grew via aggregation of smaller particles or similar sized particles by considering the fundamentals governing colloidal stability in the mother liquor at a crystallization temperature of 100 °C [30]. Application of the extended Derjaguin–Landau and Verwey–Overbeek (DLVO) theory that took the forces such as electrostatic repulsion, van der Waals attraction, steric hindrance, and solvation forces to two growth models showed that it was not likely that Silicalite-1 crystals grew via a particle-particle aggregation mechanism. He suggested that for very small particles, less than 1–2 nm, and more correctly termed clusters, aggregation with other clusters and/or growing crystals was a likely growth mechanism.

In a series studies on TPA-Silicalite-1 Kirschhock et al. [114–118] proposed the crystal growth mechanism by aggregation of intermediate nanoblocks in the synthesis solution. They suggested that the formation of a suspension of colloidal

Fig. 20. Siliceous entities suggested to occur in the TPAOH-TEOS system: (a) bicyclic pentamer; (b) pentacyclic octamer; (c) tetracyclic undecamer; (d) "trimer" in mixtures with composition $(TPAOH)_{0.36}(TEOS)(H_2O)_{6.0}$; (e) nanoslab mixtures with composition $(TPAOH)_{0.36}(TEOS)(H_2O)_{17.5}$. Reprinted with permission from [115] American Chemical Society.

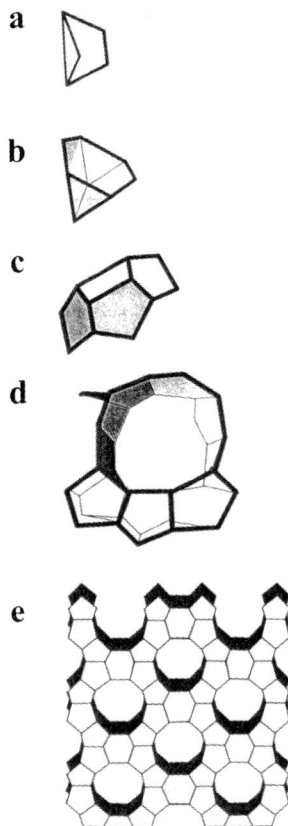

Silicalite-1 particles from tetraethyl orthosilicate (TEOS) and aqueous tetrapropylammonium hydroxide (TPAOH) in solution proceeds through a series of discrete molecular steps, which were identified with ^{29}Si NMR, X-ray scattering, gel permeation spectroscopy (GPC), and IR. At room temperature, the polycondensation process led to the selective formation of the 33 Si atoms containing precursors that is unique to the Silicalite framework connectivity (Fig. 20). These precursors condensed with each other to form larger species with dimensions of $1.3 \times 4.0 \times 4.0$ nm (Fig. 20e). These nanoslabs formed thin 2×2 sheets linked via the small sides in the b and c direction at room temperature (Fig. 21). Due to stability criteria, the thin sheets first stacked in the a direction to give mechanically more stable blocks (intermediates), which then could be built into larger particles under hydrothermal conditions. This proposed aggregation mechanism was supported by the interaction potentials of different faces of the nanoslabs estimated using extended DLVO theory.

References see page 804

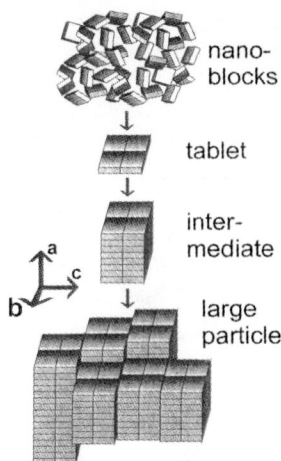

Fig. 21. Schematic representation of the formation of Silicalite-1 material from clear solutions at elevated temperatures. Reprinted with permission from [114] American Chemical Society.

Nikolakis et al. [44] studied the seeded growth of TPA-Silicalite-1 by simulations and dynamic light scattering, atomic force microscopy, and transmission electron microscopy. At high concentrations of silica in the solution, growth was observed, with a linear growth rate that was practically independent of the total silica added to the system. A population of small subcolloidal particles with size 2–3 nm was observed by DLS and TEM experiments. The growth rate of seeds was determined through modeling of a static particle in a suspension of subcolloidal particles. Good agreement with experimental results regarding the growth rate and activation energy was possible considering DLVO interactions under a constant surface charge. These results suggested that association of subcolloidal particles with the crystal, followed by fast rearrangements on the crystal surface, took place. The constant surface charge model was also verified with atomic force microscopy force measurements between a Silicalite-1 surface and a glass sphere. With this simulation they showed that only a relatively narrow subcolloidal particle size distribution (up to 4 nm in diameter) could result in reasonable growth rates and activation energy. This indicated that possible aggregates of subcolloidal particles were not directly consumed in the growth of an already existing crystal, and consequently, the smallest existing subcolloidal particles made the largest contribution to the growth.

The evolution of TPA-Silicalite-1 particle sizes has been monitored by dynamic light scattering (DLS) [25–29, 44, 114, 115], X-ray scattering (SAXS) [35–43], or neutron scattering (SANS) [43]. All authors agreed on the formation of a small species of nanoscopic dimensions (3–5 nm) with a more or less internally ordered structure resembling the final Silicalite-1 product. Although a globular or cylindrical geometry was attributed to these structures during the evaluation of the SAXS and SANS data, the exact nature of these particles, conveniently termed "nanoblocks", remains unclear. Several authors also observed the formation of larger intermediates in the crystallization process. Whereas one group favored a fractal growth pattern, others have preferred an aggregation mechanism, or even claimed the dissolution of nanoblocks into precursors as a nutrient reservoir. The fact is that the nanoblocks have been observed to be present throughout the crystallization process, which in itself rather hints at a successive aggregation mechanism.

Zeolite Beta, a wide-pore, high-silica zeolite, was first synthesized by Wadlinger et al. [193], using tetraethylammonium hydroxide (TEAOH) as a templating agent. It was reported that zeolite Beta also could be obtained from systems containing TEAOH-diethanolamine [224, 225], TEAOH-TEABr-triethanolamine [225, 226], and TEABr–NH$_4$OH [227]. Further templating agents for the synthesis of zeolite Beta such as diaza-1,4-bicyclo-[2,2,2]-octane [228], dibenzyl-1,4-diazabicyclo-[2,2,2]-octane [229], benzyldimethlamine-benzylchloride [230], and 4,4′-trimetylenedipiperidine [231], also have been reported.

The structure of zeolite Beta was revealed to be an intergrowth of two [232] or three polymorphs [233], one end-member of the structure with alternating translations of the layers being the only known real zeolite structure showing chirality. Zeolite Beta possesses 3D 12-membered ring pores with interconnected channel system. Its pore system consists of straight 12-membered ring ($\sim$7.3 × 6.4 Å) channels running in the *a* and *b* directions and a more tortuous 12-membered ring ($\sim$5.5 × 5.5 Å) channel system running parallel to the *c* direction. The unit cell of zeolite Beta consists of 64 T-atoms, and a typical composition is represented by the formula Na$_7$[Al$_7$Si$_{57}$O$_{192}$]·xH$_2$O [234].

Systematic studies of the nucleation and crystal growth rates have been reported only for reaction mixtures containing TEA$^+$ cations [227, 235–238]. Perez-Pariente et al. [235–238] investigated its crystallization kinetics and mechanism in Na$^+$, K$^+$ and TEAOH containing systems. 100 % crystalline zeolite Beta as the sole crystalline phase could be obtained from gels with Si/Al ratios smaller than 500 at temperatures below 423 K. An increase of the synthesis temperature from 373 to 393 K resulted in an increase of the Si/Al ratio of the zeolite product. The most siliceous sample of zeolite Beta synthesized in this way had a Si/Al ratio of 106. This finding was consistent with the limits established in the original patent (Si/Al = 5–100) [193]. The formation of ZSM-5 and cristobalite from the same gels at 423 K was in agreement with a general observation [18] that at higher temperatures the formation of denser phase was favored. The preferred formation of ZSM-5 over Beta at higher temperatures indicated that TEA$^+$ is not a true template for the nucleation of zeolite Beta, but functions essentially as a pore-filling agent in the crystal growth process. It was shown that in this system a higher Si/Al ratio in the gel caused a decrease of both the crystal growth rate and the number of crystals produced. This relationship was similar to that observed for typical low-silica materials, like zeolite Y [80], and opposite to the behavior of high-silica zeolites, like ZSM-5 [202]. They also reported that crystallization rate and crystal size of zeolite Beta made from Na$^+$ and K$^+$-containing gels were dependent on the total alkali content and the molar fraction of each cation. The results suggested that, since no zeolite was obtained in the absence of alkali cation, both alkali and TEA$^+$ cations were required

References see page 804

for the crystallization of zeolite Beta. On the other hand, an adequate balance between both types of cations was needed to give a minimum crystallization time, resulting from the highest apparent nucleation and crystallization rates. This effect appears to be general for high-silica zeolites, since similar results were reported for other materials, for example, ZSM-5 [239] and ZSM-12 [240]. The crystal size of the final product increased and the number of crystals decreased with the K^+ content of the gel. The size of ZSM-5 zeolite crystals made from K^+-containing gels was found to be larger than that of crystals obtained from Na^+-containing gels, even though the induction time decreased and the crystallization rate for the K^+-containing system was higher [239, 241] or similar [242, 243] to those for the Na^+-containing system.

Eapen et al. [227] carried out a systematic study on the crystallization of zeolite Beta using TEABr, in place of TEAOH, in combination with NH_4OH as an organic templating species and silica sol as a source of silica. Zeolite Beta could be crystallized in the temperature range 373–433 K by using seed crystals of zeolite Beta. The optimized gel composition was found to be within the range $Si/Al = 7.5$–29, $H_2O/SiO_2 = 18$–25, $TEABr/SiO_2 = 0.25$–0.50, $Na_2O/SiO_2 = 0.08$–0.12, and $OH^-/SiO_2 = 0.7$–1.05. The increase in the gel Si/Al ratio and synthesis temperature led to the crystallization of ZSM-12 instead of zeolite Beta.

Since the range of Si/Al ratios of zeolite Beta was established in the original patent ($Si/Al = 5$–100) [193], synthesis efforts to widen this range have had limited success. It was reported that the synthesis of a so-called pure-silica zeolite Beta from a Na^+-free system had been accomplished by using dimethyldibenzylammonium cation as the structure-directing agent [244]. Unfortunately, this method required the use of deboronated borosilicate zeolite Beta as seed crystals. Camblor et al. [245] first presented the unseeded synthesis of pure-silica zeolite Beta in a hydrothermal system containing TEA^+ and F^- ions at near neutral pH. Pure-silica zeolite Beta showed a much better resolution of the X-ray diffraction peaks. This would be a consequence not only of its large average crystal size (0.5–5 μm well-faceted truncated square bipyramid), but also of the almost complete absence of $Si–O^-$ defect groups. For pure-silica compositions worked out at high pH, TEA^+ cations had been reported to direct the crystallization towards the formation of ZSM-12 and ZSM-5 zeolites, and that even in the presence of deboronated zeolite Beta seeds, ZSM-12, ZSM-5, cristobalite, or magadiite impurities appeared in a previous study [244].

On the other hand, Borade and Clearfield [246, 247] investigated the incorporation of higher amounts of Al into the zeolite Beta framework. The effects of various reaction parameters such as Na_2O/Al_2O_3, SiO_2/Al_2O_3, $TEAOH/SiO_2$, H_2O/Al_2O_3, reaction temperature, and time on the formation of final products were investigated. A crystalline Beta zeolite with Si/Al ratio as low as 4.5 was obtained at 443 K in 44 h from a dense gel system containing 1 Na_2O : 1 Al_2O_3 : 9 SiO_2 : 1.6 TEAOH : 60 H_2O. Vaudry et al. [248] reported the synthesis of zeolite Beta with Si/Al ratio 6.7 (8.3 Al atoms per unit cell) in Na^+ and TEA^+-containing systems. Na^+ cations were needed for charge balance beyond 6 Al/unit cell. As observed by Gabelica et al. [249], the number of Al atoms in a unit cell of zeolite Beta corresponded to the number of TEA$^+$ cations needed to fill the micropore system. Crystallographic siting for TEA$^+$ in the zeolite Beta structure was not determined. As a consequence, the number of TEA$^+$ molecules per unit cell had to be evaluated by thermogravimetric or elemental analysis. Although published data about the TEA$^+$ content vary from 5.2 to 9.8 molecules per tetrahedral cell, some agreement was reached on a value between 6 and 7 TEA$^+$ ions per unit cell [235, 236, 250–252].

Zeolites are typically synthesized by hydrothermal crystallization methods at autogeneous pressure. Recently, new crystallization methods for zeolite synthesis have been developed, as they might enable us to prepare zeolites with novel structures, compositions, and convenient forms such as membranes. One such vapor-phase transport method was successfully employed for synthesis of powdery zeolites [253–255] and zeolitic membranes [256–259]. This method involves crystallization of dry aluminosilicate gel in the presence of volatile structure-directing agent and steam. However, this method could not be applied for the preparation of all types of zeolites, since many of them needed nonvolatile structure-directing agents for their crystallization. It was shown that another crystallization method, the dry gel conversion technique, was successful for the crystallization of zeolite Beta [260–262]. In this method, zeolite Beta with Si/Al ratios ranging from 15 to 365 was crystallized from dry gels containing TEAOH as a structure-directing agent. Complete conversion of gel to zeolite was obtained in 3 to 12 h at 453 K.

The physical properties and catalytic activity of zeolites would be closely related to the aluminum content and distribution in their frameworks. Suib et al. [263] first examined the distribution of aluminum in zeolites, including zeolites A, X, Y, and ZSM-5 by scanning Auger electron spectroscopy (AES). They reported that the surface Si/Al ratio was close to the bulk composition in zeolite crystals. On the other hand, von Ballmoos and Meier [264] were the first to report that aluminum zoning existed in large ZSM-5 crystals. They proved a significant enrichment of aluminum in the outer shell of the crystals and a decline in aluminum concentration from the rim to the core of a crystal by electron microprobe analysis. Chao and Chern [265] also showed a similar heterogeneous aluminum distributions in large single and twinned ZSM-5 crystals by electron microprobe analysis.

The elemental distribution in ZSM-5 has been reported by using various surface analytical techniques such as X-ray photoelectron spectroscopy (XPS) [266–268], secondary ion mass spectrometry (SIMS) [269], and combined XPS, energy dispersive X-ray analysis (EDX), and proton-induced gamma-ray emission (PIGE) [270, 271]. It was shown that not only constant aluminum concentrations throughout the ZSM-5 crystals but also increases in aluminum content from the rim to the core of the zeolites could occur.

From these results, Jacobs and Martens [272] summarized four typical types of aluminum distribution profiles, that is, an aluminum-poor core with the aluminum steadily increasing to the rim of the crystal (type I), an aluminum-rich core with the aluminum steadily decreasing to the rim of the crystal (type II), strong

References see page 804

enrichment of the aluminum in the outer shell of the crystal (type III), and homogeneous aluminum distributions (type IV). They stated that the nucleation mechanism determined the aluminum gradient. A solution nucleation mechanism should result in a type I profile, while a gel nucleation should result in a type IV profile.

Althoff et al. [273] investigated the influence of the aluminum source, the template, and the presence of additional ions on the incorporation of aluminum during the course of synthesis by measuring the spatial distribution of aluminum over ZSM-5 crystals. It was shown that by using different templates the aluminum gradient in crystals could be controlled. An explanation for the differences between the templates was given in the terms of the interaction with the template molecules. Tetrapropylammonium (TPA^+) ion as template led to strongly zoned profiles with the aluminum enriched in the outer shell of the crystals owing to the preferential interaction between TPA^+ and silicate species. 1,6-hexanediol as template or a completely inorganic reaction mixture led to homogeneous aluminum profiles with enrichment of the aluminum neither in the cores nor in the rims of the crystals. They assumed that Na^+ ion interacts strongly with aluminate species in TPA^+-free systems and an incorporation of aluminum is facilitated in the growing crystal in comparison to the TPA^+ system.

The phenomenon of isomorphous substitution is well known in the field of mineralogy [18]. By isomorphous substitution, framework atoms of crystalline compounds are replaced by atoms of other elements without changing the crystal structure. Isomorphous substitution is an important way to modify zeolite properties for practical applications and has achieved considerable interest in the field of zeolite chemistry [10, 18]. Barrer [10] discussed the thermodynamics of isomorphous substitution, and noted that the main factors governing success of isomorphous substitutions appeared to be primarily the ratio of the radii of the atoms involved. Pauling [274] formulated the main criteria for the occurrence of isomorphous substitution, which were primarily derived from crystal chemistry and geometric considerations. According to Pauling's rule of gradients, cations prefer tetrahedral coordination if the ratio of the ionic radii between M^{n+} and O^{2-} $r_M/r_O = 0.214-0.4$ and octahedral site if $r_M/r_O = 0.4-0.6$. Ions larger than 0.55 Å, therefore, exist preferentially in octahedral coordination in oxygen lattices.

The isomorphous substitution of Al^{3+} by other trivalent elements in zeolite frameworks has been studied extensively. The most widely studied metallosilicate systems have been borosilicates, gallosilicates, and ferrisilicates with ZSM-5 (MFI) type structures.

Although boron often exists in 3-fold planar coordination with oxygen, a number of reports have shown that the synthesis of a material with the ZSM-5 structure and with boron atoms incorporated in the zeolitic framework was possible [275–279]. Evidence for this substitution was given by Taramasso et al. [275] from measurements of the size of the unit cell, and by Gabelica et al. [279] from high-resolution solid state ^{11}B MAS/NMR data.

The incorporation mechanism of boron was investigated by studying the boric acid/borate equilibrium in MFI type borosilicate zeolite synthesis mixtures [280]. It was found that boron incorporation in the MFI structure required the presence of TPA$^+$, whereas aluminum did not. The proposed T-atom condensation and hydrolysis mechanism required trigonal boron and not borate ions for boron incorporation in zeolite frameworks and TPA$^+$ ions to compensate the negative charge of every incorporated framework boron site. As the maximum amount of TPA$^+$ ions in the MFI unit cell was four, it was concluded that the maximum amount of boron atoms per unit cell was equal to four.

Because gallium-bearing zeolites with the MFI structure were of special interest as catalysts for the transformation of lower alkanes to aromatics [281–285], numerous published papers have been devoted to the study of gallium state and behavior in zeolite structures. However, the knowledge about chemical processes occurring during the synthesis of gallosilicate zeolites is still incomplete. Gallium is next to aluminum in the respective group of the periodic table and could, therefore, be expected to replace aluminum in aluminosilicate zeolites. Taking into account that gallium in alkaline solutions preferentially forms tetrahedrally coordinated gallate anions, it also is conceivable that this element could substitute for silicon in tetrahedral framework positions. Recently, an excellent review on the synthesis, characterization, and catalytic application of gallium-containing microporous and mesoporous materials focusing on the isomorphous substitution of gallium into zeolite framework was published [286]. In that review Fricke et al. reported that silicate and gallate anions in the gel reacted very rapidly, and that the reactions were similar to those appearing in aluminosilicate solutions and followed the same patterns if the silica, metal, or base contents were changed.

Iron ions readily form sparingly soluble or insoluble hydroxides or oxides that are difficult to depolymerize and mobilize into the useful metallosilicate anionic species able to condense further into viable growing zeolite nuclei. It was pointed out that three factors were critical in the preparation of ferrisilicate zeolites over a wide range of Si/Fe ratios: (1) the avoidance of iron hydroxide precipitation, (2) the necessity of using low-molecular-weight silica sources, and (3) the need to suppress the formation of iron complexes with the organic amine crystal-directing agents [287]. To circumvent the problems stated above, a modification of the standard zeolite synthesis method was devised to prepare "pure" ferrisilicate zeolite [287]. Iron tends to precipitate as a rust-red colloidal ferric hydroxide at pH > 4 [288]. Once formed, ferric hydroxide is almost completely insoluble, thereby limiting the availability of FeO_2^- species for incorporation into the silicate unit during crystal growth. Complex formation between iron and silica species occurs at pH = 3–4. Therefore, initial formation of a ferrisilicate complex at low pH could avoid precipitation of rust-red iron hydroxide at elevated pH. The use of a complexing template or structure-directing agent that bound strongly to the iron species also prevented it from forming the ferrisilicate gel. A strongly coordinating neutral amine, such as pyridine, was found to limit incorporation of the iron into the gel and the final crystalline phase. In order to avoid any possible complex formation between the iron and any free amine impurity present, the TPA$^+$ was introduced after the ferrisilicate gel was formed [288].

Titanium silicalite-1 (TS-1) is the first discovered member of the group of crys-

References see page 804

talline microporous materials made of oxides of titanium and silicon. TS-1 has attracted much interest for its unique catalytic properties by virtue of the proposal that Ti assumes tetrahedral coordination in substituting for Si in framework position of crystalline silica. In oxidation reactions, with H_2O_2 as the oxidant, many organic compounds could be oxidized selectively and efficiently. Notari [289] reviewed synthesis, structure and role in oxidation catalysis of titanium silicates. TS-1 has been obtained by the hydrothermal crystallization of a gel obtained from tetraethyl orthosilicate (TEOS) and tetraethyl orthotitanate (TEOT) in the presence of tetrapropylammonium hydroxide (TPA-OH). This procedure consists of the preparation of a solution of mixed alkoxides of silicon and titanium followed by hydrolysis with alkali free solution of TPA-OH, distillation of the alcohol, and crystallization of the resulting gel at 448 K [290]. Whereas alkalis do not interfere with the synthesis of most zeolites, and in many cases are even required, they interfere with the synthesis of titanium silicates [290–293]. In the presence of alkalis, the formation of an amorphous precipitate containing Ti^{IV} is observed. Redissolution and thus incorporation of Ti^{IV} into the zeolitic structure is difficult. Samples of TS-1 prepared in the presence of Li, Na, and K in the synthesis gel produced materials that have considerable amounts of extra-framework TiO_x, poor catalytic properties, and give a high rate of H_2O_2 decomposition. It was shown that the temperature at which the reagents are mixed and the rate of hydrolysis were critical, even in the absence of alkalis. Cooling of reagents to 273 K, efficient stirring, and low rates of mixing have been recommended to prevent the precipitation of extra-framework TiO_x species [294, 295]. The use of tetrabutyl orthotitanate (TBOT) as a titanium-containing precursor was proposed to simplify the procedure and incorporate more Ti in TS-1 by reducing the rate of hydrolysis of the precursor [296, 297]. It was described that other methods for synthesis for TS-1 include the use of colloidal SiO_2 and tetrapropylammonium peroxo titanate [289] and the use of a SiO_2-TiO_2 coprecipitated dry gel that was impregnated with an aqueous solution of TPA-OH [298, 299]. TS-1 was obtained indeed by these methods, but the impurities contained in the starting materials, particularly Al^{3+}, were incorporated into a crystalline product and modify the catalytic properties.

For the synthesis of silica-based zeolites the most common mineralizer is the hydroxide anion OH^-. The replacement of OH^- by fluoride anion F^- as a mineralizer makes it possible to synthesize zeolites at pH values lower than 10, even in slightly acidic media. HF as well as salts such as NH_4F, NaF, KF, and CsF can be used as fluoride sources. At such pH values the solubility of silica increases significantly in the presence of fluoride because of the formation of hexafluorosilicate SiF_6^{2-}-species. The first clear example of the use of the fluoride was for crystallization of Silicalite-1 in slightly alkaline media [300]. The fluoride route was then extensively investigated for silica-based zeolites, alumino- and gallophosphates. Thorough reviews of zeolite synthesis potentialities and limitations in the presence of fluoride ions, with critical comparison with other conventional synthesis methods were presented by the Mulhouse groups who pioneered this technique [301–303].

The fluoride route is especially suited for synthesis of high-silica zeolites with various structures, such as MFI-, FER-, and BEA-structure type materials. This route could be used to prepare zeolites in which silicon was partly substituted by T atoms (T = B, Al, Ga, Fe, Ge, and Ti) [302, 303]. The Si/T ratio in the zeolite varied with the synthesis conditions and was larger than about 10. When the crystallization was carried out in the presence of an organic cation, for example, TPA^+ ion, as in the case of high-silica zeolite, fluoride was generally occluded in the pores of materials as a compensating negative charge, in addition to the negative framework charge, of the organic cations. In T-rich samples, the organic cation compensated partially or completely the negative charge of the framework. The temperatures of crystallization were similar to those used in the conventional syntheses but the crystallization time was generally longer.

The prepared gels, which were white before heating with iron, did not precipitate as low-soluble species, but formed soluble complexes with fluoride ions. This scheme was different from the conventional synthesis of ferrisilicate in alkaline media where the obtained gel was pale yellow [304, 305]. Up to 3.4 Fe/unit cell in the MFI framework could be introduced in the CsF-containing media [306]. The incorporation of iron in the MFI framework could depend on the solubility of the fluoride salt. On the other hand, both the induction time and crystallization time decreased with increasing pH. The pH of initial gels increased in the order $NH_4^+ < Na^+ < K^+ = Cs^+$. A thorough review of zeolite synthesis potentialities and limitations in the presence of fluoride ions, with a critical comparison with other conventional synthesis methods was presented by Guth et al. [302].

Using this method also had another effect. Axon and Klinowski [307] demonstrated that MFI zeolites synthesized in fluoride media were virtually free from structural defects, that is, internal silanol groups. Such defects were usually present in crystals prepared by conventional synthesis methods and could be removed by a longer hydrothermal treatment only as reported by Dessau et al. [308].

4.2.2.15
Conclusions

The synthesis of classical zeolites, such as zeolites A, X, Y, mordenite, ZSM-5 and Beta, is typically undertaken in hydrothermal systems involving the combination of the appropriate amounts of silicates and aluminates, and usually in basic media. As the synthesis proceeds at elevated temperature, typically in the range 100–200 °C, zeolite crystals are formed by a nucleation step, and these zeolite nuclei grow by assimilation of aluminosilicate material from solution phase. The amorphous gel phase, which formed by mixing of starting material solutions, is regarded as a reservoir of nutrients, dissolving to replenish the solution with aluminosilicate species as crystal growth occurs simultaneously. It is well known that temperature and alkalinity influence the degree of supersaturation of the synthesis solution, which is the principal driving force for nucleation. Other factors such as the nature of reactants, type of alkali cation, addition of organic compound, and pretreatment of the amorphous gel can also affect crystallization kinetics and even the type of zeolite that forms.

References see page 804

The crystallization curves are characteristically sigmoid in shape with an inflection point, which separates the primary autocatalytic, or self-accelerating, stage of crystal mass growth from the final stage of a delayed growth. The autocatalytic nucleation mechanism is suggested based on an assumption that autocatalytic nuclei, lying dormant in the amorphous gel phase, begin growing after their release from the gel as it dissolved. As the cumulative zeolite crystal surface area increases due to crystal growth, the rate of solute consumption increases, which increases the rate of gel dissolution, and results in increased rate of dormant nuclei activation. The crystal growth mechanism is a solution-mediated process. This size-independent linear crystal growth is governed by surface reaction of soluble species.

Adding seed crystals of the desired zeolite phases to synthesis batches increases the rate of crystallization and forces the solution to produce that particular phase. The additional surface area of seed crystals results in the more rapid assimilation of nutrient material from solution, reducing the supersaturation, even to the extent that nucleation of new crystal phases is prohibited. Initial breeding, which results from microcrystalline dust being washed off seed surfaces into the crystallization solution, also contributes to promotion of the crystallization rate and enhancement of the crystalline mass.

The gel aging at ambient temperature results in the shortening of the induction period and the acceleration of the crystallization process. But in some cases, the gel aging also influences the type of zeolites formed. The increase in the number of the quasi-crystalline nuclei inside the gel matrix and/or in the liquid phase during the gel aging results in an increase in the crystallization rate.

In the case of low-silica zeolites, it is suggested that four- or six-membered rings and cages of aluminosilicate tetrahedra, stabilized by alkali metal cations, dominate the synthesis chemistry. On the other hand, a templating or structure-directing mechanism is suggested in the synthesis of high-silica zeolites wherein the organic molecules, such as alkylammonium cations and amines, interact with silica anions and stabilize a certain clathrasil or zeolite structure. The crystallization process, especially of Silicalite-1 synthesis, in the homogeneous system has been monitored by several novel analysis techniques. Even though small sub-colloidal particles and aggregates of such particles have been observed, their role in nucleation and crystal growth still remains unclear. The possible growth mechanisms discussed in the literature are that an existing crystal might grow by incorporation of small particles or aggregates of such particles or by dissolution of nanoblocks into precursors as a nutrient reservoir.

References

1 R. M. Barrer, *J. Chem. Soc.* **1948**, 2158–2163.

2 R. M. Barrer, E. A. D. White, *J. Chem. Soc.* **1951**, 1267–1278.

3 R. M. Barrer, E. A. D. White, *J. Chem. Soc.* **1952**, 1561–1571.

4 R. M. Barrer, J. W. Baynham, *J. Chem. Soc.* **1956**, 2882–2909.

5 R. M. BARRER, J. W. BAYNHAM, F. W. BULTITUDE, W. M. MEIER, *J. Chem. Soc.* **1959**, 195–208.

6 R. M. BARRER, P. J. DENNY, *J. Chem. Soc.* **1961**, 971–982.

7 R. M. BARRER, P. J. DENNY, *J. Chem. Soc.* **1961**, 983–1000.

8 R. M. BARRER, D. J. MARSHALL, *J. Chem. Soc.* **1964**, 485–497.

9 R. M. BARRER, *Chem. Ind.* **1968**, 1203–1213.

10 R. M. BARRER, Hydrothermal Chemistry of Zeolites, Academic Press, London, 1982.

11 R. M. MILTON, Union Carbide Corporation, US Patent, 2,882,243, 1959.

12 D. W. BRECK, W. G. EVERSOLE, R. M. MILTON, T. B. REED, T. L. THOMAS, *J. Am. Chem. Soc.* **1956**, *78*, 5963–5972.

13 R. M. MILTON, Union Carbide Corporation, US Patent, 2,882,244, 1959.

14 D. W. BRECK, E. M. FLANIGEN, Molecular Sieves, Society of Chemical Industry, London, 1968, 47–61.

15 D. W. BRECK, Union Carbide Corporation, US Patent, 3,130,007, 1964.

16 R. M. MILTON, in: Zeolite Synthesis, M. L. OCCELLI, H. E. ROBSON (Eds), *ACS, Symp. Ser.* American Chemical Society, Washington, D. C., **1989**, *398*, 1–10.

17 D. W. BRECK, *J. Chem. Edu.* **1964**, *41*, 678–689.

18 D. W. BRECK, Zeolite Molecular Sieves: Structure, Chemistry and Uses, Wiley, New York, 1974.

19 Synthesis Commission of the International Zeolite Association, *Verified Synthesis of Zeolitic Materials* H. ROBSON (Ed.), *Microporous Mesoporous Mater.* **1998**, *22*, 495–670.

20 The home page of the Synthesis Commission of the International Zeolite Association with additional recipes, http://www.iza-online.org.

21 W. M. MEIER, in: Molecular Sieves. W. M. MEIER, J. B. UYTTERHOEVEN (Eds), *Adv. Chem. Ser.* American Chemical Society, Washington, D. C., **1973**, *121*, 10–27.

22 E. FLANIGEN, in: Molecular Sieves. W. M. MEIER, J. B. UYTTERHOEVEN (Eds), *Adv. Chem. Ser.* American Chemical Society, Washington, D. C. **1973**, *121*, 119–139.

23 E. M. FLANIGEN, in: Introduction to Zeolite Science and Practice. H. VAN BEKKUM, E. M. FLANIGEN, J. C. JANSEN (Eds), *Stud. Surf. Sci. Catal.* Elsevier, Amsterdam, **1991**, *58*, 13–34.

24 A. G. RANDOLPH, M. A. LARSON, Theory of Particulate Processes, 2nd edn, Academic Press, New York, 1988.

25 T. A. M. TWOMEY, M. MACKAY, H. P. C. E. KUIPERS, R. W. THOMPSON, *Zeolites* **1994**, *14*, 162–168.

26 A. E. PERSSON, B. J. SCHOEMAN, J. STERTE, J.-E. OTTERSTEDT, *Zeolites* **1994**, *14*, 557–567.

27 B. J. SCHOEMAN, O. REGEV, *Zeolites* **1996**, *17*, 447–456.

28 B. J. SCHOEMAN, *Zeolites* **1997**, *18*, 97–105.

29 B. J. SCHOEMAN, *Microporous Mater.* **1997**, *9*, 267–271.

30 B. J. SCHOEMAN, *Microporous Mesoporous Mater.* **1998**, *22*, 9–22.

31 J. DOUGHERTY, L. E. ITON, J. W. WHITE, *Zeolites* **1995**, *15*, 640–649.

32 C. S. CUNDY, B. M. LOWE, D. M. SINCLAIR, *J. Crystal Growth* **1990**, *100*, 189–202.

33 C. S. CUNDY, M. S. HENTY, R. J. PLAISTED, *Zeolites* **1995**, *15*, 342–352.

34 S. L. BURKETT, M. E. DAVIS, *J. Phys. Chem.* **1994**, *98*, 4647–4653.

35 O. REGEV, Y. COHEN, E. KEHAT, Y. TALMON, *Zeolites* **1994**, *14*, 314–319.

36 R. W. CORKERY, B. W. NINHAM, *Zeolites* **1997**, *18*, 379–386.

37 W. H. DOKTER, H. F. VAN GARDEREN, T. P. M. BEELEN, R. A. VAN SANTEN, W. BRAS, *Angew. Chem. Int. Ed. Engl.* **1995**, *34*, 73–75.

38 P.-P. E. A. DE MOOR, T. P. M. BEELEN, R. A. VAN SANTEN, *Microporous Mater.* **1997**, *9*, 117–130.

39 P.-P. E. A. DE MOOR, T. P. M. BEELEN, B. U. KOMANSCHEK, O. DIAT, R. A. VAN SANTEN, *J. Phys. Chem. B*, **1997**, *101*, 11077–11086.

40 P.-P. E. A. DE MOOR, T. P. M. BEELEN, R. A. VAN SANTEN, *J. Phys. Chem. B* **1999**, *103*, 1639–1650.

41 P.-P. E. A. DE MOOR, T. P. M. BEELEN, R. A. VAN SANTEN, K. TSUJI, M. E. DAVIS, *Chem. Mater.* **1999**, *11*, 36–43.

42 P.-P. E. A. DE MOOR, T. P. M. BEELEN, B. U. KOMANSCHEK, L. W. BECK, P. WAGNER, M. E. DAVIS, R. A. VAN SANTEN, *Chem. Eur. J.* **1999**, *5*, 2083–2088.

43 J. N. WATSON, L. E. ITON, R. I. KEIR, J. C. THOMAS, T. L. DOWLING, J. W. WHITE, *J. Phys. Chem. B* **1997**, *101*, 10094–10104.

44 V. NIKOLAKIS, E. KOKKOLI, M. TIRRELL, M. TSAPATSIS, D. G. VLACHOS, *Chem. Mater.* **2000**, *12*, 845–853.

45 C. J. J. DEN OUDEN, R. W. THOMPSON, *Ind. Eng. Chem. Res.* **1992**, *31*, 369–373.

46 J. C. J. BART, N. BURRIESCI, F. CARIATI, M. PETRERA, C. ZIPELLI, *Zeolites* **1983**, *3*, 226–232.

47 E. N. COKER, R. W. THOMPSON, A. G. DIXON, A. SACCO, JR., S. S. NAM, S. L. SUIB, *J. Phys. Chem.* **1993**, *97*, 6465–6469.

48 F. POLAK, A. CICHOCKI, in: Molecular Sieves. W. M. MEIER, J. B. UYTTERHOEVEN (Eds), *Adv. Chem. Ser.* American Chemical Society, Washington, D. C. **1973**, *121*, 209–216.

49 S. P. ZHDANOV, N. N. SAMULEVICH, in: Proceedings of the 5th International Conference on Zeolites L. V. C. REES (Ed.), Heyden, London, 1980, 75–84.

50 P. M. SLANGEN, J. C. JANSEN, H. VAN BEKKUM, *Microporous Mater.* **1997**, *9*, 259–265.

51 S. P. ZHDANOV, in: Molecular Sieve Zeolites. L. B. SAND, E. M. FLANIGEN (Eds), *Adv. Chem. Ser.* American Chemical Society, Washington, D. C., **1971**, *101*, 20–41.

52 G. T. KERR, *J. Phys. Chem.* **1966**, *70*, 1047–1050.

53 J. CIRIC, *J. Colloid Interface Sci.* **1968**, *28*, 315–324.

54 C. L. ANGELL, W. H. FLANK, in: Molecular Sieves-II. J. R. KATZER (Ed.), *ACS Symp. Ser.* American Chemical Society, Washington, D. C., **1977**, *40*, 194–206.

55 F. ROOZEBOOM, H. E. ROBSON, S. S. CHAN, *Zeolites* **1983**, *3*, 321–328.

56 P. K. DUTTA, D. C. SHIEH, M. PURI, *J. Phys. Chem.* **1987**, *91*, 2332–2336.

57 S. UEDA, N. KAGEYAMA, M. KOIZUMI, in: Proceedings of the 6th International Zeolite Conference, D. OLSON, A. BISIO (Eds), Butterworth, Guildford, 1984, 905–913.

58 B. D. MCNICOL, G. T. POTT, K. R. LOOS, N. MULDER, in: Molecular Sieves. W. M. MEIER, J. B. UYTTERHOEVEN (Eds), *Adv. Chem. Ser.* American Chemical Society, Washington, D.C., **1973**, *121*, 152–161.

59 T. ANTONIC, B. SUBOTIC, in: Proceedings of the 12th International Zeolite Conference, M. M. J. TREACY, B. K. MARCUS, M. E. BISHER, J. B. HIGGINS (Eds), Mater. Res. Soc. Warrendale, **1999**, *3*, 2049–2056.

60 C. W. DAVIES, A. L. JONES, *Trans. Faraday Soc.* **1955**, *51*, 812–817.

61 B. SUBOTIC, L. SEKOVANIC, *J. Crystal Growth* **1986**, *75*, 561–572.

62 B. SUBOTIC, A. GRAOVAC, *Stud. Surf. Sci. Catal.* **1985**, *24*, 199–206.

63 J. BRONIC, B. SUBOTIC, I. SMIT, LJ. A. DESPOTOVIC, in: Innovation in Zeolite Materials Science. P. J. GROBET, W. J. MORTIER, E. F. VANSANT, G. SCHULTZ-EKLOFF (Eds), *Stud. Surf. Sci. Catal.*, Elsevier, Amsterdam, **1988**, *37*, 107–114.

64 B. SUBOTIC, I. SMIT, L. SEKOVANIC, in: Proceedings of the 5th International Conference on Zeolites. L. V. C. REES (Ed.), Heyden, London, 1980, 10–19.

65 B. SUBOTIC, in: Zeolite Synthesis. M. L. OCCELLI, H. E. ROBSON (Eds), *ACS Symp. Ser.* American Chemical Society, Washington, D. C., **1989**, *398*, 110–123.

66 G. GOLEMME, A. NASTRO, J. B. NAGY, B. SUBOTIC, F. CREA, R. AIELLO, *Zeolites* **1991**, *11*, 776–783.

67 B. SUBOTIC, J. BRONIC, in: Proceedings of the 9th International Zeolite Conference, R. VON BALLMOOS, J. B. HIGGINS, M. M. J. TREACY (Eds), Butterworth-Heinemann, Boston, 1993, 321–328.

68 S. GONTHIER, L. GORA, I. GURAY, R. W. THOMPSON, *Zeolites* **1993**, *13*, 414–418.

69 C. Falamaki, M. Edrissi, M. Sohrai, *Zeolites* **1997**, *19*, 2–5.

70 W. Meise, F. E. Schwochow, in: Molecular Sieves. W. M. Meier, J. B. Uytterhoeven (Eds), *Adv. Chem. Ser.* American Chemical Society, Washington, D. C. **1973**, *121*, 169–178.

71 B. Subotic, T. Antonic, S. Bosnar, J. Bronic, M. Skreblin, in: Porous Materials in Environmentally Friendly Processes. I. Kiricsi, G. Pal-Borbely, J. B. Nagy, H. G. Karge (Eds), *Stud. Surf. Sci. Catal.* Elsevier, Amsterdam, **1999**, *125*, 157–164.

72 T. Antonic, B. Subotic, N. Stubicar, *Zeolites* **1997**, *18*, 291–300.

73 S. Mintova, N. H. Olson, V. Valtchev, T. Bein, *Science,* **1999**, *283*, 958–960.

74 L. Gora, K. Streletzky, R. W. Thompson, G. D. J. Phillies, *Zeolites* **1997**, *18*, 119–131.

75 B. J. Schoeman, J. Sterte, J.-E. Otterstedt, *J. Chem. Soc. Chem. Commun.* **1993**, 994–995.

76 B. J. Schoeman, J. Sterte, J.-E. Otterstedt, *Zeolites* **1994**, *14*, 110–116.

77 B. J. Schoeman, J. Sterte, J.-E. Otterstedt, *Zeolites* **1994**, *14*, 208–216.

78 B. J. Schoeman, J. Sterte, J.-E. Otterstedt, *Zeolites* **1994**, *14*, 568–575.

79 H. Kacirek, H. Lechert, *J. Phys. Chem.* **1975**, *79*, 1589–1593.

80 H. Kacirek, H. Lechert, *J. Phys. Chem.* **1976**, *80*, 1291–1296.

81 P. S. Singh, T. L. Dowling, J. N. Watson, J. W. White, *Phys. Chem. Chem. Phys.* **1999**, *1*, 4125–4130.

82 D. Domine, J. Quobex, Molecular Sieves, Society of Chemical Industry, London, 1968, 78–84.

83 N. N. Feoktistova, S. P. Zhdanov, W. Lutz, M. Bulow, *Zeolites* **1989**, *9*, 136–139.

84 C. S. Cundy, B. M. Lowe, D. M. Sinclair, *Faraday Discuss,* **1993**, *95*, 235–252.

85 T. Sano, S. Sugawara, Y. Kawakami, A. Iwasaki, M. Hirata, I. Kudo, M. Ito, M. Watanabe, in: Zeolites and Related Microporous Materials: State of the Art 1994. J. Weitkamp, H. G. Karge, H. Pfeifer, W. Holderich (Eds), *Stud. Surf. Sci. Catal.* Elsevier, Amsterdam, **1994**, *84*, 187–194.

86 G. S. Wiersema, R. W. Thompson, *J. Mater. Chem.* **1996**, *6*, 1693–1699.

87 A. E. Nielsen, Kinetics of Precipitation, Pergamon Press, Oxford, 1964.

88 M. W. Anderson, J. R. Agger, J. T. Thornton, N. Forsyth, *Angew. Chem. Int. Ed. Engl.* **1996**, *35*, 1210–1213.

89 V. Alfredsson, T. Ohsuna, O. Terasaki, J.-O. Bovin, *Angew. Chem. Int. Ed. Engl.* **1993**, *32*, 1210–1213.

90 T. Ohsuna, O. Terasaki, V. Alfredsson, J.-O. Bovin, D. Watanabe, S. W. Carr, M. W. Anderson, *Proc. R. Soc. London A,* **1996**, *452*, 715–740.

91 J. R. Agger, N. Pervaiz, A. K. Cheetham, M. W. Anderson, *J. Am. Chem. Soc.* **1998**, *120*, 10754–10759.

92 S. Sugiyama, S. Yamamoto, O. Matsuoka, H. Nozoye, J. Yu, G. Zhu, S. Qiu, O. Terasaki, *Microporous Mesoporous Mater.* **1999**, *28*, 1–7.

93 G. T. Kerr, *J. Phys. Chem.* **1968**, *72*, 1385–1386.

94 A. Culfaz, L. B. Sand, in: Molecular Sieves. W. M. Meier, J. B. Uytterhoeven (Eds), *Adv. Chem. Ser.* American Chemical Society, Washington, D.C. **1973**, *121*, 140–151.

95 E. Narita, *Ind. Eng. Chem. Prod. Res. Dev.* **1985**, *24*, 507–512.

96 Y. V. Mirskii, V. V. Pirozhkov, *Russ. J. Phys. Chem.* **1970**, *44*, 1508–1509.

97 J. Warzywoda, R. W. Thompson, *Zeolites* **1991**, *11*, 577–582.

98 S. Gonthier, R. W. Thompson, in: Advanced Zeolite Science and Applications. J. C. Jansen, M. Stocker, H. G. Karge, J. Weitkamp (Eds), *Stud. Surf. Sci. Catal.* Elsevier, Amsterdam, **1994**, *85*, 43–73.

99 J. Garside, R. J. Davey, *Chem. Eng. Commun.* **1980**, *4*, 393–424.

100 R. F. Strickland-Constable, *AIChE Sym. Ser.* **1972**, *68*, 1–7.

101 R. W. Thompson, A. Dyer, *Zeolites* **1985**, *5*, 302–308.

102 L.-Y. Hou, R. W. Thompson, *Zeolites* **1989**, *9*, 526–530.

103 R. D. Edelman, D. V. Kudalkar, T. Ong, J. Warzywoda, R. W. Thompson, *Zeolites* **1989**, *9*, 496–502.

104 J. Warzywoda, R. D. Edelman, R. W. Thompson, *Zeolites* **1991**, *11*, 318–324.

105 E. A. Tsokanis, R. W. Thompson, *Zeolites* **1992**, *12*, 369–373.

106 L. Gora, R. W. Thompson, *Zeolites* **1995**, *15*, 526–534.

107 L. Gora, K. Streletzky, R. W. Thompson, G. D. J. Phillies, *Zeolites* **1997**, *19*, 98–106.

108 S. Kasahara, K. Itabashi, K. Igawa, in: New Developments in Zeolite Science and Technology. Y. Murakami, A. Iijima, J. W. Ward (Eds), *Stud. Surf. Sci. Catal.* Elsevier, Amsterdam, **1986**, *28*, 185–192.

109 R. Xu, J. Zhang, W. Pang, S. Li, Report No. 421, Jilin University, Peoples Republic of China, **1983**.

110 L. Gora, R. W. Thompson, *Zeolites* **1997**, *18*, 132–141.

111 B. Subotic, A. M. Tonejc, D. Bagovic, A. Cizmek, T. Antonic, in: Zeolites and Related Microporous Materials: State of the Art 1994. J. Weitkamp, H. G. Karge, H. Pfeifer, W. Holderich (Eds), *Stud. Surf. Sci. Catal.* Elsevier, Amsterdam, **1994**, *84*, 259–266.

112 H. Lechert, in: Structure and Reactivity of Modified Zeolites. P. A. Jacobs, N. I. Jaeger, P. Jiru, V. B. Kazansky, G. Schulz-Ekloff (Eds), *Stud. Surf. Sci. Catal.* Elsevier, Amsterdam, **1984**, *18*, 107–123.

113 S. Nicolas, P. Massiani, M. Vera Pacheco, F. Fajula, F. Figueras, in: Innovation in Zeolite Materials Science. P. J. Grobet, W. J. Mortier, E. F. Vansant, G. Schulz-Ekloff (Eds), *Stud. Surf. Sci. Catal.* Elsevier, Amsterdam, **1988**, *37*, 115–122.

114 C. E. A. Kirschhock, R. Ravishankar, P. A. Jacobs, J. A. Martens, *J. Phys. Chem. B*, **1999**, *103*, 11021–11027.

115 C. E. A. Kirschhock, R. Ravishankar, L. Van Looveren, P. A. Jacobs, J. A. Martens, *J. Phys. Chem. B*, **1999**, *103*, 4972–4978.

116 C. E. A. Kirschhock, R. Ravishankar, F. Verspeurt, P. J. Grobet, P. A. Jacobs, J. A. Martens, *J. Phys. Chem. B*, **1999**, *103*, 4965–4971.

117 R. Ravishankar, C. E. A. Kirschhock, P.-P. Knops-Gerrita, E. J. P. Feijen, P. J. Grobet, P. Vanoppen, F. C. De Schryver, G. Miehe, H. Fuess, B. J. Schoeman, P. A. Jacobs, J. A. Martens, *J. Phys. Chem. B*, **1999**, *103*, 4960–4964.

118 R. Ravishankar, C. E. A. Kirschhock, B. J. Schoeman, P. Vanoppen, P. J. Grobet, S. Storck, W. F. Maier, J. A. Martens, F. C. De Schryver, P. A. Jacobs, *J. Phys. Chem. B*, **1998**, *102*, 2633–2639.

119 E. F. Freund, *J. Crystal Growth*, **1976**, *34*, 11–23.

120 B. Fahlke, P. Starke, V. Seefeld, W. Wieker, K.-P. Wendlandt, *Zeolites* **1987**, *7*, 209–213.

121 A. Katovic, B. Subotic, I. Smit, Lj. A. Despotovic, M. Curic, in: Zeolite Synthesis. M. L. Occelli, H. E. Robson (Eds), *ACS Symp. Ser.* American Chemical Society, Washington, D. C. **1989**, *398*, 124–139.

122 A. Katovic, B. Subotic, I. Smit, Lj. A. Despotovic, *Zeolites* **1989**, *9*, 45–53.

123 D. M. Ginter, G. T. Went, A. T. Bell, C. J. Radke, *Zeolites* **1992**, *12*, 733–741.

124 D. M. Ginter, A. T. Bell, C. J. Radke, *Zeolites* **1992**, *12*, 742–749.

125 B. M. Lowe, N. A. MacGilp, T. V. Whittam, in: Proceedings of the 5th International Conference on Zeolites. L. V. C. Rees (Ed.), Heyden, London, 1980, 85–93.

126 K. Hamilton, E. N. Coker, A. Sacco, Jr. A. G. Dixon, R. W. Thompson, *Zeolites* **1993**, *13*, 645–653.

127 T. Antonic, B. Subotic, V. Kaucic, R. W. Thompson, in: Porous Materials in Environmentally Friendly Processes. I. Kiricsi, G. Pal-Borbely, J. B. Nagy, H. G. Karge (Eds) *Stud.*

Surf. Sci. Catal. Elsevier, Amsterdam, **1999**, *125*, 13–20.

128 P. K. DUTTA, R. ASIAIE, in: Synthesis of Microporous Materials. M. L. OCCELLI, H. E. ROBSON (Eds), Van Nostrand Reinhold, New York, **1992**, *1*, 522–536.

129 H. BORER, W. M. MEIER, in: Molecular Sieve Zeolites. L. B. SAND, E. M. FLANIGEN (Eds), *Adv. Chem. Ser.* American Chemical Society, Washington, D. C. **1971**, *101*, 122–126.

130 J. F. CHARNELL, *J. Cryst. Growth* **1971**, *8*, 291–294.

131 G. SCOTT, A. G. DIXON, A. SACCO, JR. R. W. THOMPSON, in: Zeolites: Facts, Figures, Future. P. A. JACOBS, R. A. VAN SANTEN (Eds), *Stud. Surf. Sci. Catal.* Elsevier, Amsterdam, **1989**, *49*, 363–372.

132 M. MORRIS, A. SACCO, JR. A. G. DIXON, R. W. THOMPSON, *Zeolites* **1991**, *11*, 178–183.

133 G. SCOTT, R. W. THOMPSON, A. G. DIXON, A. SACCO, JR. *Zeolites* **1990**, *10*, 44–50.

134 G. ZHU, A. QIU, J. YU, F. GAO, F. XIAO, R. XU, Y. SAKAMOTO, O. TERASAKI, in: Proceedings of the 12th International Zeolite Conference. M. M. J. TREACY, B. K. MARCUS, M. E. BISHER, J. B. HIGGINS (Eds), Mater. Res. Soc. Warrendale, **1999**, *3*, 1863–1870.

135 M. MORRIS, A. G. DIXON, A. SACCO, JR. R. W. THOMPSON, *Zeolites* **1993**, *13*, 113–121.

136 P. D. HOPKINS, in: Zeolite Synthesis. M. L. OCCELLI, H. E. ROBSON (Eds), *ACS Symp. Ser.* American Chemical Society, Washington, D. C. **1989**, *398*, 152–160.

137 P. D. HOPKINS, in: Synthesis of Microporous Materials. M. L. OCCELLI, H. E. ROBSON (Eds), Van Nostrand Reinhold, New York, **1992**, *1*, 129–138.

138 R. H. JARMAN, M. T. MELCHIOR, *J. Chem. Soc. Chem. Commun.* **1984**, 414–416.

139 P. K. DUTTA, D. C. SHIEH, *J. Phys. Chem.* **1986**, *90*, 2331–2334.

140 M. T. MELCHIOR, in: Intrazeolite Chemistry. G. D. STUCKY, F. G. DWYER (Eds), *ACS Symp. Ser.* American Chemical Society, Washington, D. C. **1983**, *218*, 243–265.

141 M. MABILIA, R. A. PEARLSTEIN, A. J. HOPFINGER, *J. Am. Chem. Soc.* **1987**, *109*, 7960–7968.

142 R. M. BARRER, D. E. MAINWARING, *J. Chem. Soc. Dalton Trans.* **1972**, 2534–2546.

143 R. M. BARRER, R. BEAUMONT, C. COLELLA, *J. Chem. Soc. Dalton Trans,* **1974**, 934–941.

144 J. ROCHA, J. KLINOWSKI, J. M. ADAMS, *J. Chem. Soc. Faraday Trans.* **1991**, *87*, 3091–3097.

145 E. COSTA, A. LUCAS, M. A. UGUINA, J. C. RUIZ, *Ind. Eng. Chem. Res.* **1988**, *27*, 1291–1296.

146 L. V. C. REES, S. CHANDRASEKHAR, *Zeolites* **1993**, *13*, 524–533.

147 E. I. BASALDELLA, R. TORRES SANCHEZ, S. L. PEREZ DE VARGAS, D. CAPUTO, C. COLELLA, in: Proceedings of the 12th International Zeolite Conference. M. M. J. TREACY, B. K. MARCUS, M. E. BISHER, J. B. HIGGINS (Eds), Mater. Res. Soc. Warrendale, **1999**, *3*, 1663–1670.

148 M. E. DAVIS, R. F. LOBO, *Chem. Mater.* **1992**, *4*, 756–768.

149 S. I. ZONES, D. S. SANTILLI, in: Proceedings of the 9th International Zeolite Conference. R. VON BALLMOOS, J. B. HIGGINS, M. M. J. TREACY (Eds), Butterworth-Heinemann, Boston, **1993**, 171–174.

150 Y. KUBOTA, M. M. HELMKAMP, S. I. ZONES, M. E. DAVIS, *Microporous Mater.* **1996**, *6*, 213–229.

151 H. GIES, in: Advanced Zeolite Science and Applications. J. C. JANSEN, M. STOCKER, H. G. KARGE, J. WEITKAMP (Eds), *Stud. Surf. Sci. Catal.* Elsevier, Amsterdam, **1994**, *85*, 295–327.

152 H. K. BEYER, I. BELENYKAJA, in: Catalysis by Zeolites. B. IMELIK, C. NACCACHE, Y. BEN TAARIT, J. C. VEDRINE, G. COUDURIER, H. PRALIAUD (Eds), *Stud. Surf. Sci. Catal.* Elsevier, Amsterdam, **1980**, *5*, 203–210.

153 H. LECHERT, *Microporous Mesoporous Mater.* **2000**, *40*, 181–196.

154 S. MA, L. LI, R. XU, Z. YIE, in: Zeolites: Synthesis, Structure, Technology and Application. B. DRZAJ, S. HOCEVAR, S. PEJOVNIK (Eds), *Stud. Surf. Sci. Catal.* Elsevier, Amsterdam, **1985**, *24*, 191–198.

155 G. T. KOKOTAILO, J. CIRIC, in: Molecular Sieve Zeolites. L. B. SAND, E. M. FLANIGEN (Eds), *Adv. Chem. Ser.* American Chemical Society, Washington, D. C. **1971**, *101*, 109–121.

156 V. FULOP, G. BORBELY, H. K. BEYER, S. ERNST, J. WEITKAMP, *J. Chem. Soc. Faraday Trans.* **1989**, *85*, 2127–2139.

157 M. M. J. TREACY, J. M. NEWSAM, D. E. W. VAUGHAN, K. G. STROHMAIER, W. J. MORTIER, *J. Chem. Soc. Chem. Commun.* **1989**, 493–495.

158 M. AUDIER, J. M. THOMAS, J. KLINOWSKI, D. A. JEFFERSON, L. A. BURSILL, *J. Phys. Chem.* **1982**, *86*, 581–584.

159 F. DELPRATO, L. DELMOTTE, J. L. GUTH, L. HUVE, *Zeolites* **1990**, *10*, 546–552.

160 F. DOUGNIER, J. PATARIN, J. L. GUTH, D. ANGLEROT, *Zeolites* **1992**, *12*, 160–166.

161 O. TERASAKI, T. OHSUNA, V. ALFREDSSON, J.-O. BOVIN, D. WATANABE, S. W. CARR, M. W. ANDERSON, *Chem. Mater.* **1993**, *5*, 452–458.

162 S. L. BURKETT, M. E. DAVIS, *Microporous Mater.* **1993**, *1*, 265–282.

163 E. J. P. FEIJEN, K. DE VADDER, M. H. BOSSCHAERTS, J. L. LIEVENS, J. A. MARTENS, P. J. GROBET, P. A. JACOBS, *J. Am. Chem. Soc.* **1994**, *116*, 2950–2957.

164 M. W. ANDERSON, K. S. PACHIS, F. PREBIN, S. W. CARR, O. TERASAKI, T. OHSUNA, V. ALFREDSSON, *J. Chem. Soc. Chem. Commun.* **1991**, 1660–1664.

165 J. P. ARHANCET, M. E. DAVIS, *Chem. Mater.* **1991**, *3*, 567–569.

166 N. HANIF, M. W. ANDERSON, V. ALFREDSSON, O. TERASAKI, *Phys. Chem. Chem. Phys.* **2000**, *2*, 3349–3357.

167 P. CHU, F. G. DWYER, J. C. VARTULI, Mobil Oil Corporation, US Patent, 4,778,666, 1988.

168 J. C. JANSEN, A. ARAFAT, A. K. BARAKAT, H. VAN BEKKUM, in: Synthesis of Microporous Materials. M. L. OCCELLI, H. E. ROBSON (Eds), Van Nostrand Reinhold, New York, **1992**, *1*, 507–521.

169 A. ARAFAT, J. C. JANSEN, A. R. EBAID, H. VAN BEKKUM, *Zeolites* **1993**, *13*, 162–165.

170 P. M. SLANGEN, J. C. JANSEN, H. VAN BEKKUM, *Zeolites* **1997**, *18*, 63–66.

171 P. M. SLANGEN, J. C. JANSEN, H. VAN BEKKUM, G. W. HOFLAND, F. VAN DER HAM, G. J. WITKAMP, in: Proceedings of 12th International Zeolite Conference. M. M. J. TREACY, B. K. MARCUS, M. E. BISHER, J. B. HIGGINS (Eds), Mater. Res. Soc. Warrendale, **1999**, *3*, 1553–1560.

172 H. A. BENESI, *J. Catal.* **1967**, *8*, 369–374.

173 R. J. LEONARD, *Econ. Geol.* **1927**, *22*, 18–43.

174 P. K. BAJPAI, M. S. RAO, K. V. G. K. GOKHALE, *Ind. Eng. Chem. Prod. Res. Dev.* **1978**, *17*, 223–227.

175 L. B. SAND, Molecular Sieves, Society of Chemical Industry, London, 1968, 71–77.

176 W. M. MEIER, *Z. Kristallogr.* **1961**, *115*, 439–450.

177 W. L. KRANICH, Y. H. MA, L. B. SAND, A. H. WEISS, I. ZWIEBEL, in: Molecular Sieve Zeolites. L. B. SAND, E. M. FLANIGEN (Eds), *Adv. Chem. Ser.* American Chemical Society, Washington, D. C. **1971**, *101*, 502–513.

178 O. J. WITTEMORE, *Am. Mineral.* **1972**, *57*, 1146–1151.

179 S. UEDA, H. MURATA, M. KOIZUMI, *Am. Mineral.* **1980**, *65*, 1011–1018.

180 S. UEDA, T. FUKUSHIMA, M. KOIZUMI, *J. Clay Sci. Jpn.* **1982**, *22*, 18–28.

181 K. ITABASHI, T. FUKUSHIMA, K. IGAWA, *Zeolites* **1986**, *6*, 30–34.

182 P. K. BAJPAI, *Zeolites* **1986**, *6*, 2–8.

183 E. G. DEROUANE, J. G. FRIPIAT, in: Proceedings of the 6th International Zeolite Conference. D. OLSON, A. BISIO (Eds), Butterworth, Guildford, 1984, 717–726.

184 P. BODART, J. B. NAGY, E. G. DEROUANE, Z. GABELICA, in: Structure

and Reactivity of Modified Zeolites. P. A. JACOBS, N. I. JAEGER, P. JIRU, V. B. KAZANSKY, G. SCHULZ-EKLOFF (Eds), *Stud. Surf. Sci. Catal.* Elsevier, Amsterdam, **1984**, *18*, 125–132.

185 P. A. JACOBS, J. A. MARTENS, Synthesis of High-silica Aluminosilicate Zeolites, *Stud. Surf. Sci. Catal.* Elsevier, Amsterdam, **1987**, *33*, 321–329.

186 G. J. KIM, W. S. AHN, *Zeolites* **1991**, *11*, 745–750.

187 J. WARZYWODA, A. G. DIXON, R. W. THOMPSON, A. SACCO, JR. S. L. SUIB, *Zeolites* **1996**, *16*, 125–137.

188 R. MOSTOWICZ, L. B. SAND, *Zeolites* **1982**, *2*, 143–146.

189 P. K. BAJPAI, M. S. RAO, K. V. G. K. GOKHALE, *Ind. Eng. Chem. Prod. Res. Dev.* **1981**, *20*, 721–726.

190 Y. SUN, S. QIU, T. SONG, W. PANG, Y. YUE, *J. Chem. Soc. Chem. Commun.* **1993**, 1048–1050.

191 P. BODART, J. B. NAGY, Z. GABELICA, E. G. DEROUANE, *J. Chim. Phys. Phys.-Chim. Biol.* **1986**, *83*, 777–790.

192 F.-Y. DAI, M. SUZUKI, H. TAKAHASHI, Y. SATO, in: Zeolite Synthesis. M. L. OCCELLI, H. E. ROBSON (Eds), *ACS Symp. Ser.* American Chemical Society, Washington, D.C. **1989**, *398*, 244–256.

193 R. L. WADLINGER, G. T. KERR, E. J. ROSINSKI, Mobil Oil Corporation, US Patent, 3,375,205, 1968.

194 B. M. LOK, T. R. CANNAN, C. A. MESSINA, *Zeolites* **1983**, *3*, 282–291.

195 R. J. ARGAUER, G. R. LANDOLT, Mobil Oil Corporation, US Patent, 3,702,886, 1972.

196 C. J. PLANK, E. J. ROSINSKI, M. K. RUBIN, Mobil Oil Corporation, Great Britain Patent, 1,334,243, 1970.

197 P. CHU, Mobil Oil Corporation, US Patent, 3,709,979, 1973.

198 E. M. FLANIGEN, J. M. BENNETT, R. W. GROSE, J. P. COHEN, R. L. PATTON, R. M. KIRCHNER, J. V. SMITH, *Nature*, **1978**, *271*, 512–516.

199 H. GIES, B. MARLER, *Zeolites* **1992**, *12*, 42–49.

200 G. T. KOKOTAILO, S. L. LAWTON, D. H. OLSON, W. M. MEIER, *Nature*, **1978**, *272*, 437–438.

201 D. H. OLSON, G. T. KOKOTAILO, S. L. LAWTON, W. M. MEIER, *J. Phys. Chem.* **1981**, *85*, 2238–2243.

202 P. A. JACOBS, J. A. MARTENS, Synthesis of High-silica Aluminosilicate Zeolites, *Stud. Surf. Sci. Catal.* Elsevier, Amsterdam, **1987**, *33*, 47–146.

203 J. C. JANSEN, in: Introduction of Zeolite Science and Practice. H. VAN BEKKUM, E. M. FLANIGEN, J. C. JANSEN (Eds), *Stud. Surf. Sci. Catal.* Elsevier, Amsterdam, **1991**, *58*, 77–136.

204 L. D. ROLLMANN, in: Inorganic Compounds with Unusual Properties-II. R. B. KING (Ed.), *Adv. Chem. Ser.* American Chemical Society, Washington, D.C. **1979**, *173*, 387–395.

205 R. K. ILER, The Chemistry of Silica, Wiley, New York, 1979.

206 D. W. LEWIS, C. M. FREEMAN, C. R. A. CATLOW, *J. Phys. Chem.* **1995**, *99*, 11194–11202.

207 A. CHATTERJEE, *J. Mol. Catal. A: Chemical*, **1997**, *120*, 155–163.

208 A. CHATTERJEE, R. VETRIVEL, *J. Mol. Catal. A: Chemical*, **1996**, *106*, 75–81.

209 A. CHATTERJEE, R. VETRIVEL, *J. Chem. Soc. Faraday Trans.* **1995**, *91*, 4313–4319.

210 E. DE VOS BURCHART, H. VAN KONINGSVELD, B. VAN DE GRAAF, *Microporous Mater.* **1997**, *8*, 215–222.

211 K. J. CHAO, J. C. LIN, Y. WANG, G. H. LEE, *Zeolites* **1986**, *6*, 35–38.

212 R. K. McMULLAN, M. BONAMICO, G. A. JEFFREY, *J. Chem. Phys.* **1963**, *39*, 3295–3310.

213 P. K. DUTTA, M. PURI, *J. Phys. Chem.* **1987**, *91*, 4329–4333.

214 C. D. CHANG, A. T. BELL, *Catal. Lett.* **1991**, *8*, 305–316.

215 C. S. GITTELMAN, A. T. BELL, C. J. RADKE, *Microporous Mater.* **1994**, *2*, 145–158.

216 Z. GABELICA, B. J. NAGY, P. BODART, N. DEWAELE, A. NASTRO, *Zeolites* **1987**, *7*, 67–72.

217 S. L. BURKETT, M. E. DAVIS, *Chem. Mater.* **1995**, *7*, 920–928.

218 S. L. BURKETT, M. E. DAVIS, *Chem. Mater.* **1995**, *7*, 1453–1463.

219 R. GOUGEON, L. DELMOTTE, D. LE NOUEN, Z. GABELICA, *Microporous Mesoporous Mater.* **1998**, *26*, 143–151.

220 E. J. J. GROENEN, A. G. T. G.
KORTBEEK, M. MACKAY, O. SUDMEIJER,
Zeolites **1986**, *6*, 403–411.

221 R. A. VAN SANTEN, J. KEIJSPER, G.
OOMS, A. G. T. G. KORTBEEK, in: New
Developments in Zeolite Science and
Technology. Y. MURAKAMI, A. IIJIMA,
J. W. WARD (Eds), *Stud. Surf. Sci.
Catal.* Elsevier, Amsterdam, **1986**, *28*,
169–175.

222 J. J. KEIJSPER, M. F. M. POST, Zeolite
Synthesis, M. L. OCCELLI, H. E.
ROBSON (Eds), *ACS Symp. Ser.*
American Chemical Society,
Washington, D. C. **1989**, *398*, 28–48.

223 G. BOXHOORN, O. SUDMEIJER, P. H.
G. VAN KASTEREN, *J. Chem. Soc. Chem.
Commun.* **1983**, 1416–1418.

224 T. R. CANNAN, R. J. HINCHEY, UOP,
US Patent, 5,139,759, 1992.

225 U. LOHSE, B. ALTRICHTER, R. DONATH,
R. FRICKE, K. JANCKE, B. PARLITZ, E.
SCHREIER, *J. Chem. Soc. Faraday
Trans.* **1996**, *92*, 159–165.

226 M. K. RUBIN, Mobil Oil Corporation,
US Patent, 5,164,169, 1992.

227 M. J. EAPEN, K. S. N. REDDY, V. P.
SHIRALKAR, *Zeolites* **1994**, *14*, 295–302.

228 P. CAULLET, J. HAZM, J. L. GUTH, J. F.
JOLY, J. LYNCH, F. RAATZ, *Zeolites* **1992**,
12, 240–250.

229 M. K. RUBIN, Mobil Oil Corporation,
Eur. Pat. Appl. 159,847, 1985.

230 M. K. RUBIN, Mobil Oil Corporation,
Eur. Pat. Appl. 159,846, 1985.

231 B. MARLER, R. BOHME, H. GIES, in:
Proceedings of the 9th International
Zeolite Conference. R. VON BALLMOOS,
J. B. HIGGINS, M. M. J. TREACY (Eds),
Butterworth-Heinemann, Boston,
1993, 425–432.

232 J. M. NEWSAM, M. M. J. TREACY,
W. T. KOETSIER, C. B. DE GRUYTER,
Proc. R. Soc. London A, **1988**, *420*,
375–405.

233 J. B. HIGGINS, R. B. LA PIERRE, J. L.
SCHLENKER, A. C. ROHRMAN, J. D.
WOOD, G. T. KERR, W. J. ROHRBAUGH,
Zeolites **1988**, *8*, 446–452.

234 The database of the Structure
Commission of the International
Zeolite Association Website, http://
www.iza-online.org.

235 J. PEREZ-PARIENTE, J. A. MARTENS, P.
A. JACOBS, *Appl. Catal.* **1987**, *31*, 35–
64.

236 J. PEREZ-PARIENTE, J. A. MARTENS, P.
A. JACOBS, *Zeolites* **1988**, *8*, 46–53.

237 M. A. CAMBLOR, J. PEREZ-PARIENTE,
Zeolites **1991**, *11*, 202–210.

238 M. A. CAMBLOR, J. PEREZ-PARIENTE,
Zeolites **1991**, *11*, 792–797.

239 A. ERDEM, L. B. SAND, *J. Catal.* **1979**,
60, 241–256.

240 S. ERNST, P. A. JACOBS, J. A. MARTENS,
J. WEITKAMP, *Zeolites* **1987**, *7*, 458–
462.

241 A. NASTRO, C. COLELLA, R. AIELLO, in:
Zeolites: Synthesis, Structure,
Technology and Application. B. DRZAJ,
S. HOCEVAR, S. PEJOVNIK (Eds), *Stud.
Surf. Sci. Catal.* Elsevier, Amsterdam,
1985, *24*, 39–46.

242 A. NASTRO, L. B. SAND, *Zeolites* **1983**,
3, 57–62.

243 J. B. NAGY, P. BODART, H. COLLETTE,
C. FERNANDEZ, Z. GABELICA, A.
NASTRO, R. AIELLO, *J. Chem. Soc.
Faraday Trans.* **1989**, *85*, 2749–2769.

244 J. C. VAN DER WAALS, M. S. RIGUTTO,
H. VAN BEKKUM, *J. Chem. Soc. Chem.
Commun.* **1994**, 1241–1242.

245 M. A. CAMBLOR, A. CORMA, S.
VALENCIA, *Chem. Commun.* **1996**,
2365–2366.

246 R. B. BORADE, A. CLEARFIELD,
Microporous Mater. **1996**, *5*, 289–297.

247 R. B. BORADE, A. CLEARFIELD, *Chem.
Commun.* **1996**, 626–626.

248 F. VAUDRY, F. DI RENZO, P. ESPIAU,
F. FAJULA, PH. SCHULZ, *Zeolites* **1997**,
19, 253–258.

249 Z. GABELICA, N. DEWAELE, L.
MAISTRIAU, J. B. NAGY, E. G.
DEROUANE, ZEOLITE SYNTHESIS, M. L.
OCCELLI, H. E. ROBSON (Eds), *ACS
Symp. Ser.* American Chemical Society,
Washington, D.C. **1989**, *398*, 518–
543.

250 J. PEREZ-PARIENTE, J. A. MARTENS, P.
A. JACOBS, *Zeolites* **1992**, *12*, 280–286.

251 M. A. NICOLLE, F. DI RENZO, F.
FAJULA, P. ESPIAU, T. DES COURIERES,
in: Proceedings of the 9th
International Zeolite Conference.
R. VON BALLMOOS, J. B. HIGGINS,

M. M. J. TREACY (Eds), Butterworth-Heinemann, Boston, 1993, 313–320.

252 M. A. CAMBLOR, A. CORMA, J. PEREZ-PARIENTE, *Zeolites* **1993**, *13*, 82–87.

253 W. XU, J. DONG, J. LI, J. LI, F. WU, *J. Chem. Soc. Chem. Commun.* **1990**, 755–756.

254 M. H. KIM, H. X. LI, M. E. DAVIS, *Microporous Mater.* **1993**, *1*, 191–200.

255 M. MATSUKATA, N. NISHIYAMA, K. UEYAMA, *Microporous Mater.* **1993**, *1*, 219–222.

256 M. MATSUKATA, N. NISHIYAMA, K. UEYAMA, *J. Chem. Soc. Chem. Commun.* **1994**, 339–340.

257 N. NISHIYAMA, K. UEYAMA, M. MATSUKATA, in: Zeolites Related Microporous Materials: State of the Art 1994. J. WEITKAMP, H. G. KARGE, H. PFEIFER, W. HOLDERICH (Eds), *Stud. Surf. Sci. Catal.* Elsevier, Amsterdam, **1994**, *84*, 1183–1190.

258 N. NISHIYAMA, K. UEYAMA, M. MATSUKATA, *J. Chem. Soc. Chem. Commun.* **1995**, 1967–1968.

259 N. NISHIYAMA, K. UEYAMA, M. MATSUKATA, *Microporous Mater.* **1996**, *7*, 299–308.

260 P. R. HARI PRASAD RAO, M. MATSUKATA, *Chem. Commun.* **1996**, 1441–1442.

261 P. R. HARI PRASAD RAO, K. UEYAMA, M. MATSUKATA, *Appl. Catal. A: General*, **1998**, *166*, 97–103.

262 P. R. HARI, PRASAD RAO, C. A. LEON Y LEON, K. UEYAMA, M. MATSUKATA, *Microporous and Mesoporous Mater.* **1998**, *21*, 305–313.

263 S. L. SUIB, G. D. STUCKY, R. J. BLATTNER, *J. Catal.* **1980**, *65*, 174–178.

264 R. VON BALLMOOS, W. M. MEIER, *Nature*, **1981**, *289*, 782–783.

265 K. J. CHAO, J. Y. CHERN, *Zeolites* **1988**, *8*, 82–85.

266 E. G. DEROUANE, S. DETREMMERIE, Z. GABELICA, N. BLOM, *Appl. Catal.* **1981**, *1*, 201–224.

267 E. G. DEROUANE, J. P. GILSON, Z. GABELICA, C. MOUSTY-DESBUQUOIT, J. VERBIST, *J. Catal.* **1981**, *71*, 447–448.

268 A. E. HUGES, K. G. WILSHIER, B. A. SEXTON, P. SMART, *J. Catal.* **1983**, *80*, 221–227.

269 J. DWYER, F. R. FITCH, F. MACHADO, G. QIN, S. M. SMYTH, J. C. VICKERMAN, *J. Chem. Soc. Chem. Commun.* **1981**, 422–424.

270 G. DEBRAS, A. GOURGUE, J. B. NAGY, G. DECLIPPELEIR, *Zeolites* **1985**, *5*, 369–376.

271 J. B. NAGY, P. BODART, H. COLLETTE, J. EL HAGE-AL ASSWAD, Z. GABELICA, R. AIELLO, A. NASTRO, C. PELLEGRINO, *Zeolites* **1988**, *8*, 209–220.

272 P. A. JACOBS, J. A. MARTENS, Synthesis of High-silica Aluminosilicate Zeolites. *Stud. Surf. Sci. Catal.* Elsevier, Amsterdam, **1987**, *33*, 91–95.

273 R. ALTHOFF, B. SCHULZ-DOBRICK, F. SCHÜTH, K. UNGER, *Microporous Mater.* **1993**, *1*, 207–218.

274 L. PAULING, The Nature of the Chemical Bond and the Structure of Molecules and Crystals; an Introduction to Modern Structural Chemistry. Ithaca, NY, Cornel University Press, 1967.

275 M. TARAMASSO, G. PEREGO, B. NOTARI, in: Proceedings of the 5th International Conference on Zeolites. L. V. C. REES (Ed.), Heyden, London, 1980, 40–48.

276 K. F. M. G. J. SCHOLLE, A. P. M. KENTGENS, W. S. VEEMAN, P. FRENKEN, G. P. M. VAN DER VELDEN, *J. Phys. Chem.* **1984**, *88*, 5–8.

277 W. HOLDERICH, H. EICHORN, R. LEHNERT, L. MAROSI, W. MROSS, R. REINKE, W. RUPPEL, H. SCHLIMPER, in: Proceedings of the 6th International Zeolite Conference. D. OLSON, A. BISIO (Eds), Butterworth, Guildford, 1984, 545–555.

278 M. G. HOWDEN, *Zeolites* **1985**, *5*, 334–338.

279 Z. GABELICA, G. DEBRAS, J. B. NAGY, in: Catalysis on the Energy Scene. S. KALIAGUINE, A. MAHAY (Eds), *Stud. Surf. Sci. Catal.* Elsevier, Amsterdam, **1984**, *19*, 113–121.

280 R. DE RUITER, J. C. JANSEN, H. VAN BEKKUM, *Zeolites* **1992**, *12*, 56–62.

281 C. R. BAYENSE, A. J. H. P. VAN DER POL AND J. H. C. VAN HOOF, *Appl. Catal.* **1991**, *72*, 81–98.

282 G. GIANNETTO, A. MONTES, N. S. GNEP, A. FLORENTINO, P. CARTRAUD AND M. GUISNET, *J. Catal.* **1993**, *145*, 86–95.

283 Y. ONO, *Catal. Rev. Sci. Eng.* **1992**, *34*, 179–226.

284 M. GUISNET, N. S. GNEP AND F. ALARIO, *Appl. Catal.* **1992**, *89*, 1–30.

285 G. GIANNETTO, R. MONQUE AND R. GALIASSO, *Catal. Rev. Sci. Eng.* **1994**, *36*, 271–304.

286 R. FRICKE, H. KOSSLICK, G. LISCHKE, M. RICHTER, *Chem. Rev.* **2000**, *100*, 2303–2405.

287 R. SZOSTAK, V. NAIR, T. L. THOMAS, *J. Chem. Soc. Faraday Trans. 1*, **1987**, *83*, 487–494.

288 F. A. COTTON, G. WILKINSON, Advanced Inorganic Chemistry, 4th edn, Wiley, New York, **1980**.

289 B. NOTARI, *Adv. Catal.* **1996**, *41*, 253–334.

290 M. TARAMASSO, G. PEREGO, B. NOTARI, Snamprogetti S.p.A. US Patent, 4,410,501, 1983.

291 B. NOTARI, in: Innovation in Zeolite Materials Science. P. J. GROBET, W. J. MORTIER, E. F. VANSANT, G. SCHULZ-EKLOFF (Eds), *Stud. Surf. Sci. Catal.* Elsevier, Amsterdam, **1988**, *37*, 413–425.

292 B. NOTARI, in: Structure-Activity and Selectivity Relationship in Heterogeneous Catalysis. R. K. GRASSELLI, A. W. SLEIGHT (Eds), *Stud. Surf. Sci. Catal.* Elsevier, Amsterdam, **1991**, *67*, 243–256.

293 G. BELLUSSI, A. CARATI, M. G. CLERICI, A. ESPOSITO, in: Preparation of Catalysts V. G. PONCELET, P. A. JACOBS, P. GRANGE, B. DELMON (Eds), *Stud. Surf. Sci. Catal.* Elsevier, Amsterdam, **1991**, *63*, 421–429.

294 B. KRAUSHAAR–CZARNETZKI, J. H. C. VAN HOOFF, *Catal. Lett.* **1989**, *2*, 43–47.

295 A. J. H. P. VAN DER POL, J. H. C. VAN HOOFF, *Appl. Catal. A* **1992**, *92*, 93–111.

296 A. THANGARAJ, R. KUMAR, S. P. MIRAJKAR, P. RATNASAMY, *J. Catal.* **1991**, *130*, 1–8.

297 A. THANGARAJ, R. KUMAR, P. RATNASAMY, *J. Catal.* **1992**, *137*, 252–256.

298 M. PADOVAN, F. GENONI, G. LEOFANTI, G. PETRINI, G. TREZZA, A. ZECCHINA, in: Preparation of Catalysts V. G. PONCELET, P. A. JACOBS, P. GRANGE, B. DELMON (Eds), *Stud. Surf. Sci. Catal.* **1991**, *63*, 431–438.

299 M. A. UGUINA, G. OVEJERO, R. VAN GRIEKEN, D. SERRANO, M. CAMACHO, *J. Chem. Soc. Chem. Commun.* **1994**, 27–28.

300 E. M. FLANIGEN, R. L. PATTON, Union Carbide Corporation, US Patent, 4,073,865, 1978.

301 J. L. GUTH, H. KESSLER, R. WEY, in: New Development in Zeolite Science and Technology. Y. MURAKAMI, A. IIJIMA, J. W. WARD (Eds), *Stud. Surf. Sic. Catal.* Elsevier, Amsterdam, **1986**, *28*, 121–128.

302 J. L. GUTH, H. KESSLER, J. M. HIGEL, J. M. LAMBLIN, J. PATARIN, A. SEIVE, J. M. CHEZEAU, R. WEY, in: Zeolite Synthesis. M. L. OCCELLI, H. E. ROBSON (Eds), *ACS Symp. Ser.* American Chemical Society, Washington, D. C. **1989**, *398*, 176–195.

303 H. KESSLER, J. PATARIN, C. SCHOTT-DARIE, in: Advanced Zeolite Science and Applications. J. C. JANSEN, M. STOCKER, H. G. KARGE, J. WEITKAMP (Eds), *Stud. Surf. Sci. Catal.* Elsevier, Amsterdam, **1994**, *85*, 75–113.

304 R. SZOSTAK, T. L. THOMAS, *J. Catal.* **1986**, *100*, 555–557.

305 P. FEJES, J. B. NAGY, J. HALASZ, A. OSZKO, *Appl. Catal. A: General* **1998**, *175*, 89–104.

306 F. TESTA, F. CREA, R. AIELLO, J. B. NAGY, in: Porous Materials in Environmentally Friendly Processes. I. KIRICSI, G. PAL-BORBELY, J. B. NAGY, H. G. KARGE (Eds), *Stud. Surf. Sci. Catal.* Elsevier, Amsterdam, **1999**, *125*, 165–171.

307 S. A. AXON, J. KLINOWSKI, *Appl. Catal. A: General* **1992**, *81*, 27–34.

308 R. M. DESSAU, K. D. SCHMILL, G. T. KERR, G. L. WOOLERY, L. B. ALEMANY, *J. Catal.* **1987**, *104*, 484–489.

4.2.3
Synthesis of AlPO₄s and Other Crystalline Materials

Joël Patarin, Jean-Louis Paillaud and Henri Kessler

4.2.3.1
Scope

The present section is a review on the synthesis of three-dimensional open-framework phosphates (3D open framework). Since the 1980s, numerous such materials, essentially aluminophosphates and derived solids, gallo-, beryllo-, zinco- and iron phosphates, have been reported. Most of the Al,P-based solids are microporous, that is, they show adsorption properties after removal of the occluded water and/or organic molecules used in their synthesis. However, generally, the structure of the other phosphates collapses on removal of the molecules trapped during their synthesis, only a few of them show adsorption properties. The materials described in this section will be designated by their published names or acronyms and/or when known, by their structure type whose code as defined by the Structure Commission of the International Zeolite Association (IZA) is a group of three capital letters [1] which will be given in bold type, for example, **AFI** and **AEL** for the aluminophosphates AlPO₄-5 and AlPO₄-11, respectively.

Recently, in addition to the large number of 3D open-framework phosphates, numerous lamellar (2D framework) and chain-like (1D framework) alumino-, gallo- and zincophosphates have been reported. Their synthesis is very similar to that of the 3D open framework phosphates. However, their review is beyond the scope of the present section. A few mesostructured aluminophosphates that are obtained in the presence of surfactants via a liquid crystal templating mechanism have been published since the late 1990s. Some show an organized 3D framework structure, but most of them are lamellar. They will be briefly discussed at the end of the present section.

4.2.3.2
History of the Synthesis of 3D Open-Framework Phosphates

In 1982, the synthesis of a new family of molecular sieves, the aluminophosphate family, with an Al/P ratio equal to 1 was reported by researchers from Union Carbide Corporation [2]. Up to that date the incorporation of phosphorus in silica-based zeolites from an alkaline medium had been investigated by several authors, but except in the study published by Flanigen and Grose [3], the phosphorus content of the materials was rather low and most probably some was occluded as phosphate in the voids of the structures. Thus, Kühl and coworkers [4, 5] for example, reported the synthesis of the aluminosilicophosphates ZK-21 and ZK-22 of the **LTA** type, which turned out to contain trapped phosphate anions in the sodalite

References see page 867

cage besides little framework phosphorus. Flanigen and Grose [3] reported a number of phosphate-containing materials of the structure type **ANA, CHA, GIS, LTA** and **LTL** with at least 12 P atoms per 100 framework T atoms (T = Al, Si, P) as determined by microprobe analysis on the fine crystalline powder. The crystal structure of the **ANA**-type silicoaluminophosphate $Na_{13}Al_{24}Si_{13}P_{11}O_{96} \cdot 16H_2O$ has been solved by Artioli et al. [6]. It was found that phosphorus is present as framework atom in an ordered Al, Si/P distribution.

The new aluminophosphate family discovered by Wilson et al. [2] has been obtained by combining the hydrothermal conditions employed for the synthesis of the dense phases of $AlPO_4$ (acidic to neutral pH) but in the presence of organic structure-directing species as for silica-based zeolites. Indeed, the $AlPO_4$ forms equivalent to the silica dense phases quartz, cristobalite and tridymite are known. The quartz-type $AlPO_4$ berlinite has aroused much interest, in particular for its piezoelectric properties. Its hydrothermal synthesis is usually performed in strong acidic medium with an excess of H_3PO_4 [7].

By using milder conditions and in the absence of organic species, d'Yvoire [8] prepared earlier several hydrates of $AlPO_4$. They include the synthetic counterparts of the minerals variscite and metavariscite $AlPO_4 \cdot 2H_2O$ and six synthetic hydrates $AlPO_4 \cdot nH_2O$, designated H1 through H6. Some of these hydrates are microporous, e.g., H1, H2 and H3. The structure of $AlPO_4$-H1 with a 1D system of channels circumscribed by 18 TO_4 (T = Al, P) [9] is close to that of the essentially organic-free VPI-5 (**VFI**) prepared in the presence of organic species such as di-n-propylamine or tetrabutylammonium cations [10, 11]. The chemical formula of as-synthesized VPI-5 is $AlPO_4 \cdot 2.33H_2O$. By slow dehydration the corresponding porous $AlPO_4$ structure with 18-membered-ring openings (18-MR) is obtained [12], whereas by fast dehydration the aluminophosphate $AlPO_4$-8 (**AET**) with a 1D system of channels circumscribed by 14 TO_4 is formed [13, 14]. The hydrate $AlPO_4$–H2 shows ten-membered-ring channels along [001] [15], $AlPO_4$–H3 is of the structure type **APC** with a 2D system of eight-membered-ring channels [16] and $AlPO_4$–H4 is a dense phase, not a molecular sieve [17].

The discovery of the aluminophosphate molecular sieves [2] was followed by that of a large number of derived materials with the same or different structures and at least 14 elements incorporated into the framework beside Al and P. Over 100 organic species have been reported as structure-directing agents.

Soon after the discovery of $AlPO_4$-based solids, the first 3D open-framework gallophosphates were synthesized by Parise [18] using a route similar to that employed for the aluminophosphates. Later, besides this conventional route, the addition of HF to the reaction mixtures has led to numerous novel materials. Moreover, the use of a solvothermal route, i.e., from an essentially nonaqueous medium with or without fluoride, allowed other new materials or already known phases to be obtained.

The first beryllophosphates were synthesized by Harvey and Meier in 1989 [19] from mixtures containing alkaline cations and tetraethylammonium hydroxide, and the first zincophosphates were reported by Gier and Stucky in 1991 [20]. They were obtained by a low temperature route, 4 to 100 °C, in a wide range of pH values (2 to 12). Later a large number of other zincophosphates have been published by several groups and the number of new solids is still increasing (see Sect. 4.2.3.7).

In the past, several reviews on phosphate molecular sieves have been published. Those by Flanigen [21], Wilson [22] and Szostak [23] address essentially the AlPO$_4$-based solids, the others contain a large part on phosphates beside other inorganic open-framework materials [24–26].

4.2.3.3
Synthetic Methods

4.2.3.3.1 Synthesis Procedures

Three main routes are used for the synthesis of 3D open-framework phosphates, i.e., (1) the conventional hydrothermal route, (2) the nonconventional one with fluoride as a mineralizer and (3) the solvothermal route in the absence or the presence of fluoride using organic solvents instead of water.

(1) In contrast to rather strong alkaline pH values used in the conventional synthesis of silica-based zeolites, the usual pH of the reaction mixture is slightly acidic to slightly alkaline (typically, starting pH = 3–10).

(2) The pH range is essentially the same as in (1). Generally, in the presence of fluoride well formed crystals are obtained, and fluorine is found in the structures as a terminal or bridging species, but it may also be occluded in small cubic building units (double-four-rings) present in a number of structures (see Sect. 4.2.3.6.2).

(3) In the solvothermal route the reaction mixture is essentially nonaqueous, generally small quantities of water from the reactants are present. In a few cases an organic molecule may play the role of solvent and of the source of a structure-directing species resulting from its decomposition [27].

4.2.3.3.2 Chemical Parameters

A. Sources of framework atoms Some sources of framework atoms for 3D open-framework phosphates are given in Table 1. The preferred and most used aluminum sources are pseudoboehmite and aluminum isopropoxide. Pseudoboehmite reacts slowly during the preparation of the reaction mixture; thus the initial pH may be lower. The alkoxide hydrolyzes rapidly, therefore a reactive form of alumina is obtained that reacts quickly with phosphoric acid. Higher initial pH values and higher supersaturations are then obtained, and most probably different precursor species. Crystalline sources such as gibbsite are less used, they tend to lead to dense phases. The use of sodium aluminate is not preferred because Na$^+$ cations may direct to undesired sodium aluminophosphate phases, also when NH$_4^+$ cations are present the ammonium hydroxyaluminophosphate named AlPO$_4$-15 [28] is generally formed. The sources of gallium are usually the sulfate or an amorphous oxyhydroxide GaOOH·xH$_2$O.

The most used source of phosphorus oxide is phosphoric acid. All types of silicas

References see page 867

Tab. 1. Some framework T-atom sources used in the synthesis of 3-D open-framework phosphates

Aluminum	Pseudoboehmite
	Alkoxide (isopropoxide, isobutoxide)
	Amorphous aluminum hydroxide
Less used	Gibbsite
	$AlPO_4$ (dense phase or molecular sieve)
	$AlPO_4 \cdot 2H_2O$ (metavariscite)
	$AlPO_4 \cdot 1.5H_2O$ (H3 of d'Yvoire)
Gallium	Gallium sulfate
	Amorphous gallium oxyhydroxide
Phosphorus	Phosphoric acid
Less used	Polyphosphoric acid ($H_{10}P_8O_{25}$)
	Triethylphosphate
Silica	Colloidal Fumed silica
	Alkoxide Precipitated silica
Metal (Mg, Zn, Co, Mn, Fe, ...)	Water soluble salt (acetate, sulfate)
	Oxide
Other elements	Water soluble salt
	Oxide

given in Table 1 have been equally used. For the sources of metals such as Mg, Zn, Co, Mn, Fe, the acetates are preferred. As sources of other elements, soluble species or oxides are generally used.

B. Organic structure-directing species   Over 100 organic structure-directing species have been reported for the synthesis of 3D open-framework phosphates. Generally, these include quaternary ammonium ions (linear or cyclic mono-, bis-, tris- or polyquaternary), trialkylamines, dialkylamines, monoalkylamines, cyclic amines, alkylethanolamines, cyclic diamines, alkyldiamines and polyamines.

Some of the organics direct the formation of only a few structure types (1, 2 or 3), whereas others will orient to as much as about 10 different structures. On the other hand, a given structure type can be obtained from several organic species.

In a few cases, the starting organic molecule decomposes under the synthesis conditions and a decomposition product will be occluded in the structure (organic template).

Usually, the crystalline obtained solid is calcined to remove the occluded organic species and to lead to the porous solid. Solvent extraction is possible only for a very limited number of solids.

C. The solvent   Up to now the aqueous medium has been the most used medium. Water is involved in the dissolution of the gel of the framework elements. Moreover, it generally enters the structure as hydration water of inorganic cations (when present). In the materials which are prepared in the presence of large organic cations, generally only small amounts of water are occluded because little space is available beside the organic species.

As mentioned above (3), the solvothermal synthesis route using solvent-water or essentially nonaqueous systems has been developed. The organic solvents are usually alcohols such as butane-2-ol, hexanol, ethyleneglycol, triethyleneglycol and glycerol.

D. Chemical composition of the starting mixture   The overall chemical composition is usually given in terms of molar ratios of the constituents expressed as oxides (e.g., SiO_2/Al_2O_3, P_2O_5/Al_2O_3, $((CH_3)_4N)_2O/Al_2O_3$, H_2O/Al_2O_3, etc.). The molar composition may also more conveniently be expressed by taking into account the real reactants, for example, $x\,(CH_3)_4NCl{:}y\,SiO_2{:}Al(OH)_3{:}z\,H_3PO_4{:}w\,H_2O$. The starting pH will have an equilibrium value only when the mixture has reached equilibrium.

E. Preparation of the starting mixture   Generally the preparation conditions will influence the result of a synthesis through the dissolution rate of the gels, which will determine the nature and concentration of the species present at a given time in the liquid phase. This will depend on the reactivity (chemical and physical nature) of the reactants, the way and order of their addition, the efficiency of the homogenization of the mixture, etc. The lack of reproducibility or the difficulties that often arise in the scale up of the synthesis may partly result from an incomplete control of this type of factor together with that of the heating rate.

F. Aging   The reaction mixture has sometimes to be aged, with or without stirring, at a temperature that is below the crystallization temperature, generally near room temperature. During this process, chemical and structural changes of the mixture occur that influence the solid and the liquid phase, and nucleation of a given material may start during this period.

G. Seeding   The addition of seeds of the desired material to the starting mixture will generally lead, through the decrease of the crystallization time, to a pure metastable phase that will be kinetically favored with respect to other metastable phases. But sometimes it may be the only way to obtain a metastable material from a mixture whose supersaturation is high enough for crystal growth but not for nucleation (thermodynamic aspect).

Seeds may consist of finely ground crystals or may be contained in a "nucleation mixture" that has been obtained by aging above room temperature, for example.

4.2.3.3.3 Physical Parameters

A. Crystallization temperature   The crystallization temperature range for micro- and mesoporous materials is rather large, that is, from around room temperature (even ~4 °C) to about 200 °C. Some materials may crystallize over a wide temper-

References see page 867

ature range, e.g., 150 °C, but some may be produced only in a small temperature interval, e.g., 20 °C.

The most open materials are generally obtained at lower temperatures, and dense phases such as quartz-type $AlPO_4$ or $GaPO_4$ will form preferably at higher temperatures. Crystallization will be faster at higher temperatures.

The heating rate to reach the crystallization temperature may be critical, all the more if a large autoclave or a large crystallization vessel is used. Stirring of the reaction mixture will help in reaching temperature equilibrium throughout the reaction mixture. Microwave heating may be used when a high heating rate is desired, but such a technique may be only applicable to small volumes.

B. Heating time – agitation Depending on the chosen operating conditions for the synthesis of a given material, for example, the temperature, crystallization may last from a few minutes to several months. When heating is prolonged, other phases may appear from the less stable to the most stable, according to Ostwald's rule [29], for example, dense $AlPO_4$s or $GaPO_4$s. The agitation of the reaction mixture by turbostirring or tumbling of the autoclave, for example, may be preferred in order to reach and maintain a good homogeneity of temperature and reaction mixture. As a general rule, it can be considered that the crystallization time is significantly decreased when the reaction mixture is stirred.

C. Pressure Crystallization is generally carried out at autogeneous pressure. In some cases, the pressure may be increased because of the decomposition of organic species such as tetraalkylammonium hydroxides.

4.2.3.4
Aluminophosphates

A large part of the known microporous aluminophosphates and derived molecular sieves (Al, P-based) has been patented by Union Carbide Corporation between 1982 and 1985. The synthesis of a number of phases, by a different route, in particular silicoaluminophosphates, was patented by Mobil Oil Corporation in 1985. Moreover, other industrial or academic groups have published similar or new phases with their own designations.

The designations used by the UCC researchers for the $AlPO_4$ family was $AlPO_4$-*n*, where *n* is a sequence number. Later, each family with other framework T atoms beside Al and P received a specific designation (Table 2). The same sequence number is used for a given structure type regardless of the framework composition.

The materials patented by Mobil Oil Corporation were obtained in a mixed hexanol-water medium. In general they contain Si, Al, P as framework T atoms. They were designated MCM-*n*. In Table 3, the silicoaluminophosphates are reported together with their $AlPO_4$ or SAPO-*n* analogs (according to the X-ray diffraction patterns given in the Mobil patents) [30–36]. In addition, silicoaluminophos-

Tab. 2. Designations used by Union Carbide Corporation for the AlPO$_4$-based materials [22]

Designation	Framework T-atoms	
AlPO$_4$-n	Al, P	
SAPO-n	Si, Al, P	
MeAPO-n	Me, Al, P	Me = Co, Fe, Mg, Mn, Zn
MeAPSO-n	Me, Al, P, Si	
ElAPO-n	El, Al, P	El = As, B, Be, Ga, Ge, Li, Ti
ElAPSO-n	El, Al, P, Si	

phates, aluminophosphates containing iron [37], antimony, boron, germanium or vanadium [38] have also been patented.

The designations used by other industrial or academic groups for phosphate molecular sieves are collected in Table 4 [39–53].

4.2.3.4.1 AlPO$_4$-n Materials Reported by Union Carbide Corporation

In Table 5 [54–58], typical organic species for the structure types n reported by Union Carbide Corporation are given. For each structure type is also shown the pore size and the pore volume values as reported by Flanigen [21]. Pore sizes and pore volumes were determined by adsorption techniques on solids calcined at 500–600 °C to remove the occluded organic structure-directing agents and H$_2$O. The pore size values result from the adsorption of gauge molecules of known kinetic diameter. Under different synthesis conditions the same organic molecule may

Tab. 3. Silicoaluminophosphates patented by Mobil Oil Corporation [30–36] and their AlPO$_4$ or SAPO-n analogs

Designation	Analogs (IZA code)
MCM-1	AlPO$_4$–H$_3$ (**APC**)
MCM-2	SAPO-34 (**CHA**)
MCM-3	[a]
MCM-4	[a]
MCM-5	[a]
MCM-6	SAPO-5 (**AFI**)
MCM-7	SAPO-11 (**AEL**)
MCM-8	SAPO-20 (**SOD**)
MCM-9	VPI-5 (+SAPO-11) (**VFI** + **AEL**)
MCM-10	[a]
MCM-37	SAPO-8 (**AET**)

[a] Unknown.

References see page 867

Tab. 4. Designations used by other industrial and academic groups for phosphate molecular sieves

Designation	Corresponding group	T element beside P
CFAP-*n*; CFSAPO-*n*	China Fudan University	Al; Al, Si [39–42]
CHNUAP-*n*	China Hunan Norm. University	Al [43]
CNU-*n*	China Nanjiing University	Al [44]
DAF-*n*	Davy Faraday Laboratory, London	Mg, Al[a]; Zn[b]
JDF-*n*	Jilin Davy Faraday, Jilin and London	Al [45]
MIL-*n*	Material Institut Lavoisier, Versailles	Al [46], Ga[c]
Mu-*n*	Mulhouse University	Al [47,48], Ga[c]
STA-*n*	Saint Andrews, Fife	Mg, Al[a]
UCSB-*n*	University of California Santa Barbara	[a], [c]
UiO-*n*	University of Oslo	Al [49–51], Al, Co[a]; Zn[b]
ULM-*n*	University of Le Mans	Al [52,53], Ga[c]
ZYT-*n*	Mitsubishi Chemical Industries	Al, Si[a]

[a] See Sect. 4.2.3.5. [b] See Sect. 4.2.3.7. [c] See Sect. 4.2.3.6.

orient to different structure types; thus di-n-propylamine and ethylenediamine lead to eight and five different materials respectively (Table 6). On the other hand, a given structure type can be obtained from several organic species; thus, more than 25 direct the crystallization of the type **AFI** (Table 7).

4.2.3.4.2 Other Aluminophosphates Prepared in Aqueous Medium

Only a few other aluminophosphates have been reported so far (Table 8 [59–66]). $AlPO_4$-12-TAMU [59], which was obtained with tetramethylammonium as the structure-directing agent, is of the type **ATT** that has a 2-dimensional system of eight-membered-ring channels. The structures of CHNUAP-3 and -4 [43] are unknown. Their calcination leads to open solids with a good adsorption capacity for water and cyclohexane. SBM-6 [60], which was synthesized in the presence of the tris nickel complex of 1,3-diaminopropane, contains diprotonated 1,3-diaminopropane resulting from the decomposition of the complex. It should be noted that the framework shows only five- and six-coordinated aluminum, no AlO_4 tetrahedra are present. Several materials are isostructural with $AlPO_4$-EN3 [57], i.e., JDF-2 [61], Mu-10 [47], UiO-12 [62], $AlPO_4$-53 (A) [63] and CFSAPO-1 (A) [39]. It should be noted that in the synthesis of UiO-12 [62], the tetramethyl-ammonium cations decomposed into dimethylammonium cations, therefore UiO-12 is identical to Mu-10 [47]. The structure of the as-synthesized solids contains AlO_4–OH–AlO_4 units in addition to alternating AlO_4 and PO_4 tetrahedra. Calcined dehydrated $AlPO_4$-53 (B) [63] is isostructural with calcined MCS-1 [58] and UiO-12 500 (calcined at 500 °C) [62]. The removal of the organic species is accompanied by that of the bridging OH and the resulting structure (IZA structure code **AEN**) shows alternating AlO_4 and PO_4 tetrahedra only, which form a 2D eight-membered-ring channel system. Calcined $AlPO_4$-53 (C) [63] is isostructural with UiO-12 750 [62] (calcined at 750 °C); it contains a 1-dimensional system of 8-

Tab. 5. AlPO$_4$-n materials reported by Union Carbide Corporation and typical organic species used for their synthesis. Pore size and pore volume are also shown [21]

Structure type n[a]	IZA code[b]	Typical organic species	Pore size (Å)	Saturation H$_2$O pore vol./ (cm^3 g^{-1})
5	AFI	(n-C$_3$H$_7$)$_3$N	8	0.31
8	AET	(n-C$_4$H$_9$)$_2$NH	8.7 × 7.9[c]	
9	d			
11	AEL	(n-C$_3$H$_7$)$_2$NH	6	0.16
12	det.[e] [54]	H$_2$N(CH$_2$)$_2$NH$_2$		
14	det. [55]	(CH$_3$)$_3$CNH$_2$	4	0.19
15	det. [28]	NH$_4^+$		
16	AST	Quinuclidine	3	0.3
17	ERI	Quinuclidine	4.3	0.28
18	AEI	(C$_2$H$_5$)$_4$N$^+$	4.3	0.35
20	SOD	(CH$_3$)$_4$N$^+$	3	0.24
21[f]	det [56]	Pyrrolidine		
22	AWW	Quinuclidine	3.9[c]	
23[g]	d	Pyrrolidine		
24	ANA		1.6 × 4.2[c]	
25	ATV	Calc. type 21 → 25[h]	3	0.17
26	d		4.3	0.23
28	d	Calc. type 23 → 28[h]	3	0.21
31	ATO	(n-C$_3$H$_7$)$_2$NH	6.5	0.17
33[i]	ATT	(CH$_3$)$_4$N$^+$	4	0.23
34[j]	CHA	(C$_2$H$_5$)$_4$N$^+$	4.3	0.3
35[k]	LEV	Quinuclidine	4.3	0.3
36	ATS	(n-C$_3$H$_7$)$_3$N	8	0.31
37[k]	FAU	(CH$_3$)$_4$N$^+$ + (n-C$_3$H$_7$)$_4$N$^+$	8	0.35
39	ATN	(n-C$_3$H$_7$)$_2$NH	4	0.23
40	AFR	(n-C$_3$H$_7$)$_4$N$^+$	7	0.33
41	AFO	(n-C$_3$H$_7$)$_2$NH	6	0.22
42	LTA	(CH$_3$)$_4$N$^+$ + diethanolamine	4.3	0.3
43[l]	GIS	(n-C$_3$H$_7$)$_2$NH	4.3	0.3
44[m]	CHA	Cyclohexylamine	4.3	0.34
46	AFS	(n-C$_3$H$_7$)$_2$NH	7	0.28
47[n]	CHA	N,N-diethanolamine	4.3	0.3
50[o]	AFY	(n-C$_3$H$_7$)$_2$NH	6.1; 4.0 × 4.3[c]	
52	AFT	(n-C$_3$H$_7$)$_3$N + (C$_2$H$_5$)$_4$N$^+$	2.8 × 4.4[c]	
53[p]	AEN	H$_2$N(CH$_2$)$_2$NH$_2$	3.1 × 4.3; 2.7 × 5[c]	
54[q]	VFI	(n-C$_3$H$_7$)$_2$NH	12.5[c]	

Tab. 5. *(continued)*

[a] Some of the numbers missing in this table correspond to layered structures (e.g. 1,2,3,4,6). [b] When available. [c] From the reported crystal structure [1]. [d] Unknown structure. [e] det. = structure determined. [f] Precursor of $AlPO_4$-25. [g] Precursor of $AlPO_4$-28. [h] The calcination of $AlPO_4$-21 and -23 yields $AlPO_4$-25 and -28, respectively. [i] Isostructural with $AlPO_4$-12 TAMU. [j] $(C_2H_5)_4N^+$ is used in the synthesis of the silico- and metalloaluminophosphate forms. $AlPO_4$-34 is obtained with at least 5 different organic species in the presence of fluoride (Table 9). [k] Only the SAPO, MeAPO and MeAPSO forms have been reported. [l] The $AlPO_4$ form has been obtained using the solvothermal route with dimethylformamide in the presence of fluoride [27]. [m] Only the SAPO, MeAPO, MeAPSO and ElAPSO forms have been reported. [n] See footnote [k]. [o] Only the MeAPO form has been reported. [p] Isostructural with $AlPO_4$-EN3 [57], MCS-1 [58] Mu-10 [47], JDF-2, UiO-12 (Table 8) and CFSAPO-1 [39–42]. [q] Isostructural with H1 [9], MCM-9 (Table 3) and VPI-5 [11].

Tab. 6. $AlPO_4$-based materials which were obtained with di-n-propylamine and ethylenediamine

Organic species	Structure type or designation
Di-n-propylamine	11, 31, 39, 41, 43, 46, 50, 54
Ethylenediamine	12, 21 and AlPO-EN-3 [57]
	CHNUAP-3 and CHNUAP-4 [43]

Tab. 7. Organic species directing the formation of the **AFI**-type structure

Tetraethylammonium	Methyldiethanolamine
Tetrapropylammonium	Methylethanolamine
Trimethylethanolammonium	Triethanolamine
Triethylamine	2-Methylpyridine
Tri-n-propylamine	3-Methylpyridine
Triethanolamine	4-Methylpyridine
Dicyclohexylamine	Pyridine
N-methylcyclohexylamine	Piperidine
n-butyldimethylamine	N-methylpiperidine
Cyclohexylamine	3-Methylpiperidine
N,N-dimethylbenzylamine	N,N-dimethylpiperizine
N,N-diethylethanolamine	N,N'-tetraethyldiaminobutane
Aminodiethylethanolamine	1,4-diazabicyclo[2.2.2]octane
N,N-dimethylethanolamine	

Tab. 8. Other 3-D open-framework aluminophosphates with the organic structure-directing species used in their synthesis, the dimensionality of the channel system and the pore openings in terms of number of T atoms in the largest rings

Name (IZA code)	Organic structure-directing species (R)	Chemical formula*	Dimensionality of the channel system (pore openings: number of T atoms**)	Ref.
AlPO₄-12-TAMU (ATT)	Tetramethylammonium hydroxide	$[(AlPO_4)_3] \cdot R$	2 dim. (8T)	59
CHNUAP-3	Ethylenediamine	a	b	43
CHNUAP-4	Ethylenediamine	a	b	43
SBM-6	Tris nickel complex of 1,3-diaminopropane	$[Al_3(PO_4)_3OH] \cdot H_2N(CH_2)_3NH_3 \cdot H_2O$	3 dim. (10T, 8T)	60
With the AlPO₄-EN3 topology (AEN):				
AlPO₄-EN3	Ethylenediamine	$[Al_6(PO_4)_6]_4 \; R_4 \cdot 16H_2O$	2 dim. (8T)	57
MCS-1	Methylamine (+HF)c	$[Al_{24}P_{24}O_{88.8}(OH)_{14.4}]F_{2.4}(RH)_{2.4}$		58
JDF-2	Methylamine (H₂O + EG)d	$[Al_3(PO_4)_3OH] \cdot RH$		61
Mu-10	Dimethylamine	$[Al_3(PO_4)_3OH] \cdot RH$		47
UiO-12	Tetramethylammonium hydroxide	$[Al_3(PO_4)_3OH] \cdot CH_3NH_2CH_3$		62
AlPO₄-53 (A)	Methylamine	$[Al_3(PO_4)_3] \cdot R \cdot 14/8H_2O$		63
APDAB-200	1,4-diaminobutane	$[Al_4P_4O_{17}] \cdot RH_2$	3 dim. (10T)	64
UiO-26	1,3-diaminopropane	$[Al(PO_4) \; O] \cdot RH_2 \cdot H_2O$	1 dim. (10T)	51
Mu-13	1,7,10,6-tetraoxa 4,13-diaza-cyclooctadecane	$[Al_{90}(PO_4)_{90}] \cdot (RH_2)_6(OH)_{12} \cdot 11H_2O$	Cages (6T)	48
(MSO)	4-(2-aminoethyl) diethylenetriamine	$[Al_9(PO_4)_{12}] \cdot C_{24}H_{81}N_{16} \cdot 17H_2O$	3 dim. (12T)	66

*The framework composition is given in [ ] brackets.
**Approximate pore opening corresponding to 6T atoms: 3 Å; 8T atoms: 3–4 Å; 10T atoms: 4–6 Å; 12T atoms: 6–8 Å.
a Not specified. b Unknown structure. c Fluoride is incorporated in the solid as methylammonium fluoride. d JDF-2 was obtained via a solvothermal route in the presence of ethyleneglycol (EG) and a little water.

membered-ring channels [63]. APDAB-200 [64] and UiO-26 [51] were synthesized with 1,4-diaminobutane and 1,3-diaminopropane respectively, they seem to have the same structure with 10-membered-ring channels [51]. The cage-like alumino-phosphate Mu-13 [48] is isostructural with the aluminosilicate MCM-61 [65] of the **MSO** structure type [1]. The protonated 1,7,10,16-tetraoxa-4,13-diaza-cyclooctadecane ("Kriptofix 22") which is occluded in the cages can be removed by calcination at 900 °C, but the largest cage openings are only of about 3 Å, therefore it is a small-pore material. With 4-(2-aminoethyl)diethylenetriamine as the organic structure-directing species, Xu et al. [66] were successful in the synthesis of a new 12-membered-ring aluminophosphate in which the Al/P ratio is 3/4. The framework contains 12-membered-ring channels that intersect each other with 12-membered windows in three directions. The channels are filled with water and amine cations. The window contains three P=O groups which protrude into the channels and are hydrogen-bonded with two terminal N atoms of the amine. The material is stable up to 260 °C. On removal of the organic species at about 550 °C there is structure collapse.

4.2.3.4.3 Aluminophosphates Prepared in a Fluoride Medium

The materials of the types **AFI** ($AlPO_4$, MeAPO), **AEL** ($AlPO_4$, SAPO), **CHA** (SAPO, MeAPO, MeAPSO), **FAU** (SAPO) and **SOD** (MeAPO) have also been prepared in the presence of fluoride. In general, HF is added to a conventional synthesis gel. Typically, $AlPO_4$-5, for example, is obtained by heating a gel of molar composition 1 (n-C_3H_7)$_3$N:1 Al_2O_3:1 P_2O_5:1 HF:70 H_2O at 170 °C for 17 h with stirring (tumbling of the autoclave). Essentially the same conditions are used for the synthesis of SAPO-5; the starting molar SiO_2/Al_2O_3 is typically in the range 0–0.6.

Fluorine is present in the as-synthesized materials. In the $AlPO_4$-type solids with no net framework charge, fluoride neutralizes the cationic organic species, whereas in the substituted materials less fluorine is present because part of the organic cations compensate the negative framework charge. By calcination, fluorine-containing species are removed together with the decomposition products of the organic species. The resulting solid is essentially fluorine free and the adsorption properties are similar to those of the solid prepared in the absence of fluoride. In some cases the thermal stability of the material prepared in fluoride medium is higher than that of the solid produced in its absence, this is presumably due to an increased crystallinity.

The new aluminophosphates which were obtained in the presence of fluoride are reported in Table 9 [67–72]. $AlPO_4$-CJ2 is an ammonium hydroxyfluoroalumino-phosphate with eight-membered-ring channels containing NH_4^+ cations [67]. Fluorine is part of the framework as bridging and terminal species [68]. The tetragonal variant of $AlPO_4$-16 (type **AST**) is obtained in the presence of quinuclidine and HF [69] whereas the usual cubic form is produced in the absence of HF. Fluoride was found to be located in the double-four-ring units of the structure like in the **AST**-type octadecasil [1]. As can be seen in Table 9, a triclinic **CHA**-type alumino-phosphate has been obtained with at least five different organic species [70].

Tab. 9. 3-D open-framework aluminophosphates obtained in the presence of fluoride[a] with the organic species used in their synthesis, the dimensionality of the channel system and the pore openings in terms of number of T atoms in the largest rings

Material designation and IZA code when available	Organic structure-directing species (R) (+ fluoride)	Chemical formula	Dimensionality of the channel system (pore openings: number of T atoms*)	Ref.
AlPO$_4$-Cl2	Hexamethylenetetramine (decomposes → NH$_4^+$)	(NH$_4$)$_{0.88}$(H$_3$O)$_{0.12}$[AlPO$_4$(OH)$_{0.33}$F$_{0.67}$]	1 dim. (8T) 2 dim. (8T)	67, 68
Tetragonal variant of AlPO$_4$-16 (AST)	Quinuclidine	[Al$_{10}$(PO$_4$)$_{10}$]·2.2RH·F$_{1.6}$·3H$_2$O	Cages (6T, 4T)	69
Triclinic **CHA** type	1-methylimidazole (R)** Morpholine Piperidine Pyridine N,N,N',N',-tetramethyl-ethylenediamine	[Al$_3$(PO$_4$)$_3$F]·RH	3 dim. (8T)	70
AlPO$_4$-**LTA**	Tetramethylammonium + diethanolamine	b	3 dim. (8T)	71
b	1,4-diaminobutane	[Al$_3$(PO$_4$)$_3$F$_2$]·RH$_2$	1 dim. (10T) 2 dim. (8T)	72
ULM-3	1,3-diaminopropane	[Al$_3$(PO$_4$)$_3$F$_2$]·RH$_2$·H$_2$O	1 dim. (10T) 2 dim. (8T)	52
ULM-4	1,3-diaminopropane	[Al$_3$(PO$_4$)$_3$F$_2$]·RH$_2$·H$_2$O	1 dim. (10T) 1 dim. (8T)	52
ULM-6	1,3-diaminopropane	[Al$_4$(PO$_4$)$_4$F$_2$]·RH$_2$·H$_2$O	2 dim. (8T)	53
UiO-6 (**OSI**)	Tetraethylammonium hydroxide (and NaF or KF)	Na$_{0.5}$[Al$_{16}$(PO$_4$)$_{16}$F$_{1.5}$]·R K$_{1.5}$[Al$_{16}$(PO$_4$)$_{16}$F$_3$]·R$_{1.5}$	1 dim. (12T)	49
UiO-7 (**ZON**)	Tetramethylammonium hydroxide	b	2 dim. (8T)	50
MIL-27	Tris-(2 aminoethylamine)	[Al$_6$(PO$_4$)$_6$F$_3$]·RH$_3$·H$_2$O	2 dim. (8T)	46

* Approximate pore opening corresponding to 6T atoms: 3 Å; 8T atoms: 3–4 Å; 10T atoms: 4–6 Å; 12T atoms: 6–8 Å.
** The chemical formula was given for this structure-directing species.
a Generally, fluoride is added as aqueous HF. b Not specified.

Two fluorine atoms were found to bridge two Al atoms of a four-membered ring connecting two double-six rings of the **CHA**-type topology. Hydrogen bonding between the protonated amine and fluorine has been evidenced. The removal of the organic cations and HF by calcination results in the aluminophosphate AlPO$_4$-34. **LTA**-type AlPO$_4$ was synthesized for the first time by using a fluoride medium containing tetramethylammonium cations and diethanolamine (DEA) as a combination of structure-directing species [71]. A starting mixture with the molar composition 0.078 (CH$_3$)$_4$NCl:0.93 DEA:1 Al$_2$O$_3$:1 P$_2$O$_5$:0.17 HF:40 H$_2$O was heated at 170 °C for 8.5 h. SAPO, CoAPO and MeAPSO (Me = Co, Zn) samples of the type **LTA** were also obtained with the same organics (Sect. 4.2.3.5). As mentioned earlier (Sect. 4.2.3.2), the aluminosilicophosphates ZK-21 and ZK-22 of the type **LTA** which had been previously reported contain only a small amount of phosphorus (typically $0.01 \leq$ P/(P + Si + Al) ≤ 0.04). SAPO-42 too is essentially an aluminosilicate material with little phosphorus [5]. By using 1,3-diaminopropane and 1,4-diaminobutane as structure-directing species and varying the synthesis conditions, the fluoroaluminophosphates ULM-3 [52], ULM-4 [52], ULM-6 [53] and Al$_3$(PO$_4$)$_3$F$_2$·H$_3$N(CH$_2$)$_4$NH$_3$ [72] were obtained (see Table 14 for the corresponding GaPO$_4$ forms). 1,3-diaminopropane leads to three ULM-n phases with different channel systems; the frameworks are built up from the connection of PO$_4$ tetrahedra with AlO$_4$X (X = F or H$_2$O) trigonal bipyramids and AlO$_4$F$_2$ octahedra. The structure of the 1,4-diaminobutane-containing solid [72] is related to that of the ULM-3 phase. The subnetwork of the 1,4-diaminobutane cations induces a noncentric symmetry. Akporiaye et al. [50] reported two fluoroaluminophosphates designated UiO-6 [49] and UiO-7 that were prepared with tetraethyl- and tetramethylammonium cations, respectively, as the organic structure-directing species. UiO-6 (**OSI**) was made in an inorganic organic system, i.e., NaF (or KF), HF and (C$_2$H$_5$)$_4$NOII. Its structure shows a 1D system of 12-membered-ring channels; it is stable to the removal of the organic species. The largest molecule to be adsorbed by UiO-6 is 1,3,5-trimethylbenzene confirming the 12-ring channel system; the micropore volume as determined by water adsorption on the calcined solid is 0.11 cm^3 g^{-1}. UiO-7 [50] of the structure type **ZON** is isostructural with the zincoaluminophosphate ZAPO-1 ZnAl$_3$P$_4$O$_{16}$·(CH$_3$)$_4$N which was also prepared with tetramethylammonium cations [1]. It has a 2D eight-membered-ring channel system. MIL-27 [46] was obtained with tris-(2-aminoethylamine) as the organic species; the triprotonated amine and water are occluded in the structure that is built up from the connection of PO$_4$ tetrahedra, AlO$_4$ tetrahedra, AlO$_4$F trigonal bipyramids and AlO$_4$F$_2$ octahedra. The authors report no adsorption properties.

4.2.3.4.4 Aluminophosphates Prepared by Solvothermal Synthesis

A number of aluminophosphates that previously had been synthesized via the hydrothermal route have been prepared by solvothermal synthesis, i.e., AlPO$_4$-5, AlPO$_4$-11 [73], AlPO$_4$-12 [74], AlPO$_4$-17 [75], AlPO$_4$-20, -21 and -41 [76]. Generally, the organic solvent has been ethyleneglycol and the organic structure-directing molecule the same as for the hydrothermal synthesis. In addition, a number of new materials have been reported. Usually, the solvothermal synthesis procedure includes the dispersion of the alumina source in the organic solvent, then the addition of aqueous phosphoric acid (85 %). After stirring to a homogeneous mixture, the organic structure-directing species and possibly HF are added. After a possible aging, the mixture is heated in an autoclave between 100 and 200 °C under autogeneous pressure. As mentioned in Sect. 4.2.3.3.1, actually the synthesis is a two-solvent synthesis, water-organic solvent, because some water from the reactants is present, in particular from phosphoric acid. In Table 10 [77–89], the new aluminophosphates are collected with an example of a corresponding structure-directing organic molecule and an organic solvent used. The dimensionality of the channel system and the pore openings in terms of number of T atoms in the largest rings are also reported. It should be noted that generally the Al/P ratio is smaller than 1. JDF-20 has channels with large elliptical 20-membered-ring openings intersected by ten- and eight-membered-ring channels [45, 77, 78]. There are four P=O groups and two O–H–O units defining a star shape of the 20-ring, the free aperture is 6.2×7.9 Å [78]. On heating at about 300 °C, JDF-20 transforms to AlPO₄-5 [77]. AlPO₄-HAD, which is obtained with 1,6-hexanediamine is another material with large openings, it shows intersecting 12- and eight-membered-ring channels [82]. With dimethylformamide as organic structure-directing species and solvent, Vidal and coworkers [83, 84] obtained the monoclinic form of AlPO₄-sodalite. Dimethylformamide (DMF) is occluded in the cages of the solid. On the contrary, Paillaud et al. [27] observed that when HF was present in the starting mixture, DMF decomposed into CO and dimethylamine, the latter being occluded in the AlPO₄ type **GIS** formed. AlPO-CJ3 was reported by Wang et al. [86]; it was obtained in nonaqueous as well as in aqueous medium. On removal of the occluded ethanolamine at 450 °C, AlPO₄-D [87] with a 2D system of eight-membered-ring channels is formed (IZA structure code **APD**). The aluminophosphate $AlP_2O_6(OH)_2 \cdot H_3O$ named AlPO-CJ4 which has been prepared in the presence of 2-aminopyridine [88] does not contain any organic molecule. Presumably 2-aminopyridine is too large to be accommodated in the eight-membered-ring channel. Interestingly the framework features chiral propeller-like motifs formed by Al-centered octahedra with three cyclic 4-membered rings.

The first anionic aluminophosphate AlPO-CJB1 with Brönsted acidity was reported by Yan et al. [89]. The removal of the occluded organic species leads to a framework which is stable up to 600 °C and whose net negative charge is balanced by protons acting as Brönsted acid centers.

4.2.3.5
Isomorphously Substituted Aluminophosphates

The incorporation into the AlPO₄ framework of elements with different valencies modifies the properties of the materials and increases the possibilities to use them,

References see page 867

Tab. 10. 3-D open-framework aluminophosphates prepared by solvothermal synthesis

Name (IZA code)	Organic structure-directing species (R)	Organic solvent	Chemical formula	Dimensionality of the channel system (pore openings: number of T atoms *))	Ref.
JDF-20	Triethylamine	Diethyleneglycol	$[Al_5P_6O_{24}H]\cdot 2RH$	3 dim. (20T, 10T, 8T))	45, 77, 78
AlPO$_4$-JDF	Ethanolamine	Ethyleneglycol	$[Al_2(PO_4)_2]\cdot R$	1 dim. (8T)	79
CAM-1	Triethylamine	Triethyleneglycol	$[Al_6(PO_4)_6]\cdot 2R\cdot 3H_2O$	a	80
AlPO$_4$-**GIS** (**GIS**)	Dimethylformamide	Ethyleneglycol (+HF)	$[Al_2(PO_4)_2F]\cdot(CH_3)_2NH_2$	3 dim. (8T)	27
UT-6 (**CHA**)	Pyridine (+HF)	Tetraethyleneglycol	$[Al_3(PO_4)_3F]\cdot RH$	3 dim. (8T)	81
AlPO$_4$-HDA	1,6-hexanediamine	Ethyleneglycol	$[Al_4(PO_4)_5]\cdot RH_2$	2 dim. (12T, 8T)	82
AlPO$_4$-**SOD** (monoclinic)	Dimethylformamide	Dimethylformamide	$[Al_3(PO_4)_3]\cdot R\cdot H_2O$	Cages (6T, 4T)	83, 84
AlPO-DETA	Diethylenetriamine	Phenol	$[Al_2(PO_4)_3]\cdot C_4N_3H_{16}$	1 dim. (12T)	85
AlPO-CJ3	Ethanolamine	Ethyleneglycol	$[Al_2(PO_4)_2]\cdot OCH_2CH_2NH_3$	1 dim. (8T)	86
AlPO-CJ4	2-aminopyridine	2-butanol	$[AlP_2O_6(OH)_2]\cdot H_3O$	3 dim. (8T)	88
AlPO-CJB1	Hexamethylenetetramine	Ethyleneglycol	$[Al_{12}(PO_4)_{13}]\cdot(CH_2)_6N_4H_3$	1 dim. (8T)	89

* Approximate pore opening corresponding to 6T atoms: 3 Å; 8T atoms: 3–4 Å; 10T atoms: 4–6 Å; 12T atoms: 6–8 Å.

a Structure unknown.

for example, in acid or redox catalysis and as ion-exchangers [21]. As mentioned in Sect. 4.2.3.2, in the past several reviews on the AlPO₄-based molecular sieves have been published [21–26].

Between 1982 and 1985 numerous isomorphously substituted aluminophosphates have been patented by Union Carbide Corporation (UCC) [21] and Mobil Corporation [30–38] (see Tables 2 and 3). UCC designated their materials as SAPO-*n*, MeAPO-*n*, MeAPSO-*n*, ElAPO-*n* and ElAPSO-*n* where S = Si, A = Al, Me = Co, Fe, Mg, Mn, Zn and El = As, B, Be, Ga, Ge, Li, Ti. The materials patented by Mobil Corporation are essentially silicoaluminophosphates (MCM-n). In Table 11 [90–115] are reported the 3D microporous AlPO₄-based materials, generally they are isostructural with an AlPO₄-n form. In Table 12 [116–126] are listed more recent 3D microporous Me–AlPO₄ materials for which the topologies have no AlPO₄ counterparts, they have been reported by the groups of Stucky and Wright.

All these materials are prepared with procedures similar to those used for the AlPO₄s (Sect. 4.2.3.3), the additional framework elements being added to the aluminophosphate gel. However, it appears that the optimum crystallization temperature depends on composition and structure. This last point is well illustrated with the preparation of MeAPO- and MeAPSO-**FAU** materials. Indeed, by modification of the synthesis procedure used for the crystallization of SAPO-37 [91] (higher TMA and water contents of the gel), it was possible to obtain MeAPO-**FAU** and MeAPSO-**FAU** (Me = Co, Zn) [95] in a large range of gel composition and temperature. At higher temperature and lower TMA concentration, CoAPO- and SAPO-**AFR** crystallize [95].

Except for the topologies **AFY** (MeAPO-50 with Me = Co, Mg) [99], **ATN** (MAPO-39 with M = Mg) [101], **AWO** (CoAPO-21) [104], **DFO** (DAF-1) [107], **FAU** (MeAPO-**FAU** with Me = Co, Zn) [91, 95], **LTA** (CoAPO-**LTA** and CoAPSO-**LTA**) [95], **VFI** (only SAPO = MCM-9) [35] and **ZON** (ZAPO-M1) [115], the compositions for the materials given in Table 11 have been patented by UCC.

4.2.3.5.1 Silicoaluminophosphates (SAPO)

The composition of the anhydrous SAPO materials may be expressed as $R_r Si_x Al_y P_z O_2$, where R is the structure-directing species, r is typically 0–0.3 and $x + y + z = 1$ with $0 < x \leq 0.16$. Generally, P is substituted by Si and the resulting framework negative charge is compensated by a cationic organic species ($P^{5+} = Si^{4+} + R^+$). When two Si replace one Al and one P atom, no net negative framework charge is formed ($Al^{3+} + P^{5+} = 2\ Si^{4+}$).

4.2.3.5.2 Metalloaluminophosphates (MeAPO) and Aluminometallophosphates (AMePO)

In contrast to the substitution mechanism observed for the silicoaluminophosphates, in the Me-substituted aluminophosphates the metal appears to substitute for Al rather than P in a hypothetical AlPO₄ framework, $Al^{3+} = Me^{2+} + R^+$. This

References see page 867

Tab. 11. Isomorphously substituted aluminophosphates isostructural with $AlPO_4$-n forms[a].

IZA Code	SAPO	MeAPO	MeAPSO	ElAPO	ElAPSO
AEI	+		+	(As, Ga, Ge, Ti)	
AEL	+	(Co, Fe, Mg, Mn, Zn) [90]	(Co, Fe, Mg, Mn, Zn)	(As, Be, Ti)	(As, Ge, B, Ti)
AEN	[39]				
AET	[92]				
AFI	+	(Cr, Co, Fe, Mg, Mn, Zn) [93,94]	(Co, Fe, Mg, Mn, Zn)	(Be, Ga, Ge, Li, Ti)	(Be, Ga, Ti)
AFO	+	(Co, Fe, Mg, Mn, Zn)	(Co, Fe, Mg, Mn, Zn)	(As, Be, Ti)	(B)
AFR	[95,96]	(Co) [95]	(Co, Zn) [97]		
AFS	+	+	(Co, Fe, Mg, Mn, Zn) [98]		
AFY[c]		(Co, Zn, Mg, Mn) [99]			
ANA	[6]	(Co) [100]			
AST	+	+	+		+
ATN		(Mg) [101]	+		
ATO	[102]	(Co, Mg, Mn, Zn) [103]	(Co, Fe, Mg, Mn, Zn)	(Be, Ga)	
ATS[c]		(Co) [104]	(Co, Mg, Mn, Zn)		
AWO		(Co, Fe, Mg, Mn, Zn)			
CHA	[105,106][b]	(Mg) [107]	(Co, Fe, Mg, Mn, Zn)	(Be, Li)	(As, B, Be, Ga, Ge, Li, Ti)
DFO[c]		(Co, Fe, Mg) [108]			
ERI	+	(Co, Zn) [95]	(Co, Fe, Mg)	(Ga, Ge)	
FAU[c]	[95]	(Mg, Co) [110,111]	(Co, Zn) [95]		
GIS	[109]	(Co) [37,112]	[110]		
LEV[c]	+	(Co) [95]	+		
LTA	+	(Co, Mg, Mn) [113]	(Co, Zn) [95]		(As, Ge, B, Ti)
RHO[c]					
SOD	+	(Mg) [114]	(Co, Fe, Mg, Mn, Zn)	(Be, Ga, Ge, Li, Ti)	(Be, Ga)
VFI	[35]				
ZON		(Zn) [115]			

[a] Most of the materials in this table have been reported by the UCC researchers [21 and references therein]. The references given in the Table refer to crystallographic studies on specific phases. + Compositional variants also reported in ref. [21]. [b] Isostructural with ZYT-6. [c] For this material there is no $AlPO_4$ analog.

type of substitution results in a net negative framework charge. In all cases z, the mole fraction of P, is essentially equal to 0.5 and $x + y = 0.5$. In Table 11, the metal content xMe expressed as $Me_xAl_yP_zO_2$ for the MeAPO-n materials typically varies from 0.01 to 0.25 [21].

However, by using amines with a high charge/volume ratio (i.e., low C/N ratio) in the hydrothermal synthesis, Stucky and coworkers [100, 111] have prepared a series of materials formulated $R_rMe_xAl_yPO_4$ with higher degrees of substitution (Me = Co, Zn, $x + y = 1$, $x = y$ or $x < y$ or $x > y$). It is important to note that in most syntheses the organic structure-directing species play also the role of solvent beside other solvents such as ethyleneglycol. Small quantities of water from reactants (H_3PO_4, etc.) are also present. Their materials have been named AMeP-**XXX**n ($y < x$) or MeAP-**XXX**n ($x < y$) or UCSB-n. In this series **XXX** are the three letters of the framework topologies given by the IZA Structure Commission and n refers to a different organic structure-directing agent or a different metal content (except for UCSB-n for which n refers to a specific structure). Thus, they prepared ACP-**XXX**n materials (C for cobalt) with the topologies **AEI, ANA, CHA, EDI, GIS, MER, PHI, SOD** and **THO** (also AZP-**THO**3, Z for zinc) [100, 111]. Among the topologies obtained, **EDI, MER, PHI** and **THO** have no AlPO₄ counterpart (Table 12). With the same procedure, they also produced materials with lower metal contents (Me/Al $\leq$ 1) like CAP-**CHA**1-10, MAP-**CHA**1 (M for magnesium), ZAP-**CHA**1, CAP-**AEI**1, CAP-**FAU**1, CAP-**GIS**1 [100, 111], CAP-**RHO**1, MAP-**RHO**1, MnAP-**RHO**1 [113].

New structures with new topologies have also been synthesized (Table 12). The use of 1,4-diaminobutane as structure-directing agent in ethyleneglycol medium led to ACP-2 $0.25NH_4[Co_{0.75}Al_{0.25}PO_4]\cdot0.25R$ [100] while ethylenediamine in the same medium or in water gave an isostructural zincoaluminophosphate [117]. The 3D framework contains a 2D eight-ring channel system. Ethylenediamine also allowed the crystallization of ACP-1 (**ACO**) [100] and ACP-3 [100], the latter is isostructural with UiO-20 and DAF-2 (**DFT**) [118, 119]. The **ACO** topology is made of a body-centered cubic arrangement of double four-membered-ring units (D4R) occluding a water molecule. ACP-1 has a 3D eight-ring channel system.

UCSB-4 $[Al_{0.66}Co_{0.33}PO_4]\cdot0.16R$ and   UCSB-5 $[Co_{0.5}Al_{0.5}PO_4(OH)_{0.16}]\cdot$ $0.33R\cdot0.08H_2O$ [100, 116] were prepared with $(CH_3)_2N(CH_2)_2N(CH_3)_2$ and 1,3-diaminopropane as structure-directing agents in ethyleneglycol and water media, respectively. While UCSB-4 has a framework topology with 2D intersecting 8-MR channels, UCSB-5 possesses straight 10-MR channels and zigzag 8-MR channels.

Pure UCSB-6 was prepared with 1,7-diaminoheptane. With 1,8-diaminooctane or 1,9-diaminononane + di-n-propylamine or di-isopropylamine as co-solvents to increase the solubility a contaminated material was obtained [121]. The 3D 12-membered-ring channel system of UCSB-6 is similar to that found in zeolite EMC-2 of topology **EMT** [127]. In UCSB-6 the sodalite cages of EMC-2 are replaced by one-sided capped cancrinite cages. The IZA code for UCSB-6 is **SBS**. An increase of the alkyl chain length of the amine led to a new material UCSB-8 with the new

References see page 867

Tab. 12. 3-D microporous MeAlPO$_4$ materials with no AlPO$_4$ equivalents.

Name (IZA code)	Organic structure-directing species (R)	Chemical formula*	Dimensionality of the channel system (pore openings: number of T atoms)[a,b]	Ref.
UCSB-4	(CH$_3$)$_2$N(CH$_2$)$_2$N(CH$_3$)$_2$	[Co$_{0.33}$Al$_{0.66}$PO$_4$]·0.16R	2 dim. (8T)[c]	100, 116
UCSB-5	1,3-diaminopropane	[Al$_{0.5}$Co$_{0.5}$PO$_4$(OH)$_{0.16}$]·0.33R·0.08H$_2$O	2 dim. (10T, 8T)[c]	116
ACP-2	H$_2$N(CH$_2$)$_4$NH$_2$	0.25NH$_4$[Al$_{0.25}$Co$_{0.75}$PO$_4$]·0.25R [ZnAlPO$_4$]·0.5R	2 dim. (8T)[c]	100, 117
ACP-MER1-4 (MER)	Ethylenediamine or 1,2-diaminopropane or 1,3-diaminopropane (R1)	[Al$_y$Co$_y$PO$_4$]·R^d [Al$_{0.25}$Co$_{0.5}$PO$_4$]·R^d [Al$_x$Co$_{1-x}$PO$_4$]·R1	3 dim. (8T)	100, 111
ACP-PHI1 (PHI)	2,2-dimethyl-1,3-diaminopropane	[Al$_{0.25}$Co$_{0.75}$PO$_4$]·NH$_4$·R	3 dim. (8T)	100
AZP-THO3 (THO)	1,4-diaminobutane	[Al$_{0.2}$Zn$_{0.8}$PO$_4$]·0.4R	3 dim. (8T)	100
ACP-1 (ACO)	Ethylenediamine + one non specified compound	[Al$_{0.11}$Co$_{0.89}$PO$_4$]·0.5R·0.25H$_2$O	3 dim. (8T)	100
ACP-3, UiO-20 (DFT)	Ethylenediamine	[Al$_x$Co$_{1-x}$PO$_4$]·0.5–0.5x R (x = 0.15)	3 dim. (8T)	100, 118
ACP-EDI (EDI)	N-methylethylenediamine NH$_2$CH$_2$C(CH$_3$)$_2$NH$_2$	[Al$_{0.2}$Co$_{0.8}$PO$_4$]·0.4R	2 dim. (12T)	100, 120
UCSB-6 (SBS)	1,7-diaminoheptane(R**) or 1,8-diaminooctane + di-n-propylamine or 1,9-diaminononane + di-isopropylamine	[Al$_{0.45}$Co$_{0.55}$PO$_4$]·0.25R [Al$_{1-x}$Mg$_x$PO$_4$]^e [Al$_{1-x}$Mn$_x$PO$_4$]^e [Al$_{1-x}$Zn$_x$PO$_4$]^e	3 dim. (12T)	121

UCSB-8 (**SBE**)	1,9-diaminononane + di-n-propylamine (R) or 1,10-diaminodecane + di-n-propylamine or 1,9-diaminononane + 4,7,10-trioxa-1,13-tridecanediamine or ethyleneglycol bis(3-aminopropyl)ether	$[Al_{0.5}Co_{0.5}PO_4]\cdot0.25R$ $[Al_{1-x}Mn_xPO_4]^e$ $[Al_{1-x}Mg_xPO_4]^e$ $[Al_{1-x}Zn_xPO_4]^e$	3 dim. (12T, 8T)	121
UCSB-10 (**SBT**)	4,7,10-trioxa-1,13-tridecanediamine and other nonspecified polyether amines	$[Al_{1-x}Zn_xPO_4]^e$ $[Al_{1-x}Co_xPO_4]^e$ $[Al_{1-x}Mg_xPO_4]^e$	3 dim. (12T)	121
STA-1 (**SAO**)	$C_7H_{13}N-(CH_2)_7-C_7H_{13}N$ (R) or 1,3,5-tris(quinuclidiniomethyl)benzene	$[Mg_{0.18}Al_{0.82}PO_4]\cdot0.094R\cdot0.22H_2O$	3 dim. (12T)	122
STA-2 (**SAT**)	$C_7H_{13}N-(CH_2)_4-C_7H_{13}N$	$[Mg_{0.15}Al_{0.85}PO_4]\cdot0.083R\cdot0.625H_2O$	3 dim. (8T)	123
STA-5 (**BPH**)	1,3,5-tris(triethylammoniomethyl)benzene	$[Mg_{0.15}Al_{0.85}PO_4]\cdot0.064R\cdot0.94H_2O$	3 dim. (12T, 8T)	124
STA-6 (**SAS**)	1,4,8,11-tetramethyl-1,4,8,11-tetraazatetradecane	$[Mg_{0.2}Al_{0.8}PO_4]\cdot0.094R\cdot0.16H_2O^f$	1 dim. (8T)	125
STA-7 (**SAV**)	1,4,7,10,13,16-hexamethyl-1,4,7,10,13,16-hexaazacyclooctadecane	$[Co_{0.2}Al_{0.8}PO_4]\cdot0.096R\cdot0.375H_2O$ $[Mg_{0.2}Al_{0.8}PO_4]\cdot0.082R\cdot0.292H_2O$ $[Zn_{0.2}Al_{0.8}PO_4]\cdot0.108R\cdot0.417H_2O^g$	2 dim. (8T)	126

*The organic structure-directing species may be charged or neutral.
**The chemical formula was given for this structure-directing species.
[a] See the web site http://www.iza-structure.org/databases/ for a precise description of the channel systems. [b] The sizes of the pores may be slightly different depending on the composition. [c] These topologies are not yet included in the Database of Zeolite Structures (http://www.iza-structure.org/databases/). [d] These compounds contain unidentified occluded species. [e] For these compounds only the framework composition has been given and $0.4 < x < 0.5$; $0.3 < y < 0.5$. [f] Mn- and Fe-STA-6 have also been reported but no chemical composition has been given. [g] Zn-STA-7 was produced only with 1,4,8,11-tetramethyl-1,4,8,11-tetraazatetradecane as structure-directing agent.

framework topology **SBE** [121]. It can be prepared either with 1,9-diaminononane or 1,10-diaminodecane and di-isopropylamine as co-solvent. The synthesis of AlZnPO$_4$ and AlMnPO$_4$ analogs needs the presence of a polyether diamine or ethyleneglycol bis(3-aminopropyl)ether. The structure consists of an orthogonal channel system with 12-ring openings in two dimensions and 8-ring apertures in the third.

For the synthesis of UCSB-10 (**SBT**) different polyether amines with a molecular weight between 170 and 240 have been used [121]. The structure is similar to that of faujasite, and as for UCSB-6 and EMC-2 a comparison between the **FAU** [128] and **SBT** topologies is possible. The structure of UCSB-10 is obtained by replacing the sodalite cages of the **FAU** topology with the one-sided capped cancrinite cages.

Diquinuclidinium cations of the form $[C_7H_{13}N-(CH_2)_n-C_7H_{13}N]^{2+}$ have been used by Wright and coworkers in the hydrothermal preparation of two microporous magnesium aluminophosphates STA-1 [122] and STA-2 [123] of topology **SAO** and **SAT** respectively. STA-1 is obtained for $n = 7$, 8 and 9. The structure has large-pore channels, bounded by 12-membered rings. These channels are linked to form large cavities. STA-2 is prepared with the diquinuclidinium cations for which $n = 4$ or 5. This structure can be viewed in terms of cavities that are arranged in columns. Two cancrinite cages rotated by 60° with respect to each other, sandwich a double six-ring unit (D6R). Below and above this unit are cavities containing the organic cations. The cavities have six eight-membered-ring windows giving access to six identical adjacent columns.

The use of trisquaternary templates like 1,3,5-tris(triethylammoniomethyl)-benzene led to STA-5 [Mg$_{0.15}$Al$_{0.85}$PO$_4$]·0.064R·0.94H$_2$O [124] isostructural with the beryllophosphate-H of topology **BPH** [19] whose topology contains a 1D cage and channel structure with 12-membered-ring openings. Note that STA-5 was never obtained pure, MAPO-5 was always found as a coproduct.

The same group synthesized STA-6 (**SAS**) [125] and STA-7 (**SAV**) [126] in the presence of two distinct azamacrocycles. The smallest macrocycle, namely 1,4,8,11-tetramethyl-1,4,8,11-tetraazatetradecane, produced Fe-, Mn- or Mg-STA-6 [Mg$_{0.2}$Al$_{0.8}$PO$_4$]·0.094R·0.16H$_2$O and Zn-STA-7 [Zn$_{0.2}$Al$_{0.8}$PO$_4$]·0.108R·0.417H$_2$O or Co-STA-7. The use of the larger azamacrocycle 1,4,7,10,13,16-hexamethyl-1,4,7,10,13,16-hexaazacyclooctadecane led to the crystallization of Co-STA-7 [(Co$_{0.2}$Al$_{0.8}$PO$_4$)·0.096R·0.375H$_2$O] and Mg-STA-7 [(Mg$_{0.2}$Al$_{0.8}$PO$_4$)·0.082R·0.292H$_2$O]. Without divalent cations in the synthesis both azamacrocycles gave AlPO$_4$-21.

The framework of STA-6 displays a 1D 8-MR channel system. On the other hand, the framework of STA-7 possesses two distinct eight-membered-ring channel systems giving rise to a 2D 8-MR channel system. The first 8-MR channel is made up of linked cages, whereas the second is made of smaller interconnecting voids.

Only a few data are reported concerning the thermal stability of the above frameworks with a high degree of metal incorporation [100, 121]. However, it results generally in a decrease of the thermal stability of the solid compared to the one of the aluminophosphates and less substituted aluminophosphates reported in Table 11. For many of them, the removal of the occluded organic structure-directing species leads to a collapse of the structure [99, 113].

4.2.5.3
Metallosilicoaluminophosphates (MeAPSO), Element-Aluminophosphates (ElAPO) and Element-Silicoaluminophosphates (ElAPSO)

The other most studied compositional families are the MeAPSO-*n*, ElAPO-*n* and ElAPSO-*n* families (Table 11). Ternary (ElAPO), quaternary (MeAPSO and ElAPSO) and even quinary and senary framework compositions have been reported [21 and references therein]. The last two contain aluminum, phosphorus and silicon with additional combinations of divalent Me metals.

In all the materials presented above, the real incorporation of the framework heteroelements is supported by various characterization techniques. In addition to the structure resolution by X-ray techniques, gas adsorption measurements reveal the free pore volume and powder X-ray diffraction the absence of impurities. Chemical and electron microprobe analysis give the framework stoichiometry. The latter technique is of particular interest because the impurities are not taken into account. For many elements such as Al, Si, Mg, Li, B, Zn and Ga, solid state NMR spectroscopy allows to probe the tetrahedral siting. Other spectroscopic techniques (UV-visible, IR, etc.) are also useful to evidence or not the metal incorporation in the framework.

4.2.3.6
Gallophosphates and Isomorphously Substituted Materials

In the 1980s, the synthesis of microporous phosphate-based solids was extended to the synthesis of gallophosphates and then to that of metallogallophosphates (MeGaPOs, Me = Co, Zn, etc.). As for the aluminophosphates, various synthesis routes have been used to prepare these materials. Beside the conventional route, which consists in a hydrothermal treatment of a gallophosphate mixture containing water and usually an organic structure-directing species, a large number of new solids were obtained from fluoride-containing mixtures or by using a solvothermal route with an essentially nonaqueous medium.

As mentioned in Sect. 4.2.3.1, although numerous chain- and layered gallophosphates were synthesized, the present review focuses mainly on the 3D framework gallophosphates.

4.2.3.6.1 Gallophosphates Prepared with the Conventional Route
In Table 13 [129–147] are collected the different 3D gallophosphates prepared by a hydrothermal treatment, their corresponding structure-type when it is known, the organic structure-directing agent used in the synthesis and some characteristics

References see page 867

Tab. 13. 3-D open-framework gallophosphates prepared via the conventional route (aqueous medium) and the organic structure-directing species used in their synthesis

Name (IZA code) ((AlPO₄-n analog))	Organic structure directing species (R)	Chemical formula*	Dimensionality of the channel system (pore openings: number of T atoms)	Ref.
GaPO₄-12(en) ((AlPO₄-12))	Ethylenediamine	$[Ga_3(PO_4)_3 H_2O] \cdot R$	1 dim. (8T)	18, 129
GaPO₄-14 = GaPO₄-C5 ((AlPO₄-14))	Isopropylamine	$[Ga_4(PO_4)_4OH] \cdot R \cdot H_2O$	2 dim. (8T)	18, 130, 131
GaPO₄-21 = GaPO₄-C4 (AWO) ((AlPO₄-21))	Ethanolamine Ethylenediamine (R)** Isopropylamine Dimethylamine(HF)ᵃ	$[Ga_3(PO_4)_3(H_2O)] \cdot R$	1 dim. (8T)	18, 131–135
GaPO₄-25 = GaPO₄-C12 (ATV) ((AlPO₄-25))	Obtained by calcination of GaPO₄-21	/***	1 dim. (8T)	18, 131, 136
GaPO₄-a = GaPO₄-C3	Tetraethylammonium Quinuclidine (R) Tetramethylammonium Tripropylamine Diazabicyclo[2.2.2]octane Di-n-propylamine Morpholine Triethylamine	$[Ga_9(PO_4)_9OH] \cdot R$	1 dim. (8T)	18, 131, 137–139
GaPO₄-b	Tri-n-propylamine	/	/	139

GaPO$_4$-C7 ((AlPO$_4$-15))	n-Propylamine Guanidine (HF)[a]	NH$_4$ [Ga$_2$(PO$_4$)$_2$(OH)(H$_2$O)]·H$_2$O	3 dim. (8T)	131, 140–142
/	d-Co(en)$_3$ en = 1,2 diaminoethane	[Ga$_2$(PO$_4$)$_4$H$_3$]·R	2 dim. (12T)	143
Mu-8	N,N,N′,N′ tetramethyl ethylenediamine	[Ga$_{12}$(PO$_4$)$_{12}$(OH)$_4$]·2R·4H$_2$O	1 dim. (8T)	144
GaP2	1,3-diaminopropane	[Ga$_3$(PO$_4$)$_3$(H$_2$O)R]	1 dim. (10T) 2 dim. (8T)	145
GaP1 Analog of ULM-3 (see Table 14)	1,3-diaminopropane	[Ga$_3$(PO$_4$)$_3$(OH)$_2$]·R·H$_2$O	1 dim. (10T) 2 dim. (8T)	145
Mu-6	1,4,8,11-tetraazacyclotetradecane	[Ga$_{12}$P$_{16}$O$_{52}$OH$_{12}$R$_4$]	2 dim. (20T)	146
/	1,4-diaminobutane	[Ga$_4$(HPO$_4$)$_2$(PO$_4$)$_3$(OH)$_3$]·2R·yH$_2$O $y \approx 5.4$	3 dim. (20T)	147

*The organic structure-directing species may be charged or neutral.
**The chemical formula was given for this structure-directing species.
***Not specified.
[a] Presence of HF in the starting mixture.

(chemical formula and dimensionality of the channel system). The first gallophosphate molecular sieves were reported by Parise [18], however, most of these solids (GaPO$_4$-12, -14, -21 and -25) are analogs of aluminophosphates. Two other materials (GaPO$_4$-a and GaPO$_4$-b) were then patented by Wilson et al. [139]. Following these results, Xu and coworkers [131] published a series of new gallophosphates named GaPO$_4$-Cn with n ranging from 1 to 12. Among the series, some materials are also related to some AlPO$_4$s. Thus, GaPO$_4$-C7 for example is isostructural with AlPO$_4$-15 [28, 141]. Since then, only a few new 3D microporous gallophosphates were synthesized using the conventional route. Usually, the organic structure-directing species belongs to the family of linear or cyclic alkylamines or diamino-alkanes and is occluded in the porosity of the structure. Nevertheless, in some cases, alkali-metal or complex metal cations can be the structure-directing species as in Na$_3$[Ga$_5$(PO$_4$)$_4$O$_2$(OH)$_2$]·2H$_2$O [148], Rb$_2$[Ga$_4$(HPO$_4$)(PO$_4$)$_4$]·0.5H$_2$O [149] and d-Co(en)$_3$[Ga$_2$P$_4$O$_{16}$H$_3$] [143], the latter material exhibiting a chiral 3D framework.

The framework structures of these 3D microporous gallophosphates are, in general, complex because gallium can adopt mixed 4-, 5- and/or 6-fold coordination with tetrahedral, trigonal bipyramidal and octahedral environments respectively. Moreover, a few 3D gallophosphates exhibit a Ga/P molar ratio different from 1. As can be seen in Table 13, most of the solids obtained are characterized by the presence of OH groups and water molecules that are often part of the structure. Thus, for example, in the gallophosphate Mu-8 [144], OH groups bridging two gallium atoms and terminal Ga–OH$_2$ groups are present. The presence of hydroxyl groups might explain the low thermal stability of these solids compared to that of aluminosilicates and aluminophosphates.

Another characteristic of these new solids is that the organic template interacts with the inorganic framework generally via hydrogen bonds; but in some cases, it can also be integrated in the framework. This can be illustrated with the gallophosphates GaP2 [145] and Mu-6 [146]. In the former, one nitrogen atom of 1,3-diaminopropane, which is the structure-directing agent, belongs to the coordination sphere of one type of the gallium atoms leading to a gallium in 6-fold coordination (GaO$_5$N) with a Ga–N distance equal to 2.056 Å [145]. The structure of Mu-6, which is synthesized in the presence of the macrocycle 1,4,8,11-tetraazacyclotetradecane (cyclam), consists of gallophosphate chains of gallium-corner-sharing Ga$_2$P$_2$O$_4$ four-rings. Each chain is connected to the others via a gallium-cyclam complex through O–Ga–O bonds leading to a 3D framework with Ga–N bond lengths close to 2.09 Å [146]. It is interesting to note that such a structure displays a 2D channel system delimited by 20-membered rings (20-MR). Another large-pore open-framework gallophosphate was obtained by Chippindale et al. [147]. Its structure, closely related to that of the iron phosphate [NH$_3$(CH$_2$)$_3$NH$_3$]$_2$[Fe$_4$(HPO$_4$)$_2$(PO$_4$)$_3$(OH)$_3$]·γH$_2$O ($\gamma \sim 9$) [150], consists of chains of GaO$_6$ octahedra and PO$_4$ tetrahedra cross-linked by additional PO$_4$ tetrahedra to generate a 3D framework containing large tunnels delimited by 20-MRs. It is worthy of note that very recently, the equivalent fluorine-containing gallophosphate material, named ICL-1, was reported by Walton et al. [151] (see Table 14).

4.2.3.6.2 Gallophosphates Prepared from a Fluoride-containing Medium

The fluoride method was first used in the synthesis of gallophosphates in 1991. The results were exciting with the discovery of cloverite (**-CLO** structure type) [152, 153], which is the first molecular sieve with 3D 20-membered-ring channels, and the **LTA**-type gallophosphate [138]. Since then, a large number of new gallophosphates or fluorogallophosphates have been obtained in fluoride medium and in the presence of various organic structure-directing species. Thus, nowadays, more than 40 solids (1D, 2D, and 3D frameworks) have been synthesized. The characteristics of some of the 3D gallophosphates are reported in Table 14 [152–177]. Large-pore open frameworks are obtained with pore openings delimited by 12-(MIL-1 [169], TREN-GaPO [157], Mu-17 [177]), 14-(DIPYR-GaPO [158]), 16-(ULM-5 [167], ULM-16 [168]), 18-(MIL-31 [171]) and 20-(Cloverite [152]) membered rings.

For the hydrothermal synthesis of these solids, typically in the range 80–170 °C for a few hours to several days, the starting gel composition is r R:1 Ga$_2$O$_3$: 1 P$_2$O$_5$: x HF : w H$_2$O, where R is the organic species, $1 \leq r < 6$, $0.2 \leq x \leq 2$ and $40 \leq w \leq 300$. In general, the amount of R is such that the starting pH value is in the range 3–7. The preferred source of fluoride is HF. Indeed, when NH$_4$F is used as the fluoride source, the ammonium cation shows a strong directing effect and the ammonium hydroxyfluorogallophosphate (NH$_4$)$_{0.93}$(H$_3$O)$_{0.07}$GaPO$_4$(OH)$_{0.5}$F$_{0.5}$, analogous to AlPO$_4$-CJ2 [68] is obtained. Surprisingly, the *in situ* decomposition of guanidine into ammonium cations does not lead to AlPO$_4$-CJ-2 but namely to a hydroxyfluorogallophosphate related to KTiOPO$_4$(KTP) and showing small pore openings (Table 14, [160]). As for the gallophosphates prepared with the conventional route, the organic species R are mainly cyclic or linear alkylamines, or diaminoalkanes. For instance, by varying the length of the carbon chain of the α-ω diaminoalkanes, a large number of new materials belonging to the ULM-n family were synthesized. It is the case for the fluoro- or hydroxyfluorogallophosphates ULM-3, -4, -5, -6 [164–167] and MIL-31 [171]. Among the series of the gallophosphates reported in Table 14, TREN-GaPO [157], DIPYR-GaPO [158] and TMA-GaPO [159] were synthesized in the presence of two types of organic structure-directing species. The hydrothermal synthesis of these solids (water is the main solvent) is derived from the procedure developed by Kuperman et al. [178] where pyridine is used as the solvent, in the presence of HF and another amine as organic template. Thus, for TREN-GaPO, the starting mixture has the following molar composition: 0.88 TREN (tris(2-aminoethyl)amine) : Ga$_2$O$_3$: 1.3 P$_2$O$_5$: 32.1 pyridine:5.2 HF:86 H$_2$O. After a heating time of 4 days at 170 °C, this fluorogallophosphate is obtained as the main phase. Its 3D framework exhibits two parallel channel systems delimited by eight- and twelve-membered-ring openings. The most striking feature of this structure is the interplay between the organic

References see page 867

Tab. 14. 3-D open-framework gallophosphates prepared in fluoride medium and organic structure directing species used in their synthesis

Name (IZA code) ((AlPO₄ analog))	Organic structure-directing species (R)	Chemical formula*	Dimensionality of the channel system (pore openings: number of T atoms)	Ref.
Cloverite (-CLO)	Quinuclidine (R)** Quinuclidinium iodide Hexamethyleneimine 3-azabicyclo[3.2.2]nonane Piperidine $[Co(NMe_3)_2sar]^{3+a}$	$[Ga_{96}P_{96}O_{372}OH_{24}F_{24}] \cdot 24R \cdot n(H_2O)$	3 dim. (20T) 3 dim. (8T)	152–155
GaPO₄-LTA (LTA) ((AlPO₄-LTA))	Di-n-propylamine (R) Pyridine K22^b·K222^b	$[Ga_{12}(PO_4)_{12}F_3] \cdot 3R$	3 dim. (8T)	138, 155, 156
TREN-GaPO	Tris(2-aminoethyl)amine (R1) + pyridine(R2)	$[Ga_6(PO_4)_6F_4] \cdot (R1)(R2)$	1 dim. (12T) 1 dim. (8T)	157
DIPYR-GaPO	pyridine(R1) + benzylviologen (R2)	$[Ga_7(PO_4)_6F_3(OH)_2] \cdot 0.28(R1) \cdot 0.87(R3) \cdot 2H_2O$ R3 = 4,4′-dipyridyl^c	1 dim. (14T)	158
TMP-GaPO	N,N,N′,N′-tetramethylene 1,3-diaminopropane	$[Ga_4(PO_4)_3(HPO_4)F_3] \cdot R$	2 dim. (8T)	159
TMA-GaPO	Tetramethylammonium (R1) + pyridine(R2)	$[Ga_4(PO_4)_3(HPO_4)F_3] \cdot (R1,R2)$	2 dim. (8T)	159
PYR-GaPO	Pyridine(R1) + benzylviologen (R2)	$[Ga_4(PO_4)_3(HPO_4)F_3] \cdot (R1)$	2 dim. (8T)	159
GaPO₄-Cl2 ((AlPO₄-Cl2))	Hexamethylenetetramine + NH₄	$(NH_4)_{0.93}[GaPO_4F_{0.5}(OH)_{0.5}] \cdot 0.07(H_3O)$	1 dim. (8T)	68
/***	Guanidine^d	$(NH_4)[GaPO_4F]$	2 dim. (6T)	160
/	Guanidine^d	$(NH_4)_2[Ga_2(PO_4)(HPO_4)F_3]$	2 dim. (6T)	160

Material	Template	Formula	Dimensionality	Ref.
GaPO₄-tricl CHA[e] (**CHA**)[e] ((AlPO₄-34))[e]	1-methylimidazole Pyridine	$[Ga_6(PO_4)_6F_2]·2R$	3 dim. (8T)	161
GaPO₄-21 (**AWO**) ((AlPO₄-21))	Dimethylamine (R) Pyrrolidine Piperazine	$[Ga_3(PO_4)_3(OH)]·R$	1 dim. (8T)	132, 155
ULM-1	Diazabicyclo[2.2.2]octane	$[Ga_3(PO_4)(HPO_4)_2F_3(OH)]·R·0.5(H_2O)$	3 dim. (8T)	162
ULM-2	Diazabicyclo[2.2.2]octane	$[Ga_4(PO_4)_3F_2(OH)_2]·0.5R·2(H_2O)$	1 dim. (8T)	163
ULM-3	1,3-diaminopropane (R) 1,4-diaminobutane 1,5-diaminopentane	$[Ga_3(PO_4)_3F_2]·R·H_2O$	1 dim. (10T) 2 dim. (8T)	164, 165
ULM-4	1,3-diaminopropane(R) Ethylenediamine Methylamine	$[Ga_3(PO_4)_3F_2]·R·H_2O$	1 dim. (10T) 1 dim. (8T)	164, 166
ULM-5	1,6-diaminohexane	$[Ga_{16}(PO_4)_{14}(HPO_4)_2(OH)_2F_7]·4R·6(H_2O)$	1 dim. (16T) 1 dim. (8T)	167
ULM-6	1,3-diaminopropane	$[Ga_4(PO_4)_4F_2]·R·(H_2O)$	1 dim. (8T)	164
ULM-16	Cyclohexylamine	$[Ga_4(PO_4)_4F_2]·1.5R·0.5(H_2O)0.5(H_3O)$	1 dim. (16T)	168
MIL-1	1,4,7,10,13,16-hexaazacyclooctadecane	$[Ga_5(HPO_4)_2(PO_4)_4F_2]·R$	1 dim. (12T) 1 dim. (8T)	169
MIL-20	N,N,N',N'-tetramethyl ethylenediamine	$[Ga_3(PO_4)_3F]·0.5R$	3 dim. (8T)	170
MIL-31	1,9-diaminononane 1,10-diaminodecane(R)	$[Ga_9(PO_4)_9(H_2O)(OH)_2F_3]·2R·2H_2O$	1 dim. (18T)	171
GaPO₄-ZON (**ZON**) ((AlPO₄ UIO-7))	Diazabicyclo[2.2.2]octane	$[Ga_4(PO_4)_4F_2]·R$	2 dim. (8T)	172

Tab. 14. (continued)

Name (IZA code) ((AlPO₄ analog))	Organic structure-directing species (R)	Chemical formula*	Dimensionality of the channel system (pore openings: number of T atoms)	Ref.
Mu-2	4-amino-2,2,6,6-tetramethylpiperidine	$[Ga_{32}P_{32}O_{120}(OH)_{16}F_6] \cdot 6R \cdot 12H_2O$	3 dim. (8T)	173
Mu-5	Cyclam[f]	$[Ga_{20}(PO_4)_{16}F_8(OH)_4R_4]$	3 dim. (10T)	174
CYCLAM-GaPO	Cyclam[f]	$[GaPO_4(OH)_2F] \cdot GaR^g$	3 dim. (10T)	175
Mu-15	N,N,N′,N′-tetramethyl 1,3-propanediamine	$[Ga_{16}P_{16}O_{60}(OH)_2F_6O_4] \cdot 4R$	1 dim. (10T) / 1 dim. (8T)	176
Mu-17	3-(2-aminoethyl-amino)propylamine	$[Ga_{12}P_{14}O_{50}(OH)_{10}F_4] \cdot 4R$	1 dim. (12T)	177
ICL-1[h]	1,4-diaminobutane	$[Ga_4(HPO_4)_2(PO_4)_3(OH)_2F] \cdot 2R \cdot 6H_2O$	3 dim. (20T)	151

* The organic structure-directing species may be charged or neutral.
** The chemical formula was given for this structure-directing species.
*** Not specified.
[a] (N(CH₃)₃)₂sar: 1,8-bis(trimethylammonio)-3,6,10,13,16,19-hexaazabicyclo[6,6,6]-icosane. [b] K22: Kryptofix 22 (1,7,10,16-tetraoxa-4,13-diazacyclooctadecane; K222: Kryptofix 222 (4,7,13,16,21,24-hexaoxa-1,10-diazabicyclo [8.8.8] hexacosane. [c] R3 = 4,4′ dipyridyl formed by decomposition of benzylviologen dichloride. [d] Guanidinium is decomposed into ammonium cations under hydrothermal conditions. [e] Triclinic variant of chabazite. After removal of HF and the organic species by calcination GaPO₄-34 which is isostructural with AlPO₄-34 is obtained. [f] Cyclam = 1,4,8,11-tetraazacyclotetradecane. [g] The formula is that reported in Ref. [175]. [h] This sample can be obtained at room temperature.

molecules and their segregation into the two different channel systems. One channel system (8-MR) is occupied by the pyridine molecules and the other (12-MR) by the TREN molecules. A similar trend is observed for TMA-GaPO whose structure displays two types of eight-membered channel occluding the tetramethylammonium cations and the pyridine molecules, respectively. DIPYR-GaPO was prepared from a gallophosphate mixture containing also two types of organics; benzylviologen dichloride and pyridine. During the reaction, the former decomposes to form 4,4′ dipyridyl which is occluded with the pyridine in the 14-membered channels of the structure. However, the synthesis using directly 4,4′ dipyridyl in the starting mixture does not lead to DIPYR-GaPO. Therefore, it seems that its *in situ* formation together with the presence of pyridine is essential for the formation of this phase.

As for the fluoride-containing aluminophosphate system, the use of either 1-methylimidazole or pyridine yields a triclinic variant of chabazite. The removal of HF and the organic species by calcination leads to GaPO$_4$-34 which is isostructural with AlPO$_4$-34 [161]. Although F$^-$ was introduced in the starting mixture, no fluorine was found in the gallophosphate GaPO$_4$-21. Such a behavior was also observed for the hydroxygallophosphate GaPO$_4$-15 [141] as reported in Table 13. Moreover, GaPO$_4$-21 is obtained in the presence of dimethylamine, piperazine or pyrrolidine, whereas, in the absence of F$^-$, it is synthesized with isopropylamine, ethanolamine and ethylenediamine (see Table 13), the latter organic molecule leading to the fluorogallophosphate ULM-4 in fluoride medium [166]. However, as can be seen in Table 14 and as was previously observed for some aluminophosphates (see Sect. 4.2.3.4.3), in most of the 3D solids fluorine is part of the framework. Three main locations can be observed. (1) fluorine can bridge two gallium atoms increasing their coordination from 4 to 5 or 5 to 6 as it is often observed for the ULM-*n* and MIL-*n* phases. (2) It can also be present as terminal Ga–F groups. This situation is mainly encountered for the chain and layered solids but also for some 3D solids leading to interrupted frameworks. In some cases, like for the hydroxyfluorogallophosphate Mu-5, which is similar to CYCLAM-GaPO (Table 14), fluorine shares this position with OH. (3) In the third situation, F$^-$ is found located inside the small cubic building units: the so-called D4R units (double-four-ring units). The presence of such a fluorine species is unambiguously evidenced by ^{19}F MAS/NMR spectroscopy, since it leads to a signal with a chemical shift of about −70 ppm (reference CFCl$_3$). Although such an environment for F$^-$ was first observed for the clathrasil octadecasil [179], a large number of fluorogallophosphates, belonging mainly to the Mu-n family and showing this particularity were obtained. A review of the different materials characterized by the presence of this D4R-F unit is given in Table 15 [180–184]. Molecular anion-, chain-, layered- and 3D framework gallophosphates are formed. It is worthy of note that in the absence of fluoride in the starting mixture, none of these materials are obtained. For instance, in the absence of F$^-$, GaPO$_4$-a crystallizes instead of cloverite [153]. Therefore, beside its mineralizing role, F$^-$ probably plays a structure-

References see page 867

Tab. 15. Materials with a framework showing D4R units hosting a fluoride anion.

Framework atoms	Structure code (dimensionality of the structure)	Size of the largest pore opening (number of T atoms)	Ref.
Si (clathrasil)	**AST** (3D)	6	179
Si	**ISV** (3D)	12	180
Al, P	**AST** (3D)	6	181
Al, P	**LTA** (3D)	8	71
Ga, P	**LTA** (3D)	8	138
Ga, P	**-CLO** (3D)	20	153
Ga, P[a]	Mu-1 (molecular anion)[b]		182
Ga, P[a]	Mu-2 (3D)[b]	8	173
Ga, P[a]	Mu-3 (1D)[b]		183
Ga, P[a]	Mu-5 (3D)[b]	12	174
Ga, P[a]	Mu-15 (3D)[b]	10	176
Ga, P[a]	ULM-5 (3D)[b]	16	167
Ga, P[a]	ULM-18 (2D)[b]		184

[a] In these materials Ga–F terminal groups and/or bridging fluorine (Ga–F–Ga) are also present. [b] No structure code.

directing role stabilizing these small building units. The templating role of F^- can be illustrated with the synthesis of the gallophosphates Mu-5 (Table 14) and Mu-6 (Table 13). These two solids were prepared under very similar conditions with the macrocycle 1,4,8,11-tetraazacyclotetradecane (cyclam) as organic structure-directing species in the presence and the absence of fluoride respectively. Both 3D structures display the same gallium-cyclam complex. In Mu-5 this complex connects gallophosphate layers of D4Rs hosting F^-, whereas, in Mu-6 it connects chains of gallium-corner-sharing $Ga_2P_2O_4$ single four rings (S4Rs). Therefore under these experimental conditions, in the absence of fluorine, there is no formation of D4Rs.

4.2.3.6.3 Gallophosphates Prepared in Essentially Nonaqueous Medium
The synthesis conditions are quite similar to those used in aqueous medium. Usually the solvents are alcohols (ethyleneglycol, glycerol, butanol, etc.). However, the medium is not strictly nonaqueous because the phosphorus source is mainly an 85 % solution of phosphoric acid and when the synthesis is performed in fluoride medium, F^- is generally introduced in the starting mixture as an aqueous solution of hydrofluoric acid (40 %). Nevertheless, water is not the main solvent and under such conditions, the crystallization time is usually longer. The 3D framework gallophosphates that were prepared from such a quasi-nonaqueous medium are reported in Table 16 [185–189].

In some cases, the same structure-type can be obtained in aqueous and in quasi-nonaqueous medium; but sometimes with different structure-directing agents. This can be illustrated with the **LTA**-type gallophosphate. In aqueous medium, this solid is synthesized in the presence of di-n-propylamine or aza-crown ethers as or-

Tab. 16. 3-D open-framework gallophosphates prepared in quasi non-aqueous medium and the organic structure-directing species and solvents used in their synthesis

Name (IZA code) ((AlPO₄-n analog))	Organic structure-directing species (R)	Chemical formula*	Solvent	Dimensionality of the channel system (pore openings: (number of T atoms))	Ref.
GaPO₄-LTA (LTA) ((AlPO₄-LTA))	Pyridine (HF)[a]	/**	EG[b]	3 dim. (8 T)	138, 155
GaPO₄-tricl CHA[c] (CHA)[c] ((AlPO₄-34))[c]	Pyridine (HF)[a]	[Ga₆(PO₄)₆F₂]·2R	EG or GLY[b]	3 dim. (8T)	155
Cloverite (-CLO)	Piperidine (HF)[a]	/	EG	3 dim. (20 T); 3 dim. (8 T)	155, 185
GaPO₄-21 (AWO) ((AlPO₄-21))	Dimethylamine (R)*** (HF)[a] Pyrrolidine	[Ga₃(PO₄)₃(OH)]·R	EG	1 dim. (8T)	132, 155
GaPO₄-M1	Methylamine	[Ga₃(PO₄)₃]·2R·H₂O	EG	/	136
GaPO₄-M2	Methylamine	[Ga₃(PO₄)₃OH]·R	EG	2 dim. (8T)	136, 186
/	N,N,N',N' tetramethylethylenediamine	[Ga₄(PO₄)₅H]·R·H₂O	Butane,1-ol	1 dim. (16T)	187
/	1,4-diaminobutane	[Ga₄(HPO₄)(PO₄)₄]·R	EG	2 dim. (12T and (8T)	188
Pyridine-GaPO-1	Pyridine(HF)[a]	[Ga₆(PO₄)₆F₂]·2R·H₂O	PYR[b]	3 dim. (8T)	189

*The organic structure-directing species may be charged or neutral.
** Not specified.
***The chemical formula was given for this structure-directing species.
[a] Presence of HF in the starting mixture. [b] EG: ethyleneglycol; GLY: glycerol; PYR: pyridine. [c] Triclinic variant of chabazite. After removal of HF and the organic species by calcination GaPO₄-34 which is isostructural with AlPO₄-34 is obtained.

ganic templates [138, 156], whereas, when ethyleneglycol is the main solvent, this structure-type is only obtained in the presence of pyridine. It is worthy of note that the latter template leads to the formation of the triclinic **CHA**-type gallophosphate in aqueous media [161]. This structure type also crystallizes in the presence of pyridine in a concentrated nonaqueous medium (ethyleneglycol (EG) as solvent) or when the polarity of the system increases (glycerol (GLY) as solvent) [155]. Whatever the nature of the solvent, piperidine seems to be a good organic template for the hydroxyfluorogallophosphate cloverite (compare Tables 14 and 16).

Novel materials can also be obtained using this route. Mu-3, which is a 1D chain fluorogallophosphate ([183], Table 15) crystallizes only from a EG-containing mixture, whereas, in aqueous medium the hydroxyfluorogallophosphate Mu-2 ([173], Table 14) is formed.

In the presence of butane-1-ol as main solvent, Chippindale et al. [187] prepared a large-pore open-framework material with a 1D channel system delimited by 16-membered ring openings and occluding N,N,N′,N′-tetramethylethylenediamine cations. However, due to the presence of strong hydrogen bonds between P=O and P–OH groups that protrude into the large channels, the latter is divided into three smaller channels. In this interrupted framework, the Ga/P molar ratio is equal to 4/5 and gallium atoms are in tetracoordinated (GaO_4) and hexacoordinated(GaO_6) environments of oxygen atoms.

4.2.3.6.4 Isomorphously Substituted Gallophosphates

Despite the large number of known microporous gallophosphates (Tables 13–16) to date only a few metal-substituted gallophosphates (MeGaPOs) have been reported in the literature. Table 17 gives a list of the main 3D open-framework MeGaPOs [190–202]. Among the materials reported in this table, some of them have zeolite framework structures. For example, $[Me_8Ga_{16}(PO_4)_{24}]\cdot 8R$ with R = pyridine or imidazole and Me = Co, Zn, Fe or Mn adopts the laumonite framework topology [192], $[MeGa_2(PO_4)_3]\cdot R$ the sodalite topology (R = pyrrolidine, Me = Co or Zn) [192] and $[Me_8Ga_8(PO_4)_{16}]\cdot 8R$ the gismondine topology (R = pyrrolidine or guanidine, Me = Co, Zn) [192, 196]. New 3D frameworks without any zeolite counterpart like the CoGaPO-5 and CoGaPO-6 phases with structure codes **CGF** and **CGS**, respectively, are also obtained. CoGaPO-5 [192, 197] was synthesized in the presence of diazabicyclo[2.2.2]octane as organic structure-directing species, whereas quinuclidine was used to prepare CoGaPO-6 [192, 198]. Both structures are built up from MO_4 (M = Co and Ga) and PO_4 tetrahedra. The frameworks exhibit channel systems delimited by eight- and ten-membered-ring openings that intersect to form a 3D pore network occluding the organic species.

It is worthy of note that a large number of MeGaPOs have been prepared in a quasi-nonaqueous medium and mainly in the presence of ethyleneglycol as solvent. As suggested by Chippindale and Cowley [192], ethyleneglycol appears to be less likely to cause extensive decomposition of the structure-directing agent than water, the latter favoring the formation of ammonium-containing products as the major phases. This can be exemplified with the synthesis of the ammonium cobalt gallium phosphate hydrate $(NH_4)_4[Co_4Ga_8(PO_4)_{12}(H_2O)_8]$ as reported in Table 17.

This compound is obtained as a pure phase only in aqueous medium from a mixture containing pyridine, imidazole or ammonium cations as structure-directing species [193]. In the 3D framework, Co is in a hexacoordinated environment with four oxygen atoms and two H_2O molecules, such an environment being responsible for the pink color of the sample.

In the MeGaPO compounds, Me substitutes partly for gallium and the Me/Ga molar ratio is generally lower than 1. However, a new family of materials named GCP (gallium cobalt phosphate) which are analogous to the ACP (aluminum cobalt phosphate) compounds (see Sect. 4.2.3.5.2) are characterized by a Me/Ga molar ratio higher than one. Some solids of this family are isostructural with zeolites (GCP-**SOD**, GCP-**THO**, GCP-**EDI**) [111, 120, 199] but others correspond to new framework topologies like the large-pore open-framework UCSB-6 and UCSB-10 materials [121]. Both structures with the **SBS** and **SBT** structure codes, respectively, display a 3D 12-membered ring channel system.

Although Me corresponds generally to Co and Zn, at least three 3D manganese-containing frameworks have been obtained. MnGaPO-2 [192, 200] was prepared in the presence of diazabicyclo[2.2.2]octane as organic structure-directing species. In this solid Mn is in a 5-fold coordination with a distorted square-pyramidal geometry. The second solid, the first mixed-valence Mn^{II}, Mn^{III} gallium phosphate [201] was synthesized with piperazine as organic template, the latter being located in the structure at the eight-membered-ring channel intersections. In the third material [202], the manganese atoms are in hexacoordinated environments of oxygen atoms and water molecules. They form a linear $Mn_3(H_2O)_6O_8$ cluster that acts as a template, since it resides inside the tunnels built from GaO_5 trigonal bipyramids and PO_4 tetrahedra. For the synthesis of this manganese gallium phosphate, it seems that the presence of 4,4′-trimethylenedipiperidine is necessary despite the fact that the latter is not occluded in the structure.

4.2.3.6.5 Thermal Stability of the Gallophosphates

As mentioned in Sect. 4.2.3.6.1, the thermal stability of the gallophosphates is lower than that of aluminosilicates and aluminophosphates. After calcination, in order to remove the organic template, the structures generally collapse and dense phases like cristobalite or quartz-type gallophosphates crystallize. However, some materials have a better thermal stability. Thus, for the gallophosphate ULM-16 [168] the structure seems to be preserved after calcination at 800 °C. Unfortunately, no adsorption data were reported. A similar behavior is observed for the large-pore gallophosphate cloverite. Quinuclidine which is one of the organic structure-directing species (see Table 14) can be removed by heating the sample at 500 °C in air without altering the structure. After cooling down under dry atmosphere, the adsorption capacity of the calcined sample for n-hexane, xylene and mesitylene is high and equal to 0.24, 0.17 and 0.18 cm^3 g^{-1}, respectively [153]. For the **LTA**-type gallophosphate, the removal of the protonated di-n-propylamine by calcination is also possible, and leads to a porous solid with a n-hexane adsorption capacity equal

References see page 867

Tab. 17. 3-D open-framework metal gallophosphates and organic structure-directing species and solvents used in their synthesis

Name (IZA code)	Organic structure-directing species (R)	Chemical formula*	Solvent	Dimensionality of the channel system (pore openings: number of T atoms)	Ref.
MeGaPO$_4$-LAU (LAU) /**	Pyridine Imidazole	[Me$_8$Ga$_{16}$(PO$_4$)$_{24}$]·8R Me = Co, Zn, Fe, Mn	Butane,1-ol or EG[a]	1 dim. (10T)	190–192
	Pyridine Imidazole Ammonium	(NH$_4$)$_4$[Me$_4$Ga$_8$(PO$_4$)$_{12}$(H$_2$O)$_8$] Me = Co, Zn, Fe, Mn, Mg	Butane,1-ol or water	2 sets of channels	190, 193, 194
MeGaPO$_4$-SOD (SOD)	Pyrrolidine	[MeGa$_2$(PO$_4$)$_3$]·R Me = Co, Zn	EG	3 dim. (6T)	192
MeGaPO$_4$-GIS (GIS)	Pyrrolidine Guanidine	[Me$_8$Ga$_8$(PO$_4$)$_{16}$]·8R Me = Co, Zn	EG	3 dim. (8T)	192, 195, 196
CoGaPO-5 (CGF)	Diazabicyclo[2.2.2]octane	[Co$_{16}$Ga$_{20}$(PO$_4$)$_{36}$]·8R	EG	2 dim. (8T); 1 dim. (10T)	192, 197
CoGaPO-6 (CGS)	Quinuclidine	[MeGa$_3$(PO$_4$)$_4$]·R M = Co, Zn	EG	1 dim. (8T); 1 dim. (10T)	192, 198
GCP-SOD1 (SOD)	Piperazine	[GaCo$_2$(PO$_4$)$_3$]·R	EG	3 dim. (6T)	199
GCP-THO (THO)	N-methylethylenediamine (R)*** 1,3-diaminopropane	[GaCo$_4$(PO$_4$)$_5$]·2R	EG	3 dim. (8T)	199
GCP-EDI (EDI)	1,2-diaminopropane	[GaCo$_4$(PO$_4$)$_5$]·2R	water or EG	3 dim. (8T)	111, 120

GCP-2	1,4-diaminobutane	$(NH_4)[GaCo_3(PO_4)_4] \cdot R$	EG	2 dim. (8T)	199
UCSB-6GaMe (**SBS**)	$NH_2(CH_2)_nNH_2$ $7 \leq n \leq 9$	$[Me_xGa_{1-x}PO_4]^b$ Me = Co, Mg, Zn $0.4 < x < 0.5$	water + di-n-propylamine or di-isopropylamine	3 dim. (12T)	121
UCSB-10GaZn (**SBT**)	Polyether diamines	$[ZnGa(PO_4)_2]^b$	water	3 dim. (12T)	121
MnGaPO-2	Diazabicyclo[2.2.2]octane	$[Mn_4Ga_4(PO_3OH)_8(PO_4)_4] \cdot 4R$	EG	2 dim. (8T); 1 dim. (10T)	192, 200
/	Piperazine	$[Mn_2Ga_5(H_2O)(PO_4)_8] \cdot 2R$	EG + water	2 dim. (8T)	201
/	4,4'-trimethylenedipiperidine	$[Mn_3(H_2O)_6Ga_4(PO_4)_6]$	water	3 dim. (8T)	202

*The organic structure-directing species may be charged or neutral.
**Not specified.
***The chemical formula was given for this structure-directing species.
a EG: ethyleneglycol. b In Ref. [121], only the framework composition is reported.

to 14.8 wt % ($T = 25$ °C for $p/p_0 = 0.5$). This corresponds to a n-hexane pore volume of 0.22 cm^3 g^{-1}, which is close to the value usually found for zeolite A. Therefore, it can be concluded that there is no pore blockage and that the calcination does not deteriorate the structure [138]. Nevertheless, for both materials (**-CLO** and **LTA**), the structure collapses when the cooling step to room temperature is performed under moist air. This phenomenon was explained by Müller et al. [203] by the sorption of polar molecules, like water on the dangling OH groups of the framework, which would lead to a breaking of the Ga–O bond close to the Ga–OH adsorption site. It was shown by Schmidt et al. [204], that when calcined cloverite is kept at temperature above 100 °C (no rehydration), its framework remains intact and after adsorption of nonpolar molecules, like aliphatic hydrocarbons, a material similar to cloverite is obtained and can be stored at room temperature.

4.2.3.7
Zincophosphates and Beryllophosphates

4.2.3.7.1 Zincophosphates
To date, about 100 new chain-, layered- and 3D framework zincophosphates (ZnPOs) have been reported in the literature. The most representative materials are given in Table 18 [205–235]. The synthesis can be performed either by hydrothermal or solvothermal treatments of a zincophosphate mixture containing an organic and/or inorganic structure-directing species. Numerous solids have been obtained at relatively low temperatures, for example the **FAU**-type zincophosphate (analog of the zeolite faujasite) was prepared at 4 °C in the presence of tetramethylammonium and sodium cations [20, 205]. Other ZnPOs with a zeolite topology have been obtained, for example, Zn-PO$_4$-sodalite [20, 206] synthesized at 50 °C from mixtures containing sodium, Zn-PO$_4$-**ABW** [20, 221–223] prepared in the presence of ammonium or alkaline cations, but also the zincophosphate UiO-21 [214] which exhibits a chabazite framework topology and the zincophosphate ZP-4 [218], which corresponds to an intergrowth between edingtonite and thomsonite. However, most of the materials reported in Table 18 correspond to new structures with no zeolitic counterpart. It is worthy of note that after synthesis, the washing procedure appears to be critical because most of these zincophosphates transform completely into the zincophosphate tetrahydrate hopeite [236] on washing with large amounts of water [212, 232]. Surprisingly, hopeite seems to be a precursor in the crystallization of ZnPO-HEX [219], an open-framework chiral sodium zincophosphate phase (**CZP** structure type).

Depending on the synthesis conditions (amine and phosphoric acid concentrations, pH of the starting mixture, etc.), a large number of materials can be prepared with the same organic structure-directing species. This is well illustrated in Table 18 with the reaction mixtures containing triethylenetetramine, 1,3-diaminopropane, 1,4-diazabicyclo[2.2.2]octane, piperazine and guanidinium cations.

The 3D framework of ZnPOs is generally built from ZnO$_4$ and PO$_4$ tetrahedra. However, Zn can adopt 4-, 5-, and 6-coordinations with framework oxygens, water molecules or hydroxyl groups. In many cases, Zn–O–Zn linkages are present; the bridging oxygen atom being usually tri-coordinated and bonded to a phosphorus atom. For example, such a linkage is observed in the zincophosphates ZnPO/dab-D [211], UiO-21 [214], UiO-17 [215], ZnPO-W [234] and ND-1 [224]. ND-1 is the microporous solid characterized by the largest pore size. It is prepared with 1,2-diaminocyclohexane as organic template and its 3D framework displays a 1D channel system delimited by 24-membered-ring openings and occluding the trans isomer of 1,2-diaminocyclohexane only. The most striking feature of this new zincophosphate is that it contains sizable openings, even with the templates present, with a free diameter close to 8.6 Å. Therefore, the removal of the organic species by calcination at 350 °C, which leads to a collapse of the structure, is not necessary.

As was observed for some gallophosphate-based solids, the organic structure-directing species can be coordinated to the inorganic framework. This is the case for the zincophosphate $[NH_3(CH_2)_3NH_3]_2[NH_3(CH_2)_3NH_2]_2[Zn_{12}(OH_2)_2 \cdot (PO_4)_{10}] \cdot H_2O$ [209]. Its framework exhibits two channel systems delimited by eight- and ten-membered-ring openings and occluding mono- and diprotonated diaminopropane molecules, respectively. The second molecule interacts with the framework through hydrogen bonds, whereas the former acts as a ligand to one type of the zinc atoms (ZnO_3N tetrahedron with Zn–O/N distances ranging from 1.867 and 2.12 Å). Such a situation is also observed for the zincophosphates ZnPO/dab-D, solid V and products I and II (Table 18).

Rao et al. [216, 217] developed an original synthesis route to prepare new metal-phosphate materials. It consists first in synthesizing the amine phosphate and introducing it in an aqueous or nonaqueous medium containing the metal element (Zn, Co, etc.). Thus, by varying the nature of the amine phosphate, they obtained numerous 1D, 2D and 3D structures. For instance, with N-methylpiperazine phosphate (Table 18, phase III), a new 3D open-framework zincophosphate crystallizes. Its structure displays a 1D channel system with a 16-membered clover-like opening due to OH groups protruding into the aperture.

4.2.3.7.2 Thermal Stability of the Zincophosphates

Compared to the other phosphate-based materials (aluminophosphates, gallophosphates), the thermal stability of the ZnPOs is quite low and generally amorphous solids or zinc metaphosphate (β-$ZnPO_3$) and zinc pyrophosphate (α-$Zn_2P_2O_7$) are obtained after removal of the organic species by calcination. However, Harrison et al. [237] reported a series of microporous zincophosphate materials, designated $M_3Zn_4O(PO_4)_3 \cdot nH_2O$ (M = Li, Na, K, Rb, Cs), which display typical "zeolitic" dehydration/rehydration and ion-exchange reactions with thermal stabilities up to 600 °C.

4.2.3.7.3 Beryllophosphates

Only a few open-framework beryllophosphates materials are reported in the literature. Harvey and Meier [19] have synthesized five new materials from mixtures containing alkali metal cations and tetraethylammonium hydroxide at a tempera-

References see page 867

Tab. 18. Examples of 3-D open-framework zincophosphates and the organic structure-directing species used in their synthesis (syntheses performed mainly in aqueous medium)

Name (IZA code)	Organic structure-directing species (R)	Chemical formula*	Dimensionality of the channel system (pore openings: number of T atoms)	Ref.
ZnPO-X (**FAU**)	Tetramethylammonium[a] (TMA)	$Na_{67}Zn_8[Zn_{96}(PO_4)_{96}]\cdot12TMA$	3 dim. (12T)	20, 205
ZnPO-SOD (**SOD**)		$Na_6[ZnPO_4]_6\cdot8H_2O$	(6T)	20, 206
Solid III	Triethylenetetramine	$[Zn_2(PO_4)_2]\cdot0.5R$	1 dim. (8T)	207
Solid IV	Triethylenetetramine	$[Zn_3(PO_4)_2(HPO_4)]\cdot0.5R$	1 dim. (16T)	207
Solid V[b]	Triethylenetetramine	$[Zn_4(PO_4)_4R_{0.5}]$	1 dim. (8T)	207
Compound I	1,3-diaminopropane	$[Zn_4(PO_4)_4]\cdot2R$	1 dim. (8T)	208
Compound II[c]	1,3-diaminopropane	$[Zn_5(H_2O)(PO_4)_4(HPO_4)]\cdot2R$	1 dim. (8T)	208
/**	1,3-diaminopropane	$[Zn_{12}(OH_2)_2(PO_4)_{10}\ R_{12}]\cdot2R2\cdot H_2O$[d]	1 dim. (8T) / 1 dim. (10T)	209
ZnPO/dab-A	1,4-diaza-bicyclo[2.2.2]octane	$[Zn_2(HPO_4)_3]\cdot R$	1 dim. (8T)	210
ZnPO/dab-B	1,4-diaza-bicyclo[2.2.2]octane	$[Zn_4(PO_4)_2(HPO_4)_2]\cdot R\cdot3H_2O$	1 dim. (8T)	210
ZnPO/dab-C	1,4-diaza-bicyclo[2.2.2]octane	$[Zn_5(PO_4)_2(HPO_4)_4]\cdot2R\cdot H_2O$	1 dim. (8T)	211
ZnPO/dab-D[b]	1,4-diaza-bicyclo[2.2.2]octane	$[Zn_3(PO_4)_4(HPO_4)_2R]$	1 dim. (12T)	211
/	1,4-diaza-bicyclo[2.2.2]octane	$[Zn_2(HPO_4)_3]\cdot R$	2 dim. (8T)	212
/	Imidazole[e]	$[Zn_4(OH)(PO_4)_3]\cdot R$	1 dim. (8T)	213
UiO-21 (**CHA**)	1-(2-aminoethyl)piperazine	$[Zn_7(PO_4)_6]\cdot2R$	3 dim. (8T)	214
UiO-22	1-(2-aminoethyl)piperazine	$[Zn_4(PO_4)_3(OH)]\cdot R$	2 dim. (8T and 12T)	214
UiO-17	Piperazine[e]	$[Zn(H_2O)_2(PO_4)_6]\cdot2R$	2 dim. (8T)	215
Phase II[f]	Piperazine[e]	$[Zn_{3.5}(PO_4)_3(H_2O)]\cdot R$	2 dim. (8T)	216
Phase III[f]	N-methylpiperazine[e,g] / Piperazine[e]	$[Zn(HPO_4)(H_2PO_4)]\cdot0.5R$	1 dim. clover-like (16T)	216, 217
ZP-4 (**EDI/THO**)[h]	Diethylethanolamine[i]	$K[ZnPO_4]\cdot0.8H_2O$	3D(8T)	218
Crystal I Crystal II	Tetramethylammonium	$Na[ZnPO_4]\cdot H_2O$	1 dim. (12, 8 or 6T)	219, 220
ZnPO-HEX (**CZP**)[j] /(**ABW**)[k]	Ethyldiisopropylamine	β-$Li[ZnPO_4]\cdot H_2O$	1 dim. (8T)	221

	Organic SDA	Formula		References
(ABW)	1,2-diaminocyclohexane[l], Ammonium, Guanidinium	$X[ZnPO_4]$ $X = NH_4, Na, Li$	1 dim. (8T)	20, 222, 223
ND-1	1,2-diaminocyclohexane	$[Zn_3(PO_4)_2(PO_3OH)]\cdot R\cdot 2H_2O$	1 dim. (24T)	224
Product I[b]	Diethylenetriamine	$[Zn_5(PO_4)_4R]$	1 dim. (10T)	225
/[j]	Diethylenetriamine	$[Zn_4(PO_4)_3(HPO_4)]\cdot R\cdot H_2O$	2 dim. (8T)	226
Product II[b]	1,3-diaminoguanidine	$[Zn_2(PO_4)(HPO_4)R]$	2 dim. (8T)	225
DAF-3	Ethylenediamine	$[ZnPO_4]_2\cdot R$	3 dim. (8T)	227
/	Ethylenediamine	$[Zn_3(PO_4)_2(HPO_4)_{0.5}\cdot 0.5R\cdot H_2O$	1 dim. (6T), 1 dim. (8T)	228, 229
/	Guanidinium	$[Zn(HPO_4)_2]\cdot 2R$	2 dim. (12T)	230
/	Guanidinium	$[Zn_2(HPO_4)_2(H_2PO_4)]\cdot R$	1 dim. (12T)	230
/	Guanidinium	$[Zn_7(H_2O)_4(PO_4)_6]\cdot 3R\cdot H_2O$	18 ring apertures	231
/	Tetramethylammonium	$[Zn(HPO_4)(H_2PO_4)]\cdot R$	2 dim. (12T)	232, 233
ZnPO-W[m]	Trimethylamine	$[Zn_4(PO_4)_3]\cdot H\cdot H_2O$	1 dim. (6T), 1 dim. (8T)	234
ZnPO-TMA[m]	Trimethylamine	$[Zn_4(H_2O)(PO_4)_3]\cdot R$	1 dim. (8T)	234
/[m]	Ethylamine	$[Zn_4(PO_4)_3(H_2O)]\cdot R$	1 dim. (8T)	235

* The organic structure-directing species may be charged or neutral.
** Not specified.
a Presence of sodium cations in the starting mixture. b The amine coordinates to the framework via Zn–N linkages. c This zincophosphate is closely related to the mineral Thomsonite. d The monoprotonated amine R1 and diprotonated amine R2 are present in this compound; R1 acting as a ligand to Zn (ZnO_3N tetrahedron). e Amine phosphate is introduced in the starting mixture. f Synthesis performed in a butane-2-ol-water medium. g During the reaction, piperazinium cations, resulting from demethylation of the N-methylpiperazine are formed and are occluded in the 3D framework. h Intergrowths between edingtonite (EDI) and thomsonite (THO). i Diethylethanolamine acts as a pH modifier. j Chiral tetrahedral framework. k β-LiZnPO₄ is related to the other ABW-type zincophosphates but the crystallographic b parameter is doubled. l Under the experimental conditions, 1,2-diaminocyclohexane decomposed into NH_4^+. m Synthesis performed in a ethyleneglycol medium.

Tab. 19. Beryllophosphate zeolite-type structures obtained by Stucky and co-workers

Nature of the cation	Chemical formula	IZA code	Ref.
Li^+	$Li_4Be_4P_4O_{16}\cdot4H_2O$	ABW	20
Li^+	$Li_{24}Be_{24}P_{24}O_{96}\cdot40H_2O$	RHO	20
Li^+	$Li_8Cl_2Be_6P_6O_{24}$	SOD	238
Li^+	$Li_8Br_2Be_6P_6O_{24}$	SOD	239
Li^+	$Li_8(HPO_4)(BePO_4)\cdot H_2O$	LOS	240
Na^+	$Na_{96}Be_{96}P_{96}O_{384}\cdot192H_2O$	FAU	20
NH_4^+	NH_4BePO_4	ABW	222
NH_4^+	$NH_4BePO_4\cdot1/8H_2O$	MER	241

ture ranging from 100–200 °C. Four are isostructural with zeolite **RHO**, gismondine (**GIS**), edingtonite (**EDI**) and analcime (**ANA**). The fifth displays a novel framework topology (**BPH** structure type) with no aluminosilicate analog. In addition to their work on the synthesis of open-framework zincophosphates, Stucky and coworkers [20, 238–241] were successful in preparing beryllophosphates with various zeolite topologies. With the exception of the merlinoite-type beryllophosphate that was synthesized in the presence of *t*-octylamine and which contains ammonium cations in the porosity of the framework, the other materials were obtained from mixtures free of organic species. As is reported in Table 19, they are prepared mainly in the presence of sodium (**FAU**) and lithium (**ABW, RHO, SOD, LOS**) cations. It has to be noted that **ABW**-type beryllophospates can also be obtained with rubidium and cesium cations [242]. However, recently, a 3D framework beryllophosphate $[H_3N(CH_2)_3NH_3]Be_3(HPO_4)_4$ containing tetrahedral 10-rings and 12-rings was synthesized by Harrison [243] in the presence of 1,3-diaminopropane as structure-directing species.

4.2.3.8
Proposed Synthesis Mechanisms

Several mechanisms have been proposed to explain the formation of some of the microporous phosphate-based materials. For the aluminophosphates, Oliver et al. [244] have suggested that intermediates based upon 1D chains could be involved in the formation of 2D porous layers and then 3D open-framework aluminophosphate materials. Their mechanism is supported by direct experimental evidence. The parent chain $[AlP_2O_8H_2]^{2-}$ would consist of corner-sharing $Al_2P_2O_4$ four-rings bridged at the aluminum corners, two doubly bridging phosphates occurring in each four ring (see Fig. 1). This chain-like structure would be the first species to crystallize in the synthesis system. It could be isolated, for instance, in a little water-triethyleneglycol (TEG) solvent system containing triethylamine as organic template, triethylammonium cations neutralizing the negative charge of the inorganic chain. It should be noted that in such a medium, an increase of the water content in the reaction mixture leads to 2D and 3D inorganic frameworks such as the aluminophosphate JDF-20 [245]. From this parent chain, chain-to-chain transformations

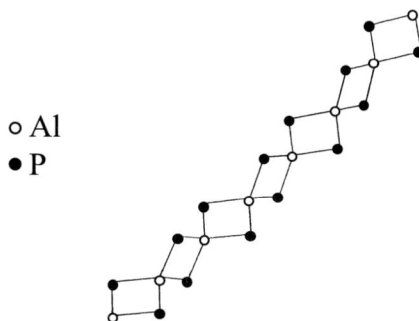

o Al
• P

would occur through hydrolysis, rotation and condensation reactions leading to the formation of secondary chains. Some of these new secondary chain structure types were also found experimentally, which reinforces the mechanism proposed. Such a chain is present in the chain-like aluminophosphate UT-2, which crystallizes from a TEG-cyclopentylamine system [246]. Depending on the synthesis conditions (pH, water content, etc.) and the type of interactions with the organic template, chain-to-layer, and chain-to-layer-to-framework transformations involving the same type of reactions (hydrolysis, rotation and condensation) would occur and allow the formation of 2D and 3D framework aluminophosphates. On the basis of this mechanism Vidal et al. [247] described the formation of the layered aluminophosphate Mu-7.

In 1995, Férey [248] proposed another mechanism for the formation of the alumino- and gallophosphate microporous solids. This mechanism is based on the fact that the structure of a large number of oxyfluorinated phosphates denoted ULM-*n* (see Table 14, Sect. 4.2.3.6.2) can be described by a limited number of secondary building units (SBU). Tetrameric, hexameric and octameric building units were identified. Thus, the structures of the ULM-3 and ULM-4 phases only differ by the arrangement of the same hexameric secondary building unit, which results from the linking of three PO_4 tetrahedra, one GaO_4F_2 octahedron and two GaO_4F trigonal bipyramids. The driving force in the mechanism proposed is the charge density matching between the inorganic and organic species (amines or ammonium cations).

The formation of the organic-inorganic hybrid material would proceed in three steps. At the beginning, IBU (initial building units, PO_4, MX*n* polyedra M = Al, Ga, X = O, F) and PBU (primary building units, phosphato complexes) would be present in the solution. Then, an oligomeric condensation resulting in the formation of SBU would occur in the solution. According to this concept, the charge density of the amine would determine the condensation, and therefore the size and the charge of the SBU formed. The two last steps would be the formation of zero-charge ammonium-SBU cation-anion pairs (MBU, molecular building units), followed by their infinite condensation. In these MBU, the inorganic SBU would bear

References see page 867

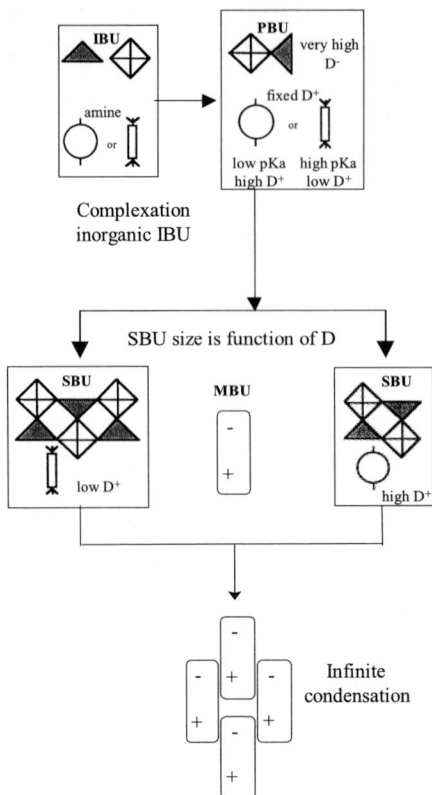

Fig. 2. Mechanism proposed for the formation of alumino- and gallophospate microporous solids with the different building units *x*BU (adapted from Ref. [248]). IBU, PBU, SBU and MBU correspond to initial, primary, secondary and molecular building units respectively (see text).

the functions for the condensation. The size and the shape of the cavities of the final solid would be determined by the topology and the steric occupancy of the MBU. A scheme of this mechanism summarizing these different steps is given in Fig. 2.

New insight in the formation of these solids was gained from *in situ* NMR experiments. Thus, Taulelle et al. [249, 250] studied the synthesis of the hydroxyfluorinated aluminophosphate CJ-2 by ^{31}P, ^{19}F and ^{27}Al solid and liquid-state NMR spectroscopy. This compound, whose formula is $(NH_4)_{0.88}(H_3O)_{0.12}$-$AlPO_4(OH)_{0.33}F_{0.67}$ [68], was obtained from a mixture containing 1,4 dia-zabicyclo[2.2.2] octane (DABCO) with the following molar composition: $1Al_2O_3/1P_2O_5/2NH_4F/1DABCO/80H_2O$. It appears that, DABCO is not incorporated in the final solid and rather acts as a pH controller. From their *in situ* liquid phase NMR study, the authors have clearly proven that, during the synthesis (from room temperature to 200 °C), a change in the coordination state of aluminum from six to five occurs. Moreover, it appears that the formation of Al–F bonds in exchange occurs during the synthesis. They concluded that an hydroxy fluorinated aluminum complex (Fig. 3a) would be formed in solution. This complex, in which aluminum exhibits 5-fold coordination, would then condense and give a cyclic

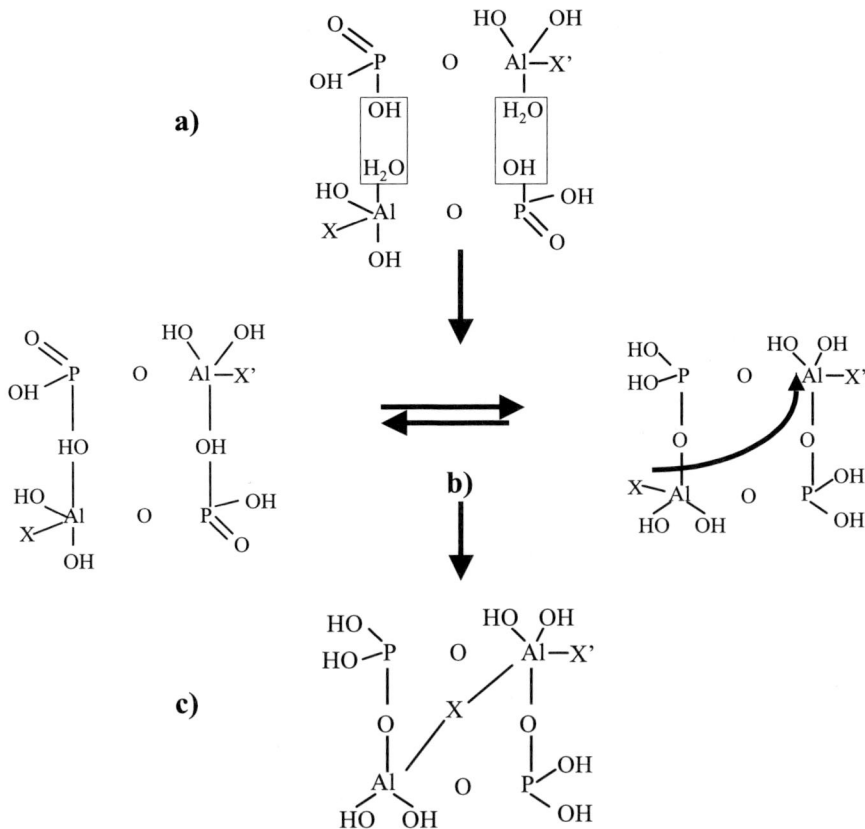

Fig. 3. Mechanism of formation of the hydroxyfluorinated aluminophosphate AlPO₄-CJ2 (adapted from Ref. [249]): (a) Hydroxfluorinated aluminum species (precursor species in solution); (b) Cyclic tetramer obtained from the precursor species; (c) Formation of the bridging bond (Al–X–Al) (X = OH/F, X′ = F/OH).

tetramer unit (Al_2P_2) (Fig. 3b), which would be the prenucleation building unit (PNBU). The structure of the hydroxyfluorinated aluminophosphate CJ-2 consists of tetramers (SBUs) with a bridging X atom (X = OH/F) between the two aluminum atoms leading to an aluminum (Al1) in 5-fold coordination (4O,1X; X = OH/F) and an aluminum (Al2) in 6-fold coordination (4O,1X (X = OH/F), 1X′(X′ = F/OH)). As no aluminum coordination state higher than 5 is observed in solution, such a bridge would be formed during nucleation and crystal growth (Fig. 3c).

In a subsequent solid-state NMR study [251], the same authors discuss the relative population and the distribution of F and OH groups within the structural building units (SBUs). They conclude that the AlPO₄-CJ-2 network is formed of

References see page 867

domains containing a mixture of different SBUs, which result from an isomerization of the PNBUs with a formation of a bridge at random with respect to OH$^-$ or F$^-$.

As reported in Sect. 4.2.3.7.1, a large number of chain-, layer- and open-framework zincophosphates can be prepared either under mild or hydrothermal conditions. In particular, Neeraj et al. [252] showed that numerous metal-phosphate phases (metal = Zn, Co, Sn, etc.) can be obtained at much lower temperatures when amine phosphates are used in the starting mixture instead of the corresponding amines. The former initially give chain or ladder structures, which, depending on the crystallization time and temperature, lead to more complex structures. Moreover, in some cases, strong structural relationships exist between the amine phosphate and the metal-phosphate phase that crystallizes. From that, and their *in situ* ^{31}P NMR and X-ray diffraction studies, the authors concluded that amine phosphates would be intermediates in the hydrothermal synthesis of open-framework metal phosphates.

From the whole of these results, it appears that several mechanisms are probably involved in the formation of the phosphate-based solids. The development of *in situ* experiments should lead in the near future to a better understanding of the genesis of these exciting solids.

4.2.3.9
Other 3D Open-Framework Metallophosphates

Besides the successful incorporation of transition metals in the open-framework aluminophosphates and gallophosphates described in the preceding sections, the synthesis of pure transition metal phosphates has been also possible. Thus, numerous molybdenum, vanadium, zirconium, titanium, iron and cobalt phosphates with a 3D open structure have been reported; a previous review has been dedicated to this subject [26 and references therein]. Also, a few indium, tin(ii) and nickel phosphates display 3D open-framework structures [26]. Since this review was published, a number of new 3D metallophosphates, essentially iron and cobalt phosphates have been reported in the literature and we will focus only on these new recently discovered materials. They are reported in Table 20 [253–265].

With the fluoride method and using diethylenetriamine as structure-directing agent Rao and coworkers [253] synthesized the open-framework iron phosphate $[Fe_5F_4(H_2PO_4)(HPO_4)_3(PO_4)_3]\cdot2R\cdot H_2O$. It possesses large elliptical voids of 24 T atoms (T = Fe, P) forming 1D channels of cross section 15×4.5 Å. Choudhury and Natarajan [254] prepared the 3D fluorooxy iron(iii) phosphate $[Fe_5(PO_4)(HPO_4)_6F_4]\cdot2R$ with ethylenediamine as structure-directing agent and acetylacetonate as a new iron source. The 3D architecture has a 1D 8-MR channel system.

In the absence of fluoride, and with ethylenediamine, Choudhury and Natarajan [255] obtained the open-framework phosphate $[Fe_2(HPO_4)_4]\cdot R$. Its structure is made of connected ladder-like chains giving rise to a 8-membered-ring 1D channel system. With the same structure-directing agent Debord et al. [256] ob-

Tab. 20. Recent 3-D open-framework metallophosphates.

Name (IZA code)	Organic structure-directing species (R)	Chemical formula*	Dimensionality of the channel system (pore openings: number of T atoms)[a]	Ref.
/	Diethylenetriamine	$[Fe_5 (H_2PO_4)(HPO_4)_3(PO_4)_3\ F_4]\cdot 2R\cdot H_2O$	1 dim. (24T)	253
/	Ethylenediamine	$[Fe_5\ (PO_4)(HPO_4)_6\ F_4]\cdot 2R$	1 dim. (8T)	254
/	Ethylenediamine	$[Fe_2(HPO_4)_4]\cdot R$	1 dim. (8T)	255
(ACO)	Ethylenediamine	$[Fe_4O(PO_4)_4]\cdot 2R\cdot H_2O$	3 dim. (8T)	256
/	Ethylenediamine	$[Fe_2(PO_4)_2]\cdot R\cdot 0.5H_2O$	b	257
/	Piperazine	$[Fe_6(HPO_4)_2(PO_4)_6]\cdot 2R\cdot 3H_2O$	1 dim. (8T)	258
/		$K[Fe_3(OH)_2(PO_4)_2]\cdot H_2O$	1 dim. (7T)	259
/	DABCOphosphate	$[Co_2(HPO_4)_3]\cdot R$	1 dim. (8T)	260
/	Diethylenetriaminephosphate	$[Co_4(PO_4)_4]\cdot 2R\cdot H_2O$	3 dim. (8T)	261
/	Diethylenetriaminephosphate	$[Co_6(PO_4)_5(HPO_4)_4]\cdot 3R\cdot H_2O$	1 dim. (16T)	261
/	Diethylenetriaminephosphate or Co(en)₃Cl₃ (en = ethylenediamine) + piperazine	$[Co_{3.5}(PO_4)_3]\cdot R$	1 dim. (12T)	262
CoPO-GIS (GIS)	Ethylenediamine	$[CoPO_4]\cdot 0.5R$	3 dim. (8T)	263
/	Piperazine	$Cs_2[Co_3(HPO_4)(PO_4)_2]\cdot H_2O$	1 dim. (16T)	264
CoPO₄-1	1,6-hexamethylenediamine	$[CoPO_4]\cdot 0.45R\cdot 0.3H_2O$	b	265
CoPO₄-2	Dipropylamine	$[CoPO_4]\cdot 0.54R\cdot 0.8H_2O$	b	265
CoPO₄-3	Dipropylamine	$[CoPO_4]\cdot 0.25R\cdot 0.75H_2O$	b	265

*The organic structure-directing species may be charged or neutral.
[a]T = Fe, Co, P. [b]The structure is not solved.

tained [$Fe_2(HPO_4)_4$]·R, which seems to be isostructural with ACP-1 [100] of topology **ACO** (see Sect. 4.2.3.5.2) with an oxygen at the center of the D4R units. More recently Kaucic and coworkers [257] also synthesized a new 3D iron phosphate of formula [$Fe_2(PO_4)_2$]·R·$0.5H_2O$ in the presence of ethylenediamine for which the thermal behavior suggests a 3D open framework, but up to now the structure remains unsolved. With piperazine Zima and Lii [258] have prepared an iron phosphate containing 8-MR channels. In an organic-free synthesis and with $Fe(C_2O_4)·2H_2O$ as iron source, Wright and coworkers [259] produced an organic-free material of composition K[$Fe_3(OH)_2(PO_4)_2$]·H_2O. The structure is isotypic with the mineral olmsteadite [260] and has a 1D 7-MR channel system.

Rao and coworkers [260–263] prepared four new open-framework cobalto-phosphates by using aminephosphates as the source of the structure-directing agent. The first, which was obtained with DABCOphosphate [$Co_2(HPO_4)_3$]·R (R = DABCO) [260], has a 1D 8-MR channel system. [$Co_4(PO_4)_4$]·$2R·H_2O$ [261] and [$Co_6(PO_4)_5(HPO_4)$]·$3R·H_2O$ [261] are obtained as a mixture when diethylene-triaminephosphate is used as the amine source. The former has a 3D 8-MR system of channels and occluded biprotonated ethylenediamine produced by the decomposition of the triamine. In contrast, for the second, diethylenetriaminepho-sphate molecules are occluded in the structure which displays elliptical 16-MR 1D channels [261]. When a mixture of the organometallic complex $Co(en)_3Cl_3$ (en = ethylenediamine) and piperazine is used as the amine source, the cobalt phosphate [$Co_{3.5}(PO_4)_3$]·R, where R is ethylenediamine is produced as a pure phase [262]. This material presents a 1D 12-MR system of channels. In ethyleneglycol, the above organometallic complex led to CoPO-**GIS** of topology **GIS** [263]. With piperazine, cesium hydroxide and oxalic acid in a water-ethyleneglycol mixture, the new organic-free $Cs_2Co_3(HPO_4)(PO_4)_2·H_2O$ material has been produced [264]. Its 3D framework is constituted of 1D 16-MR channels. Finally, Xu and coworkers [265] synthesized three new open-framework CoPOs (Table 20) which seem to have a 3D system of channels, but the structures are not yet solved.

4.2.3.10
3D Open-Framework Metal Phosphonates

Whereas inorganic materials like the molecular sieves and mesoporous materials have rigid structures with thermal and chemical stability, organic materials on the other hand are more flexible and easily tailored. The combination of inorganic and organic substructures has attracted the zeolite community; several approaches have been reported in a recent review [266].

One of the strategies used consists of the preparation of hybrid materials that exhibit selective reversible adsorption and desorption of guest molecules. To reach this goal the first method employed was the partial functionalization of the existing mesoporous materials [267–278] and molecular sieves [279–283]. Another recent strategy consists of the direct synthesis of hybrid solids such as metal organo-phosphonates [284]. Many organophosphonates are layered and structurally related to layered inorganic phosphate [285, 286]. However, several 3D metal organo-

Tab. 21. 3-D open-framework metal organophosphonates

Materials and/or chemical formula*		Pore opening	Ref.
AlMePO-α [Al$_2$(O$_3$PCH$_3$)$_3$]		36[a] (18T)[b]	287
AlMePO-β [Al$_2$(O$_3$PCH$_3$)$_3$]		36[a] (18T)[b]	288
β-Cu(O$_3$PCH$_3$)		24[a] (12T)[b]	289
Zn(O$_3$P(CH$_2$)$_2$NH$_2$)		16[a] (8T)[b]	290
M{O$_3$PCH$_2$NH(C$_2$H$_4$)$_2$NHCH$_2$PO$_3$}·H$_2$O (M = Mn, Co)		44[c]	291
Co$_3$(O$_3$P(CH$_2$)$_2$CO$_2$)·6H$_2$O		32[c]	292
Co$_2$(O$_3$PCH$_2$PO$_3$)·H$_2$O		20[c]	293
Pb$_6$(O$_3$PCH$_2$CO$_2$)$_4$		[d]	294
Mn$_3$(O$_3$PCH$_2$CO$_2$)$_2$		[d]	294
Sb$_2$O(O$_3$PCH$_2$PO$_3$)		18[c]	295
Sb[O$_3$P(CH$_2$)$_2$PO$_2$(OH)]		14[c]	295
Sb[O$_3$P(CH$_2$)$_3$PO$_2$(OH)]		24[c]	295
MIL-2	(V^{IV}O)$_2$(V^{IV}O(H$_2$O))$_4${O$_3$PCH$_2$PO$_3$}$_4$(NH$_4$)$_4$·4H$_2$O	28[c]	296
MIL-2K	K$_4$(V^{IV}O)$_2$(V^{IV}(H$_2$O))$_4${O$_3$PCH$_2$PO$_3$}$_4$·8H$_2$O	28[c]	297
MIL-3	(V^{V}O(H$_2$O))(V^{IV}O)O{O$_3$P(CH$_2$)$_2$PO$_3$}(NH$_4$)	14[c]	296
MIL-3K	K(V^{V}O(H$_2$O))(V^{IV}O)O{O$_3$P(CH$_2$)$_2$PO$_3$}	14[c]	297
MIL-5	(V^{IV}O)$_2$(OH){O$_3$P(CH$_2$)$_2$PO$_3$}·H$_2$O	16[c]	298
MIL-7(LT)	V$_2^{IV/V}$O$_3${O$_3$P(CH$_2$)$_3$PO$_3$}(NH$_4$)·3H$_2$O	16[c]	299
MIL-7(HT)	(V^{IV}O)$_2$(OH){O$_3$P(CH$_2$)$_3$PO$_3$}·3H$_2$O	16[c]	299
MIL-7K(LT)	K(V$_2^{IV}$O$_3$)(H$_2$O){O$_3$P(CH$_2$)$_3$PO$_3$}·H$_2$O	16[c]	297
MIL-7K(HT)	K(V^{IV}O)$_2$(OH){O$_3$P(CH$_2$)$_3$PO$_3$}·3H$_2$O	16[c]	297
MIL-11	Ln^{III}H[O$_3$P(CH$_2$)$_n$PO$_3$] (n = 1–3)	16, 14, 12[c,e]	300
MIL-24	(V^{IV}O(H$_2$O))(Cu^{II}(H$_2$O)){O$_3$PCH$_2$PO$_3$}	26[c]	301
MIL-26	Na[V^{IV}O(O$_3$P(CH$_2$)$_2$CO$_2$)]·2H$_2$O	26[c]	302
(V^{IV}O)$_2$(H$_2$O){O$_3$P(CH$_2$)$_3$PO$_3$}·2H$_2$O		16[c]	303
Na$_2$[(HO$_3$PCH$_2$)$_3$NH]·1.5H$_2$O		[c,d]	304

* As reported.
[a] Number of atoms forming the main channel ring including the O atoms. [b] Number of T atoms in the terminology of the preceding sections. [c] Number of atoms forming forming the large channel ring including the C, N and O atoms. [d] The structures of these materials are complex and no data about the number of atoms forming the main channel ring are given in the corresponding reference. [e] The pore openings are formed by 16 atoms for n = 3 (Ln = whole lanthanide series + Y), 14 for n = 2 (Ln = La, Ce, Pr, Nd, Eu, Gd, Yb) and 12 for n = 1 (Ln = Pr and Gd).

phosphonates have been obtained, they are reported in Table 21 [287–304]. Most of them show a 1D system of channels with small interconnecting windows.

The synthesis method of the metal organophosphonates is similar to that used for the hydrothermal synthesis of organically templated microporous metallophosphates discussed in the preceding sections but it differs in the phosphate sources. One oxygen of the phosphate group is replaced by functionalized alkyl chains or other functions (phosphonates, carboxylates, sulfonates or a mixture

References see page 867

of them). The synthesis temperatures range from 80 °C to 200 °C. Often, the organic moities of the phosphonates ensure the connectivity between the inorganic substructure of the material. This last point is very well illustrated by $VO(HPO_4)\cdot0.5H_2O$ [305], a pure phosphate, and $(V^{IV}O)_2(H_2O)\{O_3P(CH_2)_3\text{-}PO_3\}\cdot2H_2O$ [303] which are made with phosphoric acid (in a free organic medium) and ethylenediphosphonic acid as phosphorus source, respectively. In the 2D structure of $VO(HPO_4)\cdot0.5H_2O$, layers are formed by the linkage of $[V_2^{IV}O_8(H_2O)]$ dimeric units and tetrahedral $HOPO_3$ groups with terminal OH groups [301]. The same layers are present in $(V^{IV}O)_2(H_2O)\{O_3P(CH_2)_3PO_3\}_2\cdot H_2O$ but the terminal hydroxyl groups are replaced by C atoms of the di-phosphonate $H_2O_3P(CH_2)_3\text{-}PO_3H_2$ functions which ensure the connectivity between the layers to form a 3D framework.

The first reported metal phosphonate with a 3D framework is $[\beta\text{-}Cu(O_3PCH_3)]$ [289]. It was obtained with methylphosphonic acid as the phosphorus source. The structure contains 1D channels with 24-membered rings including the oxygen atoms.

The use of monophosphonates in the synthesis led to the open-framework phosphonates AlMePO-α [287], AlMePO-β [288] (Me = methyl), $\beta\text{-}Cu(O_3PCH_3)$ [289], $Zn[O_3PCH_2CH_2NH_2]$ [290], $Co_3(O_3P(CH_2)_2CO_2)\cdot6H_2O$ [292], $Pb_6(O_3PCH_2CO_2)_4$ [294], $Mn_3(O_3PCH_2CO_2)_2$ [294] and MIL-26 [302].

By replacing the monophosphonic by diphosphonic acids ($H_2O_3P\text{-}R\text{-}PO_3H_2$), it was possible to prepare other new materials. Thus, Férey and coworkers [296–303] prepared a series of solids named MIL-n for which the phosphorus sources are alkyl diphosphonic acids $H_2O_3P(CH_2)_nPO_3H_2$ with n ranging from 1 to 3. Most of these materials are vanadodiphosphonates. In all the series the vanadium may be tetravalent only or may have both V(IV) and V(V) oxidation states. The whole lanthanide series and yttrium gave MIL-11 for $n = 1$, 2 and 3. A nitrogen-containing diphosphonic acid led to $M\{O_3PCH_2NH(C_2H_4)_2NHCH_2PO_3\}\cdot H_2O$ with M = Mn, Co [291] which has ellipsoidal 44-membered-ring channels including oxygen, carbon and nitrogen atoms (pore size = 4.7×18 Å). This is the most open material listed in Table 21.

Aranda and coworkers [304] used the triphosphonate nitrilotris(methylene)-triphosphoric acid in the preparation of $Na_2[HO_3PCH_2)_3NH]\cdot1.5H_2O$, a water soluble compound. The 3D framework of the latter is very complex and possesses small channels where water molecules are located.

For most of the materials of Table 21 the pore voids are not accessible after dehydration and removal of occluded species; generally, the structures collapse below 300 °C. However, AlMePO-α and AlMePO-β [266] are thermally stable up to about 500 °C under inert atmosphere while the structures of $Pb_6(O_3PCH_2CO_2)_4$ and $Mn_3(O_3PCH_2CO_2)_2$ [294] collapse at 510 and 418 °C, respectively when the degradation of the organic moieties occurs. Reversible sorption properties have been reported only for AlMePO-α and AlMePO-β [266]. The walls of the 1D channels are lined with methyl groups covalently bonded to the inorganic framework, the resulting channel cross section is 5.8 Å. Both structures are closely related, but AlMePO-α is energetically slightly more stable than the β polymorph. Thus,

AlMePO-β may be transformed into AlMePO-α under thermal treatment with water vapor. The nitrogen adsorption/desorption isotherms at 77 K of AlMePO-α, and AlMePO-β outgassed at 450 °C are of type I as expected for a zeolite-like microporous material, the adsorption capacity is about 0.14 cm^3 g^{-1} for both products.

4.2.3.11
Organized Mesoporous Phosphate-based Materials

In addition to the microporous phosphate-based materials, organized mesoporous solids have also been reported. They are prepared in the presence of surfactants as structure-directing agents using a procedure similar to that developed by Mobil for the synthesis of the M41S materials [306]. Although numerous solids have been obtained, most of them consist of layered materials and their structures usually collapse after removal of the organic species by calcination. Nevertheless, a few aluminophosphates (Al/P) [307–314], silicoaluminophosphates (Si/Al/P) [315, 316], titanophosphates (Ti/P) [317–319] and vanadophosphates (V/P) [320–322] exhibit a hexagonal or a cubic arrangement of mesopores. Examples of these mesoporous phosphates and their characteristics (thermal stability, surface area, pore volume and pore diameter) are given in Table 22 [307–322].

4.2.3.12
Conclusion and Perspectives

The synthesis of microporous phosphate-based materials has been extremely successful in the last twenty years. However, many of these solids are not porous because the guest species occluded in the inorganic framework cannot be removed without altering the structure. It is the case for most of the zincophosphates and some gallophosphates synthesized so far.

The ability to produce novel frameworks is still quite open by varying the nature of the structure-directing species, the framework composition, the nature of the solvent and that of the mineralizing agent, or the synthesis conditions (type of heating, temperature, etc.). Indeed, the flexibility of the framework element associated with phosphorus (Al, Ga, Zn, etc.) to adopt 4-, 5-, or 6-coordination increases the number of structures that can be expected.

One goal that has now to be achieved is to develop materials with suitable properties. This could be reached by choosing organic structure-directing species on the basis of the best geometrical fit and interactions between the organic and hypothetical or known target frameworks using molecular modeling.

Acknowledgments

The authors would like to thank Dr. A. Simon, Dr. A. Matijasic, L. Sicard and L. Josien for fruitful discussions.

References see page 867

Tab. 22. Examples of organized mesoporous phosphate-based materials reported so far.

Material	Surfactant	Symmetry type	Thermally stable at 500 °C	Surface area[a] / (m² g⁻¹)	Pore volume[a] /(cm³ g⁻¹)	Pore diameter[a] / nm	Ref.
Al/P, Ga/Al/P	SDS[b]	hexagonal	/*c	630	0.4	1.7	307,308
Al/P	C$_{16}$TMA^{d,e}	hexagonal	yes	980/450^f	0.44	1.8	309, 310
Al/P	C$_{22}$TMA^{d,e}	hexagonal	yes	760	0.56	2.8	309, 311
Al/P	C$_{22}$TMA^{d,e} + TPB^g	hexagonal	yes	720	0.72	3.9	309, 311
Al/P	C$_{16}$TMA^{d,h}	hexagonal wormhole-like material	yes	480–650^i	/	1.3–3.7^i	312
Al/P	C$_{16}$TMA^{d,e,j}	cubic	no	772	/	/	313
Al/P	C$_{16}$TMA^{d,e}	hexagonal	yes	928	/	/	314, 315
Si/Al/P				830–930	/	/	
Mn/Al/P							
Si/Al/P	C$_{16}$TMA^d	hexagonal	yes	920–980	/	3.0	316
Ti/P	CnTMA^e (n = 7–17)	/	/c	740 250^k	0.79 (n = 16)	3.0 (n = 16)	317
Ti/P^l	Dodecanol + 5EO^d	/m	yes	350	/	4.5	318
Ti/P^l	CnTMA (n = 16, 18)	hexagonal	yes (350 °C)	350^n	0.18 (n = 16) 0.25 (n = 18)	2–3^n	319
V/P	C$_{14}$TMA^d	hexagonal	no	/	/	/	320
V/P	C$_{14}$TMA^d	hexagonal	/	/	/	3.0^o	321
V/P	C$_{16}$TMA^d	hexagonal	no	/	/	2.6^o	322

* Not specified.

^{a} Determined from N$_2$ adsorption-desorption isotherms. ^{b} Sodium dodecylsulfate. ^{c} The surfactant was removed by extraction. ^{d} C$_{16}$TMA: hexadecyltrimethyl-ammonium cation, C$_{22}$TMA: docosyltrimethyl-ammonium cation, C$_{14}$TMA: tetradecyltrimethylammonium cation, dodecanol + 5EO = industrial polyethyl-enoxide. ^{e} Presence of tetramethylammonium hydroxide in the starting mixture. ^{f} 980 and 450 m^2 g^{-1} were reported in references [309 and 310], respectively. ^{g} 1,3,5- tri-isopropylbenzene. ^{h} Synthesis performed in an aqueous triethanolamine medium. ^{i} Depending on the P/Al ratio in the starting mixture. ^{j} Presence of F$^-$ in the starting mixture. ^{k} Measured on the sample calcined at 540 °C. ^{l} Titanium oxo-phosphate. ^{m} Dis-ordered pore arrangement. ^{n} Determined by XRD analysis and from the pore volume. ^{o} Determined by transmission electron microscopy.

References

1 W. M. Meier, D. H. Olson, C. Baerlocher (Eds), Atlas of Zeolite Structure Types (4th revised edn), Elsevier, Amsterdam, 1996. WEB site: http://www.iza-structure.org/databases/

2 S. T. Wilson, B. M. Lok, C. A. Messina, T. R. Cannan, E. M. Flanigen, *J. Am. Chem. Soc.* **1982**, *104*, 1146–1147.

3 E. M. Flanigen, R. W. Grose, in: Molecular Sieve Zeolites-1. R. F. Gould (Ed.), *American Chemical Society Advances in Chemistry Series*, No. 101, American Chemical Society, Washington, D.C., 1971, p. 76–99.

4 G. H. Kühl, US Patent 3,355,246, 1967; 3,791,964, 1974.

5 G. H. Kühl, K. D. Schmitt, *Zeolites* **1990**, *10*, 2–7.

6 G. Artioli, J. J. Pluth, J. V. Smith, *Acta Cryst.* **1984**, *C40*, 214–217.

7 J. C. Jumar, A. Goiffon, B. Capelle, A. Zarka, J. C. Doukhan, J. Schwartzel, J. Détaint, E. Philipot, *J. Cryst. Growth* **1987**, *80*, 133–148.

8 F. d'Yvoire, *Bull. Soc. Chim. Fr.* **1961**, 1762–1776.

9 D. M. Poojary, A. Clearfield, *Zeolites* **1993**, *13*, 542–548.

10 M. E. Davis, C. Saldarriaga, C. Montes, J. M. Garcès, *Zeolites* **1988**, *8*, 362–366.

11 L. B. McCusker, C. Baerlocher, E. Jahn, M. Bülow, *Zeolites* **1991**, *11*, 308–313.

12 J. de Onate Martinez, L. B. McCusker, C. Baerlocher, *Microporous and Mesoporous Mater.* **2000**, *34*, 99–113.

13 R. M. Dessau, J. L. Schlenker, J. B. Higgins, *Zeolites* **1990**, *10*, 522–524.

14 J. W. Richardson, Jr., E. T. C. Vogt, *Zeolites* **1992**, *12*, 13–19.

15 H. X. Li, M. E. Davis, J. B. Higgins, R. M. Dessau, *J. Chem. Soc., Chem. Commun.* **1993**, 403–405.

16 J. J. Pluth, J. V. Smith, *Acta Cryst.* **1986**, *C42*, 1118–1120.

17 D. M. Poojary, K. J. Balkus, S. J. Riley, B. E. Guade, A. Clearfield, *Microporous Mater.* **1994**, *2*, 245–250.

18 J. Parise, *J. Chem. Soc., Chem. Commun.* **1985**, 605–607.

19 G. Harvey, W. M. Meier, in: Proceedings of the 8th International Zeolite Conference, Amsterdam, 1989 Zeolites: Facts, Figures, Future. P. A. Jacobs, R. A. Van Santen (Eds), *Studies in Surface Science and Catalysis*, Vol. 49A, Elsevier, Amsterdam, 1989, p. 411–420.

20 T. E. Gier, G. Stucky, *Nature* **1991**, *349*, 508–510.

21 E. M. Flanigen, in: Proceedings of the 7th International Zeolite Conference, Tokyo, 1986 New Developments in Zeolite Science and Technology. Y. Murakami, A. Iijima, J. W. Ward (Eds), Kodansha – Elsevier, Tokyo, 1986, p. 103–112.

22 S. T. Wilson, in: Introduction to Zeolite Science and Practice. H. van Bekkum, E. M. Flanigen, J. C. Jansen (Eds), *Studies in Surface Science and Catalysis*, Vol. 58, Elsevier, Amsterdam, 1991, p. 137–151.

23 R. Szostak, in: Molecular Sieves, Science and Technology. H. G. Karge, J. Weitkamp (Eds), Vol. 1, *Synthesis*, Springer, Berlin, 1998, p. 157–186.

24 R. Szostak, in: Handbook of Molecular Sieves. Van Nostrand Reinhold, New York, 1992.

25 H. Kessler, in: Comprehensive Supramolecular Chemistry. G. Alberti, T. Bein (Eds), Vol. 7, Pergamon, Oxford, 1996, p. 425–464.

26 A. K. Cheetham, G. Férey, T. Loiseau, *Angew. Chem. Int. Ed.* **1999**, *38*, 3268–3292.

27 J. L. Paillaud, B. Marler, H. Kessler, *Chem. Commun.* **1996**, 1293–1294.

28 J. J. Pluth, J. V. Smith, J. M. Bennett, J. P. Cohen, *Acta Cryst.* **1984**, *C40*, 2008–2011.

29 J. L. Guth, P. Caullet, *J. de Chimie – Physique* **1986**, *83*, 155–175.

30 E. G. Derouane, E. W. Valyocsik, R. von Ballmoos, Eur. Patent 0,146,384, **1985**, (*Chem. Abstr.* **1985**, *103*, 110 740).

31 E. G. DEROUANE, R. VON BALLMOOS, Eur. Patent 0,146,385, 1985, (*Chem. Abstr.* **1985**, *103*, 73 377).

32 E. G. DEROUANE, R. VON BALLMOOS, Eur. Patent 0,146,386, 1985, (*Chem. Abstr.* **1985**, *103*, 73 350).

33 E. G. DEROUANE, R. VON BALLMOOS, Eur. Patent 0,146,387, 1985, (*Chem. Abstr.* **1985**, *103*, 73 349).

34 E. G. DEROUANE, R. VON BALLMOOS, Eur. Patent 0,146,388, 1985, (*Chem. Abstr.* **1985**, *103*, 93 673).

35 E. G. DEROUANE, R. VON BALLMOOS, Eur. Patent 0,146,389, 1985, (*Chem. Abstr.* **1985**, *103*, 73 378).

36 E. G. DEROUANE, R. VON BALLMOOS, Eur. Patent 0,174,122, 1986, (*Chem. Abstr.* **1986**, *104*, 189 206).

37 R. VON BALLMOOS, E. G. DEROUANE, E. W. VALYOCSIK, Eur. Patent 0,163,467, 1985, (*Chem. Abstr.* **1986**, *104*, 151 798).

38 R. VON BALLMOOS, E. G. DEROUANE, Eur. Patent 0,166,520, 1986, (*Chem. Abstr.* **1986**, *104*, 151 799).

39 H. HE, Y. LONG, *J. Inclusion Phenomena* **1987**, *5*, 591–600.

40 Y. LONG, W. ZHANG, H. HE, *J. Inclusion Phenomena* **1987**, *5*, 363–372.

41 H. HE, Y. LONG, *Gaodeng Xuexiao Huaxue Xuebao* **1988**, *9*, 103–112.

42 Y. LONG, L. DONG AND H. HE, *Gaodeng Xuexiao Huaxue Xuebao* **1988**, *9*, 186–195.

43 Z. ZHAO, R. ZHAO, *Zeolites* **1993**, *13*, 634–639.

44 Q. XU, J. DONG, A. YAN, C. JIN, *Acta Phys. Chem.* **1985**, *31*, 99–108.

45 Q. HUO, R. XU, S. LI, Z. MA, J. M. THOMAS, R. H. JONES, A. M. CHIPPINIDALE, *J. Chem. Soc., Chem. Commun.* **1992**, 875–876.

46 N. SIMON, T. LOISEAU, G. FÉREY, *Solid State Sciences* **1999**, *1*, 339–349.

47 M. SOULARD, J. PATARIN, B. MARLER, *Solid State Sciences* **1999**, *1*, 37–53.

48 J. L. PAILLAUD, P. CAULLET, L. SCHREYECK, B. MARLER, *Microporous and Mesoporous Mater.* **2001**, *42*, 177–189.

49 D. E. AKPORIAYE, H. FJELLVÅG, E. N. HALVORSEN, T. HANG, A. KARLSSON, K. P. LILLERUD, *J. Chem. Soc., Chem. Commun.* **1996**, 1553–1554.

50 D. E. AKPORIAYE, H. FJELLVÅG, E. N. HALVORSEN, J. HUSTVEIT, A. KARLSSON, K. P. LILLERUD, *J. Chem. Soc., Chem. Commun.* **1996**, 601–602.

51 K. O. KONGSHAUG, H. FJELLVÅG, K. P. LILLERUD, *Microporous and Mesoporous Mater.* **2000**, *40*, 313–322.

52 F. TAULELLE, V. MUNCH, C. HUGUENARD, A. SAMOSON, T. LOISEAU, N. SIMON, J. RENAUDIN, G. FÉREY, in: Proceedings of the 12th International Zeolite Conference. M. M. J. TREACY, B. K. MARCUS, M. E. BISHER, J. B. HIGGINS (Eds), *Part IV, MRS*, Warrendale, Pennsylvania, 1999, p. 2409–2412.

53 N. SIMON, T. LOISEAU, G. FÉREY, *J. Chem. Soc., Dalton Trans.* **1999**, 1147–1151.

54 J. B. PARISE, *J. Chem. Soc., Chem. Comm.* **1984**, 1449–1450.

55 R. W. BROACH, S. T. WILSON, R. M. KIRCHNER, in: Proceedings of the 12th International Zeolite Conference. M. M. J. TREACY, B. K. MARCUS, M. E. BISHER, J. B. HIGGINS (Eds), *Part III, MRS*, Warrendale, Pennsylvania, 1999, p. 1715–1722.

56 J. M. BENNETT, J. M. COHEN, G. ARTIOLI, J. J. PLUTH, J. V. SMITH, *Inorg. Chem.* **1985**, *24*, 188–193.

57 J. B. PARISE, in: Proceedings of an International Symposium, Portorose, 1984, Zeolites, Synthesis, Structure, Technology and Application. B. DRZAJ, S. HOCEVAR, S. PEJOVNIK (Eds), *Studies in Surface Science and Catalysis*, Vol. 24, Elsevier, Amsterdam, 1985, p. 271–278.

58 A. SIMMEN, Ph. D. Thesis, E.T.H., Zurich, **1992**.

59 P. R. RUDOLF, C. SALDARRIAGA-MOLINA, A. CLEARFIELD, *J. Phys. Chem.* **1986**, *90*, 6122–6125.

60 S. NATARAJAN, J. L. P. GABRIEL, A. K. CHEETHAM, *J. Chem. Soc., Chem. Commun.* **1996**, 411–412.

61 P. I. GAI-BOYES, J. M. THOMAS, P. A. WRIGHT, R. H. JONES, S. NATARAJAN, J. CHEN, R. XU, *J. Phys. Chem.* **1992**, *96*, 8206–8209.

62 K. O. KONGSHAUG, H. FJELLWÅG, B. KLEWE, K. P. LILLERUD, *Microporous*

and Mesoporous Mater. **2000**, *39*, 333–339.

63 R. M. KIRCHNER, R. W. GROSSE-KUNSTLEVE, J. J. PLUTH, S. T. WILSON, R. W. BROACH, J. V. SMITH, *Microporous and Mesoporous Mater.* **2000**, *39*, 319–332.

64 K. MARDA, A. TUEL, S. CALDARELLI, C. BAERLOCHER, *Microporous and Mesoporous Mater.* **2000**, *39*, 465–476.

65 D. F. SHANTZ, A. BURTON, R. F. LOBO, *Microporous and Mesoporous Mater.* **1999**, *31*, 61–73.

66 Y. H. XU, B.-G. ZHANG, X.-F. CHEN, S.-H. LIU, C.-Y. DUAN, X. Z. YON, *J. Solid State Chem.* **1999**, *145*, 220–226.

67 L. YU, W. PANG, L. LI, *J. Solid State Chem.* **1990**, *87*, 241–244.

68 G. FÉREY, T. LOISEAU, P. LACORRE, F. TAULELLE, *J. Solid State Chem.* **1993**, *105*, 179–190.

69 C. SCHOTT-DARIE, J. PATARIN, P. Y. LE GOFF, H. KESSLER, E. BENAZZI, *Microporous Mater.* **1994**, *3*, 123–132.

70 H. KESSLER, J. PATARIN, C. SCHOTT-DARIE, in: Advanced Zeolite Science and Applications. J. C. JANSEN, M. STÖCKER, H. G. KARGE, J. WEITKAMP (Eds), *Studies in Surface Science and Catalysis*, Vol. 85, Elsevier, Amsterdam, p. 75–113.

71 L. SIERRA, C. DEROCHE, H. GIES, J. L. GUTH, *Microporous Mater.* **1994**, *3*, 29–38.

72 J. RENAUDIN, T. LOISEAU, F. TAULELLE, G. FÉREY, C. R. Acad. Sci. Paris **1996**, *323 II B*, 545–553.

73 Q. HUO, R. XU, *J. Chem. Soc., Chem. Commun.* **1990**, 783–784.

74 Q. GAO, S. LI, R. XU, *Mater. Letters* **1997**, *31*, 151–153.

75 Q. GAO, S. LI, R. XU, *J. Chem. Soc., Chem. Commun.* **1994**, 1465–1466.

76 R. XU, Q. HUO, W. PANG, in: Proceeedings of the 9th International Zeolite Conference. R. VON BALLMOOS, J. HIGGINS, M. M. J. TREACY (Eds), Butterworth-Heinemann, Boston, 1993, p. 271–278.

77 Q. HUO, R. XU, S. LI, Y. XU, Z. MA, Y. YUE, L. LI, in: Proceedings of the 9th International Zeolite Conference. R. VON BALLMOOS, J. HIGGINS, M. M. J. TREACY (Eds), Butterworth-

Heinemann, Boston, 1993, p. 279–286.

78 R. H. JONES, J. M. THOMAS, J. CHEN, R. XU, Q. HUO, S. LI, Z. MA, A. M. CHIPPINDALE, *J. Solid State Chem.* **1993**, *102*, 204–208.

79 Q. GAO, S. LI, R. XU, Y. YUE, *J. Mater. Chem.* **1996**, *6*, 1207–1210.

80 C. F. FENG, H. HE, J. KLINOWSKI, *J. Chem. Soc., Faraday Trans.* **1995**, *91*, 3995–3999.

81 S. OLIVER, A. KUPERMAN, G. A. OZIN, *J. Mater. Chem.* **1997**, *7*, 807–812.

82 J. YU, K. SUGIYAMA, S. ZHENG, S. QIU, J. CHEN, R. XU, Y. SAKAMOTO, O. TERASAKI, H. HIRAGA, M. LIGHT, M. B. HURSTHOUSE, J. M. THOMAS, *Chem. Mater.*, **1998**, *10*, 1208–1211.

83 L. VIDAL, J. L. PAILLAUD, Z. GABELICA, *Microporous and Mesoporous Mater.* **1998**, *24*, 189–197.

84 M. ROUX, C. MARICHAL, J. L. PAILLAUD, C. FERNANDEZ, C. BAERLOCHER, J. M. CHÉZEAU, *J. Phys. Chem.* submitted.

85 B. WEI, G. S. ZHU, J. H. YU, S. L. QIU, F. S. XIAO, O. TERASAKI, *Chem. Mater.* **1999**, *11*, 3417–3419.

86 K. X. WANG, J. H. YU, G. S. ZHU, Y. C. ZOU, R. R. XU, *Microporous and Mesoporous Mater.* **2000**, *39*, 281–289.

87 E. B. KELLER, W. M. MEIER, R. M. KIRCHNER, *Solid State Ionics* **1990**, *43*, 93–102.

88 W. F. YAN, J. H. YU, Z. SHI, R. R. XU, *Chem. Commun.* **2000**, 1431–1432.

89 W. F. YAN, J. H. YU, Z., R. R. XU, G. S. ZHU, F. S. XIAO, Y. HAN, K. SUGIYAMA, O. TERASAKI, *Chem. Mater.* **2000**, *12*, 2517–2519.

90 J. J. PLUTH, J. W. SMITH, J. W. RICHARDSON JR., *J. Phys. Chem.* **1988**, *92*, 2734–2738.

91 B. M. LOK, R. L. PATTON, R. T. GAJIK, T. R. CONNON and E. M. FLANIGEN, US Patent 4,440,871, 1984.

92 C. T. W. CHU, J. L. SCHLENKER, J. D. LEITNER, C. D. CHANG, US Patent 5,091,073, 1992.

93 K. J. CHAO, S. P. SHEN, H. S. SHEN, *J. Chem. Soc., Faraday Trans.* **1992**, *88*, 2949–2954.

94 S. RADAEV, W. JOSWIG, W. H. BAUR, *J. Mater. Chem.* **1996**, *6*, 1413–1418.

95 L. Sierra, J. Patarin, C. Deroche, H. Gies, J. L. Guth, in: Zeolites and Related Microporous Materials: State of the Art 1994. J. Weitkamp, H. G. Karge, H. Pfeifer and W. Hölderich (Eds), *Studies in Surface Science and Catalysis*, Vol. 84C, Elsevier, Amsterdam, 1994, p. 2237–2244.

96 L. B. McCusker, C. Baerlocher, *Microporous Mater.* 1996, 6, 51–54.

97 J. P. Lourenço, M. B. Ribeiro, C. Borges, J. Rocha, B. Onider, E. Garrone, Z. Gabelica, *Microporous and Mesoporous Mater.* 2000, 6, 267–278.

98 J. M. Bennett, B. K. Marcus, in: Innovation in Zeolite Materials Science. P. J. Grobet, W. J. Mortier, E. F. Vansant and G. Schulz-Ekloff (Eds), *Studies in Surface Science and Catalysis*, Vol. 37, Elsevier, Amsterdam, 1988, p. 269–279.

99 N. Novak Tusar, A. Ristic, A. Meden, V. Kaucic, *Microporous and Mesoporous Mater.* 2000, 37, 303–311 and references therein.

100 P. Feng, X. Bu, T. E. Gier, G. D. Stucky, *Nature* 1997, 398, 735–741.

101 W. H. Baur, W. Joswig, D. Kassner, A. Bienik, Kinger, J. Kornatowski, *Z. Kristallogr.* 1999, 214, 154–159.

102 W. Baur, W. Joswig, D. Kassner, J. Kornatowski, *Acta Cryst.* 1994, B50, 290–294.

103 J. W. Smith, J. J. Pluth, K. J. Andries, *Zeolites* 1993, 13, 166–169.

104 G. M. T. Cheetham, M. M. Harding, P. J. Rezkallah, V. Kaucic, N. Rajic, *Acta Cryst.* 1991, C47, 1361–1364.

105 M. Ito, Y. Shimoyama, Y. Saito, Y. Tsurita, M. Otake, *Acta Cryst.* 1985, C41, 1698–1700.

106 J. J. Pluth, J. V. Smith, *J. Phys. Chem.* 1989, 93, 6516–6520.

107 P. A. Wright, R. H. Jones, S. Natarajan, R. G. Bell, J. S. Chen, M. B. Hursthouse, J. M. Thomas, *Chem. Commun.* 1993, 633–635.

108 J. J. Pluth, J. V. Smith, J. M. Bennett, *Acta Cryst.* 1986, C42, 283–286.

109 M. Helliwall, V. Kaucic, G. M. T. Cheetham, M. M. Harding, B. M.

Karinki, P. J. Rizkallah, *Acta Cryst.* 1993, B49, 413–420.

110 E. M. Flanigen, B. H. Loh, R. L. Patton, S. T. Wilson, *Pure Appl. Chem.* 1986, 58, 1351–1358.

111 P. Feng, X. Bu, T. E. Gier, G. D. Stucky, *Microporous and Mesoporous Mater.* 1998, 23, 221–229.

112 P. A. Barrett, R. J. Jones, *Phys. Chem. Chem. Phys.* 2000, 2, 407–412.

113 P. Feng, X. Bu, T. E. Gier, G. D. Stucky, *Microporous and Mesoporous Mater.* 1998, 23, 315–322.

114 S. Han, J. V. Smith, J. J. Pluth, J. W. Richardson Jr., *J. Mineral.* 1990, 2, 787–798.

115 B. Marler, J. Patarin, L. Sierra, *Microporous and Mesoporous Mater.* 1995, 5, 151–159.

116 X. Bu, P. Feng, T. E. Gier, G. D. Stucky, *Microporous and Mesoporous Mater.* 1998, 25, 109–117.

117 A. N. Christensen, R. G. Hazell, *Acta Chem. Scand.* 1999, 53, 403–409.

118 K. O. Kongshaug, H. Fjellevag, K. P. Lillerud, *Chem. Mater.* 2000, 12, 1095–1099.

119 J. Chen, R. H. Jones, S. Natarajan, M. B. Hursthouse, J. M. Thomas, *Angew. Chem. Int. Ed.* 1994, 33, 639–640.

120 X. Bu, T. E. Gier, P. Feng, G. D. Stucky, *Chem. Mater.* 1998, 10, 2546–2551.

121 X. Bu, P. Y. Feng, G. D. Stucky, *Science* 1997, 278, 2080–2085.

122 G. W. Noble, P. A. Wright, P. Lightfoot, R. E. Morris, K. J. Hudrom, A. Kwick, *Angew. Chem. Int. Ed.* 1997, 36, 81–83.

123 G. W. Noble, P. A. Wright, A. Kvick, *J. Chem. Soc., Dalton Trans.* 1997, 4485–4490.

124 V. Patinec, P. A. Wright, R. A. Aitken, P. Lightfoot, S. D. J. Purdie, P. A. Cox, A. Kvick, G. Vaughan, *Chem. Mater.* 1999, 11, 2456–2462.

125 V. Patinec, P. A. Wright, P. Lightfoot, R. A. Aitken, P. A. Cox, *Chem. Soc., Dalton Trans.* 1999, 3909–3911.

126 P. A. Wright, M. J. Maple, A. M. Z. Slawin, V. Patinic, R. A. Aitken, S.

WELSH, P. A. COX, *J. Chem. Soc., Dalton Trans.* **2000**, 1243–1248.

127 C. BAERLOCHER, L. B. MCCUSKER, R. CHIAPPETTA, *Microporous Mater.* **1994**, 2, 269–280.

128 G. BERGERHOFF, W. H. BAUR, W. NOWACKI, *N. Jb. Miner. Mh.* **1958**, 193–200.

129 J. B. PARISE, *Inorg. Chem.* **1985**, 24, 4312–4316.

130 J. B. PARISE, *Acta Crystallogr.* **1986**, C42, 670–673.

131 R. XU, J. CHEN, S. FENG, in: Chemistry of Microporous Crystals. T. INUI, S. NAMBA, T. TATSUMI (Eds), *Studies in Surface Science and Catalysis*, Vol. 60, Elsevier, Amsterdam, 1991, p. 63–72.

132 T. LOISEAU, D. RIOU, M. LICHERON, G. FÉREY, *J. Solid State Chem.* **1994**, 111, 397–402.

133 J. B. PARISE, *Acta Crystallogr.* **1986**, C42, 144–147.

134 G. YANG, S. FENG, R. XU, H. JIEGOU, *J. Struct. Chem.* **1988**, 7, 235–240.

135 S. FENG, R. XU, *Chem. J. Chinese Univ.* **1988**, 4, 1–7.

136 Q. KAN, F. P. GLASSER, R. XU, *J. Mater. Chem.* **1993**, 3, 983–987.

137 G. YANG, S. FENG, R. XU, *J. Chem. Soc., Chem. Commun.* **1987**, 1254–1255.

138 A. MERROUCHE, J. PATARIN, M. SOULARD, H. KESSLER, D. ANGLEROT, in: Molecular Sieves. Synthesis of Microporous Materials. M. L. OCCELLI, H. E. ROBSON (Eds), Vol. I, Van Nostrand Reinhold, New York, 1992, p. 384–399.

139 S. T. WILSON, N. A. WOODWARD, E. FLANIGEN, H. EGGERT, Eur. Patent 0,226,219, 1987.

140 T. WANG, G. YANG, S. FENG, C. SHANG, R. XU, *J. Chem. Soc., Chem. Commun.* **1989**, 948–949.

141 T. LOISEAU, G. FÉREY, *Eur. J. Solid State Inorg. Chem.* **1994**, 31, 575–581.

142 S. FENG, X. XU, G. YANG, R. XU, F. P. GLASSER, *J. Chem. Soc., Dalton Trans.* **1995**, 2147–2149.

143 S. M. STALDER, A. P. WILKINSON, *Chem. Mater.* **1997**, 9, 2168–2173.

144 P. REINERT, B. MARLER, J. PATARIN, *Microporous and Mesoporous Mater.* **2000**, 39, 509–517.

145 C. BROUCA-CABARRECQ, A. MOSSET, *J. Mater. Chem.* **2000**, 10, 445–450.

146 P. REINERT, J. PATARIN, B. MARLER, *Eur. J. Solid State Inorg. Chem.* **1998**, 35, 389–403.

147 A. M. CHIPPINDALE, K. J. PEACOCK, A. R. COWLEY, *J. Solid State Chem.* **1999**, 145, 379–386.

148 M. P. ATTFIELD, R. E. MORRIS, E. GUTIERREZ-PUEBLA, A. MONGE BRAVO, A. K. CHEETHAM, *J. Chem. Soc., Chem. Commun.* **1995**, 843–844.

149 K. H. LII, *Inorg. Chem.* **1996**, 35, 7440–7442.

150 K. H. LII, Y. F. HUANG, *Chem. Commun.* **1997**, 839–840.

151 R. I. WALTON, F. MILLANGE, T. LOISEAU, D. O'HARE, G. FÉREY, *Angew. Chem. Int. Ed.* **2000**, 39, 4552–4555.

152 M. ESTERMANN, L. B. MCCUSKER, CH. BAERLOCHER, A. MERROUCHE, H. KESSLER, *Nature* **1991**, 352, 320–323.

153 A. MERROUCHE, J. PATARIN, H. KESSLER, M. SOULARD, L. DELMOTTE, J. L. GUTH, J. F. JOLY, *Zeolites* **1992**, 12, 226–232.

154 K. J. JR. BALKUS, S. J. KIM, A. M. SARGESON, in: Advanced Catalytic Materials-1996. P. W. LEDNOR, M. J. LEDOUX, D. A. NAGAKI, L. T. THOMPSON (Eds), *Materials Research Society Symposium Proceedings*, Vol. 454, Materials Research Society, Pittsburgh, 1997, p. 217–224.

155 M. M. MERTENS, C. SCHOTT-DARIE, P. REINERT, J. L. GUTH, *Microporous Mater.* **1995**, 5, 91–96.

156 P. REINERT, C. SCHOTT-DARIE, J. PATARIN, *Microporous Mater.* **1997**, 9, 107–115.

157 S. J. WEIGEL, S. C. WESTON, A. K. CHEETHAM, G. D. STUCKY, *Chem. Mater.* **1997**, 9, 1293–1295.

158 S. J. WEIGEL, R. E. MORRIS, G. D. STUCKY, A. K. CHEETHAM, *J. Mater. Chem.* **1998**, 8, 1607–1611.

159 S. J. WEIGEL, T. LOISEAU, G. FÉREY, V. MUNCH, F. TAULELLE, R. E. MORRIS, G. D. STUCKY, A. K. CHEETHAM, in: Proceedings of the 12th International Zeolite Conference. M. M. J. TREACY, B. K. BALKUS, M. E. BISHER, J. B. HIGGINS (Eds), Material Research Society, 1999, Vol. IV, p. 2453–2456.

160 T. LOISEAU, C. PAULET, N. SIMON, V. MUNCH, F. TAULELLE, G. FÉREY, *Chem. Mater.* **2000**, *12*, 1393–1399.

161 C. SCHOTT-DARIE, H. KESSLER, M. SOULARD, V. GRAMLICH, E. BENAZZI, in: Zeolites and Related Microporous Materials: State of the Art 1994. J. WEITKAMP, H. K. KARGE, H. PFEIFER, W. HÖLDERICH (Eds), *Studies Surface Science and Catalysis*, Vol. 84A, Elsevier, Amsterdam, 1994, p. 101–108.

162 T. LOISEAU, G. FÉREY, *J. Chem. Soc., Chem. Commun.* **1992**, 1197–1198.

163 T. LOISEAU, G. FÉREY, *J. Eur. J. Solid State Inorg. Chem.* **1993**, *30*, 369–381.

164 T. LOISEAU, Ph.D Université du Mans, France, 1994.

165 T. LOISEAU, R. RETOUX, P. LACORRE, G. FÉREY, *J. Solid State Chem.* **1994**, *111*, 427–436.

166 M. CAVELLEC, D. RIOU, G. FÉREY, *Eur. J. Solid State Inorg. Chem.* **1994**, *31*, 583–594.

167 T. LOISEAU, G. FÉREY, *J. Solid State Chem.* **1994**, *111*, 403–415.

168 T. LOISEAU, G. FÉREY, *J. Mater. Chem.* **1996**, *6*, 1073–1074.

169 F. SERPAGGI, T. LOISEAU, F. TAULELLE, G. FÉREY, *Microporous and Mesoporous Mater.* **1998**, *20*, 197–206.

170 T. LOISEAU, G. FÉREY, *Microporous and Mesoporous Mater.* **2000**, *35*, 606–616.

171 C. SASSOYE, T. LOISEAU, F. TAULELLE, G. FÉREY, *J. Chem. Soc., Chem. Commun.*, **2000**, 943–944.

172 A. MEDEN, R. W. GROSSE-KUNSTLEVE, CH. BAERLOCHER, L. B. MCCUSKER, *Zeitschrift für Kristallographie* **1997**, *212*, 801–807.

173 P. REINERT, B. MARLER, J. PATARIN, *J. Chem. Soc., Chem. Commun.* **1998**, 1769–1770.

174 T. WESSELS, L. B. MCCUSKER, CH. BAERLOCHER, P. REINERT, J. PATARIN, *Microporous and Mesoporous Mater.* **1998**, *23*, 67–77.

175 D. S. WRAGG, G. B. HIX, R. E. MORRIS, *J. Am. Chem. Soc.* **1998**, *120*, 6822–6823.

176 A. MATIJASIC, J. L. PAILLAUD, J. PATARIN, *J. Mater. Chem.* **2000**, *10*, 1345–1351.

177 A. MATIJASIC, V. GRAMLICH, J. PATARIN, *Solid State Science* **2000**, *accepted*.

178 A. KUPERMAN, S. NADIMI, S. OLIVER, G. A. OZIN, J. M. GARCES, M. M. OLKEN, *Nature* **1993**, *365*, 239– 241.

179 P. CAULLET, L. L. GUTH, J. HAZM, J. M. LAMBLIN, H. GIES, *Eur. J. Solid State Inorg. Chem.* **1991**, *28*, 345–361.

180 L. A. VILLAESCUSA, P. A. BARRETT, M. A. CAMBLOR, *Angew. Chem. Int. Ed.* **1999**, *38,13/14*, 1997–2000.

181 C. SCHOTT-DARIE, J. PATARIN, P. Y. LE GOFF, H. KESSLER, E. BENAZZI, *Microporous and Mesoporous Mater.* **1994**, *3*, 123–132.

182 S. KALLUS, J. PATARIN, B. MARLER, *Microporous Mater.* **1996**, *7*, 89–95.

183 P. REINERT, J. PATARIN, T. LOISEAU, G. FÉREY, H. KESSLER, *Microporous and Mesoporous Mater.* **1998**, *22*, 43–55.

184 F. TAULELLE, A. SAMOSON, T. LOISEAU, G. FÉREY, *J. Phys. Chem. B* **1998**, *102*, 8588–8598.

185 Q. HUO, R. XU, *J. Chem. Soc., Chem. Commun.* **1992**, 1391–1392.

186 F. P. GLASSER, R. A. HOWIE, Q. KAN, *Acta Crystallogr.* **1994**, *C50*, 848–850.

187 A. M. CHIPPINDALE, R. I. WALTON, C. TURNER, *J. Chem. Soc., Chem. Commun.* **1995**, 1261–1262.

188 A. M. CHIPPINDALE, A. D. LAW, *J. Solid State Chem.* **1999**, *142*, 236–240.

189 D. S. WRAGG, I. BULL, G. B. HIX, R. E. MORRIS, *J. Chem. Soc., Chem. Commun.* **1999**, 2037–2038.

190 A. M. CHIPPINDALE, R. J. WALTON, *J. Chem. Soc., Chem. Commun.* **1994**, 2453–2454.

191 A. D. BOND, A. M. CHIPPINDALE, A. R. COWLEY, J. E. READMAN, A. V. POWELL, *Zeolites* **1997**, *19*, 326–333.

192 A. M. CHIPPINDALE, A. R. COWLEY, *Microporous and Mesoporous Mater.* **1998**, *21*, 271–279.

193 A. M. CHIPPINDALE, A. R. COWLEY, R. I. WALTON, *J. Mater. Chem.* **1996**, *6*, 611–614.

194 A. R. OVERWEG, J. W. DE HAAN, P. C. M. M. MAGUSIN, R. A. VAN SANTEN, G. SANKAR, J. M. THOMAS, *Chem. Mater.* **1999**, *11*, 1680–1686.

195 A. R. COWLEY, A. M. CHIPPINDALE, *J. Chem. Soc., Chem. Commun.* **1996**, 673–674.

196 A. M. CHIPPINDALE, A. R. COWLEY, K. J. PEACOCK, *Microporous and Mesoporous Mater.* **1998**, *24*, 133–141.

197 A. M. CHIPPINDALE, A. R. COWLEY, *Zeolites* **1997**, *18*, 176–181.

198 A. R. COWLEY, A. M. CHIPPINDALE, *Microporous and Mesoporous Mater.* **1999**, *28*, 163–172.

199 P. FENG, X. BU, G. D. STUCKY, *Nature* **1997**, *388*, 735–741.

200 A. M. CHIPPINDALE, A. D. BOND, A. R. COWLEY, A. V. POWELL, *Chem. Mater.* **1997**, *9*, 2830–2835.

201 K. F. HSU, S. L. WANG, *Chem. Commun.* **2000**, 135–136.

202 K. F. HSU, S. L. WANG, *Inorg. Chem.* **2000**, *39*, 1773–1778.

203 G. MÜLLER, G. EDER-MIRTH, J. A. LERCHER, in: Zeolites: A Refined Tool For Designing Catalytic Sites. L. BONNEVIOT, S. KALLIAGUINE (Eds), Elsevier, Amsterdam, 1995, p. 71–77.

204 W. SCHMIDT, F. SCHÜTH, S. KALLUS, in: Progress in Zeolite and Microporous Materials. H. CHON, S.-K. IHM, Y. S. UH (Eds), *Studies in Surface Science and Catalysis*, Vol. 105A, Elsevier, Amsterdam, 1997, p. 771–778.

205 W. T. A. HARRISON, T. E. GIER, K. L. MORAN, J. M. NICOL, H. ECKERT, G. D. STUCKY, *Chem. Mater.* **1991**, *3*, 27–29.

206 T. M. NENOFF, W. T. A. HARRISON, T. E. GIER, G. D. STUCKY, *J. Am. Chem. Soc.* **1991**, *113*, 378–379.

207 A. CHOUDHURY, S. NATARAJAN, C. N. R. RAO, *Inorg. Chem.* **2000**, *39*, 4295–4304.

208 S. NEERAJ, S. NATARAJAN, *Chem. Mater.* **2000**, *12*, 2753–2762.

209 R. VAIDHYANATHAN, S. NATARAJAN, C. N. R. RAO, *J. Mater. Chem.* **1999**, *9*, 2789–2793.

210 W. T. A. HARRISON, T. E. MARTIN, T. E. GIER, G. D. STUCKY, *J. Mater. Chem.* **1992**, *2(2)*, 175–181.

211 W. T. A. HARRISON, T. M. NENOFF, M. M. EDDY, T. E. MARTIN, G. D. STUCKY, *J. Mater. Chem.* **1992**, *2(11)*, 1127–1134.

212 K. AHMADI, A. HARDY, J. PATARIN, L. HUVE, *Eur. J. Solid State Inorg. Chem.* **1995**, *32*, 209–223.

213 S. NATARAJAN, S. NEERAJ, C. N. R. RAO, *Solid State Sciences* **2000**, *2*, 87–98.

214 K. O. KONGSHAUG, H. FJELLVAG, K. P. LILLERUD, *Microporous and Mesoporous Mater.* **2000**, *39*, 341–350.

215 K. O. KONGSHAUG, H. FJELLVAG, K. P. LILLERUD, *J. Mat. Chem.* **1999**, *9*, 3119–3123.

216 C. N. R. RAO, S. NATARAJAN, S. NEERAJ, *J. Am. Chem. Soc.* **2000**, *122*, 2810–2817.

217 C. N. R. RAO, S. NATARAJAN, S. NEERAJ, *J. Solid State Chem.* **2000**, *152*, 302–321.

218 R. W. BROACH, R. L. BEDARD, S. G. SONG, J. J. PLUTH, A. BRAM, C. RIEKEL, H. P. WEBER, *Chem. Mater.* **1999**, *11*, 2076–2080.

219 N. RAJIC, N. Z. LOGAR, V. KAUCIC, *Zeolites* **1995**, *15*, 672–678.

220 W. T. A. HARRISON, T. E. GIER, G. D. STUCKY, R. W. BROACH, R. A. BEDARD, *Chem. Mater.* **1996**, *8*, 145–151.

221 T. R. JENSEN, *J. Chem. Soc., Dalton Trans.* **1998**, 2261–2266.

222 X. BU, P. FENG, T. E. GIER, G. D. STUCKY, *Zeolites* **1997**, *19*, 200–208.

223 Y. N. HUNT, W. T. A. HARRISON, *Microporous and Mesoporous Mater.* **1998**, *23*, 197–202.

224 G. Y. YANG, S. C. SEVOV, *J. Am. Chem. Soc.* **1999**, *121*, 8389–8390.

225 S. NEERAJ, S. NATARAJAN, C. N. R. RAO, *New J. Chem.* **1999**, 303–308.

226 S. NEERAJ, S. NATARAJAN, C. N. R. RAO, *Chem. Commun.* **1999**, 165–166.

227 R. H. JONES, J. CHEN, G. SANKAR, J. M. THOMAS, in: Zeolites and Related Microporous Materials: State of the Art 1994. J. WEITKAMP, H. G. KARGE, H. PFEIFER, W. HÖLDERICH (Eds), *Studies in Surface Science and Catalysis*, Vol. 84C, Elsevier, Amsterdam, 1994, p. 2229–2236.

228 D. CHIDAMBARAM, S. NATARAJAN, *Materials Research Bulletin* **1998**, *33*, 1275–1281.

229 S. B. HARMON, S. C. SEVOV, *Chem. Mater.* **1998**, *10*, 3020–3023.

230 W. T. A. HARRISON, M. L. F. PHILIPS, *Chem. Mater.* **1997**, *9*, 1837–1846.

231 W. T. A. HARRISON, M. L. F. PHILIPS, *Chem. Commun.* **1996**, 2771–2772.

232 M. WALLAU, J. PATARIN, I. WIDMER, P. CAULLET, J. L. GUTH, L. HUVE, *Zeolites* **1994**, *14*, 402–410.

233 W. T. A. HARRISON, L. HANNOOMAN, *Angew. Chem. Int. Ed.* **1997**, *36*, 640–641.

234 X. BU, P. FENG, G. D. STUCKY, *J. Solid State Chem.* **1996**, *125*, 243–248.

235 T. SONG, M. B. HURSTHOUSE, J. CHEN, J. XU, K. M. A. MALIK, R. H. JONES, R. XU, J. M. THOMAS, *Adv. Mater.* **1994**, *6*, *No. 9*, 679–680.

236 A. WHITAKER, *Acta Cryst.* **1975**, *B31*, 2026–2035.

237 W. T. A. HARRISON, R. W. BROACH, R. A. BEDARD, T. E. GIER, X. BU, G. D. STUCKY, *Chem. Mater.* **1996**, *8*, 691–700.

238 W. T. A. HARRISON, T. E. GIER, G. D. STUCKY, *Acta Crystallogr.* **1994**, *C50*, 471–473.

239 T. E. GIER, W. T. A. HARRISON, G. D. STUCKY, *Angew. Chem. Int. Ed.* **1991**, *30*, 1169–1171.

240 W. T. A. HARRISON, T. E. GIER, G. D. STUCKY, *Zeolites* **1993**, *13*, 242–248.

241 X. BU, T. E. GIER, G. D. STUCKY, *Microporous and Mesoporous Mater.* **1998**, *26*, 61–66.

242 V. G. NITSCH, H. SCHAEFER, *Z. Anorg. Allg. Chem.* **1975**, *417*, 11–18.

243 W. T. A. HARRISON, *Int. J. Inorg. Mater.* **2001**, *3*, 17–22.

244 S. OLIVER, A. KUPERMAN, G. A. OZIN, *Angew. Chem. Int. Ed.* **1998**, *37*, 46–62.

245 S. OLIVER, A. KUPERMAN, A. LOUGH, G. A. OZIN, J. M. GARCÈS, M. M. OLKEN, P. RUDOFF, in: Zeolites and Related Microporous Materials: State of the Art 1994. J. WEITKAMP, H. G. KARGE, H. PFEIFER, W. HÖLDERICH (Eds), *Studies in Surface Science and Catalysis*, Vol. 84A, Elsevier, Amsterdam, 1994, p. 219–225.

246 S. OLIVER, A. KUPERMAN, A. LOUGH, G. A. OZIN, *Chem. Mater.* **1996**, *8*, 2391–2398.

247 L. VIDAL, C. MARICHAL, V. GRAMLICH, J. PATARIN, Z. GABELICA, *Chem. Mater.* **1999**, *11–10*, 2728–2796.

248 G. FÉREY, *J. Fluorine Chem.* **1995**, *72*, 187–193.

249 F. TAULELLE, M. HAOUAS, C. GÉRARDIN, C. ESTOURNES, T. LOISEAU, G. FÉREY, *Colloids and Surf. A: Physicochem. Eng. Aspects* **1999**, *158*, 299–311.

250 C. IN-GÉRARDIN, M. IN, F. TAULELLE, *J. Chim. Phys.* **1995**, *92*, 1877–1880.

251 F. TAULELLE, M. PRUSKI, J. P. AMOUREUX, D. LANG, A. BAILLY, C. HUGUENARD, M. HAOUAS, C. GÉRARDIN, T. LOISEAU, G. FÉREY, *J. Am. Chem. Soc.* **1999**, *121*, 12148–12153.

252 S. NEERAJ, S. NATARAJAN, C. N. R. RAO, *Angew. Chem. Int. Ed.* **1999**, *38*, 3480–3482.

253 A. CHOUDHURY, S. NATARAJAN, C. N. R. RAO, *Chem. Commun.* **1999**, 1305–1306.

254 A. CHOUDHURY, S. NATARAJAN, *Int. J. Solid State Chem.* **2000**, *154*, 507–513.

255 A. CHOUDHURY, S. NATARAJAN, *Int. J. Inorg. Mater.* **2000**, *2*, 217–223.

256 J. R. D. DEBORD, W. M. REIFF, C. J. WARREN, R. C. HAUSHALTER, J. ZUBETIA, *Chem. Mater.* **1997**, *9*, 1994–1998.

257 N. RAJIC, R. GABROVSEK, V. KAUCIC, *Thermochimica Acta* **2000**, *359*, 119–122.

258 V. ZIMA, K. H. LII, *J. Solid State Chem.* **1998**, *139*, 326–331.

259 Z. A. D. LETHBRIDGE, P. LIGHTFOOT, R. E. MORRIS, D. S. WRAGG, P. A. WRIGHT, A. KVICK, G. VAUGHAN, *J. Solid State Chem.* **1999**, *142*, 455–460.

260 S. NATARAJAN, S. NEERAJ, C. N. R. RAO, *Solid State Sciences* **2000**, *2*, 87–98.

261 S. NATARAJAN, S. NEERAJ, A. CHOUDHURY, C. N. R. RAO, *Inorg. Chem.* **2000**, *39*, 1426–1433.

262 A. CHOUDHURY, S. NEERAJ, S. NATARAJAN, C. N. R. RAO, *Angew. Chem. Int. Ed.* **2000**, *39*, 3091–3093.

263 H.-M. YUAN, J.-S. CHEN, G.-S. ZHU, J.-Y. LI, J.-H. YU, G.-D. YANG, R.-R. XU, *Inorg. Chem.* **2000**, *39*, 1476–1479.

264 R.-K. CHIANG, C.-C. HUANG, C.-R. LIN, *J. Solid State Chem.* **2001**, *156*, 242–246.

265 J.-H. YU, Q. GAO, J.-S. CHEN, R.-R. XU, in: Progress in Zeolite and Microporous Materials. H. CHON, S.-K. IHM, Y. S. UH (Eds), *Studies in Surface Science and Catalysis*, Vol.

105A, Elsevier, Amsterdam, 1997, p. 381–388.

266 K. Maeda, F. Mizukami, *Catal. Surv. from Japan* **1999**, *3*, 119–126.

267 S. L. Burkett, S. D. Sims, S. Mann, *Chem. Commun.* **1996**, 1367–1368.

268 D. J. Macquarrie, *Chem. Commun.* **1996**, 1961–1962.

269 D. J. Macquarrie, D. B. Jackson, *Chem. Commun.* **1997**, 1781–1782.

270 C. E. Fowler, S. L. Burkett, S. Mann, *Chem. Commun.* **1997**, 1769–1770.

271 W. M. Van Rhijn, D. E. De Vos, B. F. Sels, W. D. Bossaert, P. A. Jacobs, *Chem. Commun.* **1998**, 317–318.

272 K. A. Koyano, T. Tatsumi, Y. Tanaka, S. Nakata, *J. Phys. Chem.* **1997**, *B 101*, 9436–9440.

273 T. Tatsumi, K. A. Koyano, N. Igarashi, *Chem. Commun.* **1998**, 325–326.

274 A. Cauvel, G. Renard, D. Brunel, *J. Org. Chem.* **1997**, *62*, 749–751.

275 X. Feng, G. E. Fryxell, L.-Q. Wang, A. Y. Kim, J. Liu, K. M. Kemner, *Science* **1997**, *276*, 923–926.

276 L. Mercier, T. J. Pinnavaia, *Adv. Mater.* **1997**, *9*, 500–503.

277 J. F. Diaz, K. J. Balkus Jr., F. Bedioui, V. Kurshev, L. Kevan, *Chem. Mater.* **1997**, *9*, 61–67.

278 S. O'Brien, J. Tudor, S. Barlow, M. J. Drewitt, S. Heyes, *Chem. Comm.* **1997**, 641–642.

279 A. Cauvel, D. Brunel, F. Di Renzo, P. Moreau, F. Fajula, in: Catalysis by Microporous Materials, Studies in Surface Science and Catalysis. H. K. Beyer, H. G. Karge, I. Kiricsi and J. B. Nagy (Eds), Vol. 94, Elsevier, Amsterdam, 1995, p. 286–293.

280 D. C. Calabro, US Patent 5,194,410, 1993.

281 H.-X.-Li, M. A. Camblor, M. E. Davis, *Microporous Mater.* **1994**, *3*, 117–121.

282 C. W. Jones, K. Tsuji, M. E. Davis, *Nature* **1998**, *393*, 52–54.

283 B. Adachi, J. Corker, H. Kessler, F. Lefebvre, J. M. Basset, *Microporous and Mesoporous Mater.* **1998**, *21*, 81–90.

284 A. Clearfield, in: Progress in Inorganic Chemistry, K. D. Karlin (Ed.), Vol. 47, Wiley, New-York, 1998, p. 371–384.

285 G. Cao, H. G. Hong, T. E. Mallouk, *Acc. Chem. Res.* **1992**, *25*, 420–426.

286 Y. P. Zhang, A. Clearfield, *Inorg. Chem.* **1992**, *31*, 2821–2826.

287 K. Maeda, J. Akimoto, Y. Kiyozumi, F. Mizukami, *Angew. Chem. Int. Ed. Engl.* **1995**, *34*, 1199–1201.

288 K. Maeda, J. Akimoto, Y. Kiyozumi, F. Mizukami, *J. Chem. Soc., Chem. Commun.* **1995**, 1033–1034.

289 J. Le Bideau, C. Payen, P. Palvadeau, B. Bujoli, *Inorg. Chem.* **1994**, *33*, 4885–4890.

290 S. Drumel, P. Janvier, D. Deniaud, B. Bujoli, *J. Chem. Soc., Chem. Commun.* **1995**, 1051–1052.

291 R. LaDucam, D. Rose, J. R. D. Debord, R. C. Haushalter, C. J. O'Cormor, J. Zubieta, *J. Solid State Chem.* **1996**, *123*, 408–412.

292 A. Distler, S. C. Sevov, *Chem. Commun.* **1998**, 959–960.

293 D. L. Lohse, S. C. Sevov, *Angew. Chem. Int. Ed. Engl.* **1997**, *36*, 1619–1621.

294 A. Cabeza, M. A. G. Aranda, S. Bruque, *J. Mater. Chem.* **1998**, *8*, 2479–2485.

295 B. A. Adair, G. D. de Delgado, J. M. Delgado, A. K. Cheetham, *Solid State Sciences* **2000**, *2*, 119–126.

296 D. Riou, O. Roubeau, G. Férey, *Microporous and Mesoporous Mater.* **1998**, *23*, 23–31.

297 D. Riou, P. Baltazar, G. Férey, *Solid State Sciences* **2000**, *2*, 127–134.

298 D. Riou, C. Serre, G. Férey, *J. Solid State Chem.* **1998**, *141*, 89–93.

299 D. Riou, G. Férey, *J. Mater. Chem.* **1998**, *8*, 2733–2735.

300 F. Serpaggi, G. Férey, *J. Mater. Chem.* **1998**, *8*, 2749–2755.

301 C. Paulet, C. Serre, T. Loiseau, D. Riou, G. Férey, *C. R. Acad. Sci. Paris* **1999**, *t.2 Série IIC*, 631–636.

302 M. Riou-Cavellec, M. Sanselme, G. Férey, *J. Mater. Chem.* **2000**, *10*, 745–748.

303 D. Riou, C. Serre, J. Provost, G. Férey, *J. Solid State Chem.* **2000**, *155*, 238–242.

304 H. S. Martinez-Tapia, A. Cabeza, S. Bruque, P. Pertierra, S. Garcia-Granda, M. A. G. Aranda, *J. Solid State Chem.* **2000**, *151*, 122–129.

305 M. E. LEONOWICZ, J. W. JOHNSON, J. F. BRODY, H. F. SHANNON, J. M. NEWSAM, *J. Solid State Chem.* **1985**, *56*, 370–376.

306 J. S. BECK, J. C. VARTULI, W. J. ROTH, M. E. LEONOWICZ, C. T. KRESGE, K. D. SCHMITT, C. T. W. CHU, D. H. OLSON, E. W. SHEPPARD, S. B. MCCULLEN, J. B. HIGGINS, J. L. SCHLENKER, *J. Am. Chem. Soc.* **1992**, *114*, 10834–10843.

307 B. T. HOLLAND, P. K. ISBESTER, C. F. BLANFORD, E. J. MUNSON, A. STEIN, *Mater. Res. Bull.* **1999**, *34(3)*, 471–482.

308 B. T. HOLLAND, P. K. ISBESTER, C. F. BLANFORD, E. J. MUNSON, A. STEIN, *J. Am. Chem. Soc.* **1997**, *119*, 6796–6803.

309 T. KIMURA, Y. SUGAHARA, K. KURODA, *Microporous and Mesoporous Mater.* **1998**, *22*, 115–126.

310 T. KIMURA, Y. SUGAHARA, K. KURODA, *Chem. Letters* **1997**, 983–984.

311 T. KIMURA, Y. SUGAHARA, K. KURODA, *Chem. Commun.* **1998**, 559–560.

312 S. CABRERA, J. EL HASKOURI, C. GUILLEM, A. BELTRAN-PORTER, D. BELTRAN-PORTER, S. MENDIOROZ, M. D. MARCOS, P. AMOROS, *Chem. Commun.* **1999**, 333–334.

313 J. O. PEREZ O., R. B. BORADE, A. CLEARFIELD, *J. Mol. Struct.* **1998**, *470*, 221–228.

314 D. ZHAO, Z. LUAN, L. KEVAN, *Chem. Commun.* **1997**, 1009–1010.

315 D. ZHAO, Z. LUAN, L. KEVAN, *J. Phys. Chem. B* **1997**, *101*, 6943–6948.

316 B. CHAKRABORTY, A. C. PULIKOTTIL, S. DAS, B. VISWANATHAN, *Chem. Commun.* **1997**, 911–912.

317 D. J. JONES, G. APTEL, M. BRANDHORST, M. JACQUIN, J. JIMENEZ-JIMENEZ, A. JIMENEZ-LOPEZ, P. MAIRELES-TORRES, I. PIWONSKI, E. RODRIGUEZ-CASTELLON, J. ZAJAC, J. ROZIÈRES, *J. Mater. Chem.* **2000**, *10*, 1957–1963.

318 M. THIEME, F. SCHÜTH, *Microporous and Mesoporous Mater.* **1999**, *27*, 193–200.

319 J. BLANCHARD, F. SCHÜTH, P. TRENS, M. HUDSON, *Microporous and Mesoporous Mater.* **2000**, *39*, 163–170.

320 T. ABE, A. TAGUCHI, M. IWAMOTO, *Chem. Mater.* **1995**, *77*, 1429–1431.

321 T. DOI, T. MIYAKE, *Chem. Commun.* **1996**, 1635–1636.

322 J. EL HASKOURI, M. ROCA, S. CABRERA, J. ALAMO, A. BELTRAN-PORTER, D. BELTRAN-PORTER, M. D. MARCOS, P. AMOROS, *Chem. Mater.* **1999**, *11*, 1446–1454.

4.2.4
Synthesis of Titanosilicates and Related Materials

Michael W. Anderson and João Rocha

4.2.4.1
Introduction

Zeolites and aluminophosphates consist of tetrahedrally coordinated framework atoms. However, in the field of microporous materials there is now a large class of materials that consist of structures of interlinked octahedra and tetrahedra. The archetypal material is the titanosilicate, however, many other elements can be incorporated into such frameworks. Although clays and related materials, in some way, fall into this category, they only possess two-dimensional covalent connectivity. Three-dimensional connectivity is required to impart substantial structural stability and we are only concerned with such networks in this section. Many of the structures of these materials can be seen in a recent microreview [1].

4.2.4.2
Synthesis

The synthesis of microporous titanosilicates [2–5] and other novel zeotype materials is usually carried out in Teflon-lined autoclaves under hydrothermal conditions with temperatures ranging from approximately 150 to 230 °C and times varying between a few hours and approximately 30 days. In certain cases, the materials have been prepared at relatively high temperatures. For example, single crystals of a titanosilicate analog of the mineral pharmacosiderite, $Cs_3HTi_4O_4(SiO_4)_3 \cdot 4H_2O$, have been obtained hydrothermally on a sealed gold tube at 750 °C for 40 h [6]. The pH of the synthesis gel is normally high, in the range 10–13. Seeding the gel with a small amount of the desired phase (or even with a related solid) is a common practice. The syntheses are, in general, driven by kinetics, with different metastable phases being formed with time. So far, most research has concentrated on the synthesis of microporous titanosilicates. The experience gathered in this field is guiding the efforts aimed at other systems, most notably zirconosilicates. Hence, here we shall discuss in more detail the synthesis of microporous titanosilicates, particularly ETS-10 (Engelhard titanosilicate material number 10).

The synthesis of ETS-10 was first reported by Kuznicki in 1989 [3]. The titanium source used was $TiCl_3$. A slight modification of this method afforded highly crystalline and pure ETS-10 suitable for structural elucidation [7, 8]. $TiCl_4$ is another much used (and cheaper) titanium source. For example, using this precursor Das et al. [9] reported a method for the rapid (<16 h) synthesis of ETS-10. Liu and Thomas [10] have shown that under proper conditions the thermodynamically stable TiO_2 phases of anatase and rutile can be used as suitable titanium sources for the synthesis of porous titanosilicates. A few organotitanium compounds have also been used to prepare microporous titanosilicates. The synthesis of ETS-10 has been carried out at 200 °C using a 2^3 factorial method to optimize the overall composition of reaction mixture to produce pure material within a short (18 h) time. Kinetic studies on ETS-10 and ETS-4 have been performed using the optimum compositions and the apparent activation energies were calculated as 66.78 and 14.47 kJ mol^{-1}, respectively [11]. Because they are very sensitive to moisture, these compounds are difficult to handle and require special facilities for their transfer during the gel preparation. For example, Chapman and Roe [2] used $Ti(OC_2H_5)_4$ for preparing GTS-1 while Clearfield and coworkers [12] synthesized a novel porous titanosilicate using titanium isopropoxide. Sodium silicate solutions, fumed and colloidal silica are adequate silicon sources. Organosilicon compounds, such as tetraethylorthosilicate have also been used [12]. The synthesis of ETS-10 usually (although not always) requires the presence in the parent gel of both Na^+ and K^+ ions. Other phases, such as some of the AM (Aveiro/Manchester) materials [13] are produced pure only when a single type of cation is present. Potassium fluoride is often used for preparing ETS-10 but its presence is not crucial. A comprehensive study of the hydrothermal synthesis conditions that yield pure and highly crystal-

References see page 899

line ETS-10 from $TiCl_3$ and anatase has been reported [14]. This work examines the influence on the synthesis of the following parameters: presence of fluoride, sodium and potassium ions, seeds, H_2O/Si and Si/Ti molar ratios, parent gel pH, temperature and time. A similar study has been carried out by Das et al. [15], using $TiCl_3$ as the titanium source.

Most porous titanosilicates and other related new materials, to date, can be synthesized without the addition of any organic template molecules. However, several groups have prepared ETS-10 with a range of templates: tetramethylammonium chloride, [16, 17] pyrrolidine, tetraethylammonium chloride, tetrapropylammonium bromide,1,2-diaminoethane, [17] choline chloride and the bromide salt of hexaethyl diquat-5 [18].

The framework substitution of titanium and silicon by other elements requires a judicious choice of the respective source that is usually introduced in the parent synthesis gel. The following sources of aluminum, gallium, niobium and boron for element insertion in the ETS-10 framework have been reported: $NaAlO_2$, [19] $GaCl_3$, [20] $Nb(HC_2O_4)_5$ [21] and $Na_2B_2O_4$ [22] (see also [23, 24]). In the synthesis of microporous zirconosilicates $ZrCl_4$ [25–27] and $Zr(OC_3H_7)_4$ [28] have been used as the zirconium sources. The vanadosilicate AM-6, a structural analog of ETS-10, has been prepared from $VOSO_4 \cdot 5H_2O$ [29].

4.2.4.3
ETS-10

In the early 1990s Engelhard Corporation filed a series of patents describing a small-pore material, which they named ETS-4 (Engelhard titanosilicate-4), and another wide-pore titanosilicate material named ETS-10 [3, 4]. At about the same time Chapman and Roe [2, 5] published the synthesis and X-ray diffraction pattern of a microporous titanosilicate material that apparently resembled the structure of the natural mineral zorite by comparison of the diffraction patterns. Owing to its wide-pore nature and thermal stability ETS-10 is arguably the most important octahedral/tetrahedral framework microporous titanosilicate to be synthesized to date. Consequently, it is important to understand the structure of this material in some detail.

The structure of ETS-10 was solved and reported briefly in 1994 by Anderson et al. [7] and described fully by the same group in the following year [8]. The typical crystal morphology of ETS-10 is shown in Fig. 1. These scanning electron micrographs show (1) the particle has a pseudo-4-fold symmetry along an axis indicated by the large arrow; (2) traces of faults that are perpendicular to the axis and are indicated by small arrows. This faulted structure is always present in ETS-10 and Fig. 1 represents a high-quality material.

The details of the inherent disorder and the pore structure in ETS-10 are most clearly revealed in the high-resolution electron micrograph (Fig. 2). The large 12-ring pores are clearly visible. In order to discern the disorder the reader should note the stacking of these large pores from the bottom of the micrograph to the top. It will be seen that the pores stack randomly to the right or to the left.

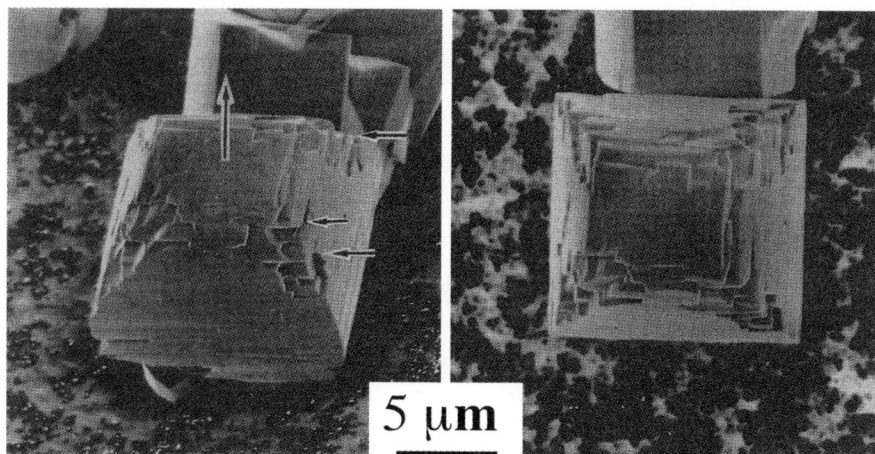

Fig. 1. Scanning electron micrographs of a high-quality ETS-10 sample. Large arrow depicts pseudo-4-fold axis, small arrows indicate faulting caused by intergrowth.

Fig. 2. High-resolution electron micrograph of ETS-10.

References see page 899

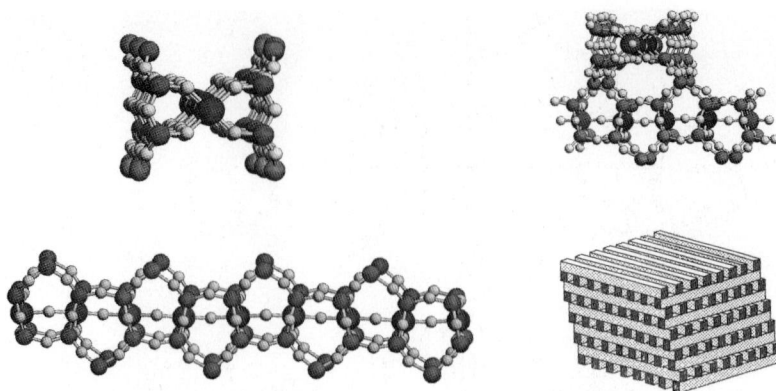

Fig. 3. Structural make-up of ETS-10: (a) and (b) show the same rod section of ETS-10 in orthogonal views – large spheres Ti, medium spheres Si and small spheres O; (c) shows how these rods lock; (d) shows one possible ordered arrangement of these orthogonal rods.

The most interesting aspect of the structure of ETS-10 is that it contains infinite –O–Ti–O–Ti–O– chains (with alternating long/short bonds [30]) which are surrounded by a silicate ring structure. These combine to make up a rod (Fig. 3a and 3b) and it is this rod nature that imparts some of the interesting optical properties described in the applications section. Adjacent layers of rods are stacked orthogonal to each other (Fig. 3c). These double layers are then stacked with a displacement of 1/4 unit cell in either the [100] direction or the [010] direction giving four possible connections (Fig. 3d). If these four possible connections are given equal probability and the stacking is random then the disordered ETS-10 structure is constructed. The pore structure of ETS-10 has 12-rings in all three dimensions; these are directly along [100] and [010] and crooked along the direction of disorder. However, there are only a handful of microporous zeolitic materials with a 3-dimensional 12-ring pore system and in this aspect ETS-10 has excellent diffusion characteristics. It should be noted that the crooked channel is not blocked nor the pore volume reduced by the disorder.

It is interesting to conjecture whether it is possible to construct ordered variants of the ETS-10 structure. The lattice energy difference between different stacking sequences is minimal, however, different sequences result in different pore architecture. A zigzag stacking of pores results in chiral symmetry and a spiral 12-ring pore in the third dimension. In order to synthesize such a material no doubt a chiral organic templating agent would be required but as yet this has not been achieved.

Mistakes in the stacking sequences (that occur frequently probably as a result of the crystal growth mechanism) result in line defects in ETS-10 that are in effect double-sized pores.

The framework formula of this basic unit is $Si_{40}Ti_8O_{104}^{16-}$, which is counter-

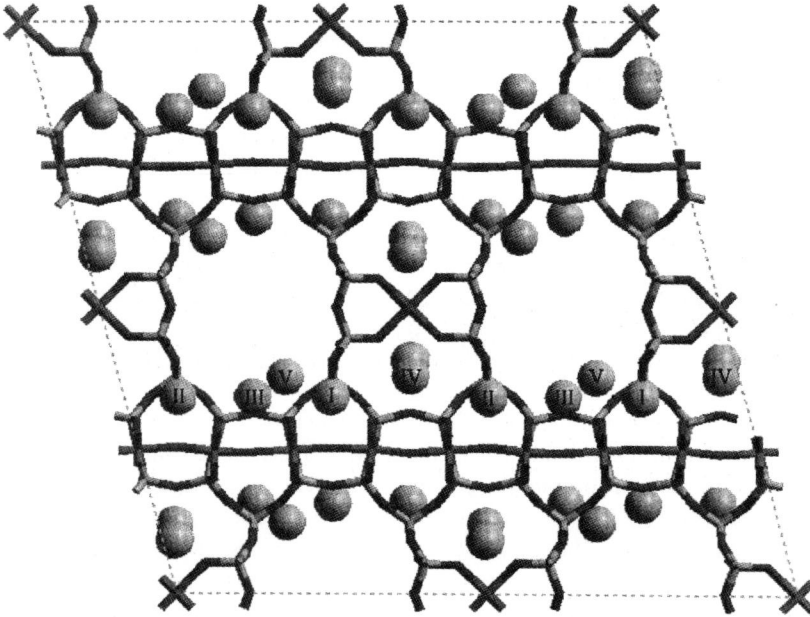

Fig. 4. Five cation sites in ETS-10 and their relationship to the framework.

balanced by 16 monovalent cations. In this original structural work it was impossible to locate the cations in ETS-10. The as-prepared material is $1/4$ K^+ and $3/4$ Na^+. The most likely location of the cations has been determined by a combination of molecular modeling and, for sodium, ^{23}Na solid-state nuclear magnetic resonance (NMR). Figure 4 shows the final structure of ETS-10 with the cations principally located in sites adjacent to the titanate chains where they balance the charge on the titanium centers.

Recently, a single-crystal study on ETS-10 has refined a superposition cell, which essentially regards the structure as a combination of all the ordered variants and confirms in essence all of the previous structural work [31].

4.2.4.4
ETS-4

Similar to ETS-10 the nominal structure of ETS-4 suggests that it could have very interesting applications. As noted above, the X-ray diffraction pattern of ETS-4 suggests strong similarities with the structure of the natural mineral zorite. In 1973 a naturally occurring alkaline titanosilicate identified as zorite was discovered [32] in trace quantities on the Siberian Tundra. The structure solution of this mineral was published [33] six years later in an article entitled "The OD structure of

References see page 899

Zorite". According to this contribution [33], the structure of this mineral is characterized by a highly disordered framework with ostensibly a 2-dimensional channel system. Two orthogonal sets of channels are defined by 12-T/O atom and 8-T atom rings (T = tetrahedral silicon; O = octahedral titanium). In reality, the disorder in zorite results in the larger 12-ring channels becoming partitioned into sections. A molecule diffusing into this 12-T/O ring channel must make detours through the 8-T ring channel system in order to pass freely. Consequently, the adsorption characteristics of zorite are far inferior to that expected for an unfaulted material. Furthermore, zorite lacks thermal stability as internal water acts as part of the structure-forming chains through the channel system. When the water is removed the framework structure collapses.

In zorite the titanium exists in two different chemical environments, titania rods, similar to ETS-10 and isolated titania semioctahedra that are exposed to the channel system. Consequently, if ETS-4 has the structure of zorite then the accessibility of the titanium centers to the channel system may be a benefit for certain applications over ETS-10, despite the stability issue. The only outstanding problem with the structure of ETS-4 is that the ^{29}Si MAS NMR spectrum does not agree with the crystal structure. The crystal structure dictates two Si environments Si(2Si, 2Ti) and Si(3Si, 1Ti) in a ratio of 2:1. The ^{29}Si NMR spectrum displays two signals but in the wrong ratio. It is still unclear at present how this discrepancy arises, however, but the most reasonable suggestion is that it is due to a large number of defect sites in ETS-4 [34, 35].

In terms of stability, the structure of zorite or ETS-4 contains structural water bound in chains along the channel system. At temperatures of 200 °C this water is lost and the structure is lost. At high temperatures, above 700 °C, there is a phase transformation into the dense phase titanosilicate structure narsarsukite [36]. Because the stoichiometry of ETS-4 and narsarsukite are different a quantity of silica must be added to ETS-4 in order to prepare pure narsarsukite at high temperatures. Controlled structural collapse of ETS-4 has recently been used to produce a molecular gate effect by closing down the pores in 0.1 Å increments for highly selective gas separation [37].

4.2.4.5
Isomorphous Framework Substitution

The insertion of hetero-atoms in the framework of microporous materials is an important process because it allows the fine tuning of their properties. Despite the fact that a considerable amount of work has been carried out on zeolites and other related systems, obtaining solid evidence for framework substitution is not a trivial task and it usually requires the combined use of several techniques. Detailed studies on framework insertion of aluminum, gallium, boron and niobium on ETS-10 have been reported, and these we now summarize.

Attempting to improve the acid characteristics of ETS-10 (and following up the early synthesis work by Kuznicki et al. [23, 24]) Rocha and Anderson and co-

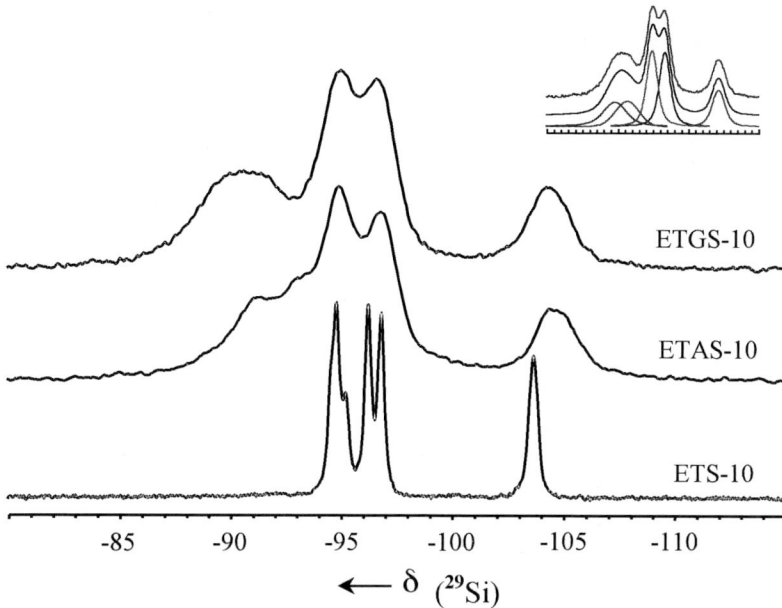

Fig. 5. ^{29}Si MAS/NMR spectra of ETS-10, ETAS-10 and ETGS-10. The inset shows a deconvolution of the ETGS-10 spectrum.

workers have incorporated aluminum [38, 39] and gallium [19, 9] into tetrahedral silicon sites, thus generating sites for zeolite-type acidity. These materials are known as ETAS-10 and ETGS-10, respectively. The optimized synthesis conditions for obtaining highly crystalline microporous ETAS-10 materials with different framework Al contents (Al/Ti molar ratio 0.1–0.48) have been reported [40]. The main evidence for the substitution of silicon by aluminum and gallium in the framework of ETS-10 is provided by solid state NMR. The ^{29}Si MAS NMR spectra of ETS-10, ETAS-10 and ETGS-10 are shown on Fig. 5. In ETS-10 there are two types of silicon chemical environments, Si(3Si, 1Ti) and Si(4Si, 0Ti), which give the two groups of resonances at $\delta = -94$ to -97 ppm and approximately -103.7 ppm, respectively. The ratio of these environments is 4:1. The spectrum reveals a further crystallographic splitting of the Si(3Si, 1Ti) site. The spectrum of ETAS-10 contains all the ETS-10 resonances plus two other peaks approximately 4 ppm downfield from the Si(3Si, 1Ti) signals that become stronger when the framework aluminum concentration increases (Fig. 6). These are ascribed to the framework incorporation of aluminum to produce Si(2Si, 1Al, 1Ti) environments. It is important to note that there is no signal approximately 4 ppm downfield from the Si(4Si, 0Ti) resonance, showing no aluminum substitution neighboring this silicon site and, hence, providing a direct proof of Al, Ti avoidance [19, 38].

References see page 899

Si(2Si,1Al,1Ti)

Si(3Si,1Ti)

Si(4Si)   Al/Ti

0.48

0.35

0.26

0.16

0.10

-85   -95   -105   -115

← δ (^{29}Si)

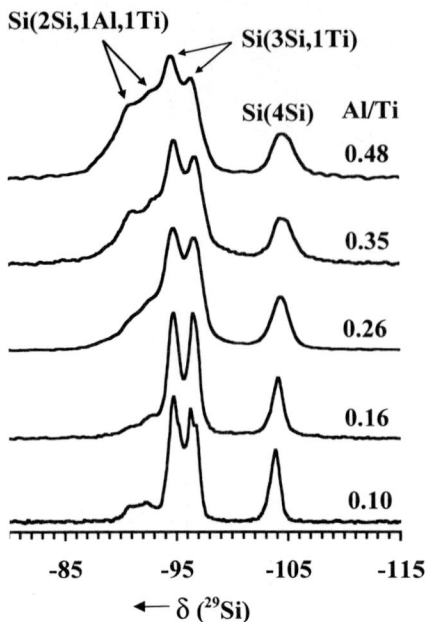

Fig. 6. ^{29}Si MAS/NMR spectra of ETAS-10 with the Al/Ti ratios depicted.

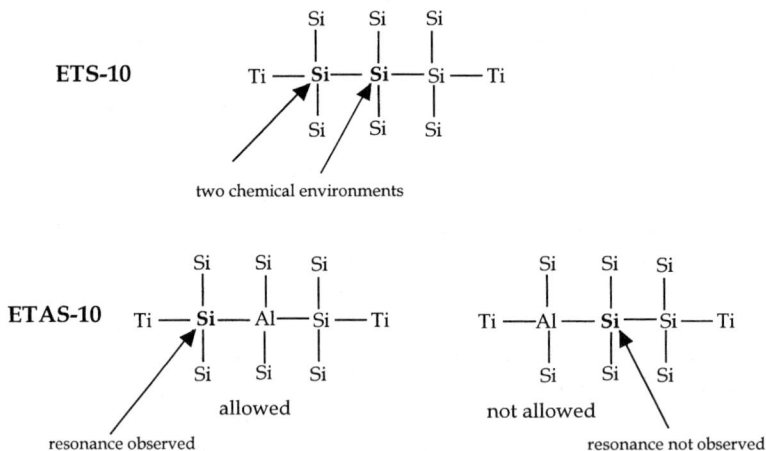

ETS-10

```
        Si   Si   Si
        |    |    |
Ti — Si— Si— Si — Ti
        |    |    |
        Si   Si   Si
```

two chemical environments

ETAS-10

```
        Si   Si   Si
        |    |    |
Ti — Si— Al— Si — Ti
        |    |    |
        Si   Si   Si
```
allowed

resonance observed

```
        Si   Si   Si
        |    |    |
Ti —Al— Si— Si — Ti
        |    |    |
        Si   Si   Si
```
not allowed

resonance not observed

The ^{27}Al MAS NMR spectrum (not shown) is also indicative of this effect since it contains a single resonance at δ approximately 60 ppm assigned to Al(4Si).

The ^{29}Si MAS NMR spectrum of gallium-substituted ETS-10 resembles the spectrum of ETAS-10: a broad peak is seen downfield from the Si(3Si, 1Ti) signals. This peak can be deconvoluted into two component signals, each one approximately 1.5 ppm downfield from the Si(2Si, 1Al, 1Ti) resonances, and which we assign to Si(2Si, 1Ga, 1Ti) as similar downfield shifts of the ^{29}Si NMR resonances have been reported for gallium-substituted zeolites. No signals are observed in the range $\delta = -99$ to -103 ppm [19, 38]. On the other hand, the ^{71}Ga MAS NMR spectrum of fully hydrated ETGS-10 (not shown) displays a broad peak at $\delta = 160$ ppm, which is characteristic of four-coordinated gallium.

Detailed analysis of spectral intensities (I_{Si}) allow framework Si/Al (or Si/Ga) and Si/Ti ratios to be calculated from Eqs (1) and (2).

$$Si/Al = \frac{\sum\limits_{n=0}^{4-m}\sum\limits_{m=0}^{4-n} I_{Si(nAl,\, mTi)}}{0.25 \sum\limits_{n=0}^{4-m}\sum\limits_{m=0}^{4-n} n I_{Si(nAl,\, mTi)}} \qquad (1)$$

$$Si/Ti = \frac{\sum\limits_{n=0}^{4-m}\sum\limits_{m=0}^{4-n} I_{Si(nAl,\, mTi)}}{0.25 \sum\limits_{n=0}^{4-m}\sum\limits_{m=0}^{4-n} m I_{Si(nAl,\, mTi)}} \qquad (2)$$

It is too early to say whether Al (Ga), Ti avoidance is a general phenomenon in titanoalumino(gallo)silicates. However, it seems logical that, when possible, Al (Ga) and Ti will not be neighbors. We have found that at relatively high Al and Ga framework concentrations slight deviations from the Al (Ga), Ti avoidance rule are observed.

The insertion of boron in ETS-10, yielding ETBS-10, has also been reported [41]. As the amount of boron in ETBS-10 increases a new ^{29}Si MAS NMR peak, attributed to Si–O–B environments, grows at δ approximately -99 ppm and (as in ETAS-10 and ETGS-10) all the resonances broaden considerably (not shown). On the other hand, the ^{11}B MAS NMR spectra (not shown) of ETBS-10 samples with different boron contents are similar and contain two groups of very sharp peaks (FWHM approximately 78 Hz) at δ approximately -1.4 and -2.9 ppm. This implies that any electric field gradients created by the electronic cloud at the ^{11}B nuclei are very small (quadrupole coupling constant estimated at 40–80 kHz) and shows that boron is in tetrahedral, rather than trigonal, coordination and replacing silicon in the framework of ETBS-10. Indeed, it has been reported that hydrated boron-substituted H-ZSM-5 zeolite gives a single sharp ^{11}B NMR peak at $\delta = -3$ ppm [42]. On the other hand, the local environment of any extra-framework boron would, in principle, be more distorted or, at least, a dispersion of boron sites would occur leading to significantly broader ^{11}B MAS NMR resonances [41].

The insertion of niobium in the framework of ETS-10, replacing titanium, has been reported to pose particular problems because NMR does not provide strong evidence for it [43]. Indeed, only a relatively small broadening of the ^{29}Si MAS/NMR resonances is observed. In addition, the ^{93}Nb MAS NMR spectrum recorded with a very fast (32 kHz) spinning rate displays a broad peak at $\delta = 100$ ppm relatively to solid Nb_2O_5 suggesting the presence of niobium in distorted octahedral coordination. However, Raman and FTIR spectroscopies provide strong evidence

References see page 899

for this framework substitution. ETS-10 gives a main strong and sharp Raman band at approximately 735 cm^{-1}, assigned to the TiO$_6$ octahedra (not shown) [43]. As the niobium content increases this peak shifts slightly and broadens and, simultaneously, a band grows at approximately 664 cm^{-1}. The latter is typical of NbO$_6$ octahedra in microporous niobosilicates. The FTIR spectrum of ETS-10 (not shown) displays bands at 446, 550 and 746 cm^{-1}, associated with the TiO$_6$ octahedra. As the niobium content of the samples increases the intensity of these bands decreases, while a new band at 918 cm^{-1} is seen. This is a further indication that niobium replaces titanium in the ETS-10 framework.

The insertion of chromium and vanadium in the framework of ETS-10 has been reported and the catalytic activity of these materials in the gas-phase oxidative dehydrogenation of cyclohexanol [44] and in the isopropanol conversion, *t*-butanol dehydration and ethanol oxidation [45] has been studied.

The great (largely overlooked) potential presented by Raman spectroscopy for studying the isomorphous framework substitutions in zeolite-type materials is particularly well illustrated in the synthetic titano-niobosilicate nenadkevichite system [46, 47]. The Raman spectra of a sample with Ti/Nb = 0.8 (Fig. 7) displays two main bands at 668 and 226 cm^{-1} associated with the NbO$_6$ octahedra. With increasing titanium content the intensity of these bands decreases (particularly of the former) while simultaneously two other strong bands grow at 764 and 290 cm^{-1}. These bands are attributed to NbO$_6$ octahedra.

The titanium for zirconium substitution in synthetic umbite has been studied in detail by powder X-ray diffraction [26]. It has been found that with increasing titanium content the unit cell shrinks in the three directions and its volume decreases systematically from approximately 985 (purely zirconeous sample) to 919 Å^3 (titaneous sample). This is not unexpected because Ti(IV) is smaller than Zr(IV). ^{29}Si MAS NMR spectroscopy provides further evidence for this framework substitution

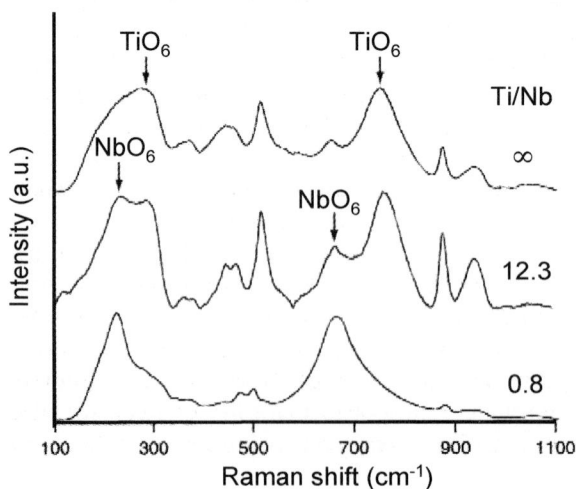

Fig. 7. Raman spectra of synthetic nenadkevichite materials. The Ti/Nb ratios are indicated.

[26]. In the structure of umbite there are three types of Si(2Si, 2Zr) sites with populations 1:1:1 [48]. The spectrum of purely zirconeous AM-2 displays a single broad peak at $\delta = -86.5$ ppm. In contrast, the spectrum of the purely titaneous sample contains a resonance at $\delta = -87.3$ ppm and two overlapping peaks at about $\delta = -86.2$ ppm and -85.9 ppm in approximately 1:1:1 intensity ratio, in accord with the crystal structure of umbite. Along the series, it is observed that as the samples zirconium content increases the peak at about $\delta = -86.5$ ppm broadens while the resonance at $\delta = -87.3$ ppm broadens and eventually disappears from the spectrum.

4.2.4.6
Other Titanosilicates

In 1994 Clearfield and coworkers [12] reported the synthesis, crystal structure and ion-exchange properties of a novel porous titanosilicate of ideal composition $Na_2Ti_2O_3SiO_4 \cdot 2H_2O$. The structure of this material has been solved from powder X-ray data by *ab initio* methods. The titanium atoms occur in clusters of four, grouped about the 4_2 axis, and are octahedrally coordinated by oxygen atoms. The silicate groups link the titanium clusters into groups of four arranged in a square of about 7.8 Å in length. These squares are linked to similar squares in the c direction by sharing corners to form a framework that encloses a tunnel. Half the Na^+ ions are located in the framework, coordinated by silicate oxygen atoms and water molecules. The remaining Na^+ ions reside in the cavity, although some of them are replaced by protons. The Na^+ ions within the tunnels are exchangeable, particularly by Cs^+ ions.

Chapman and Roe [2] have prepared a number of titanosilicate analogs of the mineral pharmacosiderite, a nonaluminosilicate molecular sieve with framework composition $KFe_4(OH)_4(AsO_4)_3$. Later, Clearfield and coworkers [49] studied by powder X-ray diffraction methods the structure of pharmacosiderite analogs with composition $HM_3Ti_4O_4(SiO_4)_3 \cdot 4H_2O$ (M = H^+, K^+, Cs^+). Harrison et al. [6] have been able to grow single crystals of the same cesium phase and solve its structure by single-crystal methods. These materials possess a most interesting structure built up from TiO_6 octahedra, sharing faces to form Ti_4O_4 cubes around the unit-cell corners and silicate tetrahedra join the titanium octahedra to form a 3D framework. Extra-framework Cs^+ species occupy sites slightly displaced from the centers of the intercage eight-ring windows, and also make $Cs-OH_2$ bonds to the water molecules that reside in the spherical cages.

Nenadkevichite is a rare mineral first found in the Lovozero region (Russia) with the composition $(Na,Ca)(Nb,Ti)Si_2O_7 \cdot H_2O$. The structure of nenadkevichite (from Saint-Hilaire, Quebec, Canada) consists of square rings of silicon tetrahedra Si_4O_{12} in the (100) plane joined together by chains of $(Nb,Ti)O_6$ octahedra in the [100] direction [50]. The pores accommodate Na^+ in two partially (0.53 and 0.54) occupied sites and water molecules in two fully occupied sites. Rocha and Anderson

References see page 899

and coworkers [46, 47] have been able to prepare a series of synthetic analogs of nenadkevichite with Ti/Nb molar ratios ranging from 0.8 to 17.1 and a purely titaneous sample.

Many porous framework titanosilicates contain Ti$-$O$-$Ti linkages that often form infinite chains. Interestingly, the materials known as AM-2, [26, 27, 51–53] AM-3 [51] and UND-1 [54] do not contain any such linkages. AM-2 is a synthetic potassium titanosilicate analog of the mineral umbite, a rare zirconosilicate found in the Khibiny alkaline massif (Russia) [48]. Although the ideal formula of umbite is $K_2ZrSi_3O_9 \cdot H_2O$, a pronounced substitution of titanium for zirconium occurs. However, the natural occurrence of purely titaneous umbite is unknown. The successful synthesis of umbite materials with different levels of titanium for zirconium substitution has been reported [26] and indicates the existence of a continuous solid solution that has not been described for any other sodium or potassium zirconosilicates. In the structure of umbite the M octahedra $(Zr,Ti)O_6$, and the T tetrahedra, SiO_4, form a 3D MT-condensed framework [48]. The M octahedron is coordinated to six T tetrahedra. In addition, to the M$-$O$-$T bonds these tetrahedra also form T$-$O$-$T links with each other. The resulting T radical has an identity period of three T tetrahedra and forms an infinite chain. Among all the known silicates and their T analogs, the umbite structure seems to be the first one to display such a MT-condensed framework. AM-3 is a sodium titanosilicate analog of the mineral penkvilksite found in Mont Saint-Hilaire, Québec (Canada) and the Kola Peninsula (Russia) with an ideal formula $Na_4Ti_2Si_8O_{22} \cdot 5H_2O$. The mineral occurs in two polytypic modifications, orthorhombic (penkvilksite-2O) and monoclinic (penkvilksite-1M) [55]. These two polytypes have been described according to the OD theory as two of the four possible maximum degree of order polytypes within a family of OD structures formed by two layers [55]. Despite the different space group symmetries the 2O (*Pnca*) and 1M ($P2_1/c$) polytypes have the same atoms, labeled in the same way, in the asymmetric unit. They differ only in the stacking of the same building blocks. Thus, the following structure description holds for both polytypes [55]. Penkvilksite contains two independent tetrahedra: the Si1-centered tetrahedra share two corners with other tetrahedra and two corners with TiO_6 octahedra; the Si2-centered tetrahedra share three corners with tetrahedra and one corner with a TiO_6 octahedron. Penkvilksite displays a new kind of connection among SiO_4 tetrahedra. Spirals of corner-sharing tetrahedra develop along [010] and have a periodicity of six tetrahedral units. The Si2-centered tetrahedra are shared between adjacent spirals, which are oriented in an alternate clockwise and counterclockwise fashion. The stacking of the layers along [001] gives rise to tetrahedral layers parallel to (100). The connection of neighboring layers of tetrahedra is due to Ti(IV) cations in octahedral coordination. AM-3 is a synthetic analog of penkvilksite-2O [51]. The synthesis of the 1M polytype has been reported recently, by Liu et al. [56, 57]. The material known as UND-1 $(Na_{2.7}K_{5.3}Ti_4Si_{12}O_{36} \cdot 4H_2O$ is the third example known of a porous framework titanosilicate containing *no* Ti$-$O$-$Ti linkages [54]. The structure of UND-1 consists of six-membered rings of SiO_4 tetrahedra and isolated TiO_6 octahedra. Each TiO_6 octahedron connects, through corner-sharing, to six SiO_4 tetrahedra on the three six-membered rings, thus forming three three-membered rings, while each SiO_4 tetrahedron connects to two isolated TiO_6 octahedra and two other SiO_4 tetrahedra of the same six-membered ring. By such connections, channels running parallel to [100] are formed with eight-membered rings containing alternative $-O-Si-O-Ti-O-$ linkages. The channel wall is covered by seven-membered rings (four SiO_4 tetrahedra and three TiO_6 octahedra) and three-membered rings. There are two cation sites in the structure. One is occupied only by K^+ and is located near the center of the seven-membered ring of the wall. The other cation site is occupied by 33 % K^+ and 67 % Na^+ and is located in the large channel near the wall.

The synthesis of a few other microporous framework titanosilicates has also been reported. For example, Chapman and Roe [5, 58] have prepared a synthetic analog of the mineral vinagradovite with framework composition $Na_8Ti_8Si_{16}O_{52}$. The structure of the mineral is composed of pyroxene chains joined to edge-sharing TiO_6 octahedra that form brookite columns. These polyhedra define one-dimensional (4 Å) channels containing zeolitic water. A number of other minerals not yet prepared in the laboratory exhibit very interesting and complexly connected frameworks (see the excellent review by Smith [59]). Verplanckite $[(Mn,Ti,Fe)_6(OH,O)_2(Si_4O_{12})_3]Ba_{12}Cl_9\{(OH,H_2O)_7\}$ has a framework with triple units of (Mn, Ti, Fe) in square-pyramidal coordination and four-rings of silicon tetrahedra [60]. The voids have a free diameter of 7 Å. Muirite, $Ba_{10}(Ca,Mn,Ti)_4Si_8O_{24}(Cl,OH,O)_{12}\cdot4H_2O$, has edge-sharing trigonal prisms and eight-rings of SiO_4 [61].

When attempting to prepare novel microporous framework titanosilicates several groups have obtained layered materials some with very interesting and unusual structures and presenting potential for being used in a number of applications such as ion exchange. One such material, known as AM-1 [8, 51] or JDF-L1, [62, 63] has the composition $Na_4Ti_2Si_8O_{22}\cdot4H_2O$. This is an unusual noncentrosymmetric tetragonal layered solid that contains five-coordinated Ti(IV) ions in the form of TiO_5 square pyramids in which each of the vertices of the base is linked to SiO_4 tetrahedra $[TiO\cdot O_4(SiO_3)_4]$ to form continuous sheet [63]. The interlamellar Na^+ ions are exchangeable, for example by protonated alkylamines. AM-4, $Na_3(Na,H)Ti_2O_2[Si_2O_6]_2\cdot2H_2O$ is yet another example of a layered titanosilicate [51, 64]. The crystal structure of AM-4 is built from TiO_6 (M) octahedra and SiO_4 (T) tetrahedra that form layers perpendicular to [001]. Each layer consists of a five-tier sandwich of T–M–T–M–T. Between the layers are Na^+ cations and water molecules. The former also exist in small cages within the layers. The major features of the structure are zigzag chains of edge-sharing TiO_6 octahedra running along the [100] that are connected together by corner-sharing pyroxene-type SiO_4 tetrahedra. Clearfield et al. [65] reported the synthesis of a layered titanosilicate, which seems to be closely related to AM-4. The same group has also carried out a considerable amount of work on the evaluation of synthetic inorganic ion exchangers for cesium and strontium removal from contaminated groundwater and wastewater using (among other materials) several microporous and layered titanosilicates, some of which possess unknown structures (see, for instance, [66] and references therein).

References see page 899

The synthesis and characterization of sodium titanosilicate JLU-1 have been reported [67]. This is a silicon-rich solid (Si/Ti = 30) prepared by templating with TMA. Although the precise structure of triclinic JLU-1 is unknown, HREM and N_2 adsorption isotherms indicate that it is a microporous material containing 6-fold Ti(IV) and with a pore size of approximately 6 Å.

4.2.4.7
Zirconosilicates

Zirconium silicates occur widely in nature and their formation under hydrothermal conditions (from approximately 300 to 550 °C) has been given considerable attention, though mainly for the solution of general geophysical and mineralogical problems (see [28] and references therein). More than 20 natural and synthetic zirconium silicates are known and for about one-third of them the crystal structures have been solved. Some of the first hydrothermal syntheses of zirconosilicates have been carried out by Maurice in 1949 [68]. Baussy et al. [69] summarize the early work in this field and report the hydrothermal synthesis of (among others) analogs of the minerals catapleiite ($Na_2ZrSi_3O_9 \cdot 2H_2O$) and elpidite ($Na_2ZrSi_6O_{15} \cdot 3H_2O$) at 350–500 °C. These materials have also been prepared by others [70]. Jale et al. [27] reported the hydrothermal synthesis of a potassium analog of elpidite at a relatively low temperature (200 °C). The characteristic feature of the structure of elpidite is the presence of double chains of tetrahedra (epididymite type) connected by zirconium atoms in octahedral coordination. The silicate chains form an anionic framework saturated by sodium ions. The double chains of tetrahedra are parallel to [100]. Two independent Na^+ cations are present. Na(2) has an octahedral coordination formed by four oxygens of the tetrahedra and two symmetrically equivalent water molecules. Na(1) occurs in a cavity formed by adjacent double chains of tetrahedra and it is bonded to seven oxygens and a water molecule [71].

Recently, an excellent example of the interesting and promising chemistry of sodium zirconosilicates has been given by Bortun et al. [28]. These workers reported on the synthesis, characterization and properties of three novel layered materials and five other zirconosilicates. In particular, a synthetic analog of the mineral gaydonnayite (ideal formula $Na_2ZrSi_3O_9 \cdot 2H_2O$) has been prepared. This material has also been synthesized by Rocha and Anderson and coworkers (AV-4) [26] and by Jale et al. [27]. The framework of gaydonnayite is composed of sinusoidal single chains of SiO_4 tetrahedra, repeating every six tetrahedra [72]. The chains are extended alternately along [011] and [01 $\bar{1}$] and are cross-linked by a ZrO_6 octahedron and two distorted NaO_6 octahedra.

Another interesting example of a microporous zirconosilicate is petarasite and its synthetic analog AV-3 [26, 73, 74]. This rare mineral ($Na_5Zr_2Si_6O_{18}(Cl,OH) \cdot 2H_2O$) possesses a very unusual structure consisting of an open 3D framework built of corner-sharing six-membered rings and ZrO_6 octahedra [73]. Elliptical channels (3.5×5.5 Å) defined by mixed six-membered rings, consisting of pairs of SiO_4 tetrahedra linked by zirconium octahedra, run parallel to the b- and c-axes. Other channels limited by six-membered silicate rings run parallel to the c-axis. The sodium, chloride and hydroxyl ions and the water molecules reside within the channels. The framework does not collapse until the release of Cl at approximately 800 °C.

The preparation of synthetic umbite materials has already been discussed in this contribution. The chemistry of other fascinating microporous framework solids is waiting to be explored. For instance, the mineral lemoynite, $(Na,K)_2CaZr_2Si_{10}O_{26} \cdot 5-6H_2O$, possesses a $ZrSi_5O_{13}$ framework with wide open channels where sodium, potassium and calcium cations and water molecules reside [75]. This framework comprises thick (7 Å) layers of hexagons of silicate groups. The sheets are bound together by six-coordinated zirconium atoms. The hexagons are tilted with respect to the layer (001) plane and the architecture of these layers is new.

The synthesis and characterization of AV-8 $(Na_{0.2}K_{1.8}ZrSi_3O_9 \cdot H_2O)$ an analog of the small-pore mineral kostylevite have been reported [76]. Kostylevite and umbite are the monoclinic and orthorhombic polymorphs of $K_2ZrSi_3O_9 \cdot H_2O$, respectively. The two minerals exhibit the same octagonal, heptagonal and hexagonal distorted tunnels and window, delimited by edges from tetrahedra and octahedra alternating in exactly the same way. The difference between the two solids resides in the fact that kostylevite is a cyclohexasilicate and umbite is a long-chain polysilicate [53]. Titanosilicate UND-1 [54] is a titaneous analog of kostylevite and, thus, also an analog of AV-8. Rocha and coworkers further report on the thermal transformations of zirconosilicates AV-3 (analog of petarasite) and AM-2 (analog of umbite) which afford analogs of the dense minerals parakeldyshite and wadeite, respectively.

4.2.4.8
Niobosilicates

Comparatively few studies are available on microporous framework niobosilicates. In this section we have already discussed the work carried out on nenadkevichite analogs containing titanium and niobium [47, 51]. Recent work suggests that other niobosilicate materials presenting potential as heterogeneous catalysts are yet to be discovered. For example, Rocha and Anderson and coworkers [77] have reported the synthesis of AM-11, a novel microporous sodium niobosilicate. Although its structure is still unknown this material has a relatively large pore volume and seems to contain NbO_6 octahedra and local silicon environments Si(4Si, 0Nb), Si(3Si, 1Nb) and Si(2Si, 2Nb). A preliminary characterization of the acid-base properties has shown that AM-11 dehydrates *ter*-butanol to isobutene with remarkably high activity and selectivity.

4.2.4.9
Stannosilicates

The hydrothermal synthesis of microporous framework stannosilicates has been pioneered by Corcoran et al. [78, 79]. As pointed out by these workers, several minerals containing SnO_6 and SiO_4 polyhedra are known and a few (dense) stannosilicate phases have been crystallized from high-temperature conditions. Two

References see page 899

microporous stannosilicates have been reported by Corcoran et al. The orthorhombic phase Sn-A has composition $Na_8Sn_3Si_{12}O_{34} \cdot nH_2O$, while Sn-B has been formulated as $Na_4SnSi_4O_{12} \cdot nH_2O$. These materials show reversible water loss and have a significant ion-exchange capacity. A third, layered, stannosilicate ($Na_4SnSi_5O_{14} \cdot nH_2O$) has also been reported [78]. Subsequently, Dyer and Jafar [80, 81] reported the synthesis, characterization and cation-exchange studies of a microporous stannosilicate with a very unusual habit and composition $Na_{13.5}Sn_{10}Si_{15}O_{36}(OH)_5 \cdot 13.5H_2O$. The ion-exchange behavior has been shown to be zeolitic in character.

Stannosilicates with the structure of the minerals umbite (AV-6) [82] and kostylevite (AV-7) [83] have been reported. Further, Rocha and coworkers [84] reported the synthesis and *ab initio* structure determination of a small-pore sodium stannosilicate ($Na_2SnSi_3O_9 \cdot 2H_2O$, AV-10). The very unusual structure of AV-10 (chiral space group $C222_1$) is composed of corner-sharing SnO_6 octahedra and SiO_4 tetrahedra, forming a 3D framework structure. The SiO_4 tetrahedra form helix chains along [001] interconnected by SnO_6 octahedra. The SnO_6 octahedra are isolated by SiO_4 tetrahedra and, thus, there are no Sn−O−Sn linkages. The zeolitic water of AV-10 is reversibly lost.

4.2.4.10
Vanadosilicates

Although vanadium has already been introduced in the framework of certain zeolites in small amounts, to the best of our knowledge, only three microporous framework vanadosilicates are known to contain stoichiometric amounts of vanadium [29, 85, 86]. Canvasite and pentagonite, dimorphs of the mineral $Ca(VO)(Si_4O_{10}) \cdot 4H_2O$, have a framework formed by silicate layers of four- and eight-membered rings of tetrahedra connected vertically by V(IV) cations, which are in a square-pyramidal coordination [85, 86]. Replacement of the VO_5 groups by two bridging oxygens would produce a tetrahedral framework topologically identical to that of zeolite gismondine. The calcium cations and the water molecules reside in the channels formed by the eight-membered rings and between the SiO_2 layers. Canvasite has channels running parallel to the *c* direction with a free diameter of only 3.3 Å in the hydrated state. Hence, both canvasite and pentagonite are likely to behave (at best) as small-pore materials. Recently, the synthesis and structural characterization of the first large-pore vanadosilicate containing octahedral vanadium and possessing a structure similar to that of titanosilicate ETS-10 has been reported. The presence of stoichiometric amounts of vanadium in the framework of AM-6 gives this material a great potential for applications as a catalyst, sorbate or functional material [29].

Together with AM-6, AM-13 and AM-14 (Si/V 10 and 4, respectively), reported recently, are the only examples of large-pore vanadosilicates [87]. Adsorption isotherms reveal that AM-13 and AM-14 adsorb 7.0 and 9.3 %g/g benzene and 5.3 and 6.7 %g/g perfluoro butylamine, respectively. Both materials are promising redox catalysts.

4.2.4.11
Other Silicates

The synthesis of microporous framework silicates of a few other metals has been reported. One such case is AV-1, the synthetic analog of the rare mineral montregianite (also known as UK-6) [88, 89]. This is an yttrium silicate, with formula $Na_4K_2Y_2Si_{16}O_{38} \cdot 10H_2O$, which possesses a very unusual structure consisting of two different types of layers alternating along the [010] direction: [90] (a) a double silicate sheet, where the single silicate sheet is of the apophyllite type with four- and eight-membered rings, and (b) an open octahedral sheet composed of $[YO_6]$ and three distinct $[NaO_4(H_2O)_2]$ octahedra. The layers are parallel to the (010) plane. The K^+ ions are ten-coordinate and the six water molecules are located within large channels formed by the planar eight-membered silicate rings. The structure of the alkali calcium silicate mineral rhodesite ($HKCa_2Si_8O_{19} \cdot 6H_2O$) [91] and its synthetic analog AV-2 [89] is closely related to that of montregianite. It consists of silicate double layers, chains of edge-sharing $[Ca(O, OH_2)_6]$ octahedra and potassium cations and additional water molecules within the pores of the silicate double layers. Rhodesite and montregianite have double silicate layers of the same topology. In fact, other minerals such as delhayelite, hydrodelhayelite and macdonaldite, all have similar double silicate sheets. In rhodesite these layers possess the maximum topological symmetry (*P2mm*) while in montregianite all symmetry has been lost. The structure of rhodesite contains two sets of octahedrally coordinated calcium ions that form single chains parallel to [001]. The octahedral chains connect adjacent silicate double layers. While Ca(2) is coordinated to six terminal oxygens that belong to six different SiO_4 tetrahedra, Ca(1) is coordinated to four terminal oxygens from four SiO_4 and two oxygens that are part of water molecules. The extra-framework potassium ions are ten-coordinated to six bridging oxygens and four water molecules.

Rocha et al. [92] reported the first microporous sodium cerium silicate (AV-5) possessing the structure of the mineral montregianite. The Ce(III)/Ce(IV) ratio in AV-5 may be controlled by oxidation/reduction in an appropriate atmosphere. This allows the fine tuning of the luminescence, adsorption and ion-exchange properties of the material. Recently, the same group reported the synthesis and *ab initio* structure determination of the first sodium potassium microporous europium and terbium silicates (AV-9) [93]. The structures of AV-9 materials and the mineral montregianite are related. However, although the tetrahedral layers of AV-9 solids and montregianite are similar this is not the case with the octahedral layers. Perhaps the most important difference between the octahedral layers of AV-9 and montregianite lies in the fact that the latter contains a single kind of Y(III), facing the pores, while AV-9 contains two kinds of Eu(III), Tb(III): one is isolated by $[NaO_4(H_2O)_2]$ octahedra while the other is facing the pores. This is expected to influence, for example, the luminescence behavior of these two Eu(III), Tb(III) centers.

References see page 899

4.2.4.12
Catalysis

For catalytic applications these novel framework materials present interesting new challenges. Apart from the possibilities for shape-selectivity and high activity through high surface area (similar to zeolites) these new materials possess a number of characteristics that make them particularly interesting as prospective heterogeneous catalysts. For ETS-10 the catalytic applications have focused on the following attributes: high cation exchange capacity, which leads to many possibilities particularly in base catalysis; facile metal loading for bifunctionality; very low acidity; possible chiral activity; photocatalytic opportunities. It should be pointed out that the titanosilicates with octahedrally coordinated titanium do not show good properties for oxidation catalysis in a similar manner to four-coordinate titanium. This is due to the ligand saturation that prevents further attachment by oxidation agents such as hydrogen peroxide.

A number of papers by Bianchi, Ragaini and coworkers [94–97] detail Fischer–Tropsch chemistry over Co and Ru exchanged ETS-10. The titanium silicate ETS-10 was found to be a suitable support for metal catalysts, having high surface area, high ion-exchange capability and no acidic function. The importance of alpha-olefin readsorption within the catalyst is discussed and the nature of this readsorption is tailored by effective control of the metal distribution inside the pores of ETS-10. The CO conversion and selectivity obtained also varies depending upon whether the active metal is introduced in the ETS-10 cages by ion exchange or simply by impregnation.

Two groups, that of Anderson, Rocha and coworkers [98] and Sivasanker and coworkers [99–101] have studied the bifunctional reforming reaction of hexane to benzene over Pt-supported basic ETS-10. The basicity of the titanosilicate can be controlled through samples exchanged with different alkali metals (M = Li, Na, K, Rb, or Cs). A distinct relationship between the intermediate electronegativity (S-int) of the different metal-exchanged ETS-10 samples and benzene yield is reported, suggesting the activation of Pt by the basicity of the exchanged metal. Typically ETS-10 samples exhibit greater aromatization activities than related $Pt-Al_2O_3$ catalysts. The very high basicity of ETS-10 in comparison with, for example zeolite X, is illustrated by Anderson and coworkers [102] in a paper that monitors the relative conversion of isopropanol to acetone. The same group has also demonstrated how these same properties are effective in aldol chemistry [103] and dehydration of t-butanol [104, 105]. In this latter reaction conversions and selectivities close to 100 % are observed at relatively modest temperatures.

One area of possible catalytic activity that has not as yet been properly explored is potential for photo-catalysis. In this respect the activity of ETS-10 for the photo-catalytic degradation of cyclohexanol, cyclododecanol, 2-hexanol, and benzyl alcohol is compared with TiO_2 particles included within small and large pore zeolitic supports suspended in acetonitrile by Fox et al. [106]. Although the activity was less than that displayed by the titanium-doped silicate TS-1 this work was reported before very much was known about the structure of ETS-10 and how it can be modified. Consequently, this field of research is ripe for exploration, particularly in the knowledge of the optical properties of these novel materials.

Acylation of alcohols with acetic acid can be carried out efficiently in the liquid phase over ETS-10 exchanged with several ions [107]. The best activity for acylation of primary alcohols is found for H, Rb and Cs-ETS-10. ETS-10 may also be used for the acylation of secondary alcohols and esterification with long-chain carboxylic acids.

The interaction of methyl and ethyl acetylene with the acidic form of ETS-10 has been studied by IR and UV-Vis spectroscopy [108]. In the first adsorption step, at room temperature, π hydrogen-bonded adducts are formed between the alkyl acetylene and ETS-10 hydroxyl groups. These hydrogen-bonded species act as precursors in the second step where oligomerization takes place, leading to formation of carbocationic double-bond conjugated systems of increasing length.

A series of chromium-containing ETS-10 samples with different Cr/Ti ratios have been prepared and their catalytic activity characterized by standard probe reactions: isopropanol conversion (to propene and acetone), t-butanol conversion (iso-butene) and ethanol oxidation (acetaldehyde and ethyl acetate) [45].

The gas-phase oxidative dehydrogenation of cyclohexanol with air using ETS-10 materials has been studied [44]. At reaction temperatures below 200 °C ETS-10 is 100 % selective to cyclohexanone and 75 % cyclohexanol conversion is achieved. The introduction of Cr, Fe and K, Cs in ETS-10 affects stability and decreases conversion and selectivity towards cyclohexanone.

An electron paramagnetic resonance (EPR) study of Ti(III) in titanosilicate molecular sieves ETS-4, ETS-10 and TS-1, TiMCM-41 has been reported [109]. Ti(III) is obtained by reduction of Ti(IV) by dry hydrogen at temperatures above 673 K. Interaction of tetrahedrally coordinated Ti(III) (in TS-1 and TiMCM-41) with O_2 and H_2O_2 results in a diamagnetic Ti(IV) hydroperoxo species. Under the same conditions, octahedrally coordinated Ti(III) (in ETS-4 and ETS-10) forms a paramagnetic titanium superoxo species. The poor activity in selective oxidation reactions of ETS materials has been attributed to the absence of formation of titanium hydroperoxy species.

The status of Ti(IV) in ETS-10 and titanium silicate TS-1 has been characterized by voltammetry [110]. Both tetrahedral and octahedral TS-1 Ti(IV) species show electrochemical response. It has been shown that the use of acid solutions allows discrimination between Ti(IV) ions in TS-1 and in ETS-10 since only the former is able to coordinate water molecules.

The adsorption and surface properties of Cu-exchanged ZSM-5 and ETS-10 with low and high degrees of ion exchange have been studied by calorimetry of adsorbed NO, CO, C_2H_4 and NH_3 probe molecules [111]. Cu(I) is the prominent species in Cu-ETS-10 and the number of Cu(I) species increased as the level of copper loading increases. Unfortunately, the structure of ETS-10 collapses at high copper loadings.

ETS-10 materials have been shown to be good supports for an enzymatic alco-

References see page 899

holysis reaction [112]. The recombinant cutinase *Fusarium solani pisi* has been immobilized by adsorption on ETS-10, ETAS-10 and vanadosilicate analog of ETS-10,
AM-6. The enzymatic activity in the alcoholysis of butyl acetate with hexanol, in
organic media (isooctane), has been measured as a function of the water content
and water activity.

Finally, ETS-10 offers the potential for chiral catalysis or asymmetric synthesis if
the pure chiral polymorph of ETS-10 could be synthesized. Along with zeolite β,
ETS-10 is currently the only known wide-pore microporous material that possesses
a chiral polymorph with a spiral channel. The as-prepared materials of both these
structures contain an equal proportion of intimately intergrown enantiomorphs
and are consequently achiral. The task of synthesizing a pure chiral polymorph is,
of course, immense. However, with some of the recently developed molecular modeling techniques for designing suitable organic templates to direct structure it may
well be possible in the next decade to solve this problem [113].

4.2.4.13
Optical Properties

Owing to the presence of stoichiometric quantities of transition metals in the
framework of these novel microporous materials they have potential for interesting
optical properties [114–118]. Again most attention to date has focused on the microporous material ETS-10. The structure of ETS-10 contains –O–Ti–O–Ti–O–
chains, with alternating long-short bonds, which are effectively isolated from
each other by a silicate sheath. In effect, therefore, ETS-10 contains monoatomic
–O–Ti–O–Ti–O– wires embedded in an insulating SiO_2 environment that leads to
a 1D quantum confinement of electrons or holes within this wire. Associated with
this charge carrier confinement within this unprecedented geometrical definition
is a bandgap blue shift. The effective reduced mass, μ, of electrons and holes
within this wire is calculated to be 1.66 $m(e) < \mu < 1.97\ m(e)$ that is consistent
with a band gap of 4.03 eV, which is quite different from that of bulk TiO_2. This
type of work underlines the possible future role that these microporous materials
may play in optoelectronic and nonlinear optical applications.

Ab initio calculations on a linear –OTiOTiO– chain embedded in an envelope of
SiO_4 tetrahedra, mimicking the structure of ETS-10, confirm that the peculiar optical properties of the solid are associated with the presence of such linear chains
[119]. The UV-Vis (ultraviolet range) and the magnetic (electron spin resonance,
ESR) properties of the chains can be modified by adsorbing Na vapors; the titanium within the chains is reduced, which effects the following redox couple:
$Ti(\text{IV}) + Na \rightarrow Ti(\text{III}) + Na^+$, thereby generating unpaired electrons within the titanate chain and also generating extra cation sites. After such reduction the mate-
rial is not air stable and is easily reoxidized, generating additional Na_2O within the
channels. Such redox couples also give strong indications of possible applications
in battery technology.

Recently, Rocha and coworkers [120, 121] doped ETS-10 with Eu^{3+} and Er^{3+} by
ion-exchange techniques and studied the luminescence properties of the resultant

materials. It has been found that only Eu^{3+}-doped ETS-10 is optically active. Upon calcination at temperatures in excess of 700 °C both materials transform into dense titanosilicate analogs of the mineral narsarsukite [$(Na, K)_2TiSi_4O_{11}$], which is made up of Si_4O_{10} chains that form tubes of rings consisting of four SiO_4 tetrahedra [122]. These tubes are linked by chains of corner-sharing TiO_6 octahedra. The cavities between the Si_4O_{10} tubes and octahedral chains contain Na^+, Eu^{3+} and Er^{3+} cations. Eu^{3+} and Er^{3+}-doped narsarsukite display very interesting luminescence properties. The latter, in particular, exhibits a high and stable room-temperature emission in the visible and in the infrared spectral regions. An efficient energy transfer between the narsarsukite skeleton and the optically active Er^{3+} centers seems to occur [121].

Another way of combining, in a given silicate, microporosity and optical activity is to prepare solids with lanthanide ions in the framework rather than in the micropores. Examples of this approach have been recently reported and they are based upon montregianite-type systems. The 14-K emission spectrum (325-nm excitation) sodium cerium silicate AV-5 shows a broad band with two overlapping peaks at 377 and 410 nm [92]. A detailed study of the photoluminescence properties of microporous europium and erbium silicates indicate that AV-9 materials present potential for applications in optoelectronics [93]. These montregianite-type systems are quite flexible in the sense that it is possible to fine tune the structure, the chemical composition and the oxidation state of the elements (for example Ce(iii)/Ce(iv)). This in turn allows the fine tuning of properties, particularly luminescence. One envisages introducing, via ion exchange, other appropriate lanthanide ions into the pores, thus enabling pumping of the lanthanide optical centers in the lattice, an effect that could be used for sensing purposes. The adsorption of molecules into the pores may also change the luminescence properties of the system and, again, there is a real potential for application in sensor devices.

The photoionization of methylphenothiazine in microporous titanosilicates M-ETS-10 (M $= N^+ + K^+$, H^+, Li^+, Na^+, K^+, Ni^+, Cu^{2+} and Co^{2+}) and Na,K-ETS-4 with UV irradiation at room temperature has been studied [123]. Methylphenothiazine cation radicals are produced in M-ETS-10 materials but none are detected by EPR or diffuse reflectance UV-Vis spectroscopy on Na,K-ETS-4. The photochemistry in ETS materials is sensitive to the metal ion, pore size and internal void space.

4.2.4.14
Adsorption Properties

Dihydrogen, dinitrogen, carbon monoxide and nitric oxide have all been adsorbed, at nominally liquid nitrogen temperature, on Na^+- and K^+-exchanged ETS-10 [124]. IR spectroscopy shows formation of $M^+-(H_2)$, $M^+-(N_2)(n)$, $M^+-(CO)(n)$ and $M^+-(NO)$ ($n = 1, 2, \ldots$; $M^+ = Na, K$) adducts prevalently involving alkali-metal cations located in the 12-membered channels. These adducts give main IR ab-

References see page 899

sorption bands in the range 4050–4150 cm^{-1} for H$_2$, 2331–2333 cm^{-1} for N$_2$, 2148–2176 cm^{-1} for CO, and 1820–1900 cm^{-1} for NO, which are assigned to the fundamental stretching mode of the diatomic molecules polarized by the electric field created by the metal ions. On Na-exchanged samples, the Na$^+$–(N$_2$) and Na$^+$–(CO) species, formed at lowest dosage, evolve into Na$^+$–(N$_2$)(n) and Na$^+$–(CO)(n) ($n = 2$, 3) species upon increasing the gas-phase pressure, This reversible "solvation" process is not observed for K-exchanged samples. This result does not find a comparable precedent for CO adsorbed on zeolites and is indicative of the unique adsorption characteristics that can be expected over these highly charged, high cation exchange capacity microporous materials. It has been reported that 30–80 % Ba-exchanged ETS-4 has high thermal stability (up to temperatures in excess of 400 °C) and effectively separates nitrogen (approximately 3.6 Å) from methane (approximately 3.8 Å) [37].

4.2.4.15
Cation Exchange

Microporous titanosilicates and related zirconosilicates are very promising materials for ion-exchange applications. As clearly shown by the group of Clearfield [6, 12, 125, 126], which has carried out most ion exchange work available on these materials, the reason for their unique selectivity is related to the correspondence of the geometrical parameters of their ion-exchange sites (channels, cavities) to the size of the selectively adsorbed ions. Because of the small size of the channels or cavities, the framework in a sense acts as a coordinating ligand to the cations.

Two structurally related titanosilicates are attracting considerable interest, Na$_2$Ti$_2$O$_3$SiO$_4$·2H$_2$O and synthetic pharmocosiderite, HM$_3$Ti$_4$O$_4$(SiO$_4$)$_3$·4H$_2$O. The former displays great affinity for Cs$^+$ [12, 125, 127] and the latter is selective for low concentrations of Cs$^+$ and Sr^{2+} in the presence of ppm levels of Na$^+$, K$^+$, Mg^{2+} and Ca^{2+} cations at neutral pH [128–130]. Dyer et al. [131] have also reported on the removal of trace ^{137}Cs and ^{89}Sr by different cationic forms of synthetic pharmacosiderite. Other recent reports on M$_2$Ti$_2$O$_3$SiO$_4$·2H$_2$O further show the importance of this material as a Cs$^+$ and Sr^{2+} exchanger [132–135].

Another family of (AM-2) materials, synthetic analogs of mineral umbite K$_2$MSi$_3$O$_9$·2H$_2$O, M = Ti, Zr, have also been the subject of much investigation by the group of Clearfield [26, 27, 51–53, 136, 137] and shown to be good exchangers for Rb$^+$, Cs$^+$ and K$^+$ cations. The ion-exchange behavior of a series of mixed Zr, Ti AM-2 materials containing different amounts of Ti and Zr has been studied. It has been found that Zr-rich silicates with large channels exhibit affinity for Rb$^+$ and Cs$^+$ cations, whereas Ti-rich compounds with much smaller channels show a preference for the K$^+$ ion [138]. These data suggest that the chemical alteration of the structure of exchangers may be a promising way for tailored changes of their selectivity.

The sorption of uranium by a series of microporous titanosilicates (ETS-10, ETS-4, layered AM-4) [64], and Na$_2$Ti$_2$O$_3$SiO$_4$·2H$_2$O has been studied by the group of Dyer [139, 140]. The difference in their ability to take up uranium has been discussed in terms of their crystal structure and the determination of their cation exchange capacity. The same group has also reported on the ability of the synthetic titanosilicate analog of mineral penkvilsite-2O (AM-3) to sorb radioactive ^{60}Co [141].

ETS-10 has been shown to be particularly selective for Pb^{2+} [142] and Cd^{2+} [143] ions. Further, the penkvilksite structure has been shown to be particularly selective for Li^+ cations [144], which gives hints of possible applications in battery technology.

4.2.4.16
Conclusions

In this section it has been shown that a new and exciting field of research into microporous zeotype solids is emerging. These stable silicate materials possess mixed octahedral-tetrahedral frameworks and display entirely new structural architectures. Although much of the work performed to date has concentrated on titanosilicates, it is clear that the novel porous frameworks may contain many other elements, particularly transition metals such as Zr, Nb, Sn, V, and Y. It is anticipated that some of the new materials will find applications in areas usually associated with zeolites. Others, however, may have potential applications in fields such as optoelectronics and nonlinear optics, batteries, magnetic materials and sensors.

References

1 J. ROCHA, M. W. ANDERSON, *Eur. J. Inorg. Chem.* **2000**, 801.

2 D. M. CHAPMAN, A. L. ROE, *Zeolites* **1990**, *10*, 730.

3 S. M. KUZNICKI, US Patent 4,853,202 1989, assigned to Engelhard.

4 S. M. KUZNICKI, US Patent 4,938,989 1990, assigned to Engelhard.

5 D. M. CHAPMAN, US Patent 5,015,453 1990 assigned to Engelhard.

6 W. T. A. HARRISON, T. E. GIER, G. D. STUCKY, *Zeolites* **1995**, *15*, 408.

7 M. W. ANDERSON, O. TERASAKI, T. OHSUNA, A. PHILIPPOU, S. P. MACKAY, A. FERREIRA, J. ROCHA, S. LIDIN, *Nature* **1994**, *367*, 347.

8 M. W. ANDERSON, O. TERASAKI, T. OHSUNA, P. J. O'MALLEY, A. PHILIPPOU, S. P. MACKAY, A. FERREIRA, J. ROCHA, S. LIDIN, *Philos. Mag. B* **1995**, *71*, 813.

9 T. K. DAS, A. J. CHANDWADKAR, S. SIVASANKER, *Chem. Commun.* **1996**, 1105.

10 X. LIU, J. K. THOMAS, *Chem. Commun.* **1996**, 1435.

11 W. J. KIM, M. C. LEE, J. C. YOO, D. T. HAYHURST, *Microporous Mesoporous Mater.* **2000**, *41*, 79.

12 D. M. POOJARY, R. A. CAHILL, A. CLEARFIELD, *Chem. Mater.* **1994**, *6*, 2364.

13 J. ROCHA, P. BRANDÃO, Z. LIN, A. P. ESCULCAS, A. FERREIRA, M. W. ANDERSON, *J. Phys. Chem.* **1996**, *100*, 14978.

14 J. ROCHA, A. FERREIRA, Z. LIN, M. W. ANDERSON, *Microporous and Mesoporous Mater.* **1998**, *23*, 1253.

15 T. K. DAS, A. J. CHANDWADKAR, A. P. BUDHKAR, A. A. BELHEKAR, S. SIVASANKER, *Microporous Mater.* **1995**, *4*, 195.

16 V. VALTCHEV, S. MINTOVA, *Zeolites* **1994**, *14*, 697.

17 V. P. VALTCHEV, *J. Chem. Soc., Chem. Commun.* **1994**, 730.

18 T. P. Das, A. J. Chandwadkar, A. P. Budhkar, S. Sivasanker, *Microporous Mater.* **1996**, *5*, 401.

19 M. W. Anderson, A. Philippou, Z. Lin, A. Ferreira, J. Rocha, *Angew. Chem. Int. Ed. Engl.* **1995**, *34*, 1003.

20 J. Rocha, Z. Lin, A. Ferreira, M. W. Anderson, *J. Chem. Soc., Chem. Commun.* **1995**, 867.

21 J. Rocha, P. Brandão, J. D. Pedrosa de Jesus, A. Philippou, M. W. Anderson, *Chem. Commun.* **1999**, 471.

22 J. Rocha, P. Brandão, M. W. Anderson, T. Ohsuna, O. Terasaki, *Chem. Commun.* **1998**, 667.

23 a) S. M. Kuznicki, K. A. Thrush, US Patent 5,244,650, 1993; b) S. M. Kuznicki, R. J. Madon, G. S. Koermer and K. A. Thrush, EPA 0405978A1, 1991.

24 S. M. Kuznicki, A. K. Thrush, WO91/18833, 1991, assigned to Engelhard.

25 J. Rocha, P. Ferreira, Z. Lin, J. R. Agger, M. W. Anderson, *Chem. Commun.* **1998**, 1268.

26 Z. Lin, J. Rocha, P. Ferreira, A. Thursfield, J. R. Agger, M. W. Anderson, *J. Phys. Chem. B.* **1999**, *103*, 957.

27 S. R. Jale, A. Ojo, F. R. Fitch, *Chem. Commun.* **1999**, 411.

28 A. I. Bortun, L. N. Bortun, A. Clearfield, *Chem. Mater.* **1997**, *9*, 1854.

29 J. Rocha, P. Brandão, Z. Lin, M. W. Anderson, V. Alfredsson, O. Terasaki, *Angew. Chem. Int. Ed. Engl.* **1997**, *36*, 100.

30 G. Sankar, R. G. Bell, J. M. Thomas, M. W. Anderson, P. A. Wright, J. Rocha and A. Ferreira, *J. Phys. Chem.* **1996**, *100*, 449.

31 X. Q. Wang, A. J. Jacobson, *Chem. Commun.* **1999**, 973.

32 A. N. Mer'kov, I. V. Bussen, E. A. Goiko, E. A. Kul'chitskaya, Y. P. Men'shikov, A. P. Nedorezova, *Zap. Vses. Mineralog. O-va* **1973**, *102*, 54.

33 P. A. Sandomirskii, N. V. Belov, *Sov. Phys.-Crystallogr.* **1979**, *24*, 686.

34 A. Philippou and M. W. Anderson *Zeolites* **1996**, *16*, 98.

35 G. Cruciani, P. DeLuca, A. Nastro, P. Pattison, *Microporous and Mesoporous Mater.* **1998**, *21*, 143.

36 M. Naderi, M. W. Anderson, *Zeolites* **1996**, *17*, 437.

37 S. M. Kuznicki, V. A. Bell, I. Petrovic, P. W. Blosser, US Patent 5,989,316 1999, assigned to Engelhard.

38 M. W. Anderson, J. Rocha, Z. Lin, A. Philippou, I. Orion, A. Ferreira, *Microporous Mater.* **1996**, *6*, 195.

39 J. Rocha, Z. Lin, A. Ferreira, M. W. Anderson, *J. Chem. Soc., Chem. Commun.* **1995**, 867.

40 Z. Lin, J. Rocha, A. Ferreira, M. W. Anderson, *Colloids Surfaces A* **2001**, *179*, 133.

41 J. Rocha, P. Brandão, M. W. Anderson, T. Ohsuna, O. Terasaki, *Chem. Commun.* **1998**, 667.

42 K. F. M. G. J. Scholle, W. S. Veeman, *Zeolites* **1985**, *5*, 118.

43 J. Rocha, P. Brandão, J. D. Pedrosa de Jesus, A. Philippou, M. W. Anderson, *Chem. Commun.* **1999**, 471.

44 A. Valente, Z. Lin, P. Brandão, I. Portugal, M. W. Anderson, J. Rocha, *J. Catal.* **2001**, *200*, 99.

45 P. Brandão, A. Philippou, A. Valente, J. Rocha, M. W. Anderson, *Phys. Chem. Chem. Phys.* **2001**, *3*, 1773.

46 J. Rocha, P. Brandão, Z. Lin, A. Kharlamov, M. W. Anderson, *Chem. Commun.* **1996**, 669.

47 J. Rocha, P. Brandão, Z. Lin, A. P. Esculcas, A. Ferreira, M. W. Anderson, *J. Phys. Chem.* **1996**, *100*, 14978.

48 G. D. Ilyushin, *Inorg. Mater.* **1993**, *29*, 853.

49 E. A. Behrens, D. M. Poojari, A. Clearfield, *Chem. Mater.* **1996**, *8*, 1236.

50 P. G. Perrault, C. Boucher, J. Vicat, E. Cannillo, G. Rossi, *Acta Crystallogr., Sect. B* **1973**, *29*, 1432.

51 Z. Lin, J. Rocha, P. Brandão, A. Ferreira, A. P. Esculcas, J. D. Pedrosa de Jesus, A. Philippou, M. W. Anderson, *J. Phys. Chem.* **1997**, *101*, 7114.

52 D. M. Poojari, A. I. Bortun, L. N. Bortun, A. Clearfield, *Inorg. Chem.* **1997**, *36*, 3072.

53 M. S. Dadachov, A. Le Bail, *Eur. J. Solid State Inorg. Chem.* **1997**, *34*, 381.

54 X. Liu, M. Shang, J. K. Thomas, *Microporous Mater.* **1997**, *10*, 273.

55 S. Merlino, M. Pasero, G. Artioli, A. P. Khomyakov, *Am. Mineral.* **1994**, *79*, 1185.

56 Y. Liu, H. Du, F. Zhou, W. Pang, *Chem. Commun.* **1997**, 1467.

57 Y. Liu, H. Du, Y. Xu, H. Ding, W. Pang, Y. Yue, *Microporous and Mesoporous Mater.* **1999**, *28*, 511.

58 R. K. Rastsvetaeve, V. I. Andrianov, *Kristallografiya* **1984**, *29*, 681.

59 J. V. Smith, *Chem. Rev.* **1988**, *88*, 149.

60 A. R. Kampf, A. A. Khan, W. H. Baur, *Acta Crystallogr., Sect. B* **1973**, *B29*, 2019.

61 L. P. Solov'eva, S. V. Borisov, V. V. Bakakin, *Sov. Phys. Crystallogr. (Engl. Transl.)* **1972**, *16*, 1035.

62 H. Du, M. Fang, J. Chen, W. Pang, *J. Mater. Chem.* **1996**, *6*, 1827.

63 M. A. Roberts, G. Sankar, J. M. Thomas, R. H. Jones, H. Du, M. Fang, J. Chen, W. Pang, R. Xu, *Nature* **1996**, *381*, 401.

64 M. S. Dadachov, J. Rocha, A. Ferreira, M. W. Anderson, *Chem. Commun.* **1997**, 2371.

65 A. Clearfield, A. I. Bortun, L. N. Bortun, R. A. Cahill, *Solvent Extr. Ion Exch.* **1997**, *15*, 285.

66 A. I. Bortun, L. N. Bortun, A. Clearfield, *Solvent Extr. Ion Exch.* **1997**, *15*, 909.

67 Y. Liu, H. Du, F.-S. Xiau, G. Zhu, W. Pang, *Chem. Mater.* **2000**, *12*, 665.

68 O. D. Maurice, *Econ. Geol.* **1949**, *44*, 721.

69 G. Baussy, R. Caruba, A. Baumer, G. Turco, *Bull. Soc. Fr. Minéral. Crystallogr.* **1974**, *97*, 433.

70 G. Y. Chao, D. H. Watkinson, *Can. Mineral.* **1974**, *12*, 316.

71 E. Cannillo, G. Rossi, L. Ungaretti, *Am. Mineral.* **1973**, *58*, 106.

72 G. Y. Chao, *Can. Mineral.* **1985**, *23*, 11.

73 S. Ghose, C. Wan, G. Y. Chao, *Can. Mineral.* **1980**, *18*, 503.

74 J. Rocha, P. Ferreira, Z. Lin, J. R. Agger, M. W. Anderson, *Chem. Commun.* **1998**, 1269.

75 Y. Le Page, G. Perrault, *Can. Mineral.* **1976**, *14*, 132.

76 P. Ferreira, A. Ferreira, J. Rocha, M. A. Soares, *Chem. Mater.* **2001**, *13*, 355.

77 J. Rocha, P. Brandão, A. Philippou, M. W. Anderson, *Chem. Commun.* **1998**, 2687.

78 E. W. Corcoran Jr., D. E. W. Vaughan, *Solid State Ionics* **1989**, *32/33*, 423.

79 E. W. Corcoran Jr., J. M. Newsam, H. E. King Jr., D. E. Vaughan, *ACS Symp. Ser.* **1989**, *398*, 603.

80 A. Dyer, J. J. Jafar, *J. Chem. Soc. Dalton Trans.* **1990**, 3239.

81 A. Dyer, J. J. Jafar, *J. Chem. Soc. Dalton Trans.* **1991**, 2639.

82 Z. Lin, J. Rocha, A. Valente, *Chem. Commun.* **1999**, 2489.

83 Z. Lin, J. Rocha, J. D. Pedrosa de Jesus, A. Ferreira, *J. Mater. Chem.* **2000**, *10*, 1353.

84 A. Ferreira, Z. Lin, J. Rocha, C. M. Morais, M. Lopes, C. Fernandez, *Inorg. Chem.*, **2001**, *40*, 3330.

85 H. T. Evans, Jr. *Amer. Mineral., Sect. B* **1973**, *58*, 412.

86 R. Rinaldi, J. J. Pluth, J. V. Smith, *Acta Crystallogr. Sect. B* **1975**, *31*, 1598.

87 P. Brandão, A. Philippou, N. Hanif, P. C. Claro, J. Rocha, M. W. Anderson, (submitted).

88 J. Rocha, P. Ferreira, Z. Lin, P. Brandão, A. Ferreira and J. D. Pedrosa de Jesus, *Chem. Commun.* **1997**, 2103.

89 J. Rocha, P. Ferreira, Z. Lin, P. Brandão, A. Ferreira and J. D. Pedrosa de Jesus, *J. Phys. Chem.* **1998**, *102*, 4739.

90 S. Ghose, P. K. S. Gupta, C. F. Campana, *Am. Mineral.* **1987**, *72*, 365.

91 K.-F. Hesse, F. Z. Liebau, *Kristallogr.* **1992**, *199*, 25.

92 J. Rocha, P. Ferreira, L. D. Carlos, A. Ferreira, *Angew. Chem. Int. Ed.* **2000**, *39*, 3276.

93 D. Ananias, A. Ferreira, J. Rocha, P. Ferreira, J. P. Rainho, C. Morais, L. D. Carlos, *J. Am. Chem. Soc.*, **2001**, *123*, 5735.

94 C. L. Bianchi, S. Vitali, V. Ragaini, in: Natural Gas Conversion, V. A.

Parmaliana, D. Sanfilippo, F. Frusteri, A. Vaccari, F. Arena (Eds) *Stud. Surf. Sci. Catal.*, Vol. 119, Elsevier, Amsterdam, 1998, p. 167.

95 C. L. Bianchi, S. Ardizzone, V. Ragaini, in: Natural Gas Conversion, V. A. Parmaliana, D. Sanfilippo, F. Frusteri, A. Vaccari, F. Arena (Eds) *Stud. Surf. Sci. Catal.*, Vol. 119, Elsevier, Amsterdam, 1998, p. 173.

96 C. L. Bianchi, V. Ragaini, *J. Catal.* 1997, *168*, 70.

97 C. L. Bianchi, R. Carli, S. Merlotti, V. Ragaini, *Catal. Lett.* 1996, *41*, 79.

98 A. Philippou, M. Naderi, N. Pervaiz, J. Rocha, M. W. Anderson, *J. Catal.* 1998, *178*, 174.

99 S. B. Waughmode, T. K. Das, R. Vetrivel, S. Sivasanker, *J. Catal.* 1999, *185*, 265.

100 T. K. Das, A. J. Chandwadkar, S. Sivasanker, in: Recent Advances in Basic and Applied Aspects of Industrial Catalysis, T. S. R. Prasada Rao, G. Murali Dhar (Eds) *Stud. Surf. Sci. Catal.*, Vol. 113, Elsevier, Amsterdam, 1998, p. 455.

101 T. K. Das, A. J. Chandwadkar, H. S. Soni, S. Sivasanker, *Catal. Lett.* 1997, *44*, 113.

102 A. Philippou, J. Rocha, M. W. Anderson, *Catal. Lett.* 1999, *57*, 151.

103 A. Philippou, M. W. Anderson, *J. Catal.* 2000, *189*, 395.

104 A. Philippou, M. Naderi, J. Rocha, M. W. Anderson, *Catal. Lett.* 1998, *53*, 221.

105 T. K. Das, A. J. Chandwadkar, H. S. Soni, S. Sivasanker, *J. Mol. Catal. A – Chemical* 1996, *107*, 199.

106 M. A. Fox, K. E. Doan, M. T. Dulay, *Research in Chemical Intermediates* 1994, *20*, 711.

107 S. B. Waughmode, V. V. Thakur, A. Sudalai, S. Sivasanker, *Tetrahedron Lett.* 2001, *42*, 3145.

108 A. Zecchina, F. X. L. I. Xamena, C. Paze, G. T. Palomino, S. Bordiga, C. O. Arean, *Phys. Chem. Chem. Phys.* 2001, *3*, 1228.

109 R. Bal, K. Chaudhari, D. Srinivas, S. Sivasanker, P. Ratnasamy, *J. Mol. Catal., A Chemical* 2000, *162*, 199.

110 S. Bodoardo, F. Geobaldo, N. Penazzi, M. Arrabito, F. Rivetti, G. Spano, C. Lamberti, A. Zecchina, *Electrochem. Commun.* 2000, *2*, 349.

111 A. Gervasini, C. Picciau, A. Auroux, *Microporous and Mesoporous Mater.* 2000, *35–36*, 457.

112 F. N. Serralha, J. M. Lopes, F. Lemos, D. M. F. Prazeres, M. R. Aires-Barros, J. Rocha, J. M. S. Cabral, F. Ramôa Ribeiro, *React. Kinet. Catal. Lett.* 2000, *69*, 217.

113 D. W. Lewis, D. J. Willock, C. R. A. Catlow, J. M. Thomas, G. J. Hutchings, *Nature* 1996, *382*, 604.

114 A. J. M. deMan, J. Sauer, *J. Phys. Chem.* 1996, *100*, 5025.

115 B. Mihailova, V. Valtchev, S. Minatova, L. Konstantinov, *Zeolites* 1996, *16*, 22.

116 Y. N. Xu, W. Y. Ching, Z. Q. Gu, *Ferroelectrics* 1997, *194*, 219.

117 E. Borello, C. Lamberti, S. Bordiga, A. Zecchina, C. O. Arean, *Appl. Phys. Lett.* 1997, *71*, 2319.

128 C. Lamberti, *Microporous and Mesoporous Mater.* 1999, *30*, 155.

129 S. Bordiga, G. T. Palomino, A. Zecchina, G. Ranghino, E. Giamello, C. Lamberti, *J. Chem. Phys.* 2000, *112*, 3859.

120 J. P. Rainho, L. D. Carlos, J. Rocha, *J. Lumin.* 2000, *87–89*, 1083.

121 J. Rocha, L. D. Carlos, J. P. Rainho, Z. Lin, P. Ferreira, R. M. Almeida, *J. Mater. Chem.* 2000, *10*, 1371.

122 Y. A. Pyatenko, Z. V. Pudvkina, *Kristallografiya* 1960, *5*, 563.

123 R. M. Krishna, A. M. Prakash, V. Kurshev, L. Kevan, *Phys. Chem. Chem. Phys.* 1999, *1*, 4119.

124 A. Zecchina, C. O. Arean, G. T. Palomino, F. Geobaldo, C. Lamberti, G. Spoto, S. Bordiga, *PCCP* 1999, *1*, 1649.

125 D. M. Poojary, A. I. Bortun, L. N. Bortun, A. Clearfield, *Inorg. Chem.* 1996, *35*, 6131.

126 E. A. Behrens, D. M. Poojary, A. Clearfield, *Chem. Mater.* 1996, *8*, 1236.

127 A. I. BORTUN, L. N. BORTUN, A.
CLEARFIELD, *Solvent Extr. Ion Exch.*
1996, *14*, 341.

128 E. A. BEHRENS, A. CLEARFIELD,
Microporous. Mater. **1997**, *11*, 65.

129 E. A. BEHRENS, P. SYLVESTER, A.
CLEARFIELD, *Environ. Sci. Technol.*
1998, *32*, 101.

130 A. M. PUZIY, *J. Radioanal. Nucl. Chem.*
1998, *237*, 73.

131 A. DYER, M. PILLINGER, S. AMIN, *J.
Mater. Chem.* **1999**, *9*, 2481.

132 P. SYLVESTER, E. A. BEHRENS, G. M.
GRAZIANO, A. CLEARFIELD, *Sep. Sci.
Technol.* **1999**, *34*, 1981.

133 A. CLEARFIELD, *Solid State Sci.* **2001**, *3*,
103.

134 A. CLEARFILED, L. N. BORTUN, A. I.
BORTUN, *React. Funct. Polym.* **2000**, *43*,
85.

135 S. SOLBRA, N. ALISON, S. WAITE, S. V.
MIKHALOVSKY, A. I. BORTUN, L. N.
BORTUN, A. CLEARFIELD, *Environ. Sci.
Technol.* **2001**, *35*, 629.

136 A. I. BORTUN, L. N. BORTUN, D. M.
POOJARY, O. XIANG, A. CLEARFIELD,
Chem. Mater. **2000**, *12*, 294.

137 V. VALTCHEV, J.-L. PAILLAUD, S.
MINTOVA, H. KESSLER, *Microporous
and Mesoporous Mater.* **1999**, *32*, 287.

138 A. CLEARFIELD, A. I. BORTUN, L. N.
BORTUN, D. M. POOJARY, S. A. KHAINA-
KOV, *J. Mol. Struct.* **1998**, *470*, 207.

139 L. ATTAR, A. DYER, R. BLACKBURN,
J. Radioanal. Nucl. Chem. **2000**, *246*,
451.

140 L. ATTAR, A. DYER, *J. Radioanal. Nucl.
Chem.* **2001**, *247*, 121.

141 U. Y. KOUDSY, A. DYER, *J. Radioanal.
Nucl. Chem.* **2001**, *241*, 209.

142 S. M. KUZNICKI, K. A. THRUSH, US
Patent 4,994,191 1991, assigned to
Engelhard.

143 M. W. ANDERSON, M. MOHSEN,
unpublished results.

144 S. M. KUZNICKI, J. S. CURRAN, X.
YANG, US Patent 5,882,624 1999,
assigned to Engelhard.

4.2.5
Modification of Crystalline Microporous Solids

Takashi Tatsumi

Although crystalline microporous solids encompass zeolites and related microporous materials, pillared clays and layered phosphates, expanded hydrotalcites,
etc., herein the attention is focused on modification of zeolites and related materials, which offers the abundant chemical diversity to the zeolites. The properties
of zeolites can be varied considerably by modifying them either during or after
the actual synthesis. The types of modifications can be classified into three
groups:

- modification of internal pores by a variety of methods, such as ion exchange and
  adsorption of organic and inorganic compounds, occasionally followed by reactions;
- modification of the framework through the incorporation or removal of elements (M) other than Si including the change in the Si/M ratio;
- and modification of the external surface including the minute change in the size
  of pore opening.

4.2.5.1
Ion Exchange and Introduction of Metals into the Pore

Most zeolites have an intrinsic ability to exchange cations. This exchange ability is a result of isomorphous substitution of a cation of trivalent or lower charges (typically Al) for Si as a tetravalent framework cation. As a consequence of this substitution, a net negative charge develops on the framework of the zeolite, which is to be neutralized by cations present within the channels and cages that constitute the microporous part of the crystalline zeolite. These cations may be any of the metals, metal complexes or alkylammonium cations and can be replaced by other cations through ion exchange. A very comprehensive review on ion exchange in zeolites has recently been given by Kühl [1]. The ion exchange capacity of zeolites with low Si/Al ratio is quite high; e.g., Na-A has an ion exchange capacity of 7.0 mmol g^{-1} on a dry basis, which is higher than that of ion exchange resins ranging from 0.5– 5.0 mmol g^{-1}.

Clays are also ion exchangers. There are similarities between zeolites and clays. Both comprise similar elements, such as Si, Al, alkali and alkaline earth metals. In the case of clay minerals, divalent cations are substituted for trivalent ones or trivalent cations for tetravalent ones. Because of their two-dimensional structure, clay minerals may undergo swelling or shrinking upon cation exchange. In contrast, zeolites do not undergo any appreciable dimensional change with ion exchange. Ion exchange capacity can arise also from unsaturated valencies occurring at the termination of the crystal edges and faces or from faults within the structure. In clay minerals up to 20 % of the exchange capacity may be accounted for by these faults [2]. However, for zeolites, the exchange capacity due to such sources is relatively small compared to the intrinsic one except for high-silica zeolites, where there is only a small intrinsic exchange capacity arising from isomorphous substitution.

Since cation exchange in zeolites results in dramatic alteration of stability, adsorption properties, catalytic activities, etc., ion exchange is an important tool for modifying zeolites. Zeolites have been used commercially as water-softening agents. Such applications of zeolites and other microporous solids as ion exchangers will be described elsewhere.

Ion exchange is usually carried out in aqueous solutions. When the zeolite crystals are immersed in an aqueous electrolyte, ion exchange between the solid phase (zeolite) and the solution proceeds. Since the anionic charges of the framework are constant, the number of the cationic charges within the zeolite pores is invariable unless salt imbibition (see below) occurs.

While the anions in aqueous solution can in principle move freely, the anions in solution are usually excluded from the zeolite channels because of the repulsion effected by the negative charges of the pore aperture; however, anions can move into the zeolite by taking stoichiometric quantities of cations with them. This phenomenon is called salt imbibition [3], which becomes significant only when the ion exchange is carried out at concentrations of 0.5 molar or higher.

Because of the crystalline microporous structure of zeolites, there are several significant characteristics in ion exchange in zeolites, distinguished from other ion-exchange systems.

Ion sieving: quaternary ammonium ions such as $N(CH_3)_4^+$ are completely excluded from zeolite A. These ions are larger than the diameters of the windows giving access to the intracrystalline pore space (cages) of the zeolite, while there is enough space within the cages. For similar reasons only partial exchange can occur, giving rise to a limitation of the maximum level of exchange, which is low compared to that expected from the framework Al content. Ion sieving of inorganic cations is observed for zeolites with narrow windows. Thus Cs^+ cannot exchange with Ag^+ in analcime while Na^+ can. In general, the rate of exchange increases with increasing temperature, since the hydration equilibrium shifts towards less hydrated ions of a smaller size.

Volume exclusion effect: there is another reason for the limited degree of ion exchange. The size of the cation may be such that the sum of the volumes of all the cations required to neutralize the anionic framework of zeolite is greater than the available space within the zeolite. This steric factor is called a volume exclusion effect.

Different exchange sites: for each zeolite structure there are different sites within the zeolite in terms of the energies of interaction associated with them, leading to a particular ion population. For example, in the hydrated Na form of zeolite A, of the 12 Na^+ ions per pseudo-unit cell, eight occupy Site I adjacent to the 6-membered ring (6-MR) openings to the β-cage (sodalite cage), three occupy Site II, near the center of the 8-MR openings to the large (α) cage and one is sited in the center of the large cavity.

The equilibrium aspects of ion exchange processes have been comprehensively studied [4]. The exchange reaction between ion A^{a+}, initially in the solution, and ion B^{b+}, initially in the zeolite, can be expressed as follows:

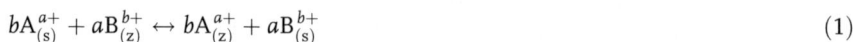

$$bA^{a+}_{(s)} + aB^{b+}_{(z)} \leftrightarrow bA^{a+}_{(z)} + aB^{b+}_{(s)} \tag{1}$$

where the subscripts (s) and (z) designate the solution and zeolite phases, respectively.

The equivalent fractions of exchanging cation in the solution and in the zeolite are defined by:

$$S_A = am_A/(am_A + bm_B) \tag{2}$$

$$Z_A = aM_A/(aM_A + bM_B) \tag{3}$$

where m_A and m_B are the molarities of the cations in solution and M_A and M_B are the equivalents of the cations in the zeolite. Obviously, $S_A + S_B = 1$ and

References see page 930

$Z_A + Z_B = 1$. The ion exchange isotherm is a plot of Z_A as a function of S_A at a given total concentration in the equilibrium solution and at a constant temperature.

The preference of the zeolite for one of two ions is expressed by the selectivity coefficient (separation factor) α_B^A defined by:

$$\alpha_B^A = Z_A^b S_B^a / (Z_B^a S_A^b) \tag{4}$$

If A^{a+} is preferred by the zeolite to B^{b+}, α_B^A is greater than unity. This is the case of Ag^+ exchange with Li^+ in zeolite A. The opposite is the case of Li^+ exchange with Na^+ in zeolite X. The selectivity series in univalent cation exchange in zeolite A was found to be:

Na > K > Rb > Li > Cs.

For zeolite X, the order of decreasing selectivity is dependent on the degree of exchange:

Cs > Rb > K > Na > Li below a level of 40 % exchange

Na > K > Rb > Cs > Li above a level of 50 % exchange

Figure 1 exemplifies the isotherms for the ion exchange in zeolite X [5]. While zeolite A is very selective to Ca^{2+} as compared to Na^+ and to a lesser extent is also selective to Mg^{2+}, in general, the selectivity is not simple for divalent cations with X and Y. These selectivities are explained in terms of degree of hydration depending on the cations and different exchange sites.

Ion exchange is usually conducted in aqueous systems. Nonaqueous solvent is rarely used due to the slow rate of exchange; the solvent can have a profound effect on the ion exchange reaction. The rate of reaction is low in dry ethanol compared to that in water. Not only the rate of ion exchange but also the actual position of the exchange equilibrium is dependent on the solvent.

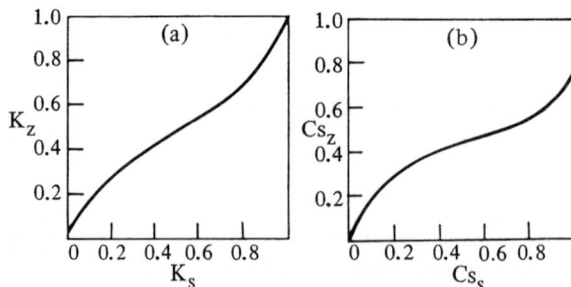

Fig. 1. Ion exchange isotherms for univalent-univalent ion exchange in zeolite X (Si/Al = 1.26) at 298 K and 0.1 M (total ion) [5]. (a) Na^+ K^+; (b) Na^+ Cs^+.

Solid-state ion exchange is a relatively new method [6]. It has advantages over the conventional ion exchange as follows: (1) handling of large volumes of salt solutions is not required; (2) the problem of the waste salt solutions can be avoided; (3) and metal cations might be introduced into narrow pore cavities in cases where ion exchange in aqueous solutions would be impeded by the large hydration shell of the cations.

Depending on the starting material, solid-state ion exchange may be expressed by two types of equations:

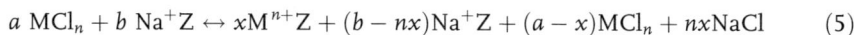

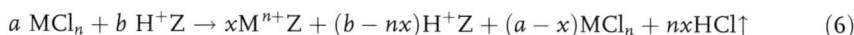

$$a\ MCl_n + b\ Na^+Z \leftrightarrow xM^{n+}Z + (b - nx)Na^+Z + (a - x)MCl_n + nxNaCl \qquad (5)$$

$$a\ MCl_n + b\ H^+Z \rightarrow xM^{n+}Z + (b - nx)H^+Z + (a - x)MCl_n + nxHCl\uparrow \qquad (6)$$

In Eq. (5), it is assumed that the reaction takes place between a chloride of the cation M and Na zeolite for the sake of simplicity. In this case, equilibrium is obtained in a similar manner to conventional ion exchange. If one starts with the hydrogen form as in the case of Eq. (6) and removes the evolved HCl in a stream of inert gas or under vacuum, the equilibrium of Eq. (6) may shift to the far right, resulting in a 100 % exchange.

Zeolites and the compound that contains the ingoing cation must interact intimately so as to cause an efficient solid-state exchange. This is attained by grinding or milling a mixture of both components. It is useful to prepare a suspension of the solids in an inert solvent and then to mix it thoroughly in order to avoid the mechanical destruction. Typically, the intimate mixture is subsequently heated to 525–625 K in a stream of inert gas or under vacuum, resulting in a maximum degree of ion exchange. Desorption of adsorbed water is required to prevent the loss of crystallinity due to the evolved HCl or HF.

Solid-state ion exchange was originally carried out to remove the acidic OH groups by way of the reaction of Y-type zeolite and NaCl. The study was extended to metals other than alkali, and alkaline earth metals and cations of metals such as transition metals [7], rare earth metals [8], and noble metals [9] are successfully incorporated into zeolites by employing this method. In solid-state exchange reactions, a trace amount of water can greatly enhance the rate of ion exchange.

Through ion exchange one may vary the cations size, or by introducing ions of different charge from those originally present the number of ions per unit cell may be changed. According to the ion and its size and valency, the cation locations often alter. Thus, within a given framework much can often be done to modify the sieving behavior. In zeolite Na-A the Na^+ ions partially block the 8-MR windows so that normal paraffins cannot penetrate the crystals. While pore-size reduction occurs upon the exchange of Na^+ by K^+ with a larger diameter, by exchanging $2Na^+$ by Ca^{2+} the windows are cleared of the cations and normal paraffins now penetrate freely (Fig. 2). As shown in Fig. 3, the increase in the adsorption does not occur in a linear fashion but rather abruptly [10]. Here the molecular sieving property is

References see page 930

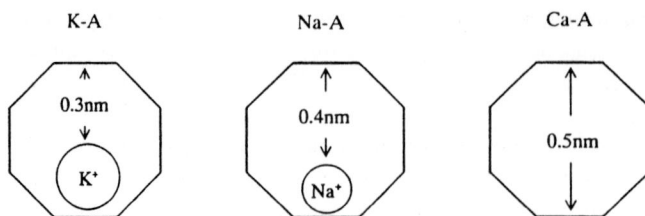

Fig. 2. Effect of ion exchange on the effective size of 8-membered ring windows of zeolite A.

determined by the location of the cations as well as their number. Branched and cyclic paraffins are too large to penetrate and so an important process has been developed for sieving normal paraffins from its mixture with other hydrocarbons.

Zeolite catalysts are mainly used for quite a number of important acid-catalyzed reactions. Ideally the total number of acid sites is equal to the total number of Al atoms on framework tetrahedral (T) sites. The overall catalytic activity of zeolites as solid acids depends on both number and property of acid sites.

In general, zeolite synthesis yields the form neutralized by sodium ions. Usually, this is also the case if template molecules (structure-directing agents) are used; the template-removal step is followed by removal of Na by ion-exchange techniques. However, many of the crystalline zeolites decompose when treated with strong acids. Therefore, the most effective and gentle methods for converting the sodium form into the hydrogen forms involve exchange of the cation by ammonium from an aqueous solution of ammonium salt. Subsequent thermal treatment of the ammonium-exchanged zeolites results in the liberation of NH_3 and the formation of

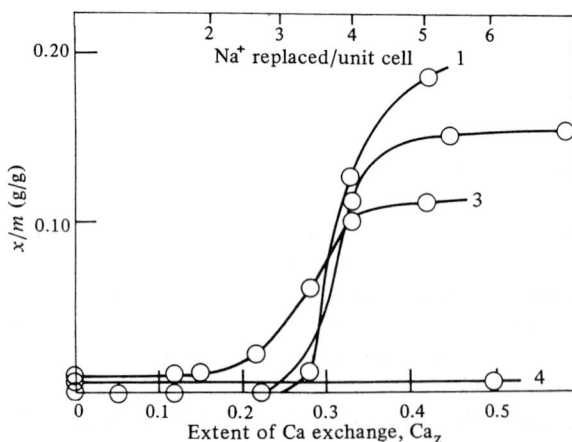

Fig. 3. Effect of calcium exchange for sodium on the sieving properties of zeolite A [10]: (1) nitrogen, 15 Torr, 77 K; (2) heptane, 45 Torr, 298 K; (3) propane, 250 Torr, 298 K; (4) isobutane, 400 Torr, 298 K.

the proton (acidic) form of the zeolites (Eq. 7).

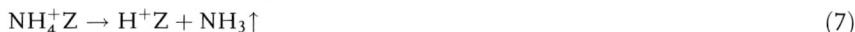

$$NH_4^+Z \rightarrow H^+Z + NH_3\uparrow \tag{7}$$

Since the acidity is adversely affected by a small amount of residual Na [11], Na must be exhaustively removed in order to obtain a highly active solid acid catalyst. Removal of Na from zeolites usually requires repeated ion-exchange steps combined with calcination in the temperature range of 823–1050 K. Only about 70 % of the Na ions are replaced by NH_4 ions during the first exchange. It is assumed that as a result of calcination Na atoms, which were not accessible for ion exchange, are redistributed over the zeolite surface and made accessible. Simultaneously, solid-state reactions occur in the zeolite and framework aluminum is removed. This phenomenon results in stabilization of the zeolite structure, as will be discussed later.

Hydrogen forms can be prepared in different ways. In general, zeolites with a Si/Al ratio of 5 or higher are resistant to acids. Direct treatment of high-silica zeolites with strong acids results in the progressive replacement of the cation by the hydronium ion. For mordenite, for example, this reaction is illustrated by the following reaction:

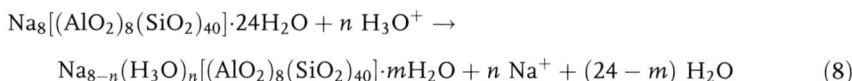

$$Na_8[(AlO_2)_8(SiO_2)_{40}]\cdot24H_2O + n\,H_3O^+ \rightarrow$$

$$Na_{8-n}(H_3O)_n[(AlO_2)_8(SiO_2)_{40}]\cdot mH_2O + n\,Na^+ + (24-m)\,H_2O \tag{8}$$

Replacement of all the original cations by the hydronium ions is possible. However, dealumination, removal of aluminum from the framework, occurs to a certain extent, which is an unavoidable side effect, as will be described below.

Elements other than hydrogen are introduced into the zeolites by means of ion exchange in order to enhance catalytic activity. By polyvalent ion exchange followed by calcination, the zeolite is changed into the protonic form:

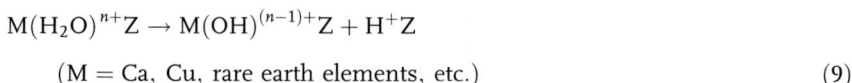

$$M(H_2O)^{n+}Z \rightarrow M(OH)^{(n-1)+}Z + H^+Z$$

$$(M = Ca,\,Cu,\,rare\ earth\ elements,\ etc.) \tag{9}$$

The first generation of zeolite FCC catalyst involved use of zeolite Y exchanged with trivalent rare earth ions and activated by calcination according to Eq. (9).

Bifunctional (metallic and acidic functions) catalysts are applied in a variety of oil refining and petrochemical processes. Zeolite supported noble metal catalysts are conveniently prepared by ion exchange using an aqueous solution of a cationic metal complex [12, 13]. For platinum and palladium, $[Pt(NH_3)_4]^{2+}$ and $[Pd(NH_3)_4]^{2+}$ are used, respectively. The ion-exchanged product is then reduced in a stream of hydrogen to produce small metal particles inside the zeolite pores ideally. This reduction is accompanied by formation of the Brønsted acid sites:

References see page 930

$$[M(NH_3)_4]^{2+}Z + H_2 \rightarrow M + 4\,NH_3 + (H^+)_2Z \quad (M = Pt\ or\ Pd) \tag{10}$$

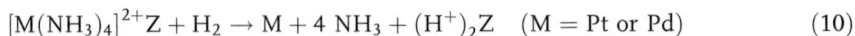

If acidic activity is undesirable, the acid sites thus formed must be neutralized before the catalytic use.

Because of the cost of noble metals, the metal loading should be low and so the metal must be well dispersed to make the exposed metal surface area as high as possible. Each of the type of metal complex, the conditions for ion exchange, the calcination procedure, and the reduction conditions has a profound effect on the dispersion of the metal. In general, the metal complex has a strong interaction with the zeolite framework. Therefore, the metal is deposited near the outer surface of the zeolite crystals, forming a thin layer of high concentration of metal. This inhomgeneity can be avoided by the addition of a competing cation such as NH_4^+.

Impregnation is an alternative to ion exchange for the introduction of other metal elements. By using a solution whose volume is equal to the pore volume of the zeolite, pore-filling impregnation can be attained. It is to be noted that ion exchange could occur using such a technique. However, different from the ion-exchange technique, not only the cation that was originally present and replaced with the cation of the added salt but also the counter anion might remain inside the pore under the conditions where salt imbibition can occur. This could result in the back-exchange with the cation present in the original zeolite. It is favorable to employ nitrates or salts of organic acids such as acetates, since these anions are easily decomposed by heating to give no residual poisonous elements. Naturally, impregnation with salts containing metal elements as anion (e.g., molybdate, tungstate, chromate, manganate, and palladate) is not suitable for the preparation of well-dispersed metal-containing catalysts. A few cationic complexes containing Mo can be utilized for metal loading by ion exchange [14].

The use of metal carbonyls can afford highly dispersed metal catalysts [13]. The reaction of protonic type zeolites with metal carbonyls is exemplified as follows:

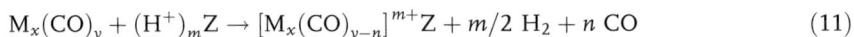

$$M_x(CO)_y + (H^+)_mZ \rightarrow [M_x(CO)_{y-n}]^{m+}Z + m/2\,H_2 + n\,CO \tag{11}$$

The subsequent treatment may produce highly dispersed metal particles.

In zeolite catalysis, emphasis has been placed on the reactions catalyzed by acids [15]. However, complete ion exchange with alkali metal cations such as K, Rb, and Cs would allow the preparation of weakly basic zeolites [16]. Oxygens as counter anions of these alkali metal cations act as basic sites. Their basicity depends on the fractional negative charge they bear and therefore, the composition of the zeolite, which can be quantified by the intermediate Sanderson electronegativity, S_{int}. It is also dependent on structural parameters such as bond length, bond angles, and Al distribution. Pt L zeolites in the alkaline cation form are active in the dehydrocyclization of hexane to benzene. The lower the S_{int} value, the more active the catalysts, showing that a high zeolite basicity results in the increased aromatization activity of the Pt sites [17].

It is to be noted that basic sites sometimes play a significant role in adsorption/ separation [16]. For example, C_8 aromatics can be separated over X and Y zeolites exchanged with alkali metal ions. The order of preference for the isomers may be related to the adsorbent intermediate electronegativity, i.e., to its basic character.

Zeolites have been considered as promising hosts to molecular clusters whose dimensions are sufficiently small that deviations from bulk behavior are observed [18]. There are two types of clusters receiving particular attention. The clusters formed from Group 12-Group 16 and Group 13 and Group 15 semiconductors have been intensively studied. The other types are alkali metal clusters.

It is expected that quantum confinement effects occur when the individual clusters of the semiconductor are smaller than the effective Bohr radius of the exciton, the bound electron–hole pair. The motion of the weakly bound electron–hole pair becomes quantized, leading to unusual optical and electronic properties. Herron et al. [19] reported the synthesis of CdS clusters located inside the sodalite cages of zeolite Y. The blueshift from 290 nm to 350 nm with the increase in CdS concentration was explained by the transition from isolated cubane-like $(CdS)_4$ to a semiconductor superlattice. It has been demonstrated that MOCVD techniques can be successfully applied to the synthesis of GaP inside the pore of zeolite Y [20]. PbI_2 clusters were incorporated into the α cages of zeolite A through the vapor phase without destroying the framework. Blueshift in absorption spectra was observed and it was concluded that superlattice reflections in the XRD pattern were produced by displacement of the cluster without atomic diffusion [21].

When alkali metal atoms are loaded into the dehydrated zeolite space, the s electrons are delocalized over many cations resident in the space, and the cationic clusters are generated. A large number of alkali-metal clusters of the type M_n^{+p} are stabilized in the nanospace of zeolite crystals [22]. The electrons of alkali atoms are shared with other alkali atoms in cages, occupying quantum levels, such as 1s and 1p, etc. In other words, the electrons are solvated by the electron trap afforded by these counter-ions. These clusters show novel properties depending on the structure of clusters, and quantum-mechanical effect on s electrons. When isolated, these clusters are color centers. At higher concentrations, aggregation of clusters leads to materials with different electronic properties, as revealed by modified EPR signal and optical spectra. These clusters are suitable candidates for the study of metal-to-insulator transitions [23]. In K-loaded K-A, ferromagnetic properties are observed depending on the average loading density of guest K atoms, although no magnetic element is contained [24]. The magnetic properties show a spin-glass behavior and are interpreted in terms of a model of itinerant electron ferromagnetism. Alkali metal clusters such as $(Na_4)^{3+}$ are formed in zeolite matrix by impregnation with NaN_3 and used in base-catalyzed reactions [25].

Research on coordination chemistry in zeolites started in the 1970s and early work was summarized by Lunsford [26]. A metal complex of the appropriate dimensions can be encapsulated in a zeolite, being viewed as a bridge between ho-

References see page 930

(a)

Free Salen I

[1,6-bis(2-hydroxyphenyl)-
2,5-diaza-1,5-hexadiene]

or N,N′-bis(salicyliden)-
ethylenediamine

Co-Salen

(b)

4

1,2-Dicyano-
benzene or
phthalodinitrile

Copper-Phthalocyanine

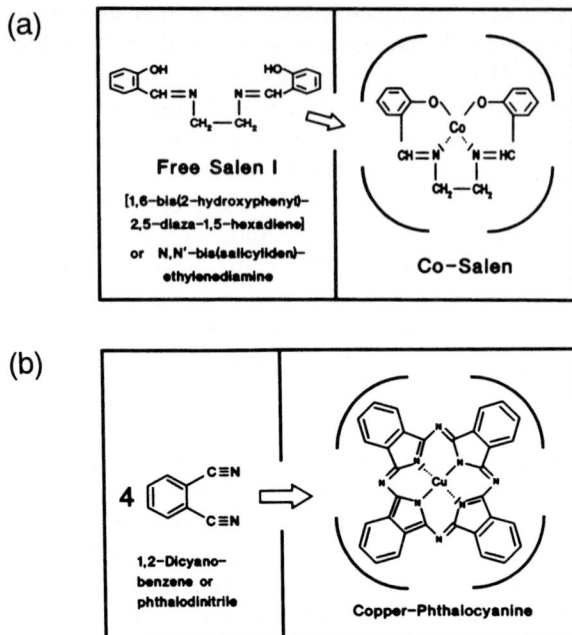

Fig. 4. Ship-in-a-bottle syntheses [27]: (a) cobalt salen; (b) copper phthalocyanine.

mogeneous and heterogeneous systems. Complexes that are smaller than the free diameters of the channels and windows have access to the cavities. On the other hand, complexes that are larger than the diameters of the windows must be synthesized *in situ*, namely, by adsorption of the ligands into the zeolites containing transition metal ions (Fig. 4a) or by synthesis of the ligands in those zeolites (Fig. 4b) [27–29]. Herron et al. [30] first referred to such zeolite guest molecules as ship-in-a-bottle complexes. Since the first report on the synthesis of a metal phthalocyanine inside zeolite Na-Y in 1977 [31], numerous examples of encapsulation of metal phthalocyanine complexes have been provided. Related porphyrin and N,N′-bis(salicylidene)ethylenediimine (SALEN) complexes have also been trapped in a zeolite cavity that has restricted apertures. These are typical examples of ship-in-a-bottle complexes and are given names like zeozymes [32] and inorganic protein [33] in regard to the biomimetic chemistry, for instance, as a model for dioxygen binding and oxygenase.

Metal carbonyl clusters encapsulated in a cavity of zeolites have been widely studied [34], forming a special subgroup of ship-in-a-bottle complexes. Even Chini-type complexes as large as $[Pt_3(CO)_3(\mu_2\text{-}CO)_3]_5^{2-}$ have been characterized in Na-Y. Such clusters are formed by the reductive carbonylation of intrazeolite metal ions, for which the presence of H_2 or H_2O as well as CO seems to be necessary.

Because the complexation of metal ions by ligands does not always proceed rapidly, excess free ligand and free noncomplexed metal as well as complexed metal might be present. There are also nonencapsulated species on the external surface of zeolite. The free ligand and outer surface species can be removed by Soxhlet extraction. Distinction between free phthalocyanine base and metallated phthalo-cyanine is made by infrared spectra; the symmetry change from D_{2h} to D_{4h} upon chelation should result in the coalescence of the split band for the free base.

Electronic spectra can be applied to the characterization of ship-in-a-bottle complexes in zeolite. Bulky complexes may be deformed by encapsulation in the re-stricted intrazeolite space. The most intense band for the phthalocyanine complex is the Q band based on the $\pi-\pi^*$ transition of the phthalocyanine ligand. The red-shift of the Q band relative to the free ligand or surface physisorbed complex was interpreted to be due to the distortion of the planar ligand to generate the saddle structure. XPS and IR spectroscopic data were also interpreted in a similar man-ner. However, it was suggested that the spectral change arose from zeolite solvent effects or protonation of peripheral nitrogens; the distortion is not clearly revealed and the subject is under dispute [32].

Zeolites can be synthesized around metal complexes, which is a versatile ap-proach to the synthesis of ship-in-a-bottle complexes [28]. This would afford the advantage that zeolite-encapsulated metal complexes are prevented from being contaminated by free ligand as well as uncomplexed metal ions. The metal com-plex must be stable under crystallization conditions. Metal phthalocyanine com-plexes and cobalticium ion Cp_2Co^+ can be encapsulated in zeolites crystallized in the presence of these complexes. The latter seemed to act as a structure-directing agent. New zeolite structure types can be derived from possible structure-directing properties of metal complexes. The use of $(CH_3)_5Cp_2Co^+$ complex resulted in the discovery of a totally new zeolite topology containing monodimensional 14-MR channels, UTD-1 [35].

4.2.5.2
Dealumination and Deboration

The chemical composition of the zeolite framework may be post-synthetically changed by the application of various techniques for dealumination [36, 37] and tetrahedral (T)-site substitution reactions.

The chemical composition of the framework is dependent on the synthesis con-ditions. For certain structures zeolites with widely varying Si/Al ratio can be syn-thesized. However, for most zeolite structures, the crystallization is possible only in a limited range of Si/Al ratio. For example, zeolite Y with the faujasite (FAU) struc-ture with a framework Si/Al ratio >3 is difficult to crystallize.

In general, the zeolite structure is susceptible to acid leaching of Al since all the Al is on the surface, mostly intracrystalline surface. This is in contrast to the clays, which contain Al not accessible to mineral acid. By treatment of high-silica zeolites with mineral acid, replacement of all the original cations by hydronium ions is possible. However, further treatment with strong acid removes the framework alu-minum ions to form hydroxyl nests.

References see page 930

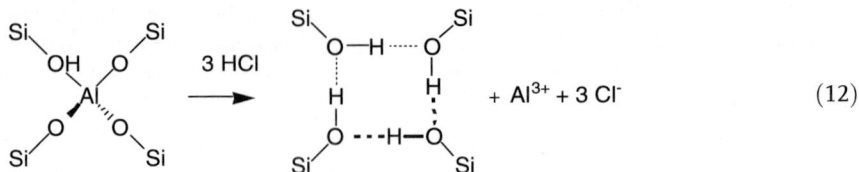

$$
\text{(Si)}_4\text{Al(OH)} + 3\,\text{HCl} \longrightarrow \cdots + \text{Al}^{3+} + 3\,\text{Cl}^- \tag{12}
$$

The crystallinity of high-silica zeolites such as mordenite, erionite, and clinoptilolite can be basically retained upon leaching with strong mineral acids. In the case of Al-rich zeolites such as type Y zeolites, however, attempts to achieve complete hydrogen exchange by this method result in structural disintegration.

It is observed that the acid strength of zeolites increases with increasing Si/Al ratio until a maximum value is reached at a Si/Al ratio of about 9 or 10 [38]. Furthermore, the acid forms of low-silica zeolites are inherently unstable. Although the enormous potential of zeolite Y as a cracking catalyst (fluid catalytic cracking: FCC) was recognized, its low stability was a serious drawback to practical use. Therefore a great effort was made and various techniques to enhance the thermal and hydrothermal stability of the zeolite Y by increasing the Si/Al of the framework have been developed, as summarized in Table 1 [37].

It was discovered that treatment of proton-exchanged Y with steam results in the stabilization of the material. By steaming NaY partially exchanged with NH_4 ions at temperatures higher than 730 K, Al species are removed from the framework, giving rise to vacancies in the lattice. Since the presence of steam is essential for dealumination, the following solid-state reaction is assumed.

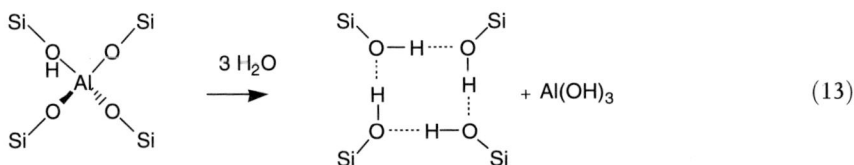

$$
\cdots + 3\,\text{H}_2\text{O} \longrightarrow \cdots + \text{Al(OH)}_3 \tag{13}
$$

Tab. 1. Summary of the effect of various modification methods on the acid activity and shape selectivity of the zeolite modified [37].

	Activity change	Pore modification
Ion exchange	Yes	Yes
Mineral acid	Yes	Sometimes
H_2O (steam)	Yes	Mesopores formed
EDTA	Yes	Mesopores formed
$SiCl_4$	Yes	Mesopores formed
$(NH_4)_2SiF_6$	Yes	–
HF	Yes	–
Organic adsorption	Yes	Yes
Coke	Yes	Yes
CVD of organometallics	Sometimes	Yes
'Al reinsertion'	(Yes)	?

The vacancies are repaired by migration of Si species originating from other parts of the crystals, resulting in the substitution of Si for Al. Thus the product should have a much higher Si/Al ratio in its framework than the starting material; however, elemental analysis of the material by ICP may show the relatively small Si/Al ratio. This is because Al removed from the framework remains as extra-(non)framework Al species, which may play an important role in acid-catalyzed reactions. The extra-framework Al species can be removed by acid washing following the steaming. By treating steam-dealuminated samples with aqueous $(NH_4)_2SiF_6$ solution, it is possible to prepare ultrastable Y samples without extra-framework Al [39].

This type of hydrothermal treatment at elevated temperatures, known as ultra-stabilization, is still the most common procedure for the industrially important dealumination of Y zeolites [40, 41]. Kerr [42] used the term "deep bed" to describe the dealumination of ammonium Y in high temperature hydrous atmosphere. McDaniel and Maher [43] employed the "shallow bed" method, obtaining a similar material having high thermal stability. The advantage of the "shallow bed" method over the "deep bed" one seems to be the control of the level of the hydrothermal environment [44]. A comparative study has revealed that the properties of steam-dealuminated products do not depend on the method but on the resultant framework Si/Al ratio.

The acid function of the stabilized Y is dependent on parameters such as the Si/Al ratio and the Na_2O content (extent of NH_4^+ exchange) of the parent zeolite, and temperature, residence time and partial pressure of steam during the hydrothermal treatment. Higher temperatures and a high degree of NH_4^+ exchange result in the higher degree of dealumination [45]. At a constant temperature, the degree of dealumination initially increases with increasing partial pressure of steam initially and eventually levels off [45].

The Si/Al ratio of the framework can be determined by the unit cell constant. Since Si has a smaller atomic radius than Al, the substitution of Si for Al leads to the decrease in the unit cell dimensions. The relationship between unit cell size and Si/Al of Y zeolite is the subject of many reports. For example,

$$N_{al} = 115.2(a_o - 24.191) \; [46] \tag{14}$$

$$N_{al} = 107.1(a_o - 24.238) \; [47] \tag{15}$$

where N_{al} is the number of framework Al atoms per unit cell (192 T atoms), and a_o is the unit cell constant in Å. The inaccuracy of the numerical values of the coefficients in Eqs (14) and (15) is due to the error made in the distinction between framework and extra-framework Al.

Since the O–T–O stretching frequencies increase with decreasing Al content in the zeolite framework, infrared spectroscopy can also be used to determine the number of framework Al atoms [47].

There have been many studies of ^{29}Si and ^{27}Al NMR spectra of zeolites [48]. While ^{29}Si magic angle spinning (MAS) NMR spectra are straightforward, ^{27}Al NMR spec-

References see page 930

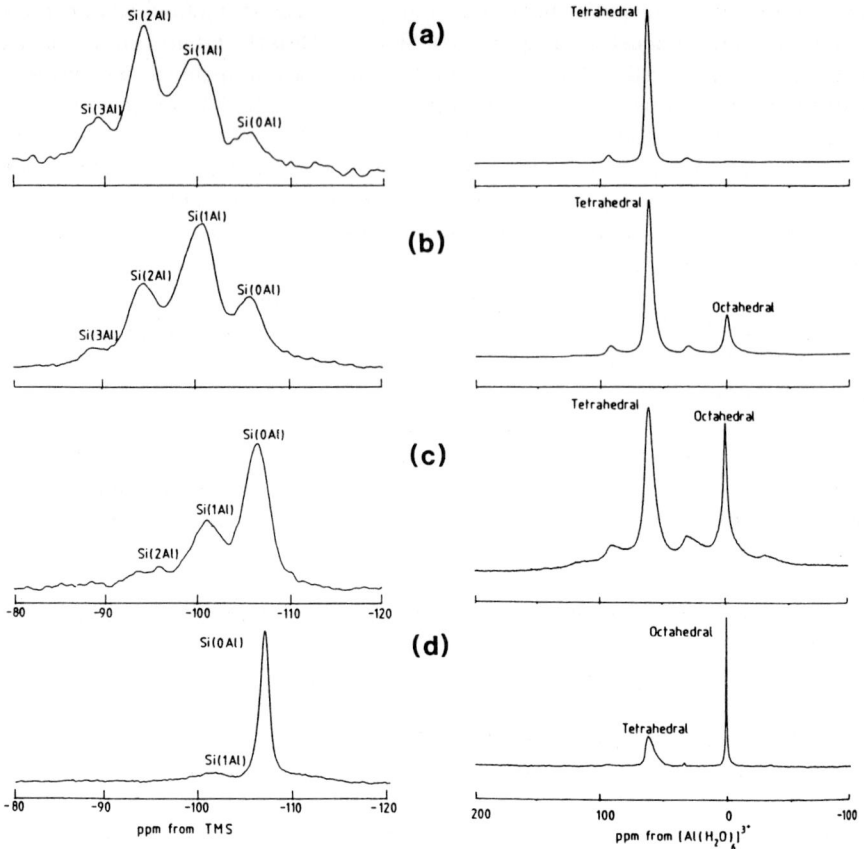

Fig. 5. High-resolution ^{29}Si (at 79.80 MHz) and ^{27}Al (at 104.22 MHz) MAS/NMR studies of the ultrastabilization of zeolite Y [49]: (a) Parent zeolite NH$_4$-Na-Y; (b) after calcining in air for 1 h at 673 K; (c) after heating to 973 K for 1 h In the presence of steam; (d) after repeated ion exchange, heating, and prolonged leaching with nitric acid.

tra are much more complex because of the quadrupolar interaction. In general, the ^{29}Si NMR spectra contain a maximum of five well-resolved peaks, corresponding to five possible distributions of Si and Al around a silicon nucleus at the center of an SiO$_4$ tetrahedron; namely, Si[4Al], Si[3Al, 1Si], Si[2Al, 2Si], Si[1Al, 3Si], Si[0Al, 4Si]. As shown in Fig. 5, the dealumination process can be followed by observing ^{29}Si and ^{27}Al NMR spectra [49]. Assuming the Lowenstein's rule, which forbids Al–O–Al linkages in the zeolite framework, the Si/Al ratio of the lattice can be calculated directly from the ^{29}Si NMR spectrum according to Eq. (16).

$$\mathrm{Si/Al} = \sum_{n=0}^{4} I_{\mathrm{Si}[n\mathrm{Al}]} \bigg/ \sum_{n=0}^{4} 0.25 n I_{\mathrm{Si}[n\mathrm{Al}]} \qquad (16)$$

This method detects the Al atoms indirectly from the effect on the Si atoms in the framework and thus detects only framework Al atoms and the Si/Al ratio for the framework. This is in contrast to the bulk chemical analysis, which include both framework and any Al occluded in the cavities or present as impurities and not an integral part of the system.

For the dealumination of zeolites, other techniques such as acid leaching [40] and EDTA extraction [40, 50] can be employed. These techniques, however, will leave lattice positions vacant that were occupied by Al. Accordingly, the stability is adversely affected unless the structural defects are remedied. Thus while mordenite [51] and beta [52] can be dealuminated by acid without structural collapse, dealumination using mineral acids cannot be applied to faujasites. Both methods produce materials with Al-deficient surfaces because the outer surface of the crystal is attacked preferentially.

Several additional methods have been developed, which include gas-phase reaction with anhydrous halides such as $SiCl_4$ at high temperatures [53, 54], and liquid-phase reaction with aqueous $(NH_4)_2SiF_6$ [55]. Both methods have an advantage over acid leaching or EDTA extraction, since Al vacancies can be filled with external Si sources.

The framework Al is replaced with Si by the reaction with $SiCl_4$ vapor at 780–830 K [53]. The stoichiometry of the dealumination of MY corresponds to Eq. (17).

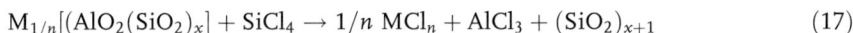

$$M_{1/n}[(AlO_2(SiO_2)_x] + SiCl_4 \rightarrow 1/n\ MCl_n + AlCl_3 + (SiO_2)_{x+1} \tag{17}$$

Starting with the Na-form, the resulting Al is present as $NaAlCl_4$ in the sample. Partial exchanging with Li enhances the volatilization of Al chloride [56]. Remaining entrapped Al chloride decomposes to aluminum oxide by treatment with water. An Al-rich surface is observed, which is ascribed to the migration of the residual Al species to the outer surface.

The use of aqueous $(NH_4)_2SiF_6$ provides a very mild method of dealuminating zeolites [55]. Zeolite Y is subjected to dealumination at <373 K.

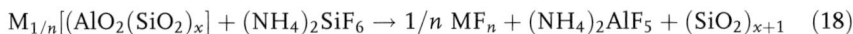

$$M_{1/n}[(AlO_2(SiO_2)_x] + (NH_4)_2SiF_6 \rightarrow 1/n\ MF_n + (NH_4)_2AlF_5 + (SiO_2)_{x+1} \tag{18}$$

The resultant $(NH_4)_2AlF_5$ can be removed by washing with hot water. Thus, the dealuminated zeolites are free of extra-framework Al. This method in general gives products also free of defects. However, if H-type zeolite is used, HF is formed, resulting in the formation of defect sites such as the hydroxyl nest [57].

XPS analysis has shown that the Si/Al ratio on the outer surface of the sample dealuminated with $(NH_4)_2SiF_6$ is much higher than the bulk Si/Al ratio. Combined with SIMS analysis, it has been concluded that the gradient of the Si/Al ratio is due both to a gradient of dealumination (diffusion controlled dealumination) and selective deposition of silica on the outer surface [58]. Silica is probably formed by hydrolysis of SiF_4 that results from the decomposition of $(NH_4)_2SiF_6$.

References see page 930

Dilute F_2 gas can be used to treat zeolites at near-ambient temperature [59]. The subsequent calcination at 823 K gives dealuminated zeolites with high crystallinity retained. The F_2-treated samples show hydrophobicity and enhanced catalytic activity for butane cracking. The activity increase is ascribed to the extra-framework F-containing Al species remaining after the calcination.

In general, the post-synthesis modification methods described above can be applied to high-silica zeolites. However, when the steaming method is applied to zeolite L, it cannot be dealuminated without suffering structural damage. In the reaction with F_2, not all zeolites show an increase in thermal stability with treatment. Both zeolites omega and L lose crystallinity when calcined after treatment [59].

The dealumination of zeolites other than Y leads to an increase in catalytic activity as well as thermal stability. The catalytic activity of mordenite [60] and ZSM-5 [61] is enhanced by hydrothermal treatment. The enhancement of activity of the high-silica zeolites only occurs under mild steaming conditions. Under severe steaming conditions, activity is decreased because of the substantially reduced amount of Al in the framework.

By means of solid-state NMR, the nature of the Al species formed during the dealumination process has been extensively studied [48]. The band at 0 ppm in the ^{27}Al NMR spectra is assigned to octahedral Al species. It is commonly accepted that octahedrally coordinated Al species belong to extra-framework Al species. Grobet and coworkers [62] have recently found octahedral Al species that are attached to the zeolite framework. They have claimed that Al–OH groups, formed by a partial hydrolysis of the framework Al–O bonds, are present only in the samples treated at low steam pressure, suggesting that the framework-bound Al–OH species are an intermediate in the dealumination process [63]. In the mild hydrothermal treatment of MCM-22, tricoordinated framework Al, which may serve as the Lewis acid site of the zeolite, is found to be formed [64].

During the dealumination treatment, pores with a diameter of approximately 20 nm are formed in the zeolite grains. This is particularly the case in the dealumination of Y by steaming and with EDTA or $SiCl_4$. In the case of $(NH_4)_2SiF_6$-treatment, there is no mesopore formation observed for thoroughly washed samples [39]. The mesopores emerge when parts of the structure collapse, associated with removal of framework Al, leaving extra-framework Si as well as extra-framework Al. The mesoporous structure of Y zeolites should be beneficial for the use as FCC catalyst, because the diffusivity of the feedstocks in the catalyst will be increased. Dealumination of mordenite by acid leaching leads to mesoporosity [65], which mitigates the problem due to the monodimensional channel system of mordenite.

The origin of these mesopores formed by the dealumination of the zeolite Y was studied by transmission electron microscope observation [66]. It was proposed that mesopores are formed along $\langle 100 \rangle$ directions in the preexisting (111) twin planes of the FAU structure; thus it was pointed out that it should be possible to control the distribution of mesopores by controlling the distribution of twins during synthesis.

Fig. 6. Adsorption isotherms on Na-Y (broken line) and dealuminated Y (Si/Al = 22) zeolite (solid lines) [53]: (a) hexane; (b) butane; (c) benzene; (d) ammonia; (e) water.

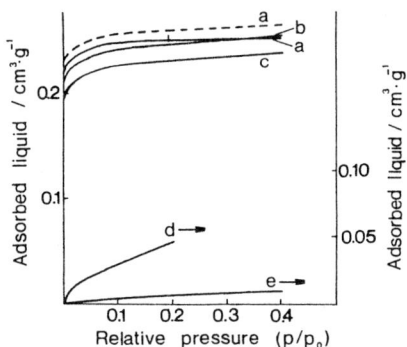

The zeolite surface in the intracrystalline pore is strongly influenced by the crystal field of zeolites, giving the zeolite solid electrolyte properties. The Al content is the predominant parameter determining the ionic character of the zeolite, its strength as an electrolyte increasing with increasing Al content. Thus the polar character of zeolite surfaces results in strong interaction with polar or polarizable molecules. However, in the high silica zeolites, the surface becomes hydrophobic [67]. High silica zeolites can be obtained by framework modification of hydrophilic zeolites or by direct synthesis. The examples of the former are siliceous types of Y, MOR, beta, MCM-22, etc.

Figure 6 shows adsorption isotherms on NaY and dealuminated Y zeolites [53]. The dealuminated Y sample has an extremely low selectivity for the adsorption of water and ammonia and a high capacity for the adsorption of hydrocarbons. The cation also affects the hydrophobicity/hydrophilicity; e.g., high silica HZSM-5 shows higher hydrophobicity than its Na^+-exchanged form [68].

Hydrophobic zeolites are effective in removing organics from waste water [69] and catalyzing organic reactions involving water as reactant, product or solvent [70]. As a quantitative measure for hydrophobicity/hydrophilicity, the Hydrophobicity Indexes have been proposed [71], which were determined from breakthrough curves using either toluene or octane and water as adsorbates.

Borosilicates, boron-containing zeolites, are most easily obtained by direct synthesis methods. Boron-beta [72] is a typical example. Borosilicates are suitable as the parent materials for post-synthesis preparation of high-silica or metal-containing zeolites, because boron can be easily extracted under mild conditions. Deboration gives materials with a large quantity of silanol nests left, which can be reoccupied with the external reagents to afford high-silica or metal-containing zeolites that are hard to be directly synthesized. Thermally stable homogeneously distributed silanol nests in MFI crystals are not obtained by dealumination procedures [73].

It has been reported that zincosilicate CIT-6 can be a useful precursor to prepare a wide range of new zeolites that have the beta topology [74]. Structure-directing

References see page 930

tetraethylammonium cations and zinc can be simultaneously removed from CIT-6 by acid extraction to form a highly hydrophobic Si-beta zeolite that has very few defects.

4.2.5.3
Insertion of Metals into Zeolitic Framework

Alumination, namely, the introduction of Al atoms into a dealuminated zeolite or a high-silica zeolite is the reaction opposite to dealumination [75]. Chang et al. [76] first demonstrated that Al could be introduced into the high-silica ZSM-5 by using $AlCl_3$ or $AlBr_3$ vapor at temperatures below 673 K. The presence of Al in the framework was confirmed by the signal due to tetrahedral aluminum in ^{27}Al NMR spectra together with the increase in Brønsted acidity. It was suggested that the alumination could proceed as the reverse reaction of dealumination, namely the substitution reaction of aluminum for silicon [77]. On the other hand, Chang et al. [76] proposed that alumination occurs by way of the reaction of $AlCl_3$ with intracrystalline defect sites such as "hydroxyl nests" as follows.

Yashima and coworkers [78] observed that aluminum could be inserted into mordenite samples that were dealuminated by treating with acid. The number of silanol groups on the defect sites was measured by the signal at −103 ppm in the ^{29}Si NMR spectra and the $C^{18}O_2$ oxygen substitution for $Si-^{16}OH$ in the hydroxyl nest [79]. The aluminum content introduced into the zeolite framework agrees with the number of hydroxyl nests, supporting that alumination occurs according to Eq. (19). Similar stoichiometry was observed for the alumination of silicalite-1.

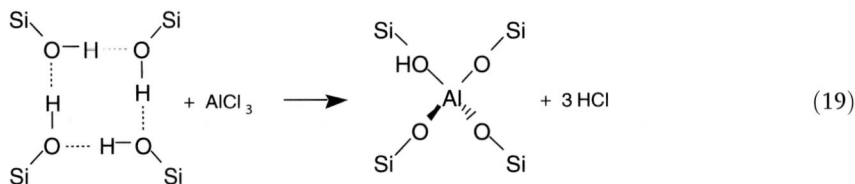

$$\text{(19)}$$

Alumination can be also carried out under liquid-phase conditions. Chang et al. [76] observed that the alumination of silicalite-1 occurs in the aqueous solution by using $(NH_4)_3AlF_6$. Aluminum atoms eliminated from the framework of zeolite Y by hydrothermal treatment can be subsequently reinserted into the framework by treatment with aqueous solution of KOH in the range of 283–303 K [80]. Aqueous NaOH is less effective than KOH in achieving reinsertion of Al. The crystallinity of the samples is largely retained in the process and depends on the concentration of the base. It has been claimed that the turnover frequency of the realuminated zeolite in m-xylene reaction is higher than that of the parent sample, as a result of the higher strength of the acid sites, caused by the higher population of the most strongly acidic $(SiO)_3-OH-Al(OSi)_3$ groupings [81].

It has been reported that the reinsertion of extra-framework Al in HZSM-5, which is dealuminated by thermal treatment at 873 K, occurs by treatment with aqueous HCl solution at 373 K [82]. Since leaching with mineral acid is one of the methods of dealuminating Al-rich zeolites such as mordenite and zeolite beta, re-alumination upon the acid treatment seems to be very puzzling. It has been proposed that the Al atom connected to the zeolite framework only by one or two remaining bonds reinserts into the framework on acid treatment according to Eq. (20) [83]. On the other hand, the reinsertion of nonframework Al species completely removed from the lattice, which is present on HZM-5 zeolite dealuminated under severe hydrothermal conditions, hardly takes place.

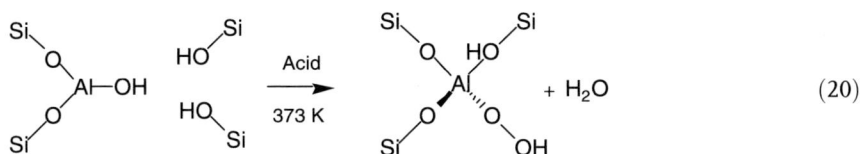

$$ (20) $$

The alumination of borosilicates is a convenient method of increasing the Al content of a high-silica material. The boron occupying the tetrahedral site is easily removable from the structure and Al can be inserted into the resultant vacant sites by low-temperature alumination. A good number of novel high-silica borosilicates have been recently prepared and Al insertion into the framework of these borosilicates offers materials having higher acidities [84].

In hydrothermal synthesis, the type and amount of T atom, other than Si, that may be incorporated into the zeolite framework are restricted due to solubility and specific chemical behavior of the T-atom precursors in the synthesis mixture. Breck [85] has reviewed the early literature where Ga, P, and Ge ions were potentially incorporated into a few zeolite structures via a primary synthesis route.

The isomorphous substitution of Ti for Si was claimed by Taramasso et al. [86]. The resulting material has the structure of silicalite-1 (pure silica MFI) with Ti in the framework positions and is named TS-1 or titanium silicalite-1. The new findings including the claim that other metals can be inserted into the zeolite framework met with skepticism. Ione et al. [87] predicted the probability of isomorphous substitution of metal ion (M^{n+}) and the stability of the M^{n+} position in the tetrahedrally surrounding oxygen atoms by using the Pauling criterion. Based on the ratio of ionic radii ρ of the cation and anion, the value for Ti and O ($\rho = 0.515$) falls out of the range ($\rho = 0.225-0.414$) for which tetrahedral coordination is expected [88]. The allowed cation would include only Al^{3+}, Mn^{4+}, Ge^{4+}, V^{5+}, Cr^{6+}, Si^{4+}, P^{5+}, Se^{6+}, and Be^{2+}. Presumably this type of estimate is surely effective, which can explain the preference of B^{3+} for trigonal coordination and the resultant instability of B^{3+} in the zeolite matrix. However, it is a very rough approximation since the completely ionic character of the T–O bond is not the case and the model assumes the atoms have a perfect round shape.

A very comprehensive review was made on TS-1 and other titanium-containing molecular sieves [89]. TS-1 proved to be a very good catalyst for liquid-phase

References see page 930

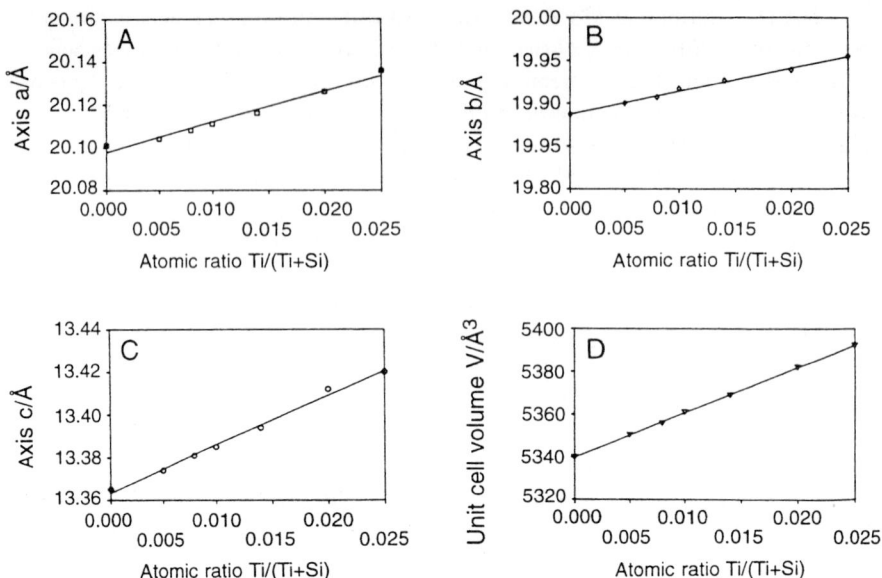

Fig. 7. Unit cell parameters of TS-1 [90].

oxidation of various organic compounds using H_2O_2 as oxidant. TS-1 is synthesized by the hydrothermal crystallization of a gel obtained from $Si(OC_2H_5)_4$ and $Ti(OC_2H_5)_4$. The incorporation of Ti into the framework of MFI structure was demonstrated by the increase in unit cell size derived from the XRD pattern, as shown in Fig. 7 [90], and the appearance of tetrahedral Ti species in UV-Vis spectra. The maximum amount of Ti that can be accommodated in the framework positions is claimed to be limited to $x = Ti/(Ti + Si)$ of 0.025. It is to be noted that, in the presence of alkalis, extra-framework Ti species are formed, giving rise to inferior catalytic properties.

It is stated that by a modified synthesis method Ti could be inserted into the lattice positions up to $x = 0.010$ [91]. However, this result is questioned [89].

After the success of TS-1 as a liquid-phase oxidation catalyst [92], great efforts have been devoted to the synthesis of titanium silicalites with different zeolitic structures. In particular, TS-1, a medium pore zeolite, shows remarkable shape selectivity. The rate of oxidation of the branched or cyclic hydrocarbons is much lower than that of the straight one [93, 94]. One of the promising titanium silicalites is Ti-beta, a large pore Ti-containing zeolite, which has been hydrothermally synthesized from gels containing tetraethylammonium hydroxide (TEAOH) as a structure-directing agent (SDA) [95]. Due to its large pore size, Ti-beta was shown to be more active than TS-1 for the oxidation of bulky substrates such as cyclic and branched molecules [96].

Ti-beta was usually obtained in very low yield [95], in contrast to TS-1. Moreover, an additional factor, namely, the presence of aluminum in the framework of Ti-beta, can contribute to the different catalytic behavior observed between Ti-beta and TS-1. In contrast to the organophilic characteristics of TS-1, the presence of Al and large concentrations of internal and external silanol groups confer a rather hydrophilic character on Ti-beta. There is therefore a strong incentive for the preparation of Ti-beta with low Al content in a better zeolite yield by using new methods. Improved methods for the synthesis of Ti-beta, e.g., use of special SDA [97], the fluoride method [98], and the dry gel conversion method [99] have been developed to obtain the Ti-beta zeolites active for oxidation with H_2O_2 in high yields. The improvement of epoxide selectivity by selectively poisoning the acid function without spoiling the oxidation activity is attained by modification by ion exchange with quaternary ammonium ions [100].

Wu et al. [101] have succeeded in the incorporation of Ti into the MWW (MCM-22) structure by adding a large amount of boric acid, which acted as a crystallization promoter. By washing the as-synthesized layered precursor material, one can obtain the material free of extra-framework Ti and of negligible acidity, which proved to be an active oxidation catalyst [102].

Another approach to substitution of metal atoms into the framework is the secondary synthesis or post-synthesis method. This is particularly effective in synthesizing metallosilicates that are difficult to be crystallized from the gels containing other metal atoms or hardly incorporate metal atoms by the direct synthesis method.

Post-synthesis modifications have been successful in preparing titanium-containing molecular sieves active in oxidation. Substitution of Ti for Al goes back to the 1980s. The reaction of zeolites with an aqueous solutions of ammonium fluoride salts of Ti or Fe under relatively mild conditions results in the formation of materials that are dealuminated and contain substantial amounts of either iron or titanium and are essentially free of defects [103]. However, no sufficient evidence for the Ti incorporation has been provided. The post-synthetic incorporation of titanium into the framework of zeolite Beta has been achieved by liquid-phase treatment employing ammonium titanyl oxalate [104].

Ti-beta has been prepared by treating Al-containing beta with a concentrated solution of perchloric or nitric acid in the presence of dissolved titanium [105]. Use of TiO_2 as the Ti source give rise to octahedral Ti as well as tetrahedral Ti. $Ti(OBu)_4$ and TiF_4 were efficient sources for the incorporation of tetrahedral Ti. In this case the extraction of Al from the framework simultaneously occurred, giving almost Al-free Ti-beta. This method was also applied to the incorporation of Ti into MOR and FAU.

Post-synthesis gas-solid isomorphous substitution methods have also been known [106]. Ti-beta essentially free of trivalent metals can be prepared from boron-beta. However, the gas phase method is not efficient for Ti incorporation and could have some disadvantage such as the deposition of TiO_2 [107].

References see page 930

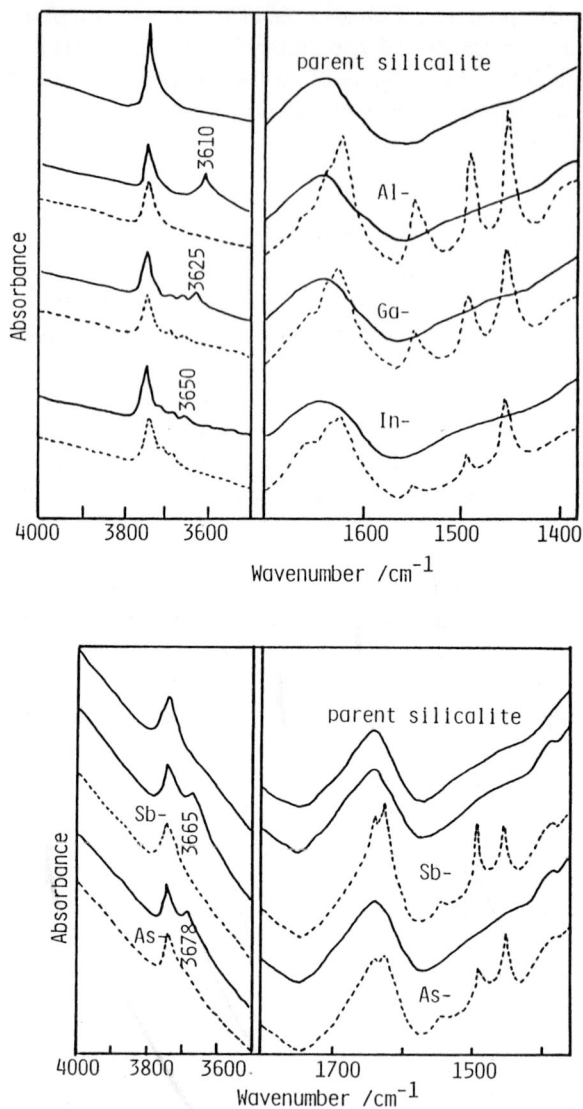

Fig. 8. IR spectra for atom-planted MFI zeolites and pyridine adsorbed on them [108].

Yashima et al. [108] proposed the "atom planting" method as the development of the alumination through the reaction of internal silanol groups with aluminum halides. By using halides of other metals, metallosilicates with the MFI structure (Ga, In, Sb, As and Ti) and with the MOR structure (Ga, Sb, and Ti) have been prepared. As shown in Fig. 8, all the materials exhibited both Brønsted and Lewis acidities. From the SiO–H stretching frequency, they concluded that the acidity of the MFI type zeolites thus prepared decreases in the order Al > Ga > In > Sb >

As. The incorporation of Ti into the MOR structure was confirmed by the appearance of the specific absorption band in the IR spectra [109].

V-beta can be prepared in a similar two-step manner. The method consists of first creating a vacant site by dealumination of the beta zeolite with nitric acid and then contacting them with an NH_4VO_3 solution [110].

4.2.5.4
Control of Pore Size and Inactivation of External Surface

Over the past three decades, a great many new synthetic zeolites have been discovered. At present (June 2001), the Structure Commission of International Zeolite Association has approved 135 framework types. In addition, there are numerous other zeolites whose structures are not yet known or are only hypothetical. The availability of a great number of these synthetic zeolites has greatly expanded the realm of shape selectivity in catalytic transformation and sorption accumulation since the diameters of the channels and cavities are different depending on the structure. Furthermore, it is possible to make alterations in the pore size of zeolites by their modification. Pore-size engineering can be achieved by a number of modification techniques such as (1) modification by a preadsorption of molecules, (2) modification by a cation exchange, (3) internal pore modification, and (4) pore-opening size modification.

Preadsorption of polar molecules such as water and amines induces a secondary selectivity factor. This influences the adsorption of a second adsorbate, which can be drastically reduced [111]. Coke deposition can also influence the diffusivity, decreasing the adsorption rates. However, these modifications are only transient because polar molecules are desorbed and coke can be burned off.

The accessibility of a zeolite structure can be accurately changed by ion exchange, as described above.

Silane (SiH_4), disilanes (Si_2H_6), and diboranes (B_2H_6) are very reactive towards the hydroxyl groups of zeolites. Vansant et al. proposed that the reaction of hydrogen zeolites with silane proceeds as follows:

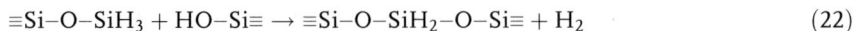

$$\equiv Si-OH + SiH_4 \rightarrow \equiv Si-O-SiH_3 + H_2 \tag{21}$$

$$\equiv Si-O-SiH_3 + HO-Si\equiv \rightarrow \equiv Si-O-SiH_2-O-Si\equiv + H_2 \tag{22}$$

Further reaction with adjacent silanol groups may also take place.

Subsequent hydrolysis of the hydrosilyl groups gives $Si(OH)_n$ groups:

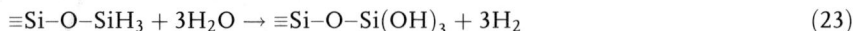

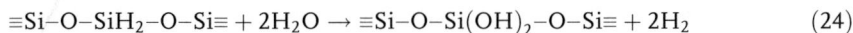

$$\equiv Si-O-SiH_3 + 3H_2O \rightarrow \equiv Si-O-Si(OH)_3 + 3H_2 \tag{23}$$

$$\equiv Si-O-SiH_2-O-Si\equiv + 2H_2O \rightarrow \equiv Si-O-Si(OH)_2-O-Si\equiv + 2H_2 \tag{24}$$

As a result, the overall reaction liberates 4 mol of H_2 per consumed SiH_4.

Boranes behave towards hydrogen zeolites in a similar manner.

References see page 930

The internal implantation of these small inorganic molecules introduces changes in the pore dimensions and electrical field within the zeolite channels [112]. Material thus obtained was mainly utilized for adsorption and separation of mixtures of rare gases, Kr and Xe [113], but it showed a great potential for shape-selective catalysis [114].

This modification method can also be applied to encapsulation of rare gas molecules in zeolites by combining the modification process following physical adsorption of the rare gas at moderate temperatures and pressures [115]. The encapsulates are stable to acids, mechanical grinding and γ-irradiation. Using small doses of modifying agent can moderate the pore-size reduction and the encapsulated gases can be released upon heating. Thus the release rate of encapsulated gases can be controlled by changing the degree of modification.

Treatment of Y zeolite with $SiCl_4$ to stabilize through dealumination is widely conducted; similar procedures applied to smaller pore zeolites leads to surface or internal deposition. When mordenite is modified with $SiCl_4$ and other metal chlorides, these metal chlorides are deposited along the interior channel or only near the channel entrance, depending on the reactivity and size of the halide molecule and on experimental conditions [116]. Impregnation of ZSM-5 with Mg salts is applied to achieve a minute reduction in the effective pore diameter to increase its shape selectivity for p-xylene formation in toluene disproportionation [117].

In order to control the pore-opening size without affecting the internal pore system of the zeolite, modifications are conducted using modifying agents with a molecular size larger than that of the zeolite pore openings and interacting only with the external surface. Employing various silicon alkoxides, Niwa and coworkers [118, 119] succeeded in enhancing the shape selectivity of zeolites. The alkoxides react with terminal silanol groups on the external surface. Upon subsequent calcination deposition of silica occurs on the external surface as an ultrathin layer, which results in narrowing of pore openings of zeolites without changing the intracrystalline structure. This method has been applied to H-MOR [118] and H-ZSM-5 [119] to lead to improved shape selectivity in cracking of C_8 alkanes (Fig. 9), alkylation of toluene with methanol, and toluene disproportionation. This technique has attracted much attention by other researchers, who observed the improved shape selectivity in xylene isomerization on ZSM-48 [120].

Similarly zeolite NaCaA was modified to be covered with a silica overlayer on the external surface, enclosing the pore-opening size precisely [121]. The rate of adsorption of lower olefins, ethylene, propylene, and 1-butene was suppressed by the modification, although the equilibrium adsorption amount was unchanged. Lower olefins thus can be separated by choosing the extent of silica deposited; the selective suppression of the adsorption of 1-butene among the mixture of ethylene, propylene, and 1-butene was characteristic of the SiNaCaA (deposited SiO_2 = 1.18 wt %).

To better understand the mechanism of the pore-opening size control by the deposition of silica on the external surface, Niwa et al. [122] used germanium alkoxide as a model reagent because the deposited GeO_2 can be distinguished from

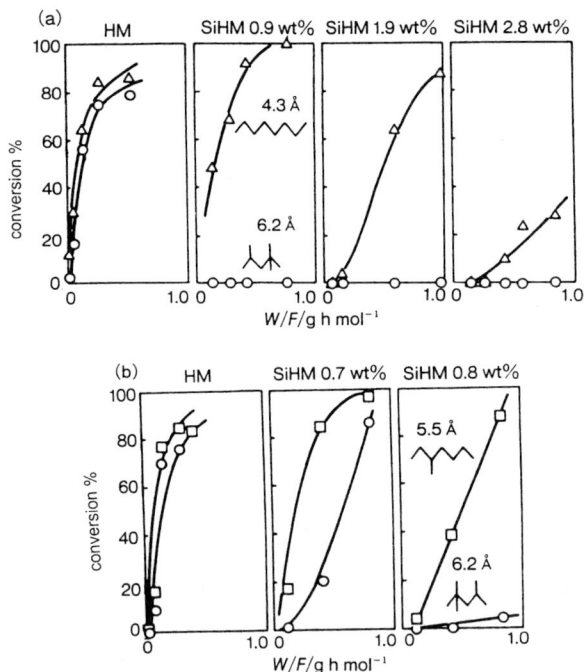

Fig. 9. Cracking of octane (□), 2,2,4-trimethylpentane (△) and 3-methylheptane (○) on HM (H-MOR) and SiHM (SiO$_2$-deposited H-MOR) at 573 K [118].

the starting zeolite. They confirmed that GeO$_2$ could be deposited on the external surface of mordenite as a thin layer, leading to the improvement in the shape selectivity. However, the shape selectivity deteriorates upon storing the catalysts under humid conditions. This is ascribed to the transformation of the ultrathin layer of GeO$_2$ into bulky particles, accompanied with local penetration of GeO$_2$ into the pore.

Zeolites exhibit a shape selectivity for various catalytic reactions. Since this is attributed to their unique pore structures, active sites on the external surface accessible to larger molecules are undesirable. For the zeolites with small crystal size, the contribution of the external surface is significant, giving rise to the decrease in shape selectivity. The inactivation of external surface is expected to result in enhancement of shape selectivity. If zeolite modifications are made by using the modification agent with a molecular size larger than that of the pore openings, it is also possible to selectively passivate the external surface.

In the pioneering work by Kaeding et al. [123], carborane-silicone polymer was used to passivate the external surface of ZSM-5, leading to the enhanced *para*-se-

References see page 930

lectivity in xylenes from toluene disproportionation. Metal chlorides (M = Si, Ge, Sn, and Ti) can be deposited to form metal compounds on external surfaces, if the operating conditions are carefully selected. However, the use of metal chlorides is limited since the liberated chloride has the potential to induce dealumination [124]. Anderson et al. [125] reported that aluminum atoms on the external surface of HZSM-5 are selectively dealuminated by the treatment with $SiCl_4$ vapor at high temperature. Monosilicic acid and trimethylphosphite also have been used for decreasing the activity of the external surface [126, 127].

Even the use of disilane causes external modification if reaction is conducted at high temperatures [128]. The incoming disilane molecules immediately react with OH groups on the external surface and at the pore mouth. The deposited molecules prevent the other disilane molecules from diffusing into the internal pores. This modification provides materials with smaller pore opening and without much loss in the internal volume, which can be advantageous for various catalytic applications.

Whether the modification occurs in internal pores, in the pore mouths, or on the external surface depends on the reactivity of the modifying agents and reaction conditions (temperature, time, partial pressure, etc.) as well as the size of modifying agents compared to the size of pore openings. External surface modification predominates when highly reactive agents are used at high temperatures. On the other hand, internal pore modification is likely to occur when less reactive reagents with a smaller size than the pore openings are employed at low temperatures. Although contradictory results are reported from different research groups on the mode of modification of the same zeolites with the same reagent, it is safe to say that the discrepancies can be attributed to the differences in the factors cited above. However, the distinction between external surface inactivation and pore size reduction is sometimes unclear, since both occur simultaneously in many cases.

The deactivation of the external surface can produce another beneficial effect. When the deactivation is due to pore-mouth plugging as a result of coke formation on the external surface of the zeolite particles, the selective removal of the active sites on the outer surface can be an effective treatment.

Coke deposition can also be considered as a technique to inactivate the external surface of zeolites. During the MTG reactions on ZSM-5, the carbonaceous deposits are formed on the external surface of the crystallites, resulting in a modification of its shape selectivity. For toluene disproportionation or toluene alkylation on ZSM-5, the para-selectivity was significantly increased by coke deposition that inactivates the acidity of the external surface, thus suppressing the secondary isomerization of p-xylene [123].

4.2.5.5
Organic Modifications

Active hydroxyl groups may be deactivated by the treatment with silylating agents:

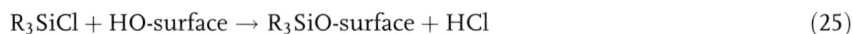

$$R_3SiCl + HO\text{-surface} \rightarrow R_3SiO\text{-surface} + HCl \tag{25}$$

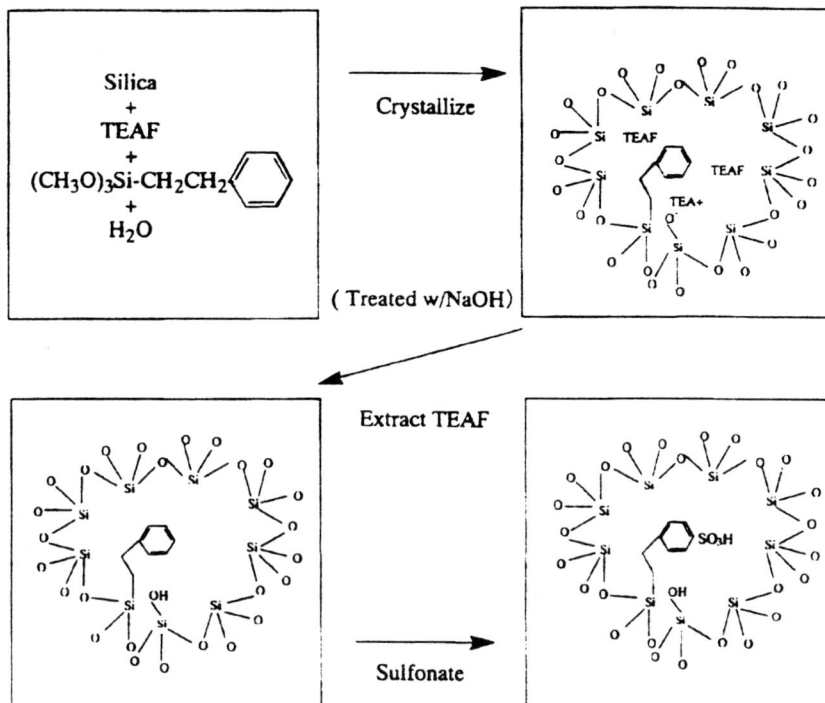

Fig. 10. Scheme of synthesis of organic-functionalized molecular sieves with the *BEA topology containing sulfophenethyl group [133].

$(CH_3)_3SiCl$ rapidly reacts with terminal silanol groups as well as hydroxyl nests in ZSM-5 [129]. Similar reactions occur by using silylating agents with alkoxy groups such as $(CH_3)_3SiOCH_3$ and $CH_3Si(OCH_3)_3$ to form siloxane bridges to the zeolite framework [130].

The silylated products show high hydrophobicity. On the other hand, they have a much lower thermal stability than the parent zeolites because organic groups are easily oxidized at elevated temperatures. However, for the catalysts or adsorbents for use at low temperatures, this type of silylation could be advantageous. It is reported that trimethylsilylation of TS-1 is effective in the enhancement of the liquid-phase oxidation using aqueous H_2O_2 [131] in a similar manner to the trimethylsilylation of mesoporous molecular sieves [132].

Organic-functionalized molecular sieves (OFMSs) have been invented, which contain intracrystalline organic functionalities covalently tethered to framework Si atoms [133]. OFMSs have been synthesized with organic moieties having acidity (Fig. 10) and basicity [134]. The catalytic tests and characterization studies have re-

References see page 930

vealed that the vast majority of the organic functional groups are contained within the micropores of the molecular sieves. The fact that they are present in the confined space within the micropores allows for their use as shape-selective catalysts. A possible disadvantage is that the use of organosilanes with pendant organic groups inevitably gives structural defects and organic groups located in micropores could spoil the microporosity. It is stated that the increase of the density of organic species within the OFMS catalysts will produce two competing effects: (1) increased activity per gram of catalyst by adding active centers, and (2) decreased activity per gram of catalyst by increasing diffusional limitations [135].

By using bis[triethoxysilyl]methane (BTESM) having a bridging methylene group between two ethoxysilanes as the Si source, zeolites with MFI and LTA structures containing an organic group as lattice (ZOL) have been synthesized [136]. Although the Si–C bond is generally strong enough to be resistant against hydrolysis, Si–CH$_2$–Si is relatively easy to be cleaved by nucleophilic substitution via possible intermediate species, Si–CH$_2^-$. The carbanion could be stabilized by the vacant d orbital of the adjacent Si atom. Supposedly, thus formed inorganic Si species and organically modified species cocrystallize to form ZOL materials. Nevertheless, the peak area of this NMR spectrum indicates that the amount of organically modified Si species in ZOL (LTA) is as high as 30 atom % of total Si.

4.2.5.6
Conclusions

The physical and chemical properties of zeolites can be changed in a controlled way by various modification techniques. Although the isomorphous substitution of metal heteroatoms for the framework silicon atoms has been briefly described above, the modification of zeolites is generally used to denote the secondary or post-synthesis treatment. Since post-synthesis modification is possible over wide-ranging conditions (temperature, solvent, atmosphere, pH, etc.) far from those for the zeolite synthesis, the modifications of zeolites present us with powerful indirect methods for manipulating the properties of zeolites. It is my firm belief that the fine tuning of the properties of zeolites will continue to be achieved by developing a variety of modification procedures as well as direct synthetic techniques.

References

1 G. H. KÜHL, in: Catalysis and Zeolites, J. WEITKAMP, L. PUPPE (Eds), Springer, Berlin, 1999, p. 81.

2 R. P. TOWNSEND, in: Introduction to Zeolite Science and Practice, H. VAN BEKKUM, E. M. FLANIGEN, J. C. JANSEN (Eds), *Studies in Surface Science and Catalysis*, Volume 58, Elsevier, Amsterdam, 1991, 359.

3 R. M. BARRER, A. J. WALKER, *Trans. Faraday So.* **1964**, *60*, 171.

4 R. P. TOWNSEND, *Pure Appl. Chem.* **1986**, *58*, 1359.

5 H. S. SHERRY, *J. Phys. Chem.* **1966**, *70*, 1158.

6 H. G. KARGE, in: Progress in Zeolite and Microporous Materials, H. CHON, S.-K. IHM, Y. S. UH (Eds), *Studies in*

Surface Science and Catalysis, Volume 105, Elsevier, Amsterdam, 1997, p. 1901.

7 A. V. KUCHEROV, A. SLINKIN, *Zeolites*, **1988**, *8*, 110.

8 H. G. KARGE, V. MARVODINOVA, Z. ZHENG, H. K. BEYER, *Appl. Catal.*, **1991**, *75*, 343.

9 J. WEITKAMP, S. ERNST, T. BOCK, A. KISS, P. KLEINSCHMIT, in: Catalysis by Microporous Materials, H. K. BEYER, H. G. KARGE, I. KIRICSI, J. B. NAGY (Eds), *Studies in Surface Science and Catalysis* **1995**, *94*, 278.

10 D. W. BRECK, W. G. EVERSOLE, R. M. MILTON, T. B. REED, T. L. THOMAS, *J. Am. Chem. Soc.* **1956**, *78*, 5963.

11 P. O. FRITZ, J. H. LUNSFORD, *J. Catal.* **1989**, *118*, 85.

12 W. M. H. SACHTLER, *Acc. Chem. Res.* **1993**, *26*, 387.

13 W. M. H. SACHTLER, Z. C. ZHANG, in: Advances in Catalysis, D. D. ELEY, H. PINES, P. B. WEISZ (Eds), Academic Press, San Diego, Vol. 39, 1993, p. 129.

14 M. TANIGUCHI, D. IMAMURA, H. ISHIGE, Y. ISHII, T. MURATA, M. HIDAI, T. TATSUMI, *J. Catal.* **1999**, *187*, 139.

15 A. CORMA, *Chem. Rev.* **1995**, *95*, 559.

16 D. BARTHOMEUF, in: Catalysis and Adsorption by Zeolites, G. ÖHLMANN, H. PFEIFER, R. FRICKE, *Studies in Surface Science and Catalysis*, Volume 65, Elsevier, Amsterdam, 1991, p. 157.

17 C. BESOUKHANOVA, J. GUIDOT, D. BARTHOMEUF, M. BREYSSE, J. R. BERNARD, *J. Chem. Soc., Faraday Trans I* **1981**, *77*, 1595.

18 N. HERRON, *J. Inclusion Phenom. Mol. Recognit. Chem.* **1995**, *21*, 283.

19 N. HERRON, Y. WANG, M. M. EDDY, G. D. STUCKY, D. E. COX, K. MOLLER, T. BEIN, *J. Am. Chem. Soc.* **1989**, *111*, 530.

20 G. D. STUCKY, J. E. MACDUGAL, *Science* **1991**, *247*, 669.

21 N. TOGASHI, Y. SAKAMOTO, T. OHSUNA, O. TERASAKI, *Mater. Sci. Eng.* **2001**, *A312*, 263.

22 N. P. BLAKE, G. D. STUCKY, *J. Inclusion Phenom. Mol. Recognit. Chem.* **1995**, *21*, 299.

23 P. P. EDWARDS, L. J. WOODALL, P. A. ANDERSON, A. R. ARMSTRONG, M. SLASKI, *Chem. Soc. Rev.* **1993**, 305.

24 Y. NOZUE, T. KODAIRA, S. OHWASHI, T. GOTO, O. TERASAKI, *Phys. Rev. B* **1993**, *48*, 12253.

25 J. A. MARTENS, P. J. GROBET, P. A. JACOBS, *Nature* **1985**, *315*, 568.

26 J. LUNSFORD, *ACS Symp. Ser.* **1977**, *40*, 473.

27 J. WEITKAMP, in: Proceedings from the Ninth International Zeolite Conference, R. VON BALLMOOS, J. B. HIGGINS, M. M. J. TREACY (Eds), Buttherworth-Heinemann, Boston, Volume 1, 1993, p. 13.

28 K. J. BALKUS, JR., A. G. GABRIELOV, *J. Inclusion Phenom. Mol. Recognit. Chem.* **1995**, *21*, 159.

29 G. SCHULZ-EKLOFF, S. ERMST, in: Preparation of Solid Catalysts, G. ERTL, K. KNÖZINGER, J. WEITKAMP (Eds), Wiley-VCH, Weinheim, 1999, p. 405.

30 N. HERRON, G. D. STUCKY, C. A. TOLMAN, *Inorg. Chim. Acta* **1985**, *100*, 135.

31 V. YU. ZAKHAROV, O. M. ZAKHAROVA, B. V. ROMANOWSKI, R. E. MARDALEISHBILI, *React. Kinet. Catal. Lett.* **1977**, *6*, 133.

32 R. PARTON, D. DE VOS, P. A. JACOBS, in: Proceedings of the NATO Advanced Study on Zeolite Microporous Solids: Synthesis, Structure and Reactivity, E. G. DEROUANE, F. LEMOS, C. NACCACHE, F. RIBEIRO (Eds), Kluwer, Dordrecht, 1992, p. 555.

33 N. HERRON, C. A. TOLMAN, G. D. STUCKY, *J. Chem. Soc, Chem. Commun.* **1986**, 1521.

34 M. ICHIKAWA, *Adv. Catal.* **1992**, *38*, 283.

35 C. C. FREYHARDT, M. TSAPATSIS, R. F. LOBO, K. J. BALKUS, JR., M. E. DAVIS, *Nature* **1996**, *381*, 295.

36 R. SZOSTAK, in: Introduction to Zeolite Science and Practice, H. VAN BEKKUM, E. M. FLANIGEN, J. C. JANSEN (Eds), Volume 58, Elsevier, Amsterdam, 1991, p. 153.

37 R. SZOSTAK, in: Molecular Sieves,

2nd edn, Blackie, London, 1991, p. 192.

38 J. A. Martens, P. A. Jacobs, *Zeolites* **1986**, *6*, 334.

39 G. W. Skeels, E. M. Flanigen, in: Zeolites: Facts, Figures, Future, P. A. Jacobs, R. A. van Santen (Eds), *Studies in Surface Science and Catalysis*, Volume 49, Elsevier, Amsterdam, 1989, p. 331.

40 G. T. Kerr, *Adv. Chem. Ser.* **1973**, *121*, 219.

41 R. M. Barrer, B. Coughlan, in: Molecular Sieves, Soc. Chem. Ind. London, 1968, p. 141.

42 G. T. Kerr, *J. Phys. Chem.* **1968**, *72*, 2594.

43 C. V. McDaniel, P. K. Maher, in: Molecular Sieves, Soc. Chem. Ind., London, 1968, p. 186.

44 G. Engelhardt, U. Lohse, V. Patzelova, M. Mägi, E. Lippmaa, *Zeolites* **1983**, *3*, 239.

45 G. Engelhardt, U. Lohse, V. Patzelova, M. Mägi, E. Lippmaa, *Zeolites* **1983**, *3*, 233.

46 D. W. Breck, E. M. Flanigen, Molecular Sieves, Soc. Chem. Ind., London, 1968, p. 47.

47 J. R. Sohn, S. J. DeCanio, J. H. Lunsford, D. J. O'Connell, *Zeolites* **1986**, *6*, 225.

48 G. Engelhardt, D. Michel, High-Resolution Solid State NMR of Silicates and Zeolites, Wiley, Chichester, 1987.

49 J. Klinowski, J. M. Thomas, C. A. Fyfe, G. C. Gobbi, *Nature* **1982**, *296*, 533.

50 H. Nakamoto, H. Takahashi, *Zeolites* **1982**, *2*, 67.

51 V. J. Frilette, M. K. Rubin, *J. Catal.* **1965**, *4*, 310.

52 M. Guisnet, P. Ayrault, C. Coutanceau, M. F. Alvarez, J. Datka, *J. Chem. Soc., Faraday Trans.* **1997**, *93*, 1661.

53 H. K. Beyer, I. Belenykaya, in: Catalysis by Zeolites, B. Imelik, C. Naccache, Y. BenTaarit, J. C. Vedrine, G. Coudurier, H. Praliaud (Eds), *Studies in Surface Science and Catalysis*, Volume 5, Elsevier, Amsterdam 1980, p. 203.

54 M. W. Anderson, J. Klinowski, *J. Chem. Soc., Faraday Trans. 1* **1986**, *82*, 1449.

55 G. W. Skeels, D. W. Breck, in: Proceedings of the Sixth International Zeolite Conference, D. H. Olson, A. Bisio (Eds), Butterworth, Guildford, 1984, p. 87.

56 B. Sulikowski, G. Borbely, H. K. Beyer, H. G. Karge, I. W. Mishin, *J. Phys. Chem.* **1989**, *93*, 3240.

57 Y. G. Shul, T. Tatsumi, H. Tominaga, *Nippon Kagaku Kai Shi*, **1989**, 429.

58 Q. L. Wang, M. Torrealba, G. Giannetto, M. Guisnet, G. Perot, M. Cahoreau, J. Caisso, *Zeolites* **1990**, *10*, 703.

59 B. M. Lok, F. P. Gortsema, C. A. Messina, H. Rastelli, T. P. J. Izod, *ACS Symp. Ser.* **1983**, *218*, 41.

60 C. Miradotos, D. Barthomeuf, *J. Chem. Soc. Chem. Commun.*, **1981**, 39.

61 R. M. Lago, W. O. Haag, R. J. Mikovsky, D. H. Olson, S. D. Hellring, K. D. Schmitt, G. T. Kerr, in: New Developments in Zeolite Science and Technology, Y. Murakami, A. Iijima, J. W. Ward (Eds), *Studies in Surface Science and Catalysis*, Volume 28, Elsevier, Amsterdam, 1986, p. 677.

62 B. H. Wouters, T.-H. Chen, P. J. Grobet, *J. Am. Chem. Soc.* **1998**, *120*, 11419.

63 B. H. Wouters, T.-H. Chen, P. J. Grobet, *J. Phys. Chem. B* **2001**, *105*, 1135.

64 D. Ma, F. Deng, R. Fy, X. Han, X. Bao, *J. Phys. Chem. B* **2001**, *105*, 1770.

65 V. R. Chumbhale, A. K. Chandwadkar, B. S. Rao, *Zeolites* **1992**, *12*, 63.

66 Y. Sasaki, T. Suzuki, Y. Takamura, A. Saji, H. Saka, *J. Catal.* **1998**, *178*, 94.

67 E. M. Flanigen, Proc. 5th Int. Conf. on Zeolites, R. V. C. Rees ed. Heyden, London, 1980, p. 760.

68 A. Jentys, G. Mirth, J. Schwank, J. A. Lercher, in: Zeolites: Facts, Figures, Future, P. A. Jacobs, R. A. van Santen (Eds), *Studies in Surface*

Science and Catalysis, Volume 49, Elsevier, Amsterdam, 1989, p. 847.

69 K. Tsutsumi, T. Kawai, T. Yanagihara, in: Zeolites and Microporous Crystals, T. Hattori, T. Yashima (Eds), *Studies in Surface Science and Catalysis*, Volume 83, Elsevier, Amsterdam, 1994, p. 217.

70 S. Namba, N. Hosokawa, T. Yashima, *J. Catal.* **1981**, *72*, 16.

71 J. Weitkamp, P. Kleinschmit, A. Kiss, C. H. Berke, in: Proceedings from the Ninth International Zeolite Conference, R. von Ballmoos, J. B. Higgins, M. M. J. Treacy (Eds), Buttherworth-Heinemann, Boston, Volume II, 1993, p. 79.

72 R. de Ruiter, K. Pamin, A. P. M. Kentgens, J. C. Jansen, H. van Bekkum, *Zeolites*, **1993**, *13*, 611.

73 R. de Ruiter, A. P. M. Kentgens, J. Grootendorst, J. C. Jansen, H. van Bekkum, *Zeolites* **1993**, *13*, 129.

74 T. Takewaki, L. W. Beck, M. E. Davis, *J. Phys. Chem. B* **1999**, *103*, 2674.

75 J. Klinowski, in Recent Advances in Zeolite Science, J. Klinowski, P. J. Barrie (Eds), *Studies in Surface Science and Catalysis*, Volume 52, Elsevier, Amsterdam, 1989, p. 39.

76 C. D. Chang, C. T.-W. Chu, J. N. Milale, R. F. Bridger, R. B. Calvert, *J. Am. Chem. Soc.* **1984**, *106*, 8143.

77 M. W. Anderson, J. Klinowski, X. Liu, *J. Chem. Soc. Chem. Commun.* **1984**, 1596.

78 K. Yamagishi, S. Namba, T. Yashima, *J. Phys. Chem.* **1991**, *95*, 872.

79 P. Wu, T. Komatsu, T. Yashima, *J. Phys. Chem.* **1995**, *99*, 10923.

80 H. Hamadan, B. Sulikowski, J. Klinowski, *J. Phys. Chem.* **1989**, *93*, 350.

81 B. Sulikowski, J. Datka, B. Gill, J. Ptaszynski, J. Klinowski, *J. Phys. Chem. B* **1997**, *101*, 6929.

82 T. Sano, R. Tadenuma, Z. B. Wang, K. Soga, *Chem. Comm.* **1997**, 1945.

83 T. Sano, Y. Uno, Z. B. Wang, C.-H. Ahn, K. Soga, *Microporous Mesoporous Mater.* **1999**, *31*, 89.

84 R. F. Lobo, M. E. Davis, *J. Am. Chem. Soc.* **1995**, *117*, 3766.

85 D. W. Breck, in *Zeolite Molecular Sieves*, Wiley, New York, 1974, p. 320.

86 M. Taramasso, G. Perego, B. Notari, US Patent 4,401,051 (1983), assigned to Enichem.

87 K. G. Ione, L. A. Vostrikova, V. M. Mastikhin, *J. Mol. Catal.* **1985**, *31*, 355.

88 M. Tielen, M. Geelen, P. A. Jacobs, *Acta Phys. Chem.* **1985**, *31*, 1.

89 B. Notari, in: Advances in Catalysis, D. D. Eley, W. O. Haag, B. C. Gates (Eds), Academic Press, San Diego, Volume 41, 1996, p. 253.

90 R. Millini, E. Previde-Massara, G. Perego, G. Bellussi, *J. Catal.* **1992**, *137*, 497.

91 A. Thangaraj, R. Kumar, S. P. Mirajkar, P. Rarnasamy, *J. Catal.* **1991**, *130*, 1.

92 G. Perego, G. Bellussi, C. Corno, M. Taramasso, F. Buomono, A. Esposito, in: New Developments in Zeolite Science and Technology, Y. Murakami, A. Iijima, J. W. Ward (Eds), *Studies in Surface Science and Catalysis*, Volume 28, Elsevier, Amsterdam, 1986, p. 129.

93 T. Tatsumi, M. Nakamura, S. Negishi, H. Tominaga, *J. Chem. Soc. Chem. Commun.* **1990**, 476.

94 T. Tatsumi, M. Nakamura, K. Yuasa, H. Tominaga, *Chem. Lett.* **1990**, 297.

95 M. A. Camblor, A. Corma, J. Perez-Pariente, *Zeolites* **1993**, *13*, 82.

96 A. Corma, P. Esteve, A. Martinez, J. Perez-Pariente, *J. Catal.* **1995**, *152*, 18.

97 J. C. van del Waal, P. Lin, M. S. Rigutto, H. van Bekkum, in: Progress in Zeolite and Microporous Materials, H. Chon, S.-K. Ihm, Y. S. Yu (Eds), *Studies in Surface Science and Catalysis*, Volume 105, Elsevier, Amsterdam, 1997, 1093.

98 M. A. Camblor, A. Corma, P. Valencia, *J. Phys. Chem. B* **1998**, *102*, 75.

99 T. Tatsumi, N. Jappar, *J. Phys. Chem. B* **1998**, *102*, 7126.

100 Y. Goa, P. Wu, T. Tatsumi, *Chem. Commun.* **2001**, 1714.

101 P. Wu, T. Tatsumi, T. Komatsu, T. Yashima, *J. Phys. Chem. B* **2001**, *105*, 2897.

102 P. Wu, T. Tatsumi, T. Komatsu, T. Yashima, *J. Catal.* **2001**, *202*, 245.

103 G. W. Skeels, E. M. Flanigen, Synthesis of Zeolites, *ACS Symp. Ser.* **1989**, *398*, 420.

104 J. S. Reddy, A. Sayari, in: Catalysis by Microporous Materials, H. K. Beyer, H. G. Karge, I. Kiricsi, J. B. Nagy (Eds), *Studies in Surface Science and Catalysis*, Volume 94, Elsevier, Amsterdam, 1995, p. 309.

105 F. Di Renzo, S. Gomez, R. Teissier, F. Fajula, in: 12th International Congress on Catalysis, A. Corma, F. V. Melo, S. Mendioroz, J. L. G. Fierro (Eds), *Studies in Surface Science and Catalysis*, Volume 130, Elsevier, Amsterdam, 2000, p. 1631.

106 M. S. Rigutto, R. de Ruiter, J. P. M. Niederer, H. van Bekkum, in: Zeolites and Related Microporous Materials: State of the Art 1994, J. Weitkamp, H. G. Karge, H. Pfeifer, W. Hölderich (Eds), *Studies in Surface Science and Catalysis*, Volume 84, Elsevier, Amsterdam, 1994, p. 2245.

107 T. Tatsumi, M. Nakamura, K. Yuasa, H. Tominaga, *Catal. Lett.* **1991**, *10*, 259.

108 T. Yashima, K. Yamagishi, S. Namba, in: Chemistry of Microporous Crystals, T. Inui, S. Namba, T. Tatsumi (Eds), *Studies in Surface Science and Catalysis*, Volume 60, Elsevier, Amsterdam, 1991, p. 171.

109 P. Wu, T. Komatsu, T. Yashima, *J. Phys. Chem.* **1996**, *100*, 10316.

110 S. Dzwigaj, M. J. Peltre, P. Massiani, A. Davidson, M. Che, T. Sen, S. Sivasanker, *Chem. Commun.* **1998**, 87.

111 R. M. Barrer, L. V. C. Rees, *Trans. Faraday Soc.* **1954**, *50*, 852.

112 A. Thijs, G. Peeters, E. F. Vansant, I. Verhaert, P. De Biervre, *J. Chem. Soc., Faraday Trans. I* **1983**, *79*, 2835.

113 E. F. Vansant, in: Innovation in Zeolite Materials Science, P. J. Grobet, W. J. Mortier, E. F. Vansant, G. Schulz-Ekloff (Eds), *Studies in Surface Science and Catalysis*, Volume 37, Elsevier, Amsterdam, 1988, p. 143.

114 Y. Yan, E. F. Vansant, *J. Phys. Chem.* **1990**, *94*, 2582.

115 Y. Yan, J. Verbiest, P. De Hulsters, E. F. Vansant, *J. Chem. Soc., Faraday. Trans. I* **1989**, *85*, 3087, 3095.

116 B. Hidalgo, M. Kato, T. Hattori, N. Niwa, Y. Murakami, *Zeolites* **1984**, *4*, 175.

117 N. Y. Chen, W. W. Kaeding, F. G. Dwyer, *J. Am. Chem. Soc.* **1979**, *101*, 6783.

118 M. Niwa, S. Morimoto, M. Kato, T. Hattori, Y. Murakami, in: Proceedings of Eighth International Congress on Catalysis, Volume 4, 1984, p. 701.

119 T. Hibino, M. Niwa, Y. Murakami, *J. Catal.* **1991**, *128*, 551.

120 P. Ratnasamy, S. K. Pokhriyal, *Appl. Catal.* **1989**, *55*, 265.

121 M. Niwa, K. Yamazaki, Y. Murakami, *Ind. Eng. Chem. Res.* **1994**, *33*, 371.

122 T. Hibino, M. Niwa, Y. Murakami, M. Sano, S. Komai, T. Hanaichi, *J. Phys. Chem.* **1989**, *93*, 7847.

123 W. W. Kaeding, C. Chu, L. B. Young, B. Weinstein, S. A. Butter, *J. Catal.* **1981**, *67*, 159.

124 S. Namba, A. Inaka, T. Yashima, *Zeolites* **1986**, *6*, 107.

125 J. R. Anderson, Y.-F. Chang, A. E. Hughes, *Catal. Lett.* **1989**, *2*, 279.

126 H. E. Bergna, M. Keane, D. H. Ralston, G. C. Sonnichesen, L. Abrams, R. D. Shannon, *J. Catal.* **1989**, *115*, 148.

127 D. R. Corbin, M. Keane, L. Abrams, R. D. Farlee, P. E. Bierstedt, T. Bein, *J. Catal.* **1990**, *124*, 268.

128 Y. Yan, J. Verbiest, E. F. Vansant, J. Philippaerts, P. De Hulsters, *Zeolites* **1990**, *10*, 137.

129 B. Kraushaar, L. J. M. van de Ven, J. W. de Haan, J. H. C. van Hooff, Innovation in Zeolite Materials Science, P. J. Grobet, W. J. Mortier, E. F. Vansant, G. Schulz-Ekloff (Eds), *Studies in Surface Science and Catalysis*, Volume 37, Elsevier, Amsterdam, 1988, p. 167.

130 T. Bein, R. F. Carver, R. D. Farlee, G. D. Stucky, *J. Am. Chem. Soc.* **1988**, *110*, 4546.

131 M. B. D'Amore, S. Schwarz, *Chem. Commun.* **1999**, 121.

132 T. Tatsumi, K. A. Koyano, N. Igarashi, *Chem. Commun.* **1998**, 325.

133 C. W. Jones, K. Tsuji, M. E. Davis, *Nature* **1998**, *393*, 52.

134 K. Tsuji, C. W. Jones, M. E. Davis, *Microporous Mesoporous Mater.* **1999**, *29*, 339.

135 C. W. Jones, M. Tsapatsis, T. Okubo, M. E. Davis, *Microporous Mesoporous Mater.* **2001**, *42*, 21.

136 K. Yamamoto, Y. Takahashi, T. Tatsumi, in: Zeolites and Mesoporous Materials at the Dawn of the 21st Century, A. Galameau, F. di Renzo, F. Fajula, J. Vedrine (Eds), *Studies in Surface Science and Catalysis*, Volume 135, Elsevier, Amsterdam, 2001, p. 299.

4.2.6
Characterization

Bodo Zibrowius and Elke Löffler

4.2.6.1
Introduction

The characterization of zeolites is obviously essential for finding new structures, but it is also of great importance in studies of zeolite applications in catalysis and adsorption and the ever-growing field of the development of new materials and devices. Generally, in systems as complex as zeolites that contain adsorbed species, any comprehensive understanding of the processes involved necessitates an adequate characterization of the specific samples being studied. Even when materials appear to have the same crystal structures and similar bulk compositions, subtle differences between the samples can determine whether a given application succeeds or not. There are, for instance, numerous examples of zeolite-based catalysts that had been previously claimed in the literature to be inactive that are, in fact, active, and vice versa. An adequate characterization with appropriate analytical techniques should (at least in principle) be able to detect the crucial differences between materials as they are synthesized, after their modification and following the preparation of the catalyst.

Seff [1], appealing for care in the preparation of zeolite samples, has highlighted another reason for sometimes apparently contradictory results: "Much of the work in zeolite science is of limited value because samples were prepared with insufficient thought, and studied, perhaps with very expensive instruments, to give results which make little sense. Most of this work is ultimately recognized as faulty by its authors and is not reported. However, anyone familiar with the zeolite literature knows that different results appear for samples that are reported as being identical, that is the work is irreproducible. The fault sometimes lies in unappreciated differences in supposedly identical initial samples. The fault often lies in experimental details which were not recognized to be important, especially to investigators unfamiliar with zeolite chemistry."

The success story of zeolites in the second half of the last century was accompanied by a tremendous development in analytical methods. In the early days of zeolite research, scientists were more or less restricted to classical chemical analysis, adsorption and ion-exchange experiments, X-ray diffraction and infrared spectroscopy [2]. Nowadays, many more spectroscopic and non-spectroscopic analytical methods (with sometimes confusing acronyms, see Table 1) are available. This wealth of characterization techniques permits many physico-chemical properties to be measured that might be directly or indirectly related to properties relevant to the application under consideration.

A full coverage of all the methods that have ever been applied to zeolites or zeolite-related solids such as microporous titanosilicates (see Sect. 4.2.4 or ref. [3]) would certainly exceed the limits of the present section. Therefore, we focus our interest on the methods most frequently used. The equipment necessary for many of these techniques (X-ray diffraction, vibrational and electronic spectroscopy, NMR and ESR spectroscopy, electron microscopy, thermal analysis and adsorption methods) is usually available in labs working on the preparation and application of zeolites. On the other hand, we have also included techniques that require highly specialized and expensive experimental setups such as neutron scattering or diffraction, and X-ray absorption spectroscopy and X-ray diffraction using synchrotron radiation. A compilation of experimental techniques and their typical fields of application is given in the next section, followed by short reviews of studies characterizing zeolites with regard to selected features, namely structure, organic template, framework composition, and acid sites. Since it is now generally accepted that an adequate picture of a zeolite can only be obtained by a combination of different techniques, particular attention will be paid to this aspect. Rather than a comprehensive review for specialists, the present section is intended to be a guide for those entering the fascinating field of zeolite research, to help them judge literature results and plan and execute their own experiments.

4.2.6.2
Experimental Methods

Zeolites differ from most of the other materials covered by the present Handbook in that they have a crystalline framework with regular cavities of molecular dimensions. Since the micropores are more or less defined by the zeolite structures, the characterization of these materials focuses on the investigation both of properties of the framework itself and of species attached to it. Nevertheless, sorption methods are indispensable for characterizing the accessible pore volume of micro- and mesopores in the zeolite sample being studied. Although the assumptions made by Brunauer, Emmet and Teller [4] in deriving their BET theory are definitely not met by the adsorption in micropores of zeolites (see Chap. 2.5 or ref. [5]), BET surface areas are widely used to characterize sorption capacities of these materials. However, not only nitrogen and noble gases, but also other molecules, in particular methane [6], n-pentane, n-hexane or benzene [7], can be used to probe the pore

Tab. 1. Acronyms for Experimental Techniques Used for the Characterization of Zeolites and Related Materials

Acronym	Description/remarks
AES	Auger electron spectroscopy
AFM	Atomic force microscopy
ATR	Attenuated total reflection, IR method
BET	Adsorption theory (isotherm) derived by Brunauer, Emmet and Teller, often used as synonym for nitrogen adsorption experiments
CIR	Cylindrical internal reflection, IR method
DAS	Dynamic angle spinning, NMR method
DOR	Double rotation, NMR method
DRIFT/DRIFTS	Diffuse reflectance infrared Fourier transform spectroscopy
DSC	Differential scanning calorimetry
DTA	Differential thermal analysis
DTG	Differential thermogravimetry
EDX/EDS	Energy dispersive X-ray (analysis)/spectroscopy
EPMA	Electron probe microanalysis, X-ray emission spectroscopy using WDX or EDX
EPR	Electron paramagnetic resonance (spectroscopy), also referred to as ESR
ESCA	Electron spectroscopy for chemical analysis, synonym for XPS
ESEEM	Electron spin-echo envelope modulation (spectroscopy), ESR method
ESR	Electron spin resonance (spectroscopy), also referred to as EPR
EXAFS	Extended X-ray absorption fine structure, XAS method
FIR	Far infrared (spectroscopy)
FR	Frequency response, sorption method
FTIR	Fourier transform infrared (spectroscopy)
HREM/HRTEM	High-resolution (transmission) electron microscopy
ICP-OES	Inductively coupled plasma optical emission spectroscopy
INS	Inelastic neutron scattering
IR	Infrared (spectroscopy)
MAS NMR	Magic-angle-spinning nuclear magnetic resonance (spectroscopy)
MQMAS	Multi-quantum magic-angle-spinning, NMR method
NIR	Near infrared (spectroscopy)
NMR	Nuclear magnetic resonance (spectroscopy)
PFG NMR	Pulsed field gradient nuclear magnetic resonance
PIGE	Proton-induced gamma-ray emission (spectroscopy)
PIXE	Proton-induced X-ray emission (spectroscopy)
QENS	Quasi-elastic neutron scattering
RBS	Rutherford backscattering

Tab. 1. *(continued)*

Acronym	Description/remarks
REDOR	Rotational echo double resonance, NMR method
SANS	Small-angle neutron scattering
SATRAS	Satellite-transition spectroscopy, NMR method
SAXS	Small-angle X-ray scattering
SEDOR	Spin-echo double resonance, NMR method
SEM	Scanning electron microscopy
STM	Scanning tunneling microscopy
TEDOR	Transferred-echo double resonance, NMR method
TEM	Transmission electron microscopy
TG	Thermogravimetry
TPD	Temperature-programmed desorption
TRAPDOR	Transfer of population in double resonance, NMR method
UPS	Ultraviolet (induced) photoelectron spectroscopy
UV-Vis	Ultraviolet-visible (spectroscopy)
WDX/WDS	Wavelength dispersive X-ray (analysis)/spectroscopy
XANES	X-ray absorption near-edge structure, XAS method
XAS	X-ray absorption spectroscopy
XAES	X-ray-induced Auger electron spectroscopy, AES method
XPS	X-ray-induced photoelectron spectroscopy
XRD	X-ray diffraction
XRF	X-ray fluorescence (spectroscopy)

system [8] and also to determine the intracrystalline mass transport coefficients and the surface barriers [9]. A very brief description of some of the most frequently used techniques for characterizing zeolites, namely X-ray powder diffraction, elemental analysis, electron microscopy, NMR and IR spectroscopy, and sorption and ion-exchange capacity measurements, can be found in a volume edited by Robson on behalf of the Structure Commission of the International Zeolite Association (IZA) [10].

The experimental methods for characterizing zeolites are very often divided into non-spectroscopic and spectroscopic techniques, and they are arranged in Table 2 in this way. Atomic absorption and emission spectroscopy are, of course, spectroscopic methods, but they are not directly applied to zeolites. Nowadays, they are the standard methods for chemical analysis of the bulk. Since some of the methods given in Table 2 (for example, calorimetry, TPD, PFG NMR and ^{129}Xe NMR spectroscopy, and temperature-programmed desorption) use probe molecules, they are strictly topics covered by Sect. 4.2.7. Others have already been discussed in detail as generic methods for the characterization of porous materials in Chap. 2, where the reader can also find the basics of these methods.

The examples of the applications of the various experimental techniques in Table 2 include pioneering papers that very often contain a detailed discussion of the essential prerequisites for the use of the applied method. Also given are very recent studies representing the state-of-the-art in the various fields. We are fully aware that dozens of other excellent papers could have been chosen, particularly in the case of very frequently applied techniques. For a more comprehensive coverage the reader is referred to the recent reviews given in Table 2 and the corresponding chapters of the Handbook. Consulting these review articles can ease the difficulties and reduce the frustration of those entering a new field. We are also aware that in a few cases some aspects of the conclusions drawn in the papers given in Table 2 are controversial. In our opinion this indicates that crucial pieces of evidence are still missing. New techniques or new combinations of techniques that have already been developed might be necessary to resolve these issues.

The application of three of the most powerful methods often used in zeolite science, namely IR and NMR spectroscopy, and X-ray diffraction, has very recently been reviewed by Karge et al. [11]. Starting with a brief description of the basics of these methods, this review compiles recent experimental results and provides an impression on the scope of problems that can be tackled by them. For applications of NMR spectroscopy in the field of catalysis in general, the reader is also referred to an excellent book edited by Bell and Pines [12]. It covers the characterization of heterogeneous catalysts, molecular diffusion and transport properties, as well as *in situ* studies of catalytic processes. Since most of the NMR active nuclei present in the zeolite framework (for example, ^{27}Al, ^{17}O, ^{69}Ga, ^{71}Ga, and ^{11}B) or as charge-balancing cations (for example, ^{23}Na, ^{39}K, and ^{133}Cs) have a nuclear spin $I > 1/2$, the enormous development in the field of solid-state NMR spectroscopy of quadrupolar nuclei in the last decade has had a great impact on the application of NMR spectroscopy in zeolite science. Examples indicating the level of sophistication that has been reached are included in Table 2. As a concise review on the recently developed special methodology for quadrupolar nuclei (for example, DOR, DAS, SATRAS, MQMAS, and TRAPDOR) we recommend a paper by Smith and van Eck [13].

During the past three decades, analytical methods using X-rays have found increasing use in the area of material characterization. Many techniques and procedures, for example EXAFS and XANES, have been developed that greatly widen the range of application. On the other hand, the versatility of the X-ray powder diffraction method has made it a technique employed in many labs all over the world. Examples of its use to elucidate and identify zeolite structures are given in Table 2 and Sect. 4.2.6.3. In addition to these papers we want to draw the reader's attention to reviews devoted to the combination of X-ray diffraction with other techniques. Whereas Cheetham and coworkers [14–16] demonstrated the advantages of com-

References see page 990

Tab. 2. Experimental Techniques for the Characterization of Zeolites and Related Materials

Technique	Target	Examples
chemical analysis (classical methods, AAS or ICP-OES) [44]	bulk composition	comparison of analytical techniques for determining the silicon and aluminum contents in zeolites [43]
thermogravimetric analysis (TG/DTG), often combined with mass spectrometry [48]	decomposition of templates or adsorbed species	template decomposition in ZSM-5 and silicalite [45]
		template decomposition in zeolite beta dependent on the calcination conditions, effect of oxygen [46]
		decomposition of adsorbed propanamines on gallium-impregnated ZSM-5 [47]
	dehydration, dehydroxylation	desorption of water from ZSM-5-type zeolites [49]
	thermal stability of occluded compounds	increased stability of alkali metal azides (LiN_3, CsN_3) in Na-Y zeolite [50]
thermal analysis (DTA, DSC), often combined with TG/DTG and mass spectrometry [48]	decomposition of templates	decomposition of tetrapropylammonium ions in MFI-type zeolites [37, 51–53]
		decomposition of tetraethylammonium in zeolite beta [54]
	coke	characterization of coke on MFI-type H-gallosilicate propane aromatization catalysts by ^{13}C MAS NMR, ESR, thermal analysis, and temperature-programmed oxidation [55]
	phase transitions	thermal stability of FAU-type [56, 57] and LTA-type [57] zeolites dependent on the kind of cations
calorimetry [48, 68]	acidity/basicity via heats of adsorption	characterization of acid sites in H-ZSM-5 zeolites by microcalorimetry and IR spectroscopy [58]
		characterization of acid sites in dealuminated Y zeolites by NH_3-TPD and calorimetric measurement of NH_3 chemisorption [59]
		detection of strong Lewis acid sites on non-framework aluminum in ZSM-5 using pyridine [60]
		strength of Brønsted acid sites in isomorphously substituted (Fe, Ga) MFI-type zeolites [61]
		comparison of acid sites in Y zeolites and mordenite [62]

Tab. 2. *(continued)*

Technique	Target	Examples
		dependence of the strength of Brønsted acid sites in Na,H-ZSM-5 on the Na$^+$ content as determined by NH$_3$ adsorption at 393 K [63]
		alkane (C$_3$–C$_6$) adsorption in acidic molecular sieves: decrease of heat of adsorption in the sequence H-MFI > H-MOR > H-FAU [64]
		characterization of mixed aluminum-gallium offretites [65] and gallosilicates with *BEA [66] and FAU structure [67]
	cations	characterization of nature and accessibility of cations (Li, Na, K) in MFI-type zeolites via adsorption of CO and N$_2$ at 195 K [69]
	pore size, pore volume	calorimetric study of the adsorption of nonpolar molecules on the molecular sieve VPI-5 [70]
		correlation between initial heats of adsorption and structural parameters of molecular sieves [71]
		determination of the topology of zeolites by adsorption microcalorimetry of organic molecules [72]
measurement of sorption isotherms [5, 80]	pore volume, pore-size distribution	evaluation of the micropore and secondary pore volume of dealuminated Y zeolites by adsorption measurements (n-hexane, n-pentane, benzene); pore-size distribution [7]
		characterization of micropore and mesopore volumes of dealuminated Y zeolites by adsorption of n-hexane [73–75]
		adsorption and diffusion of C$_6$ and C$_8$ hydrocarbons in silicalite [76]
		adsorption properties and pore-size distribution of the aluminophosphate VPI-5 [77]
		use of the t-plot-De Boer method in pore volume determinations of ZSM-5-type zeolites from nitrogen adsorption isotherms at 77 K [78]

Tab. 2. *(continued)*

Technique	Target	Examples
		reliable pore-size distribution using argon as adsorbate at 87.5 K [79]
	cations	effect of cation substitution (Li^+, K^+, Rb^+, Cs^+, Mg^{2+}, Ca^{2+}, Sr^{2+}, Ba^{2+}, Co^{2+}, Ni^{2+}, Cu^{2+}, Zn^{2+}) on the adsorption properties of xenon on Na-Y zeolite and on the ^{129}Xe chemical shift [81]
	adsorption mechanism	unusual adsorption behavior of aromatics (benzene, toluene, and *p*-xylene) on silicalite [82]
		high-resolution sorption and microcalorimetric studies of argon and nitrogen on ZSM-5 [83]
		adsorption of cyclic hydrocarbons in MFI-type zeolites [84]
sorption uptake [9]	diffusion, mass transport	intracrystalline diffusion of benzene in Na-X zeolite studied by sorption kinetics [85]
		sorption kinetics of aromatics in ZSM-5 [86]
		intracrystalline diffusion of benzene in ZSM-5 and silicalite [87]
		comparison of NMR tracer exchange and molecular uptake of benzene in pentasils [88]
		intracrystalline diffusion of benzene in MFI-type gallosilicate [89]
		sorption kinetics of *p*-ethyltoluene in Na,H-ZSM-5 [90]
		sorption kinetics of diethylbenzene isomers in MFI-type zeolites [91]
frequency response [101]	diffusion, mass transport	determination of diffusion coefficients in zeolites: krypton in mordenite [92] and krypton and xenon in zeolite A [93]
		diffusion of hydrocarbons in zeolite A [94]
		diffusion of methane in cation-exchanged A zeolites [95]
		diffusion of *p*-xylene [96, 97] and propane [98] in silicalite-1

Tab. 2. *(continued)*

Technique	Target	Examples
		comparison of the diffusivities of CO_2 in silicalite-1 and theta-1 as determined by FR, PFG NMR spectroscopy and molecular dynamic simulations [99]
		kinetics of ammonia adsorption and desorption in H-ZSM-5 [100]
		diffusion of cyclic hydrocarbons in MFI-type zeolites [84]
ion exchange [115, 116]	exchange capacity, exchange isotherms	ion-exchange properties of synthetic faujasites: univalent ions [102, 103] and divalent ions for Na^+ [102, 104]
		multicomponent ion exchange in zeolites [105–109]
		effect of hydrolysis on concurrent ion exchanges in X and Y zeolites [110, 111]
		effect of the Si/Al ratio on the ion-exchange isotherms of univalent and divalent ions in zeolite EU-1 [112]
		effect of the chemical nature of the exchange site on the ion exchange in zeolite EU-1 [113]
		Pb^{2+} exchange isotherms for Na-X zeolite at pH 5, 6, and 7 [114]
	exchange kinetics	kinetics of the exchange of divalent ions for Na^+ in zeolite A, cation diffusion coefficients [117]
		kinetics of the exchange of Ca^{2+} for Na^+ in zeolite A [118]
		cation diffusion in natural zeolite clinoptilolite [119]
TPD of probe molecules	acidic/basic sites (kind, number, strength)	combined NH_3-TPD and IR studies of Y zeolites, mordenite and ZSM-5 [120], dealuminated Y zeolites and ZSM-20 [121]
		characterization of acid sites in dealuminated Y zeolites by NH_3-TPD and calorimetric measurement of NH_3 chemisorption [59]
		characterization of acid sites in dealuminated mordenites [122] and Y zeolites [123] by TPD of ammonia and pyridine

Tab. 2. *(continued)*

Technique	Target	Examples
		combined stepwise temperature-programmed desorption of ammonia and IR spectroscopy of dealuminated beta zeolite [124], ultrastable Y zeolite, mordenite and ZSM-12 [125]
		characterization of acid sites in MeAPO-5 molecular sieves (Me = Mn, Co, Ni, Mg, Zn, Zr, Si) by NH$_3$-TPD [126]
		strength and nature of sorption sites for pyrrole in ion-exchanged (Na, K, Cs, Rb) faujasites [127]
optical microscopy	morphology	synthesis conditions for tailoring AlPO$_4$-5 crystal dimensions [128]
		different morphologies of CoAPO-5 and CoAPO-44, detection of by-products [129]
		synthesis of large optically clear silicoalumino-phosphate crystals with AFI structure [130]
	crystal growth	*in situ* observation of the growth of MFI-type zeolite crystals [131]
	twinning	evidence for crystal twinning in ZSM-5 and ZSM-8 from optical investigations using polarized light [132]
	crystal defects	*in situ* observation of the crack formation during the calcination of MFI-type zeolite crystals [133]
electron microscopy (SEM, TEM, HREM/HRTEM) [137–139]	morphology, particle size	influence of synthesis additives on the morphology of ZSM-5 crystals [134]
		synthesis conditions for tailoring AlPO$_4$-5 crystal dimensions [128]
		observation of precursors formed during the induction period of ZSM-5 synthesis using *cryo*-TEM and SAXS [135]
		particle size and morphology of aluminum-free titanium beta zeolites [136]
	crystal structure and defects [148, 149]	zeolite structures as revealed by HREM [140]
		direct determination of intergrowths in ZSM-5/ZSM-11 catalysts by HREM [141]

Tab. 2. *(continued)*

Technique	Target	Examples
		twinned Y zeolite crystals [142] and ZSM-5/ZSM-11 intergrowth structures [142, 143] studied by HREM and electron diffraction
		role of HREM in the identification and characterization of new crystalline microporous materials [144]
		direct observation of pure MEL-type zeolite [145]
		HREM and electron diffraction study of intergrowth of MFI and MEL zeolites, crystal faults in an MEL-type borosilicate [146]
		structure of the microporous titanosilicate ETS-10 [147]
	surface structure	surface structure of zeolites FAU and EMT studied by HRTEM [150]
		surface structure of zeolite L studied by HREM [151]
	metal particles	platinum, palladium and nickel particles in Na-X zeolite [152]
		platinum agglomeration in Pt-loaded K-L zeolite [153]
		iridium, rhodium and platinum nanocrystals in zeolite X [154]
AFM	surface structure	surface structure of natural heulandite [155, 156], natural [157] and synthesized mordenite [158], Linde type A zeolite [159], and SSZ-42 [160]
		growth of zeolite Y crystals [161]
		hydroxyl groups on the (100) surface of natural mordenite [157] and on the (010) surface of natural heulandite [156]
		surface structure of ZSM-5 crystals synthesized in space [162]
	arrangement of adsorbed molecules	rearrangement of tert-butyl ammonium ions on the (010) surface of clinoptilolite under the tip of the AFM [163]
		structure of an adsorbed pyridine layer on the heulandite (010) surface [164]

Tab. 2. *(continued)*

Technique	Target	Examples
electron diffraction [138, 148]	crystal structure and defects	ZSM-5/ZSM-11 intergrowth structures studied by HREM and electron diffraction [143]
		HREM and electron diffraction study of intergrowth of MFI and MEL zeolites, crystal faults in an MEL-type borosilicate [146]
		quantitative HRTEM of zeolites, quantitative measurement of electron diffraction intensities [165]
		electron diffraction structure solution of a nanocrystalline zeolite (SSZ-48) [166]
	metal particles and other occluded species	platinum, palladium and nickel particles in Na-X zeolite [152]
		MoS_2 clusters in Na-Y zeolite studied by HRTEM and electron diffraction [165]
X-ray single crystal diffraction	crystal structure	crystal structure of as-synthesized ZSM-5 [167]
		monoclinic framework structure of H-ZSM-5 [168]
		crystal structure of dehydrated Na-X zeolite [169]
		crystal structure of fully dehydrated fully Tl^+-exchanged zeolite X [170]
		crystal structure of mutinaite, the natural analog of ZSM-5 [171]
		improved description of the framework structure of cobalt-containing DAF-1 derived from micro-single-crystal diffraction employing synchrotron radiation [172]
	cation positions	cation positions in alkali-exchanged (Na^+, K^+, Rb^+, Cs^+) heulandite [173]
		cation positions in natural erionite [174]
		cation positions in fully dehydrated fully Tl^+-exchanged zeolite X [170]
		crystal structure of Pb^{2+}- and Tl^+-exchanged zeolite X containing Pb_4O_4 clusters [175]

Tab. 2. (continued)

Technique	Target	Examples
		symmetry lowering and site preference in cadmium-exchanged heulandite [176]
	phase transitions, structural changes	temperature-induced phase transition in H-ZSM-5 [177]
		preparation of a monoclinic (nearly) single crystal of H-ZSM-5 by application of uniaxial mechanical stress [178]
		in situ study of the structural changes during the calcination of CoAPSO-44 [172]
	location of adsorbed species, structure of sorption complexes [186]	*p*-xylene in H-ZSM-5 with sorbate-induced orthorhombic framework symmetry [179]
		naphtalene in H-ZSM-5 [180]
		p-dichlorobenzene [181, 182] and *p*-nitroaniline [183, 184] in MFI-type zeolites
		combined solid-state NMR and single crystal X-ray study of *p*-nitroaniline adsorbed in ZSM-5 [185]
X-ray powder diffraction [11, 194–196]	crystal structure	structure of the high-temperature form of ZSM-11 derived from Rietveld refinement combined with ^{29}Si MAS NMR [187]
		evidence from Rietveld refinement for a triple helix of water molecules in the as-synthesized aluminophosphate VPI-5 [188]
		combined X-ray and neutron diffraction and ^{29}Si MAS NMR study of siliceous ferrierite [189]
		structure determination of zeolite RUB-10 from low-resolution X-ray powder diffraction data [190]
		combined Rietveld refinement of the structure of low-silica Ca-X zeolite from neutron and X-ray diffraction data [15]
		structure of calcined $AlPO_4$-5 as determined by X-ray and neutron powder diffraction [191]

Tab. 2. *(continued)*

Technique	Target	Examples
		structure of the low-temperature form of ZSM-11 derived from ^{29}Si MAS NMR, lattice-energy minimization and Rietveld refinement [19]
		Rietveld refinement of as-synthesized AlPO$_4$-40 [192]
		NMR characterization and Rietveld refinement of the structure of rehydrated AlPO$_4$-34 [193]
	crystallinity, kinetics of crystallization	kinetics of crystallization of ZSM-5 as function of the template/Na$^+$/K$^+$ ratio [197]
		kinetics of crystallization of ZSM-5 for various alkali and alkaline earth metals [198]
		effect of alkalinity on the crystallization of silicalite-1 [199]
		kinetics of crystallization of ZSM-5 for different organic templates (tetrapropylammonium bromide, pyrrolidine), detailed kinetic analysis of nucleation and crystallization [200]
		in situ time-resolved study of the CoAPO-5 synthesis using synchroton radiation [201]
	identification and quantification of by-products	metastable phase transformations during the synthesis of ZSM-5 [197]
		zeolite phi – a mixture of chabazite and offretite [202]
		CoAPO-47 as a by-product of CoAPO-5 synthesis [201]
	phase transitions, structural changes	temperature-induced phase transition in siliceous ZSM-5/silicalite [203–205] and siliceous ZSM-11 [187]
		sorbate-induced structural changes in silicalite studied by XRD and ^{29}Si MAS NMR [206]
		in situ observation of the phase transformation of VPI to AlPO$_4$-8 [207, 208]
		combined XRD and EXAFS *in situ* observation of the thermal transformation of zinc- and cobalt-exchanged zeolite A [209, 210]

Tab. 2. *(continued)*

Technique	Target	Examples
		structural changes upon adsorption and desorption of xenon from cadmium-exchanged zeolite rho [211]
		phase transitions of MFI-type zeolites during the adsorption of *p*-xylene [212]
		temperature-induced structural changes in hydrated VPI-5 [213]
		combined NMR and synchrotron X-ray powder diffraction study of the phase transition upon K^+ ion exchange into low-silica Na-X zeolite [214]
		in situ observation of the thermal transformation of Co^{2+}- and Ni^{2+}-exchanged A, X, and Y zeolites [57]
		in situ observation of the thermal transformation of the large-pore molecular sieve MnAPO-50 into dense phases [215]
		in situ observation of the structural changes during the rehydration of $AlPO_4$-34 [193]
	crystal size	characterization of nanosized ZSM-5 crystals obtained by confined space synthesis [216]
	cation positions	combined neutron and X-ray diffraction of low-silica Ca-X zeolite, siting of the calcium ions [15]
		combined *in situ* synchrotron X-ray and neutron powder diffraction studies of lead- and cadmium-exchanged zeolite rho [217]
	location of adsorbed molecules	*p*-dichlorobenzene [17] and naphtalene [218] in MFI-type zeolites
		aromatic nitro-substituted hydrocarbons in Y zeolites [219]
		bithiophene [220] and benzene [221] in zeolite ZSM-5
		water in rehydrated $AlPO_4$-34 [193]
		residual water in alkali-metal cation-exchanged X and Y zeolites [222]

Tab. 2. *(continued)*

Technique	Target	Examples
	chemical composition	quantitative determination of the framework aluminum content of dealuminated Y zeolites based on its linear relation to the lattice constant [223]
		linear dependence of the unit cell volume of TS-1 on the titanium content of the zeolite framework [224–226]
SAXS	particle size, crystal growth	observation of precursors formed during the induction period of ZSM-5 synthesis using *cryo*-TEM and SAXS [135]
		in situ observation of the aging process in the synthesis of ZSM-5 [227]
		in situ observation of silicalite nuclei in the induction period of the synthesis [228]
		TPA-silicalite crystallization: kinetics of nucleation and growth [229]
neutron diffraction [14]	crystal structure	structure of Tl-A zeolite, ordering of silicon and aluminum [230]
		Rietveld refinements of H-Y and D-Y zeolites [231]
		combined X-ray and neutron diffraction and ^{29}Si MAS NMR study of siliceous ferrierite [189]
		combined Rietveld refinement of the structure of low-silica Ca-X zeolite from neutron and X-ray diffraction data [15]
		structure of calcined $AlPO_4$-5 as determined by X-ray and neutron powder diffraction [191]
		neutron powder diffraction study of orthorhombic and monoclinic defective silicalite, preferential siting of silicon atom vacancies in the orthorhombic framework [232]
		Rietveld refinement of titanium silicalite-1, nonrandom distribution of titanium atoms in the MFI framework [233]

Tab. 2. *(continued)*

Technique	Target	Examples
	cation positions	combined neutron and X-ray diffraction of low-silica Ca-X zeolite, siting of the calcium ions [15]
		combined MAS NMR and neutron diffraction study of the site preference in the mixed cation Li,Na-chabazite [234]
		combined *in situ* synchrotron X-ray and neutron powder diffraction studies of lead- and cadmium-exchanged zeolite rho [217]
	Brønsted acid sites	direct determination of the proton and deuteron positions in H-Y and D-Y zeolites [231]
		neutron diffraction study of acid sites in H-SAPO-37 [235]
		detection of two distinct Brønsted acid sites in the acid form of the high-silica chabazite SSZ-13 [236]
		location of Brønsted acid sites in D-ferrierite [237] and D-mordenites [238]
		identification of Brønsted acid sites in zeolite ERS-7 by combined Rietveld refinement of X-ray and neutron diffraction data [16]
	location of adsorbed species, host–guest interaction	pyridine in gallozeolite-L at 4 K [239]
		benzene in Na-Y zeolite at 4 K and room temperature [240]
		benzene in ZSM-5 [221, 241, 242]
		o-, *m-*, and *p*-xylene in Yb,Na-Y zeolites [243]
		combined neutron diffraction and ^{2}H NMR study of the interaction of benzene with acid sites in H-SAPO-37 [244]
		interaction of water with Brønsted acid sites in H-SAPO-34, formation of hydronium ions and hydrogen-bonded water molecules [245]
		aniline and *m*-dinitrobenzene in Na-Y zeolite [246]
		cyclohexane in the acid form of zeolite Y [247]

Tab. 2. *(continued)*

Technique	Target	Examples
SANS	templates	*in situ* detection of template molecules (TPA) in the crystal nuclei formed during silicalite crystallization from homogeneous solutions [228, 229]
	adsorbed molecules	aggregation of benzene in Na-Y zeolites [248]
INS [253, 254]	acid sites, interaction with probe molecules	in-plane and out-of-plane bending vibrations of hydroxyl groups in H-Y [249] and H-ZSM-5 zeolites [250]
		interaction of water with acid sites in H-mordenite [251] and H-ZSM-5 zeolite [250]
		interaction of pyrrole with Lewis acid sites [127]
		interaction of chloroform and trichloroethylene with framework oxygen in faujasites [252]
	dynamics of adsorbed molecules	translational and librational motion of ammonium ions, ammonia and water molecules in NH_4-Y zeolite [255]
QENS [260]	molecular motion, diffusion	dynamics of organic molecules adsorbed on ZSM-5 [256]
		mobility of methane in zeolite Y between 199 K and 250 K [257]
		molecular dynamics of n-pentane in Na-X zeolite [258]
		diffusion of benzene in Na-X and Na-Y zeolites [259]
IR spectroscopy [11]	framework vibrations	empirical assignment of IR bands to different framework vibrations [261, 262]
		identification of X-ray amorphous ZSM-5 zeolites [263]
		in situ observation of framework vibrations using the transmission [264] and the diffuse reflectance method [265]
	hydroxyl groups [11, 280]	formation of hydroxyl groups upon thermal treatment of NH_4-Y zeolites [266]

Tab. 2. (continued)

Technique	Target	Examples
		investigation of overtone and combination vibrations of hydroxyl groups [22, 267, 268]
		determination of different OH groups in hydrothermally dealuminated Y zeolites [269]
		vibrational lifetimes of hydroxyl groups in zeolites studied by picosecond IR pulses [270–272]
		spatial distribution of hydroxyl groups in ZSM-5 [273]
		estimation of extinction coefficients of OH vibrations in Y-type zeolites [274]
		orientation of hydroxyl groups in SAPO-5 [275]
		characterization of bridging hydroxyl groups forming a hydrogen bond based on the position of the out-of-plane bending OH vibration [276]
		in situ measurements of OH vibrations at elevated temperatures in ZSM-5 [277, 278] and mordenite [278, 279]
	acidity/basicity using probe molecules [280, 312–315]	pyridine and ammonia as probe molecules for the characterization of Lewis and Brønsted acid sites [266, 281–285]
		characterization of basic sites by adsorption of pyrrole [127, 286–289]
		characterization of Lewis acid sites by low-temperature adsorption of H_2 [290–295]
		combined NH_3-TPD and IR study on Y zeolites, mordenite and ZSM-5 [120], dealuminated Y zeolites and ZSM-20 [121]
		characterization of the acidic strength of Brønsted sites by adsorption of CO at low temperature [296–301]
		characterization of acidity of zeolites by *in situ* observation of the H/D-exchange [302, 303]

Tab. 2. *(continued)*

Technique	Target	Examples
		detection of various types of hydroxyl groups in ultrastable Y zeolites [121, 304] and ZSM-20 [121] after the adsorption of NH_3
		characterization of Lewis acid sites in dealuminated zeolites by adsorption of CO at low temperatures [285, 305, 306]
		differentiation of acid sites at the internal and external surface [307–310]
		correlation between the protonation of deuteroacetonitril measured at elevated temperatures and the cracking activity [311]
	isomorphous substitution	band at 960 cm^{-1} as a fingerprint of titanium in the framework [316, 317]
		detection of Fe(OH)Si groups in ferrisilicate pentasil zeolites [318, 319]
		shift of the framework vibration band at 1100 cm^{-1} to lower wave numbers in tin-silicate molecular sieves with MFI structure [320]
	cations	FIR vibration of A zeolites containing different cations [321, 322]
		study of FIR bands in Na,Mg-A zeolite after adsorption of probe molecules [323]
		computer modeling of vibrational spectra: no band in FIR can be unambiguously assigned to cationic vibrations at a specific site [324]
		change of FIR vibration of Na-Y and K-Y zeolites after adsorption of pyrrole: experimental evidence for an improved assignment of FIR bands [127]
		estimation of the strength of the electric field at the cation sites in A zeolites using CH_4, O_2 and N_2 [325–329]
		interaction of CO with cations in alkali-metal exchanged ZSM-5 zeolites at low temperatures [294, 330]

Tab. 2. *(continued)*

Technique	Target	Examples
		use of combination modes and overtones of metal carbonyls for the study of cations in zeolites: copper(I) carbonyls in Cu,Na-Y zeolite [331]
		low-temperature adsorption of hydrogen on acid-base pairs (sodium cations – basic oxygen anions) in faujasites [332]
		perturbation of the antisymmetric stretching T–O–T vibration by multivalent cations [333, 334]
		determination of different fundamental Cu(I)–C vibrations after adsorption of CO dependent on the zeolite type [335]
	templates	*in situ* observation of the thermal decomposition of the template in single crystals of MFI-type materials [336]
		IR and Raman spectroscopic study of the decomposition of triethyl-ammonium ions in $AlPO_4$-5 and SAPO-5 [337]
		host–guest and guest–guest interactions in $AlPO_4$-34 and SAPO-34 synthesized using morpholine as template [338]
	adsorbed molecules	diffusion coefficients of ethylbenzene and benzene in ZSM-5 [339] and of toluene in ZSM-5 single crystals [340]
		molecular orientation of *p*-xylene in silicalite-1 [28] and *p*-nitroaniline in $AlPO_4$-5 single crystals [341]
		dynamics of molecules adsorbed in zeolites (methanol [272, 342], water [272])
		in situ observation of the variation of the C–H vibration during supercritical heptane cracking over a Y-type zeolite using the CIR-IR technique [343]
		orientation of benzene, acetonitrile and pyridine molecules in SAPO-5 and GaAPO-5 [344]

Tab. 2. *(continued)*

Technique	Target	Examples
		recording the transient infrared spectrum of triplet excited duroquinone isolated in a Na-Y zeolite matrix by step-scan method [345]
Raman spectroscopy [352, 353]	framework vibrations	influence of the Si/Al ratio and cations on Raman bands of A [346, 347], X and Y zeolites [348]
		correlation of framework Raman bands with molecular sieve structure [349]
		phase transition of silicalite-1 due to the adsorption of benzene and *p*-xylene [350]
		FT-Raman spectra of aluminophosphates [351]
	templates	detection of Raman bands in the starting mixtures for the synthesis of ZSM-5 using a special pressure cell [354]
		tetrapropylammonium ions in single crystals of silicalite-1, Al and Fe MFI-type zeolites [355]
		IR and Raman spectroscopic study of the decomposition of triethylammonium ions in $AlPO_4$-5 and SAPO-5 [337]
	adsorbed molecules, host–guest interaction	interaction of pyridine with Brønsted and Lewis acid sites in faujasites [356–358]
		interaction of chloroform and trichloroethylene with framework oxygen in faujasites [252]
		host–guest interactions of *p*-chlorotoluene, toluene and chlorobenzene sorbed in completely siliceous ZSM-5 [359]
	identification of by-products	detection of anatase in Ti-containing MFI-type zeolites [360, 361]
UV-Vis/NIR spectroscopy [364, 365]	oxidation state	determination of reduction kinetics of mono- and polynuclear titanium oxide species in Na-Y zeolite [362]
		oxidation of Co^{2+} to Co^{3+} ions in presence of NO at 293 K in CoAPO-18 [363]

Tab. 2. *(continued)*

Technique	Target	Examples
	coordination, local environment	distortions of $Co^{2+}O_{4/2}$-tetrahedra in CoAPO-5 [366, 367]
		Co^{2+} ions at different sites in dehydrated X and Y zeolites [368], and mordenite [369]
		determination of changes in the local environment of Fe ions in iron silicalite due to template decomposition and interaction with adsorbates by IR, Raman, UV-Vis, EPR, XANES and EXAFS [370]
		spectroscopy (UV-Vis/NIR, ESR, XAS, XPS) and coordination chemistry of cobalt in molecular sieves [371]
	adsorbed molecules	arrangement of chromophores in $AlPO_4$-5 [372] and silicalite-1 [367, 373]
		characterization of the basicity of X and Y zeolites using adsorption of iodine [374]
		interaction of 1-butene with H-ferrierite [375]
photoluminescence spectroscopy	coordination, local environment	differentiation between isolated Cu^+ monomers and aggregated Cu^+–Cu^+ dimers in ZSM-5, mordenite, and zeolite Y [376]
		identification of three kinds of tetrahedrally coordinated V^V species in as-synthesized vanadium beta zeolite and their changes due to calcination and rehydration [377]
MAS NMR spectroscopy [11, 390–395]	chemical composition, framework ordering, isomorphous substitution	determination of the Si/Al ratio of alumino-silicates by ^{29}Si MAS NMR spectroscopy [378–380]
		ordering of silicon and aluminum in synthetic faujasites [379–382]
		evidence for boron incorporation in tetrahedral sites of zeolites from ^{11}B MAS NMR [383]
		multinuclear MAS NMR investigations of silicoaluminophosphate molecular sieves [384]
		ordering of framework atoms in MgAPO-20 [385]

Tab. 2. (continued)

Technique	Target	Examples
		multinuclear MAS NMR and IR spectroscopic study of silicon incorporation into SAPO-5, SAPO-31, and SAPO-34 [386]
		quantitative aspects of the determination of framework gallium in MFI-type gallosilicates [387]
		Mg ordering in MgAPO molecular sieves with CHA and AFI structure [388]
		characterization of mixed aluminum-gallium offretites [65] and gallosilicates with *BEA [66] and FAU structure [67]
		chemical and electronic nature of phosphorus in MgAPO-20 by ^{31}P magic-angle-turning NMR [389]
	crystallographically nonequivalent sites	crystallographically nonequivalent sites in highly siliceous zeolites [396]
		highly resolved ^{29}Si NMR spectra (20 lines) of dealuminated ZSM-5 [397]
		framework sites in aluminophosphate molecular sieves studied by ^{27}Al DOR NMR [398, 399]
		crystallographic sites in the erionite-like molecular sieves AlPO$_4$-17 and SAPO-17 [400]
		assignment of the ^{27}Al and ^{31}P NMR resonance lines to the crystallographic sites in the aluminophosphate VPI-5 [401]
		combined X-ray and neutron diffraction and ^{29}Si MAS NMR study of siliceous ferrierite [189]
		complete spectral resolution of the crystallographic sites in the triclinic CHA-like precursor of AlPO$_4$-34 [402] and the as-synthesized AlPO$_4$-14 [403] by ^{27}Al MQMAS
		interaction of aluminum at different framework sites with adsorbed water molecules in fully hydrated AlPO$_4$-11 studied by ^{27}Al{^{1}H} CPMAS NMR [404]

Tab. 2. *(continued)*

Technique	Target	Examples
		spectral resolution for crystallographically nonequivalent sites in ETS-10, improved assignment [405]
		oxygen sites in siliceous Y zeolite [406] and low-silica Na-X and Na-A zeolites [407] studied by ^{17}O DOR NMR
	framework connectivities [17, 18, 412]	two-dimensional ^{29}Si NMR investigations of framework connectivities in zeolite structures (ZSM-5 [408], ZSM-12, and KZ-2 (ZSM-22) [409])
		framework connectivities in aluminophosphate molecular sieves from two-dimensional cross-polarization and TEDOR experiments [410]
		interatomic connectivities in aluminophosphate molecular sieves using MQMAS-based methods [411]
	templates, template–framework interaction	templates in ZSM-5-type zeolites studied by ^{13}C NMR [413]
		inorganic–organic interactions during the synthesis of pure-silica zeolites studied by ^{1}H-^{29}Si CPMAS NMR [414, 415]
		^{13}C MAS NMR study of as-synthesized siliceous zeolites obtained from synthesis batches containing com-peting organic additives [416]
		quantitatively reliable spectra of occluded templates [417]
		guest–host interaction in as-made ZSM-12 studied by CPMAS NMR and REDOR [418]
	phase transitions, structural changes	sorbate-induced structural changes in ZSM-5/silicalite studied by XRD and ^{29}Si NMR [206]
		effect of template removal and rehydration on the local structure in microporous aluminophosphates and silicoaluminophosphates, reversible change of coordination of part of the framework aluminum [400, 419–422]

Tab. 2. *(continued)*

Technique	Target	Examples
		^{11}B MAS NMR study of the influence of hydration on the coordination state of boron in boron-containing MFI-type zeolites [423] and beta zeolites [424]
		temperature-induced phase transition in siliceous ZSM-5/silicalite-1 [204, 205] and siliceous ZSM-11 [187] studied by XRD and ^{29}Si NMR
		high-symmetry phases of zeolites (ZSM-39, -5, -11) at elevated temperatures [425]
		transformation of VPI-5 into AlPO$_4$-8 [426–428]
		^{27}Al DOR NMR study of the influence of hydration on the framework ordering in the aluminophosphate molecular sieves VPI-5, AlPO$_4$-5 and AlPO$_4$-8 [398, 429]
		^{27}Al DOR and MAS NMR studies of the hydration of AlPO$_4$-11 [430–432]
		variable-temperature ^{27}Al and ^{31}P MAS NMR investigations of hydrated VPI-5 [213]
		two-dimensional MAS NMR studies of AlPO$_4$-41 [422]
		^{27}Al and ^{31}P MAS NMR characterization and Rietveld refinement of the structure of rehydrated AlPO$_4$-34 [193]
	cation distribution and environment	sodium cations in dehydrated faujasite and zeolite EMT [433]
		location of Na$^+$ and Cs$^+$ cations in Cs,Na-Y zeolites studied by ^{23}Na and ^{133}Cs MAS NMR combined with X-ray powder diffraction [434]
		cation-induced transformation of boron-coordination in boron-containing ZSM-5, beta and SSZ-24 zeolites [424]
		combined MAS NMR and neutron diffraction study of the site preference in the mixed cation Li,Na-chabazite [234]

Tab. 2. *(continued)*

Technique	Target	Examples
	Brønsted acid sites [444–449]	highly resolved proton NMR spectra of hydroxyl groups in zeolites [435, 436]
		acidity of H-Y and H-ZSM-5 zeolites [437]
		geometry of Brønsted acid sites in H-Y and H-ZSM-5 zeolites [438, 439]
		acidity of dealuminated and non-dealuminated H-Y zeolite studied by broad-line ^{1}H NMR at 4 K and ^{1}H MAS NMR at 300 K [440]
		geometry and location of bridging OH groups in aluminosilicate and silicoaluminophosphate-type zeolites [441]
		determination of the quadrupolar coupling constant of the aluminum nuclei of Brønsted acid sides in dehydrated H-Y zeolite by TRAPDOR NMR [442]
		mobility of acidic protons in Brønsted sites of H-Y, H-mordenite, and H-ZSM-5 zeolites studied by ^{1}H MAS NMR at high temperature (up to 660 K) [443]
	acid sites, interaction with probe molecules [394, 447]	characterization of acid sites on dealuminated mordenite by ^{15}N MAS NMR spectroscopy of adsorbed pyridine [450]
		^{13}C MAS NMR study of the interaction between carbon monoxide and Lewis acid sites in H-ZSM-5 zeolites [451]
		proton transfer between Brønsted sites and benzene molecules in H-Y zeolite [452]
	coke	characterization of carbonaceous residues from zeolite-catalyzed reactions using ^{13}C MAS NMR [453]
		IR and ^{13}C MAS NMR study of the coke formation through the reaction of ethene over H-mordenite [454]
		^{13}C MAS NMR study of coke formation on H-ZSM-5 [455]

Tab. 2. (continued)

Technique	Target	Examples
		characterization of coke on MFI-type H-gallosilicate propane aromatization catalysts by ^{13}C MAS NMR, ESR, thermal analysis, and temperature-programmed oxidation [55]
^{129}Xe NMR spectroscopy [458–460]	pore size, pore geometry	chemical shift anisotropy related to the elliptical form of the channels in $AlPO_4$-11 and SAPO-11 [456]
		determination of void spaces in zeolites [457]
	cations	site selectivity of calcium ions in dehydrated zeolite A [461]
		application of ^{129}Xe NMR for studying mixed Y zeolites, effect of fast exchange between zeolite crystallites [462]
		effect of cation substitution (Li^+, K^+, Rb^+, Cs^+, Mg^{2+}, Ca^{2+}, Sr^{2+}, Ba^{2+}, Co^{2+}, Ni^{2+}, Cu^{2+}, Zn^{2+}) on the adsorption properties of xenon on Na-Y zeolite and on the ^{129}Xe chemical shift [81]
		polarization of xenon by di- and monovalent cations in zeolite A [463, 464]
	dispersed metals	size of small metallic particles on Pt/Na-Y zeolites [465]
		influence of calcination conditions on the formation of metal clusters in Pt/Na-Y zeolite [466]
		highly dispersed molybdenum in zeolite Y studied by EXAFS and ^{129}Xe NMR [467]
		surface heterogeneity of platinum particles on Pt/Na-Y zeolite [468]
		size and location of platinum clusters on Pt/K-L zeolite studied by ^{129}Xe NMR, XAS and xenon adsorption [469]
	adsorbed molecules	xenon occluded in Na-A zeolite, observation of five resonance lines corresponding to definite numbers of Xe atoms (1 to 5) per α-cage [470]

Tab. 2. *(continued)*

Technique	Target	Examples
		water diffusivity in Y zeolites studied by ^{129}Xe NMR [471]
		investigation of organic guest molecules in Na-Y zeolite [472]
		direct observation of mixed $Xe_n Kr_m$ clusters in Na-A zeolite by ^{129}Xe MAS NMR [463]
PFG NMR spectroscopy [484–488]	diffusion coefficients	interpretation and correlation of zeolitic diffusivities obtained from PFG NMR and sorption experiments [473]
		self-diffusion of n-paraffins [474, 475] and benzene, toluene and xylene isomers in Na-X zeolite [476]
		diffusion anisotropy in polycrystalline samples: methane in ZSM-5 [477], water in natural chabazite [478]
		selective measurement of self-diffusion coefficients of an ethane/ethene mixture adsorbed in Na-X zeolite by Fourier transform PFG NMR [479]
		self diffusion of n-alkanes in MFI-type zeolites at elevated temperatures (up to 653 K) [480]
		single-file diffusion in unidimensional channel zeolites [481–483]
	transport resistance	location of coke deposits on ZSM-5 [489]
		effect of coadsorbed benzene on the self-diffusion of methane in ZSM-5 [490]
		surface barriers on Ca,Na-A zeolites [491]
		mean distance between blockages within the channels of $AlPO_4$-5 [482]
further NMR methods applied to non-spinning samples	acid sites	acidity of H-Y zeolites before and after dealumination studied by broad-line ^{1}H NMR at 4 K and ^{1}H MAS NMR at 300 K [440]
		formation of hydroxonium ions on H-ZSM-5 zeolite [492]

Tab. 2. *(continued)*

Technique	Target	Examples
		geometry of Brønsted acid sites in H-ZSM-5 as determined by broad-line ^{1}H NMR and ^{1}H MAS NMR [493]
		acidity of H-Y zeolites; synergy between Brønsted and Lewis acid sites [494]
	acid sites, interaction with probe molecules	^{2}H NMR study of the sorption complexes of mono-, di-, and trimethylamine in H-rho zeolite [495]
		^{13}C NMR investigations of the interaction of carbon monoxide with Lewis acid sites in dealuminated Y zeolites [496] and H-ZSM-5 [497]
		geometry of the complex formed by CO adsorption on Brønsted acid sites as derived from low-temperature (4.5 K) ^{13}C NMR spectra [498]
		combined neutron diffraction and ^{2}H NMR study of the interaction of benzene with acid sites in H-SAPO-37 [244]
	adsorbed molecules, molecular motion [484, 509]	^{13}C NMR study of the molecular motion of *o*- and *p*-xylene in ZSM-5 [499] and benzene in silicalite-1 [500]
		^{2}H NMR study of the dynamics of aromatic molecules in ZSM-5 zeolites [501–504]
		^{2}H NMR study of the molecular dynamics of benzene in H-ZSM-5, Na-X, and Na-Y zeolites [505]
		^{2}H NMR study of the dynamics of ethene in silver-exchanged X-type [506] and Na-Y zeolites [507]
		^{13}C NMR study of the molecular motion of Mo(CO)$_6$ in Na-Y and H-Y zeolites [508]
	isomorphous substitution	incorporation of cobalt into aluminophosphate molecular sieves studied by spin-echo mapping ^{31}P NMR [510]
ESR spectroscopy [527, 528, ESEEM: 529]	coordination, oxidation state, isomorphous substitution	characterization of iron(III)-containing MFI-type zeolites [511]
		characterization of ruthenium species generated in H-X zeolite: interaction with probe molecules [512]

Tab. 2. *(continued)*

Technique	Target	Examples
		ESR and ESEEM spectroscopic investigations of manganese incorporation in MnAPO-11 [513, 514]
		characterization of iron in zeolites by ESR, ESEEM, and UV-Vis spectroscopy [515]
		investigation of Fe^{3+} species (lattice positions, defect sites and extra-lattice positions) in FAPO-5 and ferrisilicalite [516]
		in situ ESR monitoring of the coordination and oxidation states of copper in Cu-ZSM-5 at elevated temperatures in flowing gas mixtures: He, O_2, NO, NO_2 and H_2O [517]; CH_4 and CO [518]
		spectroscopy (UV-Vis/NIR, ESR, XAS, XPS) and coordination chemistry of cobalt in molecular sieves [371]
		incorporation and stability of Fe(III) and Fe(II) in $AlPO_4$-5 [519]
		ESR and ESEEM spectroscopic investigations of the incorporation of copper in CuAPO-5 [520], nickel in NiAPSO-34 [521], chromium in CrAPSO-5 [522], and titanium in TAPO-5, TAPO-11, TAPO-31 and TAPO-36 [523]
		changes of the oxidation state of copper ions in copper-exchanged ZSM-5 due to dehydration and rehydration/reoxidation [524]
		characterization of ruthenium-exchanged zeolites (beta, Y, and ZSM-5) by ESR spectroscopy [525]
		ESR spectroscopy of chromium in CrAPO-5 molecular sieves [526]
	Lewis acid sites using probe molecules	NO adsorption on alkaline earth (Mg^{2+}, Ca^{2+}, Sr^{2+}, and Ba^{2+}) ion-exchanged Y-type zeolites [530]
		application of different nitroxide molecules (e.g. tetramethylpiperidine-N-oxyl) as probes [531]
		NO adsorption on Na-ZSM-5 and H-ZSM-5 zeolites [532, 533]

Tab. 2. *(continued)*

Technique	Target	Examples
		adsorption-desorption behavior of NO on Lewis acid sites in Na-A zeolites [534]
		characterization of Lewis acid sites on non-framework aluminum species by adsorption of NO [306]
		adsorption of di-tert-butyl nitroxide on alkali metal (Li$^+$, Na$^+$, K$^+$, and Cs$^+$) ion-exchanged Y zeolites [535]
		application of different EPR resonance frequencies for investigation of NO adsorption on Na-A and Na-ZSM-5 zeolites [536]
	coke	coke formation through the reaction of olefines over H-mordenite: EPR study under static [537] and under on-stream conditions [538]
		characterization of coke on MFI-type H-gallosilicate propane aromatization catalysts by ^{13}C MAS NMR, ESR, thermal analysis, and temperature-programmed oxidation [55]
XPS, often in combination with XAES [549–552]	elemental analysis, type of bonding	dealumination of zeolite Y, accumulation of non-framework Al on the external crystal surface [539, 540]
		surface chemistry of zeolites [541, 542]
		nature of bonding chemistry and origin of the binding energy trends in zeolites [543–545]
		coke formation during methanol conversion over ZSM-5 [546]
		surface chemistry of the aluminophosphates VPI-5, AlPO$_4$-8 and AlPO$_4$-11 [547]
		comparative ESCA and NMR analysis of zeolites [548]
	coordination state	Ti coordination in titanium silicalite with MEL structure [553]
		Al coordination on the outer surface of dealuminated mordenites [554]
		spectroscopy (UV-Vis/NIR, ESR, XAS, XPS) and coordination chemistry of cobalt in molecular sieves [371]

Tab. 2. *(continued)*

Technique	Target	Examples
		Ti coordination in titanium ZSM-5 prepared by chemical vapor deposition [555]
	oxidation state	oxidation state of nickel in Ni-exchanged H-ZSM-5 [556]
		differentiation between internal and external metal atoms based on Auger parameters for copper in A, X and Y zeolites [557] and ZSM-5 [558, 559]
		oxidation state of copper in Cu-exchanged zeolite Y during catalyst preparation and after liquid-phase hydration of acrylonitrile [560]
AES [562]	elemental analysis	surface composition of zeolites by AES using primary electron beams of low current densities [561]
XRF	elemental analysis	comparison of analytical techniques for determining the silicon and aluminum contents in zeolites [43]
X-ray emission spectroscopy using EDX	spatially resolved elemental analysis	depth profiling of dealuminated Y zeolites by combined ESCA and EDX measurements [563]
		uniform distribution of silicon in the chabazite-like SAPO-44 [564]
		titanium distribution in aluminum-free titanium beta zeolites [136]
X-ray emission spectroscopy using WDX	spatially resolved elemental analysis	spatial distribution of aluminum over ZSM-5 crystals [273, 565, 566]
		aluminum distribution through ZSM-5 crystals synthesized in space [162]
X-ray absorption spectroscopy (EXAFS, XANES) [572, 573]	local environment (coordination, distances)	structure of CuPt particles supported in Na-Y zeolite [567]
		Ti coordination in titanium silicalite with MEL structure [553]
		combined XRD and EXAFS *in situ* observation of the thermal transformation of zinc- and cobalt-exchanged zeolite A [209, 210]
		detection of tetrahedrally coordinated framework and extra-framework gallium, and octahedral extra-framework gallium in gallosilicates with MFI structure [568]

Tab. 2. *(continued)*

Technique	Target	Examples
		Co sites in cobalt-substituted aluminophosphates [569]
		structure of TiO_6 units in the titanosilicate ETS-10 [570]
		size and location of platinum clusters on Pt/K-L zeolite studied by ^{129}Xe NMR, XAS and xenon adsorption [469]
		spectroscopy (UV-Vis/NIR, ESR, XAS, XPS) and coordination chemistry of cobalt in molecular sieves [371]
		detection of diferric (hydr)oxo-bridged binuclear clusters in Fe ZSM-5 prepared by sublimation [571]
	local environment, influence of adsorbed molecules	detection of monomeric $Ga^{+\delta}$ species in Ga/H-ZSM-5 at reaction conditions in the presence of reducing agent (H_2 or propane) [574]
		determination of changes in the local environment of Fe ions in iron silicalite due to template decomposition and interaction with adsorbates by IR, Raman, UV-Vis, EPR, XANES and EXAFS [370]
		interaction of ammonia with Ti(iv) sites in TS-1 studied by calorimetry, IR, XANES and EXAFS [575]
		changes of the oxidation state of copper ions in copper-exchanged ZSM-5 due to dehydration and rehydration/reoxidation [524]
		reversible changes in the oxygen coordination around the binuclear iron complex by *in situ* reduction with CO and reoxidation [576]
Mössbauer spectroscopy	oxidation state, coordination	chemical state of iron in MFI-type ferrisilicates [577]
		incorporation of Fe ions in AFI-type molecular sieves synthesized by microwave crystallization [519]
		valency changes of iron and tin in framework-substituted molecular sieves investigated by *in situ* Mössbauer spectroscopy [578]

Tab. 2. *(continued)*

Technique	Target	Examples
PIGE	elemental analysis	prompt nuclear and atomic reactions for elemental analysis of zeolites [579]
		chemical composition of pentasil zeolites [36]
		elemental characterization of zeolites by nuclear reactions induced by protons and deuterons [580]
PIXE	elemental analysis	prompt nuclear and atomic reactions for elemental analysis of zeolites [579]
		determination of Si/Al ratio in zeolites [581]
		gallium depth profiling in ZSM-5 galloaluminosilicate zeolites by RBS and PIXE [582]
complex impedance spectroscopy [584]	cations	detection of two distinct relaxation processes in dehydrated Na-X and Na-Y zeolites: local dipolar relaxation and long-range charge transport [583]
	protons	increased proton conductivity in zeolite H-beta in the presence of NH_3 [585]
		study of the influence of spilt-over hydrogen on the electrical properties of Pt-containing H-ZSM-5 [586]
	adsorbed or occluded species	decrease in the electrical resistance of a Pt-Y zeolite layer dependent on the amount of adsorbed n-butane [587]
		thermally activated electronic conduction in K-doped zeolite L detected by measuring the complex permittivity [588]

bining X-ray diffraction with neutron diffraction, Fyfe, Gies and coworkers [17–19] have provided numerous examples of the power of combining X-ray diffraction with solid-state NMR spectroscopy. Sankar and Thomas [20, 21] have repeatedly shown how effective the combination of X-ray absorption and X-ray diffraction techniques can be to characterize the catalysts even under operating conditions. Since NMR and X-ray absorption spectroscopy are sensitive to the local environments of the nuclei they are ideal complements to diffraction methods, which are sensitive to

References see page 990

long-range effects. Further details of the zeolite framework and occluded species can be obtained if the framework vibrations are examined by IR spectroscopic methods.

The wide use of IR spectroscopy reflects its high versatility. A variety of different experimental techniques allows all the steps of the preparation and use of catalysts to be monitored. Measurements can be carried out under conditions ranging from vacuum to *in situ* studies of chemical reactions. The range of application of IR spectroscopy has been widened by the development of special methods to overcome the inherent problems of measuring intensively colored and powdered samples. The use of diffuse reflectance introduced by Kazansky and his group [22] offered new possibilities, especially for the investigation of acid sites. Overtones and combination bands give additional information on various kinds of hydroxyl groups. *In situ* measurements can even be performed under real catalytic conditions (powder in a gas stream, detection of low-intensity bands) [23–25]. Since the necessary sample pretreatment can strongly influence the experimental results, it is a great advantage of DRIFT that combined NMR and IR investigations of hydroxyl groups can be carried out on identical sealed samples [26, 27]. The use of a microscope allows IR spectra of single crystals to be measured even in the presence of amorphous impurities and other by-products that may otherwise complicate the interpretation of the experimental data for powders. Furthermore, information on the molecular orientation of adsorbed species can be obtained using polarized radiation [28]. While other IR methods, like attenuated total reflection [29, 30], and photoacoustic detection [31–33] have been used for special applications, their usefulness in the field of zeolites is rather limited.

Multi-technique approaches as well as *in situ* studies [34, 35] have strongly contributed to the progress in the field of zeolite characterization during the last two decades. In a series of papers devoted to the characterization of ZSM-5 zeolites B.Nagy and coworkers [36–39] applied more than a dozen experimental techniques to study samples prepared by different synthesis routes and that had undergone various post-synthesis treatments. Obviously, it is not feasible to apply so much effort to every batch of zeolite synthesized, but this kind of thorough study contributes to the growth of knowledge as much as the development of new techniques. Instead of going through all the techniques mentioned in Table 2 we refer the reader to the reviews given there and to the following books: Basics and experimental results of AES, XPS, TPD, XRD, EXAFS, electron microscopy and Mössbauer spectroscopy have been compiled by Niemantsverdriet [40]. Thomas and Thomas [41] have comprehensively reviewed the techniques for characterizing catalysts and their surfaces covering most of the methods mentioned in Table 2. This book also contains a detailed description of the fundamentals of adsorption as well as of the different techniques to measure surface areas, pore volumes and pore diameters. Last, but not least, the reader is also referred to a book edited by Imelik and Védrine [42] that covers most of the techniques used for characterizing solid catalysts. For some of the techniques (EPR, Mössbauer spectroscopy, XANES and EXAFS, neutron scattering, electron microscopy and thermal methods) extensive surveys of the basics are given. The book contains numerous examples of the investigation of zeolites by XRD, neutron scattering, electron microscopy, XPS, UV-Vis, EPR, and NMR spectroscopy.

4.2.6.3
Characterization of Selected Features

4.2.6.3.1 Structure Elucidation and Identification
As soon as a zeolite batch has been synthesized it needs to be characterized with regard to particle size, crystal structure, crystallinity, and phase purity. Optical microscopy can be very helpful to gain a first impression of the quality of the material synthesized. The actual identification of the crystal structures present in the sample is usually accomplished by means of an automated X-ray powder diffractometer, which is best equipped with a sample changer to increase the throughput. Nowadays, many commercially available diffractometer systems allow the measured powder patterns to be compared directly with data contained in large data bases, for example the Powder Diffraction File of the ICDD [589]. Other valuable tools for the identification of zeolite phases present are the compilation of experimental powder patterns recorded by Lillerud [10] and the *Collection of Simulated X-ray Powder Patterns for Zeolites* by Treacy and Higgins [590] – the latter, together with the *Atlas of Zeolite Framework Types* [591], is regularly updated on the IZA Web site [592].

Of course, the vast majority of zeolites investigated are synthetic, but new natural zeolites continue to be discovered. Sometimes, zeolitic minerals are found in rather remote places: Mutinaite, a naturally occurring analog of the synthetic zeolite ZSM-5, has recently been found on Mt. Adamson, Northern Victoria Land, Antarctica [171, 593]. Zeolites of this structure type had first been synthesized more than two decades earlier [594].

Synthetic zeolites are often obtained only as rather small crystals (0.5–10 μm). However, these are still too large for their size to be determined from peak broadening in the X-ray powder patterns, a method frequently used in other fields of materials research. Provided that appropriate instrumental broadening corrections have been applied, peak broadening is a good measure for crystallite sizes up to about 300 nm [11]. Larger crystals do not give rise to any significant broadening of the peaks. Hence, in the field of zeolites the determination of crystallite size from peak broadening is limited to a few applications, mainly in regard to crystal growth.

Within the crystal size range given above, X-ray powder diffraction is not only the method of choice for routine identification, but is also the only available method to elucidate the crystal structure. Some examples of successful structure determinations using powder methods are given in Table 2. Although the methodology of this approach has been outlined by several authors, for example by McCusker and Baerlocher [194, 195], it requires a level of expertise that is in our opinion not

References see page 990

available in all groups working in the field of zeolite research. The degree of sophistication that has been reached can be appreciated from the fact that structures as complex as that of as-synthesized AlPO$_4$-40 containing 62 nonhydrogen atoms (8 P, 8 Al, 33 O, 1 N, and 12 C) in the asymmetric unit has been solved by Rietveld refinement using high-resolution powder diffraction data [192].

The advantages of combining X-ray diffraction with other methods have already been mentioned above. The combination with neutron diffraction is the most straightforward. Apart from the source of radiation, the major difference is the actual scatterer: whereas X-rays are scattered by the electrons, neutrons are scattered by the nuclei themselves. As a consequence, while for X-rays there is a strong attenuation of signal intensities at large angles, this does not occur for neutrons. Furthermore, the scattering factors of the elements for X-rays are quite different from those for neutrons. The location of light elements is one of the most important fields of application for neutron scattering. Very often, the issues that can advantageously be tackled by X-ray and neutron diffraction are complementary [14].

Textbook-like examples, where high-resolution solid state NMR spectroscopy played a fundamental role in the elucidation of zeolite structures from X-ray powder diffraction data, were reviewed by Gies et al. [17] a decade ago. For example, variable-temperature ^{29}Si MAS NMR measurements allowed a displacive phase transition in siliceous ZSM-11 zeolite to be monitored. The ^{29}Si MAS NMR spectrum at 373 K can be deconvoluted using six resonance lines with relative intensities of about 1:4:2:1:2:2. A Rietveld refinement of the data set recorded at the same temperature confirmed a proposed structure model with seven nonequivalent sites with the relative population 1:2:2:2:1:2:2 [187]. Later on, the same research group was successful in refining the structure of the low-temperature form of ZSM-11 containing 12 nonequivalent sites by combining ^{29}Si MAS NMR spectroscopy and X-ray powder diffraction with lattice energy calculations [19]. The corresponding ^{29}Si MAS NMR spectrum shows 12 well-resolved lines distributed over a range of less than 6 ppm. Solid-state ^{29}Si NMR spectroscopy yields not only the minimum number of nonequivalent sites, but allows much more detailed information on the local structure to be derived.

Soon after the first papers on the application of ^{29}Si MAS NMR to silicates [595] and zeolites [378] were published, the close correlation between the chemical shift and geometrical properties such as Si–O bond lengths and Si–O–Si bond angles α was recognized. The early work on the theoretical interpretation of ^{29}Si NMR chemical shifts in zeolites and other silicates and quantitative correlations between chemical shift and structure parameters has been reviewed by Engelhardt and Michel [596]. Since these chemical shift–structure correlations are the main reason why the combination of X-ray diffraction with solid-state NMR spectroscopy is so efficient, we discuss them in the following in more detail.

One of the empirical relationships correlates the isotropic chemical shift with the $\cos\langle\alpha\rangle/(\cos\langle\alpha\rangle - 1)$ function of the mean bond angle $\langle\alpha\rangle$. This was derived from the degree of s hybridization of the oxygen bond orbitals, which is related both to the bond angle α and the oxygen orbital electronegativity [597]. Using this linear relationship and the results of a single-crystal X-ray structure determination of monoclinic MFI framework containing 24 nonequivalent sites [168], Engelhardt and van Koningsveld [598] calculated a theoretical ^{29}Si MAS NMR spectrum. By comparing the theoretical spectrum with a highly resolved experimental one they were able to assign unambiguously the most separated lines to the various framework sites. The assignments of the other lines in the more crowded region of the spectrum was regarded by the authors as a first guideline for the spectra interpretation. A two-dimensional ^{29}Si MAS NMR study [408] has corroborated not only the firm assignments, but many of the tentative ones too, demonstrating once again the power of solid-state NMR for probing the local structure around the nucleus under study.

Using different zeolites for which high-quality structure data were available, Fyfe and coworkers [599] have critically evaluated several chemical shift–structure correlations. Both the mean Si–Si distance, that is, the distance between the target silicon atom and its first nearest neighboring silicon atoms, and the mean value of the $\cos \alpha/(\cos \alpha - 1)$ function was found to have a linear relationship with the isotropic ^{29}Si NMR chemical shift with very high correlation coefficients. Furthermore, the authors pointed out that the main limitation in the structure-chemical shift correlations is the accuracy of the X-ray data, particularly when refinements of powder data are involved. Errors of 0.1 pm in the Si–O bond lengths [596, 600, 601], 1 pm in the Si–Si distances [599] or 2° in the mean bond angles [598] correspond to comparatively large shift uncertainties on the order of 1 ppm. Very recently, Gies and coworkers [602] have demonstrated that improved structural data can be obtained from lattice-energy minimization calculations compared to those from Rietveld analysis provided that the topology of the framework is known. Even when the structural data available from X-ray analysis did not allow the lines in the ^{29}Si MAS NMR spectra to be assigned, it was possible to achieve a unique assignment from energy-minimized structures. With the above mentioned limitations with regard to the crystal size of many synthetic zeolites, the combination of X-ray powder diffraction with lattice-energy minimization and ^{29}Si MAS NMR spectroscopy offers great potential for the elucidation of structures. The structure solution of the low-temperature form of ZSM-11 is an example of the successful application of this approach [19].

It has to be mentioned that the applicability of correlations between ^{29}Si NMR chemical shifts and structure parameters is not limited to pure siliceous zeolites. Similar relationships have also been derived for sodalites and other zeolites containing Si(4Al) environments [603]. However, the spectral resolution necessary for an assignment of ^{29}Si MAS NMR lines to individual framework sites can, in general, only be obtained for siliceous zeolites.

Well-established relationships between geometrical properties such as the bond angle and NMR parameters exist not only for ^{29}Si, but also for ^{31}P and ^{27}Al [604]. For ^{27}Al, parameters describing the quadrupole interaction can be used to improve

References see page 990

the reliability. The strength of the quadrupole interaction characterized by the quadrupole coupling constant is directly related to the deviation of the AlO_4 unit from tetrahedral symmetry. This distortion can be described quantitatively by the so-called shear strain parameter ψ representing the deviations of the six O–Al–O angles from their ideal value of 109.5° [605]. Engelhardt and Veeman [401] used the linear relationship between this parameter and the quadrupole coupling to assign the ^{27}Al DOR NMR lines of the two tetrahedral sites in VPI-5. On this basis it became possible to assign the ^{31}P NMR lines unambiguously, ending a long-lasting debate.

The empirical chemical shift–structure correlations are mainly based on dense phases or zeolites containing no adsorbed molecules. The application of these relationships is likely to fail as soon as strong interactions with adsorbed or occluded species are present. Therefore, we are not surprised by the discrepancies between estimated and measured line positions as sometimes observed for as-synthesized or hydrated aluminophosphate molecular sieves, for example $AlPO_4$-14A [606] or rehydrated $AlPO_4$-34 [193]. These deviations from the empirical relationship between bond angle and chemical shift in aluminophosphates [604] are most probably due to the influence of adsorbed water molecules.

The elucidation of the structure of the titanosilicate ETS-10 is a nice example of the interplay of many different analytical methods. The structure of this material, which displays a considerable degree of disorder, has been solved using a combination of high-resolution electron microscopy, electron and powder X-ray diffraction, solid-state NMR spectroscopy, molecular modeling, and chemical analysis [147]. Whereas the information on framework ring connectivities and local disorder came from high-resolution electron microscopy, the local environment of the various silicon sites has been determined by ^{29}Si MAS NMR spectroscopy. This information allowed a trial structure to be built that was refined and, subsequently, used to simulate the HREM images and diffraction data.

Although diffraction methods are undoubtedly the major source of structural information on molecular sieves, other spectroscopic methods apart from NMR can also provide useful information on the framework structure. A systematic study of the relationship between the appearance of IR bands and structural elements of zeolites was carried out by Flanigen et al. [261]. Based on experimental data obtained for a large number of zeolites, two types of vibrations were postulated: internal vibrations of the TO_4 tetrahedra and external vibrations (for example, double rings, pore openings) that are structure sensitive. However, theoretical studies show that the framework vibrations are strongly coupled and a modification of this assignment seems to be necessary. A complete assignment of all the IR framework vibrations is not yet available, but in many cases characteristic bands can be identified. For example, when studying the synthesis of ZSM-5, IR spectroscopy can be used as a fast initial check of the quality of the zeolite material obtained [607]. The infrared spectrum of ZSM-5 has a band near 550 cm^{-1} that is not present in the spectrum of the aluminosilicate gel before zeolite synthesis. Therefore, this band is an indicator for ZSM-5 in "X-ray amorphous" samples and its intensity is a good measure for the degree of zeolite crystallization [263]. The relative intensity of the band at 550 cm^{-1} with respect to that of a band found in both the gel and the zeolite spectra near 450 cm^{-1}, I_{550}/I_{450}, can be used for this purpose [607]. A value of about 0.8 has been found for this particular intensity ratio for all pure pentasil zeolites calcined at 823 K [262]. Moreover, the progress of crystallization can be followed by the IR spectra of intermediate products.

Structural changes occurring, for instance, during template removal, adsorption processes or thermal treatments can be followed by X-ray diffraction and NMR and IR spectroscopy. Numerous examples of *ex situ* and *in situ* studies of such processes are given in Table 2. IR spectroscopy using the region of framework vibrations is a convenient method to follow structural changes. However, since framework vibrations are very often studied in the transmission mode using KBr wafers, the interpretation of these spectra can in some cases lead to erroneous conclusions as can be illustrated by the following example: The IR transmission spectrum of a SAPO-34 sample that was calcined, rehydrated and diluted in KBr shows broad and featureless framework vibration bands that are very different from those observed in the as-synthesized form. Using the diffuse reflectance mode the dehydration process can be monitored *in situ* [265]. Even partial dehydration is sufficient to make the calcined sample show the same framework vibrations as the as-synthesized one. Hence it follows that the disappearance of IR framework vibrations is not caused by a collapse of the framework structure during template decomposition, but by a distortion due to the adsorption of water. The reversibility of these structural changes for many aluminophosphate molecular sieves has been demonstrated by X-ray diffraction and/or MAS NMR spectroscopy [400, 419–422]. The influence of hydration on the structure of aluminophosphate-based materials is not always reversible [420, 608] and must be taken into consideration for any kind of investigation.

4.2.6.3.2 Templates and Their Removal

Many of the synthesis procedures developed in recent decades to produce new zeolitic materials rely on the application of organic additives as so-called templates or structure-directing agents. In a more general sense, the concept of templating and structure-directing agents also includes inorganic cations and fluoride ions [609]. In the present section, we use the term template as a synonym for an organic additive. The vast majority of successfully applied template molecules are nitrogen-based cations. The exact role of these additives during the synthesis is still under debate and might well vary from structure type to structure type. It is well known that a given template can give rise to the formation of several zeolite structures and that many different templates can produce one and the same structure. There are only a few examples where the synthesis of a certain structure seems to require the presence of a specific template. Nevertheless, some general trends can be discerned concerning the relation between molecular geometry (size, shape, rigidity) and chemical properties (basicity, hydrophobicity) of an organic additive and its ability to function as a structure-directing agent in the synthesis of zeolites, in

References see page 990

particular in the case of high-silica materials [414, 415, 610, 611]. Very recently, Balkus [612] has comprehensively reviewed the synthesis strategies for large-pore molecular sieves and has discussed the role of organic additives in the synthesis in great detail.

One approach to better understand the influence of the template on the zeolite structure is to characterize the template in the as-synthesized material as thoroughly as possible. Following the concept of structure-direction, the location of the organic molecules occluded in the cavities and channels of pore systems and the elucidation of their interactions with the inorganic framework can yield clues as to why a certain structure is obtained with a particular additive [509, 610]. For example, the experimental data thus obtained can be used to verify results of computer modeling studies on template–host interactions [613]. These computational studies aim to predict the templating ability of organic additives in the synthesis of microporous materials [614, 615]. The ultimate goal is to design a new zeolite structure with predefined properties that are advantageous to a certain application and to design a suitable template to produce it. The synthesis of a cobalt-substituted aluminophosphate with CHA structure, DAF-5 [616], using 4-piperidinopiperidine is the first example where computational design of a potential template molecule was successfully used to obtain a chosen zeolite structure.

However, there are further reasons for studying the state of the template and the organic–inorganic interactions in as-synthesized zeolites. Before a zeolite can be used in a catalytic or adsorption process it has to be activated. Whereas in the case of a template-free synthesis the activation is essentially a dehydration, the activation of template-containing zeolites necessitates the removal of the organic species. This is mostly accomplished by calcination, which is often not a trivial task. The heat of combustion generated may result in local overheating that can lead to destruction of the framework or to the removal of heteroatoms from the framework of an isomorphously substituted zeolite. The cracks observed in large MFI zeolite crystals [133] are an impressive example of undesired results of a zeolite calcination. As with the activation of template-free zeolites, self-steaming during the calcination can also result in an undesired dealumination of an aluminosilicate framework and in significant framework damage. Therefore, the calcination process should be handled and controlled carefully, especially for large crystals. Since the knowledge of the actual degradation mechanism of the template can help to avoid problems, the elucidation of these mechanisms is not only an academic task, but it allows the most appropriate procedure for the template removal to be chosen and can be crucial for a successful application of a newly synthesized zeolite.

Since alkylammonium compounds are frequently used as templating agents, their degradation in as-synthesized zeolites has been studied extensively [37, 45, 52–54, 265, 336, 337]. Parker et al. [45] investigated the thermal decomposition of several tetraalkylammonium compounds in ZSM-5 and silicalite using thermogravimetry coupled with mass spectrometry. They concluded that the tetraalkylammonium ions in the aluminum-free silicalite decompose by the Hofmann reaction to give an olefin and a trialkylamine. The trialkylamine then undergoes a series of β-elimination reactions to give di- and monoalkylamines and ultimately ammonia. In ZSM-5, the tetraalkylammonium ions associated with the acid sites are more stable and decompose at higher temperatures by sequential Hofmann reactions. These degradation mechanisms were corroborated by subsequent studies [37, 53, 336]. Identifying the volatile products, Kessler and coworkers [53] were able to show that many other reactions (for instance, oligomerizations, additions, and cyclizations) occur during the thermal decomposition of the template. Lercher and coworkers [336] studied the thermal treatment of a single crystal of ZSM-5 synthesized with tetrapropylammonium fluoride by *in situ* IR spectroscopy. The authors concluded that the rate-determining step of the template removal is the reaction to the dipropylammonium ion or dipropylamine, in support of the above degradation mechanisms. The emphasis of the ionic or nonionic mechanism is strictly dependent upon the aluminum concentration in the zeolite framework.

Whereas these studies were performed under inert atmospheres, Bourgeat-Lami et al. [54] investigated the decomposition of tetraethylammonium ions in zeolite beta by thermogravimetry coupled with mass spectrometry both under air and argon atmospheres. Again, sequential Hofmann elimination was found to be the dominating mechanism. IR and Raman spectroscopic studies of the decomposition of triethylammonium ions in $AlPO_4$-5 and SAPO-5 under air and nitrogen have recently been reported by Schnabel et al. [337]. In accordance with the above mentioned results for silicalite and ZSM-5, pronounced differences in the course of the template decomposition were found that were attributed to the interaction of the organic species with the charged framework of SAPO-5. DRIFTS is generally a very powerful and convenient technique for following the template removal *in situ* since it allows the vibrations of the template and those of the framework to be studied simultaneously. Structural changes occurring during the removal of the template can be monitored, as has been shown, for example, in a study of the decomposition of tetraethylammonium ions in as-synthesized $AlPO_4$-18 [265].

4.2.6.3.3 **Framework Composition**

Most of the methods used to determine the chemical composition of zeolites and related materials cannot distinguish between framework and non-framework contributions. The bulk composition, which can differ significantly from that of the framework, can be determined by, for example, AAS, ICP-OES, XRF, EDX, PIGE or PIXE. The potential of ^{29}Si MAS NMR spectroscopy as a reliable, standard-free method to determine the framework composition of zeolites was recognized [379, 380] as soon as the pioneering papers on silicates [595] and zeolites [378] had been published. Within a few years of this breakthrough, ^{29}Si MAS NMR spectroscopy became a standard method to determine the Si/Al ratio of the framework [391]. For Si/Al ratios of up to about 10, it is superior to other methods provided that the following conditions are fulfilled:

(1) the spectrum contains more than one line and is correctly interpreted in terms of $Q^4(mAl)$ units ($m = 0$–4),

(2) there are no Al–O–Al linkages (Loewenstein's rule is obeyed),

References see page 990

(3) only Q^4 units are present, that is, all silicon atoms are linked via oxygen to four other T-atoms.

The last condition can cause problems for dealuminated zeolites or other samples that have a relatively high concentration of defect sites. When there is a significant contribution from Q^3 units it is impossible to determine the Si/Al ratio of the framework reliably from the ^{29}Si MAS NMR spectrum alone. However, combining ^{29}Si MAS NMR with other methods, such as bulk chemical analysis, ammonium-exchange measurements and ^{27}Al and ^1H MAS NMR spectroscopy, allows even acid-leached mordenites with a high concentration of silanol groups to be characterized in great detail [617]. It should be noted that the same approach for determining the framework composition by NMR spectroscopy also works for other microporous materials, such as gallosilicates [618, 619], metal-substituted aluminophosphates (MAPOs) [385, 388], and metal-substituted titanosilicates [620].

IR spectroscopy is another method that has been used for determining the framework Si/Al ratio of aluminosilicates for many years. The presence of aluminum does not give rise to additional IR bands in the range of framework vibrations. Aluminum atoms distributed over the framework have a line-broadening effect, which is clearly visible when the spectra of Na-X, Na-Y and highly dealuminated zeolite Y are compared [621, 622]. Furthermore, the presence of aluminum lowers the wave numbers of all the stretching bands. Since the masses of aluminum and silicon atoms are approximately equal, this can only be attributed to the weaker Al–O bond strength compared to that of Si–O. This relationship has been used to estimate the Si/Al ratio in the faujasite framework [622]. For ZSM-5, the position of the most intense IR band was also found to depend linearly on the aluminum content [623, 624]. However, when the aluminum content is reduced from 5 to 0.6 atoms per unit cell the wave number only increases from 1090 cm^{-1} to 1104 cm^{-1}. Thus, determinations of Si/Al ratios based on this wave number must be regarded as rough estimates.

Isomorphous substitution into zeolite frameworks offers many perspectives for preparation of materials with novel properties. Through the incorporation of transition metals into zeolites, for example, catalysts can be produced that combine surface acidity with redox properties. Such a combination of characteristics is favorable for many catalytic reactions such as selective oxidation processes. Especially after the discovery of aluminophosphate molecular sieves [625, 626] many microporous materials were prepared via isomorphous substitution of aluminum and/or phosphorus by different elements. There are several excellent reviews [627–630] dealing with the progress in this field of zeolite research that show the versatility of physicochemical properties that can be generated in phosphate-based molecular sieves.

The incorporation of silicon into the aluminophosphate framework produces silicoaluminophosphates (SAPOs) possessing Brønsted acidity [631, 632]. The extent of isomorphous substitution that can be achieved depends on the structure type. Various mechanisms for the framework incorporation of silicon are possible

[627–629]. Only if the silicon atoms are bonded via oxygen to four adjacent aluminum atoms does the calcination of the as-synthesized sample generate one Brønsted acid site per incorporated silicon atom. This type of incorporation is sometimes referred to as monomeric substitution. The creation of pure-silica units is possible when both phosphorus and aluminum are substituted by silicon. Therefore, catalytic properties of SAPOs are determined not only by the overall silicon content, but also by the type of silicon incorporation. ^{29}Si MAS NMR is the only direct method for determining the type of silicon incorporation. Resonance lines between -89 ppm and -97 ppm are attributed to $Q^4(4Al)$ units, that is, to silicon atoms bonded via oxygen to four aluminum atoms in the aluminophosphate framework [384, 386, 400, 633–636]. Each such silicon atom should give rise to one SiOHAl group. These Brønsted acid sites can be detected by IR and ^{1}H MAS NMR spectroscopy. As regards the sensitivity for detecting silicon incorporation, IR spectroscopy, especially DRIFTS, is superior to ^{29}Si MAS NMR [637]. However, the difficulties in determining the extinction coefficient of hydroxyl groups hamper the reliable quantitative analysis for the different types of OH groups present. This problem can advantageously be solved by the application of ^{1}H MAS NMR spectroscopy [386]. Hence, the characterization of the silicon incorporation in SAPO-type materials is another nice example of the efficiency of combined IR and NMR investigations.

Other substitution mechanisms lead to $Q^4(mAl)$ units with $m < 4$. Silicon atoms surrounded in the aluminophosphate framework by four silicon atoms as next-nearest neighbors $(Q^4(0Al))$ give rise to ^{29}Si NMR signals between -106 ppm and -112 ppm [384, 633–636]. Since this is the same region as for $Q^4(0Al)$ in amorphous silica [638], even small amounts of an amorphous silica-rich by-product can falsify a quantitative analysis based on ^{29}Si MAS NMR spectra. The different mechanisms of framework substitution and their consequences for the properties of the SAPOs obtained have been reviewed comprehensively by Martens and Jacobs [629]. In this publication, the authors discuss in great detail the alternative models [635, 636, 639] for the incorporation of silicon in SAPO-37, a material that has the same topology as the widely used synthetic faujasites.

Besides the incorporation of silicon into aluminophosphate frameworks, great efforts have been made to prepare and to characterize new microporous materials containing various elements. Special attention has been paid to the incorporation of the following elements into aluminophosphate and/or (alumino)silicate frameworks: B [51, 423, 640–645], Mg [172, 385, 388, 389, 646–651], Ti [136, 224, 226, 316, 317, 360, 361, 523, 553, 555, 575, 652–685], V [377, 686–691], Fe [61, 318–320, 370, 511, 515, 516, 519, 577, 578, 692–695], Co [172, 363, 366, 367, 510, 569, 696–708], and Ga [61, 65–67, 239, 344, 387, 568, 582, 619, 709–711]. However, reports can also be found on the framework incorporation of other elements such as Cr [522, 526, 712–714], Mn [215, 513, 514, 715–717], Ni [521, 718–720], Cu [520], Zn [721, 722], Ge [723], Ru [724], In [725, 726], Sn [320, 727], and Ta [728]. Com-

References see page 990

prehensive reviews are available for some of the elements, for example, for B [729], Ti [730–734], V [377, 690, 730], and Ga [735].

For many heteroatoms it is not easy to verify that they have actually been incorporated into the crystalline frameworks because it requires a very accurate physicochemical characterization. One of the main problems is that the extent of substitution that can be achieved is often limited to a few per cent. Indeed, many claims in the literature are not supported by convincing evidence, and this is even more true of patents. The attempts to verify the incorporation of cobalt into aluminophosphates by NMR spectroscopy exemplify the pitfalls when characterizing isomorphously substituted zeolites. In several papers (for example, refs [736, 737]) it had been claimed that ^{31}P MAS NMR spectra with intense spinning sidebands over a wide frequency range can be regarded as evidence for the incorporation of paramagnetic ions into the aluminophosphate framework, while other authors [738] had claimed the decrease of the spin-lattice relaxation time to be evidence for an isomorphous substitution. However, comparison of CoAPO$_4$-11 with cobalt-impregnated samples of AlPO$_4$-11 [739] showed that ^{31}P MAS NMR spectroscopy is not an appropriate technique to prove framework substitution – the spectra obtained are almost identical. In a very detailed study, Peeters et al. [740] came to the conclusion that the incorporation of cobalt into the framework leads to some of the phosphorus becoming "invisible" in the NMR. They showed that the amount of missing phosphorus could be correlated to the Co concentration. Recently, Tuel and coworkers [510] have been able to show that conventional MAS/NMR spectroscopy is indeed unable to detect phosphorus atoms linked via oxygen to cobalt. Spin-echo mapping is necessary to record the very broad and strongly downfield shifted resonance lines, which are direct evidence for the framework siting of cobalt. There is good reason to assume that similar situations occur for the incorporation of other paramagnetic ions. The state-of-the-art in characterizing cobalt-containing molecular sieves by UV-Vis, NIR, EPR, XAS, and XPS has recently been reviewed by Verberckmoes et al. [371]. For a more general review on the incorporation of transition metal ions (V, Cr, Mn, Fe, Co) into microporous aluminophosphates, the reader is referred to a paper by Weckhuysen et al. [741].

Even if the samples are not paramagnetic, it is difficult to obtain unambiguous evidence for an isomorphous substitution, that is, the incorporation of heteroatoms into tetrahedral sites of the framework. In 1983 Taramasso et al. [316] claimed the isomorphous substitution of silicon by titanium in silicalite-1 and named the material thus obtained TS-1. This catalyst turned out to have extremely useful catalytic properties, particularly in selective oxidation reactions with H_2O_2 as the oxidant. Hence, interest in Ti-containing catalysts spread rapidly in the scientific community and many different analytical methods were used to characterize the state of titanium in the samples. Since the catalytic activity is often attributed to titanium tetrahedrally coordinated in the framework, the detection and quantification of this type of titanium is a very important issue. Some of the methods that have been applied to verify the framework incorporation are summarized in Table 3. We have compiled these data to give an idea of the range of the methods that have been used to gather valuable information on the state of the titanium atoms in the samples under study, but at the same time we do not regard all the characteristics mentioned as convincing evidence for an isomorphous substitution of titanium into the various zeolite structures. Again, we suggest that an approach employing several methods is the most reliable way to tackle such a complex problem.

An IR band at 960 cm^{-1} is widely regarded as a fingerprint of titanium in the framework [224, 316, 575, 652, 661, 664, 668, 671, 673, 678, 680]. The intensity of this band relative to that of framework bands at about 550 cm^{-1} [224, 316, 664, 678] or at about 800 cm^{-1} [680] has been used to estimate the amount of titanium in the framework. However, the origin of the band at 960 cm^{-1} is still contentious. A band in this region has been assigned to a Si–O stretching vibration in Si–O$^{\delta-}$–Ti$^{\delta+}$(IV) units [652] or SiOR groups (R = H$^+$ or TEA$^+$ [661]). Based on *ab initio* quantum-chemical calculations, an assignment to Ti–O–Si antisymmetric stretching modes has been suggested [670]. On the other hand, it has to be taken into consideration that a band in this range has also been observed for hydroxylated samples containing no heteroatoms at all [317, 674, 742]. In these cases the band has been assigned to a O$_3$Si–OH stretching vibration in hydroxylated defects [317]. For a highly dealuminated Y zeolite it has been observed that this band is not present in the framework spectra measured at 473 K [743], whereas in TS-1 the band at 960 cm^{-1} is still found in the spectra at 773 K [674]. Hence, it follows that several bands due to various species could be overlapping in the infrared range under discussion, thus hampering the characterization of titanium-containing molecular sieves. Nevertheless, an infrared band at 960 cm^{-1} can be a useful hint that heteroatoms have been incorporated – provided that its behavior during thermal treatments is also investigated. Very recently, Ricchiardi et al. [684] have presented a thorough analysis of the vibrational features of TS-1. Thin self-supported wafers of samples with different titanium contents were prepared and dehydrated under identical conditions. The intensities of the spectra were normalized using the overtone and combination framework vibrations between 1500 and 2000 cm^{-1}. Under these well-defined conditions, a quantitative correlation between the titanium content and the intensity of the band at 960 cm^{-1} was established. On the other hand, we want to point out that the usual method for investigating framework vibrations of zeolites, that is, the dilution of the zeolite in KBr and the subsequent preparation of wafers for transmission measurements, is inappropriate to study the titanium incorporation. Adsorbed water influences both the position and the intensity of bands present in the region of 960 cm^{-1} [317, 652].

Another frequently studied isomorphous substitution is that of aluminum by gallium. Considering the chemical similarities between gallium and aluminum, this approach is straightforward. For high gallium contents the incorporation of gallium into the framework can easily be verified by ^{29}Si MAS NMR spectroscopy [618]. A quantitative determination of the substitution level achieved can be very difficult, especially at lower gallium concentrations. In the ^{29}Si MAS NMR spectra lines from Q^4(1Ga) and Q^3(0Ga) units overlap. Furthermore, Bayense et al. [387]

References see page 990

Tab. 3. Characterization of Titanium-containing Zeolites by Various Analytical Methods

Structure type	Method	Characteristics used as evidence for isomorphous substitution
MFI	IR	band at 960 cm^{-1} [224, 316, 575, 652, 661, 664, 668, 671, 673, 678, 680, 684]
	Raman	band at 960 cm^{-1} [317, 360, 361, 667, 668, 673, 684]
		bands at 1125, 530 and 490 cm^{-1} (only when charge-transfer transition is excited) [683, 684]
	UV-Vis	band at 210 nm [224, 652, 656, 657, 671]
		onset at 250–260 nm [667]
	XRD	unit cell expansion [224–226, 316, 671]
	XPS	higher Ti $2p_{3/2}$ binding energy compared to anatase [553, 666]
	^{29}Si NMR	temperature of the monoclinic-orthorhombic phase transition, shoulder at −116 ppm [660]
		reduced signal intensity after Ti^{4+} to Ti^{3+} reduction with CO [682]
	XAS	EXAFS: Ti first shell coordination number of $N = 4.0$, XANES: pre-edge peak at 4967 eV indicating tetrahedral coordination [553, 575, 655, 658, 662, 665, 675, 676, 681, 684]
	neutron diffraction	refinement of site occupancy factors using five different strategies [233]
	photoluminescence	band at 495 nm in the emission spectrum [675, 685]
MEL	XRD	unit cell expansion [653]
	IR	band at 960 cm^{-1} [653, 664]
	UV-Vis	band at 240 nm [664]
*BEA	XRD	unit cell expansion [136, 677]
	IR	band at 960 cm^{-1} [661, 677]
	UV-Vis	band at 190–210 nm [669, 677]
	XAS	EXAFS: Ti first shell coordination number of $N = 4.0$, XANES: pre-edge peak at 4967 eV indicating tetrahedral coordination [669]
	^{29}Si NMR	reduced signal intensity after Ti^{4+} to Ti^{3+} reduction with CO [682]
ZSM-48	XRD	unit cell expansion [659]
	IR	band at 960 cm^{-1} [659]
AFI, AEL, ATO, ATS	ESR	^{31}P and ^{27}Al ESEEM signal of Ti(iii) after reduction with CO [523]

have shown in a very detailed study of gallium-containing MFI-type zeolites that even at a magnetic field of 14.1 T (600 MHz proton resonance frequency), no direct quantitative information on the amount of non-framework gallium can be obtained by ^{71}Ga MAS NMR spectroscopy – the resonance line can be too broad to be detected. Hence, the procedure used by Liu and Klinowski [619] to assess the amount of Q^3 defect sites in gallium-containing zeolites of the same type is rather questionable since it relies on the absence of non-framework gallium, which they deduced from the absence of the corresponding line in the ^{71}Ga MAS NMR spectrum measured at only 9.4 T. Besides NMR spectroscopy, XAS has proven to be very useful for studying the incorporation of gallium into the framework. In an EXAFS and XANES study of MFI-type gallosilicates [568], it was found that at low Si/Ga ratios a significant part of the gallium is outside the framework, both in tetrahedral and octahedral coordination. Since EXAFS cannot differentiate between tetrahedral framework and non-framework gallium, the combined use of complementary techniques is highly recommended to verify the isomorphous substitution. Very recently, Fricke et al. [735] gave an excellent review on the synthesis, characterization and catalytic application of gallium-containing microporous and mesoporous materials focusing on isomorphous substitution into zeolite frameworks. For further studies on isomorphous substitution of other elements the reader is referred to the papers compiled at the beginning of this section or given in Table 2.

Finally, it should be pointed out that catalytic performance or stability against leaching are the least convincing arguments for an isomorphous substitution. Unless a well-established, detailed reaction mechanism demands that the catalytically active heteroatoms are in the zeolite framework, the possibility must be taken into consideration that heteroatoms linked to framework T-atoms by just one or two oxygen bridges are acting as the active sites. The high dispersion of the heteroatoms that can be achieved when they are present in the synthesis gel can, in some cases, also be accomplished by treating a dealuminated zeolite with an appropriate precursor. Although commercial interests related to patents might argue otherwise, from a scientific point of view the anchoring of heteroatoms to the framework as described above should not be confused with the well-defined term isomorphous substitution, that is, a replacement of an element in a crystalline lattice by another element with similar cation radius and coordination requirements [630].

4.2.6.3.4 Acid Sites

The acidity (Brønsted and Lewis) is one of the most important properties of zeolites and is utilized in the chemical industry in large-scale plants. Apart from other factors, such as limitations of heat and mass transfer, the efficiency of an acid catalyst will be determined by the strength of the individual acid sites and their concentrations and accessibility for the reactant molecules. A comprehensive characterization of acid catalysts has to include the determination of these parameters,

References see page 990

which are *a priori* independent. The principles and methods applicable to the characterization of acid sites in zeolites are summarized in several reviews [313, 447, 448, 744–751]. Although the attention has long been focused on the acid properties of zeolites, there are also many papers on the characterization of basic sites, for example basic oxygen atoms or alkaline metal clusters [746, 752–765].

Because of the importance of Brønsted acidity in catalysis, hydroxyl groups have been studied extensively ([11] and references therein). There are several types of OH groups that can occur in molecular sieves. Bridging hydroxyl groups (strong Brønsted acid sites, for example $\equiv$Al(OH)Si$\equiv$) are formed as a result of the interaction of framework oxygen atoms with protons compensating the negative framework charge in the molecular sieve. Weakly acidic terminal OH groups occur at the external crystal surface. Interacting internal SiOH groups (hydroxyl nests) located at framework defects have been discussed in the case of ZSM-5 [296, 766–769]. SiOH groups can also exist in secondary pores generated during the dealumination of aluminosilicates [770]. In dealuminated samples, additional OH groups can be observed due to the formation of various non-framework aluminum species [121, 269, 299, 304, 771, 772]. The electrostatic field of multivalent cations can induce the dissociation of coordinatively adsorbed water molecules, leading to bridging OH groups and hydroxyls on these cations [773, 774]. Furthermore, various types of AlOH groups and POH groups can occur in aluminophosphate-based materials [775].

Among the acid sites, only hydroxyl groups can directly be investigated by IR and ^{1}H MAS NMR spectroscopy, and inelastic neutron scattering (INS). Spectroscopic investigations of Lewis acid sites necessitate the use of probe molecules. To characterize Brønsted acid catalysts without the use of probe molecules, the fundamental stretching OH vibrations have been investigated for many years using the IR transmission technique [2, 280]. However, the versatile diffuse reflection technique has increasingly become the method of choice for many investigations [22, 23, 26, 27, 265, 267, 276, 277, 318, 332, 335, 337, 386, 637, 771, 776–785]. The advantages and disadvantages of this particular technique have recently been summarized by Kustov [786]. With the development of modern line-narrowing techniques, the ^{1}H NMR spectroscopy became one of the standard methods for investigation of hydroxyl groups [444]. The major drawback of this approach is the comparably high cost of the equipment necessary. There are several reviews [394, 441, 444, 447, 448] summarizing the developments in this field and the assignments of the resonance lines observed for the various types of hydroxyl groups.

Since modern versatile IR spectrometers are nowadays available in almost every laboratory, we restrict ourselves in the following to the application of IR spectroscopy to characterize acid sites. The fundamental stretching vibrations of hydroxyl groups in molecular sieves are located in the range 3800–3200 cm^{-1}. In certain materials that contain small (6- or 8-membered) rings, for example faujasites, two different bands caused by $\equiv$Si(OH)Al$\equiv$ groups are observed: the so-called HF (high frequency) band at 3660–3600 cm^{-1}, which can be observed even when the concentration of hydroxyl groups is low, and the so-called LF (low frequency) band at about 3580–3550 cm^{-1}. It is known that the stretching vibration of the $\equiv$Si(OH)Al$\equiv$

groups depends on the Si/Al ratio of the sample, and to a lesser extent also on the type and content of cations. Many approaches that were made to explain this observation have been summarized by Karge et al. [11].

Special attention has to be paid to the assignment of bands in the range of OH stretching vibrations if the spectra are measured in the presence of adsorbed or encaged molecules. Due to the interaction of these molecules with hydroxyl groups, the OH bands shift to lower wave numbers and their integral intensities increase. The following example demonstrates that both factors may lead to misinterpretations. Bedard et al. [787] measured the spectra of as-synthesized cloverite containing quinuclidinium ions used as template. Two bands at 3674 cm^{-1} and 3163 cm^{-1} were assigned to stretching vibrations of POH and GaOH groups, respectively. The latter is strong and broad, but it falls in an IR region that could also be attributed to a NH$^+$ stretching vibration. Thibault-Starzyk et al. [788] found that the intensity of this band correlates with that of a sharp band at 1468 cm^{-1} during the template decomposition. Because the latter is characteristic of the quinuclidinium ion, the authors assigned the broad band at 3163 cm^{-1} to a vibration of the template molecule. In accordance with other papers [789, 790], two sharp bands at about 3700 cm^{-1} and 3675 cm^{-1}, which are observed after careful calcination were attributed to GaOH groups and POH groups, respectively.

In principle, the different types of OH groups present in zeolites can be distinguished by the position of the OH stretching vibration bands. In Table 4, the assignments of the fundamental OH vibrations to the various OH groups in ZSM-5-type zeolites are given. Zeolites of this particular type are of great academic and commercial interest because of their specific acid properties, in particular the low concentration and high strength of Brønsted acid sites, as well as their shape selectivity and relatively high resistance to deactivation. The data summarized in Table 4 show that it is often very difficult, and sometimes almost impossible, to identify different OH groups reliably based only on the fundamental stretching vibrations. A major improvement can be achieved by investigating the bending vibrations as well. The characteristic band positions observed in molecular sieves are given in Table 5.

Usually, the bending vibrations themselves cannot be detected by IR measurements since there is strong overlap with the vibrations of the zeolite framework in the range of 1200–400 cm^{-1}. In-plane bending and out-of-plane bending vibrations of bridging OH groups in zeolites have been directly observed by INS [249, 810, 811]. Because of the large incoherent cross section of hydrogen atoms, INS is uniquely sensitive to vibrational modes involving hydrogen atoms, whereas framework motions give rise to much weaker scattering. However, by comparison with IR spectroscopy this method is experimentally very demanding and has a lower spectral resolution [811]. Jacobs et al. [809, 813] used deuterated samples for the direct IR spectroscopic investigation of the in-plane bending modes of OD groups. After treatment with deuterium gas, new bands appeared near 870 cm^{-1} and 1550 cm^{-1} due to the formation of O–D bonds. These bands were assigned to the

References see page 990

Tab. 4. Fundamental OH Stretching Vibrations Observed in the IR Spectra of ZSM-5-type Zeolites and the Assignments Proposed

Wavenumber/cm^{-1}	Assignment
3780	terminal AlOH groups on non-framework aluminum [771]
3745–3720	terminal SiOH groups on the external surface [791–794] or on amorphous by-products (e. g. from synthesis) [791]
3738–3725	SiOH groups in the channel of the zeolite [792, 795], so-called internal silanols [296, 766, 767, 769, 794]
3725	SiOH groups (oxygen interacts with non-framework aluminum) [771]
3720	$H^+(H_2O)_x$ [777, 796]
3695–3680	OH groups from water adsorbed on Si(OH)Al groups or alkaline cations [777, 797–799]
3695–3690	OH groups of hydroxylated aluminum containing species formed during the hydrothermal treatment of ZSM-5/Al_2O_3 catalysts [800, 801]
3680–3665	AlOH groups on non-framework aluminum [771, 802, 803]
3650	terminal SiOH groups (proton interacts with non-framework aluminum) [771, 804]
3617–3600	Si(OH)Al groups [791]
3590–3560	OH groups of H_2O adsorbed on alkaline cations [777, 798]
3550	hydrogen bridged terminal AlOH and SiOH groups on zeolite-Al_2O_3-interfaces [800]
3550–3470	hydrogen-bridged SiOH groups (so-called internal silanols) [296, 766, 769, 795, 805–808]
3250	Si(OH)Al groups (proton interacts with framework oxygen) [779]

Tab. 5. Characteristic OH Vibration Band Positions of the Various Hydroxyl Groups in Zeolites and Related Materials (in cm^{-1})

Type	Stretching vibration ν_{OH}	In-plane bending vibration δ_{OH}	Out-of-plane bending vibration γ_{OH}
Si(OH)Al	3250–3660 [280, 779]	1055–990 [22, 249, 267, 809]	420–325 [249, 267, 276, 775, 809–811]
Terminal SiOH	3740–3750 [22]	795–835 [22]	–
Terminal AlOH	3770–3800 [771]	650 [778]	–
POH	3675 [775, 812]	1005–940 [775, 812]	–
MOH	3520–3610 [22]	695–955 [22]	–
H_2O (adsorbed)	3650–3740 [786]	1640 [786]	–

fundamental and the first overtone of the in-plane bending modes. Using this approach the in-plane bending modes can be studied directly. However, the use of the NIR region is more comfortable, even though the intensity of combination and overtone modes in the NIR region is about 100 times lower than that of fundamental modes. Therefore, the investigations have to be carried out using the DRIFT method, which allows weak bands to be detected.

The investigation of the overtone and combination vibrations of hydroxyl groups is valuable for the characterization of active sites. As the following examples illustrate, more detailed information can be obtained from this region than from the fundamental stretching vibrations:

(1) For the measurements of OH groups a careful dehydration of the zeolite samples is necessary, since residual water can strongly falsify the result. Adsorbed water molecules have a combination vibration band at about 5200 cm^{-1} [814, 815], and thus the absence of this band is a good measure for a complete dehydration.

(2) The investigation of hydroxyl groups is a method for verifying the incorporation of silicon in silicoaluminophosphates. The isomorphous substitution into the aluminophosphate framework leads to formation of bridging $\equiv$Si(OH)Al$\equiv$ groups. Terminal $\equiv$AlOH and $\equiv$POH groups occur at the external surface of the crystallites. Furthermore, hydroxyl groups at amorphous impurities of the sample can often not be excluded. There may be a superposition of the fundamental stretching vibration of different OH groups on aluminum [816] and silicon at about 3750 cm^{-1}. To distinguish between these two possibilities the region of combination vibrations has to be measured. Whereas a band at about 4550 cm^{-1} characterizes SiOH groups, a band at 4450 cm^{-1} is caused by AlOH groups. Using this additional information, it was possible to assign unambiguously a band at 3750 cm^{-1} observed for a SAPO-31 sample to SiOH groups [386].

In principle, the concentrations of different types of hydroxyl groups can be estimated from the intensity (integrated absorbance) of the corresponding band. To obtain the absolute data the extinction coefficient has to be determined by an independent measurement. However, the extinction coefficients for bridging OH groups reported in the literature vary significantly [274, 313]. The major advantage of ^{1}H MAS NMR over IR spectroscopy is that the signal intensities are directly proportional to the concentration of the different types of protons in the sample. Like the wave numbers of the OH stretching vibrations in IR spectroscopy, the chemical shifts of hydroxyl protons in NMR spectroscopy can also be used to distinguish between the different types of OH groups. However, the experimental data given in the literature reveal that the ^{1}H MAS NMR lines are less well resolved than the IR bands [274, 386, 817]. SAPO-34 is a convincing example: There are two bands of bridging OH groups with a difference of 25 cm^{-1} in the IR spectrum, whereas the ^{1}H MAS NMR spectrum shows only one line at 3.8 ppm [386].

The following general conclusions for an effective characterization of the Brønsted acid sites can be drawn:

References see page 990

(1) For qualitative analysis, IR spectroscopy is the most suitable method to identify different kinds of OH groups. The use of the spectral range of overtones and combination vibrations and of the diffuse reflectance method increase the potential of IR spectroscopy considerably.

(2) For a quantitative determination of the concentration of the various OH groups without the adsorption of probe molecules, ^{1}H MAS NMR is the method of choice.

(3) The wave numbers in IR spectroscopy as well as the chemical shifts in ^{1}H MAS NMR spectroscopy provide an estimate of the acid strength of the OH groups that are not influenced by an additional electrostatic interaction with the zeolite framework. More reliable information, not only on the strength, but also on the accessibility of acid sites, can be obtained by the spectroscopic investigation of the adsorption of probe molecules.

For a general discussion on the use of probe molecules for the characterization of acid sites the reader is referred to Chap. 2.12. A review on the characterization of solid acids by NMR, IR, EPR and XPS spectroscopic techniques including the use of probe molecules has recently been given by Brunner [447]. The author comes to the conclusion that these techniques, in combination with TPD, microcalorimetry and catalytic test reactions, allow a comprehensive characterization of acid sites in solids and of their interaction with adsorbed molecules.

4.2.6.3.5 Concluding Remarks

Several important areas of the characterization of microporous materials are covered only briefly or not at all in the present section. The most important such fields are the characterization of materials generated by post-synthesis modifications and the characterization of zeolites following their use as catalysts or adsorbents. In principle, all the methods mentioned earlier in the section can be applied to study modified and used zeolites. However, there are certain aspects that have to be taken into consideration when studying materials such as those that are sometimes less well defined. Below we give a few papers that describe the state-of-the-art in these fields.

For the modification of microporous materials in general, the reader is referred to chapter 4.2.5. An excellent review on the modification of zeolites and the characterization of the materials obtained in this way has also been given by Kühl [116]. His comprehensive paper covers particularly ion exchange, the introduction of metal functions into zeolites, the dealumination of aluminosilicates, and framework insertions.

The partial removal of the framework aluminum is amongst the most important procedures that aim to modify the properties of aluminosilicates. It is well known that the thermal and hydrothermal stabilities increase as the Si/Al ratio of the framework is raised [2]. Furthermore, the catalytic activity depends on the content of aluminum in the sample, both framework and non-framework. Ultrastable Y zeolite (USY), which is produced by hydrothermal dealumination of an ammonium Y zeolite [116], has enormous commercial importance as the active component of cracking catalysts. The degree of framework dealumination can be deter-

Acidity

measurement

mined ... troscopy [622]. Using
^{27}Al M ... framework aluminum
can be ... nted out, MAS NMR
under ... ency ≤ 15 kHz) is in-
sufficie ... ry recently, Kentgens
and co ... information on non-
framev ... AS NMR when high
fields (... pplied. Moreover, the
author: ... ramework aluminum
causes ... lening for part of the
framev ... d angle was found as
a long- ... s with the framework
might l

The ... lished by other treatments, ... [116, 772, 819]. The
structu ... n differ significantly
depend ... In Table 2, examples
are give ... imetry, sorption mea-
sureme

The ... th phosphorus compounds ... nodification has been
used to ... of toluene with methanol ov ... characterization of H-
ZSM-5 ... y Lischke et al. [821].
It is a ... nentary characteriza-
tion methods (ammonium exchange, TPD of ammonia, IR, and ^{27}Al and ^{31}P MAS
NMR spectroscopy). Similar findings and conclusions, in particular as regards the
formation of aluminum phosphate generated by the reaction of non-framework
aluminum with phosphoric acid, have been reported by Corma and coworkers for
modified USY zeolites [822].

The characterization of carbonaceous deposits is the most important issue in the
characterization of zeolite catalysts following their use. One of the major reasons
for their deactivation in hydrocarbon reactions is the deposition of side-products in
the zeolite pore system and the rate at which they are formed strongly affects the
utilizable time-on-stream. Elucidation of the main mechanisms for the formation
of carbonaceous deposits in the reaction under study can help to find ways to slow
down the catalyst deactivation. Furthermore, the characterization of used catalysts
can furnish valuable information for finding an appropriate procedure to regener-
ate them [823]. The large variety of side-products that do not desorb from catalysts
has caused some confusion in the literature: some authors designate as "coke"
only polyaromatics, while others include non-polyaromatic products, which often
have the same effect on the performance of the catalyst [824]. Guisnet and coworkers have very recently reviewed the organic chemistry of "coke" formation

over acid catalysts [825] and the role of acidity and pore structure in the deactivation of zeolite catalysts by these deposits [826].

The amount of carbonaceous deposits on a zeolite catalyst used, for example, in a hydrocarbon transformation, as well as its hydrogen-to-carbon ratio, can be determined by combustion and quantitative determination of the oxidation products [824, 827]. PFG NMR using appropriate probe molecules is just one of the methods [824] that can be used to gather information on the location of "coke" [484, 489]. The chemical nature of the carbonaceous deposits can be studied in more detail by spectroscopic methods (IR, UV-Vis, NMR) after the dissolution of the zeolite in HF [824]. However, the application of spectroscopic techniques does not necessarily require the catalysts to be dissolved [827]. In particular, Karge and coworkers have demonstrated that IR [454, 828], UV-Vis [455], ESR [537, 538], and NMR spectroscopy [454, 455] can be applied successfully to study the formation of "coke", even under working conditions. With the enormous progress in the development of *in situ* techniques [35], these spectroscopic methods can provide new insights into the changes occurring in a working catalyst.

Finally, we want to emphasize that the analytical instruments and methods that are nowadays available allow microporous materials to be characterized to an extent that the pioneers of zeolite research could only have dreamed of. The insight into structural details that can either be directly obtained by HRTEM and AFM or derived from X-ray powder data typifies the progress made in recent decades. However, even the application of the most advanced experimental and data analysis techniques provides no assurance that the last word has been spoken. The results have to be constantly checked and refined. For example, Lamberti et al. [226] concluded that in TS-1, "the presence of preferential substitution ... is very unlikely" and that their "experimental results seem to agree with the outcome of the quantum-chemical calculations ..., namely that Ti is homogeneously distributed on the MFI framework or may be slightly partitioned on different sites in different samples." These findings were based on a very detailed and thorough X-ray powder diffraction study using synchrotron radiation. Yet just two years later, several of these authors [233] presented neutron diffraction data that provided the "first direct evidence that Ti atoms are not equally distributed in the 12 crystallographically independent T sites in the MFI framework".

References

1 K. SEFF, in: Recent Advances and New Horizons in Zeolite Science and Technology, H. CHON, S. I. WOO, S.-E. PARK (Eds), *Studies in Surface Science and Catalysis*, Vol. 102, Elsevier, Amsterdam, 1996, p. 267–293.

2 D. W. BRECK, Zeolite Molecular Sieves, Wiley, New York, 1973, 771 pp.

3 J. ROCHA, M. W. ANDERSON, *Eur. J. Inorg. Chem.* **2000**, 801–818.

4 S. BRUNAUER, P. H. EMMETT, E. TELLER, *J. Am. Chem. Soc.* **1938**, *60*, 309–319.

5 K. S. W. SING, J. ROUQUEROL, in: Handbook of Heterogeneous Catalysis, G. ERTL, H. KNÖZINGER, J. WEITKAMP (Eds), Vol. 2, VCH, Weinheim, 1997, p. 427–439.

6 O. TALU, A. L. MYERS, *Colloids Surf., A* **2001**, *187–188*, 83–93.

7 U. LOHSE, H. STACH, H. THAMM, W. SCHIRMER, A. A. ISIRIKJAN, N. I. REGENT, M. M. DUBININ, *Z. Anorg. Allg. Chem.* **1980**, *460*, 179–190.

8 L. ABRAMS, D. R. CORBIN, *J. Inclus. Phenom. Mol. Recogn. Chem.* **1995**, *21*, 1–46.

9 M. BÜLOW, A. MICKE, *Adsorption* **1995**, *1*, 29–48.

10 H. ROBSON (Ed.), Verified Syntheses of Zeolitic Materials, 2nd edn, Elsevier, Amsterdam, 2001, 266 pp.

11 H. G. KARGE, M. HUNGER, H. K. BEYER, in: Catalysis and Zeolites: Fundamentals and Applications, J. WEITKAMP, L. PUPPE (Eds), Springer, Berlin, 1999, p. 198–326.

12 A. T. BELL, A. PINES (Eds), NMR Techniques in Catalysis, Marcel Dekker, New York, 1994, 432 pp.

13 M. E. SMITH, E. R. H. VAN ECK, *Progr. Nucl. Magn. Reson. Spectr.* **1999**, *34*, 159–201.

14 A. K. CHEETHAM, A. P. WILKINSON, *Angew. Chem.* **1992**, *104*, 1594–1608; *Angew. Chem. Int. Ed. Engl.* **1992**, *31*, 1557–1570.

15 G. VITALE, L. M. BULL, R. E. MORRIS, A. K. CHEETHAM, B. H. TOBY, C. G. COE, J. E. MacDOUGALL, *J. Phys. Chem.* **1995**, *99*, 16087–16092.

16 B. J. CAMPBELL, A. K. CHEETHAM, T. VOGT, L. CARLUCCIO, W. O. PARKER, C. FLEGO, R. MILLINI, *J. Phys. Chem. B* **2001**, *105*, 1947–1955.

17 H. GIES, B. MARLER, C. FYFE, G. KOKOTAILO, Y. FENG, D. E. COX, *J. Phys. Chem. Solids* **1991**, *52*, 1235–1241.

18 G. T. KOKOTAILO, C. A. FYFE, Y. FENG, H. GRONDEY, H. GIES, B. MARLER, D. E. COX, in: Catalysis by Microporous Materials, H. K. BEYER, H. G. KARGE, I. KIRICSI, J. B. NAGY (Eds), *Studies in Surface Science and Catalysis*, Vol. 94, Elsevier, Amsterdam, 1995, p. 78–100.

19 M. HOCHGRÄFE, B. MARLER, H. GIES, C. A. FYFE, Y. FENG, H. GRONDEY, G. T. KOKOTAILO, *Z. Kristallogr.* **1996**, *211*, 221–227.

20 G. SANKAR, J. M. THOMAS, *Top. Catal.* **1999**, *8*, 1–21.

21 G. SANKAR, J. M. THOMAS, C. R. A. CATLOW, *Top. Catal.* **2000**, *10*, 255–264.

22 L. M. KUSTOV, V. YU. BOROVKOV, V. B. KAZANSKY, *J. Catal.* **1981**, *72*, 149–159.

23 H. HOSER, A. INNOCENTI, A. RIVA, F. TRIFIRO, *Appl. Catal.* **1987**, *30*, 11–20.

24 S. A. JOHNSON, R.-M. RINKUS, T. C. DIEBOLD, V. A. MARONI, *Appl. Spectrosc.* **1988**, *42*, 1369–1375.

25 K.-H. SCHNABEL, R. FRICKE, I. GIRNUS, E. JAHN, E. LÖFFLER, B. PARLITZ, CH. PEUKER, *J. Chem. Soc., Faraday Trans.* **1991**, *87*, 3569–3574.

26 H. MIX, H. PFEIFER, B. STAUDTE, U. ZSCHERPEL, *Exper. Tech. Phys.* **1988**, *36*, 495–499.

27 CH. PEUKER, G. FINGER, E. LÖFFLER, W. PILZ, K.-H. SCHNABEL, *Z. Chem.* **1990**, *30*, 334–335.

28 F. SCHÜTH, *J. Phys. Chem.* **1992**, *96*, 7493–7496.

29 R. DE RUITER, J. C. JANSEN, J. VAN BEKKUM, *Zeolites* **1992**, *12*, 56–62.

30 C. BREMARD, G. GINESTET, J. LAUREYNS, M. LE MAIRE, *J. Am. Chem. Soc.* **1995**, *117*, 9274–9284.

31 J. PHILIPPAERTS, E. F. VANSANT, Y. A. YAN, in: Zeolites as Catalysts, Sorbents and Detergent Builders, H. G. KARGE, J. WEITKAMP (Eds), *Studies in Surface Science and Catalysis*, Vol. 46, Elsevier, Amsterdam, 1989, p. 555–566.

32 H. P. WANG, E. M. EYRING, H. HUAI, *Appl. Spectrosc.* **1991**, *45*, 883–885.

33 K. SCHRIJNEMAKERS, P. VAN DER VOORT, E. F. VANSANT, *Phys. Chem. Chem. Phys.* **1999**, *1*, 2569–2572.

34 J. M. THOMAS, G. A. SOMORJAI (Eds), *Top. Catal.* **1999**, *8*, 1–140.

35 M. HUNGER, J. WEITKAMP, *Angew. Chem.* **2001**, *113*, 3040–3059; *Angew. Chem. Int. Ed.* **2001**, *40*, 2954–2971.

36 G. DEBRAS, A. GOURGUE, J. B.NAGY, G. DE CLIPPELEIR, *Zeolites* **1985**, *5*, 369–376.

37 G. DEBRAS, A. GOURGUE, J. B.NAGY, G. DE CLIPPELEIR, *Zeolites* **1985**, *5*, 377–383.

38 G. DEBRAS, A. GOURGUE, J. B.NAGY, G. DE CLIPPELEIR, *Zeolites* **1986**, *6*, 161–168.

39 G. DEBRAS, A. GOURGUE, J. B.NAGY, G. DE CLIPPELEIR, *Zeolites* **1986**, *6*, 241–248.

40 J. W. NIEMANTSVERDRIET, Spectroscopy in Catalysis, 2nd edn, Wiley-VCH, Weinheim, 2000, 312 pp.

41 J. M. THOMAS, W. J. THOMAS, Principles and Practice of Heterogeneous Catalysis, VCH, Weinheim, 1997, 669 pp.

42 B. IMELIK, J. C. VÉDRINE (Eds), Catalyst Characterization: Physical Techniques for Solid Materials, Plenum Press, New York, 1994, 702 pp.

43 D. R. CORBIN, B. F. BURGESS, A. J. VEGA, R. D. FARLEE, *Anal. Chem.* **1987**, *59*, 2722–2728.

44 W. ZAMECHEK, in: Verified Syntheses of Zeolitic Materials, H. ROBSON (Ed.), 2nd edn, Elsevier, Amsterdam, 2001, p. 51–53.

45 L. M. PARKER, D. M. BIBBY, J. E. PATTERSON, *Zeolites* **1984**, *4*, 168–174.

46 V. KANAZIREV, G. L. PRICE, *J. Catal.* **1996**, *161*, 156–163.

47 V. KANAZIREV, K. M. DOOLEY, G. L. PRICE, *J. Catal.* **1994**, *146*, 228–236.

48 A. AUROUX, in: Catalyst Characterization: Physical Techniques for Solid Materials, B. IMELIK, J. C. VÉDRINE (Eds), Plenum Press, New York, 1994, p. 611–650.

49 B. HUNGER, M. HEUCHEL, S. MATYSIK, K. BECK, W. D. EINICKE, *Thermochim. Acta* **1995**, *269/270*, 599–611.

50 A. BERES, I. HANNUS, I. KIRICSI, *J. Thermal Analysis* **1996**, *46*, 1301–1311.

51 Z. GABELICA, J. B.NAGY, P. BODART, N. DEWAELE, A. NASTRO, *Zeolites* **1987**, *7*, 67–72.

52 M. SOULARD, S. BILGER, H. KESSLER, J. L. GUTH, *Zeolites* **1987**, *7*, 463–470.

53 S. BILGER, M. SOULARD, H. KESSLER, J. L. GUTH, *Zeolites* **1991**, *11*, 784–791.

54 E. BOURGEAT–LAMI, F. DI RENZO, F. FAJULA, P. H. MUTIN, T. DES COURIERS, *J. Phys. Chem.* **1992**, *96*, 3807–3811.

55 V. R. CHOUDHARY, C. SIVADINARAYANA, P. DEVADAS, S. D. SANSARE, P. MAGNOUX, M. GUISNET, *Microporous Mesoporous Mater.* **1998**, *21*, 91–101.

56 Z. SARBAK, *Cryst. Res. Technol.* **1996**, *31*, 601–610.

57 C. WEIDENTHALER, W. SCHMIDT, *Chem. Mater.* **2000**, *12*, 3811–3820.

58 A. AUROUX, V. BOLIS, P. WIERZCHOWSKI, P. C. GRAVELLE, J. C. VÉDRINE, *J. Chem. Soc., Faraday Trans. 1* **1979**, *75*, 2544–2555.

59 U. LOHSE, B. PARLITZ, V. PATZELOVÁ, *J. Phys. Chem.* **1989**, *93*, 3677–3683.

60 M. R. GONZALEZ, S. B. SHARMA, D. T. CHEN, J. A. DUMESIC, *Catal. Lett.* **1993**, *18*, 183–192.

61 A. AUROUX, A. TUEL, J. BANDIERA, Y. BEN TAARIT, J. M. GUIL, *Appl. Catal. A* **1993**, *93*, 181–190.

62 B. E. SPIEWAK, B. E. HANDY, S. B. SHARMA, J. A. DUMESIC, *Catal. Lett.* **1994**, *23*, 207–312.

63 M. MUSCAS, J. F. DUTEL, V. SOLINAS, A. AUROUX, Y. BEN TAARIT, *J. Mol. Catal. A* **1996**, *106*, 169–175.

64 F. EDER, M. STOCKHUBER, J. A. LERCHER, *J. Phys. Chem. B* **1997**, *101*, 5414–5419.

65 A. WÖLKER, C. HUDALLA, H. ECKERT, A. AUROUX, M. L. OCCELLI, *Solid State NMR* **1997**, *9*, 143–153.

66 M. L. OCCELLI, H. ECKERT, A. WÖLKER, A. AUROUX, *Microporous Mesoporous Mater.* **1999**, *30*, 219–232.

67 M. L. OCCELLI, G. SCHWERING, C. FILD, H. ECKERT, A. AUROUX, P. S. IYER, *Microporous Mesoporous Mater.* **2000**, *34*, 15–22.

68 A. AUROUX, *Top. Catal.* **1997**, *4*, 71–89.

69 S. SAVITZ, A. L. MYERS, R. J. GORTE, *Microporous Mesoporous Mater.* **2000**, *37*, 33–40.

70 J. JÄNCHEN, H. STACH, P. J. GROBET, J. A. MARTENS, P. A. JACOBS, *Zeolites* **1992**, *12*, 9–12.

71 H. STACH, K. FIEDLER, J. JÄNCHEN, *Pure Appl. Chem.* **1993**, *65*, 2193–2200.

72 J. M. GUIL, R. GUIL-LÓPEZ, J. A. PERDIGÓN-MELÓN, A. CORMA, *Microporous Mesoporous Mater.* **1998**, *22*, 269–279.

73 U. LOHSE, M. MILDEBRATH, *Z. Anorg. Allg. Chem.* **1981**, *476*, 126–135.

74 U. LOHSE, G. ENGELHARDT, V. PATZELOVÁ, *Zeolites* **1984**, *4*, 163–167.

75 G. WEBER, M. H. SIMONOT-GRANGE, *Zeolites* **1994**, *14*, 433–438.

76 P. WU, A. DEBEBE, Y. H. MA, *Zeolites* **1983**, *3*, 118–122.

77 M. E. Davis, C. Saldarriaga, C. Montes, J. Garces, C. Crowder, Zeolites 1988, 8, 362–366.

78 P. Voogd, J. J. F. Scholten, H. van Bekkum, Colloids Surf. 1991, 55, 163–171.

79 S. Storck, H. Bretinger, W. F. Maier, Appl. Catal. A 1998, 174, 137–146.

80 K. Meyer, P. Lorenz, B. Röhl-Kuhn, P. Klobes, Cryst. Res. Technol. 1994, 29, 903–930.

81 S.-B. Liu, B. M. Fung, T.-C. Yang, E.-C. Hong, C.-T. Chang, P.-C. Shih, F.-H. Tong, T.-L. Chen, J. Phys. Chem. 1994, 98, 4393–4401.

82 O. Talu, C. J. Guo, D. T. Hayhurst, J. Phys. Chem. 1989, 93, 7294–7298.

83 U. Müller, H. Reichert, E. Robens, K. K. Unger, Y. Grillet, F. Rouquerol, J. Rouquerol, D. Pan, A. Mersmann, Fresenius' Z. Anal. Chem. 1989, 333, 433–436.

84 L. Song, L. V. C. Rees, Microporous Mater. 2000, 35–36, 301–314.

85 M. Bülow, W. Mietk, P. Struve, P. Lorenz, J. Chem. Soc., Faraday Trans. 1 1983, 79, 2457–2466.

86 K. Beschmann, G. T. Kokotailo, L. Riekert, Chem. Eng. Process. 1987, 22, 223–229.

87 A. Zikánová, M. Bülow, H. Schlodder, Zeolites 1987, 7, 115–118.

88 Ch. Förste, J. Kärger, H. Pfeifer, L. Riekert, M. Bülow, A. Zikánová, J. Chem. Soc., Faraday Trans. 1990, 86, 881–885.

89 M. Bülow, A. Micke, J. Chem. Soc., Faraday Trans. 1994, 90, 2585–2590.

90 A. Micke, M. Bülow, M. Kočiřík, J. Phys. Chem. 1994, 98, 914–929.

91 R. Schumacher, H. G. Karge, Microporous Mesoporous Mater. 1999, 30, 307–314.

92 Y. Yasuda, J. Phys. Chem. 1982, 86, 1913–1917.

93 Y. Yasuda, G. Sugasawa, J. Catal. 1984, 88, 530–534.

94 Y. Yasuda, A. Yamamoto, J. Catal. 1985, 93, 176–181.

95 J. Cartigny, J. Giermanska-Kahn, E. Cohen De Lara, Zeolites 1994, 14, 576–581.

96 D. Shen, L. V. C. Rees, J. Chem. Soc., Faraday Trans. 1993, 89, 1063–1065.

97 D. Shen, L. V. C. Rees, J. Chem. Soc., Faraday Trans. 1995, 91, 2027–2033.

98 L. Song, L. V. C. Rees, Microporous Mater. 1996, 6, 363–374.

99 D. Shen, L. V. C. Rees, J. Chem. Soc., Faraday Trans. 1996, 92, 487–491.

100 G. Onyestyák, D. Shen, L. V. C. Rees, J. Chem. Soc., Faraday Trans. 1996, 92, 307–315.

101 Y. Yasuda, Heterogen. Chem. Rev. 1994, 1, 103–124.

102 R. M. Barrer, L. V. C. Rees, M. Shamsuzzoha, J. Inorg. Nucl. Chem. 1966, 28, 629–643.

103 H. S. Sherry, J. Phys. Chem. 1966, 70, 1158–1168.

104 H. S. Sherry, J. Phys. Chem. 1968, 72, 4086–4094.

105 K. R. Franklin, R. P. Townsend, J. Chem. Soc., Faraday Trans. 1 1985, 81, 1071–1086.

106 K. R. Franklin, R. P. Townsend, J. Chem. Soc., Faraday Trans. 1 1985, 81, 3127–3141.

107 K. R. Franklin, R. P. Townsend, J. Chem. Soc., Faraday Trans. 1 1988, 84, 687–702.

108 K. R. Franklin, R. P. Townsend, J. Chem. Soc., Faraday Trans. 1 1988, 84, 2755–2770.

109 K. R. Franklin, R. P. Townsend, Zeolites 1988, 8, 367–375.

110 R. Harjula, J. Lehto, J. H. Pothuis, A. Dyer, R. P. Townsend, J. Chem. Soc., Faraday Trans. 1993, 89, 971–976.

111 R. Harjula, A. Dyer, R. P. Townsend, J. Chem. Soc., Faraday Trans. 1993, 89, 977–981.

112 T. C. Watling, L. V. C. Rees, Zeolites 1994, 14, 687–692.

113 T. C. Watling, L. V. C. Rees, Zeolites 1994, 14, 693–696.

114 W. Shibata, K. Seff, Zeolites 1997, 19, 87–89.

115 R. P. Townsend, Pure Appl. Chem. 1986, 58, 1359–1366.

116 G. Kühl, in: Catalysis and Zeolites: Fundamentals and Applications, J. Weitkamp, L. Puppe (Eds), Springer, Berlin, 1999, p. 81–197.

117 F. Danes, F. Wolf, Z. Phys. Chem. (Leipzig) 1973, 252, 15–32.

118 D. Drummond, A. De Jonge, L. V. C. Rees, *J. Phys. Chem.* **1983**, *87*, 1967–1971.

119 A. Dyer, K. J. White, *Thermochim. Acta* **1999**, *340–341*, 341–348.

120 B. M. Lok, B. K. Marcus, C. L. Angell, *Zeolites* **1986**, *6*, 185–194.

121 H. Miessner, H. Kosslick, U. Lohse, B. Parlitz, V.-A. Tuan, *J. Phys. Chem.* **1993**, *97*, 9741–9748.

122 H. G. Karge, V. Dondur, *J. Phys. Chem.* **1990**, *94*, 765–772.

123 H. G. Karge, V. Dondur, J. Weitkamp, *J. Phys. Chem.* **1991**, *95*, 283–288.

124 G. M. Robb, W. Zhang, P. G. Smirniotis, *Microporous Mesoporous Mater.* **1998**, *20*, 307–316.

125 W. Zhang, E. C. Burckle, G. Smirniotis, *Microporous Mesoporous Mater.* **1999**, *33*, 173–185.

126 G. Lischke, B. Parlitz, U. Lohse, E. Schreier, R. Fricke, *Appl. Catal. A* **1998**, *166*, 351–361.

127 H. Förster, H. Fuess, E. Geidel, B. Hunger, H. Jobic, C. Kirschhock, O. Klepel, K. Krause, *Phys. Chem. Chem. Phys.* **1999**, *1*, 593–603.

128 G. Finger, J. Richter-Mendau, M. Bülow, J. Kornatowski, *Zeolites* **1991**, *11*, 443–448.

129 U. Lohse, R. Bertram, K. Jancke, I. Kurzawski, B. Parlitz, E. Löffler, E. Schreier, *J. Chem. Soc., Faraday Trans.* **1995**, *91*, 1163–1172.

130 D. Demuth, G. D. Stucky, K. K. Unger, F. Schüth, *Microporous Mater.* **1995**, *3*, 473–487.

131 A. Iwasaki, M. Hirata, I. Kudo, T. Sano, S. Sugawara, M. Ito, M. Watanabe, *Zeolites* **1995**, *15*, 308–314.

132 C. Weidenthaler, R. X. Fischer, R. D. Shannon, O. Medenbach, *J. Phys. Chem.* **1994**, *98*, 12687–12694.

133 E. R. Geus, H. van Bekkum, *Zeolites* **1995**, *15*, 333–341.

134 R. Mostowicz, L. B. Sand, *Zeolites* **1983**, *3*, 219–225.

135 O. Regev, Y. Cohen, E. Kehat, Y. Talmon, *Zeolites* **1994**, *14*, 314–319.

136 J. C. van der Waal, P. J. Kooyman, J. C. Jansen, H. van Bekkum, *Microporous Mesoporous Mater.* **1998**, *25*, 43–57.

137 P. Gallezot, C. Leclercq, in: Catalyst Characterization: Physical Techniques for Solid Materials, B. Imelik, J. C. Védrine (Eds), Plenum Press, New York, 1994, p. 509–558.

138 O. Terasaki, T. Ohsuna, N. Ohnishi, K. Hiraga, *Curr. Opin. Solid State Mater. Sci.* **1997**, *2*, 94–100.

139 A. K. Datye, in: Handbook of Heterogeneous Catalysis, G. Ertl, H. Knözinger, J. Weitkamp (Eds), Vol. 2, VCH, Weinheim, 1997, p. 493–512.

140 L. A. Bursill, E. A. Lodge, J. M. Thomas, *Nature* **1980**, *286*, 111–113.

141 J. M. Thomas, G. R. Millward, *J. Chem. Soc., Chem. Commun.* **1982**, 1380–1383.

142 J. M. Thomas, G. R. Millward, S. Ramdas, M. Audier, in: Intrazeolite Chemistry, G. D. Stucky, F. G. Dwyer (Eds), ACS Symposium Series, No. 218, 1983, p. 181–198.

143 G. R. Millward, S. Ramdas, J. M. Thomas, M. T. Barlow, *J. Chem. Soc., Faraday Trans. 2* **1983**, *79*, 1075–1082.

144 O. Terasaki, J. M. Thomas, G. R. Millward, D. Watanabe, *Chem. Mater.* **1989**, *1*, 158–162.

145 O. Terasaki, T. Ohsuna, H. Sakuma, D. Watanabe, Y. Nakagawa, R. C. Medrud, *Chem. Mater.* **1996**, *8*, 463–468.

146 T. Ohsuna, O. Terasaki, Y. Nakagawa, S. I. Zones, K. Hiraga, *J. Phys. Chem. B* **1997**, *101*, 9881–9885.

147 M. W. Anderson, O. Terasaki, T. Ohsuna, A. Philippou, S. P. MacKay, A. Ferreira, J. Rocha, S. Lidin, *Nature* **1994**, *367*, 347–350.

148 O. Terasaki, Y. Sakamoto, J. Yu, Y. Nozue, T. Ohsuna, N. Ohnishi, Y. Horikawa, K. Hiraga, G. Zhu, S. Qiu, R. Xu, M. Anderson, *Supramol. Sci.* **1998**, *5*, 189–195.

149 J. M. Thomas, O. Terasaki, P. L. Gai, W. Zhou, J. Gonzalez-Calbet, *Acc. Chem. Res.* **2001**, *34*, 583–594.

150 V. Alfredson, T. Ohsuna, O. Terasaki, J.-O. Bovin, *Angew. Chem.* **1993**, *105*, 1262–1264; *Angew. Chem. Int. Ed. Engl.* **1993**, *32*, 1210–1213.

151 T. Ohsuna, Y. Horikawa, K. Hiraga, O. Terasaki, *Chem. Mater.* **1998**, *10*, 688–691.

152 A. Kleine, P. L. Ryder, N. Jaeger, G. Schulz–Eckloff, *J. Chem. Soc., Faraday Trans. 1* **1986**, *82*, 205–212.

153 M. M. J. Treacy, *Microporous Mesoporous Mater.* **1999**, *28*, 271–292.

154 A. Tonscheidt, P. L. Ryder, N. I. Jaeger, G. Schulz-Ekloff, *Zeolites* **1996**, *16*, 271–274.

155 G. Binder, L. Scandella, A. Schumacher, N. Kruse, R. Prins, *Zeolites* **1996**, *16*, 2–6.

156 S. Yamamoto, S. Sugiyama, O. Matsuoka, T. Honda, Y. Banno, H. Nozoye, *Microporous Mesoporous Mater.* **1998**, *21*, 1–6.

157 S. Yamamoto, O. Matsuoka, S. Sugiyama, T. Honda, Y. Banno, H. Nozoye, *Chem. Phys. Lett.* **1996**, *260*, 208–214.

158 S. Sugiyama, S. Yamamoto, O. Matsuoka, T. Honda, H. Nozoye, S. Qiu, J. Yu, O. Terasaki, *Surf. Sci.* **1997**, *377–379*, 140–144.

159 S. Sugiyama, S. Yamamoto, O. Matsuoka, H. Nozoye, J. Yu, G. Zhu, S. Qiu, O. Terasaki, *Microporous Mesoporous Mater.* **1999**, *28*, 1–7.

160 M. W. Anderson, N. Hanif, J. R. Agger, C.-Y. Chen, S. I. Zones, in: Zeolites and Mesoporous Materials at the Dawn of the 21st Century, A. Galarneau, F. Di Renzo, F. Fajula, J. Védrine (Eds), *Studies in Surface Science and Catalysis*, Vol. 135, Elsevier, Amsterdam, 2001, p. 141.

161 M. W. Anderson, J. R. Agger, J. T. Thornton, N. Forsyth, *Angew. Chem.* **1996**, *108*, 1301–1304; *Angew. Chem. Int. Ed. Engl.* **1996**, *35*, 1210–1213.

162 E. N. Coker, J. C. Jansen, F. Di Renzo, F. Fajula, J. A. Martens, P. A. Jacobs, A. Sacco Jr., *Microporous Mesoporous Mater.* **2001**, *46*, 223–236.

163 A. L. Weisenhorn, J. E. MacDougall, S. A. C. Gould, S. D. Cox, W. S. Wise, J. Massie, P. Maivald, V. B. Elings, G. D. Stucky, P. K. Hansma, *Science* **1990**, *247*, 1330–1333.

164 M. Komiyama, T. Shimaguchi, T. Koyama, M. Gu, *J. Phys. Chem.* **1996**, *100*, 15198–15201.

165 N. Ohnishi, T. Ohsuna, Y. Sakamoto, O. Terasaki, K. Hiraga, *Microporous Mesoporous Mater.* **1998**, *21*, 581–588.

166 P. Wagner, O. Terasaki, S. Ritsch, J. G. Nery, S. I. Zones, M. E. Davis, K. Hiraga, *J. Phys. Chem. B* **1999**, *103*, 8245–8250.

167 D. H. Olson, G. T. Kokotailo, S. L. Lawton, W. M. Meier, *J. Phys. Chem.* **1981**, *85*, 2238–2243.

168 H. van Koningsveld, J. C. Jansen, H. van Bekkum, *Zeolites* **1990**, *10*, 235–242.

169 D. H. Olson, *Zeolites* **1995**, *15*, 439–443.

170 Y. Kim, Y. W. Han, K. Seff, *Zeolites* **1997**, *18*, 325–333.

171 G. Vezzalini, S. Quartieri, E. Galli, A. Alberti, G. Cruciani, Å. Kvick, *Zeolites* **1997**, *19*, 323–325.

172 G. Muncaster, G. Sankar, C. R. A. Catlow, J. M. Thomas, R. G. Bell, P. A. Wright, S. Coles, S. J. Teat, W. Clegg, W. Reeve, *Chem. Mater.* **1999**, *11*, 158–163.

173 P. Yang, T. Armbruster, *J. Solid State Chem.* **1996**, *123*, 140–149.

174 A. Alberti, A. Martucci, E. Galli, G. Vezzalini, *Zeolites* **1997**, *19*, 349–352.

175 Y. H. Yeom, Y. Kim, K. Seff, *Microporous Mesoporous Mater.* **1999**, *28*, 103–112.

176 J. Stolz, P. Yang, T. Armbruster, *Microporous Mesoporous Mater.* **2000**, *37*, 233–242.

177 H. van Koningsveld, J. C. Jansen, H. van Bekkum, *Zeolites* **1987**, *7*, 564–568.

178 H. van Koningsveld, F. Tuinstra, J. C. Jansen, H. van Bekkum, *Zeolites* **1989**, *9*, 253–256.

179 H. van Koningsveld, F. Tuinstra, H. van Bekkum, J. C. Jansen, *Acta Crystallogr. B* **1989**, *45*, 423–431.

180 H. van Koningsveld, J. C. Jansen, *Microporous Mater.* **1996**, *6*, 159–167.

181 H. van Koningsveld, J. C. Jansen, A. J. M. de Man, *Acta Crystallogr. B* **1996**, *52*, 131–139.

182 H. van Koningsveld, J. C. Jansen, H. van Bekkum, *Acta Crystallogr. B* **1996**, *52*, 140–144.

183 G. Reck, F. Marlow, J. Kornatowski, W. Hill, J. Caro, *J. Phys. Chem.* **1996**, *100*, 1698–1704.

184 H. van Koningsveld, J. H. Koegler, *Microporous Mater.* **1997**, *9*, 71–81.

185 C. A. Fyfe, D. H. Brouwer, *Microporous Mesoporous Mater.* **2000**, *39*, 291–305.

186 S. Zhen, K. Seff, *Microporous Mesoporous Mater.* **2000**, *39*, 1–18.

187 C. A. Fyfe, H. Gies, G. T. Kokotailo, C. Pasztor, H. Strobl, D. E. Cox, *J. Am. Chem. Soc.* **1989**, *111*, 2470–2474.

188 L. B. McCusker, Ch. Baerlocher, E. Jahn, M. Bülow, *Zeolites* **1991**, *11*, 308–313.

189 R. E. Morris, S. J. Weigel, N. J. Henson, L. M. Bull, M. T. Janicke, B. F. Chmelka, A. K. Cheetham, *J. Am. Chem. Soc.* **1994**, *116*, 11849–11855.

190 H. Gies, J. Rius, *Z. Kristallogr.* **1995**, *210*, 475–480.

191 A. J. Mora, A. N. Fitch, M. Cole, R. Goyal, R. H. Jones, H. Jobic, S. W. Carr, *J. Mat. Chem.* **1996**, *6*, 1831–1835.

192 V. Ramaswamy, L. B. McCusker, Ch. Baerlocher, *Microporous Mesoporous Mater.* **1999**, *31*, 1–8.

193 A. Tuel, S. Caldarelli, A. Meden, L. B. McCusker, Ch. Baerlocher, A. Ristić, N. Rajić, G. Mali, V. Kaučič, *J. Phys. Chem. B* **2000**, *104*, 5697–5705.

194 L. B. McCusker, in: Zeolites and Related Microporous Materials: State of the Art 1994, J. Weitkamp, H. G. Karge, H. Pfeifer, W. Hölderich (Eds), *Studies in Surface Science and Catalysis*, Vol. 84, Elsevier, Amsterdam, 1994, p. 341–356.

195 Ch. Baerlocher, L. B. McCusker, in: Advanced Zeolite Science and Applications, J. C. Jansen, M. Stöcker, H. G. Karge, J. Weitkamp (Eds), *Studies in Surface Science and Catalysis*, Vol. 85, Elsevier, Amsterdam, 1994, p. 391–428.

196 G. Bergeret, P. Gallezot, in: Catalyst Characterization: Physical Techniques for Solid Materials, B. Imelik, J. C. Védrine (Eds), Plenum Press, New York, 1994, p. 417–444.

197 A. Erdem, L. B. Sand, *J. Catal.* **1979**, *60*, 241–256.

198 R. Mostowicz, L. B. Sand, *Zeolites* **1982**, *2*, 143–146.

199 S. G. Fegan, B. M. Lowe, *J. Chem. Soc., Faraday Trans. 1* **1986**, *82*, 785–799.

200 S. Mintova, V. Valtchev, E. Vultcheva, S. Veleva, *Zeolites* **1992**, *12*, 210–215.

201 P. Norby, J. C. Hanson, *Catal. Today* **1998**, *39*, 301–309.

202 R. F. Lobo, M. J. Annen, M. E. Davis, *J. Chem. Soc., Faraday Trans.* **1992**, *88*, 2791–2795.

203 D. G. Hay, H. Jaeger, *J. Chem. Soc., Chem. Commun.* **1984**, 1433.

204 D. G. Hay, H. Jaeger, G. W. West, *J. Phys. Chem.* **1985**, *89*, 1070–1072.

205 J. Klinowski, T. A. Carpenter, L. F. Gladden, *Zeolites* **1987**, *7*, 73–78.

206 C. A. Fyfe, G. J. Kennedy, C. T. De Schutter, G. T. Kokotailo, *J. Chem. Soc., Chem. Commun.* **1984**, 541–542.

207 M. J. Annen, D. Young, M. E. Davis, O. B. Cavin, C. R. Hubbard, *J. Phys. Chem.* **1991**, *95*, 1380–1383.

208 H. He, P. Barnes, J. Munn, X. Turrillas, J. Klinowski, *Chem. Phys. Lett.* **1992**, *196*, 267–273.

209 L. M. Colyer, G. N. Greaves, A. J. Dent, S. W. Carr, K. K. Fox, R. H. Jones, in: Zeolites and Related Microporous Materials: State of the Art 1994, J. Weitkamp, H. G. Karge, H. Pfeifer, W. Hölderich (Eds), *Studies in Surface Science and Catalysis*, Vol. 84, Elsevier, Amsterdam, 1994, p. 387–394.

210 L. M. Colyer, G. N. Greaves, S. W. Carr, K. K. Fox, *J. Phys. Chem. B* **1997**, *101*, 10105–10114.

211 J. B. Parise, D. R. Corbin, L. Abrams, *Microporous Mater.* **1995**, *4*, 99–110.

212 B. F. Mentzen, P. Gelin, *Mat. Res. Bull.* **1995**, *30*, 373–380.

213 J. de Oñate Martínez, L. B. McCusker, Ch. Baerlocher, G. Engelhardt, *Microporous Mesoporous Mater.* **1998**, *22*, 127–134.

214 Y. Lee, S. W. Carr, J. B. Parise, *Chem. Mater.* **1998**, *10*, 2561–2570.

215 N. Novak Tušar, A. Ristić, A. Meden, V. Koučič, *Microporous Mesoporous Mater.* **2000**, *37*, 303–311.

216 C. J. H. Jacobsen, C. Madsen, T. V. W. Janssens, H. J. Jakobsen, J. Skibsted, *Microporous Mesoporous Mater.* **2000**, *39*, 393–401.

217 Y. Lee, B. A. Reisner, J. C. Hanson, G. A. Jones, J. B. Parise, D. R. Corbin, B. H. Toby, A. Freitag, J. Z. Larese, *J. Phys. Chem. B* **2001**, *105*, 7188–7199.

218 B. F. Mentzen, M. Sacerdote-Peronnet, J.-F. Bérar, F. Lefebvre, *Zeolites* **1993**, *13*, 485–492.

219 H. Klein, C. Kirschhock, H. Fuess, *J. Phys. Chem.* **1994**, *98*, 12345–12360.

220 C. Eylem, J. A. Hriljac, V. Ramamurthy, D. R. Corbin, J. B. Parise, *Chem. Mater.* **1996**, *8*, 844–849.

221 R. Goyal, A. N. Fitch, H. Jobic, *J. Phys. Chem. B* **2000**, *104*, 2878–2884.

222 C. E. A. Kirschhock, B. Hunger, J. Martens, P. A. Jacobs, *J. Phys. Chem. B* **2000**, *104*, 439–448.

223 H. Fichtner-Schmittler, U. Lohse, G. Engelhardt, V. Patzelová, *Cryst. Res. Technol.* **1984**, *19*, K1–K3.

224 A. Thangaraj, R. Kumar, S. P. Mirajkar, P. Ratnasamy, *J. Catal.* **1991**, *130*, 1–8.

225 R. Millini, E. Previde Massara, G. Perego, G. Bellussi, *J. Catal.* **1992**, *137*, 497–503.

226 C. Lamberti, S. Bordiga, A. Zecchina, A. Carati, A. N. Fitch, G. Artioli, G. Petrini, M. Salvalaggio, G. L. Marra, *J. Catal.* **1999**, *183*, 222–231.

227 J. Dougherty, L. E. Iton, J. W. White, *Zeolites* **1995**, *15*, 640–649.

228 J. N. Watson, L. E. Iton, J. W. White, *Chem. Commun.* **1996**, 2767–2768.

229 J. N. Watson, L. E. Iton, R. I. Keir, J. C. Thomas, T. L. Dowling, J. W. White, *J. Phys. Chem. B* **1997**, *101*, 10094–10104.

230 A. K. Cheetham, M. M. Eddy, D. A. Jefferson, J. M. Thomas, *Nature* **1982**, *299*, 24–26.

231 M. Czjzek, H. Jobic, A. N. Fitch, T. Vogt, *J. Phys. Chem.* **1992**, *96*, 1535–1540.

232 G. Artioli, C. Lamberti, G. L. Marra, *Acta Crystallogr. B* **2000**, *56*, 2–10.

233 C. Lamberti, S. Bordiga, A. Zecchina, G. Artioli, G. Marra, G. Spanò, *J. Am. Chem. Soc.* **2001**, *123*, 2204–2212.

234 L. J. Smith, H. Eckert, A. K. Cheetham, *J. Am. Chem. Soc.* **2000**, *122*, 1700–1708.

235 L. M. Bull, A. K. Cheetham, P. D. Hopkins, B. M. Powell, *J. Chem. Soc., Chem. Commun.* **1993**, 1196–1198.

236 L. J. Smith, A. Davidson, A. K. Cheetham, *Catal. Lett.* **1997**, *49*, 143–146.

237 A. Martucci, A. Alberti, G. Cruciani, P. Radaelli, P. Ciambelli, M. Rapacciuolo, *Microporous Mesoporous Mater.* **1999**, *30*, 95–101.

238 A. Martucci, G. Cruciani, A. Alberti, C. Ritter, P. Ciambelli, M. Rapacciuolo, *Microporous Mesoporous Mater.* **2000**, *35–36*, 405–412.

239 P. A. Wright, J. M. Thomas, A. K. Cheetham, A. K. Nowak, *Nature* **1985**, *318*, 611–614.

240 A. N. Fitch, H. Jobic, A. Renouprez, *J. Phys. Chem.* **1986**, *90*, 1311–1318.

241 J. C. Taylor, *Zeolites* **1987**, *7*, 311–318.

242 M. Sacerdote-Peronnet, B. F. Mentzen, *Mat. Res. Bull.* **1993**, *28*, 767–774.

243 M. Czjzek, H. Fuess, T. Vogt, *J. Phys. Chem.* **1991**, *95*, 5255–5261.

244 L. M. Bull, A. K. Cheetham, B. M. Powell, J. A. Ripmeester, C. I. Ratcliffe, *J. Am. Chem. Soc.* **1995**, *117*, 4328–4332.

245 L. Smith, A. K. Cheetham, R. E. Morris, L. Marchese, J. M. Thomas, P. A. Wright, J. Chen, *Science* **1996**, *271*, 799–802.

246 C. Kirschhock, H. Fuess, *Microporous Mater.* **1997**, *8*, 19–28.

247 G. Vitale, C. F. Mellot, A. K. Cheetham, *J. Phys. Chem. B* **1997**, *101*, 9886–9891.

248 A. J. Renouprez, H. Jobic, R. C. Oberthür, *Zeolites* **1985**, *5*, 222–224.

249 W. P. J. H. Jacobs, H. Jobic, J. H. M. C. van Wolput, R. A. van Santen, *Zeolites* **1992**, *12*, 315–319.

250 H. Jobic, A. Tuel, M. Krossner, J. Sauer, *J. Phys. Chem.* **1996**, *100*, 19545–19550.

251 H. Jobic, M. Czjzek, R. A. van Santen, *J. Phys. Chem.* **1992**, *96*, 1540–1542.

252 A. M. Davidson, C. F. Mellot, J. Eckert, A. K. Cheetham, *J. Phys. Chem. B* **2000**, *104*, 432–438.

253 H. Jobic, *Spectrochim. Acta A* **1992**, *48*, 293–312.

254 H. Jobic, *Physica B* **2000**, *276*, 222–225.

255 W. P. J. H. Jacobs, R. A. van Santen, H. Jobic, *J. Chem. Soc., Faraday Trans.* **1994**, *90*, 1191–1196.

256 J. H. Williams, J. Klinowski, *J. Chem. Soc., Faraday Trans.* **1992**, *88*, 1335–1338.

257 H. Jobic, M. Bee, G. J. Kearly, *J. Phys. Chem.* **1994**, *98*, 4660–4665.

258 H. Jobic, *Phys. Chem. Chem. Phys.* **1999**, *1*, 525–530.

259 H. Jobic, A. N. Fitch, J. Combet, *J. Phys. Chem. B* **2000**, *104*, 8491–8497.

260 H. Jobic, *J. Phys. IV* **2000**, *10*, Pr1-77–104.

261 E. M. Flanigen, H. Khatami, H. A. Szymanski, in: Advances in Chemistry Series, R. E. Gould (Ed.), Vol. 101, American Chemical Society, Washington D.C., 1971, p. 201–229.

262 J. C. Jansen, F. J. van der Gaag, H. van Bekkum, *Zeolites* **1984**, *4*, 369–372.

263 P. A. Jacobs, E. G. Derouane, J. Weitkamp, *J. Chem. Soc., Chem. Commun.* **1981**, 591–593.

264 W. P. J. H. Jacobs, J. H. M. C. van Wolput, R. A. van Santen, *Zeolites* **1993**, *13*, 170–182.

265 E. Löffler, Ch. Peuker, B. Zibrowius, U. Zscherpel, K.-H. Schnabel, in: Proc. 9th Int. Zeolite Conf., Montreal 1992, R. van Ballmoos, J. B. Higgins, M. M. J. Treacy (Eds), Vol. I, Butterworth–Heinemann, Stoneham, 1993, p. 521–528.

266 J. B. Uytterhoeven, L. G. Christner, W. K. Hall, *J. Phys. Chem.* **1965**, *69*, 2117–2126.

267 K. Beck, H. Pfeifer, B. Staudte, *Microporous Mater.* **1993**, *2*, 1–6.

268 E. Löffler, U. Zscherpel, Ch. Peuker, B. Staudte, *J. Mol. Struct.* **1993**, *293*, 269–272.

269 V. Patzelová, E. Drahorádová, Z. Tvarůžková, U. Lohse, *Zeolites* **1989**, *9*, 74–77.

270 J. Kubota, M. Furuki, Y. Goto, J. Kondo, A. Wada, K. Domen, C. Hirose, *Chem. Phys. Lett.* **1993**, *204*, 273–276.

271 M. J. P. Brugmans, A. W. Kleyn, A. Lagendijk, W. P. J. H. Jacobs, R. A. van Santen, *Chem. Phys. Lett.* **1994**, *217*, 117–122.

272 M. Bonn, H. J. Bakker, K. Domen, C. Hirose, A. W. Kleyn, R. A. van Santen, *Catal. Rev. – Sci. Eng.* **1998**, *40*, 127–173.

273 F. Schüth, R. Althoff, *J. Catal.* **1993**, *143*, 388–394.

274 M. A. Makarova, A. F. Ojo, K. Karim, M. Hunger, J. Dwyer, *J. Phys. Chem.* **1994**, *98*, 3619–3623.

275 F. Schüth, D. Demuth, B. Zibrowius, J. Kornatowski, G. Finger, *J. Am. Chem. Soc.* **1994**, *116*, 1090–1095.

276 L. M. Kustov, E. Loeffler, V. L. Zholobenko, V. B. Kazansky, in: Zeolites: A Refined Tool for Designing Catalytic Sites, L. Bonneviot, S. Kaliaguine (Eds), *Studies in Surface Science and Catalysis*, Vol. 97, Elsevier, Amsterdam, 1995, p. 63–70.

277 Ch. Peuker, *J. Mol. Struct.* **1995**, *349*, 317–320.

278 B. Lee, J. N. Kondo, F. Wakabayashi, K. Domen, *Bull. Chem. Soc. Jpn.* **1998**, *71*, 2149–2152.

279 J.-P. Shen, T. Sun, X.-W. Yang, D.-Z. Jiang, E.-Z. Min, *J. Phys. Chem.* **1995**, *99*, 12332–12334.

280 J. W. Ward, in: Zeolite Chemistry and Catalysis, J. A. Rabo (Ed.), ACS Monograph, Vol. 171, American Chemical Society, Washington D.C., 1976, p. 118–306.

281 J. W. Ward, *J. Catal.* **1968**, *10*, 34–46.

282 C. A. Emeis, *J. Catal.* **1993**, *141*, 347–354.

283 T. Barzetti, E. Selli, D. Moscotti, L. Forni, *J. Chem. Soc., Faraday Trans.* **1996**, *92*, 1401–1407.

284 H. Bludau, H. G. Karge, W. Niessen, *Microporous Mesoporous Mater.* **1998**, *22*, 297–308.

285 W. Daniell, N.-Y. Topsoe, H. Knözinger, *Langmuir* **2001**, *17*, 6233–6239.

286 D. Barthomeuf, *J. Phys. Chem.* **1984**, *88*, 42–45.

287 D. Barthomeuf, A. de Mallmann, in: Innovation in Zeolite Materials Science, P. J. Grobet, W. J. Mortier, E. F. Vansant, G. Schulz-Ekloff (Eds), *Studies in Surface Science and Catalysis*, Vol. 37, Amsterdam, 1987, p. 365–374.

288 C. Binet, A. Jadi, J. Lamotte, J. C. Lavalley, *J. Chem. Soc., Faraday Trans.* **1996**, *92*, 123–129.

289 D. Murphy, P. Massiani, R. Franck, D. Barthomeuf, *J. Phys. Chem.* **1996**, *100*, 6731–6738.

290 L. M. Kustov, V. Yu. Borovkov, V. B. Kazanskij, *Kinet. Katal.* **1984**, *25*, 471–477.

291 L. M. Kustov, I. B. Mishin, V. Yu. Borovkov, V. B. Kazanskij, *Kinet. Katal.* **1984**, *25*, 724–728.

292 L. M. Kustov, V. B. Kazansky, *J. Chem. Soc., Faraday Trans.* **1991**, *87*, 2675–2678.

293 K. Beck, H. Pfeifer, B. Staudte, *J. Chem. Soc., Faraday Trans.* **1993**, *89*, 3995–3998.

294 S. Bordiga, E. Garrone, C. Lamberti, A. Zecchina, C. Otero Areán, V. B. Kazansky, L. M. Kustov, *J. Chem. Soc., Faraday Trans.* **1994**, *90*, 3367–3372.

295 V. B. Kazansky, *J. Mol. Catal. A* **1999**, *141*, 83–94.

296 A. Zecchina, S. Bordiga, G. Spoto, L. Marchese, G. Petrini, G. Leofanti, M. Padovan, *J. Phys. Chem.* **1992**, *96*, 4991–4997.

297 M. A. Makarova, K. M. Al-Ghefaili, J. Dwyer, *J. Chem. Soc., Faraday Trans.* **1994**, *90*, 383–386.

298 I. Mirsojew, S. Ernst, J. Weitkamp, H. Knözinger, *Catal. Lett.* **1994**, *24*, 235–248.

299 M. Maache, A. Janin, J. C. Lavalley, E. Benazzi, *Zeolites* **1995**, *15*, 507–516.

300 M. Jiang, H. G. Karge, *J. Chem. Soc., Faraday Trans.* **1996**, *92*, 2641–2649.

301 J. Datka, G. Gil, J. Weglarski, *Microporous Mesoporous Mater.* **1998**, *21*, 75–79.

302 G. J. Kramer, R. A. van Santen, C. A. Emeis, A. K. Nowak, *Nature* **1993**, *363*, 529–531.

303 B. Lee, J. N. Kondo, F. Wakabayashi, K. Domen, *Catal. Lett.* **1999**, *59*, 51–54.

304 M. A. Makarova, J. Dwyer, *J. Phys. Chem.* **1993**, *97*, 6337–6338.

305 V. Gruver, J. J. Fripiat, *J. Phys. Chem.* **1994**, *98*, 8549–8554.

306 G. Catana, D. Baetens, T. Mommaerts, R. A. Schoonheydt, B. M. Weckhuysen, *J. Phys. Chem. B* **2001**, *105*, 4904–4911.

307 A. Corma, V. Fornés, L. Forni, F. Márquez, J. Martínez-Triguero, D. Moscotti, *J. Catal.* **1998**, *179*, 451–458.

308 J.-H. Kim, A. Ishida, M. Niwa, *React. Kinet. Catal. Lett.* **1999**, *67*, 281–287.

309 C. Otero Areán, E. Escalona Platero, M. Peñarroya Mentruit, M. Rodríguez Delgado, F. X. Llabrés i Xamena, A. García-Raso, C. Morterra, *Microporous Mesoporous Mater.* **2000**, *34*, 55–60.

310 M. Trombetta, T. Armaroli, A. G. Alejandre, J. R. Solis, G. Busca, *Appl. Catal. A* **2000**, *192*, 125–136.

311 F. Thibault-Starzyk, A. Travert, J. Saussey, J.-C. Lavelley, *Top. Catal.* **1998**, *6*, 111–118.

312 J. A. Lercher, C. Gründling, G. Eder-Mirth, *Catal. Today* **1996**, *27*, 353–376.

313 J. A. Martens, W. Souverijns, W. van Rhijn, P. A. Jacobs, in: Handbook of Heterogeneous Catalysis, G. Ertl, H. Knözinger, J. Weitkamp (Eds), Vol. 1, VCH, Weinheim, 1997, p. 324–365.

314 H. Knözinger, in: Handbook of Heterogeneous Catalysis, G. Ertl, H. Knözinger, J. Weitkamp (Eds), Vol. 2, VCH, Weinheim, 1997, p. 707–732.

315 H. Knözinger, S. Huber, *J. Chem. Soc., Faraday Trans.* **1998**, *94*, 2047–2054.

316 M. Taramasso, G. Perego, B. Notari, US Patent 4,410,501, 1983, assigned to Snamprogetti S.p.A., Milan, Italy.

317 D. Scarano, A. Zecchina, S. Bordiga, F. Geobaldo, G. Spoto, G. Petrini, G. Leofanti, M. Padovan,

G. Tozzola, *J. Chem. Soc., Faraday Trans.* **1993**, *89*, 4123–4130.

318 L. M. Kustov, V. B. Kazansky, P. Ratnasamy, *Zeolites* **1987**, *7*, 79–83.

319 J. Datka, T. Abramowicz, *J. Chem. Soc., Faraday Trans.* **1994**, *90*, 2417–2421.

320 N. K. Mal, V. Ramaswamy, S. Ganapathy, A. V. Ramaswamy, *J. Chem. Soc., Chem. Commun.* **1994**, 1933–1934.

321 M. D. Baker, J. Godber, K. Helwig, G. A. Ozin, *J. Phys. Chem.* **1988**, *92*, 6017–6024.

322 R. V. Jasra, N. V. Choudary, K. V. Rao, G. C. Pandey, S. G. T. Bhat, *Chem. Phys. Lett.* **1993**, *211*, 214–219.

323 H. Kosslick, Ch. Peuker, A. Roethe, K.-P. Roethe, *Z. Phys. Chem. (Leipzig)* **1990**, *271*, 89–92.

324 K. Krause, E. Geidel, J. Kindler, H. Förster, K. S. Smirnov, *Vib. Spectrosc.* **1996**, *12*, 45–52.

325 E. Cohen de Lara, Y. Delaval, *J. Phys. Chem.* **1974**, *78*, 2180–2181.

326 T. Yamazaki, I. Watanuki, S. Ozawa, Y. Ogino, *Langmuir* **1988**, *4*, 433–438.

327 E. Cohen de Lara, *Mol. Phys.* **1989**, *66*, 479–492.

328 F. Jousse, E. Cohen de Lara, *J. Phys. Chem.* **1996**, *100*, 233–237.

329 F. Jousse, A. V. Larin, E. Cohen de Lara, *J. Phys. Chem.* **1996**, *100*, 238–244.

330 K. Hadjiivanov, P. Massiani, H. Knözinger, *Phys. Chem. Chem. Phys.* **1999**, *1*, 3831–3838.

331 V. Yu. Borovkov, H. G. Karge, *J. Chem. Soc., Faraday Trans.* **1995**, *91*, 2035–2039.

332 V. B. Kazansky, V. Yu. Borovkov, A. Serich, H. G. Karge, *Microporous Mesoporous Mater.* **1998**, *22*, 251–259.

333 Z. Sobalik, Z. Tvarůžková, B. Wichterlová, *J. Phys. Chem. B* **1998**, *102*, 1077–1085.

334 H.-Y. Chen, X. Wang, W. M. H. Sachtler, *Phys. Chem. Chem. Phys.* **2000**, *2*, 3083–3090.

335 V. Yu. Borovkov, M. Jiang, Y. Fu, *J. Phys. Chem. B* **1999**, *103*, 5010–5019.

336 M. Nowotny, J. A. Lercher, H. Kessler, *Zeolites* **1991**, *11*, 454–459.

337 K.-H. Schnabel, G. Finger, J. Kornatowski, E. Löffler, Ch. Peuker, W. Pilz, *Microporous Mater.* **1997**, *11*, 293–302.

338 L. Marchese, A. Frache, E. Gianotti, G. Martra, M. Causa, S. Coluccia, *Microporous Mesoporous Mater.* **1999**, *30*, 145–153.

339 H. G. Karge, W. Nießen, *Catal. Today* **1991**, *8*, 451–465.

340 G. Müller, T. Narbeshuber, G. Mirth, J. A. Lercher, *J. Phys. Chem.* **1994**, *98*, 7436–7439.

341 F. Marlow, D. Demuth, G. Stucky, F. Schüth, *J. Phys. Chem.* **1995**, *99*, 1306–1310.

342 M. Bonn, H. J. Bakker, A. W. Kleyn, R. A. van Santen, *J. Phys. Chem.* **1996**, *100*, 15301–15304.

343 Z. Dardas, M. G. Süer, Y. H. Ma, W. R. Moser, *J. Catal.* **1996**, *159*, 204–211.

344 W. P. J. H. Jacobs, D. G. Demuth, S. A. Schunk, F. Schüth, *Microporous Mater.* **1997**, *10*, 95–109.

345 H. Sun, H. Frei, *J. Phys. Chem. B* **1997**, *101*, 205–209.

346 P. K. Dutta, B. Del Barco, *J. Phys. Chem.* **1985**, *89*, 1861–1865.

347 P. K. Dutta, B. Del Barco, *J. Phys. Chem.* **1988**, *92*, 354–357.

348 C. Brémard, M. Le Maire, *J. Phys. Chem.* **1993**, *97*, 9695–9702.

349 P. K. Dutta, D. C. Shieh, M. Puri, *Zeolites* **1988**, *8*, 306–309.

350 S. Ashtekar, J. J. Hastings, L. F. Gladden, *J. Chem. Soc., Faraday Trans.* **1998**, *94*, 1157–1161.

351 D. B. Akolekar, S. Bhargava, W. van Bronswijk, *Appl. Spectrosc.* **1999**, *53*, 931–937.

352 C. Brémard, D. Bougeard, *Adv. Mater.* **1995**, *7*, 10–25.

353 P.-P. Knops-Gerrits, D. E. De Vos, E. J. P. Feijen, P. A. Jacobs, *Microporous Mater.* **1997**, *8*, 3–17.

354 H.-G. Buge, Ch. Peuker, W. Pilz, E. Jahn, O. Rademacher, *Z. Phys. Chem. (Leipzig)* **1990**, *271*, 881–889.

355 C. Brémard, J. Laureyns, J. Patarin, *J. Raman Spectrosc.* **1996**, *27*, 439–445.

356 P. J. Hendra, C. Passingham, G. M. Warnes, R. Burch, D. J. Rawlence, *Chem. Phys. Lett.* **1989**, *164*, 178–184.

357 R. Burch, C. Passingham, G. M. Warnes, D. J. Rawlence, *Spectrochim. Acta A* **1990**, *46*, 243–251.

358 R. Ferwerda, J. H. van der Maas, P. J. Hendra, *Vib. Spectrosc.* **1994**, *7*, 37–47.

359 Y. Huang, E. A. Havenga, *Chem. Mater.* **2001**, *13*, 738–746.

360 G. Deo, A. M. Turek, I. E. Wachs, D. R. C. Huybrechts, P. A. Jacobs, *Zeolites* **1993**, *13*, 365–373.

361 W. Pilz, Ch. Peuker, V. A. Tuan, R. Fricke, H. Kosslick, *Ber. Bunsenges. Phys. Chem.* **1993**, *97*, 1037–1040.

362 G. Grubert, M. Wark, N. I. Jaeger, G. Schulz-Ekloff, O. P. Tkachenko, *J. Phys. Chem. B* **1998**, *102*, 1665–1671.

363 E. Gianotti, L. Marchese, G. Martra, S. Coluccia, *Catal. Today* **1999**, *54*, 547–552.

364 M. Che, F. Bozon-Verduraz, in: Handbook of Heterogeneous Catalysis, G. Ertl, H. Knözinger, J. Weitkamp (Eds), Vol. 2, VCH, Weinheim, 1997, p. 641–664.

365 B. M. Weckhuysen, R. A. Schoonheydt, in: Spectroscopy of Transition Metal Ions on Surfaces, B. M. Weckhuysen, P. Van Der Voort, G. Catana (Eds), Leuven University Press, 2000, p. 221–268.

366 I. Girnus, K. Hoffmann, F. Marlow, J. Caro, G. Döring, *Microporous Mater.* **1994**, *2*, 537–541.

367 F. Marlow, K. Hoffmann, G.-G. Lindner, I. Girnus, G. van de Goor, J. Kornatowski, J. Caro, *Microporous Mater.* **1996**, *6*, 43–49.

368 A. A. Verberckmoes, B. M. Weckhuysen, J. Pelgrims, R. A. Schoonheydt, *J. Phys. Chem.* **1995**, *99*, 15222–15228.

369 J. Dědeček, B. Wichterlová, *J. Phys. Chem. B* **1999**, *103*, 1462–1476.

370 S. Bordiga, R. Buzzoni, F. Geobaldo, C. Lamberti, E. Giamello, A. Zecchina, G. Leofanti, G. Petrini, G. Tozzola, G. Vlaic, *J. Catal.* **1996**, *158*, 486–501.

371 A. A. Verberckmoes, B. M. Weckhuysen, R. A. Schoonheydt, *Microporous Mesoporous Mater.* **1998**, *22*, 165–178.

372 J. Caro, F. Marlow, M. Wübbenhorst, *Adv. Mater.* **1994**, *6*, 413–416.

373 K. Hoffmann, F. Marlow, J. Caro, *Zeolites* **1996**, *6*, 281–286.

374 E. J. Doskocil, S. V. Bordawekar, B. G. Kaye, R. J. Davis, *J. Phys. Chem. B* **1999**, *103*, 6277–6282.

375 C. Pazè, B. Sazak, A. Zecchina, J. Dwyer, *J. Phys. Chem. B* **1999**, *103*, 9978–9986.

376 H. Yamashita, M. Matsuoka, K. Tsuji, Y. Shioya, M. Anpo, M. Che, *J. Phys. Chem.* **1996**, *100*, 397–402.

377 S. Dzwigaj, M. Matsuoka, M. Anpo, M. Che, *J. Phys. Chem. B* **2000**, *104*, 6012–6020.

378 E. Lippmaa, M. Mägi, A. Samoson, M. Tarmak, G. Engelhardt, *J. Am. Chem. Soc.* **1981**, *103*, 4992–4996.

379 G. Engelhardt, U. Lohse, E. Lippmaa, M. Tarmak, M. Mägi, *Z. Anorg. Allg. Chem.* **1981**, *482*, 49–64.

380 J. Klinowski, S. Ramdas, J. M. Thomas, C. A. Fyfe, J. S. Hartman, *J. Chem. Soc., Faraday Trans. 2* **1982**, *78*, 1025–1050.

381 S. Ramdas, J. M. Thomas, J. Klinowski, C. A. Fyfe, J. S. Hartman, *Nature* **1981**, *292*, 228–230.

382 G. Engelhardt, E. Lippmaa, M. Mägi, *J. Chem. Soc., Chem. Commun.* **1981**, 712–713.

383 Z. Gabelica, J. B.Nagy, P. Bodart, G. Debras, *Chem. Lett.* **1984**, 1059–1062.

384 C. S. Blackwell, R. L. Patton, *J. Phys. Chem.* **1988**, *92*, 3965–3970.

385 P. J. Barrie, J. Klinowski, *J. Phys. Chem.* **1989**, *93*, 5972–5974.

386 B. Zibrowius, E. Löffler, M. Hunger, *Zeolites* **1992**, *12*, 167–174.

387 C. R. Bayense, A. P. M. Kentgens, J. W. de Haan, L. J. M. van de Ven, J. H. C. van Hooff, *J. Phys. Chem.* **1992**, *96*, 775–782.

388 U. Lohse, B. Parlitz, D. Müller, E. Schreier, R. Bertram, R. Fricke, *Microporous Mater.* **1997**, *12*, 39–49.

389 A. Philippou, F. Salehirad, D.-P. Luigi, M. W. Anderson, *J. Phys. Chem. B* **1998**, *102*, 8974–8977.

390 J. Klinowski, *Progr. Nucl. Magn. Reson. Spectr.* **1984**, *16*, 237–309.

391 G. Engelhardt, D. Michel, High-Resolution Solid State NMR of Silicates and Zeolites, Wiley, Chichester, 1987, 485 pp.

392 C. A. Fyfe, K. T. Mueller, G. T. Kokotailo, in: NMR Techniques in Catalysis, A. T. Bell, A. Pines (Eds), Marcel Dekker, New York, 1994, p. 11–67.

393 H. Pfeifer, in: Solid–State NMR II: Inorganic Matter, P. Diehl, E. Fluck, H. Günther, R. Kosfeld, J. Seelig (Eds), NMR Basic Principles and Progress, Vol. 31, Springer, Berlin, 1994, p. 31–90.

394 H. Pfeifer, H. Ernst, in: Annual Reports on NMR Spectroscopy, I. Ando, G. A. Webb (Eds), Vol. 28, Academic Press, London, 1994, p. 91–187.

395 M. Stöcker, in: Advanced Zeolite Science and Applications, J. C. Jansen, M. Stöcker, H. G. Karge, J. Weitkamp (Eds), *Studies in Surface Science and Catalysis*, Vol. 85, Elsevier, Amsterdam, 1994, p. 429–507.

396 C. A. Fyfe, G. C. Gobbi, W. J. Murphy, R. S. Ozubko, D. A. Slack, *J. Am. Chem. Soc.* **1984**, *106*, 4435–4438.

397 C. A. Fyfe, J. H. O'Brien, H. Strobl, *Nature* **1987**, *326*, 281–282.

398 B. F. Chmelka, Y. Wu, R. Jelinek, M. E. Davis, A. Pines, in: Zeolite Chemistry and Catalysis, P. A. Jacobs, N. I. Jaeger, L. Kubelková, B. Wichterlová (Eds), *Studies in Surface Science and Catalysis*, Vol. 69, Elsevier, Amsterdam, 1991, p. 435–442.

399 R. Jelinek, B. F. Chmelka, Y. Wu, P. J. Grandinetti, A. Pines, P. J. Barrie, J. Klinowski, *J. Am. Chem. Soc.* **1991**, *113*, 4097–4110.

400 B. Zibrowius, U. Lohse, *Solid State NMR* **1992**, *1*, 137–148.

401 G. Engelhardt, W. Veeman, *J. Chem. Soc., Chem. Commun.* **1993**, 622–623.

402 C. Fernandez, J.-P. Amoureux, L. Delmotte, H. Kessler, *Microporous Mater.* **1996**, *6*, 125–130.

403 C. Fernandez, J.-P. Amoureux, J. M. Chezeau, L. Delmott, H. Kessler, *Microporous Mater.* **1996**, *6*, 331–340.

404 C. Fernandez, L. Delevoye, J.-P. Amoureux, D. P. Lang, M. Pruski, *J. Am. Chem. Soc.* **1997**, *119*, 6858–6862.

405 S. Ganapathy, T. K. Das, R. Vetrivel, S. S. Ray, T. Sen, S. Sivasanker, L. Delevoye, C. Fernandez, J.-P. Amoureux, *J. Am. Chem. Soc.* **1998**, *120*, 4752–4762.

406 L. M. Bull, A. K. Cheetham, T. Anupold, A. Reinhold, A. Samoson, J. Sauer, B. Bussemer, Y. Lee, S. Gann, J. Shore, A. Pines, R. Dupree, *J. Am. Chem. Soc.* **1998**, *120*, 3510–3511.

407 U.-T. Pingel, J.-P. Amoureux, T. Anupold, F. Bauer, H. Ernst, C. Fernandez, D. Freude, A. Samoson, *Chem. Phys. Lett.* **1998**, *294*, 345–350.

408 C. A. Fyfe, H. Grondey, Y. Feng, G. T. Kokotailo, *J. Am. Chem. Soc.* **1990**, *112*, 8812–8820.

409 C. A. Fyfe, Y. Feng, H. Gies, H. Grondey, G. T. Kokotailo, *J. Am. Chem. Soc.* **1990**, *112*, 3264–3270.

410 C. A. Fyfe, K. T. Mueller, H. Grondey, K. C. Wong-Moon, *J. Phys. Chem.* **1993**, *97*, 13484–13495.

411 M. Pruski, C. Fernandez, D. P. Lang, J.-P. Amoureux, *Catal. Today* **1999**, *49*, 401–409.

412 P. R. Bodart, J.-P. Amoureux, M. Pruski, A. Bailly, C. Fernandez, *Magn. Reson. Chem.* **1999**, *37*, S69–S74.

413 G. Boxhoorn, R. A. van Santen, W. A. van Erp, G. R. Hays, R. Huis, D. Clague, *J. Chem. Soc., Chem. Commun.* **1982**, 264–265.

414 S. L. Burkett, M. E. Davis, *Chem. Mater.* **1995**, *7*, 920–928.

415 S. L. Burkett, M. E. Davis, *Chem. Mater.* **1995**, *7*, 1453–1463.

416 A. Fonseca, J. B.Nagy, J. El Hage-Al Asswad, G. Demortier, R. Mostowicz, F. Crea, *Zeolites* **1995**, *15*, 131–138.

417 M. Kovalakova, B. H. Wouters, P. J. Grobet, *Microporous Mesoporous Mater.* **1998**, *22*, 193–201.

418 D. F. Shantz, C. Fild, H. Koller, R. F. Lobo, *J. Phys. Chem. B* **1999**, *103*, 10858–10865.

419 C. S. Blackwell, R. L. Patton, *J. Phys. Chem.* **1984**, *88*, 6135–6139.

420 M. GOEPPER, F. GUTH, L. DELMOTTE, J. L. GUTH, H. KESSLER, in: Zeolites: Facts, Figures, Future, P. A. JACOBS, R. A. VAN SANTEN (Eds), *Studies in Surface Science and Catalysis*, Vol. 49, Elsevier, Amsterdam, 1989, p. 857–866.

421 B. ZIBROWIUS, U. LOHSE, J. RICHTER-MENDAU, *J. Chem. Soc., Faraday Trans.* 1991, *87*, 1433–1437.

422 S. CALDARELLI, A. MEDEN, A. TUEL, *J. Phys. Chem. B* 1999, *103*, 5477–5487.

423 K. F. M. G. J. SCHOLLE, W. S. VEEMAN, *Zeolites* 1985, *5*, 118–122.

424 C. FILD, D. F. SHANTZ, R. F. LOBO, H. KOLLER, *Phys. Chem. Chem. Phys.* 2000, *2*, 3091–3098.

425 H. STROBL, C. A. FYFE, G. T. KOKOTAILO, C. T. PASZTOR, D. M. BIBBY, *J. Am. Chem. Soc.* 1987, *109*, 4733–4734.

426 L. MAISTRIAU, Z. GABELICA, E. G. DEROUANE, E. T. C. VOGT, J. VAN OENE, *Zeolites* 1991, *11*, 583–592.

427 J. A. MARTENS, E. FEIJEN, J. L. LIEVENS, P. J. GROBET, P. A. JACOBS, *J. Phys. Chem.* 1991, *95*, 10025–10031.

428 D. AKPORIAYE, M. STÖCKER, *Microporous Mater.* 1993, *1*, 423–430.

429 R. JELINEK, B. F. CHMELKA, Y. WU, M. E. DAVIS, J. G. ULAN, R. GRONSKY, A. PINES, *Catal. Lett.* 1992, *15*, 65–73.

430 P. J. BARRIE, M. E. SMITH, J. KLINOWSKI, *Chem. Phys. Lett.* 1991, *180*, 6–12.

431 M. P. J. PEETERS, J. W. DE HAAN, L. J. M. VAN DE VEN, J. H. C. VAN HOOFF, *J. Chem. Soc., Chem. Commun.* 1992, 1560–1562.

432 M. P. J. PEETERS, J. W. DE HAAN, L. J. M. VAN DE VEN, J. H. C. VAN HOOFF, *J. Phys. Chem.* 1993, *97*, 5363–5369.

433 M. HUNGER, G. ENGELHARDT, H. KOLLER, J. WEITKAMP, *Solid State NMR* 1993, *2*, 111–120.

434 H. KOLLER, B. BURGER, A. M. SCHNEIDER, G. ENGELHARDT, J. WEITKAMP, *Microporous Mater.* 1995, *5*, 219–232.

435 D. FREUDE, M. HUNGER, H. PFEIFER, G. SCHELER, J. HOFFMANN, W. SCHMITZ, *Chem. Phys. Lett.* 1984, *105*, 427–430.

436 H. PFEIFER, D. FREUDE, M. HUNGER, *Zeolites* 1985, *5*, 274–286.

437 D. FREUDE, M. HUNGER, H. PFEIFER, W. SCHWIEGER, *Chem. Phys. Lett.* 1986, *128*, 62–66.

438 D. FREUDE, J. KLINOWSKI, H. HAMDAN, *Chem. Phys. Lett.* 1988, *149*, 355–362.

439 D. FREUDE, J. KLINOWSKI, *J. Chem. Soc., Chem. Commun.* 1988, 1411–1413.

440 P. BATAMACK, C. DORÉMIEUX-MORIN, J. FRAISSARD, *Catal. Lett.* 1991, *11*, 119–128.

441 M. HUNGER, M. W. ANDERSON, A. OJO, H. PFEIFER, *Microporous Mater.* 1993, *1*, 17–32.

442 C. P. GREY, A. J. VEGA, *J. Am. Chem. Soc.* 1995, *117*, 8232–8242.

443 P. SARV, T. TUHERM, E. LIPPMAA, K. KESKINEN, A. ROOT, *J. Phys. Chem.* 1995, *99*, 13763–13768.

444 H. PFEIFER, *J. Chem. Soc., Faraday Trans. 1* 1988, *84*, 3777–3783.

445 E. BRUNNER, *J. Mol. Struct.* 1995, *355*, 61–85.

446 M. HUNGER, *Solid State NMR* 1996, *6*, 1–29.

447 E. BRUNNER, *Catal. Today* 1997, *38*, 361–376.

448 M. HUNGER, *Catal. Rev. – Sci. Eng.* 1997, *39*, 345–393.

449 E. BRUNNER, U. STEINBERG, *Progr. Nucl. Magn. Reson. Spectr.* 1998, *32*, 21–57.

450 J. A. RIPMEESTER, *J. Am. Chem. Soc.* 1983, *105*, 2925–2927.

451 D. ZSCHERPEL, E. BRUNNER, M. KOCH, H. PFEIFER, *Microporous Mater.* 1995, *4*, 141–147.

452 T. MILDNER, D. FREUDE, *J. Catal.* 1998, *178*, 309–314.

453 E. G. DEROUANE, J.-P. GILSON, J. B.NAGY, *Zeolites* 1982, *2*, 42–46.

454 J.-P. LANGE, A. GUTSZE, J. ALLGEIER, H. G. KARGE, *Appl. Catal.* 1988, *45*, 345–356.

455 H. G. KARGE, H. DARMSTADT, A. GUTSZE, H. VIETH, G. BUNTKOWSKY, in: Zeolites and Related Microporous Materials: State of the Art 1994, J. WEITKAMP, H. G. KARGE, H. PFEIFER, W. HÖLDERICH (Eds), *Studies in Surface Science and Catalysis*, Vol. 84,

Elsevier, Amsterdam, 1994, p. 1465–1474.

456 M. A. Springuel-Huet, J. Fraissard, *Chem. Phys. Lett.* **1989**, *154*, 299–302.

457 Q. Chen, M. A. Springuel-Huet, J. Fraissard, in: Catalysis and Adsorption by Zeolites, G. Öhlmann, H. Pfeifer, R. Fricke (Eds), *Studies in Surface Science and Catalysis*, Vol. 65, Elsevier, Amsterdam, 1991, p. 219–232.

458 D. Raftery, B. F. Chmelka, in: Solid-State NMR I: Methods, P. Diehl, E. Fluck, H. Günther, R. Kosfeld, J. Seelig (Eds), NMR Basic Principles and Progress, Vol. 30, Springer, Berlin, 1994, p. 111–158.

459 M.-A. Springuel–Huet, J.-L. Bonardet, J. Fraissard, *Appl. Magn. Reson.* **1995**, *8*, 427–456.

460 J.-L. Bonardet, J. Fraissard, A. Gédéon, M.-A. Springuel-Huet, *Catal. Rev. – Sci. Eng.* **1999**, *41*, 115–225.

461 C. Tsiao, D. R. Corbin, C. R. Dybowski, *J. Phys. Chem.* **1990**, *94*, 867–869.

462 R. Ryoo, C. Pak, B. F. Chmelka, *Zeolites* **1990**, *10*, 790–793.

463 A. K. Jameson, C. J. Jameson, A. C. de Dios, E. Oldfield, R. E. Gerald II, G. L. Turner, *Solid State NMR* **1995**, *4*, 1–12.

464 C. J. Jameson, H.-M. Lim, A. K. Jameson, *Solid State NMR* **1997**, *9*, 277–301.

465 L. C. de Menorval, J. P. Fraissard, T. Ito, *J. Chem. Soc., Faraday Trans. 1* **1982**, *78*, 403–410.

466 B. F. Chmelka. R. Ryoo, S.-B. Liu, L. C. de Menorval, C. J. Radke, E. E. Petersen, A. Pines, *J. Am. Chem. Soc.* **1988**, *110*, 4465–4467.

467 J. M. Coddington, R. F. Howe, Y.-S. Yong, K. Asakura, Y. Iwassawa, *J. Chem. Soc., Faraday Trans.* **1990**, *86*, 1015–1016.

468 A. Bifone, T. Pietrass, J. Kritzenberger, A. Pines, B. F. Chmelka, *Phys. Rev. Lett.* **1995**, *74*, 3277–3280.

469 S. J. Cho, W.-S. Ahn, S. B. Hong, R. Ryoo, *J. Phys. Chem.* **1996**, *100*, 4996–5003.

470 M. G. Samant, L. C. de Menorval, R. A. Dalla Betta, M. Boudart, *J. Phys. Chem.* **1988**, *92*, 3937–3938.

471 N. Bansal, C. Dybowski, *J. Magn. Reson.* **1990**, *89*, 21–27.

472 L. C. de Menorval, D. Raftery, S.-B. Liu, K. Takegoshi, R. Ryoo, A. Pines, *J. Phys. Chem.* **1990**, *94*, 27–31.

473 J. Kärger, J. Caro, *J. Chem. Soc., Faraday Trans. 1* **1977**, *73*, 1363–1376.

474 J. Kärger, H. Pfeifer, M. Rauscher, A. Walter, *J. Chem. Soc., Faraday Trans. 1* **1980**, *76*, 717–737.

475 J. Kärger, W. Heink, *J. Magn. Reson.* **1983**, *51*, 1–7.

476 A. Germanus, J. Kärger, H. Pfeifer, N. N. Samulevich, S. P. Zhdanov, *Zeolites* **1985**, *5*, 91–95.

477 B. Zibrowius, J. Caro, J. Kärger, Z. *Phys. Chem. (Leipzig)* **1988**, *269*, 1101–1106.

478 N.-K. Bär, J. Kärger, H. Pfeifer, H. Schäfer, W. Schmitz, *Microporous Mesoporous Mater.* **1998**, *22*, 289–295.

479 U. Hong, J. Kärger, H. Pfeifer, *J. Am. Chem. Soc.* **1991**, *113*, 4812–4815.

480 W. Heink, J. Kärger, H. Pfeifer, K. P. Datema, A. K. Nowak, *J. Chem. Soc., Faraday Trans.* **1992**, *88*, 3505–3509.

481 V. Gupta, S. S. Nirvarthi, A. V. McCormick, H. T. Davis, *Chem. Phys. Lett.* **1995**, *247*, 596–600.

482 V. Kukla, J. Kornatowski, D. Demuth, I. Girnus, H. Pfeifer, L. V. C. Rees, S. Schunk, K. K. Unger, J. Kärger, *Science* **1996**, *272*, 702–704.

483 K. Hahn, J. Kärger, *J. Phys. Chem. B* **1998**, *102*, 5766–5771.

484 J. Caro, M. Bülow, H. Jobic, J. Kärger, B. Zibrowius, in: Advances in Catalysis, D. D. Eley, H. Pines, P. B. Weisz (Eds), Vol. 39, Academic Press, San Diego, 1993, p. 351–414.

485 J. Kärger, H. Pfeifer, *J. Chem. Soc., Faraday Trans.* **1991**, *87*, 1989–1996.

486 J. Kärger, D. M. Ruthven, Diffusion in Zeolites and Other Microporous Solids, Wiley, New York, 1992, 585 pp.

487 J. Kärger, H. Pfeifer, in: NMR Techniques in Catalysis, A. T. Bell, A. Pines (Eds), Marcel Dekker, New York, 1994, p. 69–137.

488 F. Stallmach, J. Kärger, *Adsorption* **1999**, *5*, 117–133.

489 J. Kärger, H. Pfeifer, J. Caro, M. Bülow, H. Schlodder, R. Mostowicz, J. Völter, *Appl. Catal.* **1987**, *29*, 21–30.

490 C. Förste, A. Germanus, J. Kärger, H. Pfeifer, J. Caro, W. Pilz, A. Zikánová, *J. Chem. Soc., Faraday Trans. 1* **1987**, *83*, 2301–2309.

491 J. Kärger, H. Pfeifer, R. Seidel, B. Staudte, Th. Groß, *Zeolites* **1987**, *7*, 282–284.

492 P. Batamack, C. Dorémieux-Morin, J. Fraissard, D. Freude, *J. Phys. Chem.* **1991**, *95*, 3790–3796.

493 M. Hunger, D. Freude, D. Fenzke, H. Pfeifer, *Chem. Phys. Lett.* **1992**, *191*, 391–395.

494 P. Batamack, C. Dorémieux-Morin, R. Vincent, J. Fraissard, *Microporous Mater.* **1994**, *2*, 515–524.

495 A. J. Vega, Z. Luz, *Zeolites* **1988**, *8*, 19–26.

496 A. Michael, W. Meiler, D. Michel, H. Pfeifer, D. Hoppach, J. Delmau, *J. Chem. Soc., Faraday Trans. 1* **1986**, *82*, 3053–3067.

497 E. Brunner, H. Pfeifer, T. Wutscherk, D. Zscherpel, Z. *Phys. Chem.* **1992**, *178*, 173–183.

498 M. Koch, E. Brunner, H. Pfeifer, D. Zscherpel, *Chem. Phys. Lett.* **1994**, *228*, 501–505.

499 J. B.Nagy, E. G. Derouane, H. A. Resing, G. R. Miller, *J. Phys. Chem.* **1983**, *87*, 833–837.

500 B. Zibrowius, M. Bülow, H. Pfeifer, *Chem. Phys. Lett.* **1985**, *120*, 420–423.

501 R. Eckman, A. J. Vega, *J. Am. Chem. Soc.* **1983**, *105*, 4841–4842.

502 R. R. Eckman, A. J. Vega, *J. Phys. Chem.* **1986**, *90*, 4679–4683.

503 I. Kustanovich, D. Fraenkel, Z. Luz, S. Vega, H. Zimmermann, *J. Phys. Chem.* **1988**, *92*, 4134–4141.

504 I. Kustanovich, H. M. Vieth, Z. Luz, S. Vega, *J. Phys. Chem.* **1989**, *93*, 7427–7431.

505 B. Zibrowius, J. Caro, H. Pfeifer, *J. Chem. Soc., Faraday Trans. 1* **1988**, *84*, 2347–2356.

506 B. Boddenberg, R. Burmeister, *Zeolites* **1988**, *8*, 480–487.

507 R. Burmeister, B. Boddenberg, M. Verfürden, *Zeolites* **1989**, *9*, 318–320.

508 H. Koller, A. R. Overweg, L. J. M. van de Ven, J. W. de Haan, R. A. van Santen, *Microporous Mater.* **1997**, *11*, 9–17.

509 D. F. Shantz, R. F. Lobo, *Top. Catal.* **1999**, *9*, 1–11.

510 L. Canesson, Y. Boudeville, A. Tuel, *J. Am. Chem. Soc.* **1997**, *119*, 10754–10762.

511 G. P. Handreck, T. D. Smith, *J. Chem. Soc., Faraday Trans. 1* **1989**, *85*, 3195–3214.

512 G.-D. Lei, L. Kevan, *J. Phys. Chem.* **1991**, *95*, 4506–4514.

513 G. Brouet, X. Chen, C. W. Lee, L. Kevan, *J. Am. Chem. Soc.* **1992**, *114*, 3720–3726.

514 C. W. Lee, X. Chen, G. Brouet, L. Kevan, *J. Phys. Chem.* **1992**, *96*, 3110–3113.

515 D. Goldfarb, M. Bernardo, K. G. Strohmaier, D. E. W. Vaughan, H. Thomann, *J. Am. Chem. Soc.* **1994**, *116*, 6344–6353.

516 G. Catana, J. Pelgrims, R. A. Schoonheydt, *Zeolites* **1995**, *15*, 475–480.

517 A. V. Kucherov, J. L. Gerlock, H.-W. Jen, M. Shelef, *Zeolites* **1995**, *15*, 9–14.

518 A. V. Kucherov, J. L. Gerlock, H.-W. Jen, M. Shelef, *Zeolites* **1995**, *15*, 15–20.

519 A. Brückner, U. Lohse, H. Mehner, *Microporous Mesoporous Mater.* **1998**, *20*, 207–215.

520 T. Muñoz, Jr., A. M. Prakash, L. Kevan, K. J. Balkus, Jr., *J. Phys. Chem. B* **1998**, *102*, 1379–1386.

521 M.-A. Djieugoue, A. M. Prakash, L. Kevan, *J. Phys. Chem. B* **1999**, *103*, 804–811.

522 Z. Zhu, L. Kevan, *Phys. Chem. Chem. Phys.* **1999**, *1*, 199–206.

523 A. M. Prakash, L. Kevan, M. H. Zahedi-Niaki, S. Kaliaguine, *J. Phys. Chem. B* **1999**, *103*, 831–837.

524 G. Turnes Palomino, P. Fisicaro, S. Bordiga, A. Zecchina, E. Giamello, C. Lamberti, *J. Phys. Chem. B* **2000**, *104*, 4064–4073.

525 P. L. CARL, S. C. LARSEN, *J. Catal.* **2000**, *196*, 352–361.

526 B. V. PADLYAK, J. KORNATOWSKI, G. ZADROZNA, M. ROZWADOWSKI, A. GUTSZE, *J. Phys. Chem. A* **2000**, *104*, 11837–11843.

527 P. H. KASAI, R. J. BISHOP, in: Zeolite Chemistry and Catalysis, J. A. RABO (Ed.), ACS Monograph, Vol. 171, American Chemical Society, Washington D.C., 1976, p. 350–391.

528 M. CHE, E. GIAMELLO, in: Catalyst Characterization: Physical Techniques for Solid Materials, B. IMELIK, J. C. VÉDRINE (Eds), Plenum Press, New York, 1994, p. 131–179.

529 D. GOLDFARB, in: Spectroscopy of Transition Metal Ions on Surfaces, B. M. WECKHUYSEN, P. VAN DER VOORT, G. CATANA (Eds), Leuven University Press, 2000, p. 93–133.

530 J. H. LUNSFORD, *J. Phys. Chem.* **1970**, *74*, 1518–1522.

531 E. V. LUNINA, G. L. MARKARYAN, O. O. PARENAGO, A. V. FIONOV, *Colloids Surf., A* **1993**, *72*, 333–343.

532 A. GUTSZE, M. PLATO, H. G. KARGE, F. WITZEL, *J. Chem. Soc., Faraday Trans.* **1996**, *92*, 2495–2498.

533 B. STAUDTE, A. GUTSZE, W. BÖHLMANN, H. PFEIFER, B. PIETREWICZ, *Microporous Mesoporous Mater.* **2000**, *40*, 1–7.

534 T. RUDOLF, A. PÖPPL, W. BRUNNER, D. MICHEL, *Magn. Reson. Chem.* **1999**, *37*, S93–S99.

535 M. GUTJAHR, A. PÖPPL, W. BÖHLMANN, R. BÖTTCHER, *Colloids Surf., A* **2001**, *189*, 93–101.

536 T. RUDOLF, A. PÖPPL, W. HOFBAUER, D. MICHEL, *Phys. Chem. Chem. Phys.* **2001**, *3*, 2167–2173.

537 J.-P. LANGE, A. GUTSZE, H. G. KARGE, *J. Catal.* **1988**, *114*, 136–143.

538 H. G. KARGE, J.-P. LANGE, A. GUTSZE, M. LANIECKI, *J. Catal.* **1988**, *114*, 144–152.

539 TH. GROß, U. LOHSE, G. ENGELHARDT, K.-H. RICHTER, V. PATZELOVÁ, *Zeolites* **1984**, *4*, 25–29.

540 I. GROHMANN, TH. GROß, *J. Electron. Spectrosc. Relat. Phenom.* **1990**, *53*, 99–106.

541 T. L. BARR, M. A. LISHKA, *J. Am. Chem. Soc.* **1986**, *108*, 3178–3186.

542 T. L. BARR, L. M. CHEN, M. MOHSENIAN, M. A. LISHKA, *J. Am. Chem. Soc.* **1988**, *110*, 7962–7975.

543 Y. OKAMOTO, M. OGAWA, A. MAEZAWA, T. IMANAKA, *J. Catal.* **1988**, *112*, 427–436.

544 T. L. BARR, *Zeolites* **1990**, *10*, 760–765.

545 W. GRÜNERT, M. MUHLER, K.-P. SCHRÖDER, J. SAUER, R. SCHLÖGL, *J. Phys. Chem.* **1994**, *98*, 10920–10929.

546 B. A. SEXTON, A. E. HUGHES, D. M. BIBBY, *J. Catal.* **1988**, *109*, 126–131.

547 H. HE, K. ALBERTI, T. L. BARR, J. KLINOWSKI, *J. Phys. Chem.* **1993**, *97*, 13703–13707.

548 T. L. BARR, *Microporous Mater.* **1995**, *3*, 557–564.

549 S. KALIAGUINE, in: Recent Advances and New Horizons in Zeolite Science and Technology, H. CHON, S. I. WOO, S.-E. PARK (Eds), *Studies in Surface Science and Catalysis*, Vol. 102, Elsevier, Amsterdam, 1996, p. 191–230.

550 M. STÖCKER, *Microporous Mater.* **1996**, *6*, 235–257.

551 J. KLINOWSKI, T. L. BARR, *Acc. Chem. Res.* **1999**, *32*, 633–640.

552 W. GRÜNERT, R. SCHLÖGL, in: Molecular Sieve – Science and Technology, H. G. KARGE, J. WEITKAMP (Eds), Springer, Berlin, 2001, Vol. 4, Chap. 6, (in press).

553 D. TRONG ON, L. BONNEVIOT, A. BITTAR, A. SAYARI, S. KALIAGUINE, *J. Mol. Catal.* **1992**, *74*, 233–246.

554 M. J. REMY, M. J. GENET, P. P. NOTTE, P. F. LARDINOIS, G. PONCELET, *Microporous Mater.* **1993**, *2*, 7–15.

555 J. W. YOO, CH. W. LEE, J.-S. CHANG, S.-E. PARK, J. KO, *Catal. Lett.* **2000**, *66*, 169–173.

556 S. BADRINARAYANAN, R. I. HEGDE, I. BALAKRISHNAN, S. B. KULKARNI, P. RATNASAMY, *J. Catal.* **1981**, *71*, 439–442.

557 B. A. SEXTON, T. D. SMITH, J. V. SANDERS, *J. Electron. Spectrosc. Relat. Phenom.* **1985**, *35*, 27–43.

558 E. S. SHPIRO, W. GRÜNERT, R. W. JOYNER, G. N. BAEVA, *Catal. Lett.* **1994**, *24*, 159–169.

559 T. Liese, W. Grünert, *J. Catal.* **1997**, *172*, 34–45.

560 V. K. Kaushik, M. Ravindranathan, *Zeolites* **1992**, *12*, 415–419.

561 S. L. Suib, G. D. Stucky, R. J. Blattner, *J. Catal.* **1980**, *65*, 174–178.

562 J. C. Bertolini, J. Massardier, in: Catalyst Characterization: Physical Techniques for Solid Materials, B. Imelik, J. C. Védrine (Eds), Plenum Press, New York, 1994, p. 247–270.

563 J. N. Ness, D. J. Joyner, A. P. Chapple, *Zeolites* **1989**, *9*, 250–252.

564 U. Lohse, B. Parlitz, B. Altrichter, K. Jancke, E. Löffler, E. Schreier, F. Vogt, *J. Chem. Soc., Faraday Trans.* **1995**, *91*, 1155–1161.

565 K. Chao, J. Chern, *Zeolites* **1988**, *8*, 82–85.

566 R. Althoff, B. Schulz-Dobrick, F. Schüth, K. Unger, *Microporous Mater.* **1993**, *1*, 207–218.

567 M.-S. Tzou, M. Kusunoki, K. Asakura, H. Kuroda, G. Moretti, W. M. H. Sachtler, *J. Phys. Chem.* **1991**, *95*, 5210–5215.

568 P. Behrens, H. Kosslick, V. A. Tuan, M. Fröba, F. Neissendorfer, *Microporous Mater.* **1995**, *3*, 433–441.

569 P. A. Barrett, G. Sankar, C. R. A. Catlow, J. M. Thomas, *J. Phys. Chem.* **1996**, *100*, 8977–8985.

570 G. Sankar, R. G. Bell, J. M. Thomas, M. W. Anderson, P. A. Wright, J. Rocha, *J. Phys. Chem.* **1996**, *100*, 449–452.

571 P. Marturano, L. Drozdová, A. Kogelbauer, R. Prins, *J. Catal.* **2000**, *192*, 236–247.

572 M. Vaarkamp, D. C. Koningsberger, in: Handbook of Heterogeneous Catalysis, G. Ertl, H. Knözinger, J. Weitkamp (Eds), Vol. 2, VCH, Weinheim, 1997, p. 475–493.

573 D. C. Koningsberger, B. L. Mojet, G. E. van Dorssen, D. E. Ramaker, *Top. Catal.* **2000**, *10*, 143–155.

574 G. D. Meitzner, E. Iglesia, J. E. Baumgartner, E. S. Huang, *J. Catal.* **1993**, *140*, 209–225.

575 V. Bolis, S. Bordiga, C. Lamberti, A. Zecchina, F. Carati, F. Rivetti, G. Spanò, G. Petrini, *Microporous Mesoporous Mater.* **1999**, *30*, 67–76.

576 A. A. Battiston, J. H. Bitter, D. C. Koningsberger, *Catal. Lett.* **2000**, *66*, 75–79.

577 A. Meagher, V. Nair, R. Szostak, *Zeolites* **1988**, *8*, 3–11.

578 K. Lázár, A. J. Chandwadkar, P. Fejes, J. Čejka, A. V. Ramaswamy, *J. Radioanal. Nucl. Chem.* **2000**, *246*, 143–148.

579 G. Debras, E. G. Derouane, J.-P. Gilson, Z. Gabelica, G. Demortier, *Zeolites* **1983**, *3*, 37–42.

580 D. Decroupet, H. Meurisse, G. Demortier, *Zeolites* **1987**, *7*, 540–544.

581 M. Peisach, A. E. Pillay, C. A. Pineda, T. Themistocleous, *J. Radioanal. Nucl. Chem.* **1992**, *159*, 71–76.

582 Z. Gabelica, G. Demortier, *Nucl. Instrum. Methods Phys. Res., Sect. B* **1998**, *136–138*, 1312–1321.

583 U. Simon, U. Flesch, *J. Porous Mater.* **1999**, *6*, 33–40.

584 U. Simon, M. Franke, *Microporous Mesoporous Mater.* **2000**, *41*, 1–36.

585 U. Simon, U. Flesch, W. Maunz, R. Müller, C. Plog, *Microporous Mesoporous Mater.* **1998**, *21*, 111–116.

586 M. E. Franke, U. Simon, F. Roessner, U. Roland, *Appl. Catal. A* **2000**, *202*, 179–182.

587 K. Alberti, L. Gubicza, F. Fetting, *Chem. Ing. Tech.* **1993**, *65*, 940–943.

588 P. A. Anderson, R. G. Bell, C. R. A. Catlow, F. L. Chang, A. J. Dent, P. P. Edwards, I. Gameson, I. Hussain, A. Porch, J. M. Thomas, *Chem. Mater.* **1996**, *8*, 2114–2120.

589 Powder Diffraction File PDF1/2, International Centre for Diffraction Data, 12 Campus Boulevard, Newton Square, PA 19073–3273, U.S.A.; http://www.icdd.com.

590 M. M. J. Treacy, J. B. Higgins, Collection of Simulated XRD Powder Patterns for Zeolites, 4th edn, Elsevier, Amsterdam, 2001, 379 pp.

591 Ch. Baerlocher, W. M. Meier, D. H. Olson, Atlas of Zeolite Framework Types, 5th edn, Elsevier, Amsterdam, 2001, 302 pp.

592 http://www.iza–structure.org/databases.

593 E. Galli, G. Vezzalini, S. Quartieri, A. Alberti, M. Franzini, *Zeolites* **1997**, *19*, 318–322.

594 R. J. Argauer, G. R. Landolt, US Patent 3,702,886, 1972, assigned to Mobil Oil Corp., USA.

595 E. Lippmaa, M. Mägi, A. Samoson, G. Engelhardt, A.-R. Grimmer, *J. Am. Chem. Soc.* **1980**, *102*, 4889–4893.

596 G. Engelhardt, D. Michel, High-Resolution Solid State NMR of Silicates and Zeolites, Wiley, Chichester, 1987, p. 122–134.

597 G. Engelhardt, R. Radeglia, *Chem. Phys. Lett.* **1984**, *108*, 271–274.

598 G. Engelhardt, H. van Koningsveld, *Zeolites* **1990**, *10*, 650–656.

599 C. A. Fyfe, Y. Feng, H. Grondey, *Microporous Mater.* **1993**, *1*, 393–400.

600 A.-R. Grimmer, R. Radeglia, *Chem. Phys. Lett.* **1984**, *106*, 262–265.

601 A.-R. Grimmer, in: Nuclear Magnetic Shieldings and Molecular Structure, J. A. Tossell (Ed.), Kluwer, Dordrecht, 1993, p. 191–201.

602 M. Hochgräfe, H. Gies, C. A. Fyfe, Y. Feng, H. Grondey, *Chem. Mater.* **2000**, *12*, 336–342.

603 G. Engelhardt, S. Luger, J. Ch. Buhl, J. Felsche, *Zeolites* **1989**, *9*, 182–186.

604 D. Müller, E. Jahn, G. Ladwig, U. Haubenreisser, *Chem. Phys. Lett.* **1984**, *109*, 332–336.

605 G. Engelhardt, H. Koller, P. Sieger, W. Depmeier, A. Samoson, *Solid State NMR* **1992**, *1*, 127–135.

606 M. Goepper, J. L. Guth, *Zeolites* **1991**, *11*, 477–482.

607 G. Coudurier, C. Naccache, J. C. Védrine, *J. Chem. Soc., Chem. Commun.* **1982**, 1413–1414.

608 M. Briend, A. Shikholeslami, M.-J. Peltre, D. Delafosse, D. Barthomeuf, *J. Chem. Soc., Dalton Trans.* **1989**, 1361–1362.

609 S. T. Wilson, in: Verified Syntheses of Zeolitic Materials, H. Robson (Ed.), 2nd edn, Elsevier, Amsterdam, 2001, p. 27–31.

610 R. F. Lobo, S. I. Zones, M. E. Davis, *J. Inclusion Phenom. Mol. Recognit. Chem.* **1995**, *21*, 47–78.

611 M. E. Davis, S. I. Zones, in: Synthesis of Porous Materials: Zeolites, Clays, and Nanostructures, M. L. Occelli, H. Kessler (Eds) Marcel Dekker, New York, 1996, p. 1–34.

612 K. J. Balkus, in: Progress in Inorganic Chemistry, K. D. Karlin (Ed.), Vol. 50, Wiley, New York, 2001, p. 217–268.

613 D. W. Lewis, C. M. Freeman, C. R. A. Catlow, *J. Phys. Chem.* **1995**, *99*, 11194–11202.

614 J. M. Thomas, D. W. Lewis, *Z. Phys. Chem.* **1996**, *197*, 37–48.

615 D. W. Lewis, D. J. Willock, C. R. A. Catlow, J. M. Thomas, G. J. Hutchings, *Nature* **1996**, *382*, 604–606.

616 D. W. Lewis, G. Sankar, J. K. Wyles, J. M. Thomas, C. R. A. Catlow, D. J. Willock, *Angew. Chem.* **1997**, *109*, 2791–2793; *Angew. Chem. Int. Ed. Engl.* **1997**, *36*, 2675–2677.

617 H. Stach, J. Jänchen, H.-G. Jerschkewitz, U. Lohse, B. Parlitz, B. Zibrowius, M. Hunger, *J. Phys. Chem.* **1992**, *96*, 8473–8479.

618 G. Engelhardt, D. Michel, High-Resolution Solid State NMR of Silicates and Zeolites, Wiley, Chichester, 1987, p. 253–256.

619 X. Liu, J. Klinowski, *J. Phys. Chem.* **1992**, *96*, 3403–3408.

620 M. W. Anderson, J. Rocha, Z. Lin, A. Philippou, I. Orion, A. Ferreira, *Microporous Mater.* **1996**, *6*, 195–204.

621 H. K. Beyer, I. M. Belenykaja, F. Hange, M. Tielen, P. J. Grobet, P. A. Jacobs, *J. Chem. Soc., Faraday Trans. 1* **1985**, *81*, 2889–2901.

622 H. Fichtner-Schmittler, U. Lohse, H. Miessner, H.-E. Maneck, *Z. Phys. Chem. (Leipzig)* **1990**, *271*, 69–79.

623 E. Loeffler, Ch. Peuker, H. G. Jerschkewitz, *Catal. Today* **1988**, *3*, 415–420.

624 N. L. Spiridonova, S. E. Spiridonov, S. N. Khadzhiev, *Kinet. Katal.* **1988**, *29*, 1212–1215.

625 S. T. Wilson, B. M. Lok, E. M. Flanigen, US Patent 4,310,440, 1982, assigned to Union Carbide Corp., USA.

626 S. T. Wilson, B. M. Lok, C. A. Messina, T. R. Cannan, E. M.

FLANIGEN, *J. Am. Chem. Soc.* **1982**, *104*, 1146–1147.

627 E. M. FLANIGEN, B. M. LOK, R. L. PATTON, S. T. WILSON, *Pure Appl. Chem.* **1986**, *58*, 1351–1358.

628 E. M. FLANIGEN, R. L. PATTON, S. T. WILSON, in: Innovation in Zeolite Materials Science, P. J. GROBET, W. J. MORTIER, E. F. VANSANT, G. SCHULZ-EKLOFF (Eds), *Studies in Surface Science and Catalysis*, Vol. 37, Elsevier, Amsterdam, 1988, p. 13–27.

629 J. A. MARTENS, P. A. JACOBS, in: Advanced Zeolite Science and Application, J. C. JANSEN, M. STÖCKER, H. G. KARGE, J. WEITKAMP (Eds), *Studies in Surface Science and Catalysis*, Vol. 85, Elsevier, Amsterdam, 1994, p. 653–685.

630 J. A. MARTENS, P. A. JACOBS, in: Catalysis and Zeolites: Fundamentals and Applications, J. WEITKAMP, L. PUPPE (Eds), Springer, Berlin, 1999, p. 53–80.

631 B. M. LOK, C. A. MESSINA, R. L. PATTON, R. T. GAJEK, T. R. CANNAN, E. M. FLANIGEN, US Patent 4,440,871, 1984, assigned to Union Carbide Corp., USA.

632 B. M. LOK, C. A. MESSINA, R. L. PATTON, R. T. GAJEK, T. R. CANNAN, E. M. FLANIGEN, *J. Am. Chem. Soc.* **1984**, *106*, 6092–6093.

633 J. A. MARTENS, M. MERTENS, P. J. GROBET, P. A. JACOBS, in: Innovation in Zeolite Materials Science, P. J. GROBET, W. J. MORTIER, E. F. VANSANT, G. SCHULZ-EKLOFF (Eds), *Studies in Surface Science and Catalysis*, Vol. 37, Amsterdam, 1988, p. 97–105.

634 E. JAHN, D. MÜLLER, K. BECKER, *Zeolites* **1990**, *10*, 151–156.

635 J. A. MARTENS, P. J. GROBET, P. A. JACOBS, *J. Catal.* **1990**, *126*, 299–305.

636 P. P. MAN, M. BRIEND, M. J. PELTRE, A. LAMY, P. BEAUNIER, D. BARTHOMEUF, *Zeolites* **1991**, *11*, 563–572.

637 B. ZIBROWIUS, E. LÖFFLER, G. FINGER, E. SONNTAG, M. HUNGER, J. KORNATOWSKI, in: Catalysis and Adsorption by Zeolites, G. ÖHLMANN, H. PFEIFER, R. FRICKE (Eds), *Studies in Surface Science and Catalysis*, Vol.

65, Elsevier, Amsterdam, 1991, p. 537–548.

638 G. ENGELHARDT, D. MICHEL, High-Resolution Solid State NMR of Silicates and Zeolites, Wiley, Chichester, 1987, p. 170–175.

639 J. A. MARTENS, C. JANSSENS, P. J. GROBET, H. K. BEYER, P. A. JACOBS, in: Zeolites: Facts, Figures, Future, P. A. JACOBS, R. A. VAN SANTEN (Eds), *Studies in Surface Science and Catalysis*, Vol. 49, Elsevier, Amsterdam, 1989, p. 215–225.

640 H. KESSLER, J. M. CHEZEAU, J. L. GUTH, H. STRUB, G. COUDURIER, *Zeolites* **1987**, *7*, 360–366.

641 M. B. SAYED, A. AUROUX, J. C. VÉDRINE, *J. Catal.* **1989**, *116*, 1–10.

642 M. W. SIMON, S. S. NAM, W. XU, S. L. SUIB, J. C. EDWARD, C.-L. O'YOUNG, *J. Phys. Chem.* **1992**, *96*, 6381–6388.

643 W. Q. XU, S. L. SUIB, C.-L. O'YOUNG, *J. Catal.* **1993**, *144*, 285–295.

644 R. DE RUITER, A. P. M. KENTGENS, J. GROOTENDORST, J. C. JANSEN, H. VAN BEKKUM, *Zeolites* **1993**, *13*, 128–138.

645 S. HAN, K. D. SCHMITT, S. E. SCHRAMM, P. T. REISCHMANN, D. S. SHIHABI, C. D. CHANG, *J. Phys. Chem.* **1994**, *98*, 4118–4124.

646 P. A. WRIGHT, R. H. JONES, S. NATARAJAN, R. G. BELL, J. CHEN, M. B. HURSTHOUSE, J. M. THOMAS, *J. Chem. Soc., Chem. Commun.* **1993**, 633–635.

647 W.-L. SHEA, R. B. BORADE, A. CLEARFIELD, *J. Chem. Soc., Faraday Trans.* **1993**, *89*, 3143–3149.

648 D. B. AKOLEKAR, *J. Catal.* **1994**, *146*, 62–68.

649 S. PRASAD, D. H. BARICH, J. F. HAW, *Catal. Lett.* **1996**, *39*, 141–146.

650 S. PRASAD, J. F. HAW, *Chem. Mater.* **1996**, *8*, 861–864.

651 D. B. AKOLEKAR, R. F. HOWE, *J. Chem. Soc., Faraday Trans.* **1997**, *93*, 3263–3268.

652 M. R. BOCCUTI, K. M. RAO, A. ZECCHINA, G. LEOFANTI, G. PETRINI, in: Structure and Reactivity of Surfaces, C. MORTERRA, A. ZECCHINA, G. COSTA (Eds), *Studies in Surface Science and Catalysis*, Vol. 48, Elsevier, Amsterdam, 1989, p. 133–144.

653 J. S. Reddy, R. Kumar, P. Ratnasamy, *Appl. Catal.* **1990**, *58*, L1–L4.

654 A. Tuel, J. Diab, P. Gelin, M. Dufaux, J.-F. Dutel, Y. Ben Taarit, *J. Mol. Catal.* **1990**, *63*, 95–102.

655 P. Behrens, J. Felsche, S. Vetter, G. Schulz-Ekloff, N. I. Jaeger, W. Niemann, *J. Chem. Soc., Chem. Commun.* **1991**, 678–680.

656 A. Zecchina, G. Spoto, S. Bordiga, A. Ferrero, G. Petrini, G. Leofanti, M. Padovan, in: Zeolite Chemistry and Catalysis, P. A. Jacobs, N. I. Jaeger, L. Kubelková, B. Wichterlová (Eds), *Studies in Surface Science and Catalysis*, Vol. 69, Elsevier, Amsterdam, 1991, p. 251–258.

657 F. Geobaldo, S. Bordiga, A. Zecchina, E. Giamello, G. Leofanti, G. Petrini, *Catal. Lett.* **1992**, *16*, 109–115.

658 E. Schultz, C. Ferrini, R. Prins, *Catal. Lett.* **1992**, *14*, 221–231.

659 D. P. Serrano, H.-X. Li, M. E. Davis, *J. Chem. Soc., Chem. Commun.* **1992**, 745–747.

660 A. Tuel, Y. Ben Taarit, *J. Chem. Soc., Chem. Commun.* **1992**, 1578–1580.

661 M. A. Camblor, A. Corma, J. Pérez-Pariente, *J. Chem. Soc., Chem. Commun.* **1993**, 557–559.

662 A. Lopez, M. H. Tuilier, J. L. Guth, L. Delmotte, J. M. Popa, *J. Solid State Chem.* **1993**, *102*, 480–491.

663 S. Pei, G. W. Zajac, J. A. Kaduk, J. Faber, B. I. Boyanov, D. Duck, D. Fazzini, T. I. Morrison, D. S. Yang, *Catal. Lett.* **1993**, *21*, 333–344.

664 A. Tuel, Y. Ben Taarit, *Zeolites* **1993**, *13*, 357–364.

665 S. Bordiga, S. Coluccia, C. Lamberti, L. Marchese, A. Zecchina, F. Boscherini, F. Buffa, F. Genoni, G. Leofanti, G. Petrini, G. Vlaic, *J. Phys. Chem.* **1994**, *98*, 4125–4132.

666 I. Grohmann, W. Pilz, G. Walther, H. Kosslick, V. A. Tuan, *Surf. Interf. Anal.* **1994**, *22*, 403–406.

667 J. Klaas, K. Kulawik, G. Schulz-Ekloff, N. I. Jaeger, in: Zeolites and Related Microporous Materials: State of the Art 1994, J. Weitkamp, H. G. Karge, H. Pfeifer, W. Hölderich (Eds), *Studies in Surface Science and Catalysis*, Vol. 84, Elsevier, Amsterdam, 1994, p. 2261–2268.

668 E. Astorino, J. B. Peri, R. J. Willey, G. Busca, *J. Catal.* **1995**, *157*, 482–500.

669 R. J. Davis, Z. Liu, J. E. Tabora, W. S. Wieland, *Catal. Lett.* **1995**, *34*, 101–113.

670 A. J. M. de Man, J. Sauer, *J. Phys. Chem.* **1996**, *100*, 5025–5034.

671 A. Tuel, *Zeolites* **1996**, *16*, 108–117.

672 A. Zecchina, S. Bordiga, C. Lamberti, G. Ricchiardi, D. Scarano, G. Petrini, G. Leofanti, M. Mantegazza, *Catal. Today* **1996**, *32*, 97–106.

673 E. Duprey, P. Beaunier, M.-A. Springuel-Huet, F. Bozon-Verduraz, J. Fraissard, J.-M. Manoli, J.-M. Brégeault, *J. Catal.* **1997**, *165*, 22–32.

674 T. E. W. Nießen, Untersuchungen zur Herstellung und Charakterisierung von titan– und vanadiumhaltigen mikro- und mesoporösen Feststoffen, Shaker Verlag, Aachen, 1997, 124 pp.

675 C. Lamberti, S. Bordiga, D. Arduino, A. Zecchina, F. Geobaldo, G. Spanò, F. Genoni, G. Petrini, A. Carati, F. Villain, G. Vlaic, *J. Phys. Chem. B* **1998**, *102*, 6382–6390.

676 V. Bolis, S. Bordiga, C. Lamberti, A. Zecchina, A. Carati, F. Rivetti, G. Spanò, G. Petrini, *Langmuir* **1999**, *15*, 5753–5764.

677 A. Carati, C. Flego, E. Previde Massara, R. Millini, L. Carluccio, W. O. Parker, Jr., G. Bellussi, *Microporous Mesoporous Mater.* **1999**, *30*, 137–144.

678 Z. Fu, D. Yin, D. Yin, Q. Li, L. Zhang, Y. Zhang, *Microporous Mesoporous Mater.* **1999**, *29*, 351–359.

679 R. Bal, K. Chaudhari, D. Srinivas, S. Sivasanker, P. Ratnasamy, *J. Mol. Catal. A* **2000**, *162*, 199–207.

680 H. Gao, W. Lu, Q. Chen, *Microporous Mesoporous Mater.* **2000**, *34*, 307–315.

681 D. Gleeson, G. Sankar, C. R. A. Catlow, J. M. Thomas, G. Spanò, S. Bordiga, A. Zecchina, C. Lamberti, *Phys. Chem. Chem. Phys.* **2000**, *2*, 4812–4817.

682 A. Labouriau, K. C. Ott, J. Rau, W. L. Earl, *J. Phys. Chem. B* **2000**, *104*, 5890–5896.

683 C. Li, G. Xiong, J. Liu, P. Ying, Q. Xin, Z. Feng, *J. Phys. Chem. B* **2001**, *105*, 2993–2997.

684 G. Ricchiardi, A. Damin, S. Bordiga, C. Lamberti, G. Spanò, F. Rivetti, A. Zecchina, *J. Am. Chem. Soc.* **2001**, *123*, 11409–11419.

685 A. S. Soult, D. D. Pooré, E. I. Mayo, A. E. Stiegman, *J. Phys. Chem. B* **2001**, *105*, 2687–2693.

686 M. S. Rigutto, H. van Bekkum, *Appl. Catal.* **1991**, *68*, L1–L7.

687 G. Centi, S. Perathoner, F. Trifiro, A. Aboukais, C. F. Aissi, M. Guelton, *J. Phys. Chem.* **1992**, *96*, 2617–2629.

688 P. R. H. P. Rao, A. V. Ramaswamy, *Appl. Catal. A* **1993**, *93*, 123–130.

689 A. Tuel, Y. Ben Taarit, *Appl. Catal. A* **1993**, *102*, 201–214.

690 J. Kornatowski, B. Wichterlová, J. Jirkovský, E. Löffler, W. Pilz, *J. Chem. Soc., Faraday Trans.* **1996**, *92*, 1067–1078.

691 M. H. Zahedi-Niaki, S. M. J. Zaidi, S. Kaliaguine, *Appl. Catal. A* **2000**, *196*, 9–24.

692 R. B. Borade, *Zeolites* **1987**, *7*, 398–403.

693 R. Szostak, V. Nair, T. L. Thomas, *J. Chem. Soc., Faraday Trans. 1* **1987**, *83*, 487–494.

694 J. W. Park, H. Chon, *J. Catal.* **1992**, *133*, 159–169.

695 W. Fan, R. Li, B. Zhong, H. Du, E. Roduner, *Microporous Mesoporous Mater.* **1998**, *25*, 95–101.

696 R. A. Schoonheydt, R. de Vos, J. Pelgrims, H. Leeman, in: Zeolites: Facts, Figures, Future, P. A. Jacobs, R. A. van Santen (Eds), *Studies in Surface Science and Catalysis*, Vol. 49, Elsevier, Amsterdam, 1989, p. 559–568.

697 C. Montes, M. E. Davis, B. Murray, M. Narayana, *J. Phys. Chem.* **1990**, *94*, 6425–6430.

698 B. Kraushaar-Czarnetzki, W. G. M. Hoogervorst, R. R. Andrea, C. A. Emeis, W. H. J. Stork, *J. Chem. Soc., Faraday Trans.* **1991**, *87*, 891–895.

699 G. Nardin, L. Randaccio, V. Kaučič, N. Rajić, *Zeolites* **1991**, *11*, 192–194.

700 J. Chen, G. Sankar, J. M. Thomas, R. Xu, G. N. Greaves, D. Waller, *Chem. Mater.* **1992**, *4*, 1373–1380.

701 L. Marchese, J. Chen, J. M. Thomas, S. Coluccia, A. Zecchina, *J. Phys. Chem.* **1994**, *98*, 13350–13356.

702 V. Kurshev, L. Kevan, D. J. Parillo, C. Pereira, G. T. Kokotailo, R. J. Gorte, *J. Phys. Chem.* **1994**, *98*, 10160–10166.

703 J. Jänchen, M. P. J. Peeters, J. H. M. C. van Wolput, J. P. Wolthuizen, J. H. C. van Hooff, U. Lohse, *J. Chem. Soc., Faraday Trans.* **1994**, *90*, 1033–1039.

704 S. Thomson, V. Luca, R. Howe, *Phys. Chem. Chem. Phys.* **1999**, *1*, 615–619.

705 Q. Gao, B. M. Weckhuysen, R. A. Schoonheydt, *Microporous Mesoporous Mater.* **1999**, *27*, 75–86.

706 M. Höchtl, A. Jentys, H. Vinek, *Microporous Mesoporous Mater.* **1999**, *31*, 271–285.

707 J. Sponer, J. Čejka, B. Wichterlová, *Microporous Mesoporous Mater.* **2000**, *37*, 117–127.

708 W. B. Fan, R. A. Schoonheydt, B. M. Weckhuysen, *Phys. Chem. Chem. Phys.* **2001**, *3*, 3240–3246.

709 C. R. Bayense, J. H. C. van Hooff, J. W. de Haan, L. J. M. van de Ven, A. P. M. Kentgens, *Catal. Lett.* **1993**, *17*, 349–361.

710 C. Otero Areán, G. Turnes Palomino, F. Geobaldo, A. Zecchina, *J. Phys. Chem.* **1996**, *100*, 6678–6690.

711 C. Otero Areán, B. Bonelli, G. Turnes Palomino, A. M. Canaleta Safont, E. Garrone, *Phys. Chem. Chem. Phys.* **2001**, *3*, 1223–1227.

712 B. M. Weckhuysen, R. A. Schoonheydt, *Zeolites* **1994**, *14*, 360–366.

713 J. D. Chen, H. E. B. Lempers, R. A. Sheldon, *J. Chem. Soc., Faraday Trans.* **1996**, *92*, 1807–1813.

714 P. Brandão, A. Philippou, A. Valente, J. Rocha, M. Anderson, *Phys. Chem. Chem. Phys.* **2001**, *3*, 1773–1777.

715 J. J. PLUTH, J. V. SMITH, J. W.
RICHARDSON, JR., *J. Phys. Chem.* **1988**,
92, 2734–2738.

716 Z. OLENDER (LEVI), D. GOLDFARB, J.
BATISTA, *J. Am. Chem. Soc.* **1993**, *115*,
1106–1114.

717 H.-L. ZUBOWA, M. RICHTER, U. ROOST,
B. PARLITZ, R. FRICKE, *Catal. Lett.*
1993, *19*, 67–79.

718 X. YAN, J. W. COUVES, R. H. JONES,
C. R. A. CATLOW, G. N. GREAVES, J. S.
CHEN, J. M. THOMAS, *J. Phys. Chem.
Solids* **1991**, *52*, 1229–1234.

719 M. HELLIWELL, B. GALLOIS, B. M.
KARIUKI, V. KAUČIČ, J. R. HELLIWELL,
Acta Crystallogr. B **1993**, *49*, 420–428.

720 M. HARTMANN, N. AZUMA, L. KEVAN,
J. Phys. Chem. **1995**, *99*, 10988–10994.

721 B. MARLER, J. PATARIN, L. SIERRA,
Microporous Mater. **1995**, *5*, 151–159.

722 G. GONZÁLEZ, C. PIÑA, A. JACAS, M.
HERNÁNDEZ, A. LEYVA, *Microporous
Mesoporous Mater.* **1998**, *25*, 103–108.

723 H. KOSSLICK, V. A. TUAN, R. FRICKE,
CH. PEUKER, W. PILZ, W. STOREK, *J.
Phys. Chem.* **1993**, *97*, 5678–5684.

724 K. LATHAM, D. THOMPSETT, C. D.
WILLIAMS, C. I. ROUND, *J. Mater.
Chem.* **2000**, *10*, 1235–1240.

725 G. VORBECK, J. JÄNCHEN, B. PARLITZ,
M. SCHNEIDER, R. FRICKE, *J. Chem.
Soc., Chem. Commun.* **1994**, 123–124.

726 S. HAN, K. D. SCHMITT, C. D. CHANG,
Inorg. Chim. Acta **2000**, *304*, 297–300.

727 S. KOWALAK, M. PAWLOWSKA, L. M.
KUSTOV, in: Catalysis by Microporous
Materials, H. K. BEYER, H. G. KARGE,
I. KIRICSI, J. B.NAGY (Eds), *Studies in
Surface Science and Catalysis*, Vol. 94,
Elsevier, Amsterdam, 1995, p. 203–
210.

728 Y. S. KO, W. S. AHN, *Microporous
Mesoporous Mater.* **1999**, *30*, 283–291.

729 R. MILLINI, G. PEREGO, G. BELLUSSI,
Top. Catal. **1999**, *9*, 13–34.

730 G. BELLUSSI, M. S. RIGUTTO, in:
Advanced Zeolite Science and
Application, J. C. JANSEN, M. STÖCKER,
H. G. KARGE, J. WEITKAMP (Eds),
Studies in Surface Science and Catalysis,
Vol. 85, Elsevier, Amsterdam, 1994, p.
177–213.

731 R. MILLINI, G. PEREGO, *Gazz. Chim.
Ital.* **1996**, *126*, 133–140.

732 B. NOTARI, *Adv. Catal.* **1996**, *41*, 253–
334.

733 G. N. VAYSSILOV, *Catal. Rev. – Sci. Eng.*
1997, 39, 209–251.

734 R. J. SAXTON, *Top. Catal.* **1999**, *9*, 43–
57.

735 R. FRICKE, H. KOSSLICK, G. LISCHKE,
M. RICHTER, *Chem. Rev.* **2000**, *100*,
2303–2405.

736 D. GOLDFARB, *Zeolites* **1989**, *9*, 509–515.

737 C. MONTES, M. E. DAVIS, B. MURRAY,
M. NARAYANA, *J. Phys. Chem.* **1990**, *94*,
6425–6430.

738 S.-H. CHEN, S.-P. SHEU, K.-J. CHAO, *J.
Chem. Soc., Chem. Commun.* **1992**,
1504–1505.

739 B. KRAUSHAAR-CZARNETZKI, W. G. M.
HOOGERVORST, R. R. ANDREA, C. A.
EMEIS, W. H. J. STORK, in: Zeolite
Chemistry and Catalysis, P. A. JACOBS,
N. I. JAEGER, L. KUBELKOVÁ, B.
WICHTERLOVÁ (Eds), *Studies in Surface
Science and Catalysis*, Vol. 69, Elsevier,
Amsterdam, 1991, p. 231–240.

740 M. P. J. PEETERS, L. J. M. VAN DE VEN,
J. W. DE HAAN, J. H. C. VAN HOOFF,
Colloids Surf. A **1993**, *72*, 87–104.

741 B. M. WECKHUYSEN, R. R. RAO, J. A.
MARTENS, R. A. SCHOONHEYDT, *Eur. J.
Inorg. Chem.* **1999**, 565–577.

742 M. BERGMANN, E. LÖFFLER, M. MUHLER,
K. GENOV, M. WARK, (submitted to
Phys. Chem. Chem. Phys.).

743 E. LÖFFLER, (unpublished results).

744 J. A. RABO, G. J. GAIDA, *Catal. Rev. –
Sci. Eng.* **1989–90**, *31*, 385–430.

745 H. G. KARGE, in: Catalysis and
Adsorption by Zeolites, G. ÖHLMANN,
H. PFEIFER, R. FRICKE (Eds), *Studies
in Surface Science and Catalysis*, Vol.
65, Elsevier, Amsterdam, 1991, p. 133–
156.

746 D. BARTHOMEUF, in: Catalysis and
Adsorption by Zeolites, G. ÖHLMANN,
H. PFEIFER, R. FRICKE (Eds), *Studies
in Surface Science and Catalysis*, Vol.
65, Elsevier, Amsterdam, 1991, p.
157–169.

747 V. B. KAZANSKY, in: Advanced Zeolite
Science and Applications, J. C.
JANSEN, M. STÖCKER, H. G. KARGE, J.
WEITKAMP (Eds), *Studies in Surface
Science and Catalysis*, Vol. 85, Elsevier,
Amsterdam, 1994, p. 251–272.

748 R. A. van Santen, in: Advanced Zeolite Science and Applications, J. C. Jansen, M. Stöcker, H. G. Karge, J. Weitkamp (Eds), *Studies in Surface Science and Catalysis*, Vol. 85, Elsevier, Amsterdam, 1994, p. 273–294.

749 J. Sauer, in: Zeolites and Related Microporous Materials: State of the Art 1994, J. Weitkamp, H. G. Karge, H. Pfeifer, W. Hölderich (Eds), *Studies in Surface Science and Catalysis*, Vol. 84, Elsevier, Amsterdam, 1994, p. 2039–2057.

750 W. E. Farneth, R. J. Gorte, *Chem. Rev.* **1995**, *95*, 615–635.

751 L. Dixit, T. S. R. P. Rao, *Appl. Spectrosc. Rev.* **1996**, *31*, 369–472.

752 D. B. Akolekar, M. Huang, S. Kaliaguine, *Zeolites* **1994**, *14*, 519–522.

753 X. S. Liu, K. K. Iu, J. K. Thomas, *J. Phys. Chem.* **1994**, *98*, 7877–7884.

754 A. Philippou, M. W. Anderson, *J. Am. Chem. Soc.* **1994**, *116*, 5774–5783.

755 M. M. Huang, S. Kaliaguine, M. Muscas, A. Auroux, *J. Catal.* **1995**, *157*, 266–269.

756 E. B. Uvarova, L. M. Kustov, V. B. Kazansky, in: Catalysis by Microporous Materials, H. K. Beyer, H. G. Karge, I. Kiricsi, J. B.Nagy (Eds), *Studies in Surface Science and Catalysis*, Vol. 94, Elsevier, Amsterdam, 1995, p. 254–261.

757 D. Barthomeuf, *Catal. Rev. – Sci. Eng.* **1996**, *38*, 521–612.

758 J. C. Lavalley, *Catal. Today* **1996**, *27*, 377–401.

759 J. H. Xie, M. M. Huang, S. Kaliaguine, *Appl. Surf. Sci.* **1997**, *115*, 157–165.

760 J.-C. Lavalley, J. Lamotte, A. Travert, J. Czniewska, M. Ziolek, *J. Chem. Soc., Faraday Trans.* **1998**, *94*, 331–335.

761 E. Bosch, S. Huber, J. Weitkamp, H. Knözinger, *Phys. Chem. Chem. Phys.* **1999**, *1*, 579–584.

762 J. T. Timonen, T. T. Pakkanen, *Microporous Mesoporous Mater.* **1999**, *30*, 327–333.

763 P. Concepcion-Heydorn, C. Jia, D. Herein, N. Pfander, H. G. Karge, F. C. Jentoft, *J. Mol. Catal. A* **2000**, *162*, 227–246.

764 M. Sánchez-Sánchez, T. Blasco, *Chem. Commun.* **2000**, 491–492.

765 B. L. Su, V. Norberg, J. A. Martens, *Langmuir* **2001**, *17*, 1267–1276.

766 G. L. Woolery, L. B. Alemany, R. M. Dessau, A. W. Chester, *Zeolites* **1986**, *6*, 14–16.

767 R. M. Dessau, K. D. Schmitt, G. T. Kerr, G. L. Woolery, L. B. Alemany, *J. Catal.* **1987**, *104*, 484–489.

768 M. Hunger, J. Kärger, H. Pfeifer, J. Caro, B. Zibrowius, M. Bülow, R. Mostowicz, *J. Chem. Soc., Faraday Trans. 1* **1987**, *83*, 3459–3468.

769 R. M. Dessau, K. D. Schmitt, G. T. Kerr, G. L. Woolery, L. B. Alemany, *J. Catal.* **1988**, *109*, 472–473.

770 U. Lohse, E. Löffler, M. Hunger, J. Stöcker, V. Patzelová, *Zeolites* **1987**, *7*, 11–13.

771 E. Loeffler, U. Lohse, Ch. Peuker, G. Oehlmann, L. M. Kustov, V. L. Zholobenko, V. B. Kazansky, *Zeolites* **1990**, *10*, 266–271.

772 J. Scherzer, in: Catalytic Materials: Relationship Between Structure and Reactivity, T. E. Whyte, Jr., R. A. Dalla Betta, E. G. Derouane, R. T. K. Baker (Eds), ACS Symposium Series, Vol. 248, American Chemical Society, Washington D.C., 1984, p. 157–200.

773 A. E. Hirschler, *J. Catal.* **1963**, *2*, 428–439.

774 C. J. Plank, Proc. 3rd Int. Congr. Catal., Vol. 1, North-Holland, Amsterdam, 1964, p. 727–735.

775 E. Löffler, IR-spektroskopische Untersuchungen der aziden Zentren in ZSM-5-Zeolithen und SAPO-Molekularsieben, Shaker Verlag, Aachen, 1998, 140 pp.

776 L. Kubelková, H. Hoser, A. Riva, F. Trifiro, *Zeolites* **1983**, *3*, 244–248.

777 K.-H. Schnabel, Ch. Peuker, B. Parlitz, E. Löffler, U. Kürschner, H. Kriegsmann, *Z. Phys. Chem. (Leipzig)* **1987**, *268*, 225–234.

778 H. Mix, H. Pfeifer, B. Staudte, *Chem. Phys. Lett.* **1988**, *146*, 541–544.

779 V. L. ZHOLOBENKO, L. M. KUSTOV, V. Yu. BOROVKOV, V. B. KAZANSKY, *Zeolites* **1988**, *8*, 175–178.

780 W. R. MOSER, C.-C. CHIANG, R. W. THOMPSON, *J. Catal.* **1989**, *115*, 532–541.

781 K. A. MARTIN, R. F. ZABRANSKY, *Appl. Spectrosc.* **1991**, *45*, 68–72.

782 R. SALZER, U. FINSTER, F. ROESSNER, K.-H. STEINBERG, P. KLAEBOE, *Analyst* **1992**, *117*, 351–354.

783 J. A. MÜLLER, WM. C. CONNER, *J. Phys. Chem.* **1993**, *97*, 1451–1454.

784 K. BECK, E. BRUNNER, B. STAUDTE, Z. *Phys. Chem.* **1995**, *190*, 1–7.

785 T. V. VOSKOBOINIKOV, B. COQ, F. FAJULA, R. BROWN, G. MCDOUGALL, J. L. COUTURIER, *Microporous Mesoporous Mater.* **1998**, *24*, 89–99.

786 L. M. KUSTOV, *Top. Catal.* **1997**, *4*, 131–144.

787 R. L. BEDARD, C. L. BOWES, N. COOMBS, A. J. HOLMES, T. JIANG, S. J. KIRKBY, P. M. MACDONALD, A. M. MALEK, G. A. OZIN, S. PETROV, N. PLAVAC, R. A. RAMIK, M. R. STEELE, D. YOUNG, *J. Am. Chem. Soc.* **1993**, *115*, 2300–2313.

788 F. THIBAULT–STARZYK, A. JANIN, J.-C. LAVALLEY, *Angew. Chem.* **1997**, *109*, 1017–1019; *Angew. Chem. Int. Ed. Engl.* **1997**, *36*, 989–991.

789 T. L. BARR, J. KLINOWSKI, H. HE, K. ALBERTI, G. MÜLLER, J. A. LERCHER, *Nature* **1993**, *365*, 429–431.

790 G. MÜLLER, G. EDER-MIRTH, H. KESSLER, J. A. LERCHER, *J. Phys. Chem.* **1995**, *99*, 12327–12331.

791 P. A. JACOBS, R. VON BALLMOOS, *J. Phys. Chem.* **1982**, *86*, 3050–3052.

792 J. DATKA, T. TUŻNIK, *Zeolites* **1985**, *5*, 230–232.

793 G. QIN, L. ZHENG, Y. XIE, C. WU, *J. Catal.* **1985**, *95*, 609–612.

794 A. JENTYS, G. MIRTH, J. SCHWANK, J. A. LERCHER, in: Zeolites: Facts, Figures, Future, P. A. JACOBS, R. A. VAN SANTEN (Eds), *Studies in Surface Science and Catalysis*, Vol. 49, Elsevier, Amsterdam, 1989, p. 847–856.

795 A. JENTYS, G. RUMPLMAYR, J. A. LERCHER, *Appl. Catal.* **1989**, *53*, 299–312.

796 J. C. VÉDRINE, A. AUROUX, V. BOLIS, P. DEJAIFVE, C. NACCACHE, P. WIERZCHOWSKI, E. G. DEROUANE, J. B.NAGY, J.-P. GILSON, J. H. C. VAN HOOFF, J. P. VAN DEN BERG, J. WOLTHUIZEN, *J. Catal.* **1979**, *59*, 248–262.

797 A. ISON, R. J. GORTE, *J. Catal.* **1984**, *89*, 150–158.

798 A. JENTYS, G. WARECKA, M. DEREWINSKI, J. A. LERCHER, *J. Phys. Chem.* **1989**, *93*, 4837–4843.

799 A. JENTYS, G. WARECKA, J. A. LERCHER, *J. Mol. Catal.* **1989**, *51*, 309–327.

800 C. D. CHANG, S. D. HELLRING, J. N. MIALE, K. D. SCHMITT, P. W. BRIGANDI, E. L. WU, *J. Chem. Soc., Faraday Trans. 1* **1985**, *81*, 2215–2224.

801 A. K. GHOSH, R. A. KYDD, *Zeolites* **1990**, *10*, 766–771.

802 V. B. KAZANSKIJ, KH. M. MINACHEV, B. K. NEFEDOV, V. YU. BOROVKOV, D. A. KONDRATEV, G. D. CHUKIN, L. M. KUSTOV, T. N. BONDARENKO, L. D. KONOVALCHIKOV, *Kinet. Katal.* **1983**, *24*, 679–682.

803 E. A. STEPANOVA, V. S. KOMAROV, M. F. SINILO, L. P. SHIRINSKAYA, *Zh. Prikl. Spektrosk.* **1989**, *51*, 950–956.

804 P. N. AUKETT, S. CARTLIDGE, I. J. F. POPLETT, *Zeolites* **1986**, *6*, 169–174.

805 C. T.-W. CHU, D. CHANG, *J. Phys. Chem.* **1985**, *89*, 1569–1571.

806 K. SUZUKI, T. SANO, H. SHOJI, T. MURAKAMI, S. IKAI, S. SHIN, H. HAGIWARA, H. TAKAYA, *Chem. Lett.* **1987**, 1507–1510.

807 K. YAMAGISHI, S. NAMBA, T. YASHIMA, *J. Phys. Chem.* **1991**, *95*, 872–877.

808 A. ZECCHINA, S. BORDIGA, G. SPOTO, L. MARCHESE, G. PETRINI, G. LEOFANTI, M. PADOVAN, *J. Phys. Chem.* **1992**, *96*, 4985–4990.

809 W. P. J. H. JACOBS, J. H. M. C. VAN WOLPUT, R. A. VAN SANTEN, H. JOBIC, *Zeolites* **1994**, *14*, 117–125.

810 M. J. WAX, R. R. CAVANAGH, J. J. RUSH, G. D. STUCKY, L. ABRAMS, D. R. CORBIN, *J. Phys. Chem.* **1986**, *90*, 532–534.

811 H. JOBIC, *J. Catal.* **1991**, *131*, 289–293.

812 J. B. Peri, *Disc. Faraday Soc.* **1971**, 55–65.

813 W. P. J. H. Jacobs, J. H. M. C. van Wolput, R. A. van Santen, *Chem. Phys. Lett.* **1993**, *210*, 32–37.

814 W. Hanke, K. Möller, *Zeolites* **1984**, *4*, 244–250.

815 U. Zscherpel, E. Brunner, B. Staudte, *Z. Phys. Chem. (Leipzig)* **1990**, *271*, 931–939.

816 H. Knözinger, P. Ratnasamy, *Catal. Rev. – Sci. Eng.* **1978**, *17*, 31–70.

817 H. Pfeifer, D. Freude, J. Kärger, in: Catalysis and Adsorption by Zeolites, G. Öhlmann, H. Pfeifer, R. Fricke (Eds), *Studies in Surface Science and Catalysis*, Vol. 65, Elsevier, Amsterdam, 1991, p. 89–115.

818 J. A. van Bokhoven, A. L. Roest, D. C. Koningsberger, J. T. Miller, G. H. Nachtegaal, A. P. M. Kentgens, *J. Phys. Chem. B* **2000**, *104*, 6743–6754.

819 R. Szostak, in: Introduction to Zeolite Science and Practice, H. van Bekkum, E. M. Flanigen, P. A. Jacobs, J. C. Jansen (Eds), *Studies in Surface Science and Catalysis*, Vol. 137, Elsevier, Amsterdam, 2001, p. 261–297.

820 W. W. Keading, C. Chu, L. B. Young, B. Weinstein, S. A. Butter, *J. Catal.* **1981**, *67*, 159–174.

821 G. Lischke, R. Eckelt, H.-G. Jerschkewitz, B. Parlitz, E. Schreier, W. Storek, B. Zibrowius, G. Öhlmann, *J. Catal.* **1991**, *132*, 229–243.

822 W. Kolodziejski, V. Fornés, A. Corma, *Solid State NMR* **1993**, *2*, 121–129.

823 M. L. Occelli, M. Kalwei, A. Wölker, H. Eckert, A. Auroux, S. A. C. Gould, *J. Catal.* **2000**, *196*, 134–148.

824 M. Guisnet, in: Handbook of Heterogeneous Catalysis, G. Ertl, H. Knözinger, J. Weitkamp (Eds), Vol. 2, VCH, Weinheim, 1997, p. 626–632.

825 M. Guisnet, P. Magnoux, *Appl. Catal. A* **2001**, *212*, 83–96.

826 M. Guisnet, P. Magnoux, D. Martin, in: Catalyst Deactivation 1997, C. H. Bartholomew, G. A. Fuentes (Eds), *Studies in Surface Science and Catalysis*, Vol. 111, Elsevier, Amsterdam, 1997, p. 1–19.

827 H. G. Karge, in: Introduction to Zeolite Science and Practice, H. van Bekkum, E. M. Flanigen, P. A. Jacobs, J. C. Jansen (Eds), *Studies in Surface Science and Catalysis*, Vol. 137, Elsevier, Amsterdam, 2001, p. 707–746.

828 H. G. Karge, W. Nießen, H. Bludau, *Appl. Catal. A* **1996**, *146*, 339–349.

4.2.7
Characterization of the Pore Width of Zeolites and Related Materials by Means of Molecular Probes

Yvonne Traa and Jens Weitkamp

4.2.7.1
Introduction

With molecular probes, various properties of crystalline microporous solids can be explored. One example is the determination of surface acidity/basicity by adsorption and desorption of basic/acidic probe molecules (e.g., ammonia, pyridine, carbon dioxide, chloroform or deuterochloroform) and observing the sorption processes by infrared (IR) spectroscopy, nuclear magnetic resonance (NMR) spectroscopy, mass spectrometry or gas chromatography (see Chapter 2.11). Another possibility is the evaluation of surface hydrophobicity/hydrophilicity by sorption of mixtures of nonpolar and polar substances (see Chapter 2.10). The assessment of micropore volume and pore size can be accomplished by adsorption of xenon monitored by ^{129}Xe NMR spectroscopy (see Chapter 2.12), by adsorption of nitrogen and other small molecules, usually followed gravimetrically or volumetrically (see Chapter 2.5), or by determining the heats of adsorption by means of microcalorimetry (see Chapter 2.9).

This section also deals with the evaluation of the pore size of crystalline microporous solids with molecular probes, but only such methods will be discussed that are based on size effects, that is, where the dimensions of the probe molecules (or of the transition states/product molecules formed from them) and the pore width are similar. These methods include adsorption of molecules of different size large enough to "feel" the presence of the micropores and, therefore, allowing an assessment of the pore width, and test reactions in which the selectivities and/or conversions depend, in an unambiguous manner, on the pore width, that is, shape-selective reactions. In the early days of zeolite science, these two techniques were the most popular tools for collecting information on the approximate crystallographic pore size of zeolites with unknown structures. With the advent of more sophisticated and highly efficient crystallographic methods, a rapid determination even of complex new structures became feasible. Hence, the initial incentive for the application of methods using molecular probes has shifted [1]: Nowadays, these techniques are primarily used as quick tests for probing the effective pore width under catalytically relevant conditions and/or of molecular sieves manipulated and modified with post-synthesis methods such as chemical vapor deposition (CVD), deliberate or unwanted coking, isomorphous substitution of framework atoms and the like.

In Sect. 4.2.7.2, some general aspects will be discussed that are important for the detailed understanding of the methods covered. Section 4.2.7.3 will be devoted to adsorption, that is, the use of molecular probes without chemical reactions. Finally, in Sect. 4.2.7.4, shape-selective catalytic reactions that have been employed for characterizing the width of micropores will be reviewed.

4.2.7.2
General Aspects

4.2.7.2.1 Dimensions of Probe Molecules and Intracrystalline Cavities

For the discussion of size effects, it is vital that the dimensions be defined in an appropriate way. This problem was tackled in an excellent paper by Cook and Conner [2]. These authors stress that the hard-sphere picture underlies the thinking about adsorption, that is, adsorptive/adsorbate molecules and the adsorbent are generally both considered to be rigid structures composed of hard-sphere atoms or ions. However, in reality, the molecules and the host matrix are in continuous vibration and often quite flexible. Therefore, if results of adsorption experiments are to be explained with the hard-sphere picture, the dimensions of the probe molecules and of the intracrystalline cavities should be brought in line with this simple

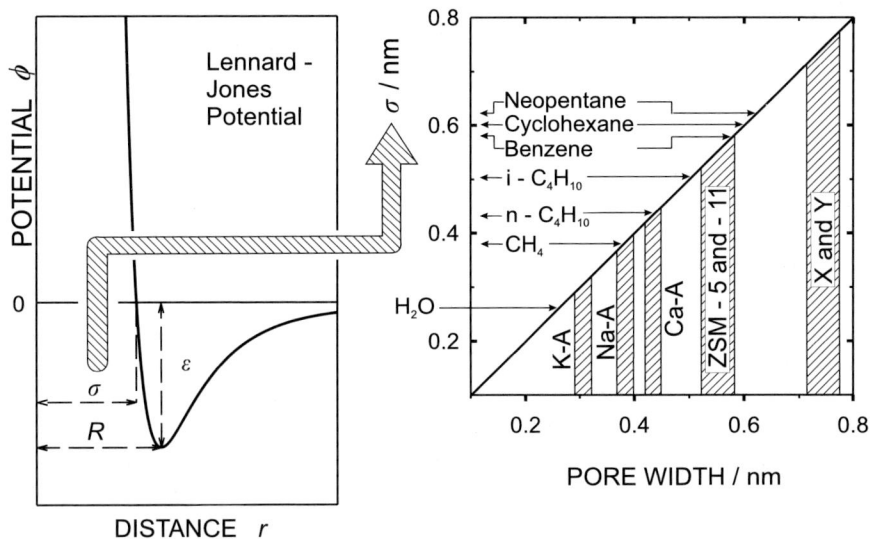

Fig. 1. Dimensions of probe molecules and intracrystalline cavities (σ is the "kinetic diameter", which appears as a parameter in the Lennard–Jones potential).

picture, that is, the model employed for interpreting the adsorption process should be consistent with the type of dimensions used for the description of the probe molecules and the intracrystalline cavities. By contrast, the "kinetic diameter" σ (from the Lennard–Jones potential, see Fig. 1) of the adsorptive and the pore size based on the ionic oxygen radius of 0.135 nm are generally used, and these dimensions do not satisfy this criterion of consistency [2]. In many instances, molecules do diffuse into pores that are considered to possess a width lower than the molecular diameter. An example is cyclohexane ($\sigma = 0.60$ nm, [3]) in zeolite ZSM-5 (0.53×0.56 and 0.51×0.55 nm, [4], see Fig. 1) [5]. For this reason, Cook and Conner [2] proposed a redefinition of framework atom sizes, based on the average physical extension of electron-density distributions in zeolite frameworks. With this modification, the maximum dimension of the ZSM-5 pores would be 0.63 nm, now permitting cyclohexane to enter its pores.

Another fact that should be taken into account is that, for cylindrical pores, the consideration of just one molecular dimension is usually insufficient. Hardly any molecule has spherical symmetry. In this context, Webster et al. [6] advanced the concept of effective minimum dimensions of molecules. MIN-1, the minimum dimension through a molecule, and MIN-2, the second minimum dimension through the same molecule perpendicular to MIN-1, determine whether or not this molecule can enter the pores of a given material. These dimensions can be calculated with molecular orbital theory and allow a more sophisticated description of the molecular behavior inside the pores.

References see page 1053

Furthermore, one should consider the flexibility of molecules and how this flexibility can affect the diffusion and adsorption inside the pores. Choudhary and Akolekar [7] proposed the shuttlecock-shuttlebox model to account for the fact that larger molecules than expected do diffuse into pores of a given width. Their model envisages the compression of alkyl groups of branched molecules similar to the compression of feathers of a shuttlecock in a shuttlebox.

In addition, one should keep in mind that the pore dimensions given in the *Atlas of Zeolite Framework Types* [4] are coarse data and subject to changes due to both the experimental conditions and the precise form of the porous material. For example, the effective pore dimension varies with temperature. The best means for probing this are catalytic tests performed at different temperatures [8]. This will be discussed in more detail in Sect. 4.2.7.4. Another parameter that can affect the pore size is the aluminum content: Framework aluminum tends to reduce the pore volume and to broaden the pore-size distribution. Steam treatment has been reported to reduce the apparent pore size [9]. Furthermore, the shape of the pore aperture can change during adsorption [3], and zeolites can undergo structural rearrangement, which might alter the pore size [10]. Wu and Ma [11] showed that the adsorption capacity for various hydrocarbons on zeolite ZSM-5 decreases as the radius of the cation increases. Thus, certain parts of the channel system can be blocked by cations, and the pore aperture can be contracted. In the extreme case, molecules are excluded from the pore system. For example, uncalcined offretite with large organic cations in its channels adsorbs neither cyclohexane nor n-hexane, whereas the calcined zeolite does [12]. In the *Atlas of Zeolite Framework Types* [4], the pore size of the species examined first is given. Especially in zeolite minerals, different cations are often present, reducing the pore size as compared to the synthetic zeolites and the pure-silica analogs. Thus, the reader should pay attention to the exact form of the species the pore size of which was determined. Another possible reason for pore blockage is amorphous material in the intracrystalline cavities [13]. Finally, preadsorption of polar molecules such as water or ammonia often affects the subsequent adsorption of other molecules by clustering around the cations, reducing the apparent pore size and adsorption capacity and eventually blocking the pores [14]. Therefore, particular attention should always be paid to the hydration state of the zeolite.

In conclusion, one should be very careful with predicting whether or not a given molecule has access to the pores of a given material merely from tabulated sizes of molecules and cavities. Only if the dimensions of the probe molecules and the dimensions of the intracrystalline cavities are chosen in a way consistent with each other and the adsorption model, meaningful predictions are possible [8].

4.2.7.2.2 Molecular Sieving

Molecular sieving is the selective adsorption of molecules into the intracrystalline void system of a molecular sieve and the exclusion of others due to their dimensions being above the critical size. One example is illustrated in Fig. 2: A gaseous mixture of n-pentane and 2-methylbutane was continuously passed over a fixed bed of calcium-exchanged zeolite A (or zeolite "5A") with an effective pore

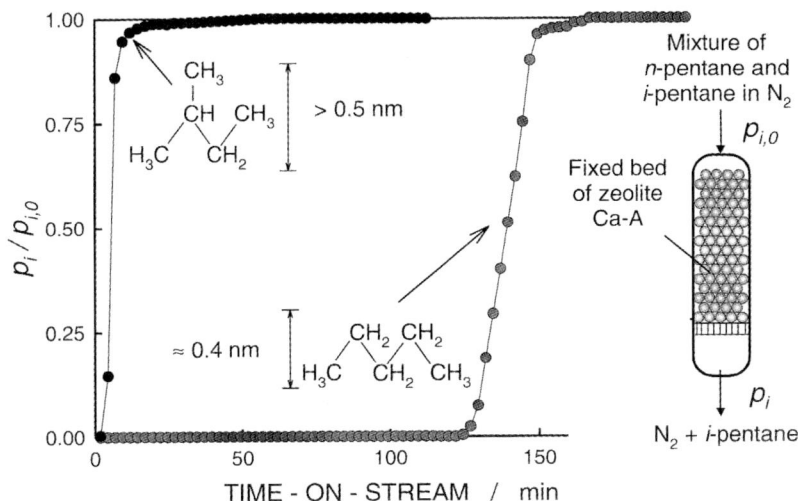

Fig. 2. Breakthrough curves for the adsorption of an n-pentane/2-methylbutane mixture over zeolite Ca-A in a fixed-bed flow-type adsorber.

diameter of approximately 0.5 nm (5 Å). This is large enough to allow for the diffusion of the n-alkane molecules through the eight-membered ring windows of zeolite A, but too small for the uptake of the branched alkane. Hence, 2-methylbutane breaks through at the adsorber outlet directly after the onset of the experiment, while the smaller n-pentane is completely adsorbed by the zeolite for about 2 h, whereupon its adsorption capacity is exhausted.

The example displayed in Fig. 2 represents the extreme case in molecular sieving, namely pore size exclusion, that is, one molecular species is so bulky that it is completely prevented from entering the intracrystalline cavities. In many instances, however, molecules do enter the pores, but their diffusion inside the pores is very slow. In such cases, very different results for the adsorption capacity will be obtained depending on whether or not the system was given enough time to reach adsorption equilibrium. Furthermore, for the purpose of pore-size characterization, one must keep in mind that adsorptive separation on zeolites can be accomplished not only by molecular sieving, but also by selective cation-adsorbate interaction, by selective sorption due to hydrophilic or hydrophobic surface properties or by selective sorption due to acidic/basic surface properties. The discussion in this section will essentially be restricted to molecular sieving, however, a clear-cut distinction between the different mechanisms of adsorptive separation is sometimes difficult.

Obviously, if adsorption at the external surface of a molecular sieve occurs to a significant extent, this will obscure the desired information on the pore size. Masuda and Hashimoto [15] demonstrated that up to 50 % of the total amount ad-

References see page 1053

sorbed can be located on the external surface, if the zeolite crystals are very small. One possibility to eliminate the undesired adsorption of the probe molecule(s) on the external surface is to add another component (e.g., as a solvent in liquid-phase adsorption) the molecules of which are so bulky that they are completely hindered from entering the pores. These bulky molecules will often cover the external surface exclusively, especially if their concentration in the bulk fluid phase is high.

Another problem arises when experimental data obtained over a large range of pore sizes are compared. If the dimensions of the molecule and the pore are approximately equal repulsive forces will favor the adsorption, reaction or production of species with a small cross section. However, Santilli et al. [16] reported that with pore sizes between 0.6 and 0.7 nm, attractive forces between the zeolite walls and hexane isomers begin to stabilize the branched isomers relative to n-hexane, resulting in an increasing preference for adsorbing branched hexanes. This is one example of selective cation-adsorbate interaction. With further increasing pore size, wall effects decrease, and the ratio of C_6 species in the pores is determined by the boiling points of the hydrocarbons. Therefore, tests for probing the pore width should only be used for an appropriate pore-size range.

Wu et al. [17] pointed out a few more pitfalls: They found that the amount of cyclohexane sorbed in large crystals of zeolite ZSM-5 was much lower after 2 h of equilibration than that sorbed in smaller crystals of the same zeolite. This observation was explained by the increasing time required to reach adsorption equilibrium with increasing crystallite size [17]. Another fact that could hinder the adsorption is imperfections of the large crystals that are invisible with conventional characterization techniques. Therefore, when interpreting adsorption results, one should always take into account the crystallite size of the samples. Another factor that can severely affect the results are impurities in the adsorptives [17]. Even at low concentrations, impurities in an adsorptive can lead to erroneous results due to selective adsorption of the impurity, especially if its molecular dimensions are small.

4.2.7.3
Adsorption of Probe Molecules with Different Size

Probing the pore width of microporous materials by adsorptives of different molecular size has been a popular method since the beginning of zeolite science. Consequently, a vast amount of results are scattered in the scientific and patent literature. No comprehensive discussion of these results is aimed at in the present review. Rather, a few selected and instructive examples will be presented.

4.2.7.3.1 Characterization of Various Zeolites in Comparison
Adsorption of an appropriate set of components with different molecular dimensions is a widely accepted procedure for characterizing the pore sizes of zeolites. A very thorough study on several 8-, 10- and 12-membered-ring zeolites was made by Wu et al. [17], which can be looked upon as a basis for a large part of the later work. During static adsorption at room temperature, substantial amounts of

n-hexane were sorbed by all zeolite samples, since its critical dimensions are smaller than, or equal to, their crystallographic pore openings. The largest adsorptive, mesitylene (1,3,5-trimethylbenzene), was sorbed only in the pores of the 12-membered-ring zeolites. The authors concluded that the ability to sorb benzene and cyclohexane classifies zeolite ZSM-23 (MTT) (in parentheses, the structural code of the International Zeolite Association is given [4]) and ZSM-48 as medium-pore zeolites. However, having only one-dimensional channels, zeolites ZSM-23 and ZSM-48 adsorbed less benzene or cyclohexane than zeolites ZSM-5 (MFI) and ZSM-11 (MEL), which have intersecting 10-membered-ring channels. The lower adsorption capacity of the 12-membered-ring zeolite ZSM-12 (MTW) as compared to zeolite Y (FAU) with its very spacious pore system was proposed to be a consequence of its denser structure with one-dimensional channels (like the large-pore zeolite mordenite (MOR); the 8-membered-ring pore channels of mordenite do not adsorb the hydrocarbons used). The lesser amount of mesitylene sorbed by ZSM-12 as compared to mordenite was said to be consistent with its smaller crystallographic pore opening, impeding the diffusion of the large mesitylene molecules into its intracrystalline voids [17].

Wu et al. [17] also applied a dynamic adsorption method using a thermogravimetric analyzer. These experiments were carried out at 373 K; equilibrium was judged to have been achieved when the weight gain was less than 17 ng s^{-1}. ZSM-23 and ZSM-48 exhibited a comparable capacity ratio of 3-methylpentane to n-hexane as ZSM-5 and ZSM-11, but a considerably lower ratio of uptake rate for these two hydrocarbons, reflecting again the unidimensionality of the channels in ZSM-23 and ZSM-48. However, the capacity of cyclohexane sorbed by ZSM-48 was larger and the rate of uptake faster than on ZSM-23, consistent with the greater eccentricity of the pore openings and the channels of ZSM-23.

In a further evaluation of their experimental data, Wu et al. [17] determined effective pore sizes of the zeolites by using the smaller cycloalkanes as a measure of the minor axis and the larger aromatics as a measure of the major axis of the effective pore opening. Just one example is given here for illustration: Since the uptake rate of cyclohexane (about 0.5 × 0.6 nm size) in ZSM-23 was very low, the minor axis of the ZSM-23 channels was concluded to be about 0.45 nm, approaching that of the smallest dimension of cyclohexane. Its ability to sorb *p*-xylene (about 0.4 × 0.6 nm size) but not readily *o*-xylene (about 0.4 × 0.7 nm size) indicates that its major axis appears to be only 0.65 nm. Table 1 gives an overview of which molecules can enter the pores of important zeolites. When using this table, the reader should always be aware that the adsorption behavior is dependent on the experimental conditions and on changes in the pore size, which can be caused by ion exchange and so forth (see Sect. 4.2.7.2.1).

Guil et al. [20] chose similar probe molecules, that is, n-hexane, toluene, *m*-xylene and mesitylene, for probing the pore size of the new zeolite ITQ-4 (IFR) in comparison to several medium- and large-pore zeolites. These authors defined so-called packing densities as the amount of adsorbate that fills up the micropores during

References see page 1053

Tab. 1. Molecules (a) adsorbed in; (ae) adsorbed very slowly and in small amounts; or (e) excluded from the pores of various zeolites [16–22].

Zeolite	n-Hexane (0.39 × 0.43 × 0.91)[1]	3-Methylpentane (0.46 × 0.58 × 0.86)	Benzene (0.34 × 0.62 × 0.69)	p-Xylene (0.37 × 0.62 × 0.86)	Cyclohexane (0.47 × 0.62 × 0.69)	2,2-Dimethylbutane (0.59 × 0.62 × 0.67)	o-Xylene (0.41 × 0.69 × 0.75)	Mesitylene (0.37 × 0.78 × 0.85)
Ca,Na-A (LTA)[2]	a	e	e	e	e	e	e	e
H-ZSM-23 (MTT)	a	–[3]	a	a	ae	e	ae	e
ZSM-35 (FER)	a	–[3]	–[3]	–[3]	a	–[3]	–[3]	e
H-ZSM-48[4]	a	–[3]	a	–[3]	a	–[3]	–[3]	e
ZSM-22 (TON)	a	–[3]	–[3]	–[3]	a	–[3]	–[3]	–[3]
H-ZSM-11 (MEL)	a	a	a	a	a	a	a	e
H-ZSM-5 (MFI)	a	a	a	a	a	a	a	e
EU-1 (EUO)	a	–[3]	–[3]	–[3]	a	–[3]	–[3]	–[3]
Na-MCM-22 (MWW)	a	–[3]	a	–[3]	a	–[3]	a	a
H-ZSM-12 (MTW)	a	a	a	–[3]	a	a	–[3]	a/e[5]
H-Mordenite (MOR)	a	a	a	–[3]	a	a	–[3]	a
CIT-1 (CON)	a	–[3]	–[3]	–[3]	–[3]	–[3]	–[3]	a
L (LTL)	a	a	–[3]	–[3]	a	a	–[3]	a[6]
SSZ-24 (AFI)	a	–[3]	–[3]	–[3]	a	a	–[3]	a/e[6,5]
H-Beta (BEA)	a	a	–[3]	–[3]	a	a	–[3]	a
EMT (EMT)	a	–[3]	–[3]	–[3]	a	–[3]	–[3]	–[3]
H-Y (FAU)	a	–[3]	a	–[3]	a	a	–[3]	a

[1] Molecular dimensions in nm estimated from Courtald space-filling models; the two smaller parameters define the critical dimensions [17]. [2] In parentheses, the structural code of the International Zeolite Association is given [4]. [3] This adsorption experiment was not carried out in Refs [16–22]. [4] No three-letter code has been assigned to zeolite ZSM-48 by the IZA Structure Commission yet. [5] The virgule indicates that different results were reported by different authors. [6] In this experiment, 1,3,5-triisopropylbenzene (kinetic diameter ≈ 0.85 nm [21]) was adsorbed instead of mesitylene.

single-component adsorption per micropore volume (as determined by nitrogen sorption). The micropore volume of zeolite ITQ-4 was found to be nearly as large as that of zeolite Beta (BEA), suggesting the existence of wide channels and large cavities. However, its n-hexane packing density, which is very sensitive to the presence of cavities, was between those of zeolites SSZ-24 (AFI) and ZSM-12 (MTW), which both possess unidirectional 12-membered-ring channels. The packing density of *m*-xylene in ITQ-4 was 4.55 mmol g^{-1}. From this, the authors concluded that ITQ-4 possesses 12-membered-ring channels, since in 10-membered-ring channels, into which *m*-xylene can penetrate, the packing density was much lower. Mesitylene did not have access to the micropores of ITQ-4 [20].

Santilli et al. [16] studied the competitive dynamic adsorption of n-hexane, 3-methylpentane and 2,2-dimethylbutane at 130 °C on several small-, medium-, large- and extra-large-pore zeolites. Except for zeolites erionite (ERI) and ZSM-23 (MTT), the three hydrocarbons were sorbed on all zeolites used. Zeolite erionite sorbed only n-hexane, ZSM-23 only n-hexane and 3-methylpentane.

Lourenço et al. [23] recorded adsorption isotherms of cyclohexane on zeolites H-mordenite, H-ZSM-5, H-SAPO-40 (AFR) and H-Y ($n_{Si}/n_{Al} = 4.5$). They proposed that the lowest rate of cyclohexane adsorption and the low total amount adsorbed on H-mordenite were in line with the presence of the unidirectional channel system in this zeolite, in contrast to the high adsorption rate on H-ZSM-5, that has a 3D channel system facilitating diffusion of cyclohexane, despite its pore diameter being smaller than that of H-mordenite. On zeolite H-Y with its large cavities, cyclohexane was adsorbed rapidly and in large amounts. In H-SAPO-40, only the main 12-membered-ring channels were accessible to cyclohexane. The high sorption rate and capacity indicated that its structure involves a larger void volume and fewer constraints than that of H-mordenite.

Lobo et al. [21] used adsorption experiments to characterize the new extra-large-pore zeolite UTD-1 (DON): The adsorption capacities of n-hexane, 2,2-dimethyl-butane and cyclohexane were similar for zeolites SSZ-24 (AFI), L (LTL) and UTD-1, but considerably larger for zeolites Y and VPI-5 (VFI). This was explained by the former having 1D pore systems, the latter having large cavities or very large 18-membered-ring channels (VFI). The pore size of the different zeolites was assessed with 1,3,5-triisopropylbenzene: This large molecule (kinetic diameter ≈ 0.85 nm) had no access to the channels of SSZ-24 (diameter 0.73 nm), adsorbed very slowly in the pores of zeolite Y (diameter 0.74 nm), but had easy access to the pores of the 14-membered-ring zeolite UTD-1 and to VPI-5. (Strangely enough and un-commented by the authors, 1,3,5-triisopropylbenzene was adsorbed on zeolite L, in which the diameter of the 12-membered-ring channels is 0.71 nm.)

Yuen et al. [24] employed linear dialkylazodicarboxylate chromophores with different sizes of the alkyl groups as probes for pore-size characterization. The uptake of the azo compounds from an isooctane solution was followed in a facile way with a UV/Vis spectrophotometer. The method demonstrated no uptake of the chromophores for small-pore zeolites, somewhat hindered uptake for medium-pore zeo-

References see page 1053

lites, and variable uptake rates for large-pore zeolites, because the solvent could compete with the azo compound. For medium-pore zeolites, the uptake of the chromophores was inversely related to the crystallite size. The technique allows for a quick probing of the pore system of novel zeolite materials. In this study, the structure of zeolite SSZ-25 (MWW) was examined in comparison to the known structures of various zeolites, such as ZSM-5, ZSM-22, ZSM-23, Y, Beta and other SSZ zeolites. Water was found to compete with the azo compounds in the adsorption experiments, therefore, it should carefully be excluded from the system. As expected, the presence of coke reduced the uptake of the chromophores. Though a large amount of adsorption isotherms and other useful data were collected in this study, the authors themselves stressed that the analysis of the data is hardly foolproof and that additional data is needed, for example concerning crystallite size, zeolite composition and so on, before this method can possibly be used as a standard test.

A vast amount of adsorption data was collected by Breck [25]: Adsorption capacities of most of the common gases and hydrocarbon vapors on various zeolites and zeolite minerals are listed in this book. Where a marked effect of the cation on the sieving characteristic is evident (see Sect. 4.2.7.2.1), adsorption capacities for different cation-exchanged forms of the zeolites are given. For example, the potassium form of zeolite A ("3A") has a smaller, the calcium form ("5A") a larger pore size than the sodium form of zeolite A ("4A"). It is concluded that on zeolite Na-A, for example, molecules with a kinetic diameter larger than 0.36 nm are not adsorbed (see Table 2). Breck's book continues to be very useful as a quick reference on whether a molecule can be adsorbed on a specific zeolite or not, though it does not contain information on the more recent zeolites.

For the characterization of small-, medium- and large-pore zeolites by adsorption, Otake [19] proposed the so-called R_{CN} index, defined as the ratio of the adsorption capacity of cyclohexane to that of n-hexane, as a simple and versatile tool. Whereas catalytic test reactions (see Sect. 4.2.7.4) are only applicable to medium-pore *or* large-pore zeolites, but not to both at the same time, the R_{CN} index was claimed to be linearly dependent on the average diameter of the largest pore over a wide range of pore sizes of zeolites with eight- to twelve-membered rings. A further advantage of this index is that it eliminates the dependence of the amount

Tab. 2. Adsorption properties of various zeolites [25].

Zeolite	Does not adsorb molecules with a kinetic diameter
Na-A (LTA)	>0.36 nm at 77 K, >0.40 nm at 300 K
Ca-A (LTA)	>0.43 nm at 300 K, >0.44 nm at 420 K
Erionite (ERI) (mineral)	>0.43 nm at 400 K
Chabazite (CHA) (mineral)	>0.43 nm at 400 K
Mordenite (MOR) (mineral)	>0.48 nm at 300 K
Ca-X (FAU)	>0.78 nm at 300 K
Na-X (FAU)	>0.80 nm at 298 K
Na,K-L (LTL)	>0.81 nm at 323 K

sorbed on the pore volume and that it is expected to reflect only the size and shape of the zeolite pore. It was demonstrated that the R_{CN} index could also detect subtle changes in the pore size due to ion exchange: The R_{CN} value decreased with increasing cation radius of cation-exchanged Beta zeolites. However, Na-Beta and Ca-Beta had different R_{CN} values, even though the cation radii of Na and Ca are the same. Similar inconsistencies appeared when comparing different zeolites. In summary, though the proposed index is an interesting tool for characterizing zeolite pore sizes, it is in an early stage of development. More data and a refinement are needed, before one can judge its general applicability and usefulness.

4.2.7.3.2 Various Methods for Pore-size Characterization by Adsorption

Besides the relatively simple methods for pore-size characterization by adsorption that were presented in Sect. 4.2.7.3.1, a wealth of other methods have been proposed. Some of them will be described in this section.

First, methods similar to the R_{CN} index (see Sect. 4.2.7.3.1, last paragraph) were introduced, the difference being that the time dependence of the adsorption process is taken into consideration: In a study on ZSM-5 and ZSM-11 zeolites, Harrison et al. [26] defined the so-called pore constraint index as the ratio of the amount of n-hexane adsorbed after 15 min to the amount of 2,3-dimethylbutane (or cyclohexane) adsorbed after 15 min. This index is based on the fact that, under the experimental conditions chosen, molecules with a kinetic diameter less than the dimensions of the zeolite pore openings, such as n-hexane, were rapidly adsorbed and approached their sorption capacity within 15 min. In contrast, molecules such as 2,3-dimethylbutane and cyclohexane, the diameters of which are close to the channel dimensions, were only slowly adsorbed. In such cases, the amounts sorbed after 15 min are not true sorption capacities, but represent relative rates of adsorption. Such data can, however, still yield valuable information on the zeolite pore structure: A plot of sorbate uptake on ZSM-5 after 15 min versus the kinetic diameter of the sorbate molecule revealed a sharp cut-off in sorbate uptake at a molecular diameter of approximately 0.58 to 0.60 nm, which agrees well with the pore dimensions of ZSM-5. By contrast, on zeolite ZSM-11, even 2,3-dimethylbutane and cyclohexane approached their adsorption capacities within 15 min. Thus, the decreasing value of the pore constraint index on going from silicalite-1 over ZSM-5 and ZSM-5/ZSM-11 intermediates to ZSM-11 was accounted for in terms of decreasing channel tortuosity allowing for a more rapid rate of diffusion of larger molecules into the zeolite pores. Therefore, this test facilitates an estimation of pore sizes within a small range, which could also be useful for the assessment of pore size changes by CVD. In addition, the method reveals an important advantage of adsorption tests over catalytic test reactions of being independent of the n_{Si}/n_{Al} ratio of the zeolite, that is, the pure-silica analog silicalite-1 can be tested as well.

A similar adsorption test was proposed in the patent literature [27]: Zeolites selective for the production of p-dialkylbenzenes were characterized by determining relative rates of adsorption of p- and o-xylene. Highly selective zeolites, such as

References see page 1053

ZSM-5, typically had an adsorption capacity of $m_{p\text{-xylene}}/m_{zeolite} \geq 0.01$. The equilibrium adsorption capacity was determined gravimetrically at 120 °C; *p*-xylene was used preferentially since it is the xylene isomer that reaches equilibrium within the shortest time, but *m*- or *o*-xylene or isomer mixtures can also be used. The time for *o*-xylene sorption for 30 % of the *p*-xylene capacity was longer than 10 min with selective zeolites. This sorption time, the so-called $t_{0.3}$, could be extended considerably by precoking of the catalyst. Hence, this adsorption test is useful for characterizing pore-size reduction by coking and modification with oxides of antimony, phosphorus, boron or magnesium or CVD. $t_{0.3}$ was found to be a direct measure of the diffusivity [28] and of the pore tortuosity [29]. A similar parameter, which was also claimed to be a measure of the pore tortuosity, is the relative *o*-xylene adsorption velocity, $V_{ROA} \equiv (m_{o\text{-xylene, adsorbed at 180 min}})/(m_{p\text{-xylene, adsorbed at infinite time}})$ [29]. Baeck et al. [30] showed that the rate constant for n-hexane sorption on Mg-ZSM-22 decreased strongly with increasing deposition time of silicon alkoxide. Therefore, the rate constant is also a measure for the contraction of the pore opening. However, the adsorption capacities of n-hexane were unchanged by CVD, suggesting that only the pore opening at the external surface of the crystallites was reduced, whereas the channel diameters in the interior of the crystals were not.

Santilli [31] developed the so-called pore-probe technique that allows for the measurement of the absolute steady-state concentrations of molecules within the pores of zeolites even at temperatures near to or at typical reaction conditions, that is, the adsorption capacity during the reaction can be determined. To achieve this, the catalyst/adsorbent is exposed to the feed flow until the product stream reaches steady state. One then switches to a nitrogen flow, and the catalyst is cooled quickly in order to avoid further reaction. The material desorbed is collected and analyzed by gas chromatography using an external standard. Blank runs using adsorbents without pores are employed to measure the steady-state concentrations of all feed components in the dead space and on the external surface. By comparing the amount of C_6 hydrocarbons in the pores of H-ZSM-5 and Na-ZSM-5, the influence of an acid-catalyzed reaction on the adsorption capacity could be determined [31]. In addition, the change of the adsorption capacity with temperature can be calculated.

Choudhary et al. [32] studied zeolite adsorption properties by injecting different compounds into a gas chromatographic column packed with zeolite particles at temperatures close to those employed in catalytic processes. The flowing carrier gas eluted the adsorbate, and the peak was recorded. The retention time data were corrected for the voids present in the zeolite column and connecting tubes. A sharp decrease in the retention time (or sorption) of the sorbate (during the passage of the sorbate pulse) on the zeolite was observed with an increase in the steric hindrance due to chain branching. The method appears to be useful for determining whether a molecule can enter the channels or not. Thielmann et al. [33] used a similar technique that combines thermal desorption with inverse gas chromatography. A peak during thermodesorption indicates that the respective molecule had access to the pore system of the porous material. In addition, this method seems to permit a discrimination between adsorption in micropores, mesopores and on the external surface [33]. Anunziata and Pierella [34] also performed desorption experiments, but they used a conventional TPD (temperature-programmed desorption) equipment with a flame ionization detector. During desorption of *m*-xylene or mesitylene from zeolite ZSM-11, they observed one peak at low temperature, mostly due to the desorption of *m*-xylene and mesitylene adsorbed on the external surface, and one peak at higher temperature, mostly due to molecules that partially penetrated the pores of the zeolite. Naphthalene did not have access to the channels and was desorbed from the external surface in one peak at relatively low temperatures.

Denayer et al. [35] considered the fact that most industrial zeolite processes are performed at elevated hydrocarbon pressures or in the liquid phase, but adsorption of organic molecules is commonly studied at low partial pressures in the gas phase. In the liquid phase, the adsorbent is saturated with adsorbate molecules, which approaches the industrial conditions in a much more realistic way than the zero coverage limit that is usually studied in the gas phase. In the liquid phase, nonidealities, such as adsorbate–adsorbate interactions and surface heterogeneity, become more important. Therefore, the situation is complex, and liquid-phase adsorption properties cannot be extrapolated easily from gas-phase adsorption isotherms. However, data on competitive adsorption of organic molecules with different polarity and carbon number at temperatures relevant to catalysis and elevated pressures are lacking in the literature. Denayer et al. [35] used a classical HPLC (high-pressure liquid chromatography) set-up with pellets or crystals of the adsorbent in the HPLC column and a refractometer as detector, which allowed to work at temperatures exceeding the boiling temperatures of the fluids employed as mobile phases. The authors found that the competitive adsorption between n-alkanes depends strongly on the pore size of the zeolite. In the large-pore zeolite Y, adsorption was governed by chain segment interactions rather than interaction with complete molecules. Thus, whereas the adsorption constants depend strongly on the carbon number of the n-alkanes *in the gas phase*, short-chain and long-chain n-alkanes adsorb in a nonselective way *in the liquid phase*, as they are built of the same carbon chain segments. Selective adsorption occurred only in the presence of aromatics or polar molecules. The molecular weight of the molecules had only a minor influence. However, when the amount of adsorbed molecules decreased, long-chain n-alkanes started being adsorbed preferentially over short-chain n-alkanes as they then had more free space to align themselves with the zeolite surface. In the smaller-pore system of ZSM-5, n-alkanes were stretched along the pore axis and had a stronger interaction with the zeolite. Sorbate–sorbate interactions were reduced, and a stronger competitive adsorption behavior between short-chain and long-chain n-alkanes was observed. However, short-chain n-alkanes (up to n-heptane) had access to the two channel types of ZSM-5, whereas longer n-alkanes had only restricted access to the pore system. In addition, it was observed that the polarity of the zeolite determines to what extent long-chain n-alkanes are adsorbed preferentially over shorter n-alkanes. The work of Denayer et al. [35] shows, in an impressive way, the lack of realistic data on adsorption properties of porous mate-

References see page 1053

rials, and is an appeal to perform competitive adsorption experiments in the liquid phase and/or under elevated pressure.

An unusual method of pore-size characterization was applied by Yoon et al. [36]: A series of brightly colored charge-transfer complexes was assembled *in situ* by intercalation of various aromatic donors with different pyridinium acceptors in a number of large-pore zeolites. Upon the deliberate introduction of water vapor into these variously colored zeolites, the diffuse reflectance spectra underwent pronounced spectral shifts of the charge-transfer bands, the magnitude of the shifts being uniquely dependent on the molecular size/shape of the donors and acceptors as well as the dimensions of the zeolite cavities. For example, the bathochromic shift of the charge-transfer band of complexes of methylviologen with 1,4-dimethoxybenzene decreased rapidly with increasing pore size from zeolite mordenite (MOR) over zeolite L (LTL) to zeolite Omega (MAZ). In zeolite Y (FAU) with its spacious supercages, no shift at all was observed. Thus, intermolecular charge-transfer complexes provide a method for probing the pore width of large-pore zeolites, which is especially useful because the other techniques proposed up to now mostly employ molecular probes more appropriate for the characterization of medium-pore zeolites. Another useful spectroscopic method is FT-IR spectroscopy. With this technique, Trombetta et al. [37] showed, for example, that o-xylene and pivalonitrile do not reach the internal zeolitic OH groups of zeolite ZSM-5. However, FT-IR spectroscopy is more appropriate for the characterization of acidic and basic sites of zeolites than of their pore size and will not be dealt with further here.

Another widely applied method for pore-size characterization is microcalorimetry: Derouane et al. [38] found that the heats of adsorption of methane increase considerably with decreasing pore size of the zeolitic adsorbents employed. These results were extended by many researchers, but are not within the scope of this section.

An approach including much modeling work was made by Webster et al. [39]: They used smaller probe molecules (O_2, N_2, CO, Ar, CH_4) to characterize the eight-membered-ring channels and larger probe molecules or atoms (C_2H_6, C_3H_8, n-C_4H_{10}, i-C_4H_{10}, Xe, SF_6) for the twelve-membered-ring channels of zeolite mordenite. All these molecular probes had access to the channel system, but by using such a broad variety, the channels could be well assessed by gradually increasing the size of the molecules. The authors defined three different interaction types between the probe molecules and the pores: (1) strong interaction with two opposing walls, (2) strong interaction with one wall and weak interaction with the opposite wall and (3) interaction with one wall only. Since the larger probes were excluded from the small channels, they could provide information on the large channel and the aperture of the small channels only, yielding no information on the small channels directly. By contrast, the smaller molecules could access regions that the larger molecules could not. The data sets were analyzed by the new multiple equilibrium analysis (MEA) method, providing adsorption capacities, accessible surface areas, pore volumes as well as enthalpies and entropies of adsorption. The big advantage of this technique is its accuracy and the possibility to determine *accessible* surface areas (e.g., 265 m^2 g^{-1} for Na-MOR and 346 m^2 g^{-1} for H-MOR), whereas with the BET model values of 458 and 462 m^2 g^{-1} were found for the Na$^+$ form and the H$^+$ form, respectively.

Much modeling work was done regarding adsorption and diffusion of molecular probes in crystalline microporous solids. In combination with experimental data on known structures, computer modeling can provide information on unknown structures and help to predict experiments. Deka and Vetrivel [40] demonstrated, for instance, the efficiency of the force field energy minimization technique for the study of adsorption of large molecules inside the micropores of zeolites. These techniques are, however, beyond the scope of this section.

4.2.7.3.3 Molecular Probes for Zeolites with Different Pore Sizes

Small-pore zeolites For the characterization of small-pore zeolites, most of the molecular probes discussed so far cannot be employed because they are too large to enter the pores. Kerr [41] used n-hexane, water and cyclohexane for characterizing zeolite ZK-5 (KFI). Water and n-hexane were adsorbed in large amounts, whereas only a small amount of cyclohexane was adsorbed. Eder and Lercher [42] reported that propane, n-butane, n-pentane and n-hexane had access to the pores of ZK-5 but no isoalkanes.

den Exter et al. [43] studied the adsorption of several small molecules on aluminum-free deca-dodecasil 3R (DDR) and also observed that linear alkanes and alkenes could enter the pores, whereas isobutane was only adsorbed in mesopores or on the external surface. Stewart et al. [44] reported, however, that even n-butane was excluded (at 273 K) from the pores of zeolite Sigma-1 (DDR), the aluminosilicate analog of deca-dodecasil 3R. This could be due to pore-size narrowing by the cations balancing the negative framework charges.

No doubt the most thoroughly characterized small-pore zeolite is zeolite A (LTA) [8, 13, 25, 45, 46]: This zeolite can be used for the separation of, for instance, n-butanol from iso-butanol [45], n-butane from isobutane [13, 45] and oxygen from nitrogen [25, 45, 46]. With nitrogen ($\sigma = 0.364$ nm) and oxygen ($\sigma = 0.346$ nm), subtle changes of the pore size of zeolite A were probed: On zeolite K-A with its small pore size ("3A") neither oxygen nor nitrogen were adsorbed at -196 °C, whereas on Ca-A with its considerably larger pore size ("5A") both molecules were adsorbed at this low temperature [45]. The pore size of Na-A ("4A") lies in between the pore sizes of Ca-A and K-A, and on this zeolite the sieving characteristics changed with temperature: At -196 °C, oxygen but no substantial amount of nitrogen was adsorbed. At -75 °C or even higher temperatures, nitrogen was adsorbed in larger quantities than oxygen [45]. This preferential adsorption is, however, not due to the molecular size, because both molecules have access to the channel system at this temperature, but due to chemical properties of the zeolite and the probe molecules. If the pore size of zeolite Na-A was slightly decreased by CVD, nitrogen was no longer adsorbed [46].

References see page 1053

Medium-pore zeolites As an excellent catalyst for the skeletal isomerization of butenes, zeolite H-ferrierite (FER) has attracted significant interest. Adsorption was used to characterize its pore system containing two perpendicularly intersecting channel systems, one consisting of ten-membered rings, the other of eight-membered rings. van Well et al. [47] found that the pore volumes occupied by n-heptane and n-hexane were significantly lower than the ones occupied by n-pentane, n-butane and propane. From this, the authors supposed that the complete pore structure of FER is not accessible to n-heptane and n-hexane. By comparing ^{13}C NMR spectra of molecules adsorbed on H-ferrierite and Na-ZSM-22 (TON), a zeolite with 1D ten-membered-ring channels, they concluded that n-hexane adsorbs exclusively in the ten-membered-ring channels of H-ferrierite. Similar conclusions were drawn by Eder and Lercher [42]: As detected by IR spectroscopy, only about 90 % of the acid sites of zeolite H-FER were accessible to n-hexane. Isobutane and isopentane were adsorbed only in small amounts and slowly. The authors pointed out that the additional interaction between the acid site and the sorbed hydrocarbon can lead to synergistic and antagonistic effects depending on whether or not it re-directs the probe molecules to positions different from those preferred in the absence of acid sites, which should always be kept in mind.

The largest amount of work in this field was devoted to ZSM-5 (MFI), the most important medium-pore zeolite. ZSM-5 is the cheapest medium-pore zeolite, and, therefore, it is important to know for possible applications of this zeolite whether certain molecules can access its pore system or not. Table 1 gave a first account of this subject. Table 3 is designed to extend the picture for zeolite ZSM-5. When using this table, the reader should always be aware of the fact that it depends, *inter alia*, on the experimental conditions whether molecules have access to the pores or not.

Multicomponent adsorption experiments on zeolite ZSM-5 in the gas phase were performed by Klemm and Emig [58]: During adsorption of the three xylene isomers on the zeolite in a recycle adsorber, an excess of *m*-xylene and a lack of *p*-xylene were observed at the adsorber outlet in the initial stage of the experiment. These effects disappeared when approaching steady state and were explained by a higher diffusion coefficient of *p*-xylene. Namba et al. [53] performed competitive adsorption experiments of xylene isomers in a batch adsorber in the liquid phase using, as solvent, 1,3,5-triisopropylbenzene the molecular dimension of which was too large to enter the pores of the zeolite. This technique may eliminate the effect of the external surface, because the solvent, the concentration of which is much higher than that of the adsorptives, covers the external zeolite surface. At 283 K, only *p*-xylene was adsorbed on zeolite H-ZSM-5, whereas at higher temperatures all three isomers were adsorbed, even though the *p*-xylene adsorption capacities were much higher than those of *o*- and *m*-xylene. Dessau [48] obtained similar results with xylenes at the same experimental conditions. During competitive adsorption of n-alkanes and aromatics on ZSM-5, he observed a marked preference for the n-alkanes, in distinct contrast to adsorption studies on faujasites. In counterdiffusion studies, in which *p*-xylene was adsorbed initially, it was rapidly displaced upon addition of n-nonane, because this hydrocarbon has a higher molecu-

Tab. 3. Adsorption properties of zeolite MFI [15, 17, 48–60].

Compound	Adsorption in MFI zeolites	Ref.
n-Hexane	Yes	17, 50, 51, 54, 56, 59
3-Methylpentane	Yes	17, 50, 51, 59
Benzene	Yes	15, 17, 48, 50, 54, 55
Toluene	Yes	15, 51, 55
p-Xylene	Yes	15, 17, 48, 49, 51, 53–58
Cyclohexane	Yes	17, 50
Ethylbenzene	Yes	52, 54
1-Methylbutylbenzene	Yes	48
Methylnonanes	Yes	49
1,4-Diisopropylbenzene	Yes	53
2,3-Dimethylbutane	Yes	51
1,2,4-Trimethylbenzene	Yes	17, 56
Cresol isomers	Yes	53
2-Methylnaphthalene	Yes	60
m-Xylene	Yes[1]	49, 50, 52–54, 56
Neopentane	Yes[2]/no[3]	48, 49
2,2-Dimethylbutane	Yes[4]/no[5]	17, 50, 59
o-Xylene	Yes[2,4]/no[3,6]	17, 48, 49, 54, 56, 57
1,3-Diisopropylbenzene	No	53
1-Methylnaphthalene	No	60
cis-Decalin	No	48
trans-Decalin	No	48
Tetramethylsilane	No	48
Cyclooctane	No	48
1-Ethylpropylbenzene	No	48
Mesitylene	No	15, 17, 48, 51

[1] Not adsorbed during competitive adsorption of all xylene isomers at 283 K in the liquid phase [53]. [2] Adsorption at room temperature in the liquid phase [48]. [3] Adsorption at 333 K in the gas phase [49]. [4] Adsorption at 373 K in the gas phase [17]. [5] Adsorption at room temperature [50] or at 323 K [59] in the gas phase. [6] Adsorption at 303 K in the liquid phase [57].

lar weight. The author argued that this behavior of ZSM-5 might be due to the fact that, unlike A and Y zeolites, it contains no large cavities in which n-alkanes can coil around themselves. The coiling of high molecular weight alkanes inside A and Y zeolites should result in an additional entropy loss upon sorption, thereby reversing the normal order of preferential adsorption of the higher molecular weight component.

Weitkamp et al. [59, 60] performed competitive adsorption experiments in a flow-type fixed-bed adsorber in the gas phase. During adsorption of an n-hexane/3-methylpentane/2,2-dimethylbutane mixture over Na-ZSM-5, 2,2-dimethylbutane broke through immediately (see Fig. 3), that is, this hexane isomer could not enter

References see page 1053

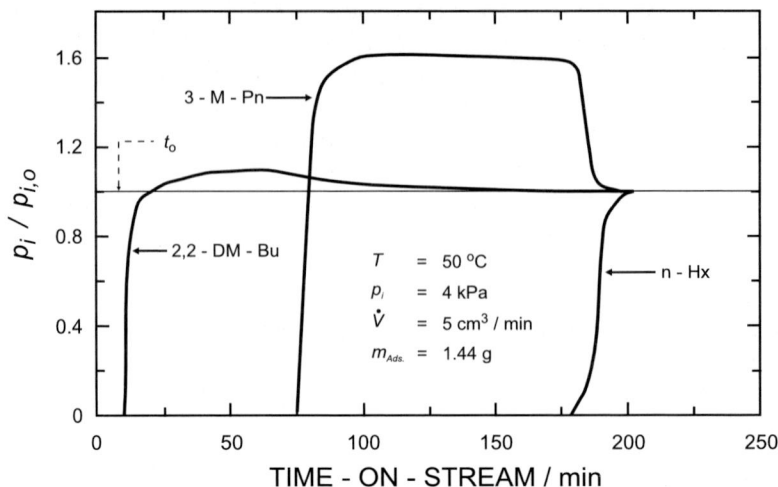

Fig. 3. Breakthrough curves for the adsorption of an n-hexane/
3-methylpentane/2,2-dimethylbutane mixture over zeolite
Na-ZSM-5 (n-Hx = n-hexane, 3-M-Pn = 3-methylpentane,
2,2-DM-Bu = 2,2-dimethylbutane).

the zeolite pores. 3-Methylpentane and n-hexane had access to the pore system, but n-hexane was adsorbed preferentially. Therefore, 3-methylpentane reached partial pressures p_i at the adsorber outlet that were higher than the partial pressure $p_{i,0}$ at the adsorber inlet, because it was displaced from the zeolite pores by n-hexane [59]. During adsorption of a 2-methylnaphthalene/1-methylnaphthalene mixture on zeolite H-ZSM-5, the larger 1-methylnaphthalene was not adsorbed at 100 °C, whereas 2-methylnaphthalene could enter the zeolite pores [60]. On zeolite H-ZSM-12, both isomers were adsorbed, the larger isomer breaking through first. On zeolite Na-Y, the interaction between the zeolite and 1-methylnaphthalene was larger than the one with 2-methylnaphthalene. Therefore, the smaller 2-methylnaphthalene was displaced by 1-methylnaphthalene.

Another medium-pore zeolite that was characterized by adsorption is zeolite MCM-22 (MWW) [18, 22] and its pure-silica analog ITQ-1 (MWW) [61]: Corma et al. [22] determined adsorption isotherms of different probe molecules in the gas phase. The uptake of *m*-xylene was about half the value for toluene. The value for *o*-xylene was much lower and approximately equal to that of 1,2,4-trimethylbenzene. These results prompted the authors to suggest the existence of micropores and cavities of two different sizes, i.e. narrower micropores, to which only toluene has access, and wider micropores or cavities that are penetrated by toluene as well as *m*-xylene. The much lower uptake of *o*-xylene and 1,2,4-trimethylbenzene was ascribed to adsorption at the entrances of the micropores and cavities. In another series of experiments, toluene adsorption isotherms were recorded after preadsorption of other adsorptives and an intermediate outgassing. The toluene uptake was lower when preadsorption had been performed, the effect being less important for preadsorbed molecules of larger size [22]. Ravishankar et al. [18] found that the adsorption capacities of n-hexane, cyclohexane, *m*-xylene and mesitylene on zeolite MCM-22 were virtually independent of the aluminum content. MCM-22 adsorbed moderate amounts of mesitylene, even though this hydrocarbon should not be able to enter the ten-membered-ring pore openings. Therefore, the authors suggested that some of the large twelve-membered-ring cages should be accessible from the external surface, either through defect centers or through the presence of some twelve-membered-ring cages at the external surface [18].

Large-pore and extra-large-pore zeolites For large-pore zeolites, suitable choices of probe molecules for pore-size characterization are tertiary alkylamines with alkyl groups of varying bulkiness or with perfluorinated alkyl groups [62]. Adsorption of the critically sized molecule perfluorotriethylamine, $(C_2F_5)_3N$, showed that the effective aperture of the dehydrated mineral faujasite is substantially smaller than that of synthetic zeolite Y, since its adsorption capacity was much smaller. Perfluorotriethylamine was not adsorbed at all on Ca-X and Ca-Y zeolites. Li-X, Na-X and Cs-X adsorbed $(C_2F_5)_2NC_3F_7$ ($\sigma = 0.77$ nm), $(C_3H_7)_3N$ ($\sigma = 0.81$ nm) and $(C_4H_9)_3N$ ($\sigma = 0.81$ nm), but not $(C_4F_9)_3N$ ($\sigma = 1.02$ nm), whereas Ca-X and Ba-X only adsorbed $(C_2F_5)_2NC_3F_7$. From this, the authors concluded that the effective pore diameter of Na-X is about 0.9 to 1.0 nm and that of Ba-X or Ca-X 0.8 to 0.9 nm [62].

Davis et al. [63] distinguished extra-large-pore zeolites from large-pore zeolites by adsorption experiments with 1,3,5-triisopropylbenzene; this molecule was too bulky to enter the pore system of faujasite with reasonable uptake rates, but did have easy access to the channels of VPI-5.

4.2.7.4
Catalytic Test Reactions

Catalytic test reactions for probing the pore width of porous materials have much in common with adsorption tests with the same objective. Adsorption, of course, does occur during these catalytic tests as well, but, in addition, a chemical reaction takes place in a shape-selective manner. It is, therefore, crucial for a design and understanding of test reactions for probing the pore dimensions, to address the fundamentals of shape-selective catalysis.

4.2.7.4.1 Shape-selective Catalysis in Microporous Materials
Different definitions have been in use for shape-selective catalysis. Of these, the most straightforward and useful one is as follows [1]: Shape-selective catalysis encompasses all effects in which the selectivity of a reaction depends, in an unambiguous manner, on the pore width or pore architecture of the porous solid. A comprehensive review on the fundamentals of shape-selective catalysis has been

References see page 1053

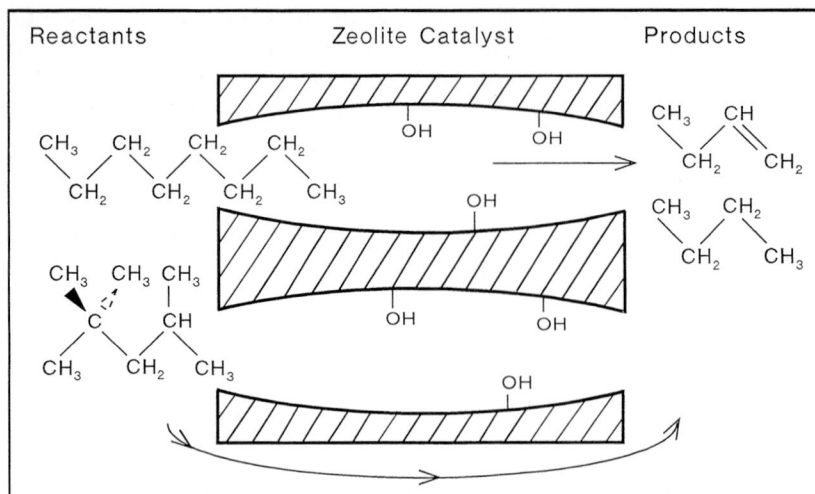

Fig. 4. Selective cracking of n-octane in the presence of 2,2,4-trimethylpentane as an example for reactant shape selectivity.

published recently [64]. Three types of shape selectivity are usually discerned, namely reactant, product and restricted transition state shape selectivity [65].

Both reactant and product shape selectivity have their origins in mass transfer effects, that is, the hindered diffusion of reactant and product molecules, respectively, in the pores of the zeolite catalyst. Therefore, we refer to those two shape-selectivity effects as mass transfer shape selectivity. As one example for the reactant type of mass transfer shape selectivity, the competitive cracking of n-octane and 2,2,4-trimethylpentane is depicted in Fig. 4. This situation can indeed be understood as molecular sieving combined with catalytic conversion: 2,2,4-trimethylpentane is too bulky to enter the pores of the zeolite and is, therefore, hindered from reaching the catalytically active sites in the pores of the zeolite. This molecule can only be converted at catalytic sites located on the external surface of the zeolite crystallites, or it can leave the reactor without being converted. By contrast, the slender n-octane does have access to the pores of the zeolite, where it is readily converted. The net effect, which can be detected at the reactor outlet, is the selective conversion of n-octane.

As an example of the product type of mass transfer shape selectivity, the acid-catalyzed ethylation of toluene has been chosen in Fig. 5. Product shape selectivity is the reverse of reactant shape selectivity: Both reactants are small enough to enter the zeolite pores, but of the potential products (*o*-, *m*- and *p*-ethyltoluene), only the slim *p*-ethyltoluene is small enough to leave the pore system. The two bulkier eth-yltoluene isomers, even though they may form in relatively spacious intracrystalline cages or at channel intersections, are unable to escape from the pores and do

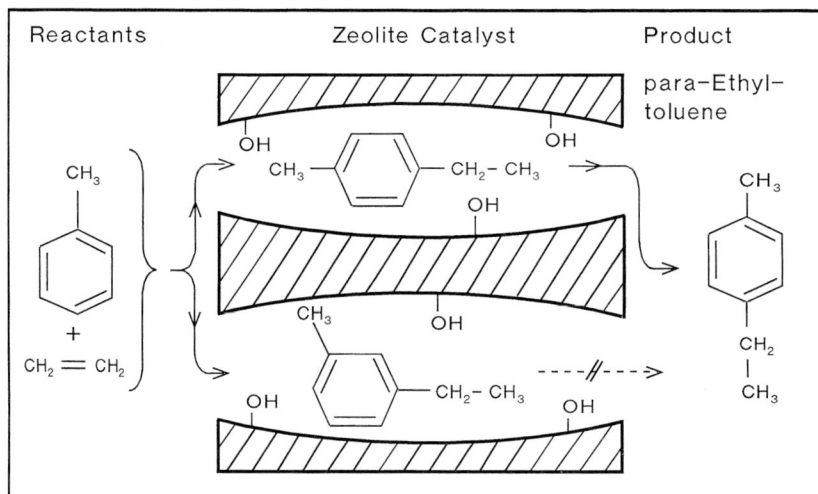

Fig. 5. Selective formation of *p*-ethyltoluene in the alkylation of toluene with ethylene as an example for product shape selectivity.

not occur in the reactor effluent. Ultimately, these entrapped products may be catalytically transformed into smaller molecules (e.g., isomerized into *p*-ethyltoluene), which are able to leave the pores, or into coke, which is deposited inside the pores. Webster et al. [8] distinguished two types of product shape selectivity: Preferential-diffusion product shape selectivity appears, if two or more reaction products formed within the confines of the structure have effective diffusivities that differ sufficiently from each other to allow one product to preferentially diffuse out of the structure. Often, the smaller diffusivity of the larger isomer allows for isomerization to the diffusion-preferred (smaller) product. Size-exclusion product shape selectivity arises from an actual confinement of a reaction product within the structure of the solid.

A typical example for restricted transition state shape selectivity is depicted in Fig. 6. *m*-Xylene can undergo acid-catalyzed isomerization into *p*-xylene (and *o*-xylene, which is omitted from Fig. 6 for clarity), and transalkylation into toluene and one of the trimethylbenzene isomers. It is evident that transalkylation is a bimolecular reaction and, as such, it necessarily proceeds via bulkier transition states and intermediates than the monomolecular isomerization. In a zeolite with the appropriate pore width, there will be just enough space for the accommodation of the transition states and intermediates for the monomolecular reaction, but no room for the formation of the bulky transition states and intermediates of the bi-

References see page 1053

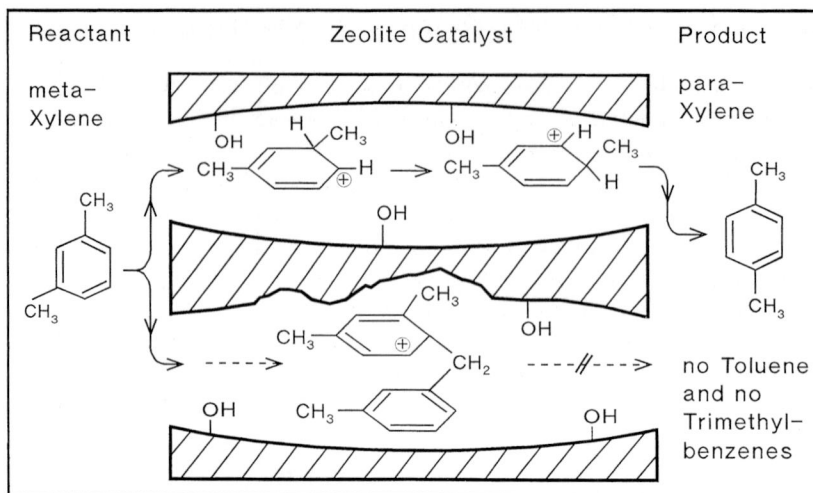

Fig. 6. Suppression of trimethylbenzene plus toluene formation during *m*-xylene isomerization as an example for restricted transition state shape selectivity.

molecular reaction, the net effect being a complete suppression of the latter reaction. As opposed to mass transfer shape selectivity, restricted transition state shape selectivity is due to intrinsic chemical effects, which emerge from the limited space around the intracrystalline active sites. As long as one assumes that there is no catalytic contribution from the external surface, the influence of the crystallite size is as follows: If mass transfer shape selectivity is operative, the length of the intracrystalline diffusion path and, hence, the measurable selectivity effects will decrease with decreasing crystallite size. If restricted transition state shape selectivity is operative, the measurable selectivity will be independent of the crystallite size. If, on the other hand, there is a significant contribution of the nonselective external surface, the influence of the crystallite size on the measurable catalytic selectivity will be more complex. The selectivity will decrease with decreasing crystallite size irrespective of whether the reaction is controlled by mass transfer or restricted transition state shape selectivity. There will only be a gradual difference that will be difficult to assess.

Several catalytic test reactions for the characterization of the effective pore widths of zeolites have been proposed so far; the results of the best-known test reactions were expressed in terms of the quantitative criteria constraint index (CI), refined or modified constraint index (CI^*) and spaciousness index (SI). Extended review articles on these indices and other test reactions have already been published [1, 66, 67]. This section will, in large part, directly refer to the paper by Weitkamp and Ernst [1]. In addition, recent advances in test reactions for probing the pore width will be discussed.

4.2.7.4.2 **Test Reactions for Monofunctional Acidic Molecular Sieves**

Competitive cracking of n-hexane and 3-methylpentane – the constraint index,
CI The method of characterizing the effective pore width of zeolites by catalytic
tests was first employed by researchers at Mobil Oil Corp. They introduced the
constraint index [68] which is based on the competitive cracking of an equimolar
mixture of n-hexane and 3-methylpentane on the monofunctional acidic form of
the zeolite. As long as the catalyst pores are sufficiently spacious, branched alkanes
are cracked at higher rates than their unbranched isomers. The opposite holds for
medium-pore zeolites, such as H-ZSM-5. Based on this shape-selectivity effect, the
constraint index was defined at Mobil as the ratio of first-order rate constants (k) of
the cracking of n-hexane and 3-methylpentane:

$$CI \equiv \frac{k_{\text{n-Hx}}}{k_{\text{3-M-Pn}}} \tag{1}$$

The constraint index has been used routinely in Mobil's patents for about two decades. In the scientific literature, precise experimental conditions for its determination were given (reaction temperature between 290 and 510 °C, liquid hourly
space velocity ($LHSV$) between 0.1 and 1 h^{-1}, 10 vol.-% of each reactant in helium
as carrier gas, mass of catalyst around 1 g, overall conversion 10 to 60 %, fixed-bed
reactor at atmospheric pressure) [68]. The reactor effluent is to be analyzed after
20 min time-on-stream. With the performance equation for integral fixed-bed reactors [69], Eq. (1) can be rewritten as

$$CI \equiv \frac{\log(1 - X_{\text{n-Hx}})}{\log(1 - X_{\text{3-M-Pn}})} \quad (X = \text{conversion}) \tag{2}$$

According to Frilette et al. [68], the constraint index allows for a classification of
zeolites into large-pore (twelve-membered-ring), medium-pore (ten-membered-
ring) and small-pore (eight-membered-ring) molecular sieves:

$CI < 1$:   large-pore materials;

$1 \leq CI \leq 12$:   medium-pore materials;

$12 < CI$:   small-pore materials.

Constraint indices taken from the literature are summarized in Table 4. By and
large, a correct classification of eight-, ten- and twelve-membered-ring zeolites can
be achieved. Exceptions are zeolites ZSM-12 (MTW) and MCM-22 (MWW). ZSM-
12 possesses twelve-membered-ring pores that are, however, strongly puckered
and, hence, narrowed, and on the basis of its constraint index it would be classified
as a material with ten-membered-ring pores. Zeolite MCM-22 disposes of large
intracrystalline cages that are only accessible via ten-membered-ring windows. In
this case, the constraint index of 1.5 can be looked upon as an averaged value

References see page 1053

Tab. 4. Constraint indices for selected zeolites. Data from the patent literature [70, 71], values in parentheses from the open literature [72, 73].

Zeolite	CI	Classification after CI	True ring size
Erionite (ERI)	38	small pores	8-ring
ZSM-23 (MTT)	9.1		10-ring
ZSM-22 (TON)	7.3 (7.4)		10-ring
ZSM-5 (MFI)	6–8.3 (4.6)		10-ring
ZSM-11 (MEL)	5–8.7	medium pores	10-ring
ZSM-50 (EUO)	2.1		10-ring
MCM-22 (MWW)	1.5		10-ring
ZSM-12 (MTW)	2.3		12-ring
Mordenite (MOR)	0.5 (1.0)		12-ring
Beta (BEA)	0.6–2.0	large pores	12-ring
X or Y (FAU)	0.4 (0.2)		12-ring

characterizing the mean space available in the ten-membered-ring pores and in the large cavities.

Values for the constraint indices of more recent zeolite structures are given in Ref. [74], revealing even more shortcomings of the constraint index test. One example is that zeolites with 14-membered rings and pore openings greater than 0.8 nm cannot be distinguished from large-pore zeolites, simply because the degree of absence of spatial constraints cannot vary. Another example is that some zeolites with large internal cavities and pores composed of eight- or nine-membered rings do not generate constraint indices higher than zeolites with ten-membered-ring pores, probably again because the constraint index test averages the space in the pores and the cavities.

Other advantages and disadvantages of the constraint index have been discussed in detail in Ref. [1]. The reaction can suffer from a relatively fast catalyst deactivation, if large-pore zeolites are used, which makes a reliable determination of *CI* difficult, and a (sometimes pronounced) temperature dependence for medium-pore zeolites. The reason for the latter has been investigated in much detail by Haag et al. [75, 76]. These authors showed, in a convincing manner, that two basically different cracking mechanisms can be operative, namely the classical bimolecular chain-type mechanism involving tri-coordinated alkylcarbenium ions and a monomolecular mechanism via nonclassical penta-coordinated alkylcarbonium ions. The latter has a higher activation energy, hence its contribution increases with increasing reaction temperature, and, due to its requiring less space, the constraint index decreases with increasing temperature. However, Macedonia and Maginn [77] recently used Monte Carlo integration methods employing a classical molecular mechanics force field to predict values for the constraint indices of 12 different zeolites. These authors concluded that it is not necessary to invoke such a change in mechanism to explain decreasing constraint indices with increasing temperature. Their calculations indicated that the impact of confinement on the bimolecular transition state decreases with increasing temperature, effecting a decrease of the constraint index. This was said to be caused by a competition between energetic confinement effects that dominate at lower temperatures and entropic effects that become dominant at high temperature.

As to the nature of the shape-selectivity effects, Haag et al. [78] demonstrated in a most impressive study with H-ZSM-5 samples of equal concentration of acidic sites but different crystal sizes (0.05 to 2.7 μm) that neither the measurable rate of cracking of n-hexane nor that of the bulkier 3-methylpentane depend on the length of the intracrystalline diffusion paths. From this finding, the selectivity effects encountered in the competitive cracking of n-hexane and 3-methylpentane have to be interpreted in terms of intrinsic chemical effects (i.e., restricted transition state shape selectivity) rather than by mass transport effects. Haag et al. [78] suggested that the rate-controlling step in the chain-type mechanism of acid-catalyzed alkane cracking via carbocations is the chain-propagating hydride transfer between a cracked alkylcarbenium ion and a feed molecule, requiring significantly more space for the transition state if the feed alkane is branched, the net effect being a significant inhibition of cracking of 3-methylpentane. The simulations performed by Macedonia and Maginn [77] indicated, however, that for zeolites the pores of which are too small to accommodate the bimolecular transition state, such as ZSM-23 (MTT) and ferrierite (FER), the monomolecular mechanism dominates, with the measured constraint index attributed to reactant shape selectivity. Only for zeolites the pores of which are large enough for the bimolecular transition state but small enough for confinement effects was the bimolecular reaction predominant, and the selectivity was based on restricted transition state shape selectivity.

Recently, Baeck et al. [30] demonstrated that the constraint index is useful for probing subtle changes of the pore size as effected by CVD. The constraint index determined on zeolite Mg-ZSM-22 (TON) increased from about 9.9 to about 13.3 after 3 h of deposition of tetraethyl orthosilicate, showing the decrease of the pore opening by CVD.

In conclusion, with the constraint index, the idea of probing the pore width by catalytic test reactions was introduced into zeolite science. It no doubt fostered the search for alternative and improved catalytic test reactions. In spite of its shortcomings, enumerated above, the constraint index test has been widely used, and ample data is available in the literature.

Isomerization and disproportionation of _m_-xylene On acidic catalysts, _m_-xylene can undergo isomerization into _o_- and _p_-xylene and disproportionation (or transalkylation) into 1,2,3-, 1,2,4- or 1,3,5-trimethylbenzene and toluene (see Fig. 7). While it is obvious that disproportionation is necessarily a reaction involving a bimolecular transition state, there is some ambiguity as to whether the acid-catalyzed isomerization of xylenes proceeds via a monomolecular or a bimolecular pathway [79, 80].

References see page 1053

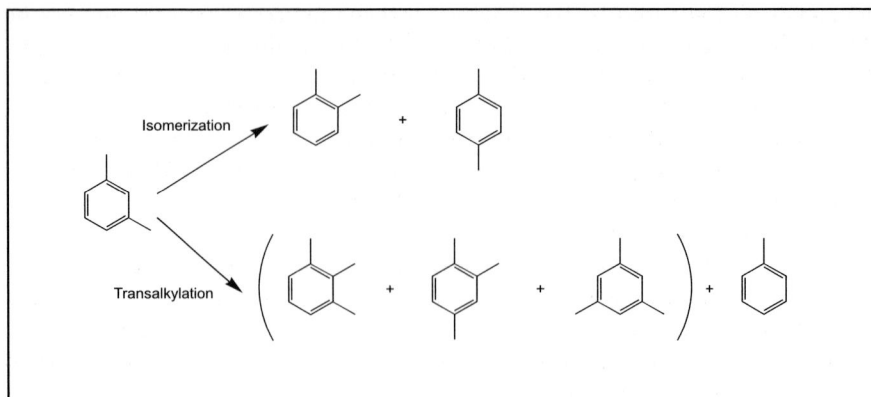

Fig. 7. Principal reaction pathways of *m*-xylene over acidic catalysts.

The use of *m*-xylene conversion for the characterization of the effective pore width of zeolites was first proposed by Gnep et al. [81]. These authors identified three selectivity criteria that may furnish valuable information on the effective pore width: (1) the relative rates of formation of *o*- and *p*-xylene, (2) the ratio of rates of disproportionation and isomerization and (3) the distribution of the trimethylbenzene isomers formed in the disproportionation reaction.

Criterion (1) is based on the finding that, in the absence of shape selectivity, *o*- and *p*-xylene are formed at virtually the same rate. With decreasing pore width, however, the formation of *p*-xylene is increasingly favored over the formation of the bulkier *o*-xylene [81]. This effect is best interpreted in terms of product shape selectivity, that is, progressively hindered diffusion of *o*-xylene molecules, as the pores are getting narrower.

Criterion (2) is a quantitative expression of the observation that, with decreasing pore width, isomerization of *m*-xylene is more and more favored over its disproportionation. At 350 °C, the bimolecular disproportionation is completely suppressed in the medium-pore zeolite ZSM-5, whereas isomerization (which is likely to proceed via the monomolecular mechanism at these high temperatures) does not experience significant hindrance [81]. Later, Olson and Haag [28] demonstrated that, if zeolite catalysts with different pore systems are used, a linear relationship results between the ratio of rate constants of disproportionation and isomerization and the dimensions of the largest zeolite cavities. One of the conclusions of this study was that the suppression of disproportionation is due to restricted transition state shape selectivity rather than to mass transfer effects.

Finally, criterion (3) was deduced from the observation that the disproportionation of *m*-xylene in the spacious cages of zeolite Y and in the much narrower channels of mordenite (MOR) resulted in significantly different contributions of the trimethylbenzene isomers: in the product obtained on H-mordenite, about 95 % of the trimethylbenzene consisted of the 1,2,4-isomer as compared to 24 % of the 1,3,5-isomer beside 74 % of the 1,2,4-isomer in the product from H-Y (on both zeolites, only negligible amounts of 1,2,3-trimethylbenzene were formed, and this was attributed to its low concentration in thermodynamic equilibrium) [81]. To interpret the hindered formation of 1,3,5-trimethylbenzene from *m*-xylene in mordenite, the authors mainly invoked restricted transition state shape selectivity.

m-Xylene isomerization as a catalytic test reaction was later adopted by a number of other groups [1, 64] and the criteria proposed by Gnep et al. [81] were refined and applied to a broader structural variety of molecular sieves. Dewing [82] introduced the so-called *R* value, defined as the ratio of rate constants observed experimentally for the formation of *o*- and *p*-xylene from *m*-xylene under conditions of diffusional limitations. Joensen et al. [83] coined the term shape selectivity index (*SSI*), which is defined as the yield ratio of *p*- and *o*-xylene observed on a shape-selective zeolite and extrapolated to zero conversion minus the same ratio obtained under identical experimental conditions on a catalyst with sufficiently large pores, that is, in the complete absence of shape selectivity in *m*-xylene isomerization. Neither Dewing's *R* value nor the shape selectivity index of Joensen et al. received much attention by others.

A critical evaluation of the published data [1] revealed that a comparison of the rates of isomerization and disproportionation, if applied quantitatively to a larger number of zeolites, does not allow for a reliable ranking of large-pore molecular sieves. No correlation at all exists between criterion (2) and the window or pore size [1]. Therefore, criterion (2) will not be dealt with further.

According to Martens et al. [84] there seems to be, however, valuable information on the pore architecture in the distribution of the trimethylbenzenes formed from *m*-xylene in large-pore zeolites. If disproportionation occurs at all in 10-membered-ring zeolites, it leads to 1,2,4-trimethylbenzene exclusively, probably because the diphenylmethane-type carbocation intermediate can be best accommodated in their narrow pores. An exception is zeolite ZSM-50 (EUO) in which 1,2,3- and 1,3,5-trimethylbenzene are formed as well. This can be rationalized if one assumes that the disproportionation takes place, at least in part, in the spacious side pockets running perpendicular to the ten-membered-ring channels. For the zeolites with the largest pores or cages, the distribution of the trimethylbenzenes is close to equilibrium. Hence, mechanistic, that is, kinetic conclusions from these data should be drawn with great care, and, therefore, criterion (3) is not considered further here.

With respect to criterion (1), it has been concluded [1] that it allows for a safe discrimination between ten- ($S_{\text{p-xylene}}/S_{\text{o-xylene}} \geq 2$, S = selectivity) and twelve-membered-ring ($S_{\text{p-xylene}}/S_{\text{o-xylene}} \approx 1.0$ to 1.5) zeolites. However, later work revealed that the situation is more complex, the main problem being that the experimental conditions for the test are not exactly defined. For instance, the selectivity ratios $S_{\text{p-xylene}}/S_{\text{o-xylene}}$ for zeolite MCM-22 (MWW) reported in the literature differ greatly (see Table 5). Whereas Corma et al. [86–88] and Adair et al. [89] determined their experimental data in dependence of time-on-stream and extrapolated to zero time-on-stream in order to account for the deactivation of the catalysts due to

References see page 1053

Tab. 5. Results from *m*-xylene reactions over zeolite H-MCM-22 (MWW) [85–89]; *TOS* = time-on-stream, *X* = conversion, *S* = selectivity.

Crystallite size/ μm	n_{Si}/n_{Al}	TOS/ h	T/K	$\dfrac{\dot{n}_{H_2}}{\dot{n}_{m\text{-}xylene}}$	$X_{m\text{-}xylene}/$ %	$\dfrac{S_{p\text{-}xylene}}{S_{o\text{-}xylene}}$	Ref.
1–2	14	2	573	4.0	58.1	1.03	85
1–2	14	2	573	4.0	39.7	1.02	85
1–2	14	2	573	4.0	20.2	1.06	85
1–2	14	2	573	4.0	9.5	1.07	85
1–2	14	2	573	4.0	4.1	0.87	85
0.5	11	0	623	0	not given	2.0	86–88
submicrometer	14	0	590	0	not given	2.6	89

coking and their different decay rates, Ravishankar et al. [85] obtained the selectivity ratio after 2 h time-on-stream. However, since in the latter experiments hydrogen was present in the feed (as opposed to the former), deactivation might still be absent after 2 h time-on-stream. Possibly, the presence of hydrogen can account for the low $S_{p\text{-}xylene}/S_{o\text{-}xylene}$ ratio observed in Ref. [85]. The selectivity ratio seems to be independent of the *m*-xylene conversion. The differences in the $S_{p\text{-}xylene}/S_{o\text{-}xylene}$ ratio cannot be attributed to the different crystal sizes, since the ratio is based on diffusion effects and is expected to increase with increasing pore size. Because the deviations shown in Table 5 are in the same range as the differences in the $S_{p\text{-}xylene}/S_{o\text{-}xylene}$ ratio of ten- and twelve-membered-ring zeolites, it is vital that for comparisons of different zeolites the ratios be determined under identical experimental conditions.

If one ignores Ref. [85], zeolite MCM-22 can be correctly identified as medium-pore zeolite, its selectivity ratio being somewhat higher than that of large-pore zeolites, but lower than that of typical medium-pore zeolites (e.g., $S_{p\text{-}xylene}/S_{o\text{-}xylene} \approx 2.9$ for ZSM-5) [88] because of its large intracrystalline cavities. Regarding the group of zeolites that give access to the intrazeolitic pore system only via ten-membered-ring channels, but have much space inside the pores due to large cavities (MWW), large side pockets (EUO) or twelve-membered-ring bridges (NES), MCM-22 (MWW) and EU-1 (EUO) have selectivity ratios in the range of medium-pore zeolites (2.6 and 3.9, respectively), but the ratio of NU-87 (NES) is considerably lower (1.6, typical for large-pore zeolites) [89]. The authors could not explain this fact. Corma et al. [90] also reported values typical for twelve-membered-ring zeolites for zeolite NU-87. However, zeolite CIT-1 (CON), having ten- and twelve-membered-ring pore openings, was classified as medium-pore zeolite by Corma et al. [90], but as large-pore zeolite by Adair et al. [89]. More experimental data on the characteristics of more recently discovered zeolites in the *m*-xylene test reaction can be found in Refs [23, 89–92].

The data discussed so far may lead one to conclude that the *m*-xylene test reaction is of limited use for probing the pore width of zeolites. However, Jones et al. [91] recently reported that the reactions of *m*-xylene could give, under proper experimental conditions, information enabling the characterization of medium-, large-

and extra-large-pore zeolites with one and the same test reaction, whereas all other test reactions proposed so far are only appropriate for medium- *or* large-pore zeolites. These authors found that large- and extra-large-pore zeolites with 1D channel systems, such as CIT-5 (CFI), SSZ-24 (AFI), SSZ-31 and UTD-1 (DON), have an $S_{\text{p-xylene}}/S_{\text{o-xylene}}$ ratio < 1, whereas multidimensional zeolites such as zeolite Y (FAU), Beta (BEA) display ratios > 1 (note, however, that zeolite L (LTL) with uni-dimensional pores has a selectivity ratio > 1 as well). In addition, the authors stress that it is imperative that all zeolites be compared under exactly the same experimental conditions, for example, similar conversion levels and even the same flow rate (sufficiently high to ensure the absence of significant external diffusion effects on reaction selectivity) are important. Besides, all zeolites should have roughly the same crystal size.

In conclusion, the *m*-xylene test reaction has found widespread application for probing the pore width of zeolites, probably because it is easy to handle (the number of products that have to be considered is small). Its usefulness and reliability, however, continue to be a matter of debate. If properly handled, the test reaction might allow for a meaningful classification of several classes of zeolite materials. However, a safe ranking of zeolites with respect to their pore width within these classes is certainly difficult and less safe.

Other test reactions Csicsery [93, 94] recognized pronounced shape-selectivity effects during the reaction of 1-ethyl-2-methylbenzene in acidic zeolites. A wealth of information on the pore size of the catalyst can be deduced from the product distributions. Of course, all three criteria advanced for *m*-xylene conversion (see previous section) are applicable to Csicsery's test as well. The latter furnishes, however, additional information, *inter alia* because two types of alkyl groups of different bulkiness can be transferred. Because of the larger size of 1-ethyl-2-methylbenzene and the intermediates, transition states and products derived from it, this test might be much more appropriate for probing the pores of twelve-membered-ring zeolites than the *m*-xylene test.

The disproportionation of ethylbenzene into benzene and the isomeric diethyl-benzenes was originally proposed by Karge et al. [95] as a test reaction to collect rapid information on the number of strong Brønsted acid sites in zeolites. Later, the reaction was found to be suitable for probing the pore width as well. From comparative catalytic experiments with a variety of ten- and twelve-membered-ring zeolites [96, 97], the following criteria were established: (1) twelve-membered-ring zeolites exhibit an induction period whilst ten-membered-ring zeolites do not; (2) after the induction period, there is very little or no deactivation in twelve-membered-ring zeolites, whereas in all ten-membered-ring zeolites, there is considerable deactivation from the very beginning of the catalytic experiment; (3) on twelve-membered-ring zeolites, the distribution of the diethylbenzenes is approximately 5 mol-% ortho-, 62 mol-% meta- and 33 mol-% para-isomer; on ten-membered-ring zeolites, the ortho-isomer is often completely absent or it appears as a very

References see page 1053

minor component, especially at the onset of the catalytic experiment. The combined application of the above criteria allows for a safe discrimination between medium- and large-pore zeolites (for the structures examined so far), but not yet for a ranking of the zeolites according to their effective pore width. However, Das et al. [98] demonstrated that the *p*-diethylbenzene selectivity increased from 33.4 % on H-ZSM-5 to 99.4 % on silylated H-ZSM-5 and attributed this effect to the pore-size regulation achieved by vapor deposition of the bulky tetraethyl orthosilicate followed by its calcination at 813 K. In a later study [99], the authors employed reaction mixtures containing 80 % *m*-xylene and 20 % ethylbenzene. With this mixture, essentially two reactions occurred on ZSM-5 zeolites at 678 K, namely *m*-xylene isomerization and ethylbenzene dealkylation to benzene and ethylene. With increasing silylation time, the *m*-xylene conversion decreased much more strongly than the ethylbenzene conversion. After a silylation period of 210 min, the conversion of *m*-xylene was zero, whereas the ethylbenzene conversion was still about 20 %. This example shows that such test reactions can be useful for probing subtle differences in the pore size. However, they need to be designed, defined and refined for a small pore-size range.

Kim et al. [100] proposed a test reaction particularly suitable for probing the effective pore width of large-pore zeolites. They chose a very bulky reactant, namely *m*-diisopropylbenzene, which has no access to ten-membered-ring pores, for the conversion with propene. Under the reaction conditions chosen, isomerization of *m*-diisopropylbenzene to its para-isomer and alkylation to the three isomeric triisopropylbenzenes occurred. It is the rationale in the design of this catalytic test that 1,3,5-triisopropylbenzene is a bulkier product than its 1,2,4-isomer. A problem associated with this test reaction is the deactivation of the catalyst due to its direct exposure to an alkene under a substantial partial pressure. Nonetheless, with a sophisticated sampling strategy, from the selectivity ratio of the 1,3,5- to the 1,2,4-isomer, conclusions could be drawn on the effective pore width of the investigated materials, which are in general agreement with their spaciousness indices (see Sect. 4.2.7.4.3). To minimize coke deposition in the catalyst pores, Singh et al. [101] lately proposed a modification of the test, that is, the use of isopropanol instead of propene as alkylating agent and of a large excess of *m*-diisopropylbenzene and a solvent. Unfortunately, the authors restricted their studies to zeolites Y, mordenite and Beta, so that the complete and critical evaluation of this test reaction for probing extra-large-pore zeolites is still lacking.

The alkylation of biphenyl with propene has been suggested as a test reaction for examining the pore size of pillared clays, large-pore zeolites and related microporous materials [102]. The major selectivity criterion discussed is the content of *o*-isopropylbiphenyl in the monoalkylated product fraction. The reaction could have a much broader potential for characterizing large- and even extra-large-pore molecular sieves, but more systematic work on the influence of the pore size and geometry of the catalyst on the distribution of the mono-, di- and, desirably, trialkylated products is needed.

An index analogous to the constraint index, yet designed for large-pore zeolites, was proposed by Miller [103] and Doblin et al. [104]. Miller introduced the C_8 selectivity index (C_8 *S.I.*), defined as

$$C_8 \ S.I. \equiv \frac{\ln(1 - X_{\text{n-Oc}})}{\ln(1 - X_{2,2,4\text{-TM-Pn}})} \tag{3}$$

as determined during competitive cracking of a 1:1 mixture of n-octane (n-Oc) and 2,2,4-trimethylpentane (2,2,4-TM-Pn) [103]. The rationale of this test is that n-octane can easily enter the pores, whereas 2,2,4-trimethylpentane cannot. This test is also suitable for bifunctional zeolites, since the effect of the metal on the index is small. Doblin et al. [104] proposed a more general reaction constraint parameter, namely the selectivity ratio (*SR*), defined as

$$SR \equiv \frac{\log(1 - X_{\text{less branched isomer}})}{\log(1 - X_{\text{more highly branched isomer}})} \tag{4}$$

as determined during competitive cracking of a 1:1 mixture of two alkanes with the same number of carbon atoms. With this definition, Mobil's constraint index becomes a special case of the selectivity ratio. (Note that, in the original papers of Miller [103] and Doblin et al. [104], the indices were formally defined in a slightly different manner that, from a chemical reaction engineering viewpoint, needed some re-definition.) Compared to the constraint index, the selectivity ratio is a more flexible index, since probe molecules can be chosen from a variety of alkanes to more closely match the catalyst pore size to be characterized. This is an interesting approach, but more experimental data collected on a much broader variety of zeolites including extra-large-pore materials is needed, before the usefulness of *SR* can be assessed.

Flego and Perego [105] recently proposed the aldol condensation of acetone as a more unusual test for characterizing both the acid-site density and the pore dimension of small-, medium- and large-pore zeolites in their acidic forms. The products can be readily analyzed by UV-Vis spectroscopy. Medium-pore zeolites were shown to give phorone as the final product, whereas large-pore zeolites tend to favor the formation of isophorone. However, no quantitative criterion was coined, hence this test reaction appears to need more refinement, before it can be recommended for a more general application.

The last test reaction for acidic zeolites that has to be discussed is the "methanol conversion test": The acid-catalyzed methanol-to-hydrocarbons (MTH) reaction is known to proceed in a stepwise manner [106, 107]: With increasing severity, dimethyl ether, olefins and a mixture of aromatics plus alkanes are successively formed. By suitable adjustment of the reaction conditions, either the yield of light olefins or that of aromatics plus alkanes can be maximized, the corresponding process variants being referred to as methanol-to-olefins (MTO) and methanol-to-gasoline (MTG), respectively. Large amounts of water are necessarily formed in the MTH reaction, and the risk exists that this brings about an undesired steaming of the zeolites with a concomitant framework dealumination. To minimize these

References see page 1053

effects, the zeolites to be tested are to be used in a form with a sufficiently high n_{Si}/n_{Al} ratio, yet a minimum amount of aluminum must be present in the framework since it generates the Brønsted acidity and, hence, the catalytic activity.

The "methanol conversion test" relies on the finding that the selectivity of the MTH reaction is strongly influenced by the zeolite pore geometry. Zeolite structures with eight-membered-ring pore apertures, such as erionite (ERI) or chabazite (CHA) are capable of converting methanol selectively to light olefins. Aromatics, if formed at the catalytic sites, would be trapped inside the large cages existing in most of these small-pore zeolites. On medium-pore zeolites, such as ZSM-5, aromatics do occur in the product, but due to the space limitations inside the pores, the aromatics distribution terminates at around C_{10} with a maximum at around C_8. In zeolite mordenite (MOR), the bulky polymethylbenzenes tend to be the main products, whereas in ZSM-12 (MTW) with its smaller effective pore diameter, a broader aromatics distribution has been observed [106]. These and other selectivity effects in the MTH reaction were exploited by Yuen et al. [108] for the characterization of AFI and CHA molecular sieves. Webster et al. [8] used the results from the test performed at different temperatures with zeolite H-ZSM-5 to assess the change of the effective pore dimension with temperature. The authors concluded that the effective channel size of H-ZSM-5 is between 0.662 and 0.727 nm at 300 °C, the MIN-2 (see Sect. 4.2.7.2.1) dimensions of the reaction products *p*-xylene (which is formed) and *o*-xylene (which is not formed), and at least 0.764 nm at 370 °C, the MIN-2 dimension of the reaction product 1,2,3-trimethylbenzene. Upon increasing the temperature, the dimensions of the channel intersections increase as well, namely from 0.817 by 0.888 nm to 0.908 by 0.909 nm [8].

A similar test reaction as with the methanol conversion which is based on mass transfer shape selectivity, but reactant shape selectivity instead of product shape selectivity, was used by Bendoraitis et al. [109]: These authors determined the catalytic pore sizes of ZSM-5, ZSM-23 (MTT) and mordenite on the basis of the sizes of molecules converted during the dewaxing of waxy distillate feeds to 0.55 by 0.70, 0.45 by 0.65 and 0.90 by 1.0 nm, respectively. As these examples show, this kind of test reaction can provide much useful information. However, the evaluation of the data is troublesome, since many products have to be considered and no criterion for an easy quantitative analysis has been defined. In addition, the crystal size of the materials examined should be kept constant, which is often difficult.

4.2.7.4.3 Test Reactions for Bifunctional Molecular Sieves

Since the early 1980s, the potential of catalytic reactions occurring on bifunctional forms of zeolites (i.e., the Brønsted acid form modified by a small amount, i.e., typically 0.1 to 1 wt.-%, of a hydrogenation/dehydrogenation component, usually platinum or palladium) for characterizing the pore size and pore architecture of zeolites has been explored. Particular emphasis was placed on the isomerization and hydrocracking of long-chain n-alkanes, such as n-decane, and to hydrocracking of cycloalkanes, for example, butylcyclohexane. These reactions have in common that they are conducted under hydrogen, which is activated by the metal component of the catalyst. One very favorable consequence of the presence of activated hydrogen is that coke formation and the concomitant catalyst deactivation are absent or very slow. This not only makes the experiments much easier (neither the conversion nor the selectivities vary with time-on-stream), but also eliminates the risk of progressively narrowing the pores by the deposition of carbonaceous residues formed in the test reaction.

The mechanisms of hydrocarbon conversion over bifunctional zeolite catalysts have been extensively discussed (see, e.g., Refs [1, 110–113]) and are beyond the scope of this section. In brief, the saturated hydrocarbon reactant is first dehydrogenated on the noble metal to the corresponding olefins that, in turn, are protonated at Brønsted acid sites. The resulting carbenium ions, while adsorbed at acid sites, undergo skeletal rearrangements and β-scissions. Finally, the product carbocations are desorbed from the acid sites as alkenes, which are hydrogenated on the noble metal. At low conversion, for example, monobranched cations, formed from n-alkanes in the first rearrangement step, will usually be readily desorbed and hydrogenated to isoalkanes, and these appear as the sole products. Upon raising the conversion, the monobranched alkylcarbenium ions undergo another rearrangement. The resulting dibranched carbenium ions desorb and appear in the product as dibranched isoalkanes. Upon further increasing the conversion, the dibranched cations rearrange once more into tribranched ones. These can undergo the very rapid type A β-scission [113]. Its salient feature is that β-scission starts from a tertiary and leads again to a tertiary carbenium ion. Tertiary carbenium ions are much more stable than secondary or primary carbenium ions. Indeed, the mechanism of hydrocracking of long-chain n-alkanes in the absence of shape selectivity (i.e., using a zeolite or any other catalyst with sufficiently spacious pores) proceeds via this route, that is, several skeletal rearrangements followed by type A β-scission; besides, there is a smaller contribution of type B β-scission [113] of dibranched carbenium ions, involving tertiary and the less stable secondary carbenium ions.

Isomerization and hydrocracking of long-chain n-alkanes – the refined or modified constraint index, *CI** Whereas hydrocracking of n-alkanes proceeds via highly branched and, hence, relatively bulky alkylcarbenium ions in large- and extra-large-pore zeolites, the tribranched precursor ions of the favorable type A β-scission cannot form under the steric constraints imposed in medium-pore zeolites. The system is then forced into alternative routes via less bulky intermediates, that is, the narrower the pores, the higher will be the contributions of β-scissions involving less stable secondary or even primary carbenium ions. This shift in the hydrocracking mechanism with decreasing pore width brings about a large number of selectivity changes in the hydrocracked product.

Striking shape-selectivity effects not only occur in hydrocracking, but also in skeletal isomerization, which precedes hydrocracking. If n-decane is converted on a bifunctional catalyst with sufficiently spacious pores, such as zeolite Pt/Ca-Y, the product mixture obtained at low conversion consists of all possible isodecanes with one branching, that is, 2-, 3-, 4- and 5-methylnonane, 3- and 4-ethyloctane as well

References see page 1053

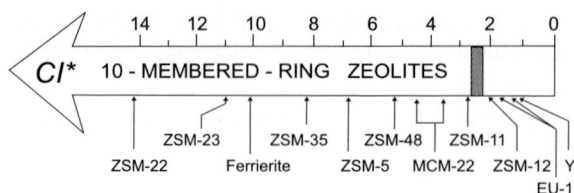

Fig. 8. Refined or modified constraint indices for various zeolites. Data from Refs [88, 117, 118].

as 4-propylheptane [114]. As the catalyst pores are getting narrower, the bulkier isomers can no longer be formed inside the pores or cannot escape from there. On Pt/H-ZSM-5, for example, neither the propyl nor the ethyl isomers are formed [113, 115].

Another shape-selectivity effect concerns the formation of the four isomeric methylnonanes. While in the absence of shape selectivity and at low conversions, that is, in the kinetic regime, 2- and 5-methylnonane form about half as fast as 3- and 4-methylnonane (which can be readily understood in terms of the branching mechanism via protonated cyclopropanes, see Ref. [114]), 2-methylnonane is the kinetically preferred isomer in ten-membered-ring zeolites. Based on the observation that, with decreasing pore width of the zeolite, the amount of 2-methylnonane formed from n-decane at low conversions increases relative to the other methylnonanes, the refined or modified constraint index, CI^*, was defined as

$$CI^* \equiv Y_{\text{2-methylnonane}}/Y_{\text{5-methylnonane}} \quad (\text{at } Y_{\text{isodecanes}} \approx 5 \text{ \%}) \ (Y = \text{yield}) \tag{5}$$

in the isomerization of n-decane [116]. It should be noted that the reaction mechanisms, on which Mobil's original constraint index (see Sect. 4.2.7.4.2) are based, are entirely different. The only feature both indices have in common is that their numerical values increase with decreasing pore size of the zeolite [64].

CI^* values taken from the literature are presented in Fig. 8. It is evident that the CI^* values for ten-membered-ring zeolites extend over a relatively broad range, namely from approximately 2.7 to 15; hence, this is the range where the modified constraint index is particularly useful. On the other hand, only a very narrow range, namely from approximately 1 to 2.3, is available for 12-membered-ring zeolites. It should, however, be kept in mind when using this or other indices, that the accuracy of indices determined from catalytic tests is limited. For example, for ZSM-5, CI^* values of 4.5 and 7.0 have been determined [1].

While the modified constraint index is now widely employed for characterizing ten-membered-ring molecular sieves, the precise origin and nature of the shape selectivity effects on which it is based have not yet been fully elucidated. From the finding that the diffusion coefficients of the four isomeric methylnonanes in H-ZSM-5 and H-ZSM-11 at 80 °C, that is, at a *subcatalytic temperature*, were practically identical [115], the strongly preferred formation of 2-methylnonane was attributed to restricted transition state shape selectivity rather than to mass transfer shape selectivity. Essentially the same conclusion was drawn by Martens and Jacobs [119] from n-decane isomerization studies with Pt/H-ZSM-5 of different crystal size. Increasing the crystal size from 1 to approximately 15 μm brought about only a minor increase in CI^* from 7.5 to 9.4, and this was considered to be essentially consistent with the earlier interpretation of the favored formation of 2-methylnonane in 10-membered-ring zeolites.

Webb and Grest [120] recently used molecular dynamics simulations to estimate the self-diffusion coefficients of n-decane and the methylnonane isomers at what they considered a *catalytically relevant temperature* of 327 °C (actually, catalytic studies on the isomerization of long-chain n-alkanes such as n-decane are often performed at temperatures around 250 °C). These authors arrived at the conclusion that the diffusion coefficients of the methylnonane isomers were clearly different in all medium-pore zeolites studied. The diffusion coefficients were minimal for different isomers in the various zeolites. This was explained by effects related to the commensurability between the molecular structure and the voids present at intersections of the main channels. Molecular motion of the methylnonane isomers was regarded to be dominated by a competition amongst CH_3 groups to place themselves at energetically favorable positions along the channel, and diffusion down the channel was accomplished by jumps between successive pocket mouths. This means that molecules that are incommensurate with the void system diffuse faster, because these molecules are less efficiently locked to intersection positions. The relevant result was that, for five of the seven zeolites included in the study, the order of CI^* values was the same as their order of the ratios of diffusion coefficients for 2-methylnonane and 5-methylnonane. From this, Webb and Grest [120] concluded that the shape selectivity effect underlying the modified constraint index CI^* does have to do, at least in part, with different mass transfer rates of the isomeric methylbranched isomers.

The modified constraint index exploits the selectivity occurring in the first step of the n-alkane reaction network. Numerous additional shape-selectivity effects are encountered in the consecutive reactions, namely the formation of dibranched isomers and hydrocracking. After a careful inspection of all these effects, Martens et al. [116, 119] defined a total of eight quantitative criteria. For a thorough discussion of these criteria and their usefulness for characterizing the pore width of microporous materials, the reader is referred to the pertinent literature [1, 116, 119]. While the modified constraint index has been well accepted for characterizing the pore width of medium-pore zeolites, the seven other criteria did not reach the same popularity.

In certain cases, for instance, if zeolites with very large pore sizes or void dimensions are to be characterized, it can be advantageous to use a probe molecule with more than ten carbon atoms. For example, an attempt was undertaken to detect differences between the pore systems of zeolite Y (FAU) and zeolite ZSM-20 (FAU/EMT), which is an intergrowth of zeolites FAU and EMT [121]. For this purpose, n-tetradecane was selected as feed hydrocarbon with the rationale that there are three

References see page 1053

isomers with a propyl and one with a butyl side chain among the monobranched isomers and that such bulky isomers may be suitable for detecting subtle differences in the pore systems. Indeed, such differences were found: 5-butyldecane was lacking in the product obtained on Pd/H-ZSM-20 at low conversion ($X_{n\text{-tetradecane}} \approx$ 2 %), while it did appear at the same conversion on zeolite Pt/Ca-Y. Moreover, the three isomeric propylundecanes appeared in significantly different distributions on both bifunctional catalysts [121].

Martens and coworkers [122, 123] went one step further and suggested the use of n-heptadecane as probe molecule for the characterization of large-pore zeolites. In this case, however, the information was acquired from the selectivity of hydrocracking, rather than from feed isomers of different bulkiness as in the test with n-tetradecane proposed by Weitkamp et al. [121].

Very recently, Raichle et al. [124] demonstrated that the selectivity for ethane, propane and n-butane (the mixture of these alkanes is a high-quality feedstock to steamcrackers for the manufacture of ethene and propene) during the hydroconversion of methylcyclohexane on acidic zeolites increases significantly with increasing spatial constraints inside the zeolite pores and correlates well with the modified constraint index. This example shows that the modified constraint index data are capable of providing valuable quantitative information concerning the shape-selective performance of zeolites with ten-membered-ring pores in a completely different reaction. This is particularly noteworthy in view of the fact that the CI^* data have been collected on *bifunctional* forms of the zeolites, whereas Raichle et al. [124] employed their zeolite catalysts in their *monofunctional* acidic forms.

In conclusion, shape-selective test reactions based on the isomerization and hydrocracking of long-chain n-alkanes are now well established for characterizing the pore width of medium-pore zeolites. While a variety of quantitative criteria have been proposed, by far the most popular one is the modified constraint index CI^*, which is based on the selectivity of n-decane isomerization at low conversion on a bifunctional form of the zeolite. Among the features that render the determination of CI^* so straightforward is the lack of catalyst deactivation. As expected, CI^* is particularly sensitive in a certain range of pore widths only, that is, where the pore width strongly influences the selectivity of isomerization and hydrocracking. As clearly shown in Fig. 8, this is the region of ten-membered-ring zeolites. By contrast, CI^* is of little use in the range of twelve-membered-ring zeolites. Fortunately, there is another index based on a different test reaction that is ideally complementary to CI^*. This is the spaciousness index SI.

Hydrocracking of C_{10} cycloalkanes such as butylcyclohexane – the spaciousness index, SI Isomerization and hydrocracking of cycloalkanes follow essentially the same mechanistic rules as the reactions of alkanes. The cycloalkane undergoes several steps of skeletal isomerization until a structure is reached from which the favorable type A β-scission can start. In comparison with alkanes, however, the carbon-carbon bond rupture *inside* the cycloalkane ring proceeds more sluggishly, which is best interpreted in terms of an unfavorable orbital orientation in the transition state of β-scission of cyclic carbenium ions [125, 126]. Therefore, hydrocracking of C_{10} cycloalkanes over bifunctional catalysts gives, irrespective of which isomer (e.g., pentylcyclopentane, butylcyclohexane, diethylcyclohexanes or tetra-methylcyclohexanes) is used as feed hydrocarbon, entirely different carbon number distributions than hydrocracking of a C_{10} alkane. It has been shown that, during hydrocracking of C_{10} cycloalkanes, methylcyclopentane and isobutane are formed almost exclusively *in the absence of spatial constraints* [127, 128]. This can be readily accounted for by the following sequence of steps: Starting from, for example, butylcyclohexane and excluding *endocyclic* carbon-carbon bond rupture, several rearrangement, ring-contraction and ring-enlargement steps take place until one of three possible tribranched carbenium ions (cyclopentanes with one methyl and one isobutyl side chain or with one isopentyl group) is reached, which can undergo *exocyclic* type A β-scission. Type A β-scission of all these three carbocations gives isobutane and methylcyclohexane as hydrocracked products. Small amounts of byproducts (propane, n-butane, pentanes and cycloalkanes other than methyl-cyclopentane) are in line with some contribution of *exocyclic* type B β-scissions [1].

Again, the tribranched intermediates for the type A β-scission are rather bulky. Therefore, it has been predicted on the basis of carbocation chemistry that, as the pores become narrower, hydrocracking of C_{10} cycloalkanes proceeds much less selectively and a broader variety of hydrocracked products is formed, and this has indeed been found experimentally [128]. A thorough investigation of all features of shape-selective hydrocracking of C_{10} cycloalkanes (preferentially butylcyclohexane, pentylcyclopentane is, from a chemical viewpoint, equally well or even better suited, though usually not readily available) revealed that the yield ratio of iso-butane and n-butane in the hydrocracked products is a most valuable indicator of the effective pore width of the zeolite. Since this ratio increases with increasing space inside the pores, it was named the spaciousness index, *SI* [129, 130].

SI values for a number of microporous materials are given in Fig. 9. It is evident that the spaciousness index is particularly valuable for ranking twelve-membered-ring molecular sieves for which it covers a wide range from approximately 3 (ZSM-12) to more than 20 (faujasite). By contrast, all ten-membered-ring zeolites have essentially the same spaciousness index around 1, so the index is of little or no use in this range of pore widths. Nor does it seem to be sensitive for extra-large-pore zeolites, since the *SI* value for amorphous SiO_2–Al_2O_3, which is probably meso-porous or even macroporous, has been found to be essentially equal to that for zeolite Y [64].

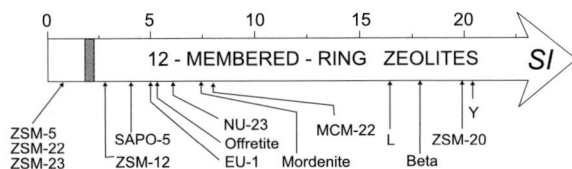

Fig. 9. Spaciousness indices for various zeolites. Data from Ref. [130].

References see page 1053

In routine applications, the spaciousness index offers several advantages: (1) As in all hydrocarbon reactions carried out on bifunctional catalysts, there is no deactivation. (2) The analysis for isobutane and n-butane is very easy and can be done quickly. (3) The *SI* value is in a very broad range independent of the butylcyclohexane conversion, the yield of hydrocracked products, the reaction temperature and other parameters. Hence, there is no prescribed conversion or yield, and a tedious search for the appropriate experimental conditions is superfluous.

Evidence has been obtained [1] that the spaciousness index is based on restricted transition state shape selectivity rather than on mass-transfer effects, at least for 12-membered-ring zeolites including ZSM-12. By using ball-shaped ZSM-12 crystallites with a diameter of approximately 0.5 μm and rod-like crystallites of the same zeolite with 11 μm by 1.5 μm, no significant influence on *SI* could be detected [130]. Nor was there any significant influence on *SI* when the n_{Si}/n_{Al} ratio was varied from 70 to 300. The independence of *SI* of the conversion implies that there is no interconversion of isobutane and n-butane under the conditions applied. This was confirmed experimentally for two zeolites, namely Pd/H-ZSM-5 and Pd/H-L [131]. Since the risk of undesired hydrogenolysis reactions on the metal is somewhat higher for Pt than for Pd, the use of the latter metal, usually in an amount around 0.25 wt.-% located inside the zeolite pores, has been recommended [1]. It has furthermore been demonstrated for zeolite EU-1 (EUO) that the amount of noble metal introduced into the pores does not influence the selectivity of the shape-selective hydrocracking of butylcyclohexane [130], at least not within a reasonable range, hence there is no risk of diminishing the effective pore width by introducing the noble metal required for making the catalyst bifunctional.

In conclusion, for the characterization of large-pore molecular sieves, the spaciousness index is the method of choice. The carbocation intermediates that govern the selectivity of hydrocracking of C_{10} cycloalkanes seem to be ideally suited for exploring the space available in the whole range of twelve-membered-ring materials [1].

4.2.7.5
Conclusions

Among the salient features of zeolites and zeolite analogs are their strictly uniform pore shapes and pore widths in the same range as molecular dimensions. Obviously, these properties can be exploited for characterizing zeolitic pores, especially their effective widths, by means of molecular probes. The available techniques rely on adsorption or shape-selective catalysis. Instructive examples for both methods are presented in this review with emphasis on their respective advantages, drawbacks and pitfalls.

Acknowledgments

The authors thank Mr. R. A. Rakoczy for helpful discussions. Financial support by Deutsche Forschungsgemeinschaft, Fonds der Chemischen Industrie and Max

Buchner-Forschungsstiftung is gratefully acknowledged. Y.T. thanks the Ministerium für Wissenschaft, Forschung und Kunst Baden-Württemberg for financial support through the Margarete von Wrangell-Habilitationsprogramm für Frauen.

References

1 J. WEITKAMP, S. ERNST, *Catal. Today* **1994**, *19*, 107–150.

2 M. COOK, W. C. CONNER, in: Proceedings of the 12th International Zeolite Conference. M. M. J. TREACY, B. K. MARCUS, M. E. BISHER, J. B. HIGGINS (Eds), Vol. 1, Materials Research Society, Warrendale, Pennsylvania, 1999, pp. 409–414.

3 D. W. BRECK, Zeolite Molecular Sieves. 2nd Edn, Robert E. Krieger Publishing Company, Malabar, Florida, 1984, p. 633–645.

4 CH. BAERLOCHER, W. M. MEIER, D. H. OLSON, Atlas of Zeolite Framework Types, 5th Revised Edn., Elsevier, Amsterdam, 302 pp.

5 I. D. HARRISON, H. F. LEACH, D. A. WHAN, in Proceedings of the Sixth International Zeolite Conference. D. OLSON, A. BISIO (Eds), Butterworths, Guildford, 1984, p. 479–488.

6 C. E. WEBSTER, R. S. DRAGO, M. C. ZERNER, *J. Am. Chem. Soc.* **1998**, *120*, 5509–5516.

7 V. R. CHOUDHARY, D. B. AKOLEKAR, *J. Catal.* **1989**, *117*, 542–548.

8 C. E. WEBSTER, R. S. DRAGO, M. C. ZERNER, *J. Phys. Chem. B* **1999**, *103*, 1242–1249.

9 US Patent 3,224,167, Dec. 21, 1965, assigned to Union Carbide Corp. (Inv.: R. A. J. SNYDER).

10 W. C. CONNER, R. VINCENT, P. MAN, J. FRAISSARD, *Catal. Lett.* **1990**, *4*, 75–84.

11 P. WU, Y. H. MA, in: Proceedings of the Sixth International Zeolite Conference. D. OLSON, A. BISIO (Eds), Butterworths, Guildford, 1984, pp. 251–260.

12 R. AIELLO, R. M. BARRER, J. A. DAVIES, I. S. KERR, *Trans., Faraday Soc.* **1970**, *66*, 1610–1617.

13 Y. NISHIMURA, H. TAKAHASHI, *Kolloid Z. Z. Polym.* **1971**, *245*, 415–419.

14 D. W. BRECK, W. G. EVERSOLE, R. M. MILTON, T. B. REED, T. L. THOMAS, *J. Am. Chem. Soc.* **1956**, *78*, 5963–5971.

15 T. MASUDA, K. HASHIMOTO, in: Zeolites and Microporous Crystals. T. HATTORI, T. YASHIMA (Eds), *Studies in Surface Science and Catalysis*, Vol. 83, Kodansha, Tokyo, Elsevier, Amsterdam, 1994, pp. 225–232.

16 D. S. SANTILLI, T. V. HARRIS, S. I. ZONES, *Microporous Mater.* **1993**, *1*, 329–341.

17 E. L. WU, G. R. LANDOLT, A. W. CHESTER, in: New Developments in Zeolite Science and Technology. Y. MURAKAMI, A. IIJIMA, J. W. WARD (Eds), *Studies in Surface Science and Catalysis*, Vol. 28, Kodansha, Tokyo, Elsevier, Amsterdam, 1986, pp. 547–554.

18 R. RAVISHANKAR, P. N. JOSHI, S. S. TAMHANKAR, S. SIVASANKER, V. P. SHIRALKAR, *Ads. Sci. Technol.* **1998**, *16*, 607–621.

19 M. OTAKE, *J. Catal.* **1993**, *142*, 303–311.

20 J. M. GUIL, R. GUIL-LÓPEZ, J. A. PERDIGÓN-MELÓN, in: Proceedings of the 12th International Congress on Catalysis. A. CORMA, F. V. MELO, S. MENDIOROZ, J. L. G. FIERRO (Eds), *Studies in Surface Science and Catalysis*, Vol. 130, Part C, Elsevier, Amsterdam, 2000, pp. 2927–2932.

21 R. F. LOBO, M. TSAPATSIS, C. C. FREYHARDT, S. KHODABANDEH, P. WAGNER, C.-Y. CHEN, K. J. BALKUS, JR., S. I. ZONES, M. E. DAVIS, *J. Am. Chem. Soc.* **1997**, *119*, 8474–8484.

22 A. CORMA, C. CORELL, J. PÉREZ-PARIENTE, J. M. GUIL, R. GUIL-LÓPEZ, S. NICOLOPOULOS, J. GONZALEZ CALBET, M. VALLET-REGI, *Zeolites* **1996**, *16*, 7–14.

23 J. P. LOURENÇO, M. F. RIBEIRO, F. RAMÔA RIBEIRO, J. ROCHA, Z.

GABELICA, *Appl. Catal. A: General* **1996**, *148*, 167–180.

24 L.-T. YUEN, J. S. GEILFUSS, S. I. ZONES, *Microporous Mater.* **1997**, *12*, 229–249.

25 D. W. BRECK, Zeolite Molecular Sieves. 2nd edn, Robert E. Krieger Publishing Company, Malabar, Florida, 1984, pp. 596–628.

26 I. D. HARRISON, H. F. LEACH, D. A. WHAU, in: Proceedings of the Sixth International Zeolite Conference. D. OLSON, A. BISIO (Eds), Butterworths, Guildford, 1984, pp. 479–488.

27 US Patent 4,117,026, Sept. 26, 1978, assigned to Mobil Oil Corp. (Invs.: W. O. HAAG, D. H. OLSON).

28 D. H. OLSON, W. O. HAAG, in: Catalytic Materials: Relationship Between Structure and Reactivity. T. E. WHYTE, JR., R. A. DALLA BETTA, E. G. DEROUANE, R. T. K. BAKER (Eds), *ACS Symposium Series*, Vol. 248, American Chemical Society, Washington, D.C., 1984, pp. 275–307.

29 S. NAMBA, J.-H. KIM, T. YASHIMA, in: Zeolites and Microporous Crystals. T. HATTORI, T. YASHIMA (Eds), *Studies in Surface Science and Catalysis*, Vol. 83, Kodansha, Tokyo, Elsevier, Amsterdam, 1994, pp. 279–286.

30 S. H. BAECK, K. M. LEE, W. Y. LEE, *Catal. Lett.* **1998**, *52*, 221–225.

31 D. S. SANTILLI, *J. Catal.* **1986**, *99*, 335–341.

32 V. R. CHOUDHARY, A. P. SINGH, R. KUMAR, *J. Catal.* **1991**, *129*, 293–296.

33 F. THIELMANN, D. A. BUTLER, D. R. WILLIAMS, E. BAUMGARTEN, in: Nanoporous Materials II. A. SAYARI, M. JARONIEC, T. J. PINNAVAIA (Eds), *Studies in Surface Science and Catalysis*, Vol. 129, Elsevier, Amsterdam, 2000, pp. 633–638.

34 O. A. ANUNZIATA, L. B. PIERELLA, in: Catalysis by Microporous Materials. H. K. BEYER, H. G. KARGE, I. KIRICSI, J. B.NAGY (Eds), *Studies in Surface Science and Catalysis*, Vol. 94, Elsevier, Amsterdam, 1995, pp. 574–581.

35 J. F. DENAYER, A. BOUYERMAOUEN, G. V. BARON, *Ind. Eng. Chem. Res.* **1998**, *37*, 3691–3698.

36 K. B. YOON, T. J. HUH, J. K. KOCHI, *J. Phys. Chem.* **1995**, *99*, 7042–7053.

37 M. TROMBETTA, T. ARMAROLI, A. G. ALEJANDRE, J. R. SOLIS, G. BUSCA, *Appl. Catal. A: General* **2000**, *192*, 125–136.

38 E. G. DEROUANE, J.-M. ANDRE, A. A. LUCAS, *J. Catal.* **1988**, *110*, 58–73.

39 C. E. WEBSTER, A. COTTONE, III, R. S. DRAGO, *J. Am. Chem. Soc.* **1999**, *121*, 12127–12139.

40 R. C. DEKA, R. VETRIVEL, *J. Catal.* **1998**, *174*, 88–97.

41 G. T. KERR, *Inorg. Chem.* **1966**, *5*, 1539–1541.

42 F. EDER, J. A. LERCHER, *J. Phys. Chem. B* **1997**, *101*, 1273–1278.

43 M. J. DEN EXTER, J. C. JANSEN, H. VAN BEKKUM, in: Zeolites and Related Microporous Materials: State of the Art 1994. J. WEITKAMP, H. G. KARGE, H. PFEIFER, W. HÖLDERICH (Eds), *Studies in Surface Science and Catalysis*, Vol. 84, Part B, Elsevier, Amsterdam, 1994, pp. 1159–1166.

44 A. STEWART, D. W. JOHNSON, M. D. SHANNON, in: Innovation in Zeolite Materials Science. P. J. GROBET, W. J. MORTIER, E. F. VANSANT, G. SCHULZ-EKLOFF (Eds), *Studies in Surface Science and Catalysis*, Vol. 37, Elsevier, Amsterdam, 1988, pp. 57–64.

45 US Patent 2,882,243, April 14, 1959, assigned to Union Carbide Corp. (Inv.: R. M. MILTON).

46 M. NIWA, K. YAMAZAKI, Y. MURAKAMI, *Chem. Lett.* **1989**, 441–442.

47 W. J. M. VAN WELL, X. COTTIN, J. W. DE HAAN, B. SMIT, J. H. C. VAN HOOFF, R. A. VAN SANTEN, in: Proceedings of the 12th International Zeolite Conference. M. M. J. TREACY, B. K. MARCUS, M. E. BISHER, J. B. HIGGINS (Eds), Vol. 1, Materials Research Society, Warrendale, Pennsylvania, 1999, pp. 91–96.

48 R. M. DESSAU, in: Adsorption and Ion Exchange with Synthetic Zeolites. W. H. FLANK (Ed.), *ACS Symposium Series*, Vol. 135, American Chemical

Society, Washington, D.C., 1980, pp. 123–135.

49 P. A. JACOBS, H. K. BEYER, J. VALYON, *Zeolites* **1981**, *1*, 161–168.

50 K. FOGER, J. V. SANDERS, D. SEDDON, *Zeolites* **1984**, *4*, 337–345.

51 J. R. ANDERSON, K. FOGER, T. MOLE, R. A. RAJADHYASKA, J. V. SANDERS, *J. Catal.* **1979**, *58*, 114–130.

52 P. WU, A. DEBEBE, Y. H. MA, *Zeolites* **1983**, *3*, 118–122.

53 S. NAMBA, Y. KANAI, H. SHOJI, T. YASHIMA, *Zeolites* **1984**, *4*, 77–80.

54 H. KARSLI, A. ÇULFAZ, H. YÜCEL, *Zeolites* **1992**, *12*, 728–732.

55 O. TALU, C.-J. GUO, D. T. HAYHURST, *J. Phys. Chem.* **1989**, *93*, 7294–7298.

56 Y. H. MA, T. D. TANG, L. B. SAND, L. Y. HOU, in: New Developments in Zeolite Science and Technology. Y. MURAKAMI, A. IIJIMA, J. W. WARD (Eds), *Studies in Surface Science and Catalysis*, Vol. 28, Kodansha, Tokyo, Elsevier, Amsterdam, 1986, pp. 531–538.

57 A. KURGANOV, S. MARMÉ, K. UNGER, in: Zeolites and Related Microporous Materials: State of the Art 1994. J. WEITKAMP, H. G. KARGE, H. PFEIFER, W. HÖLDERICH (Eds) *Studies in Surface Science and Catalysis*, Vol. 84, Part B, Elsevier, Amsterdam, 1994, pp. 1299–1306.

58 E. KLEMM, G. EMIG, in: Proceedings of the 12th International Zeolite Conference. M. M. J. TREACY, B. K. MARCUS, M. E. BISHER, J. B. HIGGINS (Eds), Vol. 1, Materials Research Society, Warrendale, Pennsylvania, 1999, pp. 235–241.

59 J. WEITKAMP, S. ERNST, M. SCHWARK, (unpublished results).

60 J. WEITKAMP, M. SCHWARK, S. ERNST, *Chem.-Ing.-Tech.* **1989**, *61*, 887–888.

61 H. DU, M. KALYANARAMAN, M. A. CAMBLOR, D. H. OLSON, *Microporous Mesoporous Mater.* **2000**, *40*, 305–312.

62 D. W. BRECK, E. M. FLANIGEN, in: Molecular Sieves. Society of Chemical Industry, London, 1968, pp. 47–61.

63 M. E. DAVIS, C. SALDARRIAGA, C. MONTES, J. CARCES, C. CROWDER, *Nature* **1988**, *331*, 698–699.

64 J. WEITKAMP, S. ERNST, L. PUPPE, in: Catalysis and Zeolites. J. WEITKAMP, L. PUPPE (Eds), Springer-Verlag, Berlin, 1999, pp. 327–376.

65 S. M. CSICSERY, in: Zeolite Chemistry and Catalysis. J. A. RABO (Ed.), *ACS Monograph*, Vol. 171, American Chemical Society, Washington, D.C., 1976, pp. 680–713.

66 L. FORNI, *Catal. Today* **1998**, *41*, 221–228.

67 J. WEITKAMP, S. ERNST, *Catal. Today* **1988**, *3*, 451–457.

68 V. J. FRILETTE, W. O. HAAG, R. M. LAGO, *J. Catal.* **1981**, *67*, 218–222.

69 O. LEVENSPIEL, Chemical Reaction Engineering. 2nd Edn, John Wiley, New York, 1972, pp. 484–485.

70 US Patent 4,686,316, Aug. 11, 1981, assigned to Mobil Oil Corp. (Inv.: R. A. MORRISON).

71 US Patent 5,234,872, Aug. 10, 1993, assigned to Mobil Oil Corp. (Invs.: M. R. APELIAN, T. F. DEGNAN, A. S. FUNG).

72 S. ERNST, R. KUMAR, M. NEUBER, J. WEITKAMP, in: Characterization of Porous Solids. K. K. UNGER, J. ROUQUEROL, K. S. W. SING, H. KRAL (Eds), *Studies in Surface Science and Catalysis*, Vol. 39, Elsevier, Amsterdam, 1988, pp. 531–540.

73 S. ERNST, J. WEITKAMP, J. A. MARTENS, P. A. JACOBS, *Appl. Catal.* **1989**, *48*, 137–148.

74 S. I. ZONES, T. V. HARRIS, *Microporous Mesoporous Mater.* **2000**, *35–36*, 31–46.

75 W. O. HAAG, R. M. DESSAU, in: Proceedings of the 8th International Congress on Catalysis. Vol. 2, Verlag Chemie, Weinheim, 1984, pp. 305–316.

76 W. O. HAAG, R. M. DESSAU, R. M. LAGO, in: Chemistry of Microporous Crystals. T. INUI, S. NAMBA, T. TATSUMI (Eds), *Studies in Surface Science and Catalysis*, Vol. 60, Elsevier, Amsterdam, 1991, pp. 255–265.

77 M. D. MACEDONIA, E. J. MAGINN, *AIChE J.* **2000**, *46*, 2504–2517.

78 W. O. HAAG, R. M. LAGO, P. B. WEISZ, *Faraday Discuss. Chem. Soc.* **1982**, *72*, 317–330.

79 S. M. CSICSERY, *J. Org. Chem.* **1969**, *34*, 3338–3342.

80 A. CORMA, E. SASTRE, *J. Catal.* **1991**, *129*, 177–185.

81 N. S. GNEP, J. TEJADA, M. GUISNET, *Bull. Soc. Chim. Fr.* **1982**, *I*, 5–11.

82 J. DEWING, *J. Mol. Catal.* **1984**, *27*, 25–33.

83 F. JOENSEN, N. BLOM, N. J. TAPP, E. G. DEROUANE, C. FERNANDEZ, in: Zeolites: Facts, Figures, Future. P. A. JACOBS, R. A. VAN SANTEN (Eds), *Studies in Surface Science and Catalysis*, Vol. 49, Part B, Elsevier, Amsterdam, 1989, pp. 1131–1140.

84 J. A. MARTENS, J. PEREZ-PARIENTE, E. SASTRE, A. CORMA, P. A. JACOBS, *Appl. Catal.* **1988**, *45*, 85–101.

85 R. RAVISHANKAR, D. BHATTACHARYA, N. E. JACOB, S. SIVASANKER, *Microporous Mater.* **1995**, *4*, 83–93.

86 A. CORMA, *Microporous Mesoporous Mater.* **1998**, *21*, 487–495.

87 A. CORMA, C. CORELL, A. MARTÍNEZ, J. PÉREZ-PARIENTE, in: Zeolites and Related Microporous Materials: State of the Art 1994. J. WEITKAMP, H. G. KARGE, H. PFEIFER, W. HÖLDERICH (Eds), *Studies in Surface Science and Catalysis*, Vol. 84, Part A, Elsevier, Amsterdam, 1994, pp. 859–866.

88 A. CORMA, C. CORELL, F. LLOPIS, A. MARTÍNEZ, J. PÉREZ-PARIENTE, *Appl. Catal. A: General* **1994**, *115*, 121–134.

89 B. ADAIR, C.-Y. CHEN, K.-T. WAN, M. E. DAVIS, *Microporous Mater.* **1996**, *7*, 261–270.

90 A. CORMA, A. CHICA, J. M. GUIL, F. J. LLOPIS, G. MABILON, J. A. PERDIGÓN-MELÓN, S. VALENCIA, *J. Catal.* **2000**, *189*, 382–394.

91 C. W. JONES, S. I. ZONES, M. E. DAVIS, *Appl. Catal. A: General* **1999**, *181*, 289–303.

92 C. W. JONES, S. I. ZONES, M. E. DAVIS, *Microporous Mesoporous Mater.* **1999**, *28*, 471–481.

93 S. M. CSICSERY, *J. Catal.* **1970**, *19*, 394–397.

94 S. M. CSICSERY, *J. Catal.* **1987**, *108*, 433–443.

95 H. G. KARGE, J. LADEBECK, Z. SARBAK, K. HATADA, *Zeolites* **1982**, *2*, 94–102.

96 J. WEITKAMP, S. ERNST, P. A. JACOBS, H. G. KARGE, *Erdöl, Kohle – Erdgas – Petrochem.* **1986**, *39*, 13–18.

97 H. G. KARGE, Y. WADA, J. WEITKAMP, S. ERNST, U. GIRRBACH, H. K. BEYER, in: Catalysis on the Energy Scene. S. KALIAGUINE, A. MAHAY (Eds), *Studies in Surface Science and Catalysis*, Vol. 19, Elsevier, Amsterdam, 1984, pp. 101–111.

98 J. DAS, Y. S. BHAT, A. B. HALGERI, *Ind. Eng. Chem. Res.* **1993**, *32*, 2525–2529.

99 A. B. HALGERI, Y. S. BHAT, in: Zeolites and Microporous Crystals. T. HATTORI, T. YASHIMA (Eds), *Studies in Surface Science and Catalysis*, Vol. 83, Kodansha, Tokyo, Elsevier, Amsterdam, 1994, pp. 163–170.

100 M.-H. KIM, C.-Y. CHEN, M. E. DAVIS, in: Selectivity in Catalysis. M. E. DAVIS, S. L. SUIB (Eds), *ACS Symposium Series*, Vol. 517, American Chemical Society, Washington, D.C., 1993, pp. 222–232.

101 P. S. SINGH, R. A. SHAIKH, R. BANDYOPADHYAY, B. S. RAO, in: Recent Advances in Basic and Applied Aspects of Industrial Catalysis. T. S. R. P. RAO, G. M. DHAR (Eds), *Studies in Surface Science and Catalysis*, Vol. 113, Elsevier, Amsterdam, 1998, pp. 473–478.

102 J.-R. BUTRUILLE, T. J. PINNAVAIA, *Catal. Lett.* **1992**, *12*, 187–192.

103 S. J. MILLER, *Microporous Mater.* **1994**, *2*, 439–449.

104 C. DOBLIN, J. F. MATHEWS, T. W. TURNEY, *Catal. Lett.* **1994**, *23*, 151–160.

105 C. FLEGO, C. PEREGO, *Appl. Catal. A: General* **2000**, *192*, 317–329.

106 C. D. CHANG, in: Shape-Selective Catalysis. C. SONG, J. M. GARCÉS, Y. SUGI (Eds), *ACS Symposium Series*, Vol. 738, American Chemical Society, Washington, D.C., 2000, pp. 96–114.

107 M. Stöcker, *Microporous Mesoporous Mater.* **1999**, *29*, 3–48.

108 L.-T. Yuen, S. I. Zones, T. V. Harris, E. J. Gallegos, A. Auroux, *Microporous Mater.* **1994**, *2*, 105–117.

109 J. G. Bendoraitis, A. W. Chester, F. G. Dwyer, W. E. Garwood, in: New Developments in Zeolite Science and Technology. Y. Murakami, A. Iijima, J. W. Ward (Eds), *Studies in Surface Science and Catalysis*, Vol. 28, Kodansha, Tokyo, Elsevier, Amsterdam, 1986, pp. 669–675.

110 J. Weitkamp, in: Hydrocracking and Hydrotreating. J. W. Ward, S. A. Qader (Eds), *ACS Symposium Series*, Vol. 20, American Chemical Society, Washington, D.C., 1975, pp. 1–27.

111 J. Weitkamp, *Erdöl, Kohle – Erdgas – Petrochem.* **1978**, *31*, 13–21.

112 P. B. Weisz, in: Advances in Catalysis. D. D. Eley, P. W. Selwood, P. B. Weisz (Eds), Vol. 13, Academic Press, New York, 1962, pp. 137–190.

113 J. Weitkamp, P. A. Jacobs, J. A. Martens, *Appl. Catal.* **1983**, *8*, 123–141.

114 J. Weitkamp, *Ind. Eng. Chem., Prod. Res. Dev.* **1982**, *21*, 550–558.

115 P. A. Jacobs, J. A. Martens, J. Weitkamp, H. K. Beyer, *Faraday Discuss. Chem. Soc.* **1982**, *72*, 353–369.

116 J. A. Martens, M. Tielen, P. A. Jacobs, J. Weitkamp, *Zeolites* **1984**, *4*, 98–107.

117 P. A. Jacobs, J. A. Martens, *Pure Appl. Chem.* **1986**, *58*, 1329–1338.

118 W. Souverijns, L. Rombouts, J. A. Martens, P. A. Jacobs, *Microporous Mater.* **1995**, *4*, 123–130.

119 J. A. Martens, P. A. Jacobs, *Zeolites* **1986**, *6*, 334–348.

120 E. B. Webb III, G. S. Grest, *Catal. Lett.* **1998**, *56*, 95–104.

121 J. Weitkamp, S. Ernst, V. Cortés-Corberán, G. T. Kokotailo, in: 7th International Zeolite Conference, Preprints of Poster Papers. Japan Association of Zeolite, Tokyo, 1986, pp. 239–240.

122 J. A. Martens, G. Vanbutsele, P. A. Jacobs, in: Proceedings from the 9th International Zeolite Conference. R. von Ballmoos, J. B. Higgins, M. M. J. Treacy (Eds), Vol. 2, Butterworth-Heinemann, Stoneham, 1993, pp. 355–362.

123 E. J. P. Feijen, J. A. Martens, P. A. Jacobs, in: 11th International Congress on Catalysis – 40th Anniversary. J. W. Hightower, W. N. Delgass, E. Iglesia, A. T. Bell (Eds), *Studies in Surface Science and Catalysis*, Vol. 101, Part B, Elsevier, Amsterdam, 1996, pp. 721–729.

124 A. Raichle, H. Scharl, Y. Traa, J. Weitkamp, in: Zeolites and Mesoporous Materials at the Dawn of the 21st Century. A. Galarneau, F. Di Renzo, F. Fajula, J. Védrine (Eds), *Studies in Surface Science and Catalysis*, Vol. 135, Elsevier, Amsterdam, 2001, p. 302 and full paper No. 26-P-10 on accompanying CD.

125 D. M. Brouwer, H. Hogeveen, *Rec. Trav. Chim.* **1970**, *89*, 211–224.

126 J. Weitkamp, S. Ernst, in: Guidelines for Mastering the Properties of Molecular Sieves. D. Barthomeuf, E. G. Derouane, W. Hölderich (Eds), *NATO ASI Series B*, Vol. 221, Plenum Press, New York, 1990, pp. 343–354.

127 J. Weitkamp, S. Ernst, H. G. Karge, *Erdöl, Kohle – Erdgas – Petrochem.* **1984**, *37*, 457–462.

128 S. Ernst, J. Weitkamp, in: Proceedings International Symposium on Zeolite Catalysis. Siófok, Hungary, May 13–16, Acta Phys. Chem., 1985, pp. 457–466.

129 J. Weitkamp, S. Ernst, R. Kumar, *Appl. Catal.* **1986**, *27*, 207–210.

130 J. Weitkamp, S. Ernst, C. Y. Chen, in: Zeolites: Facts, Figures, Future. P. A. Jacobs, R. A. van Santen (Eds), *Studies in Surface Science and Catalysis*, Vol. 49, Part B, Elsevier, Amsterdam, 1989, pp. 1115–1129.

131 J. Weitkamp, C. Y. Chen, S. Ernst, in: Successful Design of Catalysts. T. Inui (Ed.), *Studies in Surface Science and Catalysis*, Vol. 44, Elsevier, Amsterdam, 1988, pp. 343–350.

4.2.8
Application of Microporous Materials as Ion-Exchangers

Wolfgang Schmidt

Ion exchange in solids requires an anionic or cationic "backbone" of the respective exchanger phase, which can be amorphous inorganic or polymeric solids, e.g. amorphous silicates or resins, as well as crystalline materials, e.g. layered silicates or zeolites. Only a few of these materials exhibit microporosity and for ion exchange applications of microporous materials on a larger scale primarily zeolites are used. Nevertheless, microporous resins and ion exchangers of other compositions with micropores are also well established and used for specific applications [e.g. 1–3]. Due to their predominant role in large-scale applications, the present section will focus mainly on zeolites as ion exchangers. However, the general terms and theoretical bases can be transferred to other ion-exchange systems as well.

4.2.8.1
Ion-Exchange Properties of Zeolites in Aqueous Solutions

The ability to exchange cations is probably one of the most apparent attributes of zeolites besides their adsorptive and catalytic properties. Therefore, this feature has been investigated for several decades and many reports and reviews may be found in the literature. For those who need a more detailed view on the theoretical background of ion exchange in zeolites, or on the exchange properties of various cations and zeolites, the reviews and papers of Breck, Sherry, Kühl and Townsend may serve well as starting points [4–7, 8 or 9].

Zeolite frameworks consist basically of silicon coordinated tetrahedrally by oxygen atoms. These SiO_4 tetrahedra are connected via shared oxygen corners, thus forming the zeolite structures. The ion-exchange properties of zeolites are caused by bi- or trivalent metal atoms that substitute silicon atoms in the silicon oxide frameworks on the tetrahedra positions (T-sites). The hetero-atomic metal cations are typically tri- or bivalent, thus creating negative net-charges to the zeolite framework due to their coordination by four oxygen atoms. Usually, the heteroatom is the trivalent aluminum, which leads to one negative framework charge per aluminum incorporated within the framework as shown schematically in Fig. 1. In the following, the term "zeolite" will be used in its narrow sense, that is for aluminosilicate materials. The negative framework charges are compensated by cationic species that are not framework atoms. One of the most common ones is the Na^+ cation, which is usually found in as-synthesized zeolites. Depending on the synthesis conditions, other cations, such as for example K^+, NH_4^+, Ca^{2+}, Mg^{2+} or positively charged organic molecules, such as for example, tetraalkylammonium ions might also be present in the as-synthesized materials. Whatever cationic species is present, it has to be located in the channels, cavities and/or cages of the re-

a)

b)

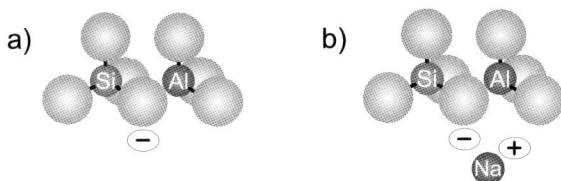

Fig. 1. (a) Negative framework charge due to a trivalent aluminum hetero-atom coordinated by four oxygen atoms in silicon oxide; (b) charge compensation by a sodium cation.

spective zeolite structure. Not only cationic species are located in the channels and cavities of the as-synthesized zeolite but also solvent molecules, which are usually water molecules. The general formula for an as-synthesized sodium containing zeolite can thus be written as:

$$Na_x(AlO_2)_x(SiO_2)_{y-x} \cdot nH_2O \quad \text{or} \quad Na_xAl_xSi_{y-x}O_{2y} \cdot nH_2O,$$

where y defines the number of tetrahedra per unit cell, and n the number of water molecules present in that unit cell. According to Loewenstein's rule Si/Al ratios of less than one are avoided for aluminum coordinated tetrahedrally by oxygen [10], therefore the lowest Si/Al ratio generally possible is 1 although exceptions are known. The compositions of zeolites range from those with Si/Al ratios of 1 in the low-silica zeolites A and X to those with infinite Si/Al ratios in purely siliceous forms, such as for example in silicalite-1 [11]. Since the amount of cations in a zeolite depends on the number of aluminum atoms in the framework, the ion-exchange capacity of different zeolites is also very variable.

During an ion exchange the charge-compensating cation is replaced by another one, whereby the net-charge within the zeolite remains zero. Multivalent cations thus replace the respective number of monovalent cations in the zeolite. This re-action can be expressed by:

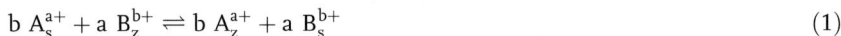

$$b \; A_s^{a+} + a \; B_z^{b+} \rightleftharpoons b \; A_z^{a+} + a \; B_s^{b+} \tag{1}$$

with z and s denoting the respective species in the zeolite and in the solution. The thermodynamic equilibrium constant is then given by the equation [12–14]:

$$K_a = \frac{a_{A_z}^b a_{B_s}^a}{a_{A_s}^b a_{B_z}^a} = \frac{f_A^b Z_A^b m_B^a \gamma_B^a}{f_B^a Z_B^a m_A^b \gamma_A^b}, \tag{2}$$

where a_{A_z}, a_{B_s}, etc., denote the respective ion activities in the solution and the zeolite in the first term, f_A and f_B the ion activity coefficients of the ions A^{a+} and B^{b+} in the zeolite, m_A and m_B the molalities of the cations in the solution, γ_A

References see page 1093

and γ_B the molal single-ion activity coefficient in the solution, and Z_A and Z_B the equivalent fractions of the cations in the zeolite, which are defined as

$$Z_A = \frac{a \cdot n_{A_z}}{a \cdot n_{A_z} + b \cdot n_{B_z}} \quad \text{and} \quad Z_B = \frac{b \cdot n_{B_z}}{a \cdot n_{A_z} + b \cdot n_{B_z}}, \tag{3a, 3b}$$

with a and b the valences of the cations A^{a+} and B^{b+}, and n_{A_z} and n_{B_z} the number of the respective cations in a unit volume of the zeolite. Thus, Z_A and Z_B are the fractions of the negative charges in the zeolite that are compensated by the cations A^{a+} and B^{b+}. To avoid calculations with the zeolite volume one may also write

$$Z_A = \frac{a \cdot m_{A_z}}{a \cdot m_{A_z} + b \cdot m_{B_z}} \quad \text{and} \quad Z_B = \frac{b \cdot m_{B_z}}{a \cdot m_{A_z} + b \cdot m_{B_z}}, \tag{4a, 4b}$$

with m_{A_z} and m_{B_z} being the molalities of the cations in the zeolite (moles of cations in the zeolite per kg of zeolite, mol kg^{-1}). The expression in the denominator of Eqs (4a and b) is the total ion-exchange capacity, Q, of the zeolite:

$$Q = a \cdot m_{A_z} + b \cdot m_{B_z} \tag{5}$$

The so-called corrected rational selectivity coefficient K_c calculates as [12–14]:

$$K_c = K'_c \frac{\gamma_B^a}{\gamma_A^b} = K_a \frac{f_B^a}{f_A^b}, \tag{6}$$

with

$$K'_c = \frac{m_B^a Z_A^b}{m_A^b Z_B^a}, \tag{7}$$

the rational selectivity coefficient. Similar to Eqs (4a and b), equivalent fractions of the cations in the solution can be calculated according to

$$S_A = \frac{a \cdot M_{A_s}}{a \cdot M_{A_s} + b \cdot M_{B_s}} \quad \text{and} \quad S_B = \frac{b \cdot M_{B_s}}{a \cdot M_{A_s} + b \cdot M_{B_s}}, \tag{8a, 8b}$$

with M_{A_s} and M_{B_s} the number of cations per unit volume of the solution (=molarity in mol l^{-1}). S_A and S_B being the fractions of the negative (anionic) charges in the solution compensated by the respective cations. Since the total number of cations in the system is constant it is sufficient to give only two distinct equivalent fractions to describe the ion-exchange equilibrium at a given temperature. Usually, Z_A and S_A are used for ion-exchange isotherms, as shown in Fig. 2. The plot of the equivalent fraction of the entering cation in the zeolite, Z_M, against the equivalent fractions of the cations in the solution, S_M, at a given temperature gives the ion-

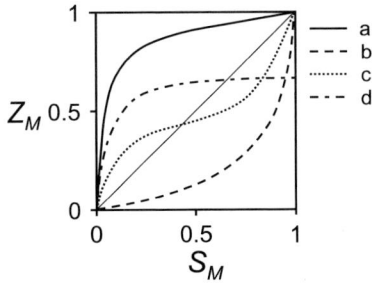

Fig. 2. Isotherms typically found for a binary ion exchange with (a) selectivity (preference) of the zeolite for the entering cation over the entire range; (b) selectivity for the leaving cation over the entire range; (c) selectivity reversal with increasing fraction of the entering cation in the zeolite; (d) incomplete exchange of an initially preferred cation [15].

exchange isotherm. The total normality of all cations has to be constant over the whole range. Figure 2 shows four typical ion-exchange isotherms that are frequently observed for binary ion exchanges in zeolites [15]. Isotherm a) is usually found when the zeolite has a preference for the entering cation while isotherm b) is observed when the zeolite prefers the leaving cation over the entering one. Isotherm c) with a sigmoidal shape is found when different exchange sites are present in the zeolite which have different preferences for the leaving cation. At low degrees of exchange, cations from sites with a preference for the entering cation are exchanged, while at higher degrees of exchange the cations are exchanged from sites with a preference for the leaving cation (=selectivity reversal). When the exchange cannot proceed completely, e.g. for steric reasons, an isotherm of type d) is found. The preference of a zeolite for one specific cation over another one can be expressed by the separation factor α_B^A:

$$\alpha_B^A = \frac{Z_A S_B}{S_A Z_B} \tag{9}$$

where $\alpha_B^A > 1$, if A^{a+} is preferred over B^{b+} by the zeolite [4].

The water molecules in the zeolite and the anion in the solution are neglected in that equation, however, in some cases salt imbibition has to be considered, as shown by Barrer and Walker [16] for high salt concentrations and by Uyama et al. [17] for nonaqueous systems. Since the cations are coordinated by different numbers of water molecules and since the cations are of different size, the amount of water within the zeolite might change during the exchange, which leads to a net-transfer of water molecules as well. Sherry and Walton [12] proposed a method to evaluate the standard free energies, standard enthalpies and standard entropies, which was originally developed by Gaines and Thomas [18] for the ion exchange in clay minerals. By neglecting salt imbibition and assuming the water activity to be

References see page 1093

unity they used the relation

$$\ln K_a = (b - a) + \int_0^1 \ln K_c \, dZ_A \tag{10}$$

to calculate K_a. The integral is evaluated from plots of log K_c versus the zeolite composition Z_A, often referred to as Kielland plots. The change of free energy associated with the ion exchange according to Eq. (1) is then given by the standard free enthalpy per equivalent of exchange

$$\Delta G^0 = -\frac{RT}{ab} \ln K_a \tag{11}$$

and the enthalpy and entropy can be calculated by the relations

$$\Delta H^0 = RT^2 \frac{d \ln K_a}{dT} \tag{12}$$

and

$$\Delta S^0 = \frac{\Delta H^0 - \Delta G^0}{T} \tag{13}$$

Theoretically, the maximum degree of exchange is determined by the net-charge of the zeolite framework, e.g. the number of aluminum atoms per zeolite unit cell, and the number of positive charges of the respective cation. The negative charges of the framework and positive charges from the cations must compensate to zero. However, there are some parameters that can significantly influence the maximum degree of exchange:

The size of the cation might be too large to fit into the zeolite pores and the cation is thus excluded from the zeolite pore system. This sieving effect is enhanced by water molecules that usually coordinate the cations in aqueous solutions. Some cations, even if they are small enough to fit into the zeolite channels, are coordinated by water molecules in several coordination spheres, which exclude them from the zeolite channels or specific cavities if the water molecules cannot be stripped from the cation. The smaller the cations and the higher their positive charge, the more difficult is the removal of water molecules coordinating these cations. Elevated temperatures lead to an enhanced stripping of water molecules from the cations and therefore are advantageous for systems where cations are excluded from specific sites due to their large hydration shells.

In some zeolites cations are located in cages that are accessible only via cage windows confined by a ring of six tetrahedra, denoted as a 6 T-ring or a 6-membered ring (6MR). The access of a specific cation to these cages could be limited due to its size, thus leading to an incomplete exchange. Even if the entering cation could pass through the 6MR, the coordination of the cation to be exchanged within the cage might be energetically more favorable, which also leads to an incomplete exchange.

For large voluminous cations, for example with several coordination shells of water, the space inside the zeolite channels and cavities might be restricted, because the space needed for a complete charge compensation by these cations exceeds the available volume within the pores. In this case either some of the water molecules must be stripped from the cations – the cations may be coordinated by framework oxygen atoms – or the number of exchanged cations is less than the theoretically possible maximum number.

A factor that apparently increases the degree of exchange into the zeolite above the theoretically maximum value is salt imbibition, i.e. the transfer of not only cations into the zeolite but also of anions associated with the excess incorporation of cations needed for the charge compensation of the anions. This factor is usually neglected, but at very high salt concentration in the exchange solution or at exchanges from salt melts salt imbibition might occur to significant extents.

A similar effect as from salt imbibition might occur when a salt or metal hydroxide/oxide layer is formed on the surface of the zeolite crystals during the ion exchange. The amount of metal cations removed from the solution is thus higher than expected from the maximum number of exchangeable cations (over-exchange).

Finally, the resistance of the respective zeolite against exchange solution plays a crucial role. Especially low silica zeolites, such as zeolite A, X, or P, may suffer significant damages by alkaline or acidic exchange solutions, e.g. in some transition metal salt solutions. The zeolite may be damaged by dealumination and partial or complete structure collapse; it may even dissolve partly or completely. A drying or calcination step after the ion exchange enhances the probability of obtaining an amorphous material, since a damaged zeolite structure is thermally less stable than an unaffected structure. This fact is often neglected and occasionally a more or less completely decomposed (amorphous) zeolite is used rather than an ion-exchanged zeolite for subsequent catalytic experiments or other investigations (especially for iron-, cobalt-, nickel- and/or copper-exchanged low-silica zeolites).

In general, ion-exchange isotherms and ion-exchange kinetics can be modeled for bi- and/or tri-component ion exchanges. The theoretical methods had been developed in the middle of the last century and then improved successively. The procedures and models for calculating ion-exchange isotherms and the kinetics of the ion exchange are described extensively in the literature and will not be discussed here. For further reading, the papers of Sherry, Barrer, Townsend, and Rees are recommended [8, 16, 19–28].

4.2.8.2
Aqueous Ion Exchange of Zeolites

Basically, every zeolite has ion-exchange properties provided the framework is not purely siliceous. In that case only a few of the slightly acidic protons from terminal

References see page 1093

silanol groups on the surface and in defect sites of the crystals might be exchanged by cations. However, if bi- or trivalent hetero-atoms are incorporated within the zeolite framework the cations needed for the charge balance can be exchanged by other cations. It would exceed the scope of the present book by far to focus on all zeolites that might be used as ion exchangers and only a few systems will be presented here. The zeolites that will be discussed in the following are those that are used most frequently as ion exchangers.

4.2.8.2.1 Ion Exchange in Zeolite A

Zeolite A in its hydrated form has the general composition $Na_{96}Al_{96}Si_{96}O_{384}$· 216 H_2O. This formula represents the composition of one unit cell of zeolite A which crystallizes in the cubic space group $Fm\bar{3}c$ with $a = 24.6$ Å [29–34] and the unit cell is formed by strictly alternating SiO_4 and AlO_4 tetrahedra, which leads to a Si/Al ratio of exactly one. The zeolite framework of structure type LTA [34] can be considered as a packing of truncated octahedra (cubooctahedra), formed by SiO_4 and AlO_4 tetrahedra. The silicon and aluminum atoms connected via four coordinating oxygen atoms are often denoted as T-atoms. Pore and channels in zeolites are often described by the number of T-atoms forming the respective cavity or channel. Thus, a 4-membered ring (4MR) is build by four T-atoms connected by shared oxygen atoms. In zeolite A such 4MR form a larger cage, denoted as β-cage. Six 4MRs that are connected to four other 4MR of the same β-cage are connected via oxygen bridges, thus forming eight 6-membered rings (6MR). Each of the thus-formed β-cages has a diameter of 6.6 Å and is connected with six further β-cages by double 4-membered rings (D4R), as shown in Fig. 3. Eight β-cages are positioned on the edges of a cube forming larger cavities 11.8 Å in diameter. These cavities are denoted as α-cages, which is not consistent with the latest IUPAC terminology. However, that expression is still quite common and will be used here. The α-cages are accessible via 8-membered rings (8MR) with openings of about 4.1 Å in diameter. During an ion exchange the β-cages are accessible only via 6MR and can be occupied merely by cations that are small enough to fit through these windows of about 2.2 Å diameter.

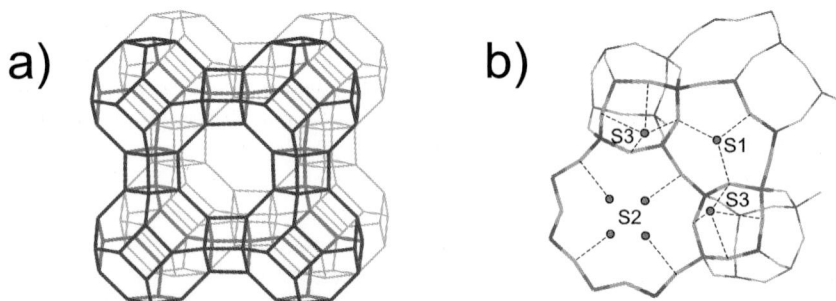

Fig. 3. (a) Structure of zeolite A (oxygen atoms are omitted, the lines connect neighboring T-atoms); (b) cation sites in zeolite A (spheres indicate sites, dotted lines coordination with framework oxygen atoms).

Several cation sites have been identified in zeolite A. Some of them (S1, S2, S3 in Fig. 3b) are energetically favored and occupied preferentially by the cations, as long as no steric hindrances occur (e.g. two cations too close to each other). If more cations are required for the charge balance, the remaining cations occupy the less-favored sites, by which some sites are only statistically occupied [31]. One of the energetically most favored sites (S1) is located on top of the centers of the 6MRs, slightly shifted out of plane to the center of the α-cages [29, 31–33]. Cations on these sites are coordinated by three next-neighbor oxygen atoms from the 6MRs. A second, slightly less favored cation site (S3) is found on top of each of the 4MRs, this time shifted slightly further to the center of the large cavity. Cations on these sites are coordinated by four oxygen atoms from the 4MRs [31, 33]. Four further likely cation sites (S2) are in each of the 8MRs forming the windows of the α-cages (cube faces). However, usually not all four of these sites are occupied simultaneously, because that would lead to steric hindrance in the ring opening [29, 31]. Furthermore, cations may be found in the β-cages on top of the 4MR and 6MR windows and in the center of the β-cage, as well as in the α-cages in front of the 8MR and in the center of the α-cage [35, 36]. Generally, cations are most frequently found on the sites S1, S2, and S3, nevertheless, the other sites may be of marked importance in some special cases [36]. Some of the sites can also be occupied by water molecules and, due to changes of the water coordination, the presence or absence of water molecules might shift the cation positions (e.g. in the direction to the framework where the cations get coordinated by more framework oxygen atoms). To check the probable position of a specific cation in zeolite A, the compilation of extra-framework sites by Mortier [36] is recommended.

The ion-exchange properties of zeolite A have been studied extensively throughout the last four decades, especially with respect to its application as water softener in detergents (builders). However, Na-A is capable of exchanging with very different cations and the exchange of Na^+ by some selected mono- and bivalent cations will be discussed here.

Figure 4a shows the ion-exchange isotherms of the monovalent cations Li^+, K^+, Rb^+, and Cs^+ with Na-A reported by Barrer et al. [37] and Fig. 4b the isotherms of Tl^+ and Ag^+ with Na-A measured by Sherry and Walton [12]. Barrer et al., as well as Sherry and Walton, used a Na-A that contained excess aluminum that they assigned to residual $NaAlO_2$ species in the β-cages. The ion-exchange capacity of the respective zeolites was thus higher than expected for the genuine zeolite framework. For the alkali metal ions Barrer et al. [37] found overall affinities of the ions for zeolite A of $Na^+ > K^+ > Rb^+ > Li^+ > Cs^+$. With the exception of Li^+, the affinity decreases as the ion radius increases and with all cations except Cs^+ a complete ion exchange could be achieved. The large Cs^+ cations cannot replace all Na^+ cations in zeolite A due to space limitations (steric hindrance) in the α-cages; depending on the exchange conditions, similar problems may arise for the exchange with Rb^+ [37]. From the data of the isotherms in Fig. 4a and from calorimetric experiments Barrer et al. derived ΔG^0, ΔH^0, and ΔS^0 of the exchange reactions,

References see page 1093

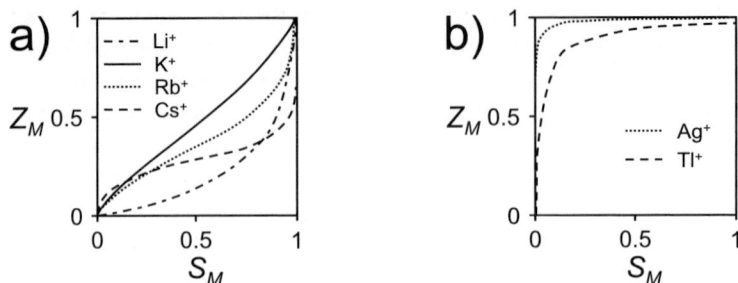

Fig. 4. Ion-exchange isotherms of Na-A with (a) M = Li$^+$, K$^+$, Rb$^+$, and Cs$^+$ (adjusted to total normality 0.1 N, 298 K) [37]; (b) M = Ag$^+$ and Tl$^+$ (total normality 0.1 N, 298 K) [12].

which are listed in Table 1. They revealed that for exchanges with Li$^+$, K$^+$, Rb$^+$, and Cs$^+$ ΔG^0 is always positive, indicating that these cations have a smaller affinity to zeolite A than Na$^+$. The standard heats of exchange ΔH^0 and the corresponding entropies ΔS^0 both have negative sign for the exchanges of K$^+$, Rb$^+$, and Cs$^+$. The entropy term, $T\Delta S^0$, is always larger than the enthalpy, ΔH^0, thus always resulting in positive standard free energies, ΔG^0. For the exchange of Na$^+$ by Li$^+$, ΔH^0 and ΔS^0 are positive with the entropy term being smaller than the heat term, resulting in a positive standard free energy. The displacement of a Na$^+$ ion by a larger uni-valent cation in the zeolite is always accompanied with a decrease in energy and entropy and with a decrease of the water content of the zeolite.

From Fig. 4b it is obvious that zeolite A has a pronounced Ag$^+$ selectivity and all Na$^+$ can be replaced by Ag$^+$, while Tl$^+$ can only replace 97 % of the Na$^+$. However, Tl$^+$ also exchanges readily with Na-A. The ion-exchange experiments with Tl$^+$ and

Tab. 1. Standard free energies (ΔG^0), enthalpies (ΔH^0), and entropies (ΔS^0, here multiplied with the temperature) calculated from ion exchange isotherms of zeolite A (energies in kJ per g equiv of exchanger).

Exchange	Ref.	ΔG^0 (298 K)/kJ (g equiv)$^{-1}$	ΔH^0 (298 K)/kJ (g equiv)$^{-1}$	$T\Delta S^0$ (298 K)/kJ (g equiv)$^{-1}$
Na$^+$/Li$^+$	37	5.45 $\pm$0.21	9.47 $\pm$0.21	4.02
Na$^+$/K$^+$	37	0.59 $\pm$0.25	−10.01 $\pm$0.25	−10.6
Na$^+$/Rb$^+$	37	2.85 $\pm$0.17	−10.68 $\pm$0.63	−13.53
Na$^+$/Cs$^+$	37	8.30 $\pm$0.21	−15.92 $\pm$1.26	−24.22
Na$^+$/Ag$^+$	12	−16.45 $\pm$0.17	−11.64 $\pm$1.88	4.86 $\pm$0.75
Na$^+$/Tl$^+$	12	−9.71 $\pm$0.08		
Na$^+$/Mg^{2+}	39	3.26	18.8	15.50
Na$^+$/Ca^{2+}	37	−0.59 $\pm$0.08	8.80 $\pm$0.21	9.39
Na$^+$/Ca^{2+}	12	−3.07 $\pm$1.26	11.30 $\pm$1.26	14.33 $\pm$1.25
Na$^+$/Ca^{2+}	39	−2.68	12.2	14.90
Na$^+$/Sr^{2+}	12	−4.23 $\pm$0.04	2.09 $\pm$0.84	6.38 $\pm$2.47
Na$^+$/Ba^{2+}	12	−4.89 $\pm$0.08	0.00 $\pm$0.84	4.86 $\pm$0.98

References see page 1093

Fig. 5. Ion-exchange isotherms of Na-A with M = Ca^{2+}, Sr^{2+}, Ba^{2+} [12], and Mg^{2+} [39] (total normality 0.1 N, 298 K).

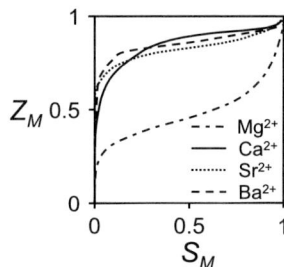

Ag^+ can be used to evaluate the amount of Na^+ on sites in the smaller β-cages. The aperture of the 6MR to the $NaAlO_2$ containing β-cages (2.2–2.5 Å) is large enough to allow the smaller Ag^+ (ion size 2.52 Å) to pass through, while the Tl^+ (ion size 2.80 Å) is excluded from those cages at 298 K [12].

The exchange properties of zeolite A for bivalent alkaline earth metals are shown by the exchange isotherms in Fig. 5. The preference of Na-A for Ca^{2+}, Sr^{2+}, and Ba^{2+} over Mg^{2+} is obvious. These three bivalent cations exchange readily into the zeolite. From the data shown in Figs 4b and 5 Rees, Sherry and Walton calculated the standard free enthalpies, ΔG^0, enthalpies, ΔH^0, and entropies, ΔS^0, of the exchange reactions, as described in Sect. 4.2.8.1. Values derived for room-temperature experiments are given in Table 1.

Sherry and Walton emphasize that the standard enthalpies of exchange are positive for the exchanges of the alkaline earth cations, while the standard entropies are large enough to result in negative standard free enthalpies, and that it is thus the entropy function that causes the preference of the zeolite for the alkaline earth cations over the sodium cation. Comparison with other studies led to the conclusion that incorrect determinations of the concentrations, e.g. due to nonequilibrium, may change the results drastically, since the Kielland plots then may show a maximum. The integration according to Eq. (10) then apparently leads to wrong values of K_a and, thus, the values for ΔG^0, ΔH^0, and ΔS^0 derived from K_a are also incorrect.

The entropic contributions were explained to be two-fold, in the aqueous phase each exchanged alkaline earth cation is replaced by two sodium cations. By this exchange, water molecules are released in the case of Ca^{2+} and Sr^{2+}, because both are coordinated by more water molecules than two Na^+. The larger Ba^{2+} disturbs the local water structure and the release of Ba^{2+} from the water phase leads to a decrease of the entropy. In the zeolite phase, one alkaline earth cation replaces two sodium cations. Taking all these contributions into account, Sherry and Walton calculated the zeolite and solution-phase contributions ΔS_Z^0 and ΔS_S^0 to the entropy of the exchange reaction ΔS^0 [12] using the relation

$$\Delta S_Z^0 = \Delta S^0 - \Delta S_S^0 = \Delta S^0 - \left(S_{Na^+}^{hyd} - \tfrac{1}{2} S_{A^{2+}}^{hyd} \right) \tag{14}$$

Tab. 2. Contributions to entropy of exchange reaction of the zeolite and the solution phase ($T = 298$ K) [12].

	Ag^+	Ca^{2+}	Sr^{2+}	Ba^{2+}
$T\Delta S_S^0/kJ$ (g equiv)$^{-1}$	1.73	5.22	4.26	−2.44
$T\Delta S_Z^0/kJ$ (g equiv)$^{-1}$	3.13	9.12	2.12	7.30

The data for ΔS^0 were taken from the experiment and the standard entropies of hydration were taken from Rosseinsky [38]. The thus calculated values are given in Table 2. The data show an increase of the entropy in the zeolite when the sodium cations are exchanged by the alkaline earth and silver cations, which is greater than the contribution of the solution phase to the total entropy of the reaction. The increase of the entropy in the zeolite by exchanging the sodium cation by the monovalent silver cation is somewhat unexpected. The authors suggest a direct coordination of the Ag^+ by oxygen atoms from the zeolite framework. The release of the water molecules previously coordinating the Ag^+ then should increase the entropy. The significant increase in entropy in the zeolite associated with the exchange of Ba^{2+} cannot be explained with coordination by framework oxygen and release of water molecules because the enthalpy of exchange is zero. The authors suppose a net-transfer of water molecules during the ion exchange from the zeolite to the solution phase that would lead to an increase of the entropy. This assumption is supported by the low water content of Ba^{2+}-exchanged zeolite A [12].

The influence of the concentration and temperature of the exchange solutions is shown for the exchange of Na^+ by Ca^{2+} and Mg^{2+} in the ion-exchange isotherms in Fig. 6 [39]. The preference of Na-A for Ca^{2+} over Na^+ is much more pronounced than that for Mg^{2+} over Na^+. Decreasing total normalities lead to higher loading of the zeolite with the bivalent cations that are even more enhanced when the temperature is increased. The values for ΔG^0, ΔH^0, and ΔS^0 that Rees calculated from data of the ion-exchange isotherms of these systems agree well with the data of Sherry and Walton, as shown in Table 1. Both exchanges are endothermic ($\Delta H^0 > 0$) which explains the enhanced ion exchange at elevated temperature. In comparison with Ca^{2+} the exchange of Mg^{2+} in zeolite A is associated with a larger endothermic enthalpy (see Table 1), which can be explained by the energy that is needed to strip water molecules from the cations. This removal of water molecules from the coordinating water shell of the cations is necessary because the hydration shells of the bivalent cations are too large to fit into the pore openings of the zeolite. The energy needed to remove water molecules from the hydration shell of the smaller Mg^{2+} is higher than that for the larger Ca^{2+}. The preference of the zeolite for Ca^{2+} over Mg^{2+} and Na^+, which has been found in the binary systems, is also obvious in the ternary system [39]. At no point more than 40 % of the charges of the zeolite are compensated by Mg^{2+}. The properties of that ion-exchange system have been investigated very thoroughly due to its importance as a detergent builder [e.g. 24, 39, 40]. Kinetic experiments on the uptake of an ion into the zeolite with time enable the calculation of diffusion coefficients. Kinetic experiments

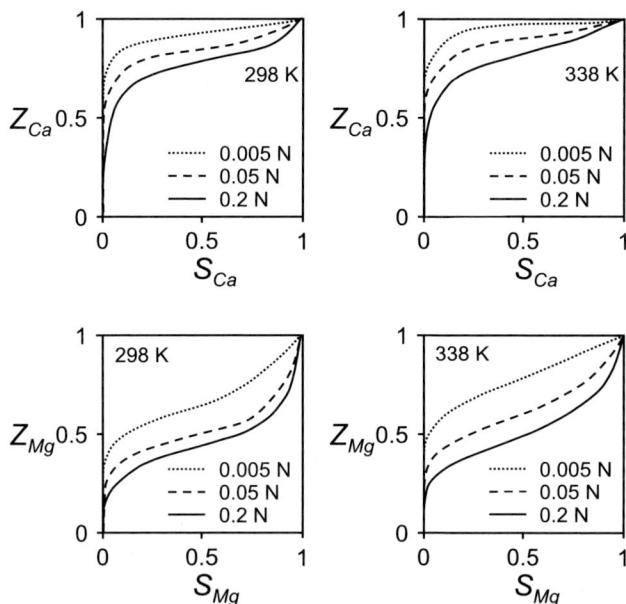

Fig. 6. Ion-exchange isotherms of Na-A with Ca^{2+} and Mg^{2+} at different total normalities at 298 K and 338 K [39].

are usually performed by measuring the ion content in the solutions with the time of exchange. Occasionally, zeolites containing radiotracers like $^{22}Na^+$ are used [24]. The radiotracer, which is easy to detect, is released into the solution during the exchange. Following the exchange kinetics of Ca^{2+} and Mg^{2+} in Na-A, Rees [39] found that the uptake of Ca^{2+} in zeolite A is ten times faster than that of Mg^{2+} from binary solutions.

The main application of zeolite A as ion exchanger is as a water softener in detergents, where the removal of Ca^{2+} and, if possible, of Mg^{2+} is of predominant importance. Nevertheless, zeolite A has been used for the exchange of several other metal cations [e.g. 41–51]. Figure 7 shows the ion-exchange isotherms of Zn^{2+}, Cd^{2+}, Co^{2+}, and Ni^{2+} with Na-A. Zn^{2+} and Cd^{2+} readily exchange into the zeolite while the uptake of Co^{2+} and Ni^{2+} is limited at 298 K. The major parameter for this different behavior seems to be the size of the hydrated cation. The exchange of Co^{2+} and Ni^{2+} into Na-A is enhanced at elevated temperature [43, 48, 49], which can be explained by water stripping from the hydration shell of the transition metal cations. A transition-metal-exchanged zeolite A contains significantly more water molecules than the genuine Na-A, which is explained by a closer packing of the water molecules due to a more ordered arrangement around the bivalent transition metal cations [43]. The stability of the zeolite A crystals is often affected by the exchange with transition-metal-containing exchange solutions [41, 48, 49, 51]. The

References see page 1093

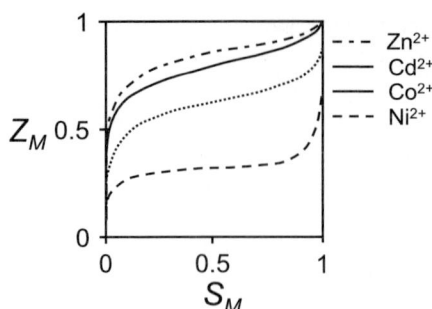

Fig. 7. Ion-exchange isotherms of Na-A with M = Zn^{2+}, Cd^{2+}, Co^{2+}, and Ni^{2+} (total normality 0.1 N, 298 K) [43].

destabilization of the crystal structure may be caused by several effects. Transition metals in aqueous solutions tend to autoprotolysis by which the solution gets slightly acidic. The autoprotolysis may even take place inside the exchanged zeolite [47, 52, 53]. The structure of the aluminum-rich zeolite A is susceptible to acidic solutions and structure damage arise from dealumination and partial dissolution of the zeolite. This effect is accompanied by a further destabilization of the dehydrated Ni^{2+}-exchanged zeolite A due to changes of coordination of the transition metal cations during the water removal. In contrast to cations like Co^{2+}, where the crystal field stabilization does not differ much between its tetrahedrally and octahedrally coordinated form, the crystal field stabilization energies for tetrahedrally and octahedrally coordinated Ni^{2+} differ significantly. An octahedral coordination is much more preferred by Ni^{2+} than a tetrahedral one. During the removal of water molecules, the Ni^{2+} cations are successively coordinated by oxygen atoms from the crystal structure, but at a distinct stage of water removal there exists no further possibility for octahedral coordination of the Ni^{2+} ion, resulting in a destabilization of the whole system [43, 45, 46]. In a completely dehydrated zeolite the Ni^{2+} cations are coordinated by only three oxygen atoms from 6MR close to site S1 [45]. In order to avoid severe damage to the zeolite structure by transition-metal-containing solutions, the pH of the exchange solution should not be too low and the contact time of the zeolite with the exchange solutions should be kept as short as possible. Often (but not in every case) only a few minutes are sufficient to reach the exchange equilibrium [e.g. 48, 49].

As shown by the above examples, rather comprehensive pictures of the ion-exchange into a specific zeolite can be drawn from carefully performed ion-exchange experiments. In the following sections the ion-exchange with other zeolites will be reviewed in less detail, the reader is referred to the specified literature for a more comprehensive discussion.

4.2.8.2.2 Ion Exchange in Zeolite X and Zeolite Y

The mineral faujasite crystallizes in cubic symmetry, $Fd\bar{3}m$, and has two isotypic synthetic forms, zeolite X and zeolite Y. The structure type, FAU [54], and thus the framework topology is identical for these three materials, while their chemical compositions vary. The general composition of the materials per unit cell is (Ca^{2+}, Mg^{2+}, Na_2^+)$_{0.5x}Al_xSi_{192-x}O_{384} \cdot 240$ H_2O with x usually ranging from 96 to 48.

Generally, a synthetic material of structure type FAU is called zeolite X when its framework has a Si/Al ratio in the range of 1.0–1.4 and zeolite Y when the ratio is in the range of 2.0–3.0. In the intermediate range a transition form with mixed contributions of both materials exists [55–57]. The notation is not arbitrary but is based on structural changes. The lattice parameters a of synthetic sodium fauja-sites decrease linearly with decreasing amounts of aluminum in the unit cell, un-less a Si/Al ratio of 1.4 is reached. Above that ratio, the lattice parameter increases instantly in a step, after which it again decreases constantly with decreasing amount of aluminum. At a Si/Al ratio of about 2.0 a second abrupt step to slightly larger values of the lattice parameter is observed, followed by a third linear region of decreasing lattice parameter with further decreasing aluminum content [57]. The reason for these two steps are probably changes in the ordering of the silicon and aluminum species in the faujasite framework. The lattice parameter a, which can range from 24.26 Å, e.g. in siliceous faujasite [58], up to 25.10 Å, e.g. in hy-drated Na-X [59], depends not only on the aluminum content but also on the type and amount of ions serving as counterbalance for the negative charges of the zeo-lite framework. Faujasites with higher Si/Al ratios also exist, but they are usually obtained by dealumination of low-silica faujasites.

The structure is also formed by β-cages as that of zeolite A, but in faujasite-type zeolites the β-cages are connected by double 6-membered rings (D6R). The β-cages are connected in the same topology as carbon atoms in the diamond structure, and thus ten β-cages form a large cavity with a diameter of about 9.0 Å. These large cavities, which are often referred to as faujasite cages, supercages, or α-cages (al-though they are not cages according to the IUPAC terminology), are accessible by four 12-membered rings (12MR). The structure of the faujasite cage and a view into one of the four 12MR is shown in Fig. 8a. A large number of studies per-formed on cation sites in FAU-type zeolites have been summarized by Mortier [36], here only the most prominent sites will be discussed. The six sites that are most frequently found in faujasites are denoted S1 to S4, where a distinction is made in two cases when a cation is positioned either in front of or behind two specific 6MRs. As shown in Fig. 8b, site S1 is positioned in the center of the D6Rs where the cation is coordinated octahedrally by six oxygen atoms from the respec-tive D6R, while site S1′ is opposite the D6Rs inside the β-cages where each cation is coordinated by three oxygen atoms from the 6MR next to it. Site S2 is located in front of 6MRs in the large faujasite cages, while site S2′ is located on the opposite side of the same 6MRs inside the β-cages. Site 3 is in front of a 4-membered ring (4MR) of a β-cage inside the faujasite cage and cations on this site are coordinated by the two nearest oxygen atoms from the 4MR. Site 4 is located in the center of a 12MR forming the window to the faujasite cages. Sometimes, deviations from these exact positions are reported [e.g. 59], but these positions serve very well as starting points for the search of cations in a faujasite. As for zeolite A some of the sites may also be occupied by water molecules. Further known water sites were summarized by Mortier [36].

References see page 1093

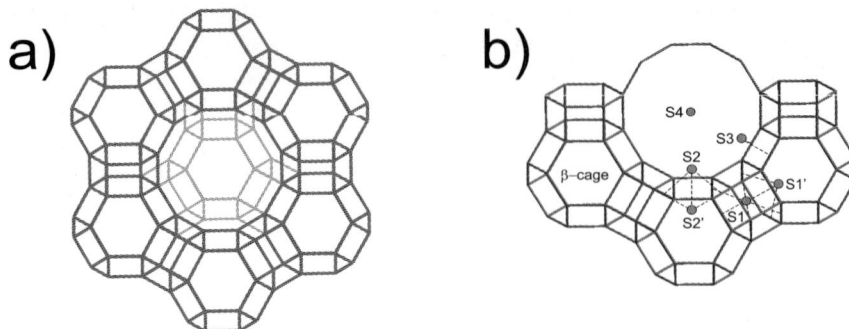

Fig. 8. (a) Structure of zeolite X; (b) cation sites in zeolite X; spheres indicate sites, dotted lines the coordination with framework oxygen atoms (oxygen atoms are omitted in both figures, the lines connect neighboring T-atoms).

Ion-exchange isotherms of the zeolites Na-X and Na-Y with the monovalent alkali metal cations are shown in Fig. 9 [60]. The ion-exchange equilibria were reached in less than one hour as proved by sodium isotope-exchange experiments. The exchange isotherms of K^+, Rb^+, and Cs^+ with zeolite X have a sigmoidal shape, indicating selectivity reversals with higher loading of the zeolite, which is due to the fact that the cations are located on at least three different sites in zeolite X. From the 85 monovalent cations 37 are located in the large faujasite cages (α-cages) where they are coordinated most probably by water molecules and are only loosely bound to the zeolite framework. These sites cannot be determined exactly by X-ray techniques, the cations may float in the zeolite cavities. K^+, Rb^+, and Cs^+ cations on these sites are preferred by the zeolite over Na^+ and Li^+ due to the smaller hydrated ionic radii of these cations [61] and coulombic interactions between these hydrated cations and the anionic sites of the zeolite framework; according to Pearson [62], a soft base like the anionic zeolite framework prefers

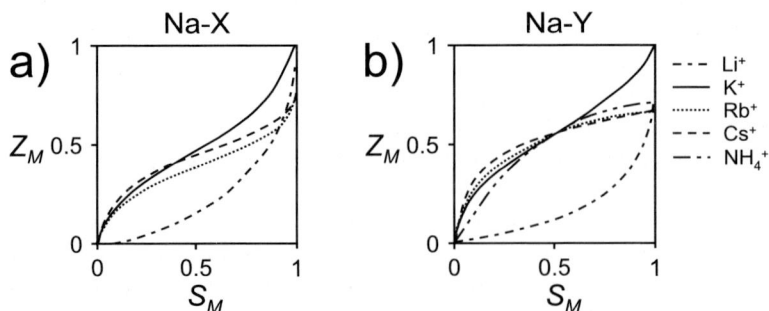

Fig. 9. Ion-exchange isotherms of (a) Na-X; (b) Na-Y with $M = Li^+$, K^+, Rb^+, Cs^+, and NH_4^+ (total normality 0.1 N, 298 K, anhydrous Na-X: $Na_{85}Al_{85}Si_{107}O_{384}$, anhydrous Na-Y: $Na_{50}Al_{50}Si_{142}O_{384}$) [60].

Fig. 10. Ion-exchange isotherms of Na-X and Na-Y with $M = Ag^+$ and Tl^+ (total normality 0.1 N, 298 K, anhydrous Na-X: $Na_{85}Al_{85}Si_{107}O_{384}$, anhydrous Na-Y: $Na_{50}Al_{50}Si_{142}O_{384}$) [60].

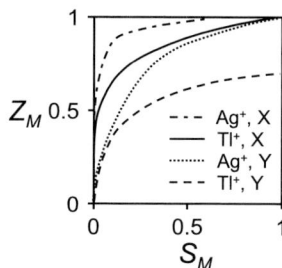

bonding with soft acids like the larger alkali metals. The 32 cations on site S2 are probably replaced next, there are no water molecules between the cations and the framework oxygen atoms as proved by X-ray structure analysis. The close distance between the cations on that position and the oxygen atoms indicates a direct coordination. The selectivity for the cations on these sites is different as for those of the hydrated cations floating in the faujasite cages. It is the net result of the free enthalpy change due to coulombic interactions of the partially dehydrated cations with the negatively charged zeolite framework and the free-enthalpy change of the partial dehydration of the cations. Finally, the 16 cations located either in the hexagonal prisms (site S1) or in the β-cages (site S1′) are replaced, provided the cation is small enough to fit through the 6MR of the β-cages. Cs^+ and Rb^+ are generally too large to pass through the 6MR windows, therefore, they cannot completely replace Na^+. As indicated by the isotherm in Fig. 9, all alkali metal cations including Na^+ are preferred over Li^+ on all sites due to the low polarizability of the hard acid Li^+. Consistently, the soft acids Ag^+ and Tl^+ with low polarizability are easily exchanged into Na-X due to strong binding on all sites, as shown in Fig. 10.

Zeolite Y has a lower ion-exchange capacity and the internal molality is less than in zeolite X. The internal molality in the zeolites changes for the different cations because the amount of water present in the zeolite depends on the type of cation balancing the negative charges of the zeolite. Nevertheless, it is always less in zeolite Y than in zeolite X; for Na-X with 264 water molecules per unit cell as used by Sherry [60], the internal molality is about 19, for Na-Y with a similar water content the internal molality is about 11. The lower number of negative charges of the framework and lower internal molality probably lead to less ion binding, in a way that there seems to be no site heterogeneity in zeolite Y. No sigmoidal shape of the exchange-isotherm characteristic for site heterogeneity is observed for the exchange of alkali metal and ammonium cations with Na-Y as shown in Fig. 9.

As in zeolite X, Ag^+ and Tl^+ are preferred by zeolite Y, as indicated by the isotherms in Fig. 10. From the isotherms in Figs. 9 and 10 it is obvious that Rb^+, Cs^+, NH_4^+, and Tl^+ are too large to fit into the β-cages; the isotherms are terminated at the point $Z_M = 0.68$ and $S_M = 1.0$, which corresponds to the amount of the 16 Na^+ cations on sites S1 or S1′ per unit cell. Only Li^+, Na^+, K^+, and Ag^+ are small enough to enter the β-cages of the zeolite Y sample. Thus, the β-cages of the zeolite

References see page 1093

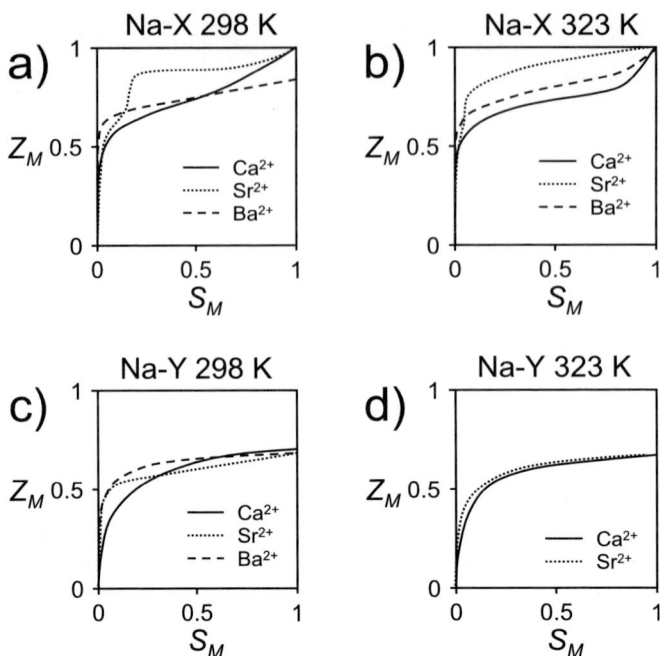

Fig. 11. Ion-exchange isotherms of Na-X at (a) 298 K; (b) 323 K and of Na-Y at (c) 298 K; (d) 323 K with M = Ca^{2+}, Sr^{2+}, and Ba^{2+} (total normality 0.1 N, anhydrous Na-X: $Na_{85}Al_{85}Si_{107}O_{384}$, anhydrous Na-Y: $Na_{51}Al_{51}Si_{141}O_{384}$) [63].

Y have smaller 6MR openings than those of the zeolite X sample. The incorporation of aluminum into the zeolite framework leads to a significant expansion of the crystal lattice.

The alkaline earth cations Ca^{2+}, Sr^{2+}, and Ba^{2+} exchange easily with the Na^{+} cations in the large faujasite cages of Na-X and Na-Y, as shown by the exchange isotherms in Fig. 11. However, the 16 Na^{+} cations located on site S1 in the D6Rs or on site S1′ in front of these D6Rs are difficult to exchange by these alkaline earth cations in both zeolites despite the fact that the zeolites initially favor the exchange of Na^{+} by the divalent cations [63]. In zeolite X, which has slightly larger 6MRs than zeolite Y, Ca^{2+}, and Sr^{2+} can replace Na^{+} on those sites at room temperature; however, the replacement of Na^{+} by Ca^{2+} or Sr^{2+} on these sites is rather slow. Ba^{2+} cannot replace Na^{+} on site S1 or S1′ at all at room temperature. An elevated temperature of 323 K (stripping of water from hydration shell) and a very long exchange time of one week enable a complete exchange on these sites in zeolite A. In zeolite Y none of the three alkaline earth cations can replace the 16 Na^{+} cations on sites S1 or S1′, not even at elevated temperature. The hydrated (298 K) or partially dehydrated (323 K) bivalent cations are too large to pass through the 6MR windows of the β-cages. The enthalpy for the exchange of these three cations for Na^{+} has either a positive or a small negative value. It is the entropy of the re-

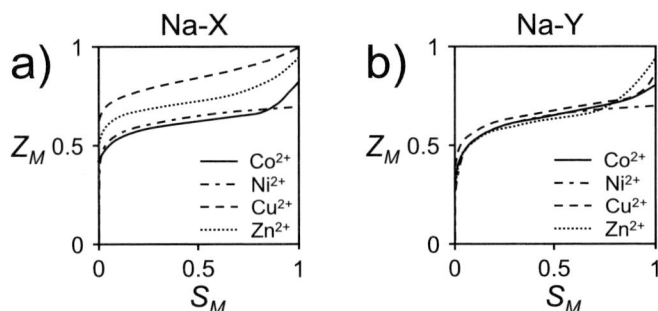

Fig. 12. Ion-exchange isotherms of (a) Na-X; (b) Na-Y with M = Co^{2+}, Ni^{2+}, Cu^{2+}, and Zn^{2+} (total normality 0.01 N, 298 K, Na-X: $Na_{85}Al_{85}Si_{107}O_{384} \cdot 255H_2O$, Na-Y: $Na_{54}Al_{54}Si_{138}O_{384} \cdot 241H_2O$) [66].

action that is responsible for the initial selectivity of the faujasite zeolites for Ca^{2+}, Sr^{2+}, and Ba^{2+} [63]. From the corrected selectivity coefficients Sherry concluded that there is no siting of Sr^{2+} and Ba^{2+} in the large faujasite cages of zeolite Y, while in all other cases at least partial siting of the alkaline earth cations is observed.

The exchange of transition metal cations into faujasite zeolites has been studied mainly with respect to catalysis and for applications in water purification. Many studies dealing with these subjects were published but only a rather limited number of studies on the ion-exchange process itself. However, the studies on transition metal exchange on zeolite X and zeolite Y show significant differences between the exchange of alkali or alkaline earth metal cations and transition metal cations [64–71]. Despite their small sizes, some transition metal cations cannot be exchanged completely into the faujasite zeolites at 298 K, as has been observed in several studies [e.g. 64–67]. Figure 12 shows the ion-exchange isotherms of the bivalent transition metal cations Co^{2+}, Ni^{2+}, Cu^{2+}, and Zn^{2+} with zeolite X and zeolite Y at 298 K. All cations are preferred by the zeolites at a loading of the zeolites of $Z_M < 0.5$–0.6, at higher loading the isotherms level out in a plateau that ends in a rather steep final increase of the loading close to $S_M = 1$ except for Ni^{2+}, for which no sigmoidal shape of the isotherm is observed. The ease of exchange into both zeolites increases in the order $Ni^{2+} < Co^{2+} < Zn^{2+} < Cu^{2+}$. At 298 K an almost complete exchange is only observed for Cu^{2+} in Na-X. The other cations exchange only partially with Na-X and Na-Y. At first glance, one might argue that the cations are excluded from the β-cages, similar to the larger alkaline earth metal cations, especially with respect to the fact that at slightly elevated temperatures (303–218 K) a more or less complete exchange of Na^+ by Co^{2+}, Zn^{2+} in Na-X and Na-Y or Cu^{2+} in Na-X is observed [66, 72, 73]. However, a closer look shows that this explanation for the restricted exchange at room temperature does not hold for the exchange of the transition metal cations. The radii of the hydrated transition

References see page 1093

metal cations and those of Ca^{2+}, Sr^{2+}, and Ba^{2+} are rather similar (4.0–4.3 Å [61]) but the Pauling radii of the transition metal cations of 0.70–0.75 Å are much smaller than those of the alkaline earth cations. Obviously, the theoretical exchange levels of 0.82 (Na-X) and 0.68 (Na-Y), which correspond to the restricted access to the β-cages of the faujasite zeolites used by Maes and Cremers [66], are not matched by most of the exchange isotherms; the numbers found are in some cases below these values but often also exceed them. Furthermore, Cu^{2+} and Ni^{2+} have almost identical hydration energies [38], thus differences of the sizes of the hydrated cations or energies of dehydration cannot explain the significant differences of the exchange behavior of the different transition metal cations sufficiently. Maes and Cremers [66] showed that the β-cages of zeolite Y are occupied by transition metal cations even at moderate temperatures and that transition metal cations may pass through the 6MR of the β-cages even at low transition metal loading [64]. They assume that differences of the transition metal exchange may be caused by different coordination of the respective cations by oxygen atoms from the zeolite framework; e.g., Ni^{2+} strictly prefers octahedral coordination due to crystal field stabilization energies while other cations, such as Co^{2+}, may accept both octahedral and tetrahedral coordination. Thus, different siting of the transition metal cations in the hydrated zeolites may significantly affect the ion-exchange process. An aspect that cannot be neglected is the protonation, especially after repeated exchanges, of the zeolites in the slightly acidic transition metal solutions; one may consider the transition metal exchange as a ternary exchange process. Due to protonation, the maximum degree of exchange of the zeolites by transition metals may be reduced significantly [66] and, as mentioned before, low-silica zeolites, such as zeolite X, may suffer severe damage from the acidic exchange solutions, while the structure of the more siliceous zeolite Y is more resistant to an acidic environment [48, 49, 51, 64].

During dehydration the transition metal cations successively get coordinated by framework oxygen and, finally, after complete dehydration, they migrate on sites with favorable coordination (e.g. sites S1, S1', S2, S2', S3) [44, 74–76]. On some of these sites, for example on site S1 or on site S1', they get locked during the dehydration at elevated temperature, which means they cannot be re-exchanged at room temperature [77–80].

4.2.8.2.3 Ion Exchange in Zeolite P

Zeolite P belongs to a family of zeolites that are related to the mineral gismondine. The framework of zeolite P is extremely flexible and, depending on the content of water molecules and on the composition of the zeolite, structural changes are observed. Subtle changes of the zeolite framework are observed not only due to its framework composition but also to its extra-framework composition resulting in distortions of the zeolite framework. It is usually obtained either in a pseudo-cubic form, denoted as zeolite P_c [81], a tetragonal form [82], denoted as zeolite P_t, or an orthorhombic form, denoted as zeolite P_o [83]. For low-silica zeolite P, a monoclinic crystal structure was determined [84]. The general gismondine-type structure topology of zeolite P is shown in Fig. 13a. It consists of two connected double-

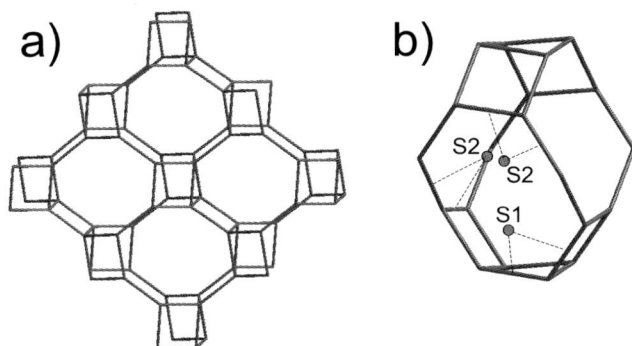

Fig. 13. (a) Structure of the pseudo-cubic modification of zeolite P; (b) cation sites in zeolite P; spheres indicate sites, dotted lines the coordination with framework oxygen atoms (oxygen atoms are omitted in both figures, the lines connect neighboring T-atoms) [81].

crankshaft chains parallel to [100] and [010] forming a two-dimensional pore system. The pore system consists of intersecting elliptical channels with sizes of about 3.1×4.5 Å and 2.8×4.8 Å that are confined by 8MR [85].

The position of the cation sites depends on the respective structure modification. Figure 13b shows the two sites S1 and S2 found by Baerlocher and Meier [81] in the pseudo-cubic zeolite P_c. Cations on both sites are coordinated by water molecules and two oxygen atoms from the zeolite framework. They are located in the cavities that are formed at the intersections of the two 8MR channel systems. Their position in front of the 8MR windows is slightly shifted into the cavities. The water sites, which were also described by Baerlocher and Meier [81] are omitted from Fig. 13b. Further sites had been described for the other forms of zeolite P [82–84, 86] and for gismondine [36]. However, as can be seen from Fig. 13b for the pseudo-cubic Na-P, there are only a few cation positions in one specific form of zeolite P. As mentioned, the framework of zeolite P is very flexible and undergoes reversible structural changes either as a result of reducing the content of water in the crystals or due to ion exchange. These structural changes significantly affect the thermodynamics of the ion exchange of zeolite P [87]. Figure 14 shows the ion-exchange isotherms of alkali metal cations with Na-P at 298 K. The Na^+-containing zeolite has a high selectivity for K^+, Rb^+, and Cs^+, while Na^+ is significantly preferred over Li^+. In contrast to K^+, Rb^+, and Cs^+, which completely replace the Na^+ in the zeolite, Li^+ exchanges Na^+ only to a maximum content of $Z_{Li} = 0.5$. During the successive incorporation of Li^+, K^+, Rb^+, and Cs^+ the symmetry of the zeolite changes from cubic (Na-P_c) to tetragonal (Na-P_t) [87]. This reversible transition is due to the loss of water during the exchange and to the site occupation of the entering cations.

Zeolite P is not only very selective for alkali metal cations, except Li^+, but also for

References see page 1093

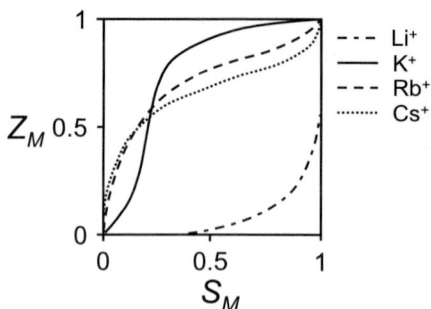

Fig. 14. Ion-exchange isotherms of Na-P with M = Li+, K+, Rb+ and Cs+ (total normality 0.1 N, 298 K, Na-P: $Na_{34.8}Al_{35.4}Si_{60.6}O_{192}\cdot71.4H_2O$) [87].

alkaline earth metal cations as shown in Fig. 15. Ca^{2+}, Sr^{2+}, and Ba^{2+} completely exchange Na^+ and are preferred at any composition of the zeolite [87, 88]. Despite the fact that zeolite P has similar sizes of its cavity openings as zeolite A, it has a much higher selectivity for Ba^{2+}, which is almost completely removed from the solution by Na-P up to a loading of the zeolite of $Z_{Ba} = 0.5$. Similar to the exchanges with alkali metal cations, phase transitions from cubic to tetragonal symmetry are observed with successive loading of the zeolite with the entering cation.

The ion-exchange isotherms of other metal cations at 298 K are shown in Fig. 16. All exchanges proceed quite quickly and the zeolite has significantly different preferences for three metal cations Pb^{2+}, Zn^{2+}, and Ni^{2+}. It clearly prefers Pb^{2+} over Na^+ and, at high loading of the zeolite, Pb^{2+} can replace basically all the sodium from zeolite P; the maximum equilibrium fraction of Pb^{2+} in the zeolite is 0.95. However, Moirou et al. [89] observed an almost complete and irreversible destruction of the zeolite structure after that exchange. Nevertheless, the amorphous Pb^{2+}-containing material could be completely re-exchanged with Na^+. In contrast, Ni^{2+} is only preferred by Na-P at very low loading of the transition metal, the maximum equilibrium fraction of Ni^{2+} in the zeolite is only 0.27. The selectivity of Na-P for Zn^{2+} is in between those for the other two metal cations. The maximum equilibrium fraction of Zn^{2+} in the zeolite is 0.76, indicating that more exchange sites are accessible for Zn^{2+} than for Ni^{2+}, but still not as much as for Pb^{2+}. The crystallinity of zeolite P exchanged with Zn^{2+} and Ni^{2+} seems to be slightly reduced, but a complete destruction of the zeolite framework, as happened after the exchange with Pb^{2+}, was not observed. Since all three cations in their hydrated form are too large to pass through the windows of the cavities of zeolite P, size-

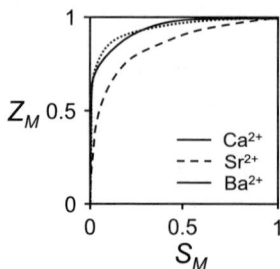

Fig. 15. Ion-exchange isotherms of Na-P with M = Ca^{2+}, Sr^{2+}, and Ba^{2+} (total normality 0.1 N, 298 K for Sr^{2+} and Ba^{2+}, 303 K for Ca^{2+}, anhydrous Na-P: $Na_{34.8}Al_{35.4}Si_{60.6}O_{192}$ for Sr^{2+} and Ba^{2+}, $Na_{46.8}Al_{46.8}Si_{49.2}O_{192}$ for Ca^{2+}) [87, 88].

Fig. 16. Ion-exchange isotherms of Na-P with M = Pb^{2+}, Zn^{2+}, and Ni^{2+} (total normality 0.015 N for Pb^{2+}, 0.014 N for Ni^{2+}, and 0.012 N for Zn^{2+}, 298 K, anhydrous Na-P: $Na_{26.6}K_{2.1}Ca_{1.4}Al_{29.4}Si_{65.5}O_{192}$) [89].

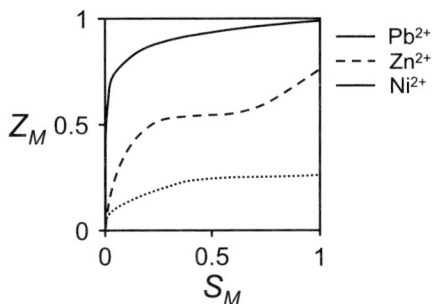

exclusion effects alone cannot explain the different selectivity of the zeolite for the specific heavy metal cations. Moirou et al. [89] assume a similar site specificity of the metal cations as observed for the faujasite-type zeolites. The cations in the zeolite, even if they are partly dehydrated, are rather coordinated by water molecules than by oxygen atoms from the zeolite framework that is not expected to be favorable for every heavy metal cation. Furthermore, dehydration of the cations, which seems to be necessary to fit them into the restricted space of the cavities, might lead to less-advantageous coordination. At higher loading, the restricted space in the pore system may require Ni^{2+} to be coordinated by less than six water molecules which is energetically less favored. Thus, crystal field stabilization energies are supposed to affect the selectivity of the zeolite for a specific metal cation in zeolite P as well [89]. Apart from the structure collapse of Pb^{2+}-exchanged zeolite P, structural transition of zeolite P, as reported for the exchange of alkali metal and/or alkaline earth metal cations, were not observed by Moirou et al. [89] for the exchange with Pb^{2+}, Zn^{2+}, and Ni^{2+}.

4.2.8.2.4 Ion Exchange in ZSM-5

ZSM-5 is a high-silica zeolite of structure type MFI [90] crystallizing in the space group *Pnma* [91, 92]. The structure of ZSM-5 is formed by chains of 5-membered rings (5MR), denoted as pentasil chains, which are connected in two dimensions. They can be considered to form so-called pentasil layers along [100] by connecting the chains via 5MRs as shown in Fig. 17a. The pentasil layers are connected by 4MRs and 6MRs along [100] by which the three-dimensional structure of ZSM-5 is formed, as illustrated in Fig. 17b. The lattice parameters of ZSM-5 are $a = 20.07$ Å, $b = 19.92$ Å and $c = 13.42$ Å [34]. The structure has 12 crystallographic T-sites, which can be occupied either by silicon or aluminum atoms. In contrast to the zeolites described before, ZSM-5 has no cavities but intersecting channels. There exist two types of channels, straight ones along [010] and sinusoidal channels along [100]. The channels are formed by 10-membered rings (10MR) and are slightly elliptical with diameters of 5.1×5.5 Å (straight channels) and 5.3×5.6 Å (sinusoidal channels) [34, 92]. At the channel intersections larger voids of 8.7–9.0 Å are formed by the two channels. The general composition of a unit cell is

References see page 1093

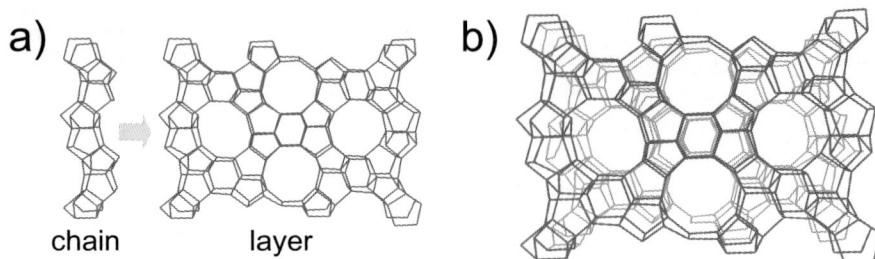

Fig. 17. (a) Formation of the structure of ZSM-5 from chains of 5MR; (b) view into the straight channels of ZSM-5 along [010] (oxygen atoms are omitted in both figures, the lines connect neighboring T-atoms).

$Na_xAl_xSi_{96-x}O_{192}\cdot 16\,H_2O$ with $x < 27$. Usually ZSM-5 has a rather high Si/Al ratio and is thus called a high-silica zeolite. Its aluminum-free form is called silicalite-1 and is accessible directly via a hydrothermal reaction [93]. Since no framework aluminum is present in silicalite-1, it has basically no ion-exchange properties. In ZSM-5 usually only a few aluminum atoms per ZSM-5 unit cell are present, which results in a rather low ion-exchange capacity of ZSM-5. The sites of the framework aluminum atoms as well as those of the charge balancing cations in ZSM-5 are not as well known as for the low-silica zeolites A, X, Y, and P. The reason for this is that specific sites that may also exist in ZSM-5 are occupied by only a few cations, which makes them difficult to detect in diffraction experiments. Therefore, the framework structure of ZSM-5 is well established, but the location of its cation sites is still under investigation. Nevertheless, there has been some progress and specific sites have been determined, of which a short selection will be presented below. The nomenclature for the sites in ZSM-5 is not strictly handled by the authors and therefore, the notations of the sites will be arbitrary and differ from those given by the respective authors who all used their own notations.

The cation positions that were reported for ZSM-5 on the basis of crystallographic data are shown in Fig. 18. The two views show identical parts of the ZSM-5 structure rotated by 90° and allow location of the cations in the straight and sinusoidal channels. Two sites for Cs^+ were suggested by Lin et al. [94] in the straight channels of Cs-ZSM-5, the first one (Cs 1) close to the center of the channel on top of a 6MR in the hydrated zeolite, the second one (Cs 2) in the dehydrated zeolite near the wall where it is coordinated by oxygen atoms from a 6MR. The occupation of a Ni^{2+} site (Ni 1) in the sinusoidal site on top of a 4MR was reported by Liu et al. [95] who investigated cation sites in dehydrated Ni(II)-ZSM-5. They proposed another site (Ni 2) in a small cage in the wall of the sinusoidal channels as shown in Fig. 18. Huddersman and Rees [96] investigated Tl-exchanged ZSM-5 and found three Tl^+ sites in or close to the intersections of the straight and sinusoidal channels of the hydrated zeolite. According to the authors the most reliably located site due to its occupancy is almost in the center of the channel intersection (Tl 3), while the other two sites (Tl 1 and Tl 2) were found closer toward

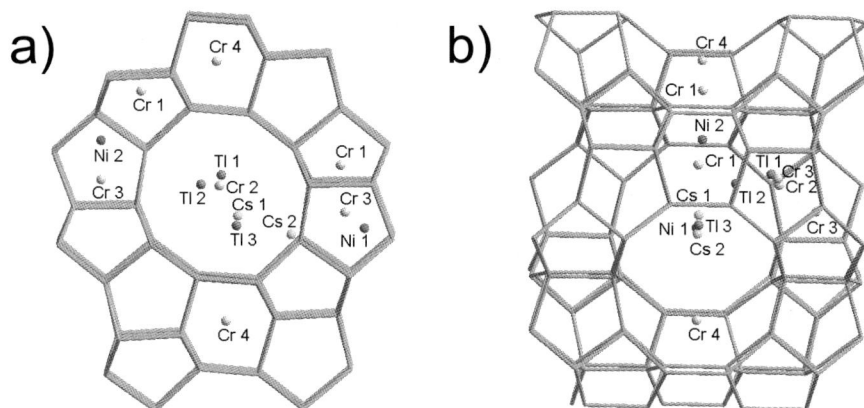

Fig. 18. Cation sites in ZSM-5 according to the notation in the text, (a) view along a straight channel; (b) view along a sinusoidal channel (oxygen atoms are omitted in both views, the lines connect neighboring T-atoms).

the back of the intersection near a 6MR of the framework. All Tl^+ cations are predominantly coordinated by water molecules on these three positions. The site Tl 3 is very similar to site Cs 1, which was reported by Lin et al. [94], the distance between these two sites is only 0.51 Å. Rachapudi et al. [97] published four sites (Cr 1–Cr 4) for Cr^{3+} cations, all of them accessible either in the straight or sinusoidal channels. The occupation of site Cr 3 was only observed in one sample, while Cr^{3+} cations on sites Cr 1, Cr 2, and Cr 4 were found more frequently. Site Cr 3 is located in a cage formed by eight 5MRs (pentasil unit). Lamberti et al. [98] proposed two distinct sites in the straight channels of Cu(I)-ZSM-5 on the base of spectroscopic data (not shown in Fig. 18). One site (Cu 1) is located on top of a T-atom directly at the intersection of the two channels where the Cu^+ is coordinated by three oxygen atoms of that TO_4 unit. The second site (Cu 2) is close by on the opposite side of the respective channel, and the cation is coordinated by two oxygen atoms close to the site and one slightly further away, but also belonging to the same TO_4 unit. Stokes et al. [99] supposed similar positions for Pd^+ in Pd(I)-ZSM-5 on the basis of EPR studies.

As shown by these few examples, the position on which a specific cation is found in ZSM-5 depends strongly on the type of cation and on the degree of hydration of both the zeolite and the cation. However, some sites, e.g. Cs 1/Tl 3 or Cr 2/Tl 1, seem to be preferred by some cations. The fact that cations in dehydrated ZSM-5 seem to be preferentially located on sites close to 4MRs has been taken as an indication for aluminum atoms in the 4MRs. Further cation sites may be found in the literature but discussing every published extra-framework site would be beyond the scope of the present section. In ion-exchanged and hydrated ZSM-5 materials par-

References see page 1093

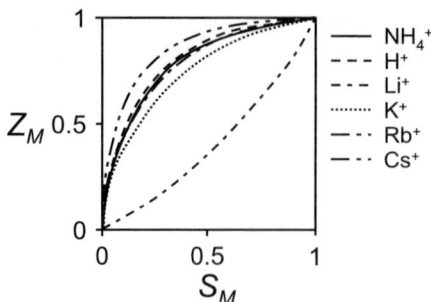

Fig. 19. Ion-exchange isotherms of Na-ZSM-5 with M = NH_4^+, H^+, Li^+, K^+, Rb^+, and Cs^+ (total normality 0.05 N, 298 K, anhydrous Na-ZSM-5: $Na_{2.4}Al_{2.4}Si_{93.6}O_{192}$) [100].

ticular water sites have also been found that are of importance for the coordination of the respective metal cations in the zeolite [e.g. 94, 96].

The synthesis of ZSM-5 is usually performed by using a structure-directing agent, called a template, which is most frequently a tetraalkylammonium cation, an amine, or an alcohol. A positively charged template molecule may also balance some of the negative framework charges. The positions of these template cations are also known for several materials but usually larger template cations cannot be exchanged by a simple ion exchange in aqueous solutions. They have to be removed by calcination, a process that will not be covered in this section.

The environment for cations in ZSM-5 is different from those in the zeolites A, X, Y, and P. ZSM-5 contains only a relative small amount of aluminum and, thus, only a few extra-framework cations for charge compensation are needed. Due to the low amount of cations, the distances between neighboring cations could thus be quite large. In contrast to the low-silica zeolites, the siliceous ZSM-5 is less susceptible to acidic environments (e.g. transition-metal solutions) and exchanges in acidic solutions are possible without too severe damage to the zeolite structure. Even a direct exchange of sodium cations by protons is possible. The ion-exchange isotherms for the exchange of Na^+ by monovalent cations in ZSM-5 are shown in Fig. 19. At 298 K, as well as at 333 K, the selectivity of ZSM-5 for these cations decreases in the series $Cs^+ > H^+ = NH_4^+ = Rb^+ > K^+ > Na^+ > Li^+$, as shown by Matthews and Rees on the base of calculations of the ΔG^0 of these exchange reactions [100]. All the monovalent cations are able to replace the Na^+ in the zeolite, but larger cations that are only weakly hydrated are preferred by ZSM-5, as predicted by Sherry [101] for high-silica zeolites. Interestingly, the exchange of Na^+ by NH_4^+ is enhanced at room temperature compared to exchanges at elevated temperature [102, 103], which is surprising since on other zeolites like faujasites the NH_4^+ selectivity increases with temperature.

Cations with higher charges, such as the alkaline earth cations Ca^{2+}, Sr^{2+}, or Ba^{2+} or the rare earth cation La^{3+}, can only replace a small amount of the Na^+ in ZSM-5, as shown in Fig. 20. The hydration shells of these cations are too large and only partially dehydrated cations may enter the channels of ZSM-5. The larger alkaline earth cations are less strongly hydrated and thus the selectivity of ZSM-5 for the cations increases in the order $Ca^{2+} < Sr^{2+} < Ba^{2+}$.

Fig. 20. Ion-exchange isotherms of Na-ZSM-5 with M $= Ca^{2+}$, Sr^{2+}, Ba^{2+}, and La^{3+} (total normality 0.05 N, 298 K, anhydrous Na-ZSM-5: $Na_{2.4}Al_{2.4}Si_{93.6}O_{192}$) [100].

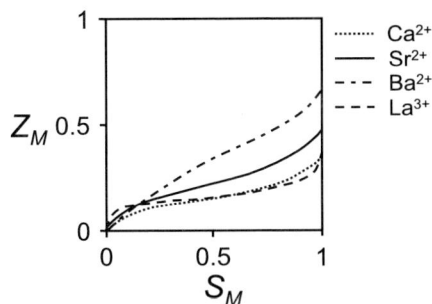

At elevated temperature the amount of alkaline earth metal cations in ZSM-5 can be increased due to enhanced water stripping from the cations [100]. The large, but trivalent, rare-earth cation La^{3+} is also strongly hydrated and, thus, is excluded from the zeolite channels as well. It exchanges only a small fraction of the Na^{+} present in ZSM-5. Another problem arising with trivalent cations in high-silica zeolites is the spacing of the exchange sites, the distances may be too large to allow a charge compensation by one multivalent cation. Chu and Dwyer [103] explained the selectivity of ZSM-5 for large cations by the weak anionic field strength of that zeolite. Due to this weak field, the cation selectivity is mainly controlled by the size of the cation. In order to enter the zeolite channels, all cations have to strip water molecules from their hydration shells and, thus, the weakly hydrated cations, such as the large alkali-metal cations, may enter preferentially the channels of ZSM-5. The bivalent alkaline-earth-metal cations are more strongly hydrated and, thus, the selectivity of the zeolite for those cations is reduced.

In contrast to the alkaline-earth cations, bivalent transition-metal cations exchange much easier with Na^{+} in ZSM-5. As shown in Fig. 21, Zn^{2+}, Ni^{2+}, and Cu^{2+} completely replace Na^{+} in ZSM-5. Zn^{2+} and Cu^{2+} are slightly preferred over Na^{+} at any degree of exchange while Ni^{2+} is less preferred, the selectivity sequence is $Cu^{2+} > Zn^{2+} > Ni^{2+}$. However, a complete exchange by Ni^{2+} is possible, which could not be achieved with the low-silica zeolites. The selectivity for these transition metal cations is not much affected by the aluminum content of the ZSM-5

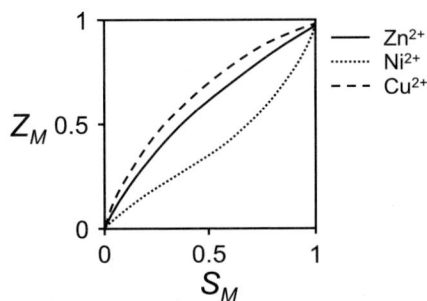

Fig. 21. Ion-exchange isotherms of Na-ZSM-5 with M $= Zn^{2+}$, Ni^{2+}, and Cu^{2+} (total normality 0.1 N, 298 K, anhydrous Na-ZSM-5: $Na_{4.6}Al_{4.6}Si_{91.4}O_{192}$) [103].

References see page 1093

framework. Chu and Dwyer [103] observed no changes of the selectivity of ZSM-5 samples with Si/Al ratios ranging from 20 to 103.

Depending on the preparation conditions, the aluminum is not distributed homogeneously over the whole ZSM-5 crystals but enriched on the outer zones of the crystals (zoning). Therefore, the exchange takes place in the outer regions of the crystals and on its external surface. Handreck and Smith [104] showed that up to 25 % of all exchange sites of ZSM-5 may be on the external surface of the crystals. The cations in hydrated ZSM-5 are most probably coordinated by water molecules that are stripped from the cations on dehydration of the zeolite. The coordination of the cations changes during the removal of the water molecules and coordinative preferences may direct different cations onto specific sites.

4.2.8.3
Nonaqueous Ion Exchange on Zeolites

Ion exchange from aqueous solutions is the most commonly applied method; however, ion exchange from nonaqueous systems is also possible. A method that is occasionally used, even if not very frequently, is the exchange from solvents others than water. Polar organics, either pure or mixed with water, have been used [105–108] as well as more exotic solvents like nonaqueous ammonia solutions [109]. However, the success of the ion exchange depends strongly on the solvent used, and possibly no ion exchange at all might take place, as experienced by Ho et al. [110]. The solvent molecule has to fit into the zeolite pores, otherwise an ion exchange is not likely. The rational selectivity coefficients, K_{Na}^M, for the exchange of Na^+ by some cations (M^{m+}) in various solvents are listed in Table 3 [111].

The solvent has not only a significant influence on the selectivity of the zeolite for a specific cation but also on the exchange kinetics, as shown in Table 4. The exchange in organic solvents is a slower process than in water and the exchange rate decreases with decreasing dielectric constant. In some systems, e.g. exchange of Na^+ by K^+ or NH_4^+ in zeolite A (solvent ethanol) [105] or K^+ in zeolite X (see

Tab. 3. Dependency of the rational selectivity coefficient K_{Na}^M for the ion exchange of Na^+ by $M = Li^+$, K^+, Ca^{2+}, and Ag^+ in zeolite X in different solvents; $\varepsilon =$ dielectric constant of solvent (303 K, salt concentration 0.05 M) [111].

Solvent	ε	K_{Na}^M			
		Li^+	K^+	Ca^{2+}	Ag^+
water	78.7	8.4×10^{-2}	1.10	0.44	45.9
methanol	31.7	4.2×10^{-2}	0.96	4.5×10^{-3}	18.1
ethanol	23.6	3.9×10^{-2}	0.70	1.1×10^{-3}	6.34
n-propanol	19.8	1.5×10^{-2}	0.50		
i-propanol	17.9	1.3×10^{-2}	0.19		
i-butanol	17.1	1.1×10^{-2}	0.19		

Tab. 4. Data on the exchange kinetics of zeolite X in different solvents; $t_{1/2}$ = time required to attain 50 % exchange equilibrium (303 K, salt concentration 0.05 M) [111].

Cation	Solvent	$t_{1/2}$/min
K^+	water	18
K^+	methanol	19
K^+	ethanol	48
Ag^+	water	40
Ag^+	methanol	120
Ag^+	ethanol	295

nonaqueous solvents in Table 4), a selectivity reversal is observed if organic solvents are used instead of water.

While ion exchanges from solutions others than aqueous ones are rather rarely applied solid-state ion exchange is a more frequently used technique [112–121]. The zeolite is contacted with a salt or oxide of the cation to be exchanged and then heated to temperatures in the range of 470–1100 K. The ion exchange proceeds either from a salt melt covering the zeolite particles or via a real solid-state reaction between the two solid components. The advantage of this method is its easy applicability and the fact that no large amounts of exchange solutions have to be handled. However, the zeolites have to be washed after the exchange to remove the residual salts and oxides and salt imbibition into the zeolite occurs easily from the salt melts covering the zeolite crystals. Some zeolites can accommodate remarkable amounts of salt inside their pore systems, e.g. zeolite A occludes ten $NaNO_3$ per unit cell when it is brought in contact with a $NaNO_3$ melt at about 600 K [122]. Thus, the ion-exchange capacity of the zeolite is drastically changed due to the occluded salt. However, depending on the location of the salt inside the zeolite, the surplus cations in the zeolite either may be completely or partially re-exchanged by other cations. If the salt is occluded in smaller cavities or cages, the re-exchange depends on the size of the respective cations. If they cannot pass the windows to that small cage, they are excluded and no complete re-exchange is possible [105].

4.2.8.4
Application of Zeolites as Ion Exchanger

4.2.8.4.1 Detergent Builders
The ability of detergents to clean textiles is mainly based on the ability of surfactants (tensides) to interact with hydrophobic surfaces. Dirt particles are removed from the surfaces and included in micelles, which are formed due to the polar (hydrophilic)/nonpolar (lipophilic) functionality of the surfactant molecules. The dirt particles remain in the aqueous suspension due to repulsive forces between the micelles. Unfortunately, the surfactants used form complexes with Ca^{2+} and

References see page 1093

Tab. 5. Development of the compositions of builders in detergents [123].

Year	Detergent composition
1907	sodium silicate + soda
1933	sodium diphosphate
1946	sodium tripolyphosphate
1976	zeolite A + sodium tripolyphosphate
1983	zeolite A + soda + cobuilder (e.g., polycarboxylates, citrate, phosphonate)
1994	zeolite A + silicates + cobuilder (e.g., polycarboxylates, citrate, phosphonate)
1994	zeolite P

Mg^{2+} ions, a process that significantly reduces the cleaning activity of the surfactant. These alkaline earth cations do not only react with the surfactants but also contribute to incrustations on the textiles, e.g. as carbonates. Therefore, it is advantageous to remove them from washing solutions, which is achieved by so-called builders, preferentially during the first one to two minutes of the washing process. The builders act as water softeners and as supports for the surfactants. The composition of the detergents changed during the last 100 years as shown in Table 5 [123]. The first builders used silicates (sodium waterglass) as ion exchangers that were replaced by phosphates in the mid-1930s. For ecological reasons, alternatives for the phosphates had to be found in the mid-1970s. Due to its high ion-exchange capacity and high selectivity for Ca^{2+}, zeolite A was a perfect substitute for the phosphates [124–128], which were successively reduced in the builder compositions over time. The exchange of Na^+ by Ca^{2+} in zeolites usually proceeds rapidly even at room temperature, which is especially important for the US market where washing times are usually shorter and the washing temperatures rather low compared to Europe. In the 1990s special silicates were re-employed and zeolite P appeared to be a suitable alternative to zeolite A. Zeolite P had been considered as a builder in the 1970s but was not used due to its higher production costs (longer crystallization times) and lower exchange capacities of the early zeolite P samples compared to zeolite A. Since 1994 new types of zeolite P with higher aluminum contents have been on the market that are suitable substitutes for zeolite A. However, zeolite A is still the dominant compound in zeolite-containing builders. The builder makes 35–45 % of washing powders of which 15–25 % are zeolites in phosphate-free detergents. The composition of phosphate-free builder systems is given in Table 6.

Tab. 6. Composition of phosphate-free builders [123].

Builder/cobuilder	Function	Fraction in formulation
zeolite A	ion exchanger	15–25 %
polycarboxylate	dispersing agent	2–6 %
citrate	complexing agent	0–10 %
sodium carbonate (soda)	alkali provider	5–15 %

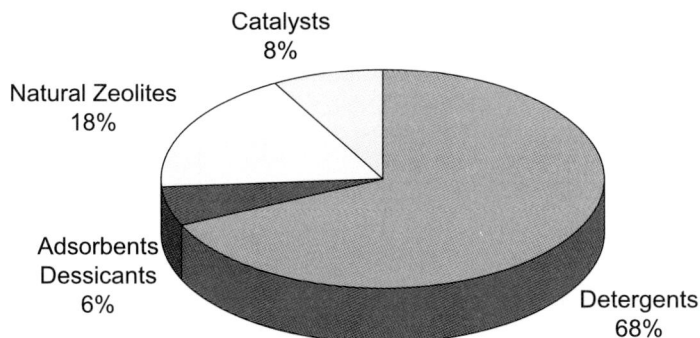

Fig. 22. Zeolite market in 1996 [130].

At room temperature zeolite A is capable of binding 160–175 mg CaO per gram of zeolite and at 333 K 180 g CaO per gram of zeolite [123, 129]. The Mg^{2+} exchange capacity of zeolite A depends more strongly on the temperature than the Ca^{2+}-exchange capacity. At room temperature zeolite A binds about 20–40 mg MgO per gram of zeolite which increases to 50 mg MgO per gram of zeolite at 363 K [123]. The ion-exchange kinetics is faster at elevated temperature, due to the enhanced stripping of water molecules from the cations.

Within the last decades, during which phosphates have been successively replaced in detergents due to ecological considerations, the application of zeolite A as detergent builder became the largest market segment for zeolites, as shown in Fig. 22 [130]. The development of the zeolite A production continuously increased until the mid-1990s when silicates and zeolite P reappeared on the market, as shown in Fig. 23.

There are several facts that make zeolite P interesting as a substitute for zeolite A. Its average particle size of about 1 μm in commercial products is smaller than

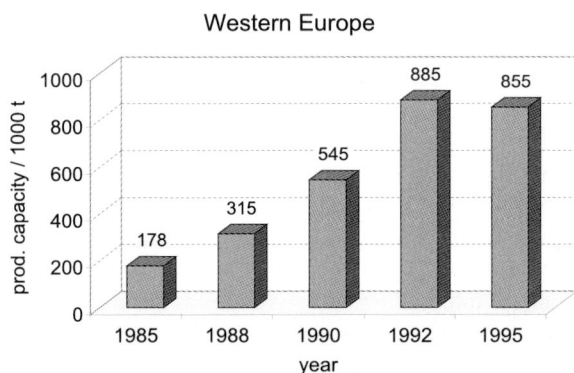

Fig. 23. Production capacity for zeolite A in Western Europe [123].

References see page 1093

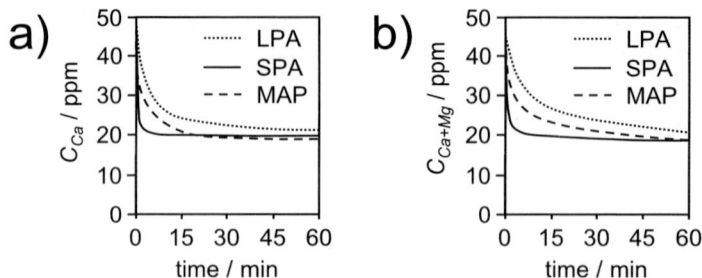

Fig. 24. (a) Calcium depletion (303 K, 0.0025 N); (b) hardness ion depletion (Ca/Mg = 2, 303 K, 0.0025 N) of solutions by large-particle zeolite A (LPA, particle size: 8 μm, crystallite size: 2.5–3.0 μm), small-particle zeolite A (SPA, particle size: 3.4 μm, crystallite size: 1.0–1.5 μm), and maximum aluminum zeolite P (MAP, particle size: 2 μm, crystallite size: 0.1–0.2 μm) [88].

that of zeolite A of 2.5–3.5 μm. Smaller particles contribute less to the textile incrustation and provide a larger external surface area for the deposition of surfactants. Thus, they are advantageous for detergent builders. The ion-exchange capacity of the high-alumina modifications of zeolite P are about 140–165 mg CaO per gram of zeolite at room temperature and 180 mg CaO per gram of zeolite at 333 K [123, 129], while the Mg^{2+}-exchange capacities of both zeolites are almost the same [123, 129]. The ion-exchange kinetics of the two zeolites is the subject of controversial discussions. While some authors report a faster uptake of Ca^{2+} and Mg^{2+} in maximum alumina zeolite P (MAP) [123, 129], Borgstedt et al. [88] observed faster uptake of the cations in small-particle zeolite A, as shown in Fig. 24. Factors like particle size, dimensionality and accessibility of the pore system play a crucial role for the ion exchange and have to be considered when Ca^{2+} and Mg^{2+} depletion from solutions are discussed with respect to kinetics. The exchange kinetics of Ca^{2+} and Mg^{2+} are significantly dependent on the size of the zeolite particles: the smaller the particles the faster the ion exchange. For zeolite particles of similar size, one should expect a more rapid exchange for zeolites with a 3D pore system, as in zeolite A, than for a 2D pore system, as in zeolite P. As shown in Fig. 24, the fastest exchange kinetics is observed for small zeolite A particles of even slightly larger size than that of the zeolite P that was used for the investigation. However, zeolite P seems to bind the hardness cations slightly stronger than zeolite A and less of them are released from Ca^{2+}- and Mg^{2+}-loaded zeolite P [123], and the bleach stability of a builder is increased when zeolite P is used instead of zeolite A [123, 129]. At present, zeolite A is still the most used zeolite in builder systems. The integration of zeolite P into builder systems depends strongly on the actual price for the zeolite production and as long as the prices for zeolite A does not increase significantly, there is only little stimulus for the use of zeolite P. Alternatively, crystalline and/or amorphous silicates are used complementary to zeolite A in builders, with their easy solubility making them especially attractive for soluble builders [123]. Their application in detergents is increasing but their share in builder formulations is still rather low compared to zeolite A.

4.2.8.4.2 Waste Water Treatment and Deposition of Heavy Metal Waste

The regeneration of water is of predominant importance not only with respect to drinking water but also to environmental considerations. High ammonium concentrations, as are quite frequently found in municipal waters, are not only poisonous but also contribute significantly to the eutrophication of lakes and rivers, and the toxicity of heavy metal cations is well known. Especially copper cations are toxic for mammals [131], whereas copper, cadmium, and nickel are among the most toxic metals for higher plants and microorganisms [131–133]. Lead is not as acutely toxic as copper, cadmium and nickel, but it easily accumulates in organisms [132]. In order to avoid health risks and environmental pollution, both ammonium and heavy metal cations have to be removed from waste waters. Ion exchangers like organic resins, and especially zeolites, are used both for the removal of ammonium and/or metal cations from municipal, agricultural, and industrial sewage. For this kind of applications synthetic zeolites [134–136] as well as natural zeolites, especially clinoptilolite, are applied [137–142]. Clinoptilolite is able to remove not only considerable amounts of NH_4^+ cations from aqueous solutions [137, 141–145] but also, to a lesser amount but still efficiently, heavy metal cations like Zn^{2+}, Pb^{2+}, Cu^{2+}, and Cd^{2+} [138–141, 143, 146–148]. Other natural zeolites that have been investigated for the cleaning of municipal water are mordenite and ferrierite, which both show similar exchange activities as clinoptilolite, however, clinoptilolite is found to be superior to these two minerals [138, 139, 141]. As documented by Semmens and Martin [140], as well as by Loizidou and Townsend [138, 139], the capacity and selectivity of clinoptilolite for metal cations depends strongly on the pretreatment of the zeolite, i.e. conditioning of the zeolite with K^+, NH_4^+, and/or Ca^{2+} solutions.

Natural zeolites usually occur in changing compositions and with various amounts of impurities. As an alternative, the well-defined synthetic zeolites may be used. Very high alumina zeolites, such as zeolite A, can be used for the regeneration of waste waters, since they easily remove heavy metal cations from aqueous solutions (see Fig. 7). However, they are not very stable in solution with higher metal concentrations [48] and hydrolyze in natural waters [134]. At neutral pH the half-lives of zeolite A were shown to be typically in the range of 1–2 month. Synthetic zeolites with slightly higher silicon contents, such as zeolite Y, are advantageous with respect to long term stability in aqueous solutions, even when their lower exchange capacity is taken into account. Keane [136] reported that the cation balancing the framework charge in the parent zeolite Y, i.e. Li^+, Na^+, K^+, Rb^+, or Cs^+, significantly affects the extent of removal of Cu^{2+} and Ni^{2+} as one would expect from the preference of the zeolite for specific cations. The advantage of using ion exchangers as zeolites or resins for the regeneration of waste waters compared

References see page 1093

with other methods, like chemical precipitation or reverse osmosis, is the easy application of the solid ion exchangers as the stationary phase in exchange columns. Furthermore, ion exchangers usually can be regenerated when their maximum capacity is reached and reused after the regeneration. The recovery of the exchanged cations is of special importance when zeolites are applied for the regeneration of fission product containing waste water. Natural zeolites, e.g. clinoptilolite, erionite, or mordenite, as well as synthetic zeolites, e.g. zeolite Y, efficiently remove radioactive elements, such as ^{137}Cs, ^{60}Co, ^{239}Np, or ^{235}U, from aqueous solutions [149–152]. The radioactive cations can be easily recovered from the respective zeolites and then further processed for their deposition, e.g. precipitation and storage as salts.

4.2.8.4.3 Preparation of Zeolite Catalysts by Ion Exchange

Cations in zeolites can act as active sites in catalytic reactions. Therefore, ion-exchanged zeolites are quite frequently used as catalysts in industrial processes [153]. Protonated zeolites, such as H-X, H-Y, or H-ZSM-5, are highly acidic catalysts due their ability to easily release protons that are located in the zeolite pores as charge-compensating cations. This type of catalyst is most frequently used in petrochemical processes, such as fluid catalytic cracking or isomerization reactions [154, 155], but can also be applied for the production of fine chemicals [156]. The direct protonation with mineral acids, by which the charge-compensating cation (usually Na$^+$) in the zeolite is directly exchanged against H$^+$, is possible. However, since the framework structures of zeolites easily get damaged in acidic solutions, the exchange with ammonium cations followed by a calcination step is preferred. During the thermal treatment the ammonium cations decompose according to the reaction

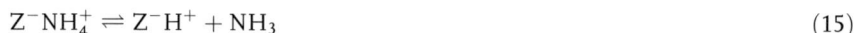

$$Z^-NH_4^+ \rightleftharpoons Z^-H^+ + NH_3 \qquad (15)$$

resulting in the protonated zeolite (Z$^-$H$^+$). The zeolite catalysts may be used as crystalline powders in fluidized-bed reactors (zeolite Y in FCC process) or as extrudates or beads. In the latter cases the protonation process is implemented in the processing of the zeolite (synthesis – filtration – drying/calcination – extrudation/pelletization (with binders, e.g. clay) – ion exchange (ammonium exchange) – drying/calcination). Zeolite Y as a FCC catalyst is usually not only applied in its protonated form, but also exchanged with rare-earth cations by which the FCC catalyst is stabilized.

Slightly basic catalysts are obtained by the incorporation of excess alkali metal cations in the zeolite, e.g. by salt impregnation (salt imbibition) with nitrates, which then form oxide species in the zeolite when the salt is decomposed by calcination.

Transition-metal-exchanged zeolites are also used as catalysts [156, 157]. The metal cations are deposited in a zeolite by ion exchange, and the obtained transition metal containing zeolite then may be used directly as a catalyst, or after the reduction of the metal. In the latter case the zeolite acts as a support for highly dispersed metal particles that are then the catalytically active components, e.g. for reduction and/or oxidation reactions [157].

4.2.8.4.4 **Preparation of Adsorbents by Ion Exchange**

Most zeolites can accommodate large amounts of gas molecules in their pores and/or cavities after the water molecules have been removed from the pores. The size of the pore opening of the zeolites determines the size of the gas molecule that can enter the zeolite pores. By using zeolites with different pore sizes, specific gases can be selectively removed from gas mixtures. This size exclusion of the gas molecules can be used for gas separations or selective adsorption processes, e.g. drying and purification of air or petrochemicals. Zeolites with different pore sizes due to their structures are used. However, a rather minute tuning of the pore size of a specific zeolite is also possible. Depending on the occupied sites and the size of the cations in the zeolite, the pore openings are blocked more or less by these cations. Thus, the pore sizes can be adjusted by exchanging the zeolite with different cations. Commercially, there are different forms of zeolite A available that contain K^+, denoted as zeolite 3A (pore opening 3.0 Å), Na^+, denoted as zeolite 4A (pore opening 3.8 Å), or Ca^{2+}, denoted as zeolite 5A (pore opening 4.3 Å), resulting in different pore openings. In the same way, the pore openings of other zeolites can be adjusted to the specific requirements of a process. Li-X is very suitable for the production of oxygen from air with purities of 90–95 %, while zeolite 5A is used for the separation of linear and branched paraffins. Zeolite 3A is applied as a water adsorbent, e.g. as a desiccating agent in sealed insulation glass, or as a drying agent for organic chemicals [130].

4.2.8.5
Conclusions

The ability to exchange cations is one of the most striking features of zeolites, beside their adsorptive and catalytic properties. The ion exchange in zeolites is governed not only by the interaction of cations with a negatively charged solid surface but also by the accessibility of the exchange sites. Cations that are too large to fit into the channels or through specific cage or cavity windows are excluded, resulting in an ion-sieving effect. Thus, the selectivity of a zeolite for a specific ion is not only dependent on its charge but also on the size of the ion and that of its hydration shell. Thermodynamic effects also affect the ion exchange behavior of a zeolite and heats of hydration/dehydration and entropy can play a major role. Furthermore, the restricted space, resulting in a restricted amount of large cations that can be accommodated in the cages and/or cavities, and crystal field stabilization forces have to be considered in some cases.

References see page 1093

These properties make zeolites very suitable for applications as water softeners and in waste water treatment. Furthermore, zeolites are widely applied in catalysis and separation techniques due to the possibility to fine tune the catalytic and adsorptive properties by a directed ion exchange. Their microporosity makes zeolites very special with respect to their ion exchange properties and there exist only a few other materials, such as microporous resins, with both the ability to exchange ions and to exclude larger ions from exchange sites due to pore-size limitations (ion-sieving effect). However, noncrystalline materials, such as resins, generally have a rather broad pore size distribution and exactly defined pore and/or window openings due to structural features are only found in crystalline materials like zeolites. Zeolites are thus especially suited for applications where minute changes of the pore openings result in high selectivities for specific cations or where well-defined pore openings are of importance.

4.2.8.6
Symbols and Abbreviations

α_B^A	separation factor
a, b, c	lattice parameters of the zeolite structure
a, b	valences of cations A^{a+} and B^{b+}
a_{As}	ion activity of cation A^{a+} in solution
a_{Az}	ion activity of cation A^{a+} in zeolite
ΔG^0	standard free enthalpy change per equivalent of exchange
ΔH^0	enthalpy change per equivalent of exchange
ΔS^0	entropy change per equivalent of exchange
DnR	double n ring of T-atoms ($n = 4, 6$)
ε	dielectric constant of solvent
f_A, f_B	ion activity coefficients of A^{a+} and B^{b+} in the zeolite
γ_A, γ_B	molal ion activity A^{a+} and B^{b+} in the solution
g equiv.	gram equivalent (gram of substance (e.g. exchanger) that provides one equivalent (mole) of charges; g (mole charges)$^{-1}$)
K_a	thermodynamic equilibrium constant
K_c	corrected rational selectivity coefficient
K_c'	rational selectivity coefficient
K_{Na}^M	rational selectivity coefficient for the exchange of Na^+ by a cation M^{m+}
M	molarity of cations in solution
m	molality of cations in solution or in the zeolite
n_{Az}, n_{Bz}	number of cations in a unit volume of a zeolite
nMR	n-membered ring of T-atoms ($n = 4, 5, 6, 8, 10, 12$)
Q	total exchange capacity of a zeolite
S_A, S_B	equivalent fraction of cations A^{a+} and B^{b+} in solution
T	temperature (absolute)
$t_{1/2}$	time required to achieve 50 % exchange equilibrium
Z_A, Z_B	equivalent fraction of cations A^{a+} and B^{b+} in zeolite

References

1 G. B. McGarvey, J. B. Moffat, *J. Catal.* **1991**, *128*, 69–83.

2 D. Casparovicova, M. Kralik, M. Hronec, *Collect. Czech. Chem. C.* **1999**, *64*, 502–514.

3 G. Chessa, G. Marangoni, N. Stevanato, A. Vasori, *React. Polym.* **1991**, *14*, 143–150.

4 D. W. Breck, Zeolite Molecular Sieves. Wiley, New York, 1974, pp. 529–592.

5 H. Sherry, *Zeolites* **1993**, *13*, 377–383.

6 G. H. Kühl, in: Catalysis and Zeolites: Fundamentals and Applications, J. Weitkamp, L. Puppe (Eds), Springer, Berlin, 1999, pp. 81–196.

7 R. P. Townsend, in: Introduction to Zeolite Science and Practice, H. van Bekkum, E. M. Flanigen, J. C. Jansen (Eds), *Studies in Surface Science and Catalysis*, Vol. 58, Elsevier, Amsterdam, 1991, pp. 359–390.

8 R. P. Townsend, *Pure Appl. Chem.* **1986**, *58*, 1359–1366.

9 R. P. Townsend, in: New Developments in Zeolite Science and Technology. Y. Murakami, A. Iijima, J. W. Ward (Eds), *Studies in Surface Science and Catalysis*, Vol. 28, Elsevier, Amsterdam, 1986, pp. 273–282.

10 W. Loewenstein, *Am. Mineral.* **1954**, *39*, 92–96.

11 H. Robson, K. P. Lillerud, Verified Synthesis of Zeolitic Materials, Elsevier, Amsterdam, 2001, pp. 201–202.

12 H. S. Sherry, H. F. Walton, *J. Phys. Chem.* **1967**, *71*, 1457–1465.

13 R. M. Barrer, B. M. Munday, *J. Chem. Soc.* **1971**, 2904–2909.

14 R. M. Barrer, B. M. Munday, *J. Chem. Soc.* **1971**, 2909–2914.

15 D. W. Breck, Zeolite Molecular Sieves, John Wiley & Sons, New York, 1974, p. 532.

16 R. M. Barrer, A. J. Walker, *Trans. Faraday Soc.* **1964**, *60*, 171–184.

17 H. Uyama, Y. Kanzaki, O. Matsumoto, *Mat. Res. Bull.* **1987**, *22*, 157–164.

18 G. L. Gaines, H. C. Thomas, *J. Chem. Phys.* **1953**, *21*, 714–718.

19 L. M. Brown, H. S. Sherry, F. J. Krambeck, *J. Phys. Chem.* **1971**, *75*, 3846–3855.

20 L. M. Brown, H. S. Sherry, *J. Phys. Chem.* **1971**, *75*, 3855–3863.

21 P. Fletcher, R. P. Townsend, *J. Chem. Soc., Faraday Trans. 2* **1981**, *77*, 955–963.

22 P. Fletcher, R. P. Townsend, *J. Chem. Soc., Faraday Trans. 2* **1981**, *77*, 965–980.

23 P. Fletcher, R. P. Townsend, *J. Chem. Soc., Faraday Trans. 2* **1981**, *77*, 2077–2089.

24 D. Drummond, A. De Jonge, L. V. C. Rees, *J. Phys. Chem.* **1983**, *87*, 1967–1971.

25 R. P. Townsend, P. Fletcher, M. Loizidou, in: Proceedings of the 6th International Zeolite Conference. D. Olson, A. Bisio (Eds), Butterworth, Guildford, 1983, pp. 110–121.

26 R. M. Barrer, *J. Chem. Soc., Faraday Trans. 2* **1984**, *80*, 629–640.

27 K. R. Franklin, R. P. Townsend, *J. Chem. Soc., Faraday Trans. 1* **1985**, *81*, 3127–3141.

28 K. R. Franklin, R. P. Townsend, *Zeolites* **1988**, *8*, 367–375.

29 T. B. Reed, D. W. Breck, *J. Am. Chem. Soc.* **1956**, *78*, 5972–5977.

30 L. Broussard, D. P. Shoemaker, *J. Am. Chem. Soc.* **1960**, *82*, 1041–1051.

31 R. Y. Yanagida, A. A. Amaro, K. Seff, *J. Phys. Chem.* **1973**, *77*, 805–809.

32 V. Gramlich, W. M. Meier, *Z. Kristallogr.* **1971**, *133*, 134–149.

33 T. Ikeda, F. Izumi, T. Kodaira, T. Kamiyama, *Chem. Mater.* **1998**, *10*, 3996–4004.

34 Ch. Baerlocher, W. M. Meier, D. H. Olson, Atlas of Zeolite Framework Types, Elsevier, Amsterdam, 2001, pp. 168–169.

35 H. S. Lee, W. V. Cruz, K. Seff, *J. Phys. Chem.* **1982**, *86*, 3562–3569.

36 W. J. Mortier, Compilation of Extra Framework Sites in Zeolites, Butterworth, Guildford, 1982, pp. 1–67.

37 R. M. BARRER, L. V. C. REES, D. J. WARD, *Proc. Roy. Soc.* **1963**, *A273*, 180–197.

38 D. R. ROSSEINSKY, *Chem. Rev.* **1965**, *65*, 467–490.

39 L. V. C. REES, in: Zeolites as Catalysts, Sorbents and Detergent Builders, H. KARGE, J. WEITKAMP (Eds), *Studies in Surface Science and Catalysis*, Vol. 46, Elsevier, Amsterdam, 1989, pp. 661–672.

40 K. R. FRANKLIN, R. P. TOWNSEND, *J. Chem. Soc., Faraday Trans. 1* **1985**, *81*, 1071–1086.

41 D. W. BRECK, W. G. EVERSOLE, R. M. MILTON, T. B. REED, T. L. THOMAS, *J. Am. Chem. Soc.* **1956**, *78*, 5963–5971.

42 R. M. BARRER, W. M. MEIER, *Trans. Faraday Soc.* **1958**, *54*, 1074–1085.

43 I. J. GAL, O. JANKOVIC, S. MALCIC, P. RADOVANOV, M. TODOROVIC, *Trans. Faraday Soc.* **1971**, *67*, 999–1008.

44 T. A. EGERTON, A. HAGAN, F. S. STONE, J. C. VICKERMANN, *J. Chem. Soc., Faraday Trans.* **1972**, *14*, 732–735.

45 P. RILEY, K. SEFF, *Inorg. Chem.* **1974**, *13*, 1355–1360.

46 N. H. HEO, W. CRUZ-PATALINGHUG, K. SEFF, *J. Phys. Chem.* **1986**, *90*, 3931–3935.

47 R. BHAT, G. P. BABU, A. N. BHAT, *J. Chem. Soc., Faraday Trans.* **1995**, *91*, 3983–3986.

48 W. SCHMIDT, C. WEIDENTHALER, in: Applied Mineralogy in Research, Economy, Technology, Ecology and Culture, D. RAMMLMAIR, J. MEDERER, TH. OBERTHÜR, R. B. HEIMANN, H. PENTINGHAUS (Eds), Balkema, Rotterdam, 2000, pp. 241–244.

49 C. WEIDENTHALER, W. SCHMIDT, *Chem. Mater.* **2000**, *12*, 3811–3820.

50 C. WEIDENTHALER, W. SCHMIDT, *Z. Kristallogr.* **2001**, *216*, 105–111.

51 W. SCHMIDT, C. WEIDENTHALER, in: Zeolites and Mesoporous Materials at the Dawn of the 21st Century, A. GALARNEAU, F. DI RENZO, F. FAJULA, J. VERDINE (Eds), *Studies in Surface Science and Catalysis*, Vol. 135, Elsevier, Amsterdam, 2001, p. 206.

52 D. BAE, K. SEFF, *Microporous Mesoporous Mater.* **1999**, *33*, 265–280.

53 P. E. RILEY, K. SEFF, *J. Phys. Chem.* **1975**, *79*, 1594–1601.

54 CH. BAERLOCHER, W. M. MEIER, D. H. OLSON, Atlas of Zeolite Framework Types, Elsevier, Amsterdam, 2001, pp. 132–133.

55 D. W. BRECK, E. M. FLANIGEN, Molecular Sieves, Soc. Chem. Ind., London, 1968, pp. 47–60.

56 G. H. KÜHL, *Zeolites* **1985**, *5*, 4–6.

57 E. DEMPSEY, G. H. KÜHL, D. H. OLSON, *J. Phys. Chem.* **1969**, *73*, 387–390.

58 J. J. HRILJAK, M. M. EDDY, A. K. CHEETHAM, J. A. DONOHUE, G. J. RAY, *J. Solid State Chem.* **1993**, *106*, 66–72.

59 D. H. OLSON, *Zeolites* **1995**, *15*, 439–443.

60 H. SHERRY, *J. Phys. Chem.* **1966**, *70*, 1158–1168.

61 E. R. NIGHTINGALE, *J. Phys. Chem.* **1959**, *63*, 1381–1387.

62 R. G. PEARSON, *J. Am. Chem. Soc.* **1963**, *85*, 3533–3539.

63 H. SHERRY, *J. Phys. Chem.* **1968**, *72*, 4086–4094.

64 A. MAES, A. CREMERS, in: Molecular Sieves, W. M. MEIER, J. B. UYTTER-HOEVEN (Eds), *Adv. Chem. Ser.*, Vol. 121, Am. Chem. Soc., Washington, 1973, pp. 230–239.

65 E. GALLEI, D. EISENBACH, A. AHMED, *J. Catal.* **1974**, *33*, 62–67.

66 A. MAES, A. CREMERS, *J. Chem. Soc., Faraday Trans. 1* **1975**, *71*, 265–277.

67 I. J. GAL, P. RADOVANOV, *J. Chem. Soc., Faraday Trans. 1* **1975**, *71*, 1671–1677.

68 M. A. KEANE, *Microporous Mater.* **1994**, *3*, 93–108.

69 M. A. KEANE, *Microporous Mater.* **1995**, *3*, 385–394.

70 M. A. KEANE, *Microporous Mater.* **1995**, *4*, 359–368.

71 M. A. KEANE, *Colloids Surf. A* **1998**, *138*, 10–20.

72 A. DYER, P. P. TOWNSEND, *J. Inorg. Nucl. Chem.* **1973**, *35*, 2993–2999.

73 F. WOLF, D. CEACAREANU, K. PILCHOWSKI, *Z. Phys. Chem. – Leipzig* **1973**, *252*, 50–64.

74 H. HOSER, S. KRZYZANOWSKI, F. TRIFIRO, *J. Chem. Soc., Faraday Trans. 1* **1975**, *71*, 665–669.

75 B. Wichterlova, P. Jiru, A. Curinova, *Z. Phys. Chem. Neue Fol.* **1974**, *88*, 180–192.

76 D. Bae, K. Seff, *Microporous Mesoporous Mater.* **1999**, *33*, 265–280.

77 P. Gallezot, B. Imelik, *J. Phys. Chem.* **1973**, *77*, 652–656.

78 P. P. Lai, L. V. C. Rees, *J. Chem. Soc., Faraday Trans. 1* **1976**, *72*, 1809–1817.

79 P. P. Lai, L. V. C. Rees, *J. Chem. Soc., Faraday Trans. 1* **1976**, *72*, 1818–1826.

80 P. P. Lai, L. V. C. Rees, *J. Chem. Soc., Faraday Trans. 1* **1976**, *72*, 1827–1839.

81 Ch. Baerlocher, W. M. Meier, *Z. Kristallogr.* **1972**, *135*, 339–354.

82 U. Håkansson, L. Fälth, S. Hansen, *Acta Crystallogr. C* **1990**, *46*, 1363–1364.

83 S. Hansen, U. Håkansson, L. Fälth, *Acta Crystallogr. C* **1990**, *46*, 1361–1362.

84 B. R. Albert, A. K. Cheetham, J. A. Stuart, C. J. Adams, *Microporous Mesoporous Mater.* **1998**, *21*, 133–142.

85 Ch. Baerlocher, W. M. Meier, D. H. Olson, Atlas of Zeolite Framework Types, Elsevier, Amsterdam, 2001, pp. 138–139.

86 A. Alberti, G. Vezzalini, *Acta Crystallogr. B* **1979**, *35*, 2866–2869.

87 R. M. Barrer, B. M. Munday, *J. Chem. Soc. A* **1971**, 2909–2914.

88 E. Borgstedt, H. Sherry, J. P. Slobogin, in: Progress in Zeolite and Microporous Materials, H. Chon, S.-K. Ihm, Y. S. Uh (Eds), *Studies in Surface Science and Catalysis*, Vol. 105, Elsevier, Amsterdam, 1997, pp. 1659–1666.

89 A. Moirou, A. Vaxevanidou, G. E. Christidis, I. Paspaliaris, *Clay. Clay. Miner.* **2000**, *48*, 563–571.

90 Ch. Baerlocher, W. M. Meier, D. H. Olson, Atlas of Zeolite Framework Types, Elsevier, Amsterdam, 2001, pp. 184–185.

91 G. T. Kokotailo, S. L. Lawton, S. L. Olson, W. M. Meier, *Nature* **1978**, *272*, 437–438.

92 S. L. Olson, G. T. Kokotailo, S. L. Lawton, W. M. Meier, *J. Phys. Chem.* **1981**, *85*, 2238–2243.

93 E. M. Flanigen, J. M. Bennet, R. W. Grose, J. P. Cohen, R. L. Patton, R. M. Kirchner, J. V. Smith, *Nature* **1978**, *271*, 512–516.

94 J.-C. Lin, K.-J. Chao, Y. Wang, *Zeolites* **1991**, *11*, 376–379.

95 Z. Liu, W. Zhang, Q. Yu, G. Lü, W. Li, S. Wang, Y. Zhang, B. Lin, in: New Developments in Zeolite Science and Technology, Y. Murakami, A. Iijima, J. W. Ward (Eds), *Studies in Surface Science and Catalysis*, Vol. 28, Elsevier, Amsterdam, 1986, pp. 415–422.

96 K. D. Huddersman, L. V. C. Rees, *Zeolites* **1991**, *11*, 270–276.

97 R. Rachapudi, P. Chintawar, H. L. Greene, *J. Catal.* **1999**, *185*, 58–72.

98 C. Lamberti, S. Bordiga, M. Salvalaggio, G. Spoto, A. Zecchina, F. Geobaldo, G. Vlaic, M. Bellatreccia, *J. Phys. Chem. B* **1997**, *101*, 344–360.

99 L. S. Stokes, D. M. Murphy, R. D. Farley, C. C. Rowlands, S. Bailey, *Phys. Chem. Chem. Phys.* **1999**, *1*, 621–628.

100 D. P. Matthews, L. V. C. Rees, *Chem. Age India* **1986**, *37*, 353–357.

101 H. S. Sherry, in: Ion Exchange, Vol. 2, J. A. Marinsky (Ed.), E. Arnold Ltd., London, 1969, p. 89.

102 E. F. Vansant, J. B. Uytterhoeven, in: Molecular Sieve Zeolites – I, R. F. Gould (Ed.), *Adv. Chem. Ser.*, Vol. 101, Am. Chem. Soc., Washington, 1971, pp. 426–435.

103 P. Chu, F. G. Dwyer, Intrazeolite Chemistry, G. D. Stucky, F. G. Dwyer (Eds), *ACS Symp. Ser.*, Vol. 218, 1983, pp. 59–78.

104 G. P. Handreck, T. D. Smith, *J. Chem. Soc., Faraday Trans. 1* **1989**, *85*, 645–654.

105 D. W. Breck, Zeolite Molecular Sieves, Wiley, New York, 1974, pp. 580–585.

106 D. Bae, K. Seff, *Zeolites* **1996**, *17*, 444–446.

107 S. A. Sayed, *Zeolites* **1996**, *17*, 361–364.

108 A. M. Tolmachev, I. V. Baranova, *Zh. Fiz. Khim.* **1977**, *51*, 2343–2345.

109 H. Uyama, Y. Kanzaki, O. Matsumoto, *Mat. Res. Bull.* **1987**, *22*, 157–164.

110 K. Ho, H. S. Lee, T. Sun, K. Seff, *Zeolites* **1995**, *15*, 377–381.

111 P. C. Huang, A. Mizany, J. L. Pauley, *J. Phys. Chem.* **1964**, *68*, 2575–2578.

112 B. Wichterlová, S. Beran, L. Kubelková, J. Nováková, A. Smiešková, R. Šebík, in: Zeolites as Catalysts, Sorbents and Detergent Builders, H. G. Karge, J. Weitkamp, *Studies in Surface Science and Catalysis*, Vol. 46, Elsevier, Amsterdam, 1989, pp. 347–353.

113 Z. Li, K. C. Xie, R. C. T. Slade, *Appl. Catal., A* **2001**, *209*, 107–115.

114 E. M. El-Malki, R. A. van Santen, W. M. H. Sachtler, *J. Catal.* **2000**, *196*, 212–223.

115 E. Rojasova, A. Smieskova, P. Hudec, Z. Zidek, *Collect. Czech. Chem. C* **2000**, *65*, 1506–1514.

116 G. Kinger, A. Lugstein, R. Swagera, M. Ebel, A. Jentys, H. Vinek, *Microporous Mesoporous Mater.* **2000**, *39*, 307–313.

117 E. M. El-Malki, D. Werst, P. E. Doan, W. M. H. Sachtler, *J. Phys. Chem. B* **2000**, *104*, 5924–5931.

118 Y. Li, J. N. Armor, *Appl. Catal., A* **1999**, *188*, 211–217.

119 M. Rauscher, K. Kesore, R. Monnig, W. Schwieger, A. Tissler, T. Turek, *Appl. Catal., A* **1999**, *184*, 249–256.

120 Y. Huang, R. M. Paroli, A. H. Delgado, T. A. Richardson, *Spectrochim. Acta A* **1998**, *54*, 1347–1354.

121 H. Förster, U. Hatje, *Solid State Ionics* **1997**, *101–103*, 425–430.

122 M. Liquornik, Y. Marcus, *J. Phys. Chem.* **1968**, *72*, 2885–2889.

123 H. Upadek, B. Kottwitz, B. Schreck, *Tenside Surfact. Det.* **1996**, *33*, 385–392.

124 T. Mukaiyima, H. Nishio, O. Okumura, in: New Developments in Zeolite Science and Technology, Murakami, A. Iijima, J. W. Ward (Eds), *Studies in Surface Science and Catalysis*, Vol. 28, 1986, 1017–1023.

125 C. P. Kurzendörfer, M. Liphard, W. von Rybinski, M. Schwuger, *Colloid Polym. Sci.* **1987**, *265*, 542–547.

126 M. J. Schwuger, M. Liphard, in: Zeolites as Catalysts, Sorbents and Detergent Builders, H. G. Karge, J. Weitkamp (Eds), *Studies in Surface Science and Catalysis*, Vol. 46, Elsevier, Amsterdam, 1989, pp. 673–690.

127 W. Leonhard, B.-M. Max, in: Zeolites as Catalysts, Sorbents and Detergent Builders, H. G. Karge, J. Weitkamp (Eds), in *Studies in Surface Science and Catalysis*, Vol. 46, Elsevier, Amsterdam, 1989, pp. 691–699.

128 H. Upadek, P. Krings, in: Zeolites as Catalysts, Sorbents and Detergent Builders, H. G. Karge, J. Weitkamp (Eds), *Studies in Surface Science and Catalysis*, Vol. 46, Elsevier, Amsterdam, 1989, pp. 701–709.

129 C. J. Adams, A. Araya, S. W. Carr, A. P. Chapple, K. R. Franklin, P. Graham, A. R. Minihan, T. J. Osinga, J. A. Stuart, in: Progress in Zeolite and Microporous Materials, H. Chon, S.-K. Ihm, Y. S. Uh (Eds), *Studies in Surface Science and Catalysis*, Vol. 105, Elsevier, Amsterdam, 1997, pp. 1667–1674.

130 A. Pfenninger, in: Molecular Sieves: Science and Technology, Vol. 2, H. G. Karge, J. Weitkamp (Eds), *Structures and Structure Determination*, Springer, Berlin, 1999, pp. 163–198.

131 H. J. M. Bowen, The Environmental Chemistry of the Elements, Academic Press, London, 1979, pp. 1–333.

132 A. Kabata-Pendias, H. Pendias, Trace Elements in Soils and Plants, CRC Press, Boca Raton, 1984, pp. 1–315.

133 R. E. Train, Quality Criteria for Water, Castle House, London, 1979, pp. 1–256.

134 H. E. Allen, S. H. Cho, T. A. Neubecker, *Water Res.* **1983**, *17*, 1871–1879.

135 S. Ahmed, S. Chughtai, M. A. Keane, *Sep. Purif. Technol.* **1998**, *13*, 57–64.

136 M. A. Keane, *Colloids Surf., A* **1998**, *138*, 11–20.

137 M. Vokáčová, Z. Matějka, J. Eliàšek, *Acta Hydroch. Hydrob.* **1986**, *14*, 605–611.

138 M. Loizidou, R. P. Townsend, *J. Chem. Soc., Dalton Trans.* **1987**, 1911–1916.

139 M. Loizidou, R. P. Townsend, *Zeolites* **1987**, *7*, 153–159.

140 M. J. Semmens, W. P. Martin, *Water Res.* **1988**, *22*, 537–542.

141 S.-J. Kang, K. Wada, *Appl. Clay Sci.* **1988**, *3*, 281–290.

142 J. Olah, J. Papp, A. Meszaros-Kis, G. V. Mucsi, D. Kallo, in: Zeolites as Catalysts, Sorbents and Detergent Builders, H. G. Karge, J. Weitkamp (Eds), *Studies in Surface Science and Catalysis*, Vol. 46, Elsevier, Amsterdam, 1989, p. 712–719.

143 R. Pode, G. Burtica, A. Iovi, E. Popovici, in: Porous Materials in Environmentally Friendly Processes, I. Kiricsi, G. Pal-Borbély, J. B. Nagy and H. Karge (Eds), *Studies in Surface Science and Catalysis*, Vol. 125, Elsevier, Amsterdam, 1999, 769–776.

144 K. Metropoulos, E. Maliou, M. Loizidou, N. Spyrellis, *J. Environ. Sci. Heal. A* **1993**, *28*, 1507–1518.

145 A. Haralambous, E. Maliou, M. Malamis, *Water Sci. Technol.* **1992**, *25*, 139–145.

146 E. Maliou, M. Malamis, P. O. Sakellarides, *Water Sci. Technol.* **1992**, *25*, 133–138.

147 H. Minamisiwa, H. Yamanka, N. Arai, T. Okutani, *Nippon Kagaku Kaishi* **1991**, 1605–1611.

148 A. Takasaka, H. Inaba, Y. Masuda, *Nippon Kagaku Kaishi* **1991**, 618–622.

149 L. A. Bray, H. Fullam, in: Molecular Sieve Zeolites – I, E. M. Flanigen, L. B. Sand (Eds), *Adv. Chem. Ser.*, Vol. 101, Am. Chem. Soc., Washington, 1971, pp. 450–455.

150 M. Foldesova, P. Lukac, J. Majling, V. Tomkova, *J. Radioanal. Nucl. Chem.* **1996**, *212*, 293–302.

151 M. T. Olguin, M. Solache, J. L. Iturbe, P. Bosch, S. Bulbulian, *Sep. Sci. Technol.* **1996**, *31*, 2021–2044.

152 D. A. White, A. Nattkemper, R. Rautiu, *Nucl. Technol.* **1999**, *127*, 212–217.

153 P. M. M. Blauwhoff, J. W. Gosselink, E. P. Kieffer, S. T. Sie, W. J. H. Stork, in: Catalysis and Zeolites: Fundamentals and Applications, J. Weitkamp, L. Puppe (Eds), Springer, Berlin, 1999, p. 437–538.

154 R. von Ballmoos, D. H. Harris, J. S. Magee, in: Handbook of Heterogeneous Catalysis, G. Ertl, H. Knözinger, J. Weitkamp (Eds), Wiley-VCH, Weinheim, 1997, pp. 1955–1986.

155 I. E. Maxwell, W. J. H. Stork, in: Introduction to Zeolite Science and Practice, H. van Bekkum, E. M. Flanigen, J. C. Jansen (Eds), *Studies in Surface Science and Catalysis*, Vol. 58, Elsevier, Amsterdam, 1991, pp. 571–630.

156 W. F. Hölderich, H. van Bekkum, in: Introduction to Zeolite Science and Practice, H. van Bekkum, E. M. Flanigen, J. C. Jansen (Eds), *Studies in Surface Science and Catalysis*, Vol. 58, Elsevier, Amsterdam, 1991, pp. 631–726.

157 F. J. Janssen, in: Handbook of Heterogeneous Catalysis, G. Ertl, H. Knözinger, J. Weitkamp (Eds), Wiley-VCH, Weinheim, 1997, pp. 1633–1668.

4.2.9
Application of Microporous Solids as Catalysts

Johannes A. Lercher and Andreas Jentys

4.2.9.1
Introduction

Molecular sieves are widely used as solid catalysts or catalyst components in areas ranging from petroleum refining [1, 4] to the synthesis of intermediates and fine chemicals [5, 10]. The reasons for the widespread use are related to the possibility to tailor the microporous material with respect to the concentration and nature of catalytically active sites and their immediate environment [11–13]. This allows control of the nature of catalytically active sites and the access of molecules to these sites [14]. Such strict control of access is only found in enzymes and similar biological catalysts, but has the inherent disadvantage that it restricts the size of molecules that can be reacted in molecular sieves. In this respect it should, however, be emphasized that molecules are able to sorb and react at the pore mouth of molecular sieves even when the molecules are too bulky to enter the zeolite pores [15, 16]. Lately, it has also been shown that molecular sieves can be delaminated into crystalline sheets [17–19]. Such sheets can be loosely arranged or pillared thus enhancing the space for molecules to react, while maintaining the well-defined crystalline environment of the catalytically active site.

Up to now, out of the large number of possible structures only a few materials have been the basis for widespread development of catalytic materials. These structures include the large pore materials such as faujasite, mordenite, and zeolite Beta, medium pore zeolites such as ZSM-5 and ZSM-11 and narrow-pore materials such as ferrierite and ZSM-22 as well as materials substituted with transition metal cations such as Ti, Co or Cr in the framework. The reasons for this relatively low number of zeolites lie in practical difficulties of industrial upscaling of catalytic synthesis processes and the traditional usage of the materials in petroleum refining.

However, in explorative experiments a large number of reactions has been investigated over a growing number of microporous materials. This has led to an excitingly high level of understanding of the catalytic chemistry and the structure–activity relationships. Numerous reviews on the use of molecular sieves as catalysts for chemical synthesis and petroleum processing have appeared over the last decades [1–13, 20, 21] and underline the importance of this class of catalysts.

The growth in the use of molecular sieves as catalysts as compared with non-functionalized macro- and mesoporous oxides has been stimulated by several factors: (1) The high surface density of active sites (in comparison with oxides) results in very active catalysts. (2) The defined pore structure allows excluding reactants from being converted and/or products to be formed or transported out of the pores due to too large a size [12]. (3) The active site and the environment of that site can be designed on the atomic level, for example, by ion exchange [22] or chemical functionalization of the framework [23]. (4) The crystalline nature of the material allows tailoring of the chemical properties of molecular sieves in a more subtle way compared to macro- and mesoporous oxides.

Many of the advantages relate to the fact that the acid/base surface properties of the dense/macroporous oxides, phosphates or sulfates depend on the way the bulk material is terminated [24]. For molecular sieves the entire pore surface and, thus, most acidic and basic sites are an integral part of the crystal structure. In other words, the acid and base sites result from charge imbalances of the framework atoms. This allows subtly and reproducibly designing and modifying acid/base sites as well as redox properties of the molecular sieves. Additionally, the controlled transport to the active site permits influence of the strength of interaction between the substrate and the catalyst and, hence, opens another tool to adjust the surface chemical properties. Overall, the rate and selectivity of catalyzed reactions over molecular sieve catalysts are influenced by factors that are concerned with the specific interface chemistry (chemical induced selectivity) and the constraints induced by the steric limitations (shape selectivity) [25, 26].

This advantage, however, also induces limitations that can only partly be overcome by adjustment of the mesoscopic properties of the molecular sieve. The most prominent concerns the minimum diameter of the molecular-sieve channels restricting the conversion of large organic molecules. A remedy for this problem can be to create a secondary meso/macro pore structure, which maximizes the entry zone to the microporous region [27]. Another limitation lies in the combination of the presence of larger molecules and of strongly basic or acidic functional groups in the pores that cause desorption of products frequently to be rate limiting or even to be impossible without the help of a coreactant (adsorption-assisted desorption). This is especially important for reactions such as condensation, oligomerization and nucleophilic substitution, which lead to larger products or to products exceeding the reactants in polarity. Finally, the catalytic chemistry inside a zeolite can be seen to occur in a microscopically small tubular reactor in which the active sites are distributed over the whole reactor. It can be intuitively understood that it is difficult to avoid sequential reactions of the same type in such an environment.

Up to now the progress towards the detailed understanding of the chemistry of the catalytic transformations has been limited to reactions practiced on a relatively large scale. This will also be the backbone of the present contribution. It should, however, be emphasized that the current rapid progress in high throughput experimentation techniques will provide a large number of data that will allow development of a broad base of structure activity relationships [28, 29]. By this it will be finally possible to reach a generic understanding of the reactivity of functional groups of the reacting molecules and their way of interaction with active sites in the molecular sieve catalysts. In this way, high throughput experimentation will also stimulate the rational design of new catalysts and/or processes. At this point it should be emphasized that much of the progress during the last years has only been possible through the progress in theoretical methods describing the solid material, the interaction and the catalytic transformation [30–33].

A significant number of excellent review articles on the use of molecular sieves for chemical synthesis and petroleum refining exists as outlined above (see also [34, 44]). Most of these, however, address the problem from the viewpoint of the organic conversion and how a molecular sieve changes activity and selectivity for a particular reaction as compared to a macroporous oxide. The present section, in contrast, aims to correlate the solid-state properties of the molecular sieve and the reaction it catalyzes. First, the focus will be on outlining the chemical requirements for functional groups in molecular sieves to catalyze a particular reaction and how these groups can be manipulated to optimize activity and selectivity. Then

References see page 1145

we will discuss how varying the pore structure can selectively influence the transport and sorption of reactants and products in the pores and/or the formation of different transition state complexes.

4.2.9.2
Direct Influences of the Proximity of Molecule and Zeolites Channel

Many of the unique properties of the zeolites as sorbents and catalysts result from the proximity of the molecule and the surrounding zeolite wall. Given the radii of oxygen and Si or Al the molecules may interact by undirected short-range dispersion forces (physisorption), which can dominate over specific chemical interactions for weakly polar and larger molecules. Such situations have been well demonstrated for, but are not confined to, alkanes in molecular sieves.

Adsorption of an alkane in the microporous material leads to two energetic components for the interactions, one that is related to the directional binding of the proton or the metal cation and another that results from the London dispersion forces a molecule experiences upon sorption in the micropores of the zeolite. The direct interaction results from localized chemical bonds that range in strength from dipole induced hydrogen-bridge bonding with alkanes to the formation of a cation with strong bases such as ammonia. The interaction resulting from the London dispersion forces summarizes all forms of nonlocalized interactions similar in nature to the bonds that give rise to the nonideal behavior of a real gas. The partition between these two forces is not known. It is possible, however, for some cases to compare microporous materials with and without Brønsted acid sites. Figure 1 shows the dependence of the heat of sorption of n-alkanes on their chain length. The initial heat of sorption increases linearly with the chain length for ZSM-5 and zeolite Y [45, 46]. This is attributed to the linear increase of the London dispersion forces with the size of the alkane. Two observations are notable, (1) the

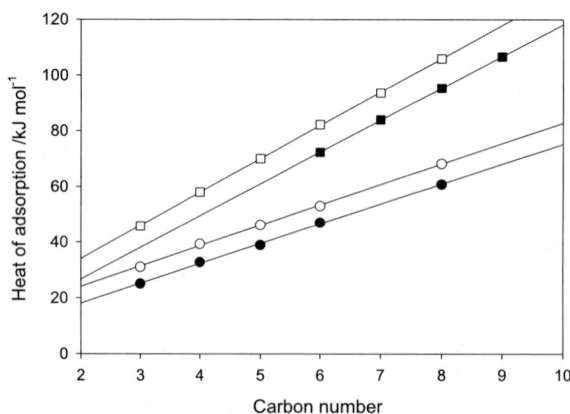

Fig. 1. Heat of sorption of n-alkanes on (□) H-ZSM-5; (■) ZSM-5 (siliceous); (○) H-Y; (●) Y (siliceous).

increment with which the heat of sorption increases per carbon atom is larger for the medium pore ZSM-5 material than for the large pore zeolite Y and (2) the presence of strong Brønsted acid sites increases the heat of sorption by 7–10 kJ mol^{-1}. The larger contribution of the latter term for ZSM-5 suggests that it was the stronger acid of the two studied. The larger increment of the heat of adsorption per carbon atom for this material, however, suggests that the narrower pore interacts more strongly with the hydrocarbon than the wider [45]. It is impressive to see that the contribution of the Brønsted acid sites to the overall heat of adsorption is small compared to the contribution of the dispersion forces. Thus, it can be concluded that the interaction between the alkane and the zeolite is governed by physisorption.

The sorption in the pores also reduces the degrees of freedom for the sorbed molecules, which is manifested in the decrease of the entropy. This is mainly due to the reduction of at least one translation degree of freedom in the adsorbed state compared to the gas phase. Intuitively it is well understood that a stronger bonding should lead to a more pronounced reduction of the translational degrees of freedom and consequently of the entropy of the sorbate. Indeed, it can be seen that with increasing chain length of the alkane (increasing heat of sorption) also the entropy in the sorbed state decreases, following a linear relation. It is interesting to note, however, that for a given heat of adsorption, the loss of entropy decreases with the pore size of the molecular sieve [45]. The influence of the pore size decreases with the size of the molecule leading to an intercept of the lines between adsorption enthalpies of 20–30 kJ mol^{-1}. In this context, it should be mentioned that Gorte and coworkers [47] recently reported a notable influence of the localized bonding for small molecules such as methane and acetonitrile.

For catalysis, these observations suggest that the different rates of reactions found with zeolites of varying pore size for reactions such as alkane cracking may be strongly related to the concentration of the reactants in the pores of the zeolites and hence are a direct result of the variations in the dispersion forces. Indeed, it has been reported that the true energy of activation of monomolecular cracking of alkanes is not only identical for a large number of alkanes in one molecular sieve, but also that with different molecular sieves such as zeolites Y or H-ZSM-5 the same energy of activation is found [48]. The constancy of the true activation energy results from the compensation between apparent energy of activation and the heat of adsorption. It should be stressed, however, that the sorption entropy will subtly depend upon the coverage (ordering in the pores) and that consequently such correlations can only be expected for low coverages.

Focusing on the more chemical variation of the molecular sieve a number of reports suggest that both the adsorption behavior and the polarization of the molecule can be influenced by the exchangeable cations. Corma et al. [49, 50] describe, for example, for n-heptane cracking over La^{3+} and Cr^{3+} partially ion-exchanged Y zeolites a linear relation between ln A and E_{app} (pre-exponential factor and apparent energy of activation). The results show that the ion exchange with trivalent

References see page 1145

cations is able to change the true energy of activation and/or the heat of adsorption by approximately 20 kJ mol^{-1}. This is tentatively ascribed to the strong polarization effect of the transition metal cation upon the alkane. Also, Crickmore [51] describes such a compensation for hydrocarbons in analysis of gas oil cracking over La-Y zeolites using data from Pachovsky and Wojciechowski [52]. Similarly, compensation behavior was also suggested for hydrocracking on n-alkanes [53].

Such strong effects are, however, not confined to molecules that are fully taken up into the pores of the zeolites, but also appear to play a role with molecules that cannot fully penetrate the zeolites pore. Such molecules may adsorb and react at the pore mouth [54, 55] Under such circumstances, however, the relation between entropy and enthalpy is rather complex leading to rather strong coverage-dependent effects [56]. These effects were named confinement and nest effect by Derouane and Chang [57], who revived older concepts of relating the physisorption to the fit between the molecule and the sorbing environment.

4.2.9.3
Molecular Sieves as Solid Acids and Bases

4.2.9.3.1 Nature and Origin of Strong Brønsted Acid Sites in Molecular Sieves
Molecular sieves consist of a three-dimensional network of metal-oxygen tetrahedra (and to a lesser extent also octahedra) that provides a regularly sized micropore structure, in which the acid and base sites are structurally included [58–60]. Acid sites result from the imbalance between the metal and the oxygen formal charge in this smallest building unit. This is seen most clearly in the case of zeolites, which comprise of a 3D network of Si–O tetrahedra. Formally, the 4+ charge on the Si cation and the 2− charge on the oxygen anion leads to neutral tetrahedra, as every oxygen belongs to two of such units. Thus, a framework consisting of only Si–O tetrahedra is neutral and may possess acid properties only through defects in such a structure. If a Si^{4+} cation is substituted by Al^{3+} the formal charge on that tetrahedron changes from neutral to −1. This negative charge is balanced by a metal cation or a proton forming a Lewis or a Brønsted acid site, respectively [2, 61, 62]. Note that the bare, negatively charged tetrahedron is then the corresponding base. Depending upon the charge on the metal cation/proton and the oxygen, the acidic or basic properties of the molecular sieves will be dominating and consequently it will be called a solid acid or base [2, 63]. Note that these acid or base properties are not a simple function of the chemical composition, but that also the structure (framework density and local strain) has a major impact.

More possibilities emerge, when also metal-oxygen tetrahedra with metal cations of formal charges different from 4+ or 3+ are incorporated such as in substituted metal aluminophosphates (see Fig. 2). Depending on the combinations of the metal cation in the framework, neutral frameworks and frameworks with cation or anion exchange sites are conceptually conceivable [64], but up to now only cation exchange molecular sieves have been found.

The differences in the charge of these primary building units determine the theoretically highest concentration of the acid (and base) sites in the framework.

Fig. 2. Possibilities of tailoring the zeolite by replacing framework Si^{4+} or Al^{3+} with difference metal cations, i.e. P^{5+}.

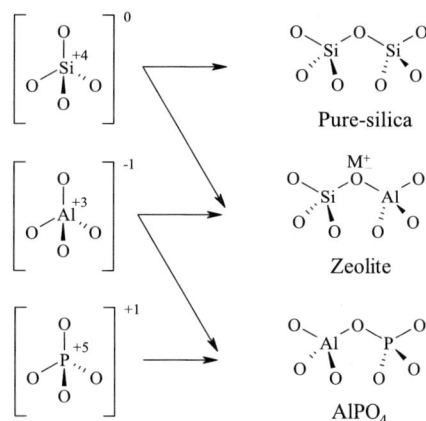

The strength of the acid (and base) sites depends primarily on the overall chemical composition of the framework, but also on more subtle factors such as the nature of the metal cations at exchange sites and the crystal structure [65–68]. Especially, the type of exchangeable cation incorporated influences the polarizability and the real charge of the framework oxygen atoms. Thus, adjusting the combination of chemical composition, (partial) cation exchange and the molecular sieve structure can lead to a wide variety of chemical properties. These relationships between the chemical composition, the crystal structure and the acid/base properties of molecular sieves have been studied extensively and were thoroughly reviewed (see, e.g., Refs [6, 69, 74]).

However, it should be emphasized that the measurement and quantification of the acid/base strength of zeolites is complex and that it is difficult to directly compare the acid/base strength of a solid with that of a liquid. This results from the fact that the stabilization of carbocations and carbanions in a microporous solid differs from that in strongly polar acid and base solutions. For zeolites, it can be stated that the concentration of aluminum in the framework is directly proportional to the concentration of acid sites and the polarity of the framework and to a first approximation indirectly proportional to the strength of acid sites [72]. For a given chemical composition of the zeolite, the polarity of the lattice increases with decreasing framework density [75].

Replacement of bulk chemicals such as H_2SO_4 and $AlCl_3$ with zeolites is a very attractive option that is only limited by the significantly lower proton (acid site) concentration of molecular sieves in comparison to liquid mineral acids. For example, 1 g of H_2SO_4 has 2×10^{-2} moles of protons, whereas 1 g of zeolite HY, with Si/Al ratio of five, contains 3×10^{-3} moles of protons. Note that this is a crude approximation of the acidic protons available for catalysis, because it assumes that with both materials all hydroxyl groups are available and catalytically

References see page 1145

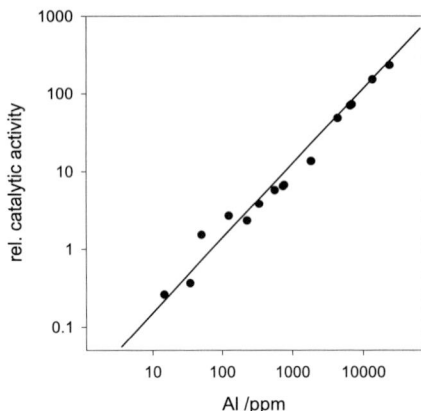

Fig. 3. Hexane cracking activity versus Si/Al ratio for H-ZSM-5.

active. Additionally, 1 g of H_2SO_4 occupies far less volume (i.e., 0.5 cm^3) than the equivalent mass of zeolites (4–6 cm^3) [34].

4.2.9.3.2 Evidence for Uniformity and Distribution of Acid Sites in Molecular Sieves
The Brønsted acid sites of high silica zeolites such as ZSM-5 exhibit uniform acid strength and uniform catalytic activity. This has been unequivocally demonstrated for two acid-catalyzed reactions, i.e., protolytic cracking and *m*-xylene isomerization. Haag and coworkers [76] demonstrated that the rate of n-hexane cracking is directly proportional to the concentration of aluminum in the framework. As outlined, n-hexane cracking is monomolecular under the reaction conditions chosen and catalyzed by strong Brønsted acid sites only (see Fig. 3). Interestingly, also the rate of n-hexene cracking, which is a reaction that should be catalyzed by much weaker acid sites, also varies linearly with the concentration of strong Brønsted acid sites. However, such relationships strongly suggest, but do not prove, that only one type of sites exists in HZSM-5. A distribution of strong sites may exist and lead to linear relations between site concentrations and catalytic activity as long as the distribution of site strength does not change with the concentration of aluminum in the zeolites framework.

The presence of Brønsted acid sites of equal strength in one particular sample of H-ZSM-5 has been demonstrated by Mirth et al. [77] for *m*-xylene isomerization using *in situ* IR spectroscopy to determine the concentration of sorbed reactants (shown in Fig. 4). Through a variation in the partial pressure of *m*-xylene the reaction rate was shown to be directly proportional to the concentration of acid sites covered indicating that all acid sites of ZSM-5 converted *m*-xylene to *o*- and *p*-xylene with the same rate per proton. It should be noted that due to the operating conditions (steric limitations in the accessibility, diffusional limitations, etc.) not all acid sites may actually participate in the reaction. The aspects of the shape selectivity that also influence activity in a complex way will be discussed in a later section.

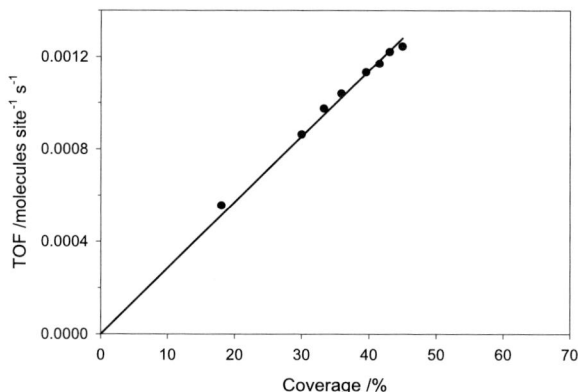

Fig. 4. Reaction rate for isomerization of *m*-xylene to *o*- and *p*-xylene as function of the coverage of acid sites.

It should be emphasized at this point that such uniform properties of acid sites are usually restricted to high-silica zeolites or, in general, to molecular sieves with a low density of acid sites. With materials of higher acid-site concentrations (such as zeolites Y and X), sites with distinct differences in the acid strength have been observed [78]. These differences were attributed mainly to the existence of neighboring acid sites that produce a local situation not unlike to that of H_2SO_4, in which two protons with differing acid strength exist.

A large number of papers, however, report very broad distributions of acid sites in all zeolitic materials (see e.g., [79–84]). Such observations stem partly from the fact that not only structurally defined strong Brønsted acid sites are present in the materials investigated. The higher the concentration of defects, the more they will influence directly (by being acid sites of medium strength) and indirectly (by altering the site strength of the strong Brønsted acid sites) the properties of the materials investigated. An excellent example of this is the catalytic activity for the protolytic cracking of n-hexane of partially alkali exchanged "clean" and "mildly steamed" H-ZSM-5 as shown in Fig. 5. [76]. It is obvious that in the latter case few very active Brønsted acid sites exist, while over two thirds of the acid sites show markedly lower activity.

The presence of neighboring acid sites appears to be important, when bimolecular reaction steps are involved in the reaction network. Hydride transfer reactions in alkane/alkene transformations depend in a nonlinear fashion upon the varying concentration of acid sites. Post and coworkers [85] showed elegantly that the rates of these bimolecular reactions depend upon the square of the concentration of the acid sites, while the rates of the monomolecular reactions (e.g., protolytic cracking [86]) depended linearly on the proton concentration. The observation should not be interpreted that two protonated species have to react, but rather that a higher concentration of acid sites also results in a higher concentration of re-

References see page 1145

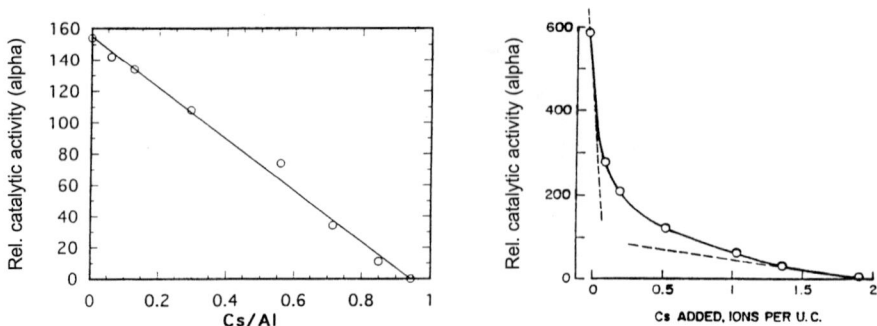

Fig. 5. Effect of Cs$^+$ exchange on the n-hexane cracking activity for H-ZSM-5 (Si/Al = 35) (a) as-prepared; (b) mildly steamed.

actants in the pores, which in turn favors the bimolecular reactions. Note that Lewis acid sites present in the zeolite may also play a significant role in enhancing the concentration of the reactant in the pores [87]. The direct role of Lewis acid sites in hydrocarbon conversions is less understood. Karge et al. [88] have shown that La^{3+} ion-exchanged zeolites that do not contain hydroxyl groups are catalytically inactive for ethylbenzene disproportionation suggesting that protons are indispensable for carbon–carbon bond rearrangement reactions. On the other hand, several reactions have been reported (the absence of hydroxyl groups is not certain in all those cases) that are well catalyzed by trivalent metal cation-exchanged zeolites [89]. The role of the metal cation in these instances appears to be more that of mediating the acid strength and modifying the adsorption strength than being the active site itself.

4.2.9.4
Acid-catalyzed Reactions

4.2.9.4.1 Activation of Alkanes and Alkenes by Formation of Carbocations
The early success of molecular sieve catalysis was triggered by the dramatic improvement in activity and selectivity for catalytic cracking of vacuum gas oil achieved by using faujasite-based catalysts instead of the previously used amorphous SiO$_2$/Al$_2$O$_3$. These catalysts exceeded the catalytic activity of amorphous SiO$_2$/Al$_2$O$_3$ catalysts (amorphous silica-alumina, ASA) by a factor of about 100 [90]. The major components to be cracked in the industrial process are alkanes, alkenes and aromatic molecules with an aliphatic substitution. Because of the industrial importance this has been a particularly active field with a very high number of papers being published. However, despite the advancements in the field, the mechanism of catalytic cracking is not completely settled and more understanding appears to be necessary to be able to fully describe all the reaction steps [91]. Excellent reviews of the state of understanding exist and the reader is referred to those contributions for detailed information (e.g., [92–94]). Details of the complex cracking chemistry

CH$_2$=CHCH$_2$CH$_3$ (g)

+

D
|
O
Al   Si

CH$_2$=CHCH$_2$CH$_3$
(interacting with wall)

+

D
|
O
Al   Si

CH$_2$=CHCH$_2$CH$_3$
|
D
|
O
Al   Si

alkyl-BAS complex

CH$_2$=CHCH$_2$CH$_3$
|
D
|
O
Al   Si

π-BAS complex

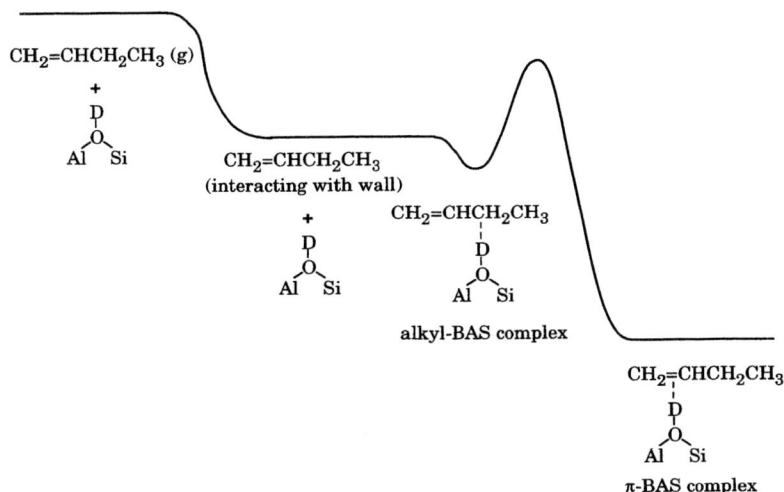

Scheme 1. Energy diagram of 1-butene adsorption on D-ZSM-5.

will be discussed in chapter 4.2.9.4.2 while the generation of the carbocations and radicals to activate the alkanes and the principal routes to cleave the molecules will be discussed in detail here, as these elementary steps are of general importance for the chemistry in zeolite pores.

The activation of an alkene appears to be a rather facile reaction. If an olefin is contacted with a zeolite at ambient temperature an alkoxy group is immediately formed. These alkoxy groups tend to be rather susceptible to alkylation by another olefin resulting in a rapidly growing alkoxy chain [95]. At low temperatures (150 K) alkenes are adsorbed in zeolites such as ZSM-5 via dispersion induced bonding on the saturated part of the molecule [96]. Only upon warming to temperatures above 170 K does the carbon–carbon bond interact via hydrogen bonding with the Brønsted acidic hydroxyl groups (see Scheme 1). Upon further heating it may be expected that the hydrogen-bonded alkene would be protonated and form an alkoxy group. It is interesting to note that for mordenite and ZSM-5 dimerization of iso-butene molecules took place before the protonation step occurs. The dimer alkoxy species formed were found to be restricted by the pore size of the zeolites. On mordenite 2,4,4-trimethyl-2-pentoxy species were identified [97], while on ZSM-5, the bulky tert-alkoxy groups can not be accommodated inside the pores and, therefore, the reaction resulted in the formation of 2,4,4-trimethyl-1-pentoxy species [98]. In addition, the double bond migration from 1-butene to 2-butene was shown to proceed without proton transfer from the Brønsted acidic groups below 230 K [99].

The activation of an alkane molecule is significantly more difficult. In principle, three routes have been postulated for zeolites, i.e., (1) hydride abstraction to form a carbenium ion, (2) the addition of a proton to form a carbonium ion and (3) radical

References see page 1145

cleavage of a C–H (C–C) bond leading to a hydrogen-deficient radical hydrocarbon fragment. These elementary reactions steps are endothermic with increasing energies of activations in the sequence of mentioning [94].

Hydride transfer and hydride abstraction occur preferentially under mild conditions. In the initial step the carbenium is formed by abstraction of a hydride either by a Lewis acid site [100] or more likely by another carbenium ion present in the zeolite [92, 101]. Such carbenium ions are formed, for example, by protonation of trace olefins present in the alkane feed. It has been shown empirically that hydride transfer reactions are favored in zeolite catalysts containing rare earth element cations such as La^{3+} or extra framework alumina and which have a low Si/Al ratio [102, 103]. The catalytic activity of zeolites for alkane activation via this route seldom coincides with the acid strength of the material. For hydride abstraction via metal cations the stabilization of the metal-hydride and the route of release of the abstracted hydrogen are most important. The difficulty to describe this reaction pathway completely and the resulting controversial discussion in the literature stem from the fact that only a few of such sites suffice to start hydride transfer reactions among hydrocarbon molecules. Thus, it is difficult to conclude unequivocally whether or not a small number of molecules has been dehydrogenated in the course of the reaction and these molecules have started a chain reaction in which hydride transfer takes place between carbenium ions and alkanes.

The rate of hydride transfer between carbenium ions (alkoxy groups in the ground state) and alkanes increases with the concentration of Brønsted acid sites in the zeolites following second-order kinetics [104]. The unexpectedly high order is the result of two factors, the weakening of the carbon–oxygen bond in the alkoxy group and the increased concentration of alkane molecules in the zeolites pores. Theoretical calculations [105] show clearly that in addition to the chemical effects described above, the transition state between the incoming alkane and the alkoxy group (carbenium ion) is also rather bulky. This is in perfect agreement with the observation that large pore zeolites such as zeolite Y show greater tendency for hydride transfer than medium-pore zeolites such as H-ZSM-5 [104].

The direct protonation of an alkane by a proton in liquid super acids at low temperatures was known from the work of Olah for some time, before Haag and Dessau [106] postulated in a landmark paper that also in zeolites alkanes may be directly protonated to form a "surface-stabilized" carbonium ions (see also ref. [107]). The proposal is based on the interpretation of the decay pattern of the carbonium ions formed as transition states in zeolites similar to that depicted in Scheme 2. Because the reactive decay of the carbonium ion leads to breaking of the C–C or C–H bonds at the insertion point of the proton, the reaction has been named protolytic cracking. The typical products are alkanes, hydrogen and primary carbenium ions, which can isomerize or desorb as alkenes. These products correspond to those observed from the collapse of a $C_6H_{15}^+$ ion in a mass spectrometer [108].

In contrast to the super acid chemistry in the liquid phase, the formation of the carbonium ions appears with zeolites only to be significant at higher temperatures (above 723 K). Also, in contrast to the stable carbonium ions in liquid super acids,

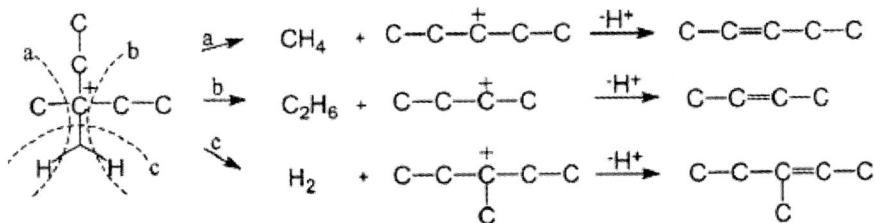

Scheme 2. Schematic reaction pathways for carbonium ion decay (protolytic cracking) of a protonated 3-methylpentane together with principle transition states for dehydrogenation and cracking.

theoretical studies suggest that carbonium ions in zeolites exist only in a transition state. Thus, it can be concluded that the catalytic manifestation only at high temperatures is a consequence of the thermodynamic equilibrium between the ground state (a hydrogen-bonded alkane molecule) and the transition state. Various recent *ab initio* calculations show that (1) the transition state for the various decay routes (dehydrogenation, cracking) differ markedly, i.e., that the reaction pathways are controlled by individual carbonium ions [109] and (2) that the framework of the zeolite exerts some stabilization to the carbonium ion [110].

Using similar product distributions, McVicker et al. [111] suggested that the alkane activation occurs via the formation and the subsequent decay of a radical cation. A similar proposal has also been made by Corma [112]. The strongest argument for such a route is the fact that the formation of radical cations is not related to the acidic properties of the zeolite, but rather to the presence of extra-framework aluminum-containing species. In this context, experiments by Narbeshuber et al. [113] should be mentioned, which show that a noncatalytic dehydrogenation occurs at short times on stream and that this route is related to the presence of extra framework alumina.

4.2.9.4.2 Reactions Involving Carbon–Carbon Bond Scission in Aliphatic Compounds
Overall, three major mechanistic pathways can be distinguished for catalytic alkane and alkene cracking on zeolites, i.e., protolytic cracking of carbonium ions, β-cracking of carbenium ions and cracking that is preceded by some oligomerization of reactants and/or products (oligomerization cracking). The importance of a particular reaction route depends upon the reaction conditions and the zeolite used.

Monomolecular (protolytic) cracking dominates at low conversions, high reaction temperatures, low reactant pressures and with medium and small pore zeolites that have a low concentration of Brønsted acid sites. All these conditions favor a low concentration of reactants in the pores and impede hydride transfer. The decay of the carbonium ion into an alkane and a smaller carbonium ion, which is the core elementary step in protolytic cracking, has already been discussed in detail in

References see page 1145

the section above with respect to the activation of alkane molecules. Thus, only two aspects should be mentioned here. First, results with n-alkanes in various zeolites suggest that the true energy of activation of cleaving the carbon–carbon bond is equal for all bonds in the molecule and that the selectivity of the cleavage is primarily related to entropic effects [114]. Second, like many other reactions discussed later in this section, protolytic cracking depends subtly upon physisorption, the presence of polarizing oxide species in the pores and the concentration of strong Brønsted acid sites. These factors may also be interdependent making the interpretation of kinetic data complex [48].

Cracking of a carbenium ion leads to the formation of an alkene and a smaller carbenium ion. The carbon–carbon bond cleaved is located in β-position to the carbon atom bearing the positive charge (β-cracking). When the smaller carbenium ion donates the proton back to the zeolite, an olefin is formed. Alternatively, the carbenium ion can abstract a hydride ion from an alkane leading to the desorption of an alkane and the formation of a new carbenium ion. Thus, for alkanes that cracking route proceeds via a chain mechanism [94]. The overall process is governed by the stability of the carbenium ions in the initial and final states of the reaction. As shown also by theoretical calculations, the difference in the stability of the carbenium ions determines the true energies of activation for the elementary step of the carbon–carbon bond cleavage. Thus, when only considering the initial state, cracking of a carbenium ion with the positive charge at a tertiary carbon will be faster than cracking of a carbenium ion with the positive charge at a secondary or primary carbon atom. Similarly, the ease of reaction decreases in the sequence tertiary > secondary > primary carbenium ion formed. Note that this is related to the influence of the Polyani relationship between the initial and final state upon the true energy of activation. The energy of activation usually increases with increasing energy level of the final state. Therefore, the rate for reactions starting from a tertiary carbenium ion and ending with a tertiary carbenium ion (Type A) is faster than the reaction starting from and ending with a secondary carbenium ion (Type C). The possible reaction pathways for the reaction are depicted in Scheme 3. Note that these rather simple assumptions for the mechanism lead to a surprisingly good prediction of the cracking selectivity observed [115].

Especially at higher reactant partial pressures this reaction route is gradually replaced by a cracking mechanism that has substantial oligomerization preceding the cracking process. Clear evidence for such a route comes from labeling studies that show complete scrambling of carbon-labeled olefinic products of cracking [116]. The importance of the mechanism increases with higher partial pressure, higher conversion and lower reaction temperature. While that route is important for accounting for the product distribution in some cases, the basic chemistry related to cracking is identical with that observed in β-cracking.

4.2.9.4.3 Reactions Involving Formation of Carbon–Carbon Bonds in Aliphatic Compounds

Following the principles of microscopic reversibility, molecular sieves are also frequently used catalysts for the reverse reaction, the alkylation of alkanes, alkenes

β-scission	Ions involved	Example
A	tert. → tert.	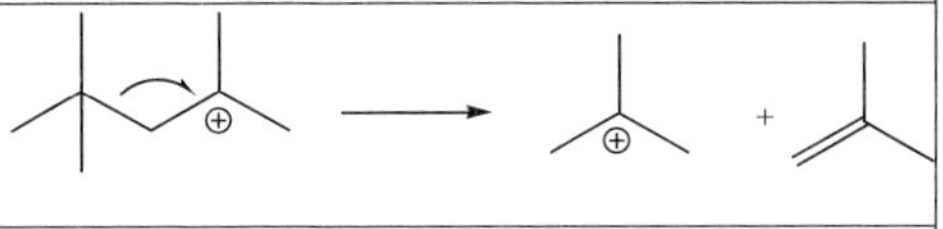
B₁	sec. → tert.	
B₂	tert. → sec.	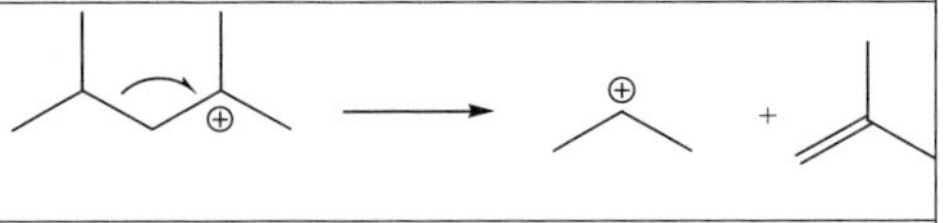
C	sec. → sec.	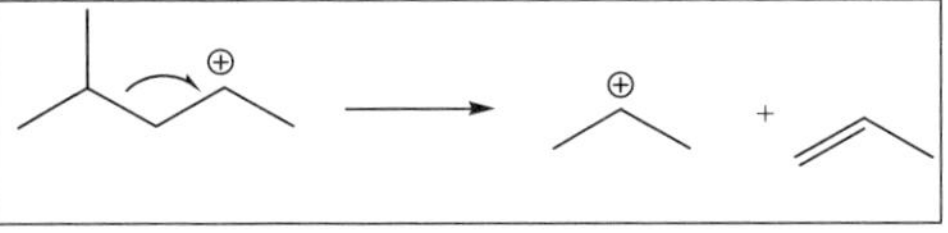

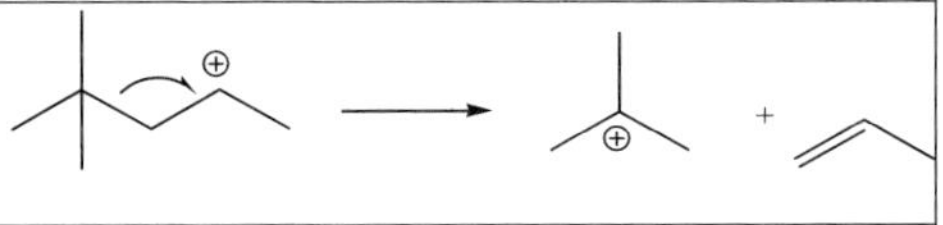

Scheme 3. β-scission mechanism for secondary and tertiary alkylcarbenium ions.

and aromatic molecules with alkenes. Owing to the lower temperature of reaction the desorption step is often difficult in these reactions, especially as the product of the reaction is more strongly adsorbed than the reactants.

The most demanding reaction is the alkylation of isobutane with light alkenes leading to the formation of a complex mixture of branched alkanes, which is an excellent blending component for gasoline [117, 118]. The reaction is commercially carried out using hydrofluoric and sulfuric acid as acid catalysts. While large pore zeolites, amongst other solid acids, have been examined for a long time, a process has not been yet commercialized.

The main reaction cycle occurring during alkylation consists of olefin addition to a tertiary butoxy group/carbenium ion, isomerization via methyl and hydride shift reactions of the resulting C_8 alkoxy group and hydride transfer from isobutane to the C_8 alkoxy group, which then desorbs as the corresponding isooctane (see Scheme 4). The reaction is initiated by the sorption of n-butene on a Brønsted acid site followed by olefin addition and hydride transfer (i.e., a n-butene dimerization followed by hydride transfer from isobutane) or by adsorption of n-butene on the Brønsted acid site, followed by hydride transfer producing an n-butane molecule and a tertiary butoxy-group.

The lifetime of the catalyst is determined by the relative rates of olefin addition and hydride transfer [119]. In order to control the rate of olefin addition the olefin concentration in the reactor is minimized using back-mixed reactors such as stirred-tank reactors operated at high conversions. The rate of hydride transfer is intrinsically influenced by the stability of the carbenium ion (in turn controlled by the chemical composition of the molecular sieve) and the steric hindrance through the micropore environment. The slower the hydride transfer (relative to the olefin

References see page 1145

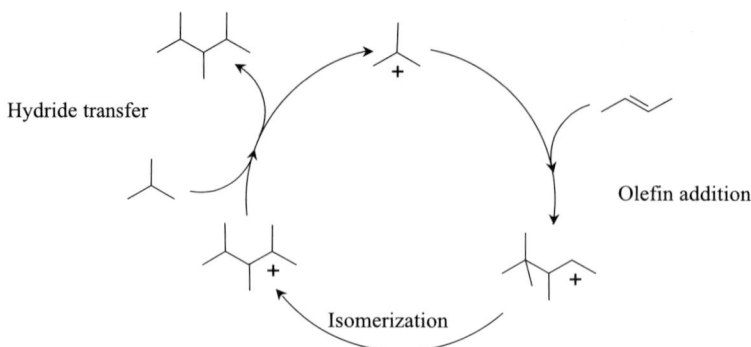

Scheme 4. Principle catalytic cycle in the alkylation of iso-butane with n-butene.

addition) the more likely multiple alkylation, leading to C_{12} or C_{16} carbenium ions, becomes. The hydride transfer to these molecules is increasingly slow; therefore, they cannot desorb and, thus, will block the particular site. Note that the rate of hydride transfer will also influence the isomer distribution, as methyl shift reactions will lead to product distributions closer to the chemical equilibrium [120].

Typical side reactions are oligomerization over weak Brønsted acid sites, producing larger olefins, which participate in alkylation and lead to the formation of larger alkylate molecules (increasing the deactivation rate) and cracking via β-scission, which are both responsible for a wider distribution in alkanes. In principle, most large pore zeolites show some activity for alkylation, but materials with a higher concentration of Brønsted acid sites have been found to be better [121, 122].

4.2.9.4.4 Reactions Involving Carbon–Carbon Bond Rearrangements

Paraffin isomerization [123, 124] was one of the first successful industrial applications of zeolite catalysts after cracking and has been studied extensively as a consequence. Because the associated processes will be discussed in a chapter 6.9 of this book, only the principle mechanistic features are addressed here. As with cracking, the catalytic activity for isomerization is strongly related to the concentration of strong Brønsted sites. Many of the catalysts used are of bifunctional nature and, hence, contain an additional metallic function, which helps to dehydrogenate the alkanes and facilitates the formation of the carbenium ions. As isomerization and β-cracking share the same intermediates (alkoxy groups in the zeolite pores) both reactions are concluded to be interlinked. It is generally accepted that a higher lifetime of the carbenium ion, usually associated with the stronger acid site creating it, leads to more pronounced cracking and less isomerization (see Scheme 5). While this is true at first sight it should be emphasized that the lifetime of the carbenium ions in monofunctional acid-catalyzed skeletal isomerization of alkanes will depend also to a large degree upon the hydride transfer characteristics of the reactants involved and the molecular sieves utilized.

Conceptually [125], two types of skeletal isomerization of alkanes can be seen, one that results in a positional shift of the branching and the other that increases

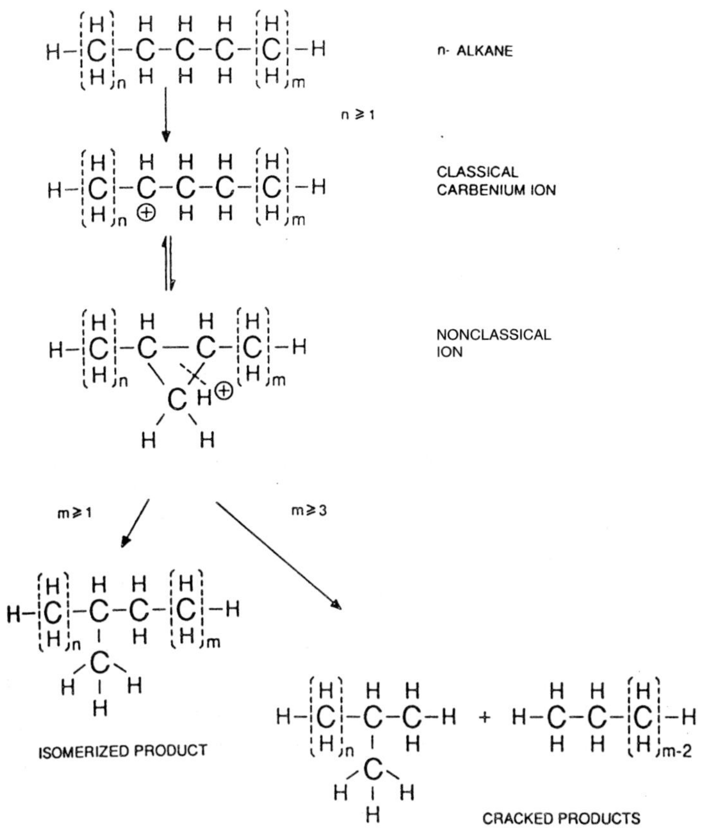

H–C–C–C–C–C–H   n- ALKANE

↓   n ≥ 1

H–C–C–C–C–C–H   CLASSICAL
⊕   CARBENIUM ION

↕

NONCLASSICAL
ION

m ≥ 1   m ≥ 3

ISOMERIZED PRODUCT

H–C–C–C–H + H–C–C–C–H   CRACKED PRODUCTS

Scheme 5. Reaction pathways for isomerization and cracking of alkanes.

or reduces the number of branches (for more extended reviews please consult, e.g., refs [126–129]). Both processes combine (facile) hydride shift reactions and the formation of a cyclopropyl ring, but differ insofar as in the type A isomerization only ring closure and opening occurs, while in the type B reaction a corner-to-corner proton jump also occurs (see Scheme 6). The transition state complex for this latter reaction appears to be significantly higher than that for reaction pathway A. Thus, changes in the branching will only occur under more severe conditions than the methyl shift isomerization. The pore size and shape of zeolites has a marked influence on the isomerization and subsequent cracking, which can be attributed to the strain to form and break the ring in the zeolite pores. Note that this has also served to classify zeolite pore structures [130].

However, molecular sieves can also catalyze skeletal isomerization of significantly more complex molecules such as the isomerization of tetrahydrodicyclopentadiene into adamantane, which is an example of a very complex rearrangement that is

References see page 1145

Type A:

Type B:

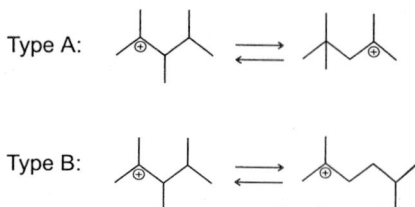

Scheme 6. Classification of skeletal rearrangements of alkylcarbenium ions.

commercially carried out over strong Lewis acids with a hydrogen-transfer initiator. The reaction can be catalyzed by rare earth exchanged zeolite Y (Scheme 7) in a $H_2/$ HCl atmosphere at 523 K. Selectivities to adamantane of up to 50 % were reported in presence of a noble metal, such as Pt, capable of catalyzing hydrogenation [131]. Initially acid-catalyzed endo- to exo-isomerization of tetrahydrodicyclopentadiene occurs followed by a series of 1,2 alkyl shifts involving secondary and tertiary carbenium ions that lead finally to the formation of adamantane [132]. A possible mechanistic route of adamantane formation from tetrahydro-dicyclopentadiene is discussed in detail in Refs. [133, 134].

Re-Y
H_2 / HCl
523 K

Scheme 7. Skeletal isomerization of tetrahydrodicyclopentadiene to adamantane.

Zeolites also catalyze the double-bond and skeletal isomerization of molecules having polar functional groups without conversion of the functional group. Suitable zeolites, such as H-ZSM-5, have to be apolar (hydrophobic) and, thus, have only a low concentration of Brønsted acid sites. With such catalysts, the interactions of the polar functional groups with zeolite pore walls are weak [135] and, hence, the activation of these groups is minimal. An example of double-bond shift isomerization is the transformation of 2-ethyl propenal into *trans*-2-methyl-2-butenal over Ce,B-ZSM-5 [136] (Scheme 8). An example of a skeletal isomerization is the allylic rearrangement of 1,4 diacetoxybutene over ZSM-5 while retaining the functional group intact (see Scheme 9).

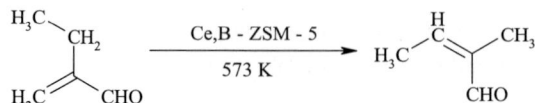

H_3C
CH_2
H_2C CHO

Ce,B - ZSM - 5
573 K

H
H_3C—C CH_3
CHO

Scheme 8. Isomerization of 2-ethyl propenal to *trans*-2-methyl-2-butenal.

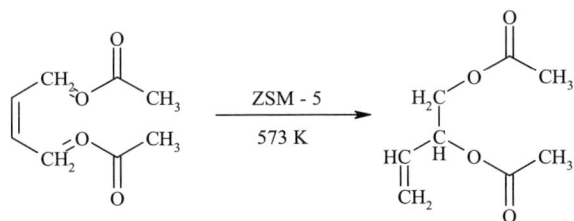

Scheme 9. Skeletal isomerization of 1,4-diacetoxybutene over ZSM-5.

4.2.9.4.5 Positional Isomerization Reactions Involving Hetero-atoms

The weaker strength of carbon–hetero-atom bonds compared to a carbon–carbon bond allows mild reaction conditions. Successful examples include the isomerization of halogenated aromatic molecules such as chlorophenols, chlorothiophene, bromothiophene and iodothiophene over ZSM-5 zeolites [137]. The optimum reaction temperature for the last three molecules gradually decreases from 300 °C to 100 °C in parallel with the increasingly weaker carbon–halogen bond.

4.2.9.4.6 Skeletal Isomerization Reactions Involving Hetero-atoms

This group of reactions involves a wide variety of organic transformations, in which the hetero-atom forms a new functional group. In general, these reactions take place over high-silica materials, because the interactions of the zeolite pore walls have to be as weak as possible in order to achieve reasonable product yields.

B-ZSM-5 and Fe-ZSM-5 have been found to be suitable catalysts for the transformation of aldehydes to corresponding ketones of the same molecular weight [138]. The iron-based catalysts appear to be more active than the boron-containing ones and water (usually added in such reactions to delay catalyst deactivation) does not have to be added. Stronger acid sites or higher concentrations of acid sites are negative for the yield.

The opening of epoxide rings can be catalyzed by Brønsted and Lewis acid forms of pentasil zeolites suggesting that both types of sites can catalyze the breaking of one of the carbon–oxygen bonds in the epoxide ring. Ti-silicalite, for example, catalyzes the reaction of (substituted) styrene oxides in acetone or methanol below 373 K resulting in phenylacetaldehydes with yields exceeding 80 % (see reaction Scheme 10) [139]. The reaction is also successfully catalyzed in the gas phase over mildly acidic B-, Fe- and Al-containing ZSM-5 catalysts [140].

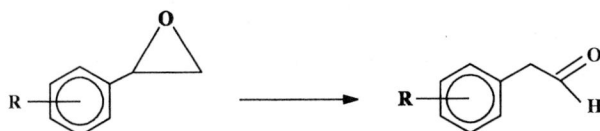

R = alkyl, aryl, arylalkyl, halogen, haloalkyl, alkoxy, alkylthio

Scheme 10. Reaction of (substituted) styrene oxide to phenylacetaldehyde.

References see page 1145

Such reactions can also be catalyzed with acceptable selectivity in the presence of other functional groups emphasizing that the zeolite reacts specifically, provided the concentration of acid sites is sufficiently low. Examples include the transformation of 1,5 dioxospiro-(2,6)-octane to 4-formyltetrahydropyrane (see Scheme 11) and the rearrangement of the very reactive isophoron-oxides to the corresponding ketoaldehyde and the α-diketone (see Scheme 12). The former reaction has been catalyzed successfully in the gas phase at 573 K using B-ZSM-5 and H-ZSM-5 as catalysts [141]. Catalysts for the latter reaction, which is carried out in solvents such as toluene and benzene are dealuminated zeolites Y and mordenite [142, 143].

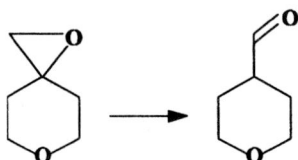

Scheme 11. Transformation of 1,5-dioxospiro-(2,6)-octane to 4-formyltetrahydropane.

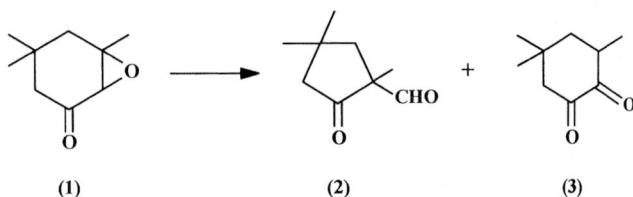

(1)　　　　　　　　　　(2)　　　　　　　　　(3)

Scheme 12. Isomerization of isophoronoxide.

Because of its economic importance, the catalytic transformation of cyclohexanone oxime into ε-caprolactam has been explored for many zeolite systems (Scheme 13). While a large number of materials is active and selective the most successful zeolites have mild acidity and catalyze the rearrangement on the outside of the crystals. Suitable zeolites have been found to be ZSM-5 type materials, which are preferentially silylated [144]. Sato et al. [145] observed that the catalytic activity and selectivity to ε-caprolactam increased in parallel with the Si/Al ratio of ZSM-5 and were directly proportional to the concentration of weak acid sites on the external surface of H-ZSM-5. Interestingly, amorphous silica that also contained a high concentration of such SiOH groups also gave a very high initial conversion, but deactivated rapidly due to coking. Also a number of other zeolites having weak

Scheme 13. Beckman rearrangement of cyclohexanone oxime into ε-caprolactam.

Brønsted acid sites on the outside of the zeolite crystals, such as B-ZSM-5, Ti-ZSM-5 and SAPO-11 are excellent catalysts, which show over 90 % yield at high space velocities. Diluents such as water, ethanol and acetonitrile help to maintain the activity of the catalyst. Highly crystalline zeolite samples were shown to be more selective and more active indicating that the regularity and/or the low density of such weakly acidic silanols are essential for high lactam selectivity. As with many of the reactions discussed in this section, limited desorption of the product from stronger acid sites appears to be the main cause for the deactivation of the catalyst. For an in-depth treatment of the subject the reader is referred to a recent review by Tatsumi [146].

Typical examples for benzamine rearrangement are the conversion of aniline and 1,3-diaminobenzene with ammonia to 2-methylpyridine and 2-amino-6-methyl-pyridine (Scheme 14). Although several acidic oxides were found to be active, the best results were obtained with H-ZSM-5. It was more active than amorphous SiO_2/Al_2O_3 (48 % conversion compared to 29 %), but showed a similar selectivity (83 % over H-ZSM-5 and 98 % over SiO_2/Al_2O_3) [147, 148]. The reaction seems to proceed via the addition of ammonia to a protonated aminobenzene (probably present in the form of a cyclic enamine). After an enamine-amine isomerization, the ring is opened via a reverse aldol-type reaction. Upon addition of the amino group to the imine double bond the ring closes again. After elimination of ammonia from the resulting aminal the final product is obtained [149, 150]. Note that this potentially provides a new simple route for the production of amino-pyridines replacing the current complicated industrial process [151].

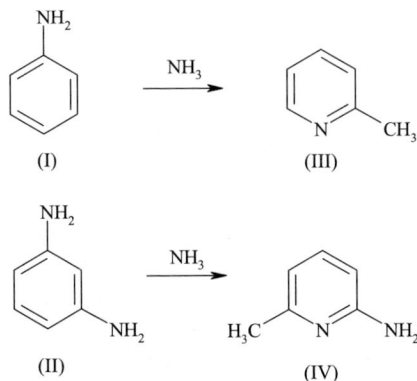

Scheme 14. Conversion of aniline and 1,3-diaminobenzene with ammonia to 2-methylpyridine and 2-amino-6-methylpyridine.

4.2.9.4.7 Nucleophilic Substitution and Addition Reactions

Conventionally, nucleophilic substitutions (and addition reactions) follow two mechanistic pathways, i.e., the two-step nucleophilic substitution (S_{N1}) and the

References see page 1145

one-step process (S$_{N2}$). In the former route, the highly polar intermediate species or, in the limiting case, the carbocation is stabilized by the catalyst, while in the latter a transition state comprising the substituent and the leaving group is formed in a concerted fashion.

Direct evidence for the presence of carbonium and carbenium ions of the mostly polar species in the molecular sieve pores is scarce. Experiments point to such species only in the presence of very strong acid sites provided relatively basic reactant molecules are used [60]. Even in such cases the interpretation of the experimental data does not seem to be unequivocal [152]. Most results suggest that the true cation exists only in the transition state resulting in a quite complex reaction coordinate. The course of the reaction is determined by the chemical nature of the leaving and the substituting group, the acid/base properties of the molecular sieve, the influence of co-reactants and the availability of space for the reaction to take place. The majority of the nucleophilic substitutions involve the replacement of an -OH group with an -NH$_2$, -NR$_2$, -S, -SH, -SR, -OR or another functional group.

One of the major problems is that in many cases the resulting product interacts more strongly with the molecular sieve than the reactant. This leads to the situation that many reactions are desorption controlled and require either a reactant to desorb (adsorption-assisted desorption) or a gaseous/liquid cocatalyst that also facilitates the desorption of the products without participating in the reaction. Note that for liquid phase reactions the solvent can assume the role of the cocatalyst.

A. Formation of ethers from alcohols Etherification, conceptually one of the simplest reactions to catalyze, occurs over most zeolites. Molecular sieves, however, have too low an acid-site density to make them interesting for commercial applications. Usually, organic resins such as amberlyst are used for that purpose [153]. On the other hand, etherification is experimentally and theoretically well studied and understood. Dimethylether formation from methanol is an excellent example to show how the reaction conditions influence the reaction mechanism, i.e., whether the reaction proceeds along the S$_{N1}$ or the S$_{N2}$ pathway [154–157]. Temperature-programmed reaction studies of methanol conversion over H-ZSM-5 suggest that three reaction routes to form dimethylether exist, i.e. via an alkoxonium cation and via two alkoxy pathways [65]. At low temperatures the reaction proceeds via an Eley–Rideal type mechanism. In the transition state one methanol molecule forms a methoxonium ion, water leaves the molecule and simultaneously another weakly sorbed methanol binds to the methyl group forming protonated dimethylether (see Scheme 15). The protonated dimethylether donates the proton back to the zeolite and desorbs.

As the reaction temperature increases, part of the methanol molecules are transformed into methoxy groups replacing the proton in bridging (SiOHAl) and terminal (SiOH) hydroxyl groups. These methoxy groups react with weakly associated methanol to form dimethylether under the simultaneous restitution of the hydroxyl group. While the methoxy group is covalently bound to the zeolite framework, its reactivity increases with the acid strength of the hydroxyl group it replaces [158, 159]. Thus, methoxy groups at bridging hydroxyl groups produce dimethyl-

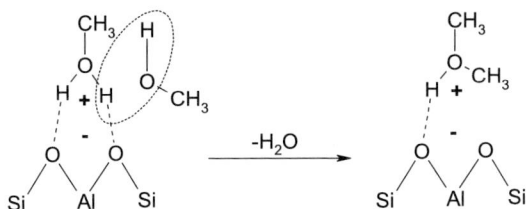

Scheme 15. Formation of dimethylether from methanol.

ether at lower temperatures than methoxy groups at terminal hydroxyl groups. Comparison of the chemistry over various zeolites indicates that the formation and reactivity of a specific type of methoxy group is connected in a complex way to the polarizability of the framework and the overall acid/base properties. Methoxy groups at bridging sites are more easily formed and consumed on FAU type materials than on MFI type materials [160].

Theoretical calculations by Blazowski and van Santen [161] suggest that the pathway to form dimethylether via a methoxonium ion is energetically favored over the pathway via methoxy groups. The data agree with the observed strong temperature dependence of the reaction mechanism as reported [87]. The *ab initio* calculations used show that the S_{N2} reaction involves a complex transition state. For a successful reaction, four elementary steps have to proceed in a synchronous manner, i.e., (1) formation of a methoxonium ion by proton donation from the zeolite, (2) cleavage of water from the methoxonium ion and formation of a methylcarbenium ion, (3) binding of the methylcarbenium ion to the second methanol molecule to form protonated dimethylether and (4) donation of the proton back to the zeolite. Note that according to the calculations all (steps) must occur in a concerted manner, as the protonated species are only found to be stable in the transition state. The transition entropy of such a complex must be quite low leading in turn to low reaction rates. Stabilization of the methoxonium ion by the catalyst would lead to a less complex transition state and, hence, one might expect the intrinsic rates of the reaction to be higher. Methanol sorbed on organic resins and heteropoly acids appears to form methoxonium ions [162].

B. Reactions of alcohols with ammonia   Amination of alcohols follows a mechanistic pathway similar to etherification [163]. Both, purely Brønsted acidic and metal-exchanged zeolites are active for the reaction [164]. As for the ether formation the mechanism varies as a function of the active site and the reaction temperature used. For the purely Brønsted acidic zeolites the basicity of the reactants dictates that at the start of the reaction ammonia is present in the molecular sieve as ammonium ions. The alkyl group reacts with the ammonium ion forming water as the leaving group and an alkylammonium ion [165]. The alkylammonium ions, however, cannot desorb under typical reaction conditions ($T = 625$ K).

References see page 1145

Even at reaction temperatures above 625 K, the alkylammonium ions are unable to desorb from the acid sites [166]. The alkylamines released into the gas phase stem either from a further nucleophilic substitution in which the alkyl group of the alkylammonium ion is scavenged by weakly adsorbed ammonia or the desorption of the alkylammonium ion is aided by the simultaneous formation of an ammonium ion (see Scheme 16). Both reaction pathways have been found to be important under typical reaction conditions. The acid strength of the zeolites should be as high as possible in order to assure that all the acid sites are covered by ammonia and amines, thus, preventing the formation of ethers and higher hydrocarbons from the alcohol over free acid sites [167, 169]. The highly desired selectivity to mono- and dialkylamines is achieved with surface modification of the zeolites by coating with silica overlayers [170]. In the resulting materials trimethylamines can be formed but cannot leave the pores and react via disproportionation to the smaller mono- and dimethylamines.

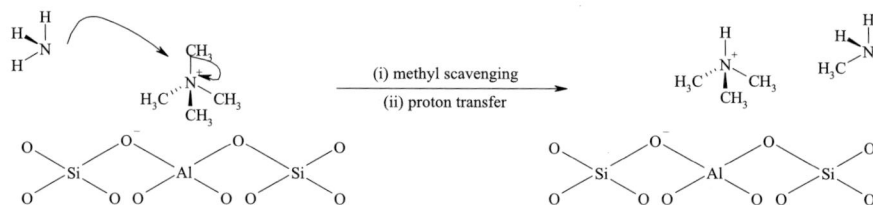

Scheme 16. Proposed mechanism for the removal of methyl amines by scavenging with ammonia.

C. Formation of alkylamines by addition of ammonia to olefins Catalysis for the addition of alkenes and ammonia requires the alkene to be activated by the Brønsted acid site of the zeolite, which is only possible when the acid sites are not completely blocked by ammonium ions. Examples include the amination of ethene with H-Y, H-erionite and H-mordenite at reaction temperatures between 500 and 550 K [171]. At higher temperatures, nitriles tend to be formed. The reaction of isobutene with ammonia is successfully catalyzed with B-ZSM-5 [172]. In addition to moderate to weak acid sites of the catalyst, mild reaction temperatures and high pressures of alkenes are favorable [173]. It is important to note that dienes can successfully be aminated with ammonia, but that diamines are not formed and the chemistry follows the Markownikow rule. Thus, the route does not allow the synthesis of 1,4-butyldiamine from butadiene.

D. Synthesis of anilines via nucleophilic substitution Nucleophilic substitution of an aromatic ring is difficult to achieve, as the π-electrons will repel the electron rich functional group of the incoming molecule and it is difficult for the aromatic ring to accommodate the additional electrons. The substitution becomes more facile when a strongly electron-withdrawing group is replaced by a more electron-donating one. Examples of this case are the reaction of chlorobenzene with ammonia to form aniline and HCl or the reaction of phenol with ammonia to give aniline and water (see Scheme 17) [174, 175]. The rate-determining step seems to be the release of the electron-withdrawing group [76, 176]. Phenol and ammonia are converted to aniline with high yields using H-ZSM-5 and Na-ZSM-5 as catalysts [177, 178]. The reaction is also catalyzed by zeolites such as Y, X, mordenite and amorphous silica-alumina. Such acidic catalysts, however, also catalyze the formation of larger amines, such as diphenylamine. These latter catalysts also show a markedly shorter lifetime, which appears to be related to the formation of larger byproducts that are suppressed in the case of medium pore materials.

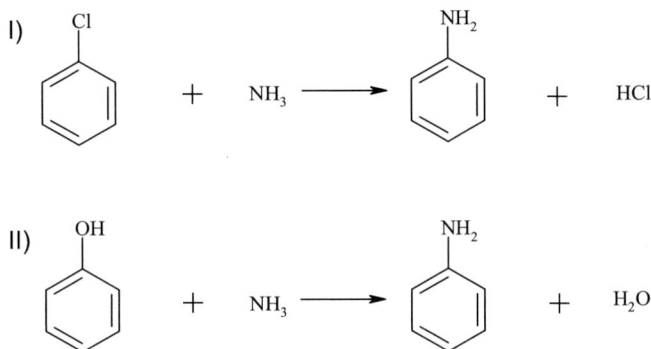

I)

II)

Scheme 17. Reactions of chlorobenzene with ammonia (I) and phenol with ammonia (II).

Similarly, the chlorine atom in chlorobenzene can be substituted by an NH_2 group using ammonia as reactant and Cu-exchanged medium pore zeolites such as Cu-ZSM-5 [179] as catalyst. The activity depends upon the concentration of exchanged Cu sites and, hence, upon a high concentration of aluminum in the zeolite framework high catalytic activity is observed. The selectivity to aniline is only satisfactory with the medium pore zeolites.

E. Nucleophilic substitution in ring systems A more complex example than the above described substitution and addition reaction is the transformation of oxygen-containing heterocycles into nitrogen or sulfur-containing heterocyclic compounds [180]. The structure of the molecular sieve has been found to be less important (provided there is enough space within in the zeolite pores to accommodate the reactants and products), compared to the acid strength and the nature of the acid site. Hatada et al. [181] reported the conversion of γ-butyrolactone into 2-pyrrolidinone over a series of metal-exchanged Y zeolites (Scheme 18). For alkali-metal- and alkaline-earth-metal-exchanged FAU a direct dependence of the yield of 2-pyrrolidinone upon the strength of the electrostatic field of the cation has been observed. This indicates that the strength of the coordination of γ-butyrolactone to the metal cation is the most important parameter influencing the catalytic conversion. For transition-metal (Co, Ni, Cu and Zn)-exchanged zeolite Y a correlation between the

References see page 1145

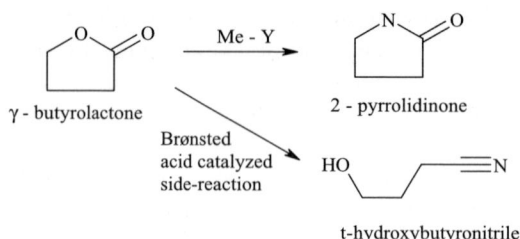

Scheme 18. Transformation of γ-butyrolactone into 2-pyrrolidinone over Me-Y.

cation field strength and the activity was not observed. It is speculated that this is due to the nonspherical nature of transition metal cation orbitals, especially of their partially filled d-orbitals, and thus the simple electrostatic model is not applicable [182]. The reaction proceeds via initial polarization of the carbonyl group of γ-butyrolactone by the metal cation. In the next step, ammonia binds to the carbon atom of the polarized carbonyl group forming an acid amide, which then rapidly dehydrates under ring closure. The stronger the electrostatic field of the metal cation the stronger the interaction between the carbonyl group and the metal cation leading to a more polarized C=O bond, which is then more reactive towards ammonia.

In contrast to the catalysis involving only carbon–carbon bond formation and breaking, these results clearly show that Lewis acid sites can also act as catalytically active sites for nucleophilic substitutions. If catalysts without Brønsted acid sites are used (i.e., zeolites exchanged with monovalent cations) the competitive side reaction leading to t-hydroxybutyronitrile via protonation of the acid amide can be completely suppressed (see Scheme 18).

F. Addition and elimination reactions of carbonyl compounds   The polar nature of the carbonyl group allows for addition of nucleophiles at the carbon atom. Molecular sieves catalyze these reactions by enhancing the polarity of the carbonyl group through interactions between the Brønsted or Lewis acid sites and the oxygen of the carbonyl group. If the nucleophile retains a proton, water can be easily eliminated and the overall reaction leads to the replacement of the oxygen by another nucleophile (see Scheme 19 and refs [183, 184]). The reactions involve the addition of H_2O, ROH, RSH, HCN and HSO_3^- to the carbonyl group yielding the corresponding hydrates, (semi)acetals, cyanhydrines etc.

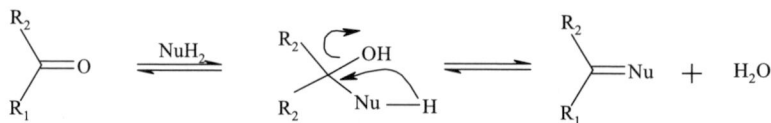

Scheme 19.   General mechanism of nucleophilic addition with subsequent elimination of water.

Acetal and ketal formation from aldehydes, respectively ketones, and alcohols occurs over mordenite and other acidic zeolites [185] slightly above ambient temperatures in the liquid phase. The reaction is not confined to simple alcohols, diols can also be converted (e.g., cyclohexanone reacts with ethylglycol to 1,4-dioxaspiro(4,5)decane [2]). Note that it is likely that desorption controls the rate of such reactions, as the product molecules are larger than the reactants and, hence, have a higher adsorption constant.

The reaction of acetonyl acetone to dimethylfuran, catalyzed by H-ZSM-5, is an example of an intramolecular addition reaction involving two carbonyl groups, followed by a β-elimination of water (Scheme 20). The Brønsted acid site of the zeolite protonates one of the carbonyl groups, while the oxygen atom of the second C=O group binds to the positively charged carbon atom of the protonated carbonyl group. The use of the rather hydrophobic zeolite H-ZSM-5 facilitates the elimination of water after the ring-closure reaction.

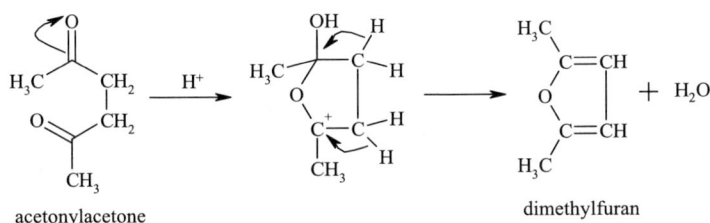

acetonylacetone

dimethylfuran

Scheme 20. Reaction of acetonylacetone to dimethylfurane.

The products of the ketone or aldehyde conversion with ammonia and amines depend upon the availability of a proton at the nitrogen atom. If hydrogen is present in that position, e.g., in the reaction of benzaldehyde with NH_3, the addition of ammonia to the carbonyl group is followed by a rapid elimination of water. The so-formed benzylidineimine subsequently dehydrogenates to form benzonitrile in the presence of transition metal ions, such as Co, Cr, Cu, Zn or Mn, (Scheme 21) [74]. If the proton at the nitrogen is lacking, e.g., in the reaction of diethylamine with cyclohexanone, the formation of a C=N bond is prevented during the dehydration step and instead a ring C=C bond is formed. The zeolites used in this case are large-pore zeolites such as Ca-X or H-mordenite. Note that with these catalysts drying agents have to be added and it appears to be likely that large-pore hydrophobic zeolites would be a better choice as catalyst.

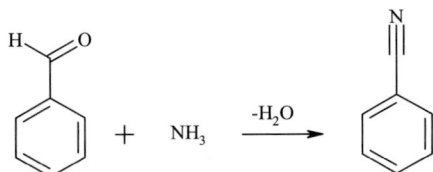

Scheme 21. Reaction of benzylaldehyde with ammonia to benzonitrile.

References see page 1145

Acidic 10- and 12-membered ring zeolites (mordenite, ZSM-5, ZSM-11) can also be used to catalyze the condensation of alkenes with aldehydes to form unsaturated alcohols, acetals, etc. (Prins reaction) [186]. Chang et al. [187] showed that this reaction involves in the initial step the activation of the aldehyde by a Brønsted acid site to generate an electrophilic species. The condensation (for example with isobutene) leads to a primary alcohol with a positive charge at the tertiary carbon atom. Elimination of water and addition of further aldehyde molecules may lead to a broad variety of products. Some of these reactions can be effectively blocked by choosing zeolites with the appropriate pore size [188, 189].

4.2.9.4.8 Cyclization Reactions

A. Diels–Alder cycloadditions   Diels–Alder-related reactions are facile to catalyze, but the zeolite has the important function to concentrate the reactants [190] and to limit the extent of addition reactions through spatial constraints [191]. This is noticeable in the highly selective cyclodimerization of butadiene to vinlycyclohexene over Cu^I-Y zeolites [191]. With analogous homogeneous catalysts a mixture of oligomers is formed. The authors speculate that the zeolite induces the selectivity by inducing constraints on the transition state, which are obviously much larger for trimers and tetramers.

Another intriguing example of zeolites catalyzing Diels–Alder chemistry is the conversion of cyclopropene to tricyclohexane over zeolites such as Na-A or K-A [192] with yields exceeding 95 %. Again, zeolites with larger void volumes catalyze oligomerization, suggesting that the sequential reactions are impeded by the zeolite pores.

B. Cyclocondensation with ammonia   Aldehydes and/or ketones form together with ammonia pyridine and alkylated pyridines as major compounds in high selectivity. A mixture of ammonia, acrolein and butanal forms over HF treated B-ZSM-5 β-ethylpyridine with over 70 % selectivity. Similarly, pyridine and methyl-pyridine are formed from a mixture of formaldehyde, acetaldehyde and ammonia [193]. The pores of the zeolite limit the formation of higher-substituted pyridines. While the mechanistic details of the reactions are unclear, it has been shown by isotope labeling experiments that the used molecules condense without significantly repeated bond-breaking and bond-making reactions [194]. The reaction can also be realized with alcohols as starting material in the presence of oxygen over B-ZSM-5 or Fe-ZSM-5 as catalysts [195]. Mild reaction temperatures are needed in order to avoid formation of ethane.

C. Fischer indole synthesis   The synthesis of substituted indoles involves the initial condensation of a phenylhydrazine and 3-heptanone to phenylhydrazone (see Scheme 22). Phenylhydrazone undergoes (Brønsted or Lewis) acid-catalyzed tautomerization to give the enhydrazine tautomer, which further rearranges and then eliminates ammonia to form the indole. Two products are possible, the bulky 2-ethyl-3-propyl-indole and the more linear 2-butyl-3-methyl-indole. In the homoge-

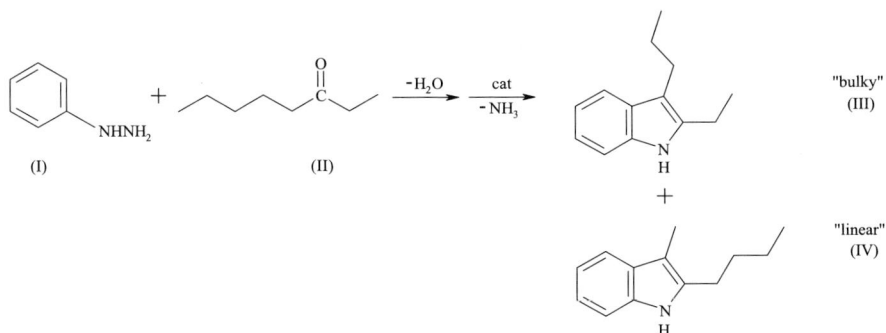

Scheme 22. The Fischer indole reaction of phenylhydrazine (I)
with 3-actanone (II) giving two indole products.

neous catalysis, the selectivity towards one of the two products is controlled by the
acid strength of the catalyst. The role of the zeolite in controlling the selectivity in
the heterogeneously catalyzed process is not resolved. Rigutto et al. [196] suggest
that the selectivity is related to constraints in the transition state. In a later study,
however, Kunkeler et al. [197, 198] found that the outer zeolite surface (or the pore
mouth) must be the site of catalytic activity and not the interior of the pores.

4.2.9.4.9 Electrophilic Substitution on the Aromatic Ring

The reactions are characterized by the attack of a species with a positive partial
charge, a positively charged species or a radical (i.e, species that are electron defi-
cient) on an aromatic ring, preferably on the carbon atom with the highest negative
charge. A broad variety of such electrophilic species has been reported to exist in
the pores of molecular sieves (see Fig. 6, after Ref. [21]). The generation of such
species occurs via several pathways, amongst which protonation, hydride abstrac-

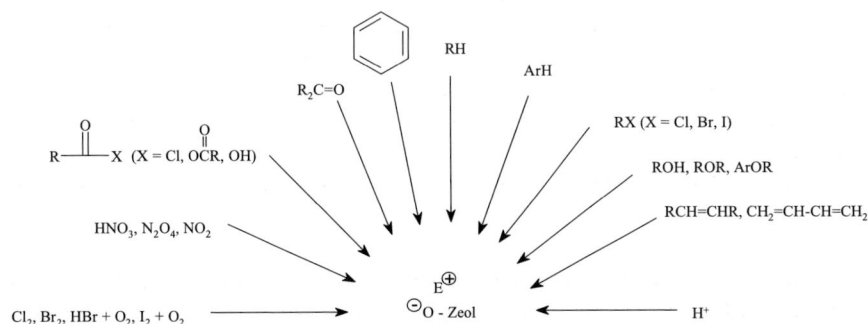

Fig. 6. Range of electrophilic agents employed in electrophilic
aromatic substitution over zeolite catalysts.

References see page 1145

tion and cleavage of polar groups are the most important ones. A general mechanism can be visualized as depicted in Scheme 23.

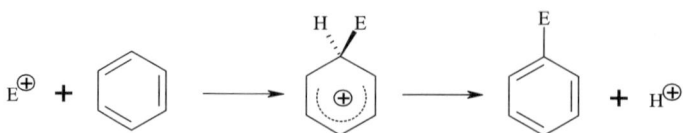

Scheme 23.   General scheme for electrophilic aromatic substitution.

In the first step, coordinative bonding between the π-electrons of the aromatic ring and the electrophile occurs frequently. Recent spectroscopic evidence for such an intermediate was reported for the methylation of toluene [199]. The aromatic ring must only be weakly held by the zeolite in order not to decrease the availability of the π-electrons. Subsequently, one carbon atom of the ring interacts with the electrophile prior to the actual substitution. In the presence of a substituent on the ring, the carbon atom position at which the interaction with the electrophile occurs will depend on the inductive effects induced by the ring substituent. For electron-donating substituents the preferred carbon atoms to accept the electrophile are those in ortho- and para-position to the substituent group [200].

The overall reactivity of the aromatic ring depends upon the nature of the substituent. Electron-donating properties of the substituents increase the availability of π-electrons at the aromatic ring, while the electron-withdrawal properties reduce it. In that respect alkyl-, hydroxyl-, alkoxy-, or amine groups increase the reactivity, while the presence of halogen or nitro groups reduces it. The reactivity of heterocycles also depends upon whether or not the ring has a π-electron excess. This results in pyrrole and thiophene being more reactive than benzene, while pyridine is less reactive [92]. The examples discussed here with respect to molecular sieve properties and the necessary adjustment of the reaction conditions are alkylation, acylation, nitration and chlorination.

Friedel–Crafts-type alkylation of benzene by alkenes involves the initial formation of a framework-associated carbenium ion, formed by protonation of the sorbed olefin. The chemisorbed alkene is covalently bound to the zeolite in the form of an alkoxy group and the carbenium ion formed exists only in the transition state [201]. As would be expected from conventional Friedel–Crafts alkylation, the reaction rate over acidic molecular sieves increases with the degree of substitution of the aromatic ring (tetramethyl > trimethyl > dimethyl > methyl > unsubstituted benzene). The spatial restrictions induced by the pore size and geometry frequently inhibit the formation of large multisubstituted products (see also the section on shape selectivity).

For similar alkylation reactions modified faujasites need lower temperatures to catalyze the reaction with the same rate (under otherwise identical reaction conditions) than amorphous silica-alumina catalysts [202]. The difference is explained with the higher site strength and density in the zeolite catalysts. The fact that the original Friedel–Crafts catalyst (promoted Lewis acid – AlCl$_3$–HCl) is reactive at yet lower temperatures than modified faujasites suggests that a microporous material with higher acid strength could lower the operating temperatures. In general, a suitable catalyst should have high acid-site strength and sorb the substituting molecule strongly. A good example of this is the alkylation of benzene with propene for which the reaction rate over divalent cation exchanged Y zeolites was found to decrease in the order Mg ~ Ca > Sr > Ba, in accordance with the decreasing acid strength of the materials [203].

Today, the alkylation of benzene or toluene with light olefins is generally performed with zeolite catalysts [204]. The zeolites frequently used (ZSM-5, Beta, MCM-22, mordenite) have rather high Si/Al ratios and exhibit shape-selective properties to various extents. The importance of the latter property depends critically on the specific alkylation process. While for alkylation of toluene with ethene the shape selectivity to para-substituted products is most important, single alkylation and catalyst stability dominates for cumene synthesis [205]. For the latter process a new dealumination technique transformed mordenite into three-dimensionally accessible fragments [206].

The alkylation of toluene with methanol over H-ZSM-5 proceeds at low temperatures via a protonated methanol species in the transition state [207] and weakly coadsorbed toluene as classically predicted for Friedel–Crafts alkylation. The reaction rate is directly proportional to the concentration of the chemisorbed methanol (in the presence of excess toluene), as shown in Fig. 7 [208]. Alkylation leads preferentially to ortho- and para- substituted products, which rapidly isomerize in the zeolite pores. Specific reaction conditions and tailoring of the catalyst pore structure can be employed so that para-substituted products are preferentially produced [209, 210]. The reasons for this selectivity and the methods for optimizing the catalyst performance will be discussed in a later section. The catalysis appears to be

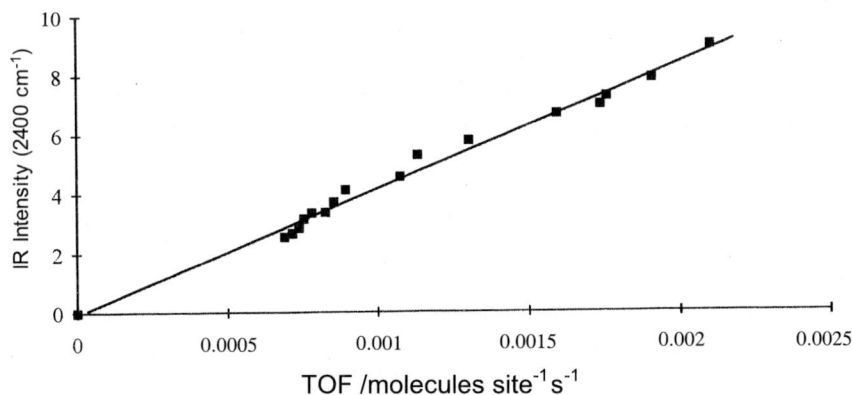

Fig. 7. Reaction rate of methylation of toluene as a function of the concentration of chemisorbed methanol.

References see page 1145

completely controlled by the Brønsted acid sites with the role of the Lewis acid sites being marginal [211].

In contrast, Brønsted and Lewis acid sites are claimed to participate in the catalysis of the alkylation of the more active phenol [212] and of aniline [213]. The Brønsted acid sites activate the alkylating agent by protonation, whereas the Lewis acid sites can activate the alkylating agent and phenol by coordination, and/or phenol by deprotonation. If activation of the alkylating agent and the phenol occurs on the same Lewis acid site, the predominant product will be the ortho-substituted isomer. For optimum results it is important that the concentration of both reactants in the pores of the zeolite is well balanced. The chemical and the shape selectivity of aniline depend strongly upon the reaction temperature. With respect to the chemical selectivity, Ione et al. [214] showed that the N-methylation of aniline with methanol drastically decreased with the reaction temperature, in line with the increasing tendency to Hoffman elimination of eventually formed amines with temperature. The *p*-selectivity of toluidine went through a pronounced maximum showing significant formation of *m*-toluidine at low and high reaction temperatures.

Acylation is currently carried out industrially with stoichiometric amounts of metal chlorides or mineral acids. Zeolites can replace liquid acids in this two-step process consisting of esterifcation and the Fries rearrangement. Several possible starting compounds for acylation, such as acid halides, carboxylic acids and acid anhydrides exist. The type of acid site (i.e., Brønsted or Lewis) in the molecular sieve has to be adjusted for the acylating agent. A Lewis acid, such as La^{3+} in the zeolite, will not activate a carboxylic acid to give an acylium ion, but will rather form a carboxylate anion. In contrast, Brønsted-acidic hydroxyl groups will readily help to generate an acylium ion. When using an acid chloride as the acylating agent, Lewis acid sites are a better choice than Brønsted acid sites, since they assist heterolytic dissociation by forming strong bonds to the halogen anion. An example of the need for Brønsted acid sites is the acylation of phenol with acetic acid to yield 2-hydroxyacetophenone [215] with H-ZSM-5 as catalyst (Scheme 24). The latter situation is exemplified by the para-directed acylation of toluene with several aliphatic acid chlorides over Zn-Y [216]. Other examples are the formation of anthraquinone from benzene and phthalic anhydride or from phthalic anhydride alone over NaCe and NaZn exchanged zeolite Y, respectively [34, 217].

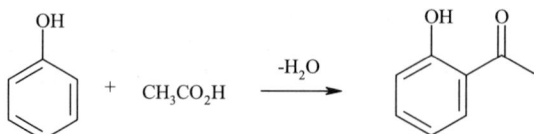

Scheme 24. Acylation of phenole with acetic acid to 2-hydroxyacetophenone.

Acylation of heterocycles, such as thiophene, seems to require a lower acid strength of the catalyst, which is best met by B-ZSM-5. The reaction of thiophene with acetic anhydride to 2-acetylthiophene proceeds with 99 % selectivity at 24 %

conversion and the conversion of pyrrole with acetic anhydride to 2-acetylpyrrole with 98 % selectivity at 41 % conversion [34].

Nitration of aromatic compounds requires very strong acid sites to stabilize the NO_2^+ cation, which is an important intermediate in liquid-phase nitration. Several nitrating agents such as HNO_3, NO_2 and N_2O_4 have been successfully applied using mainly dealuminated mordenites or faujasites and elevated pressures [218, 219]. The stability of the zeolites is a major problem, given the highly acidic reaction medium. A combination of high crystallinity and sufficient extra-zeolite surface area (the presence of extra-framework material) was found to be beneficial for stabilizing the catalysts.

In the halogenation of aromatic molecules the role of the molecular sieve is to polarize the Cl_2 or Br_2 molecule in order to enable it to attack the aromatic nucleus. The polarization is aided by an alkali or an alkaline earth cation [220]. With high-silica zeolites addition of Cl_2 to benzene dominates over the substitution leading to chlorocyclohexane as dominating product. With materials containing a high concentration of aluminum and/or less electronegative cations (e.g., K) direct formation of chlorobenzenes via electrophilic substitution has been observed [221]. The degree of chlorination and the mechanistic issues involved in the positional selectivity of the halogenation are not unequivocally related to properties of the zeolitic materials. In part, this is related to dynamic changes in the zeolite properties through the presence of the highly reactive medium.

4.2.9.5
Reactions Catalyzed by Basic Sites

In contrast to the situation found with acid-catalyzed reactions, the role of the zeolite is less well defined for base-catalyzed reactions. This is related to the fact that so-called basic zeolites contain alkali cations acting as (weak) Lewis acid sites. Thus, most of the catalytic chemistry described in this chapter involves Lewis acid and base sites. It should also be stressed that for acid/base-catalyzed reactions both sites are involved in the reaction pathway. For many of the acid-catalyzed reactions, however, the importance of the acid sites dominates so strongly that attention is paid only to the acidic function [222].

We speak, therefore, of base-catalyzed reactions if the strength of the base sites is high enough to stabilize anionic or polarized species with a marked negative charge and if these species are part of the catalytic cycle [223]. Interpretation of catalytic results with respect to the role of acid and base sites remains, however, always ambiguous as the stabilizing effect of the metal cation (for zeolites usually an alkali metal cation) is difficult to assess.

The second problem connected with defining the catalytically active site for base-catalyzed reactions relates to the observation that acid sites, irrespective of whether they are of Lewis or Brønsted nature, are always a minority species. The majority of the accessible atoms in a molecular sieve pore consists, in contrast, of oxygen.

References see page 1145

Consequently, it is straightforward to characterize the minority species (a large variety of methods have been developed in that respect (e.g., [224]), while characterizing the majority species, i.e., base sites, of the catalyst still is challenging. Thus, evidence on the location of the base sites in the molecular sieve channels is ambiguous [225]. The main question in this respect is, whether the base sites are localized (e.g., next to the alkali cation) or whether all oxygens of the molecular sieve framework act as base sites [226].

In base-catalyzed reactions relatively low rates are achievable compared to acid-catalyzed reactions and in many cases minor traces of acidic protons may change the selectivity of a reaction dramatically [110]. In order to overcome this problem, catalysts are frequently prepared with a slight excess of the alkali cations. Very strong basic sites have been created by supporting metallic sodium in the zeolite pores [189, 227]. Recently, the method of using an excess of alkali-metal cations has been expanded to load zeolites systematically with alkali metal oxides. This approach causes that the zeolites is used more as support than as base catalyst [228]. The oxidic nanophase particles in the zeolites are created by thermal decomposition of the corresponding alkali-metal acetate, nitrate or hydroxide [125, 122]. In contrast to the situation found with solid acids, dense oxides (with high surface areas) are excellent catalysts and the use of basic molecular sieves might be advantageous only if shape selectivity is needed for a particular reaction.

In general, the action of the basic catalysts can be grouped into two categories. The first function is related to the more facile proton abstraction from functional groups of reactant molecules by the high electrostatic field in the pores and the negatively charged oxygen in the molecular sieve framework. This action leads either to a stabilized carbanion or to a polarized functional group of the reacting molecule. Conceptually, it is obvious that the former represents a stabilized intermediate (i.e., it is characteristic of a two-step process), while the latter will probably form the carbanion only in a transition state (concerted one-step mechanism). The second function is governed by the polarization of a polar group in the reactant via electron pair donor–electron pair acceptor interaction with the alkali metal cation and the oxygen. Hydride transfer, which is frequently part of the catalytic sequence in base-catalyzed reactions, is more a consequence of the close vicinity of the sorbed molecules than of being induced by the basic nature of the zeolite.

Dehydrogenation of alcohols occurs over basic zeolites in the presence and in the absence of oxygen [229]. In contrast, dehydration is the prevailing reaction over acid zeolites [230]. Higher reaction temperatures are required for dehydrogenation than for dehydration, due to the higher energy of activation for the former reaction [231]. The catalytic activity is related to the concentration and the type of alkali cation, i.e., with increasing size of the alkali cation and increasing level of exchange the rate/selectivity to dehydrogenated products increases [126]. The reaction is concluded to proceed via abstraction of a proton from the hydroxyl group of the alcohol by forming an alkoxy group on the alkali cation. Detailed spectroscopic measurements and theoretical calculations suggest that the formation of the alkoxy groups does not occur at ambient temperature and is an activated process, which is preceded by an adsorption mode in which the alcohol is adsorbed on an acid-base pair site [232]. After hydride abstraction, the proton and the hydride ion recombine to hydrogen desorbing instantaneously. The ketone/aldehyde formed is frequently stabilized by strong interactions and may be difficult to remove to close the catalytic cycle [233].

Poisoning experiments with pyridine (to block the acid sites) and with phenol (to block the base sites) indeed show that dehydration requires strong acid sites for catalysis, whilst dehydrogenation requires strong basic sites [234, 235]. In general, higher concentrations of aluminum in the zeolites and (for a given alkali metal ion) higher concentrations of alkali metal ions would lead to a stronger basic zeolite [236]. This is also well established through spectroscopic probing. In this context it is, however, interesting to note that Hathaway and Davis [228] reported Cs-exchanged Y type zeolites to be an order of magnitude more active than the corresponding X-type materials for the catalytic dehydrogenation of isopropanol. This indicates that other factors, such as the stabilization and further oxidation of the ketones and aldehydes, may influence the catalytic behavior.

Basic zeolites are able to catalyze double-bond isomerization of olefins [189]. Although this can also be achieved with acidic zeolites, the lower reactivity of basic zeolites towards hydrocarbons (i.e., the complete absence of skeletal isomerization) leads to higher yields [237]. A good example of this is the double-bond isomerization of 1-octene over potassium-loaded NaY. It is claimed that high yields can be achieved in this way and that the impregnation of the zeolite with an excess of alkali cations is important to obtain a good catalyst [238].

Aldol condensations are catalyzed by acid and basic zeolites. In the base-catalyzed route the anionic species is generated by the interaction of the basic site with the hydrogen in the β-position to the carbonyl group. The β-carbon atom (bearing a negative partial charge) then forms a new C–C bond with the carbon atom of the carbonyl group of another aldehyde molecule generating a larger β-hydroxy carbonyl compound. Subsequent dehydration leads to the formation of an α,β-unsaturated aldehyde. Successful examples include the synthesis of crotonaldehyde from acetaldehyde over SAPO-34 [239] and the conversion of acetone into diacetonaldehyde, mesityloxide and subsequent products over various alkali-exchanged and alkali-oxide-loaded large-pore zeolites [240].

The pore size of the zeolite influences the product distribution via suppression of the formation of the bulkier products. The condensation of acetone over Na-X and Na-L type zeolites is an example of this shape selectivity. As outlined above acetone is converted to diacetonalcohol and mesityloxide, which may further react to isophorone. The product ratio of mesityloxide to the bulkier isophorone was 0.75 for zeolite X and 1.87 for zeolite L [241, 242].

A special case of an aldol condensation is the side-chain alkylation of alkyl-aromatics over basic zeolites such as alkali-containing faujasites [243, 244]. The reaction requires the complete absence of protons in the zeolite, since these would catalyze ring alkylation with a much higher rate [245]. The most well-studied example is the side-chain alkylation of toluene with methanol over a variety of alkali-

References see page 1145

containing zeolites [243]. Note that alkenes can also be used as alkylating agents for this reaction, but they require a higher base strength, i.e., the presence of metallic Na [246]. The alkali-exchanged zeolite has several roles. First, it polarizes the methyl group of toluene, which may lead to a carbanion structure in the transition state [243, 247]. Then, it adsorbs toluene well, so that it is present under reaction conditions with sufficient concentration [248]. Finally, it catalyzes the formation of formaldehyde from methanol. [243, 249]. The negatively charged carbon at the toluene carbanion forms a C–C bond with the positively charged carbon atom of chemisorbed formaldehyde resulting in an intermediate that rapidly eliminates water and yields styrene (see Scheme 25). The reaction rate seems to be determined by the availability of toluene (which is more readily stabilized in the faujasite pores by the larger alkali cations than methanol) and formaldehyde. Indeed, addition of an extra dehydrogenating function by the addition of ZnO to the zeolite leads to a drastic improvement in the activity [250].

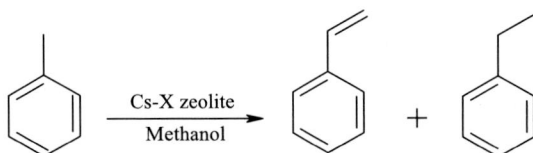

Scheme 25. Side-chain alkylation of toluene with methanol over basic zeolites.

The stability of carbanions follows the opposite sequence to that of carbenium ions, i.e., carbanions at primary carbon atoms are more stable than those at secondary or tertiary carbon atoms [251]. Thus, one would expect that it might be possible to convert methane and ethane with methanol. Unfortunately, activation and/or proton abstraction from alkanes seems not to be possible to a significant extent, as attempts to react methanol with methane or ethane have up till now failed. Presumably, one needs to couple such experiments with oxidative dehydrogenation [252] in order to achieve feasible conversions.

In a rather specific example, which has been reported recently [253], a strongly basic catalyst was used to produce 4-methyl-thiazole in a simplified reaction sequence (replacing a five-step synthesis with a two-step synthesis). The catalysts (Cs-loaded ZSM-5 and zeolite β) proved to be effective for the conversion of acetone and methylamine into the corresponding imine. In the second step this imine is converted with SO_2 into 4-methyl thiazole (Scheme 26). Using Cs sulfate as the Cs source resulted in the best catalyst and given the acidity and basicity of the reactants, one can speculate that sulfate species may also prevail in the rate-determining step.

The Meerwein–Ponndorf–Verley reaction (the corresponding reaction is the Oppenauer oxidation) is conventionally seen as a base-catalyzed reduction of a complex aldehyde by a secondary alcohol, e.g., isopropanol. The reaction is catalyzed by alkali-metal-exchanged zeolites and the product distribution is influenced by the strength of the base sites and/or by spatial constraints in the zeolite pores. An ex-

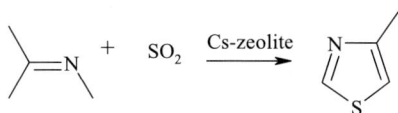

Scheme 26. Base-catalyzed conversion of acetone into an imine and subsequent reaction to 4-methyl thiazole.

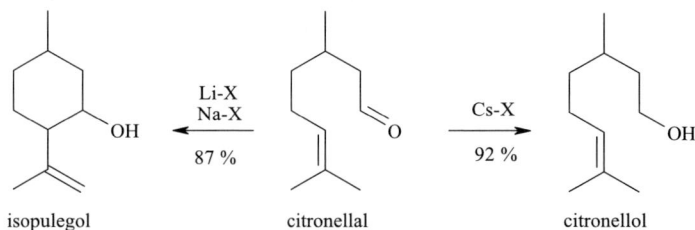

Scheme 27. Selective reduction of citronellal to isopulegol over Li or Na-X or to citronellol over Cs-X

ample is the reduction of citronellal (I) with isopropanol (Scheme 27), which gives 86 % isopulegol and 14 % citronellol at 87 % conversion with Li or Na-X as catalysts, while with Cs exchanged faujasites 99 % citronellol is produced at 77 % conversion [254]. This change in selectivity is attributed to the steric hindrance induced by the larger Cs^+ ions, but the influence of the increasing base strength cannot be ruled out. As with other base-catalyzed reactions the role of the catalyst in this example is also 2-fold, i.e., the basic oxygen helps to abstract a proton from the hydroxyl group of the alcohol, while the metal cation stabilizes the resulting alkoxy species and polarizes the carbonyl group of the aldehyde. If both molecules are adlineated, hydride transfer from the alkoxy group to the polarized aldehyde group takes place, inverting the nature of the two reactants. The remaining steps are the reverse reactions of the activation.

A recent example, the stereoselective reduction of 4-tert-butylcyclohexanone to cis-4-tert-butylcyclohexanol with secondary alcohols over zeolite β (95 % selectivity at 33 % conversion) [255] shows, however, that not so much the basic character of the molecular sieve, but the presence of Lewis acid sites is an indispensable prerequisite for an active catalyst. As long as metal cations (in the form of extra-framework clusters of aluminum oxide) are present, the zeolite is active and selective.

References see page 1145

4.2.9.6
Oxidation with Molecular Sieve Catalysts

4.2.9.6.1 Molecular Sieves as Catalysts for Oxidation Reactions

For (selective) oxidation reactions molecular-sieve-based catalysts must contain redox functionality, which can be realized by (1) isomorphous substitution of metal ions into framework positions, (2) ion exchange with metal cations and (3) grafting metal complexes onto the surface of (mesoporous) molecular sieves [256, 257]. The use of molecular sieves offers potential routes to develop a solvent-free technology for liquid-phase oxidation reactions [258]. Several problems, however, hamper the widespread use of molecular sieve as oxidation catalysts. The transition-metal cations incorporated in the zeolite framework are difficult to stabilize and may be leached out of the molecular-sieve framework. This leads not only to loss of the catalytically active species, but also causes the catalysis by homogeneous species in the fluid phase. In the case of a highly stable incorporation of the transition-metal cations, these ions frequently cannot change their oxidation state. Another difficulty with transition-metal ions that are exchanged into the molecular sieve is partial hydrolysis and the formation of oxide clusters that catalyze preferentially total oxidation. Here, the most important molecular sieves for selective oxidation are described with the aim, however, to focus on the structural aspects and typical catalyzed reactions to illustrate the principle rather than to give complete compilation of the possibilities.

4.2.9.6.2 Ti-containing Molecular Sieves

Ti-silicalite (TS-1), hydrothermally synthesized with Si/Ti ratios between 30 and 50, was the first commercially used zeolite-based partial oxidation catalyst. TS-1 has been found to be an excellent catalyst for the oxidation of small molecules with H_2O_2. The reaction is typically carried out in aqueous solutions containing $\sim 30\%$ of H_2O_2 under mild reaction conditions (20–100 °C). Product distributions can be enhanced by the shape selectivity resulting from the structural properties of the zeolite framework, while undesired (mostly acid-catalyzed) side reactions are suppressed by the low concentration of aluminum, which results in a negligible concentration of Brønsted acid sites.

According to experimental [259, 263] and theoretical studies [264–266] the active sites in TS-1 are isolated and uniformly distributed titanyl centers (see Fig. 8). Each Ti^{4+} is isolated from other titanyl centers by O–Si–O–Si–O units. The isolation of the Ti^{4+} centers is required to obtain high selectivities with TS-1, which may be related to a significantly reduced rate for H_2O_2 decomposition [40, 259]. The high affinity of TS-1 for H_2O_2 leads to the formation of surface titanium peroxo-

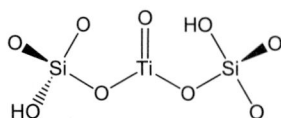

Fig. 8. Possible geometry of active sites in TS-1.

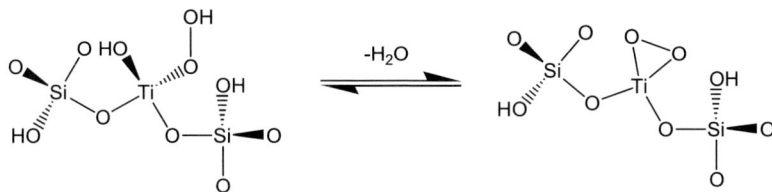

Fig. 9. Hydroxylated and dehydrated peroxotitanate species.

compounds, which can be either present in a hydrated and dehydrated form (see Fig. 9).

A general scheme of selective oxidation reactions with H_2O_2 over TS-1 is given in Fig. 10. The hydroxylation of aromatic molecules is of particular interest. TS-1 is used commercially to produce catechol and hydroquinone from phenol. The reaction is highly sensitive to the presence of extra-framework titania, which catalyzes the decomposition of H_2O_2 and the formation of quinines and coupling products. Note that this reaction can also serve as a test reaction for the fraction of titanium incorporated into the MFI framework [259]. The selectivity to hydroquinone is typically higher over TS-1 due to the shape selectivity imposed by the zeolite structure. Phenol seems to be only very weakly coordinated to the Ti site, which might be one of the reasons why the subsequent oxidation of hydroquinone to quinone is

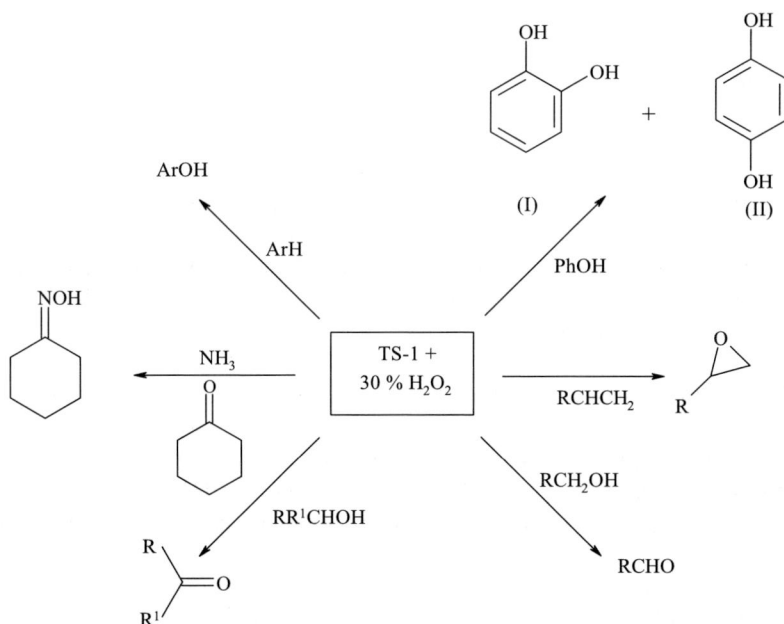

Fig. 10. Oxidation reaction over TS-1 with H_2O_2.

References see page 1145

Tab. 1. TS-1 catalyzed epoxidations with 60 % aqueous H_2O_2, Olefin/H_2O_2 molar ratio = 5:1, solvent CH_3OH.

Olefin	T/K	H_2O_2 conversion	Time/min	Selectivity to epoxides/%
Propene	313	90	72	94
1-Hexene	298	88	70	90
1-Octene	318	81	90	92
Allyl-chloride	318	98	30	92
Cyclohexene	298	10	90	–

relatively slow. For epoxidation the rate strongly depends on the nucleophilicity and the structure of the olefins (see Table 1 after Ref. [267]).

Even relatively unreactive electron-poor olefins such as allyl-chloride are epoxidized under mild conditions (~45 °C) [267], while cyclohexene is completely unreactive due to the steric restrictions of the TS-1 pores [267]. Methanol is the most effective solvent, which is assumed to be a result of the coordination of the alcohol to the titanium hydroperoxide species forming 5-membered peroxyacid-like species that facilitate the oxygen transfer. The negative effect of the presence of Brønsted acid sites for the product selectivity in propene epoxidation becomes evident when comparing the catalytic properties of Al and Ga containing TS-1 with that of purely siliceous TS-1, where the selectivity decreased from 98 % epoxide formation over TS-1 to only 6.5 % over Ti-Ga-silicalites [259]. The favored product in the latter reaction was 1-methoxypropan-2-ol formed via the acid-catalyzed addition of methanol to the epoxide.

Although shape selectivity is one of the main assets of TS-1, the microporous nature restricts its use to small molecules. For example, TS-1 is not suitable for oxidation reactions using tert-butyl hydroperoxide (TBHP), because the steric constraints of the pores limit the transition state of the oxygen transfer. To overcome the pore-size restrictions several Ti-containing molecular sieves with larger pores such as Ti-Beta and TAPSO-5 have been synthesized. This permits epoxidation of more bulky olefins (e.g., cyclohexene, cyclododecene, norborene). A list of selected examples of titanium-containing molecular sieves is given in Table 2 (after Ref. [268]).

Within the larger pores of Ti-Al-Beta the same activity was observed for the epoxidation of cyclohexene and 1-hexene than with TS-1. The presence of Brønsted acid sites, however, leads to the formation of side products via acid-catalyzed ring opening of the epoxide. Neutralizing the acid sites by ion exchange with alkali or alkaline-earth metal ions increases the selectivity to the epoxide [269].

4.2.9.6.3 Fe-containing Molecular Sieves

Iron-containing MFI zeolites are an unusual class of oxidation catalysts, because the catalytic activity of Fe-ZSM-5 differs markedly from that of bulk Fe_2O_3 [270]. Fe-ZSM-5 is unable to activate O_2, but shows high activity for the decomposition of N_2O [271]. This process is claimed to lead to the formation of a new form of sur-

Tab. 2. Titanium-containing molecular sieves in chronological order.

Material	Pore size/Å	Template	Year	Ref.
Ti-silicalite (TS-1)	5.6 × 4.4	$(Pr)_4NOH$	1983	336
Ti-silicalite (TS-2)	5.5 × 5.1	$(Bu)_4NOH$	1990	337
Ti-Y	7.4 × 7.4	Post-synthesis modification	1990	338
Ti-Al-Beta	7.6 × 6.4	$(Et)_4NOH$	1992	339
Ti-ZSM-48	5.4 × 5.4	$H_2N(CH_2)_8NH_2$	1992	340
Ti-Al-mordenite	6.7 × 7.0	None	1993	341
Ti-APSO-5	7.3 × 7.3	$C_6H_{11}NH_2$	1994	342
Ti-MCM-41	>40	$C_{16}H_{33}(CH_3)_3NOH$	1994	343, 344
Ti-HMS	>40	$C_{12}H_{25}NH_2$	1994	344

face oxygen, the so called α-oxygen [272]. At low temperatures (below 573 K) a stoichiometric surface reaction occurs on the so called α-sites until all sites are occupied with α-oxygen $(N_2O + (*)_\alpha \leftrightarrow (O)_\alpha + N_2)$.

Above 573 K α-oxygen is thermally unstable, recombines and desorbs as O_2. For the location of Fe in ZSM-5 three principle positions exists: (1) in tetrahedral positions, (2) isolated ions or small complexes inside the channels and (3) highly dispersed oxide particles. As this material was also found to be one of the most stable catalysts for the reduction of NO_x with hydrocarbons in the presence of O_2 [273, 274] the structure of this material was studied in great detail [275–278]. EXAFS revealed the presence of diferric hydroxo-bridged binuclear clusters in Fe-ZSM-5 (shown in Fig. 11), whose structures resemble that of the methane monooxygenase enzyme [278]. From spectroscopic characterization using IR, UV-Vis and EPR and catalytic experiments it was concluded that highly dispersed extra-framework iron oxide species appear to host the active α-sites [279].

Examples for the activity of Fe-ZSM-5 for the oxidation of benzene to N_2O are compiled in Table 3 after refs. [272, 280]. In contrast to Fe_2O_3 in the presence of oxygen (catalyzing complete oxidation), Fe-ZSM-5 shows an exceptional selectivity to phenol, when N_2O is used as reducing agent. This allowed the development of a new process for the production of phenol, the so-called AlphOx process [281].

Similar to the microporous Fe-ZSM-5, mesoporous iron-containing silicas with MCM-41-type structure [282, 283] can be used to oxidize large hydrocarbons to oxygenates with N_2O. For the oxidation of cyclododecane at 673 K with N_2O 20 % conversion with a yield of 5 % to cyclododecene was reported. The majority of

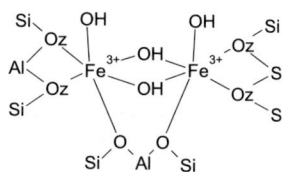

Fig. 11. Structure of binuclear Fe cluster in Fe-ZSM-5 compensating two framework charges.

References see page 1145

Tab. 3. Comparison of N_2O and O_2 in the oxidation of benzene to phenol.

Catalyst	Oxidant	Reaction temperature/K	Benzene conversion/%	Phenol selectivity/%
Fe-ZSM-5 (0.055 wt % Fe)	N_2O	623	27	98
Fe_2O_3	N_2O	623	5.5	0
Fe-ZSM-5 (0.055 wt % Fe)	O_2	773	0.3	0
Fe_2O_3	O_2	623	24.5	0

products, however, resulted from total oxidation or led to the formation of coke on the catalyst surface [283]. As outlined by the authors low selectivity observed for these catalysts most probably results from the different Fe-species present in the two catalysts. For Fe-ZSM-5 binuclear or clustered species were identified as reactive sites, while in the mesoporous catalysts Fe was found to be mainly present in an isolated form. Additionally, the differences in the acidity between ZSM-5 and MCM-41, which is significantly lower for the mesoporous material, might influence the activation of the hydrocarbons and, thus, alter the catalytic activity of the catalyst.

4.2.9.6.4 Vanadium-containing Molecular Sieves

Vanadium silicates (with MFI and MEL structure) with typical Si/V ratios between 40 and 160 are active oxidation catalysts for gas- and liquid-phase oxidation reactions [284]. Similar to the titanium silicates, the catalytically active vanadium species are supposed to be associated to the framework. Only one type of V species appears to be present [285] most probably as isolated species [286]. A possible structure of the V sites is shown in Fig. 12 [285]. The Si–O–V bonds are longer than Si–O–Ti bonds and, therefore, V seems to be more exposed. The redox properties of vanadium-containing molecular sieves are affiliated with the changes in the oxidation state of V between +4 and +5. Vanadium bound to the framework has been claimed not be extracted [287], but that view has been challenged frequently. Other methods for preparing vanadium-containing zeolites, e.g., impreg-

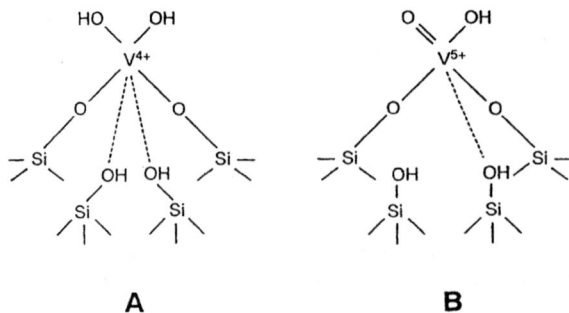

Fig. 12. Environment of vanadium in V-ZSM-11 (a) as-synthesized; (b) calcined.

nation or ion exchange, were found not to lead to active catalysts for oxidation reactions [288].

Vanadium silicates are highly selective for the oxidative dehydrogenation of propane to propene and of methanol to formaldehyde using O_2 or N_2O as oxidants [288]. V-ZSM-11 (VS-2) catalyzes the selective oxidation of C_6–C_8 n-paraffins, cyclohexane, ring hydroxylation and side-chain oxidation of aromatic hydrocarbons, hydroxylation of phenol and the sulfoxidation of thioethers in liquid-phase using H_2O_2 as oxidant [289].

It appears that oxidations with VS-2 tend to be deeper than with TS-1. While VS-1, TS-1 and TS-2 oxidize the secondary carbon atoms of n-paraffins to the secondary alcohol or ketone, only VS-2 oxidizes the primary carbon atom, thus, catalyzing the formation of the primary alcohol or aldehydes [285, 289]. Furthermore, while both the V- and Ti-containing zeolites catalyze ring hydroxylation reactions of aromatic hydrocarbons such as toluene, only VS-2 can also catalyze the oxidation of the alkyl side chain, e.g., toluene to benzyl alcohol and benzaldehyde [284, 289].

4.2.9.6.5 Other Metal-substituted Molecular Sieves

Co and Cr can be incorporated into the framework of aluminophosphates with AFI and AEL structure during hydrothermal synthesis [290]. The oxidation state of the metal ions in the framework depends on the pre-treatment procedures. While the tetrahedral coordination of the metal is stable during gas-phase reactions, it is possible that leaching from the framework and the formation of highly active complexes occurs.

In addition to the oxidation potential of $CoAPO_4$, the materials also contain moderately strong acid sites. The valency of Co suggests that it is primarily substituted on positions normally occupied by Al^{3+} in the $AlPO_4$-5 structure. If Co is present in the oxidation state +3 the framework is neutral, while Co present in the oxidation state +2 leads to the formation of Brønsted acid sites in the aluminophosphate [291]. This allows, in principle, to utilize the redox and acid/base functionality in a sequential manner during a redox cycle. $CoAPO_4$-5 has been used to oxidize cyclo- and n-hexane in the presence of acetic acid to cyclohexylacetate and hexyl-2-acetate, respectively. During the reaction the active sites are regenerated by oxidation of Co^{2+} to Co^{3+} by the acetic acid [292]. Similarly, for Co-, Fe- and Mn-substituted aluminophosphates with AFI, AEL, AEI and ATS structure, changes in the oxidation state of the transition metal and the specific end-on sorption geometry of the reactant inside the pores were identified to be the reason for the functionalization of the terminal CH_3 and penultimate CH_2 groups during the oxidation of linear alkanes by air [293].

Another example of the use of CoAPO-5 is the autooxidation of p-cresol to p-hydroxybenzaldehyde in methanolic sodium hydroxide solution [294]. Incorporation of Cr^{VI} into silicalite or aluminophosphates with AEL and AFI structure leads to materials that catalyze a variety of oxidation reactions with tert-butyl hydroperoxide (TBHP) or O_2 [295]. In the as synthesized materials Cr is present as

References see page 1145

isomorphously substituted Cr^{III}, which is oxidized to hexavalent chromyl $(CrO_2)^{II}$ attached to the framework by two metal–oxygen bonds. Cr^{VI} typically catalyzes oxidations via the oxo-metal mechanism in which $Cr^{VI}=O$ is the active oxidant [268]. As CrAPO-5 was found to catalyze reactions typical of oxo-metal oxidants, i.e., benzylic and allylic oxidations and (cyclo)alkane oxidations, the catalytic cycle is assumed to involve a reoxidation of Cr^{IV} to $Cr^{VI}=O$ by TBHP or O_2 [295]. However, a significant leaching of the active Cr from CrAPO-5 was observed during the reactions [296], which was verified by removing the aluminophosphate from the reactant by filtration of the solution or by using bulky molecules for the reaction, which cannot enter into pores of the molecular sieve [297]. In both cases the activity was partially restored, which clearly indicates that chromium is leached from the catalysts and that the observed catalytic activity results from a homogeneous catalytic reaction.

4.2.9.7
Physical Aspects of Molecular Sieve Catalysis for Chemical Synthesis

4.2.9.7.1 General Aspects
In the previous section the chemical functionalities of molecular sieves were discussed and it was shown how these functionalities can be incorporated in the solids. It was, however, the spatial constraints of molecular-sieve pores that initially made zeolites attractive for organic synthesis. It should be noted at this point that discrimination between chemical and structural aspects works well at a conceptual level, but faces quite severe limitations as soon as one tries to separate the contributions of the two effects. This is due to the fact that the chemical properties of a particular molecular sieve are interconnected with its framework density. In general terms, the strength of the acid sites in a molecular sieve with a given chemical composition will increase as the framework density increases and the pore size decreases [298]. Similarly, the polarizability of the framework seems to be higher for frameworks exhibiting a higher density. Theoretical calculations of the group of Mortier [299] suggest that these interconnections might be quite complex and difficult to predict. It is unclear at present, if such properties are determined by localized structural effects, i.e., by local bond angles and the way the tetrahedra are connected on a microscopic level [300] or if global properties, i.e., the average distance of the framework T atoms dominate [301, 302].

We would like to emphasize that the global properties of molecular sieves, often described in terms of hydrophobicity and hydrophilicity [303] or framework polarity [304] will markedly influence the chemical preference for the sorption of molecules. This may lead to quite different relative concentrations of reactants in the zeolite pores (well demonstrated and discussed for the case of TS-1 in [259]) and in the intracrystalline void space.

4.2.9.7.2 Shape Selectivity
Shape selectivity can be induced by differences in the diffusivities of the reactants and/or the products or by steric constraints of the transition state. A schematic

Reactant selectivity

Product selectivity

Restricted transition state-type selectivity

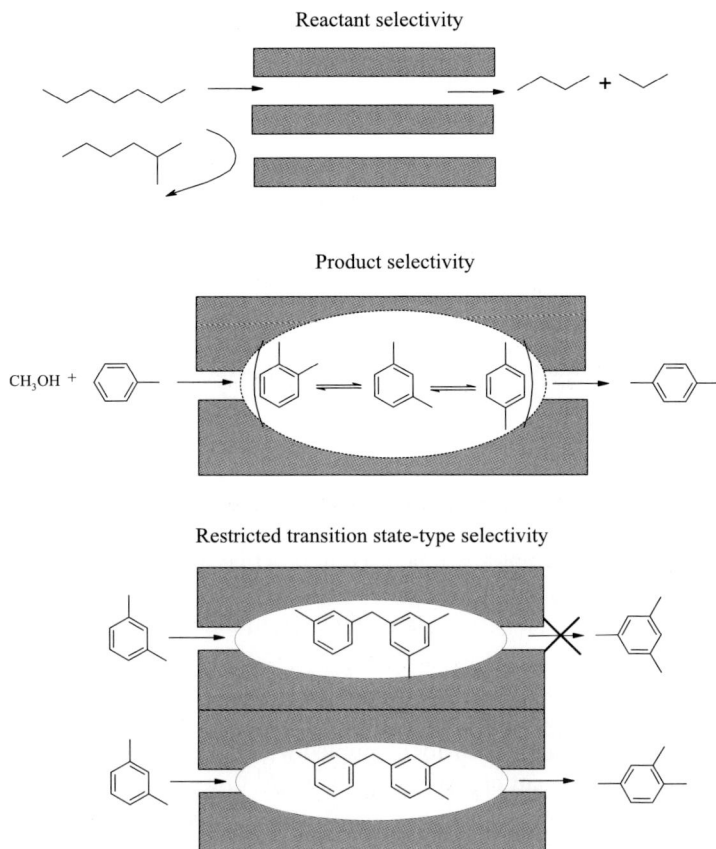

Fig. 13. Schematic representation of the three types of shape selectivity.

representation of the three types of shape selectivity, i.e., the limitations of the access of some of the reactants to the pore system (reactant selectivity), the limitation of the diffusion of some of the products out of the pores (product selectivity) and constraints in forming certain transition states (transition-state selectivity) are shown in Fig. 13. Differentiation between the latter two is difficult as the kinetic results may be disguised when the overall rate is influenced by the rates of diffusion. *In situ* IR and NMR spectroscopy have contributed much to our understanding of these complex phenomena. The aspects of shape selectivity have been extensively discussed and excellent reviews exist [11, 305–307]. The examples given here should only illustrate what can be achieved by employing a zeolite and why the pathway of a particular reaction is influenced.

The first example of shape selectivity was reported by Weisz and Frilette [308] for the dehydration of an n-butanol/iso-butanol mixture over LTA type zeolites. Because of its larger minimum kinetic diameter iso-butanol is excluded from enter-

References see page 1145

ing the zeolite pores, while n-butanol is easily dehydrated to butene. This demonstrated for the first time that the catalytically active sites are inside the zeolite pores and that the pores are able to realize a well-defined cutoff point with respect to the minimum kinetic diameter of the reacting molecules. This principle of size exclusion was then frequently used in hydrocarbon processing to remove linear hydrocarbons from a mixture of hydrocarbons (e.g., selectoforming [309]). Note that a complete separation of one group of molecules is not necessary and usually a large difference in the diffusivities of the molecules will suffice. The relatively high apparent energies of activation for configurational diffusion (diffusion through micropores of similar dimensions to the diffusing molecules [310]) often masks the presence of diffusion control. The high values also indicate that the (relative) rates of transport may change dramatically as the (reaction) temperatures are changed.

One of the most discussed cases of shape selectivity involving transition-state selectivity or product-diffusional constraints is the production of *p*-xylene over chemically modified ZSM-5 zeolites [311]. Several processes exist that utilize the shape selectivity of these zeolites, for example the alkylation of toluene with methanol [312], xylene isomerization [313, 314] and selective toluene disproportionation [315]. The first two of these examples will be used to describe in detail the principal possibilities to tailor the reaction pathway by shape selectivity.

Toluene alkylation by methanol occurs via methoxonium ions (presumably stable only in the transition state [31]) at low temperatures and via methoxy groups at high temperatures. Initially toluene is alkylated preferentially in *o*- and *p*-position of toluene, but all three isomers appear to be primary products (as shown by *in situ* IR spectroscopy [316]). High para-selectivity is coupled with rapid isomerization of the xylenes. The diffusion constant of *p*-xylene is about 10^3 times higher than that of *m*-xylene [317]. In an idealized model one would, therefore, expect to find the xylenes in their equilibrium concentrations in the MFI pores. *In situ* IR spectroscopy showed that this is not the case and that preferred sorption of *m*- and *o*-xylene and trimethylbenzenes occurs. However, this is not a result of the higher sorption enthalpies of any of the reactants or products on H-ZSM-5 [254]. Combining the rate constants for the isomerization of the individual xylene isomers and the concentrations of the products in the ZSM-5 pores, it was shown that the selectivity to *p*-xylene is high, when the rate of internal isomerization exceeds drastically the overall rate of alkylation. Thus, the results indicate that the secondary isomerization is important for the shape-selective production of *p*-xylene. However, the rates of isomerization are apparently not fast enough with respect to the diffusion to establish equilibrium in the pores. That suggests that the two steps (isomerization and diffusion) are in subtle balance and are at least so close as not to allow the assumption of a quasi-equilibrium under steady-state operation [318].

The isomerization of *m*-xylene is a good example of transition-state selectivity [319]. Irrespective of the temperature and coverage, (a particular sample of) ZSM-5 showed a product ratio of *p*- to *o*-xylene of 2:1. In the zeolite pores, only *m*-xylene was found to be sorbed in appreciable quantities. Thus, the reaction rate appears not to be influenced by the preferred retention of one of the products. The selectivity must consequently be governed by the differences in the transition states

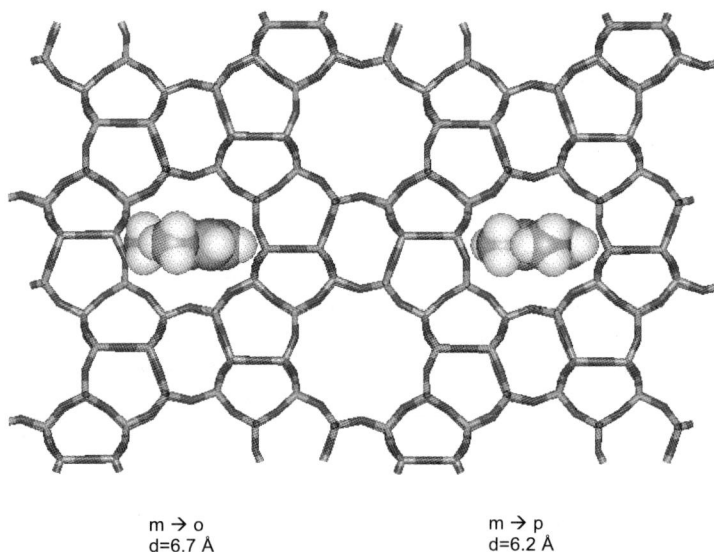

m → o   m → p
d=6.7 Å   d=6.2 Å

Fig. 14. Model of the transition state complexes in *m*-xylene isomerization.

of the two products. The constant selectivity with varying reaction temperature indicates an identical apparent energy of activation for the formation of *p*- and *o*-xylene. Therefore, the different selectivities must be caused by differences in the transition entropy. Molecular modeling indeed shows that the transition state for the *m–p* xylene transition has a smaller minimum kinetic diameter than that of the m-o xylene transition (see Fig. 14). Considering the identical energies of activation and the identical heats of adsorption of all isomers the larger kinetic diameter of the transition state of the m-o xylene transformation is concluded to require more specific configurations in the ZSM-5 channels, i.e., lower transition entropy. Note that similar effects have been observed for hydrocarbon conversions [320].

Blocking the pore mouth and reducing the diffusivities of the xylenes, for example, by chemical vapor deposition of ethoxysilanes or by selectively coking the channels does not change the overall catalytic pathway for toluene methylation, but enhances the *p*-selectivity [209, 321–324]. However, zeolites prepared in this way deactivate and this has to be balanced with higher reaction temperatures. The higher reaction temperatures are required to open new reaction channels (dealkylation, transalkylation, and disproportionation) to allow desorption of products from the pores as the longer residence times lead to polymethylated benzenes, which are unable to leave the zeolite pores and would eventually block all acid sites [258].

Shape selectivity is not confined to reactions of hydrocarbons in the absence of polar functional groups. MFI-type materials have been reported to catalyze the

References see page 1145

isomerization of cresols, chlorotoluenes, toluonitriles and toluidines [325]. In the isomerization of aniline derivatives the reaction temperatures have to be relatively mild as under severe reaction temperatures isomerization to methylpyridine would occur [326]. For dimethylanilines it could be shown that only the isomers with the smallest minimum kinetic diameter reacted (reactant selectivity), and that those with a larger kinetic diameter did not form (product selectivity). The isomerization is concluded to occur via a 1,2-methyl shift, which is interpreted to indicate transition-state selectivity [327].

While many of the shape-selective processes reported concern small and medium pore zeolites, the synthesis of fine chemicals and intermediates requires larger-pore molecular sieves. In this respect shape-selective conversions have also been reported for MOR- and BEA-type catalysts. For example 4,4′-diisopropylbiphenyl, an intermediate for liquid crystals, can be produced from propene and biphenyl in high yields over dealuminated mordenite [328]. The parent mordenite and other 12-membered ring zeolites such as H-Y, H-L or H-offretite gave poor results. The active mordenite was obtained by severe acid leaching, resulting in a catalyst with a Si/Al ratio of 1300 and large mesopores. The post synthesis modification converted the mono-dimensional channel structure of MOR into a 3D structure, in which the micropores are connected by micro and mesopores generated in the leaching procedure. It is concluded that the mesopores are indispensable for efficient mass transport of the bulky molecules and that dealumination reduces coke formation and unselective alkylation, i.e., in the para-meta positions. The recently commercialized process for the synthesis of cumene and *p*-diisopropylbenzene from propene and benzene uses the same catalysts [329].

2-Methylnapthalene (2-MN) and 2,6-dimethylnapthalene (2,6-DMN) can be selectively produced by isomerization and disproportionation of 1-methylnapthalene (1-MN) or by direct alkylation of naphthalene. The observed reactivity for the isomerization of 1-MN and 2-MN over H-ZSM-5, H-ZSM-11, H-beta, H-mordenite and H-Y led Weitkamp and Neuber [330] to propose that product shape selectivity dominates, while Fraenkel et al. [331] suggest that cavities at the external surface containing strong Brønsted acid sites are responsible for the selectivity. The idea has been followed up by Derouane [332], proposing nest-like structures to be important. However, Neuber and Weitkamp [333] showed that even the more bulky molecules can enter the zeolite pores. While the role of the external and internal acid sites is not completely resolved, the combination of silylation and poisoning experiments for xylene isomerization [210] suggest that external acid sites do not markedly contribute to the reaction pathway for such large molecules.

In recent years the role of pore-mouth catalysis has been strongly debated in relation to the hydroisomerization of alkanes [334, 335]. It has been shown that product patterns could be explained by invoking only the acid sites at the entrance of small pore materials such as erionite, ZSM-22 and Theta-1. The isomer distribution suggests that branching of n-alkanes occurs in approximately the distance between two pore openings. Sorption of branched alkanes on these materials revealed that the acid sites involved in such catalysis, however, are located inside the crystalline material [56].

4.2.9.8
Conclusions and Outlook

The examples above document some of the impressive progress that has been made over the last decades in utilizing zeolites for chemical catalysis. This progress was and is based on the gradual development of the understanding of chemical and physical processes and the material properties involved in the catalytic reactions. Rapidly advancing experimental capabilities and the progress in theoretical chemistry aided the speed of development greatly.

What is crucial at this point, however, is to minimize problems, by better definition of the materials studied and using better-documented procedures for modifying materials. Molecular sieves are fragile materials, whose properties vary drastically with subtle variations in preparation or handling procedures. The crystallinity of the material offers a unique chance over amorphous mixed oxides, as it allows controlling the active sites and their environments. However, defects in these materials may mask and even impede the catalytic chemistry, which is generic to a particular material. Defining small particles from the zeolite layers or better defining the outer surface of the material may produce materials in which we can utilize the crystalline aspects, but are not defined by the limitations of the pore size.

Much of the other challenges resemble those in other areas of catalysis. It will be important to understand how to activate molecules (saturated hydrocarbons in particular) and to remove products at low reaction temperatures. High selectivity with respect to steric production of chemicals will be needed and concepts utilizing the crystalline nature of molecular sieves itself to prepare highly stereoselective catalysts might evolve. Highly atom-efficient processes may be generated using the control of the molecular sieves upon the transport on a molecular level using creative approaches such as the reactions from two sides of molecular-sieve membranes. The challenges to develop catalysts along these lines are rather severe; the review of the progress over the last three decades, nevertheless, gives good reason to be optimistic for future exciting developments.

References

1 K. P. DE JONG, *Catal. Today* 1996, *29*, 171.
2 C. T. O'CONNER, E. VAN STEEN, M. E. DRY, in: Recent Advances in and New Horizons in Zeolite Science and Technology, H. CHON S. I. WOO, S.-E. PARK (Eds), *Studies in Surface Science and Catalysis*, Vol. 102, Elsevier, Amsterdam, 1996, p. 323.
3 T. F. DEGNAN, *Topics Catal.* 2000, *13*, 349.
4 A. NISHIJIMA, T. KAMEOKA, T. SATO, N. MATSUBAYASHI, Y. NISHIMURA, *Catal. Today* 1998, *45*, 261.

5 W. F. HÖLDERICH, J. RÖSELER, G. HEITMANN, A. T. LIEBENS, *Catal. Today* 1997, *37*, 353.
6 K. TANABE, W. F. HÖLDERICH, *Appl. Catal. A* 1999, *181*, 399.
7 B. M. CHOUDARY, M. L. KANTAM, P. L. SANTHI, *Catal. Today* 2000, *57*, 17.
8 M. G. CLERICI, *Topics Catal.* 2000, *13*, 373.
9 Y. ONO, T. BABA, *Catal. Today* 1997, *38*, 321.
10 W. F. HÖLDERICH, *Catal. Today* 2000, *62*, 115.

11 S. M. Csicsery, *Zeolites* **1984**, *4*, 202.

12 J. Weitkamp, *Solid State Ionics* **2000**, *131*, 175.

13 A. Corma, H. Garcia, *J. Chem. Soc., Dalton Trans.* **2000**, 1381.

14 F. R. Ribeiro, F. Alvarez, C. Henriques, F. Lemos, J. M. Lopes, M. F. Ribeiro, *J. Mol. Catal. A* **1995**, *96*, 245.

15 J. A. M. Arroyo, G. G. Martens, G. F. Froment, G. B. Marin, P. A. Jacobs, J. A. Martens, *App. Catal. A,* **2000**, *192*, 9.

16 P. Andy, D. Martin, M. Guisnet, R. G. Bell, C. R. A. Catlow, *J. Phys. Chem. B* **2000**, *104*, 4827.

17 A. Corma, V. Fornes, S. B. Pergher, T. L. M. Maesen, J. G. Buglass, *Nature* **1998**, *396*, 353.

18 A. Corma, V. Fornes, J. Martinez-Triguero, S. B. Pergher, *J. Catal.* **1999**, *186*, 57.

19 A. Corma, V. Fornes, J. M. Guil, S. Pergher, T. L. M. Maesen, J. G. Buglass, *Microporous Mesoporous Materials* **2000**, *38*, 301.

20 P. B. Venuto, P. S. Landis, *Adv. Catal.* **1968**, *18*, 259.

21 P. B. Venuto, *Microp. Mater.* **1994**, *2*, 297.

22 A. V. Kucherov, A. A. Slinkin, *J. Mol. Catal.* **1994**, *90*, 323.

23 P. Fink, J. Datka, *J. Chem. Soc., Faraday Trans. I* **1989**, *85*, 3079.

24 A. Zecchina, C. Lamberti, S. Bordiga, *Catal. Today* **1998**, *41*, 169.

25 E. B. Webb, G. S. Grest, *Catal. Letters* **1998**, *56*, 95.

26 C. E. Webster, R. S. Drago, M. C. Zerner, *J. Am. Chem. Soc.* **1998**, *120*, 5509.

27 R. A. Beyerlein, C. Choifeng, J. B. Hall, B. J. Huggins, G. J. Ray, *Catal. Lett.* **1997**, *4*, 27.

28 J. P. Ramirez, R. J. Berger, G. Mul, F. Kapteijn, J. A. Moulijn, *Catal. Today* **2000**, *60*, 93.

29 A. Holzwarth, P. Denton, H. Zanthoff, C. Mirodatos, *Catal. Today* **2001**, *67*, 309.

30 J. Sauer, P. Ugliengo, E. Garrone, V. R. Saunders, *Chem. Rev.* **1994**, *94*, 2095.

31 R. A. van Santen, G. J. Kramer, *Chem. Rev.* **1995**, *95*, 637.

32 R. A. van Santen, *Catal. Today* **1999**, *50*, 511.

33 J. W. Andzelm, A. E. Alvarado-Swaisgood, F. U. Axe, M. W. Doyle, G. Fitzgerald, C. M. Freeman, A. M. Gorman, J.-R. Hill, C. M. Kölmel, S. M. Levine, P. W. Saxe, K. Stark, L. Subramanian, M. A. van Daelen, E. Wimmer, J. M. Newsam, *Catal. Today* **1999**, *50*, 451.

34 W. A. Hölderich, H. van Bekkum, in: Introduction to Zeolite Science and Practice, H. van Bekkum, E. M. Flanigan, J. C. Jansen (Eds), *Studies in Surface Science and Catalysis*, Vol. 58, Elsevier, Amsterdam, 1991, p. 631.

35 H. Hattori, in: Heterogeneous Catalysis and Fine Chemicals III, M. Guisnet, J. Barbier, J. Barrault, C. Bouchoule, D. Duprez (Eds), *Studies in Surface Science and Catalysis*, Vol. 78, Elsevier, Amsterdam, 1993, p. 35.

36 J. Cornier, J. M. Popa, M. Gubelmann, *Actual. Chim.* **1992**, *6*, 405.

37 W. A. Hölderich, *NATO ASI Ser. C* **1992**, *352*, 579.

38 W. A. Hölderich, in: Structure-Activity and Selectivity Relationships in Heterogeneous Catalysis, R. K. Grasselli, A. W. Sleight (Eds), *Studies in Surface Science and Catalysis*, Vol. 67, Elsevier, Amsterdam, 1991, p. 257

39 R. A. Sheldon, *Chemtech* **1991**, *21*, 566.

40 R. A. Sheldon, in: Dioxygen Activation and Homogeneous Catalytic Oxidation, L. I. Simandi (Ed.), *Studies in Surface Science and Catalysis*, Vol. 66, Elsevier, Amsterdam, 1991, p. 573.

41 H. van Bekkum, H. W. Kouwenhoven, *Recl. Trav. Chim. Pays-Bas* **1989**, *108*, 283.

42 J. M. Thomas, C. R. Theocharis, *Mod. Synth. Methods,* **1989**, *5*, 249.

43 K. Smith, *Bull. Soc. Chim. Fr.* **1989**, *2*, 272.

44 M. Butters, Solid Supports in Catalytic Organic Synthesis, K. Smith (Ed.), Horwood, Chichester, UK, 1992, p. 130.

45 F. Eder, M. Stockenhuber, J. A. Lercher, *J. Phys. Chem. B* **1997**, *101*, 5414.

46 F. Eder, J. A. Lercher, *J. Phys. Chem.* **1997**, *101*, 1273.

47 L. Yang, K. Trafford, O. Kresnawahjuesa, J. Sepa, R. J. Gorte, D. White, *J. Phys. Chem. B* **2001**, *105*, 1935.

48 J. A. van Bokhoven, M. Tromp, D. C. Koningsberger, J. T. Miller, J. A. Z. Pieterse, J. A. Lercher, B. A. Williams, H. H. Kung, *J. Catal.* **2001**, *202*, 129.

49 A. Corma, F. Llopis, J. B. Monton, S. Weller, *J. Catal.* **1993**, *142*, 97.

50 A. Corma, V. Fornes, J. B. Monton, A. V. Orchilles, *Appl. Catal.* **1984**, *12*, 105.

51 P. J. Crickmore, *Can. J. Chem. Eng.* **1989**, *67*, 392.

52 R. A. Pachovsky, B. W. Wojciechowski, *Can. J. Chem. Eng.* **1975**, *53*, 659.

53 J. Hoffmann, E. Petzold, *Z. Chem.* **1988**, *28*, 168.

54 J. A. M. Arroyo, G. G. Martens, G. F. Froment, G. B. Marin, P. A. Jacobs, J. A. Martens, *App. Catal. A* **2000**, *192*, 9.

55 M. C. Claude, J. A. Martens, *J. Catal.* **2000**, *190*, 39.

56 J. A. Z. Pieterse, S. Veefkind-Reyes, K. Seshan, J. A. Lercher, *J. Phys. Chem. B* **2000**, *104*, 5715.

57 E. G. Derouane, C. D. Chang, *Microporous Mesoporous Mater.* **2000**, *6*, 425.

58 D. W. Breck, Zeolite Molecular Sieves, Structure, Chemistry and Use, J. Wiley & Sons, New York, 1974, p. 29.

59 K. Tanabe, Solid Acids and Bases, Kodansha, Tokyo, 1970, p. 73.

60 V. B. Kaszansky, in: Advanced Zeolite Science and Applications, J. C. Jansen, M. Stöcker, H. G. Karge, J. Weitkamp (Eds), *Studies in Surface Science and Catalysis*, Vol. 85, Elsevier, Amsterdam, 1994, p. 251 and references cited therein.

61 G. H. Kühl, *J. Phys. Chem. Solids* **1977**, *38*, 1259.

62 J. Ward, *J. Catal.* **1967**, *9*, 231.

63 K. Tanabe, M. Misono, Y. Ono, H. Hattori, in: New Solid Acids and Bases, K. Tanabe, M. Misono, Y. Ono, H. Hattori (Eds) *Studies in Surface Science and Catalysis*, Vol. 51, Elsevier, Amsterdam, 1989, p. 142.

64 E. M. Flanigen, L. Patton, S. T. Wilson, in: Innovation in Zeolite Materials Science, P. J. Grobet, W. J. Mortier, E. F. Vansant, G. Schulz-Ekloff (Eds), *Studies in Surface Science and Catalysis*, Vol. 37, Elsevier, Amsterdam, 1988, p. 13.

65 W. J. Mortier, *J. Catal.* **1978**, *55*, 138 and references cited therein.

66 K. P. Schröder, J. Sauer, M. Leslie, C. R. A. Catlow, J. M. Thomas, *J. Chem. Phys. Lett.* **1992**, *188*, 320.

67 L. Uytterhoeven, W. J. Mortier, P. Geerlings, *J. Phys. Chem. Solids.* **1989**, *50*, 479.

68 W. J. Mortier, in: Theoretical Aspects of Heterogeneous Catalysis, J. B. Moffat (Ed.), Van Nostrand Reinhold Catalysis Series, New York, 1990, p. 135.

69 D. Barthomeuf, Catalysis by Zeolites, Elsevier, Amsterdam, 1980, p. 56.

70 D. Barthomeuf, in: Catalysis by Acids and Bases, B. Imelik, C. Naccache, G. Coudurier, Y. Ben Taarit, J. C. Vedrine (Eds), *Studies in Surface Science and Catalysis*, Vol. 20, Elsevier, Amsterdam, 1985, p. 75.

71 D. Barthomeuf, in: Catalysis in Adsorption by Zeolites, G. Öhlmann, H. Pfeifer, R. Fricke (Eds), *Studies in Surface Science and Catalysis*, Vol. 65, Elsevier, Amsterdam, 1991, p. 157.

72 J. A. Rabo, G. C. Gajda, *Catal. Rev.* **1989**, *31*, 385.

73 R. A. van Santen, in: Advanced Zeolite Science and Applications, J. C. Jansen, M. Stöcker, H. G. Karge, J. Weitkamp (Eds), *Studies in Surface Science and Catalysis*, Vol. 85, Elsevier, Amsterdam, 1994, p. 273.

74 G. Sastre, V. Fornes, A. Corma, *J. Phys. Chem. B* **2000**, *104*, 4349.

75 K. A. van Genechten, W. J. Mortier, *Zeolites* **1988**, *8*, 273.

76 R. M. Lago, W. O. Haag, R. J. Mikovsky, D. H. Olson, S. D. Hellring, K. D. Schmitt, G. T.

KERR, Proc. of the 7th International Zeolite Conference, Amsterdam, 1986, p. 677.

77 G. MIRTH, J. CEJKA, J. A. LERCHER, *J. Catal.*, **1993**, *139*, 24.

78 P. A. JACOBS, J. UYTTERHOEVEN, *J. Chem. Soc., Faraday Trans. I* **1973**, *69*, 359.

79 P. D. HOPKINS, J. T. MILLER, B. L. MEYERS, G. J. RAY, R. T. ROGINSKI, M. A. KUEHNE, H. H. KUNG, *App. Catal. A* **1996**, *136*, 29.

80 B. HUNGER, H. MIESSNER, M. VONSZOMBATHELY, E. GEIDEL, *J. Chem. Soc., Faraday Trans.* **1996**, *92*, 499.

81 J. K. LEE, H. K. RHEE, *Catal. Today* **1997**, *38*, 235.

82 F. ARENA, R. DARIO, A. PARMALIANA, *App. Catal. A* **1998**, *170*, 127.

83 C. COSTA, J. M. LOPES, F. LEMOS, F. R. RIBEIRO, *J. Mol. Catal. A* **1999**, *144*, 221.

84 T. L. M. MAESEN, E. P. HERTZENBERG, *J. Catal.* **1999**, *182*, 270.

85 A. H. F. WIELERS, N. VAARKAMP, M. F. M. POST, *J. Catal.* **1991**, *127*, 51.

86 W. O. HAAG, R. M. DESSAU, R. M. LAGO, in: Chemistry of Microporous Crystals, T. INUI, S. NAMBA, T. TATSUMI (Eds), *Studies in Surface Science and Catalysis*, Vol. 60, Elsevier, Amsterdam, 1990, p. 255.

87 R. M. LAGO, W. O. HAAG, R. J. MIKOVSKY, D. H. OLSON, S. D. HELLRING, K. D. SCHMITT, G. T. KERR, in: New Developments in Zeolite Science and Technology, Y. MURAKAMI, A. IIJIMA, J. W. WARD (Eds), *Studies in Surface Science and Catalysis*, Vol. 28, Elsevier, Amsterdam, 1986, p. 677.

88 H. G. KARGE, G. BORBÉLY, H. K. BEYER, G. ONYESTYAK, Proceedings of the 9th International Congress on Catalysis, M. J. PHILIPS, M. TERNAN (Eds), 1988, Vol I, p. 396.

89 R. CARVAJAL, P.-J. CHU, J. LUNSFORD, *J. Catal.* **1990**, *125*, 123.

90 C. J. PLANK, E. J. ROSINSKI, *Chem. Eng. Progr. Symp. Ser.* **1967**, *73*, 26.

91 A. CORMA, B. W. WOJCIECHOWSKI, Catalytic Cracking, Chemistry and Kinetics, Marcel Dekker, New York 1986.

92 F. C. JENTOFT, B. C. GATES, *Topics Catal.* **1997**, *4*, 1.

93 R. D. CORTRIGHT, J. A. DUMESIC, R. J. MADON, *Topics Catal.* **1997**, *4*, 15.

94 A. CORMA, A. V. ORCHILLÉS, *Microporous Mesoporous Mater.* **2000**, *35–36*, 21.

95 M. BORONAT, P. VIRUELA, A. CORMA, *J. Phys. Chem. A* **1998**, *102*, 982.

96 J. N. KONDO, F. WAKABAYASHI, K. DOMEN, *J. Phys. Chem B* **1998**, *102*, 2259.

97 H. ISHIKAWA, E. YODA, J. N. KONDO, F. WAKABAYASHI, K. DOMEN, *J. Phys. Chem. B* **1999**, *103*, 5681.

98 J. N. KONDO, H. ISHIKAWA, E. YODA, F. WAKABAYASHI, K. DOMEN, *J. Phys. Chem. B* **1999**, *103*, 8538.

99 J. N. KONDO, L. SHAO, F. WAKABAYASHI, K. DOMEN, *J. Phys. Chem B* **1997**, *101*, 9314.

100 A. BRAIT, A. KOOPMANS, K. SESHAN, H. WEINSTABL, A. ECKER, J. A. LERCHER, *Ind. Eng. Chem. Res.* **1998**, *37*, 873.

101 K. A. CUMMING, B. W. WOJCIECHOWSKI, *Catal. Rev. Sci. Eng.* **1996**, *38*, 101.

102 V. B. KAZANSKY, M. V. FRASH, R. A. VAN SANTEN, *Appl. Catal.* **1996**, *146*, 225.

103 B. A. WILLIAMS, S. M. BABITZ, J. T. MILLER, R. Q. SNURR, H. H. KUNG, *Appl. Catal. A* **1999**, *177*, 161.

104 A. F. H. WIELERS, M. VAARKAMP, M. F. M. POST, *J. Catal.* **1991**, *127*, 51.

105 S. R. BLASZKOWSKI, R. A. VAN SANTEN, *Topics Catal.* **1997**, *4*, 145.

106 W. O. HAAG, R. M. DESSAU, Proceedings of the 8th International Congress on Catalysis, Dechema, Frankfurt am Main, 1984, p 305.

107 S. KOTREL, H. KNÖZINGER, B. C. GATES, *Microporous Mesoporous Mat.* **2000**, *35*, 11.

108 R. HOURIET, G. PARISOD, T. GAUMANN, *J. Am. Chem. Soc.* **1997**, *99*, 3599.

109 S. P. BATES, R. A. VAN SANTEN, *Adv. Catal.* **1998**, *42*, 1.

110 S. A. ZYGMUNT, L. A. CURTISS, P. ZAPOL, L. E. ITON, *J. Phys. Chem. B.* **2000**, *104*, 1944.

111 G. B. McVicker, G. M. Kramer, J. J. Zemiak, *J. Catal.* **1983**, *83*, 286.

112 A. Corma, in: Zeolites: Facts, Figures, Future, P. A. Jacobs, R. A. van Santen (Eds), *Studies in Surface Science and Catalysis*, Vol. 49, Elsevier, Amsterdam, 1989, p. 49.

113 T. Narbeshuber, A. Brait, K. Seshan, J. A. Lercher, *J. Catal.* **1997**, *172*, 127.

114 T. Narbeshuber, H. Vinek, J. A. Lercher, *J. Catal.* **1997**, *157*, 388.

115 B. C. Gates, J. R. Katzer, G. C. A. Schuit, Chemistry of Catalytic Processes, Mc Graw-Hill New York, 1979, p. 1.

116 D. W. Werst, P. Han, S. C. Choure, E. I. Vinokur, L. Xu, A. D. Trifunac, L. A. Eriksson, *J. Phys. Chem. B* **1999**, *103*, 9219.

117 A. Corma, A. Martinez, *Catal. Rev.-Sci. Eng.* **1993**, *35(4)*, 483.

118 J. Weitkamp, Y. Traa, in: Handbook of Heterogeneous Catalysis, G. Ertl, H. Knözinger, J. Weitkamp (Eds), Wiley-VCH, Weinheim, 1997, p. 2039.

119 M. F. Simpson, J. Wie, S. Sundaresan, *Ind. Eng. Chem. Res.* **1996**, *35*, 3861.

120 G. S. Nivarthy, K. Seshan, J. A. Lercher, *Microporous Mesoporous Mater.* **1998**, *22*, 379.

121 A. Corma, A. Martinez, C. Martinez, *J. Catal.* **1994**, *146*, 185.

122 K. P. de Jong, C. M. A. M. Mesters, D. G. R. Peferoen, P. T. M. van Brugge, C. de Groot, *Chem. Eng. Sci.* **1996**, *51*, 2053.

123 W. O. Haag, R. M. Lago, US Patent 4,374,296 (1983), assigned to Mobil Oil Corp.

124 V. R. Choudhary, *Zeolites* **1978**, *7*, 272.

125 D. M. Brouwer, H. Hogeveen, *Recul. Trav. Chim. Pays-Bas.* **1970**, *89*, 211.

126 H. Pines, The Chemistry of Hydrocarbon Conversions, Academic Press, New York, 1981.

127 A. Corma, *Chem. Rev.* **1995**, *95*, 559.

128 J. A. Martens, P. A. Jacobs, in: Theoretical Aspects of Heterogeneous Catalysis, J. B. Moffat (Ed.), Van Nostrand Reinhold, New York, 1990.

129 D. S. Santili, B. C. Gates, in: Handbook of Heterogeneous Catalysis, G. Ertl, H. Knözinger, J. Weitkamp (Eds), Wiley-VCH, Weinheim, 1997, p. 123.

130 J. A. Martens, P. A. Jacobs, *Zeolites* **1986**, *6*, 334.

131 K. Honna, M. Sugimoto, N. Shimizu, K. Kurisaki, *Chem. Lett.* **1986**, 315.

132 G. C. Lau, W. F. Maier, *Langmuir* **1987**, *3*, 164.

133 R. P. Kirchen, T. S. Sorenson, S. M. Whitworth, *Can. J. Chem.* **1993**, *71*, 2016.

134 H. W. Whitlock Jr., M. W. Siefken, *J. Am. Chem. Soc.* **1968**, *90*, 4929.

135 G. Mirth, A. Kogelbauer, J. A. Lercher, Proceedings of the 9th International Zeolite Conference, R. von Ballmoos, J. B. Higgins, M. M. J. Treacy (Eds), Butterworth-Heinemann, 1992, Vol II, p. 251.

136 R. Fischer, W. A. Hölderich, W. D. Mross, H. M. Weitz, European Patent 0,167,021, 1986, assigned to BASF AG.

137 U. Dettmeier, K. Eichler, K. Kühlein, E. I. Leupold, H. Litterer, *Angew. Chem. Int. Ed. Engl.* **1987**, *26*, 468.

138 W. Hölderich, F. Merger, W. D. Mross, R. Fischer, European Patent 0,162,387, 1985, assigned to BASF AG.

139 C. Neri, F. Buonomi, European Patent 0,100,117, 1983, assigned to ANIC S. p. A.

140 M. Orisaku, K. Sano, Japanese Patent 61,112,040, 1986, assigned to Mitsubishi Gas Chem. Co. Inc.

141 N. Götz, W. Hölderich, L. Hupfer, US Patent 4,968,831, 1988, assigned to BASF AG.

142 C. Meyer, W. Laufer, W. F. Hölderich, *Catal. Lett.* **1998**, *53*, 131.

143 R. A. Sheldon, J. A. Elings, S. K. Lee, H. E. B. Lempers, R. S. Downing, *J. Mol. Catal. A* **1998**, *134*, 129.

144 H. Sato, K. Hirose, M. Kitamura, H. Tojima, N. Ishii, European Patent 236,092, 1987, assigned to Sumitomo Chemical Company.

145 H. Sato, K. Hirose, Y. Nakamura, *Chem. Lett.* **1993**, 1987.

146 T. Tatsumi, in: Fine Chemicals through Heterogeneous Catalysis, R. A. Sheldon, H. van Bekkum (Eds), VCH-Wiley, Weinheim, 2001, p 185.

147 C. D. Chang, P. D. Perkins, *Zeolites* **1983**, *3*, 298.

148 H. le Blanc, L. Puppe, K. Wedemeyer, German Patent 3,332,687, 1985, assigned to Bayer AG.

149 Th. Stamm, H. W. Kouwenhoven, R. Prins, in: Heterogeneous Catalysis and Fine Chemicals III, M. Guisnet, J. Barbier, J. Barrault, C. Bouchoule, D. Duprez (Eds), *Studies in Surface Science and Catalysis*, Vol. 78, Elsevier, Amsterdam, 1993, p. 543.

150 R. Prins, *Catal. Today* **1997**, *37*, 103.

151 J. D. Roberts, M. C. Caserio, Basic Principles of Organic Chemistry, W. A. Benjamin Inc., New York, 1965.

152 J. A. Lercher, R. A. van Santen, H. Vinek, *Catal. Lett.* **1994**, *27*, 91.

153 F. Stevens, US Patent 4,521,635, 1983, assigned to Atlantic Richfield Co.

154 T. R. Forester, R. F. Howe, *J. Am. Chem. Soc.* **1987**, *109*, 5076.

155 W. W. Kaeding, S. A. Butter, *J. Catal.* **1980**, *61*, 155.

156 G. Mirth, J. A. Lercher, in: Natural Gas Conversion, A. Holmen, K.-J. Jens, S. Kolboe (Eds), *Studies in Surface Science and Catalysis*, Vol. 61, Elsevier, Amsterdam, 1991, p. 437.

157 J. A. Lercher, G. Mirth, M. Stockenhuber, T. Narbeshuber, A. Kogelbauer, in: Acid Base Catalysis II, H. Hattori, M. Misono, Y. Ono (Eds), *Studies in Surface Science and Catalysis*, Vol. 90, Elsevier, Amsterdam, 1994, p. 147.

158 L. Kubelkova, J. Novakova, K. Nedomova, *J. Catal.* **1990**, *124*, 441.

159 L. Kubelkova, J. Novakova, P. Jiru, in: Structure and Reactivity of Modified Zeolites, P. A. Jacobs, N. I. Jäger, P. Jiru, V. B. Kazansky, G. Schulz-Ekloff (Eds), *Studies in Surface Science and Catalysis*, Vol. 18, Elsevier, Amsterdam, 1984, p. 217.

160 V. Bosacek, *J. Phys. Chem.* **1993**, *97*, 10732.

161 S. R. Blaszkowski, R. A. van Santen, *J. Am. Chem. Soc.* **1996**, *118*, 5152.

162 M. Misono, *Catal. Rev. Sci. Eng.* **1987**, *29*, 269.

163 Ch. Gründling, G. Eder-Mirth, J. A. Lercher, *Res. Chem. Int.* **1997**, *23*, 25.

164 D. R. Corbin, S. Schwarz, G. C. Sonnichsen, *Catal. Today* **1997**, *37*, 71.

165 C. Grundling, G. Eder-Mirth, J. A. Lercher, *J. Catal.* **1996**, *160*, 299.

166 D. J. Parrillo, R. J. Gorte, W. E. Farneth, *J. Am. Chem. Soc.* **1993**, *115*, 12441.

167 M. Deeba, European Patent 0,180,983, 1985, assigned to Air Products and Chemicals Inc.

168 L. Abrams, D. R. Corbin, R. D. Shannon, European Patent 0,251,597, 1987, assigned to Du Pont.

169 A. Kogelbauer, Ch. Gründling, J. A. Lercher, *J. Phys. Chem.* **1996**, *100*, 1852.

170 Ch. Gründling, G. Eder-Mirth, J. A. Lercher, *J. Catal.* **1996**, *160*, 299.

171 M. Deeba, M. E. Ford and T. A. Johnson, *J. Chem. Soc., Chem. Comm.* **1987**, *8*, 562.

172 V. Taglieber, W. Hölderich, R. Kummer, W. D. Mross, German Patent, DE 3,327,000 A1, 1983, assigned to BASF AG.

173 M. Lequite, F. Figueras, C. Moreau, S. Hub, *J. Catal.* **1996**, *163*, 255.

174 K. Hatada, Y. Ono, T. Keii, *Adv. Chem. Ser.* **1973**, *121*, 45.

175 D. S. Jones, US Patent 3,231,616, 1986, assigned to Socony Mobil Oil Co. Inc.

176 M. G. Warawdekar, R. A. Rajadhyaksha, *Zeolites* **1987**, *7*, 574.

177 C. D. Chang, P. D. Perkins, *Zeolites* **1983**, *3*, 298.

178 C. D. Chang, P. D. Perkins, European Patent 082,613, 1986, assigned to Mobil Oil Corp.

179 M. H. W. Burgers, H. van Bekkum, *J. Catal.* **1994**, *148*, 68.

180 Y. Ono, *Heterocycles* **1981**, *16*, 10.

181 K. Hatada, M. Shimada, Y. Ono, T. Keil, *J. Catal.* **1975**, *37*, 166.

182 F. A. COTTON, G. WILKINSON, Advanced Inorganic Chemistry, John Wiley & Sons, New York, 4th Edn, 1980, 1396 pp.

183 R. M. DESSAU, *Zeolites* **1990**, *10*, 205.

184 D. P. ROELOFSEN, H. VAN BEKKUM, *Recueil* **1972**, *91*, 605.

185 P. VENUTO, P. S. LANDIS, *Adv. Catal.* **1968**, *18*, 346.

186 J. MARCH, Advanced Organic Chemistry, John Wiley & Sons, New York, 4th Edn, 1992, 1995 pp.

187 C. D. CHANG, W. H. LANG, W. K. BELL, Catalysis of Organic Reactions, W. R. MOSER (Ed.), Marcel Dekker, New York, 1981, p. 73.

188 L. KUBELKOVA, J. CEJKA, J. NOVAKOVA, V. BOSACEK, I. JIRKA, P. JIRU, in: Zeolites: Facts, Figures, Future, P. A. JACOBS, R. A. VAN SANTEN (Eds), *Studies in Surface Science and Catalysis*, Vol. 49, Elsevier, Amsterdam, 1989, p. 1203.

189 L. R. MARTENS, W. J. VERMEIREN, D. R. HUYBRECHTS, P. J. GROBET, P. A. JACOBS, in: New Developments in Zeolite Science and Technology, Y. MURAKAMI, A. IIJIMA, J. WARD (Eds), *Studies in Surface Science and Catalysis*, Vol. 28, Elsevier, Amsterdam, 1986, p. 935.

190 R. M. DESSAU, *J. Chem. Soc., Chem. Commun.* **1986**, 1167.

191 I. E. MAXWELL, R. S. DOWNING, J. J. DE BOER, S. A. VAN LANGEN, *J. Catal.* **1980**, *61*, 485.

192 A. J. SCHIPPERIJN, J. LUKAS, *Recl. Trav. Chim. Pays-Bas* **1983**, *102*, 537.

193 W. F. HÖLDERICH, N. GÖTZ, G. FOUQUET, European Patent 263,464, 1988, assigned to BASF AG.

194 F. J. VAN DER GAAG, R. J. O. ADRIAANSENS, H. VAN BEKKUM, P. C. VAN GEEM, in: Recent Advances in Zeolite Science, J. KLINOWSKY, P. J. BARRIE (Eds), *Studies in Surface Science and Catalysis*, Vol. 52, Elsevier, Amsterdam, 1989, p. 283.

195 F. J. VAN DER GAAG, F. LOUTER, J. C. OUDEJANS, H. VAN BEKKUM, *Appl. Cat.* **1986**, *26*, 191.

196 M. S. RIGUTTO, H. J. A. DE VRIES, S. R. MAGILL, A. J. HOEFNAGEL, H. VAN BEKKUM, in: Heterogeneous Catalysis and Fine Chemicals III, M. GUISNET, J. BARBIER, J. BARRAULT, C. BOUCHOULE, D. DUPREZ (Eds), *Studies in Surface Science and Catalysis*, Vol. 78, Elsevier, Amsterdam, 1993, p. 66.

197 P. J. KUNKELER, M. S. RIGUTTO, R. S. DOWNING, H. J. A. DE VRIES, H. VAN BEKKUM, in: Progress in Zeolite and Microporous Materials, H. CHON, S.-K. IHM, Y. S. UH (Eds), *Studies in Surface Science and Catalysis*, Vol. 105, Elsevier, Amsterdam, 1997, p. 1269.

198 P. J. KUNKELER, D. MOESKOPS, H. VAN BEKKUM, *Microporous Mater.* **1997**, *11*, 313.

199 G. MIRTH, J. A. LERCHER, *J. Phys. Chem.* **1991**, *95*, 3736.

200 P. SYKES, Mechanism in Organic Chemistry, Longman Inc., New York, 5th edn, 1981, 397 pp.

201 T. XU, D. H. BARICH, P. D. TORRES, J. B. NICHOLAS, J. F. HAW, *J. Am. Chem. Soc.* **1997**, *119*, 396.

202 W. F. BOURGOYNE, D. D. DIXON, J. P. CASEY, *Chemtech* **1989**, 690.

203 KH. M. MINACHEV, YA. I. ISAKOV, V. J. GARIN, *Chem. Abs.* **1966**, *64*, 19452.

204 J. S. BECK, W. O. HAAG, in: Handbook of Heterogeneous Catalysis, G. ERTL, H. KNÖZINGER, J. WEITKAMP (Eds), Wiley-VCH, Weinheim, 1997, p. 2123.

205 G. MEIMA, *Cattech*, **1998**, *3*, 5.

206 G. S. LEE, J. J. MAJ, S. C. ROCKE, J. M. GARCES, *Catal. Lett.* **1989**, *2*, 243.

207 G. MIRTH, J. A. LERCHER, *J. Catal.* **1991**, *132*, 244.

208 G. MIRTH, J. CEJKA, E. NUSTERER, J. A. LERCHER, in: Zeolites and Related Microporous Materials, State of the Art 1994, J. WEITKAMP, H. G. KARGE, H. PFEIFER, W. HÖLDERICH (Eds), *Studies in Surface Science and Catalysis*, Vol. 84, Elsevier, Amsterdam, 1994, p. 287.

209 G. MIRTH, J. CEJKA, J. KRTIL, J. A. LERCHER, in: Catalyst Deactivation, B. DELMON, G. F. FROMET (Eds), *Studies in Surface Science and Catalysis*, Vol. 88, Elsevier, Amsterdam, 1994, p. 241.

210 J. CEJKA, N. ZILKOVA, B. WICHTERLOVA, G. EDER-MIRTH, J. A. LERCHER, *Zeolites*, **1996**, *17*, 265.

211 H. Vinek, M. Derewinski, G. Mirth, J. A. Lercher, *Appl. Catal.* **1991**, *68*, 277.

212 E. Santacesaria, D. Grasso, D. Gelosa, S. Carra., *Appl. Catal.* **1990**, *64*, 83.

213 P. Y. Chen, M. C. Chen, H. Y. Chu, N. S. Chang, T. K. Chuang, in: New Developments in Zeolite Science and Technology, Y. Murakami, A. Iijima, J. Ward (Eds), *Studies in Surface Science and Catalysis*, Vol. 28, Elsevier, Amsterdam, 1986, p. 739.

214 K. G. Ione, O. V. Kikhtyanin, in: Zeolites: Facts, Figures, Future, P. A. Jacobs, R. A. van Santen (Eds), *Studies in Surface Science and Catalysis*, Vol. 49, Elsevier, Amsterdam, 1989, p. 1073.

215 I. Neves, F. R. Ribeiro, J. P. Bodibo, Y. Pouilloux, M. Gubelmann, P. Magnoux, M. Giusnet, G. Perot, Proceedings of the 9th International Zeolite Conference, R. von Ballmoos, J. B. Higgins, M. M. J. Treacy (Eds), Butterworth-Heinemann, 1992, Vol. II, p. 543.

216 H. van Bekkum, A. J. Hoefnagel, M. A. van Koten, E. A. Gunnewegh, A. H. G. Vogt, H. W. Kouwenhoven, in: Zeolites and Microporous Crystals, T. Hattori, T. Yashima (Eds), *Studies in Surface Science and Catalysis*, Vol. 83, Elsevier, Amsterdam, 1993, p. 379.

217 G. Friedhoven, O. Immel, H. H. Schwarz, German Patent 2,633,458, 1978, assigned to Bayer AG.

218 L. E. Bertea, H. W. Kouwenhoven, R. Prins, in: Zeolites and Related Microporous Materials, State of the Art 1994, J. Weitkamp, H. G. Karge, H. Pfeifer, W. Hölderich (Eds), *Studies in Surface Science and Catalysis*, Vol. 84, Elsevier, Amsterdam, 1994, p. 1973.

219 A. Kogelbauer, H. W. Kouwenhoven, in: Fine Chemicals through Heterogeneous Catalysis, R. A. Sheldon, H. van Bekkum (Eds), Wiley-VCH, Weinheim, 2001, p. 123.

220 Th. M. Wortel, D. Oudijn, C. J. Vleugal, D. P. Roelofsen, H. van Bekkum, *J. Catal.* **1979**, *60*, 110.

221 A. P. Singh, S. B. Kumar, A. Paul, A. Raj, *J. Catal.* **1994**, *147*, 360.

222 H. Noller, W. Kladnig, *Catal. Rev.-Sci. Eng.* **1976**, *13*, 149.

223 J. Weitkamp, M. Hunger, U. Rymsa, *Microporous Mesoporous Mater.* **2001**, *48*, 255.

224 J. A. Lercher, Ch. Gründling, G. Eder-Mirth., *Catal. Today* **1996**, *27*, 353.

225 M. Lasperas, H. Cambon, D. Brunel, I. Rodriguez, P. Geneste, *Microporous Mater.* **1993**, *1*, 343.

226 D. Barthomeuf, *J. Phys. Chem.* **1984**, *88*, 42.

227 L. R. M. Martens, W. J. M. Vermeiren, P. J. Grobet, P. A. Jacobs, in: Preparation of Catalysts IV, B. Delmon, P. Grange, P. A. Jacobs, G. Poncelet (Eds), *Studies in Surface Science and Catalysis*, Vol. 31, Elsevier, Amsterdam, 1987, p. 531.

228 P. E. Hathaway, M. E. Davis, *J. Catal.* **1989**, *116*, 263.

229 D. E. Bryant, W. L. Kranich, *J. Catal.* **1967**, *8*, 8.

230 H. Pines, J. Manassen, *Adv. Catal.* **1966**, *16*, 49.

231 J. B. Nagy, J. P. Lange, A. Gourgue, P. Bodart, Z. Gabelica, in: Catalysis by Acids and Bases, B. Imelik, C. Naccache, G. Coudurier, Y. Ben Taarit, J. C. Vedrine (Eds), *Studies in Surface Science and Catalysis*, Vol. 20, Elsevier, Amsterdam, 1985, p. 127.

232 M. Rep, E. A. Palomares, G. Eder-Mirth, J. G. van Ommen, N. Rösch, J. A. Lercher, *J. Phys. Chem. B* **2000**, *104*, 8624.

233 H. Noller, G. Ritter, *J. Chem. Soc., Farad. Trans.* **1984**, *180*, 275.

234 T. Yashima, H. Suzuki, N. Hara, *J. Catal.* **1974**, *33*, 486.

235 P. A. Jacobs, M. Tielen, J. B. Uytterhoeven, *J. Catal.* **1977**, *50*, 98.

236 L. R. M. Martens, P. J. Grobet, P. A. Jacobs, *Nature* **1985**, *315*, 568.

237 J. C. Kim, H. X. Li, C. Y. Chen, M. E. Davies, *Microporous Mater.* **1994**, *2*, 413.

238 T. F. Brownscombe, US Patent, 4,992,613, 1991, assigned to Shell Oil Co.

239 W. F. HOLDERICH, Proceedings of the International Symposium on Acid/Base Catalysis, Sapporo, 1988, p. 2.

240 L. R. MARTENS, W. J. VERMEIREN, D. R. HUYBRECHTS, P. J. GROBET, P. A. JACOBS, Proceedings of the 9th International Congress on Catalysis, M. J. Philips, M. Ternan (Eds), 1988, Vol I, p. 420.

241 C. B. DARTT, M. E. DAVIS, *Catal. Today*, **1994**, *19*, 151.

242 P. CHU, G. W. KUEHL, US Patent 4,605,787, 1986, assigned to Mobil Oil Corp.

243 A. E. PALOMARES, G. EDER-MIRTH, M. REP, J. A. LERCHER, *J. Catal.* **1998**, *180*, 56.

244 B. KUMARI VASANTHY, M. PALANICHAMY, V. KRISHNASAMY, *Appl. Catal. A* **1996**, *148*, 51.

245 YU. N. SIDORENKO, P. N. GALICH, *Petrol Chem.* **1991**, *31*, 57.

246 R. J. DAVIS, E. J. DOSKOCIL, S. BORDAWEKAR, *Catal. Today* **2000**, *67*, 241.

247 H. ITOH, A. MIRANOTO, Y. MORAKANII, *J. Catal.* **1980**, *64*, 284.

248 A. E. PALOMARES, G. EDER-MIRTH, J. A. LERCHER, *J. Catal.* **1998**, *180*, 56.

249 M. HUNGER, U. SCHENK, M. SEILER, J. WEITKAMP, *J. Mol. Catal. A* **2000**, *156*, 153.

250 D. ARCHIER, G. COUDURIER, C. NACCACHE, Proceedings of the 9th International Zeolite Conference, R. VON BALLMOOS, J. B. HIGGINS, M. M. J. TREACY (Eds), Butterworth-Heinemann, 1992, Vol. II, p. 525.

251 J. MARCH, Advanced Organic Chemistry, 4th edn, John Wiley & Sons, New York, 1992, p. 172.

252 J. HABER, in: New Developments in Selective Oxidation by Heterogeneous Catalysts, P. RUIZ, B. DELMON (Eds), *Studies in Surface Science and Catalysis*, Vol. 72, Elsevier, Amsterdam, 1992, p. 279.

253 B. S. BESTY, F. P. GORTSEMA, G. WILDMAN, J. J. SHARKEY, US Patent 5,231,187, 1993, assigned to Merk Co.

254 J. SHABTAI, R. LAZAR, E. BIRON, *J. Mol. Catal.* **1984**, *27*, 35.

255 E. J. CREYGHTON, S. D. GANESHIE, R. S. DOWNING, H. VAN BEKKUM, *J. Chem. Soc., Chem Comm.* **1995**, 1859.

256 J. S. RAFELT, J. H. CLARK, *Catal. Today* **2000**, *57*, 33.

257 U. SCHUCHARDT, D. CARDOSO, R. SERCHELI, R. PEREIRA, R. S. DE CRUZ, M. C. GUERREIRO, D. MANDELLI, E. V. SPINACE, E. L. PIRES, *Appl. Catal. A* **2001**, *211*, 1.

258 J. M. THOMAS, R. RAJA, G. SANKAR, B. F. JOHNSON AND D. W. LEWIS, *Chem. Eur. J.* **2001**, *7*, 2973.

259 B. NOTARI, in: Innovation in Zeolite Materials Science, P. J. GROBET, W. J. MORTIER, E. F. VANSANT, G. SCHULZ-EKLOFF (Eds), *Studies in Surface Science and Catalysis*, Vol. 37, Elsevier, Amsterdam, 1988, p. 413.

260 J. M. THOMAS AND G. SANKAR, *Acc. Chem. Res.* **2001**, *34*, 571.

261 G. N. VAYSSILOV, *Catal. Rev. Sci. Eng.* **1997**, *39*, 209.

262 B. NOTARI, *Adv. Catal.* **1996**, *41*, 253.

263 V. BOLIS, S. BORDIGA, C. LAMBERTI, A. ZECCHINA, A. CARATI, F. RIVETTI, G. SPANO, G. PETRINI, *Microporous Mesoporous Mater.* **1999**, *30*, 67.

264 A. JENTYS, C. R. A. CATLOW, *Catal. Lett.* **1993**, *22*, 251.

265 A. J. M. DEMAN, J. SAUER, *J. Phys. Chem.* **1996**, *100*, 5025.

266 G. TOZZOLA, M. A. MANTEGAZZA, G. RANGHINO, G. PETRINI, S. BORDIGA, G. RICCHIARDI, C. LAMBERTI, R. ZULIAN, A. ZECCHINA, *J. Catal.* **1998**, *179*, 64.

267 M. G. CLERICI, P. INGALLANA, *J. Catal.* **1993**, *140*, 71.

268 R. A. SHELDON, I. W. C. E. ARENDS, H. E. B. LEMPERS, *Collect. Czech. Chem. Commun.* **1998**, *63*, 1724.

269 T. SATO, J. DAKKA, R. A. SHELDON, in: Zeolites and Related Microporous Materials, State of the Art 1994, J. WEITKAMP, H. G. KARGE, H. PFEIFER, W. HÖLDERICH (Eds), *Studies in Surface Science and Catalysis*, Vol. 84, Elsevier, Amsterdam, 1994, p. 1853.

270 G. I. PANOV, A. K. URIARTE, M. A. RODKIN, V. I. SOBOLEV, *Catal. Today* **1998**, *41*, 365.

271 V. SOBOLEV, G. PANOV, A. KHARITONOV, V. ROMANNIKOV, A. VOLODIN, K. KONE, *J. Catal.* **1993**, *139*, 435.

272 G. Panov, A. Kharitonov, V. Sobolev, *Appl. Catal.* **1993**, *98*, 1.

273 X. Feng, W. K. Hall, *J. Catal.* **1997**, *166*, 368.

274 H. Y. Chen, W. M. H. Sachtler, *Catal. Today* **1998**, *42*, 73.

275 R. Joyner, M. Stockenhuber, *J. Phys. Chem. B* **1999**, *103*, 5963.

276 T. V. Voskoboinikov, H. Y. Chen, W. M. H. Sachtler, *Appl. Catal. B* **1998**, *19*, 279.

277 E. M. El-Malki, R. A. van Santen, W. M. H. Sachtler, *J. Phys. Chem. B* **1999**, *103*, 4611.

278 P. Marturano, L. Drozdova, A. Kogelbauer, R. Prins, *J. Catal.* **2000**, *192*, 236.

279 A. Ribera, I. Arends, S. de Vries, J. Perez-Ramirez, R. A. Sheldon, *J. Catal.* **2000**, *195*, 287.

280 G. Panov, *Cattech* **2000**, *4*, 18.

281 A. K. Uriate, M. A. Rodkin, M. J. Gross, A. Kharitonov, G. Panov, in: Third World Congress on Oxidation Catalysis, R. K. Grasselli, S. T. Oyama, A. M. Gaffney, J. E. Lyons (Eds), *Studies in Surface Science and Catalysis*, Vol. 110, Elsevier, Amsterdam, 1997, p. 857.

282 M. Stockenhuber, M. J. Hudson, R. W. Joyner, *J. Phys. Chem. B* **2000**, *104*, 3370.

283 M. Stockenhuber, R. W. Joyner, J. M. Dixon, M. J. Hudson, G. Grubert, *Microporous Mesoporous Materials* **2001**, *44*, 367.

284 P. R. Rao, A. V. Ramaswamy, *Appl. Catal. A* **1993**, *93*, 123.

285 P. R. Hari Prasad Rao, A. V. Ramaswamy, P. Ratnasamy, *J. Catal.* **1992**, *137*, 225.

286 P. R. Hari Prasad Rao, A. V. Ramaswamy, P. Ratnasamy, *J. Catal.* **1993**, *141*, 604.

287 M. S. Rigutto, H. van Bekkum, *Appl. Catal.* **1991**, *68*, L1.

288 L. W. Zatorski, G. Centi, J. Lopez Nieto, F. Trifiro, G. Bellussi, V. Fattore, in: Zeolites: Facts, Figures, Future, P. A. Jacobs, R. A. van Santen (Eds), *Studies in Surface Science and Catalysis*, Vol. 49, Elsevier, Amsterdam, 1989, p. 1243.

289 P. R. Hari Prasad Rao, A. V. Ramaswamy, *J. Chem. Soc., Chem Commun.* **1992**, 1245.

290 E. M. Flanigen, B. M. Lok, R. L. Patton, S. T. Wilson, in: New Developments in Zeolite Science and Technology, Y. Murakami, A. Iijima, J. Ward (Eds), *Studies in Surface Science and Catalysis*, Vol. 28, Elsevier, Amsterdam, 1986, p. 103.

291 E. M. Flanigen, B. M. Lok, R. L. Patton, S. T. Wilson, in: Innovation in Zeolite Materials Science, P. J. Grobet, W. J. Mortier, E. F. Vansant, G. Schulz-Ekloff (Eds), *Studies in Surface Science and Catalysis*, Vol. 37, Elsevier, Amsterdam, 1988, p. 13.

292 B. Kraushaar-Czarnetzky, W. G. M. Hoogervorst, W. H. J. Stork, in: Zeolites and Related Microporous Materials, State of the Art 1994, J. Weitkamp, H. G. Karge, H. Pfeifer, W. Hölderich (Eds), *Studies in Surface Science and Catalysis*, Vol. 84, Elsevier, Amsterdam, 1994, p. 1896.

293 J. M. Thomas, R. Raja, G. Sanker, R. G. Bell, *Nature* **1999**, *398*, 227.

294 J. Dakka, R. A. Sheldon, Netherlands Patent 9,200,968, 1992, assigned to DSM.

295 R. A. Sheldon, *J. Mol. Catal. A* **1996**, *107*, 75.

296 H. E. B. Lempers, R. A. Sheldon, *Appl. Catal. A*, **1996**, *143*, 137.

297 H. E. B. Lempers, R. A. Sheldon, *J. Catal.* **1998**, *175*, 62.

298 W. J. Mortier, R. A. Schoon-heydt, *Progr. Solid State Chem.* **1985**, *16*, 1.

299 R. Heidler, G. O. A. Janssens, W. J. Mortier, R. A. Schoonheydt, *Microporous Mater.* **1997**, *12*, 1.

300 J. Datka, M. Boczar, B. Gil, *Langmuir* **1993**, *9*, 2496.

301 K. van Genechten, W. J. Mortier, *Zeolites* **1988**, *8*, 273.

302 L. Uytterhoeven, W. J. Mortier, P. J. Geerlings, *J. Phys. Chem.* **1989**, *50*, 479.

303 J. Weitkamp, P. Kleinschmitt, A. Kiss, C. H. Berke, Proceedings of the

9th International Zeolite Conference, R. von Ballmoos, J. B. Higgins, M. M. J. Treacy (Eds), Butterworth-Heinemann, 1992, Vol. II, p. 79.

304 D. Barthomeuf, *J. Phys. Chem.* **1992**, *83*, 249.

305 A. Corma, *Chem. Rev.* **1995**, *95*, 559.

306 S. I. Zones, T. V. Harris, *Microporous Mesoporous Mater.* **2000**, *35–36*, 31.

307 C. R. Marcilly, *Topics Catal.* **2000**, *13*, 357.

308 P. B. Weisz, V. J. Frilette, *J. Phys. Chem.* **1960**, *64*, 382.

309 N. Y. Chen, J. Maziuk, A. B. Schwartz, P. B. Weisz, *Oil Gas Journal* **1968**, *66*, 154.

310 J. Kärger, W. Keller, H. Pfeifer, S. Ernst, J. Weitkamp, *Microporous Mater.* **1995**, *3*, 401.

311 W. W. Kaeding, C. Chu, L. B. Young, B. Weinstein, S. A. Butter, *J. Catal.* **1981**, *67*, 159.

312 J. H. Kim, T. Kunieda, M. Niwa, *J. Catal.* **1998**, *173*, 433.

313 S. Morin, P. Ayrault, N. S. Gnep, M. Guisnet, *App. Catal. A* **1998**, *166*, 281.

314 C. W. Jones, S. I. Zones, M. E. Davis, *Appl. Catal. A* **1999**, *181*, 289.

315 W. W. Kaeding, C. Chu, L. B. Young, B, S. A. Butter, *J. Catal.* **1981**, *69*, 392.

316 G. Mirth, J. A. Lercher, *J. Catal.* **1994**, *147*, 199.

317 G. Mirth, J. Cejka, J. A. Lercher, *J. Catal.* **1993**, *139*, 24.

318 M. Boudart, G. Djega-Mariadassou, Kinetics of Heterogeneous Catalytic Reactions, Princeton University Press, 1984.

319 G. Mirth, J. Cejka, E. Nuster, J. A. Lercher, in: Zeolites and Microporous Crystals, T. Hattori, T. Yashima (Eds), *Studies in Surface Science and Catalysis*, Vol. 83, Elsevier, Amsterdam, 1993, p. 287.

320 T. F. Narbeshuber, H. Vinek, J. A. Lercher, *J. Catal.* **1995**, *157*, 388.

321 T. Hibino, M. Niwa, Y. Murakami, *J. Catal.* **1991**, *128*, 551.

322 J. H. Kim, A. Ishida, M. Okajima, M. Niva, *J. Catal.* **1996**, *161*, 387.

323 H. P. Röger, M. Krämer, K. P. Möller, C. T. O'Connor, *Microporous Mesoporous Mater.* **1998**, *21*, 607.

324 F. Bauer, W. H. Chen, Q. Zhao, A. Freyer, S. B. Liu, *Microporous Mesoporous Mater.* **2001**, *47*, 67.

325 F. J. Weigert, *J. Org. Chem.*, **1987**, *52*, 3296.

326 C. D. Chang, P. D. Perkins, *Zeolites* **1983**, *3*, 298.

327 F. J. Weigert, *J. Org. Chem.* **1986**, *51*, 2653.

328 G. S. Lee, J. J. Maj, S. C. Rocke, J. M. Garcés, *Catal. Lett.* **1989**, *2*, 243.

329 G. R. Meina, M. J. M. van der Aalst, M. S. U. Samson, J. M. Garces, J. D. Lee, Proceedings of the 9th International Zeolite Conference, R. von Ballmoos, J. B. Higgins, M. M. J. Treacy (Eds), Butterworth-Heinemann, 1992, Vol. II, p. 327.

330 J. Weitkamp, M. Neuber, in: Chemistry of Microporous Crystals, T. Inui, S. Namba, T. Tatsumi (Eds), *Studies in Surface Science and Catalysis*, Vol. 60, Elsevier, Amsterdam, 1990, p. 291.

331 D. Fraenkel, M. Cherniavsky, B. Ittah, M. Levy, *J. Catal.* **1986**, *101*, 273.

332 E. Derouane, *J. Catal.* **1986**, *100*, 541.

333 M. Neuber, J. Weitkamp, in: Zeolites: Facts, Figures, Future, P. A. Jacobs, R. A. van Santen (Eds), *Studies in Surface Science and Catalysis*, Vol. 49, Elsevier, Amsterdam, 1989, p. 425.

334 W. Souverijns, J. A. Martens, G. F. Froment, P. A. Jacobs, *J. Catal.* **1998**, *174*, 177.

335 J. A. M. Arroyo, G. G. Martens, G. F. Froment, G. B. Marin, P. A. Jacobs, J. A. Martens, *Appl. Catal. A* **2000**, *192*, 9.

336 M. Taramasso, G. Perego, B. Notari, US Patent 4,410,501, 1983, assigned to Snam Progetti.

337 J. S. Reddy, R. Kumar, P. Ratnasamy, *Appl. Catal.* **1990**, *58*, L1.

338 H. W. Kouwenhoven, C. Ferrini, in: New Developments in Selective Oxidation, G. Centi, F. Trifiro (Eds), *Studies in Surface Science and Catalysis*,

Vol. 55, Elsevier, Amsterdam, 1990, p. 53.

339 M. A. Camblor, A. Corma, A. Martiney, J. Perez-Pariente, *J. Chem. Soc., Chem. Commun.* **1992**, 589.

340 D. P. Serrano, H. X. Li, M. E. Davis, *J. Chem. Soc., Chem. Commun.* **1992**, 745.

341 G. J. Kim, B. R. Cho, J. H. Kim, *Catal. Lett.* **1993**, *22*, 259.

342 A. Tuel, *Zeolites* **1995**, *15*, 228.

343 A. Corma, M. T. Navarro, J. Perez-Pariente, *J. Chem. Soc., Chem. Commun.* **1994**, 147.

344 P. T. Tanev, M. Chibwe, T. Pinnavaia, *Nature* **1994**, *386*, 239.

4.2.10
Applications of Natural Zeolites

Carmine Colella

4.2.10.1
Introduction and Historical Remarks

4.2.10.1.1 The Early Applications

Natural zeolites of sedimentary origin (see Sect. 4.2.10.2), namely *zeolitic tuffs*, have been employed as multipurpose raw material for millennia [1]. Our ancestors appreciated their softness and easy workability and used them to obtain objects of various nature. Examples of this practice are the tools of the Neolithic age found in Gherla (Transylvania, Romania) made of clinoptilolite-rich tuffs [2], the bas-reliefs carved into clinoptilolite-rich tuff by Zapotec Indians, approximately 2000 BC, at Mitla, Oaxaca, Mexico [3], and, more recently, the memorial sculptures of *Matres* ("Mothers") occasionally carved in chabazite-rich Campanian ignimbrite in the Caserta arca (north of Naples, Italy) in the 8th century BC [4]. However, in consideration of some additional attractive properties of tuff, such as lightness, reasonable resistance to erosion and capacity to control the humidity level and the thermal range, the early usage of this material was especially connected with constructions: the *Oya-ishi* (stone of Oya) of Utsunomiya City, Tochigi Prefecture, Japan, which contains clinoptilolite and mordenite, has been employed in the building industry as dimension stone since the 5th century AD [1]; the chabazite-rich Rhenish tuff, locally called *trass*, has been used as a building stone in Germany and neighboring countries since the Roman times [1], whereas the Neapolitan yellow tuff, containing phillipsite and in minor extent chabazite, has been used for the same purpose by the Greek colonizers and later by the Romans in the Neapolitan area starting from about 2700 years ago [5].

The successful use of the tuff as dimension stone in central and southern Italy was favored by the introduction of a new type of hydraulic mortar, made of lime and a volcanic unlithified material called *pozzolana*. According to the famous Roman architect Vitruvius, blends of this mortar with crushed and graded tuff (most probably zeolitic) gave rise to a durable concrete [6]. This concrete was used for the construction of large hydraulic works (e.g., harbors, aqueducts, dams, bridges) and other large constructions, such as amphitheaters, temples, spas, and villas, which

Fig. 1. Castel dell'Ovo, Naples, Italy, constructed of Neapolitan yellow tuff.

are often still preserved nowadays [7]. Similar examples of use of zeolitic tuff for producing hydraulic mortars are reported in Germany, where this type of mortar was used in Roman times for the construction of the Trier–Cologne aqueduct [8].

The Romans used the tuff blocks for all kinds of construction and to meet both architectural and structural requirements. The building techniques developed by the Romans were utilized almost unchanged during the first millennium and beyond. The use of zeolitic tuff in the Middle Ages became so extensive in Italy that entire villages were constructed using this stone, for instance all the houses in the villages of Pitigliano, Sovana, and Bagnoregio and in the city of Orvieto in central Italy are made of chabazite-rich tuffs [9]. Contrary to Roman usage to cover the tuff structures with plaster, beginning in the Middle Ages, tuff was used as facing stone in recognition of its architectural value. Examples of the use of bare blocks for monumental buildings are numerous throughout Naples and includes castles (Castel dell'Ovo, 12th century, see Fig. 1; Castel Nuovo, 13th century; Castel S. Elmo, 16th century) and churches (for example, S. Domenico Maggiore, 13th century, and Santa Chiara, 14th century, see Fig. 2).

Most likely zeolitic tuff played a role in ancient times also in agriculture, for instance in the *Phlegraean* (from the Greek $\varphi\lambda o\gamma\varepsilon\rho o\varsigma$ = burning) *Fields*, near Naples, Italy, and more generally in the Campania Plain, a mostly volcanic area, rich in tuff deposits, located NW of Naples. The ancient Romans called this area *Campania Felix* (=fertile), in recognition of the remarkable fertility of its soils, composed of potassium-rich, partly zeolitized, trachytic rocks, apart from the ordinary soil constituents.

References see page 1183

Fig. 2. Santa Chiara church, Naples, Italy, constructed of Neapolitan yellow tuff.

4.2.10.1.2 The Discovery of Zeolites

Despite these ancient unconscious utilizations, natural zeolites were discovered only two hundred and fifty years ago, in 1756, by Axel F. Cronstedt, a Swedish mineralogist, who described and classified two samples of an unknown zeolite mineral (possibly stilbite, but there is no direct evidence of this), coming from Lapland, northern Sweden, and Iceland, respectively [10]. He also coined the word *zeolite*, from the Greek words ζειν (to boil) and λιϑος (stone), in allusion to their peculiar feature to froth when heated with a blowpipe.

We know now that Cronstedt's species and other similar minerals are ubiquitous constituents in the vugs and cavities of basalts and other traprock formations. They appear usually as beautiful assemblages of well-formed crystals of millimeter to centimeter size and are said to have originated in a hydrothermal environment from late-stage magmatic solutions. They are therefore called *hydrothermal* zeolites [11].

After Cronstedt's discovery, many other hydrothermal zeolites were found in nature in the decades to come, so much so that about twenty species were reported in mineralogy books one century later [12] (by comparison, some 60 species are presently listed by the International Mineralogical Association (IMA) [13]).

4.2.10.1.3 Early Studies in Adsorption and Molecular Sieving

Zeolite minerals were regarded for years as mineralogical curiosities. The investigation of their adsorption properties started only at about the middle of the 19th century and was focused mainly on chabazite. Damour in a series of studies performed from 1840 (see, for instance, ref. [14]) observed that a series of zeolites, such as stilbite, harmotome, heulandite, brewsterite, faujasite, chabazite, gmelinite, christianite (phillipsite), analcime, levyne, scolecite, mesotype (natrolite) and thompsonite, could be reversibly dehydrated with no apparent structural damage. Friedel [15], at the end of that century, showed that the space set free in a zeolite by dehydration (the species investigated were chabazite and harmotome) could be occupied by other molecules, such as ammonia, carbon dioxide, hydrogen sulfide, hydrogen, ethanol, silicon fluoride and silicon chloride. The phenomenon of adsorption was further investigated by Grandjean [16], who studied the interaction at high temperature between some zeolites, mainly chabazite, and a series of vapors, such as iodine, bromine, calomel, sulfur and cinnabar. Seeliger and Lapkamp [17], studying the adsorption isotherms at 273 K of several gases (namely, Ne + He, N_2, O_2, CH_4, NO, CO_2, C_2H_2, H_2, NH_3) on chabazite, pointed out a direct relationship between sorption capacity and critical temperature of the adsorbed gas. Weigel and Steinhof [18] in 1924 found that chabazite when dehydrated readily adsorbed the vapors of water, methanol and ethanol and formic acid, whereas acetone, ether and benzene were largely excluded. The significance of these results was pointed out by McBain [19], who was able to calculate that chabazite could sorb molecules with a molecular volume smaller than about $76\,\text{Å}^3$ and therefore that its pores should have been of less than 5 Å in diameter. A few years later Schmidt [20], Sameshima [21] and Baba [22], studying the adsorption of a series of gases of different molecular dimensions on chabazite, confirmed that adsorption was related to the dimensions of pores and cavities in the structure of the dehydrated chabazite and fixed the pore size of this zeolite at about 4.4 Å. All these papers contributed to introduce the concept of "molecular sieving", a definition coined by McBain and reported in a review book, which appeared in 1932, whose Chapter V is devoted to studies on "Sorption by chabasite, other zeolites and permeable crystals" [23].

The above results stimulated further studies on adsorption by natural zeolites. Dozen of papers appeared, in fact, in the chemical literature until the middle of the 1950s and also patents on gas separations processes of industrial significance were granted. Much of this work was made by Barrer and coworkers in the UK (mainly in London) and concerned especially chabazite and mordenite, which appeared to have the most interesting prospects of application of the zeolites known at that time. A detailed review of the earlier Barrer work in this scientific area can be found in refs [24, 25].

4.2.10.1.4 Early Studies in Ion Exchange

Contrary to adsorption and molecular sieving, ion-exchange properties of natural zeolites received scarce attention by researchers. The scientific bases of ion exchange were investigated and disclosed in the middle of the 19th century, when Thompson [26] and Way [27] reported that a column filled by a soil retained ammonium from a solution containing a fertilizer, releasing calcium. The reversibility of the ion exchange was demonstrated a few years later by Eichhorn [28], who studied the behavior of chabazite and natrolite when put into contact with dilute solutions of sodium, ammonium or calcium chloride and sodium or ammonium

References see page 1183

carbonate. However, it was not until one century later that a systematic study of natural zeolites as ion exchangers commenced. In fact, natural zeolites were for a long time considered less attractive than *permutites*, a class of partially crystalline synthetic aluminosilicate exchangers widely employed, especially for water softening, in the early decades of the 20th century [29]. Later, attention was especially attracted by *ion-exchange resins*, which, introduced in 1935 [30], became in a few years the leading product for solving problems related to water softening and deionization.

Nevertheless, some fragmentary studies on cation exchange in natural zeolites were performed in the early decades of the 20th century. Gans [29] noted that chabazite, heulandite and stilbite underwent exchange more readily than did analcime and natrolite. Zoch [31] made measurements on several zeolites, finding the ease of entry of the ammonium ion in the order chabazite and stilbite > heulandite > harmotome > scolecite. Clark and Steiger, Steiger, and Hey (see references in [32]) also made observations on a variety of crystalline zeolites.

The first systematic investigations on the ion-exchange properties of natural zeolites started in the Barrer's laboratory in the London and Aberdeen Universities at the end of the 1940s. Research was focused on chabazite and mordenite, but the main purpose of the investigation was, at the beginning, the preparation of modified molecular sieve sorbents, in order to obtain gradations in molecular sieve action and, therefore, new possibilities to separate molecular mixtures [32]. Research on the fundamentals of cation exchange equilibria and kinetics in zeolite minerals began, however, a few years later in the same laboratory under the leadership of the same scientist [33].

4.2.10.1.5 The Discovery of the Sedimentary Zeolites

Research on natural zeolites up to the early 1950s was important for the discovery of their unique properties, but the exiguity of the hydrothermal species and the lack of any method that could economically extract them from the basaltic rocks in which they were contained precluded any investigation directed to the development of large-scale industrial processes. This was also why, in a period of about fifteen years around 1950, great efforts were made by the chemical industry to obtain synthetic analogs of natural zeolites through low-temperature hydrothermal treatment of largely amorphous silico-aluminate magmas in alkaline environments [34]. In the same period efforts were also made to find in nature other occurrences of zeolite minerals, not so exiguous and therefore attractive from industrial and commercial points of view [35].

Evidence for the existence of such deposits had been reported in the literature [36]. In the United States, for instance, the chabazite from the well-known Bowie deposit, in Arizona, [37] had already been described in 1875. Many other occurrences of this type were since reported, but nobody could imagine at that time the extension of such formations and their possible commercial importance.

Starting from 1950 the geological world became aware that many enormous sedimentary deposits of volcanic origin, widespread all around the world, were rich in zeolites. The presence of zeolites was initially proved in the Green Tuff of Yokotemachi, Akita Prefecture, Japan [38]; in low-grade metamorphic rocks of Southland, New Zealand [39]; in the Neapolitan yellow tuff, around Naples, Italy [40] and later on in hundreds of locations in the United States, Japan, Eastern Europe, Russia and many other countries around the world [36].

All these occurrences originated mostly from reaction of volcaniclastic material, after deposition, with lacustrine, marine or groundwaters [41], giving rise to a variety of fine-grained zeolites, such as clinoptilolite, mordenite, phillipsite, chabazite and others. These rocks are now called *tuffs* and the zeolites composing these rocks, having sizes from less than 0.1 μm to a few micrometers, are said to be *sedimentary*.

The discovery of the sedimentary zeolites changed completely the view on this family of minerals. In fact, the large availability of zeolite-rich rocks at low price stimulated the search of possible applications, taking advantage of their attractive chemical and physical properties.

4.2.10.2
The Sedimentary Zeolites

4.2.10.2.1 Zeolite Types and Occurrence

Out of some sixty natural zeolite types, reported in a recent classification of the International Mineralogical Association [13], roughly ten types have been found with certainty in sedimentary occurrences. They are listed in Table 1, together with the structural code of identification [42] and the frequency of occurrence.

Based on the abundance and/or the interest of the zeolite types, the deposits that are exploited for applications or have chances for future mining activities are those containing clinoptilolite, mordenite, phillipsite, chabazite and, to a minor extent, erionite.

Clinoptilolite-containing deposits are practically ubiquitous [43, 44]. Reserves of clinoptilolite-rich tuffs have been found in many areas all around the world, for instance, in Europe (Bulgaria, Greece, Hungary, Italy, Romania, Slovakia, Slovenia, Turkey, Yugoslavia), in Russia and several states of the former Soviet Union (Georgia, Ukraine, Azerbaijan), in Asia (China, Iran, Japan, Korea), in Africa (South Africa), in Australia and New Zealand and lastly in many countries of the Americas, such as Argentina, Cuba, Mexico and the United States.

Mordenite-rich tuff deposits are definitely less frequent than the clinoptilolite-bearing formations. The main occurrences are reported to be in eastern Europe (Bulgaria, Hungary, Slovakia, Yugoslavia), Japan, Russia, the United States and New Zealand [43, 44].

Phillipsite and chabazite are not frequent as predominant phases in sedimentary deposits, although their presence is reported in several deposits in the United States and in a few other countries in the world [35, 43, 44]. On the contrary these

References see page 1183

Tab. 1. Zeolites found in sedimentary rocks[1].

Name	Structural code[2]	Abundance[3]
Analcime-wairakite[4]	ANA	+++
Chabazite	CHA	++
Clinoptilolite-heulandite[4]	HEU	+++++
Erionite	ERI	++
Faujasite	FAU	+
Ferrierite	FER	+
Laumontite	LAU	++
Natrolite	NAT	++
Mordenite	MOR	++
Phillipsite	PHI	+++

[1] Estimates from data of Hay [41], Mumpton [35], Hawkins [43], Gottardi and Galli [44]. [2] According to the Structure Commission of the International Zeolite Association [42]. [3] Estimated occurrence of worldwide sedimentary zeolite deposits. Abundance ranks from +++++ (very frequent) to + (rare). [4] Analcime and wairakite and clinoptilolite and heulandite are two couples of isostructural species that should have the same name. Nevertheless, for historical reasons and long-established usage the original denominations are maintained in the official nomenclature [13].

two zeolites are very common in joint occurrences especially in Italy, and, in minor extent, in Germany and in the Canary Islands (Spain) [45]. Phillipsite occurs very frequently also in deep-sea sediments [44].

Erionite-rich tuff deposits are somewhat common in the United States [46], but uncommon in other areas of the world, with some exceptions, for instance Turkey, where the presence of erionite in some locations (Cappadocia) has been associated with the incidence of mesothelioma [47].

4.2.10.2.2 Chemistry and Mineralogy

Table 2 summarizes the chemical and structural features of the main zeolite types found in the sedimentary zeolite deposits. Reported ranges of the Si/Al ratios are restricted to those more frequently exhibited by the various types [49]. Cation exchange capacities (CEC) have been calculated from the idealized formulae [13], taking in consideration the specific Si/Al ratios.

Inspecting Table 2 points out that the most common sedimentary zeolites are medium- (chabazite and phillipsite) to high-silicon types (clinoptilolite, erionite and mordenite), which is beneficial for some applications in the field of cation exchange [49], and that most of them (chabazite, clinoptilolite and mordenite) are medium- to large-pore zeolites, they are therefore potentially useful for applications based on adsorption.

Mineralogy of clinoptilolite-rich tuffs is rather uniform: smectite, quartz, cristobalite and some unreacted glass occur very frequently as ancillary phases of clinoptilolite crystals (Fig. 3a). Calcite, muscovite and feldspar are other common

Tab. 2. Chemical and structural features of some sedimentary zeolites.

Zeolite	Chemistry[1]			Structural features[2]	
	Si/Al	Cations	CEC/(meq g^{-1})	Window size/Å	Void volume
Chabazite	2.2–2.6	Ca,Na,K	3.3–3.6	3.8 × 3.8	0.48
Clinoptilolite	4.0–5.2	Ca,Na,K	2.2–2.6	3.0 × 7.6	0.34
Erionite	~3.5	Na,K,Mg,Ca[3]	~2.8	3.6 × 5.1	0.36
Mordenite	~4.9	Na,Ca	~2.3	6.5 × 7.0	0.26
Phillipsite	2.4–2.7	K,Na,Ca	3.3–3.6	3.6	0.30

[1] From Gottardi and Galli [44] and from the Report on Nomenclature for Zeolite Minerals [13]. The most common extra-framework cations are reported in order of abundance. [2] Window size is that of the largest channel [42]; void volume is expressed as cm^3 of liquid water per cm^3 of crystal [48]. [3] Substantial amounts of any of the four cations may be present.

impurities. Mordenite may be present at level of traces or in moderate amount. Clinoptilolite contents in the parent rocks are commonly over 50 %, but contents over 80 % are not rare. On the basis of the data of Table 2, CEC values in the range 1.2–2.0 meq g^{-1} must be expected. It has to be observed, however, that clinoptilolite exhibits usually exchange capabilities lower than those expected, because some extra-framework cations may not be available for the exchange [50, 51].

The mineral composition of the mordenite-rich tuffs is similar to the clinoptilolite-rich rocks. Apart from mordenite (Fig. 3b) and clinoptilolite (sometimes present at a modest level, but frequently in a comparable amount to mordenite), cristobalite, quartz, feldspar, clay minerals and unreacted glass are the ordinary coexisting phases. Mordenite percentage in the rock is usually high. Contents as high as 80 % are not unusual. The resultant CEC is normally greater than 1 meq g^{-1} with maximum values in the range 1.8–2.0 meq g^{-1}.

Mineral assemblages, containing phillipsite and/or chabazite, include also minor amounts of analcime, K-feldspar, biotite, pyroxene and some unreacted amorphous phases. In some chabazite-rich occurrences calcite may also be present. The Italian formations show two distinct typologies: phillipsite is predominant over chabazite (Fig. 3c) in the so-called *Neapolitan yellow tuff*, NW of Naples; chabazite (Fig. 3d), on the contrary, is the prevailing phase in many deposits, cropping out over large areas in central-southern Italy, for instance in the formation of *Campanian Ignimbrite* [52]. Total zeolite content averages 50–70 %. Resulting CEC values range between 1.8 meq g^{-1} and 2.5 meq g^{-1}.

Erionite (Fig. 2e) occurs usually in association with other zeolites, such as clinoptilolite, chabazite, phillipsite, apart from other minor phases, such as clay minerals and K-feldspar. Its content in the tuff may be as low as 20 % or less, but monomineralic beds of erionite have also been reported [46].

References see page 1183

a)

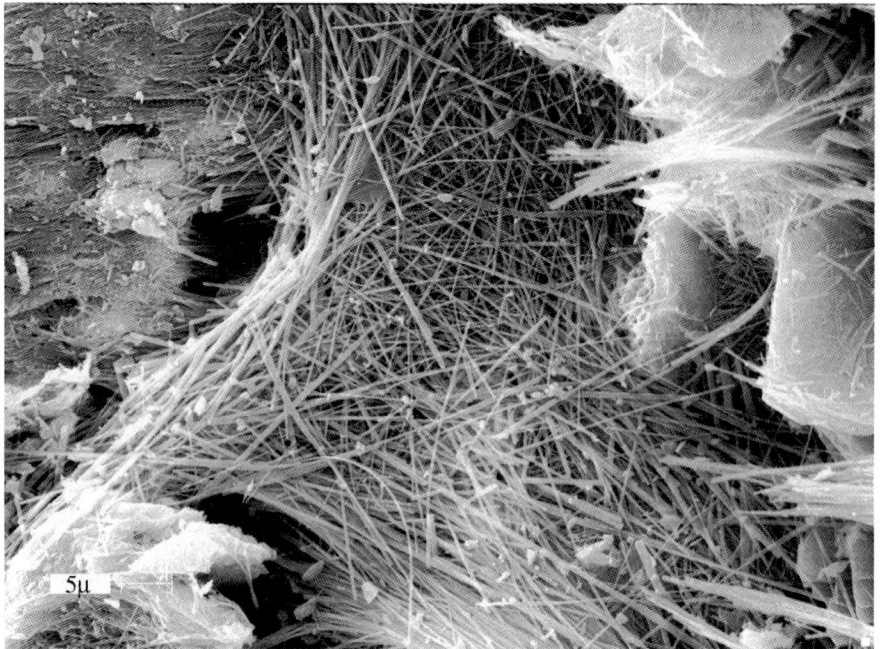

b)

c)

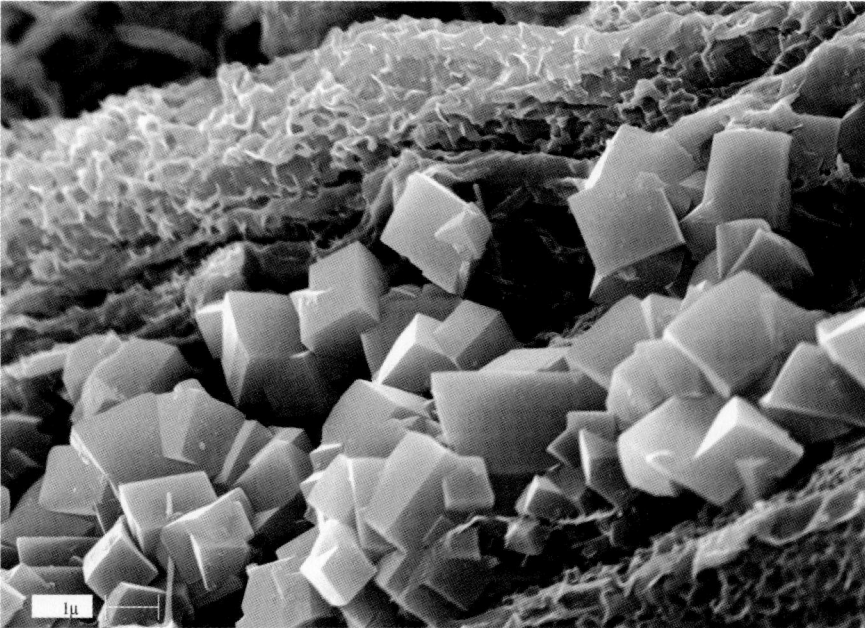

d)

e)

Fig. 3. (a) Scanning electron micrograph of a tuff sample from Pentalofos, Thrace, Greece, showing plate-shaped crystals of clinoptilolite (micrograph taken by B. de Vito); (b) scanning electron micrograph of a tuff sample from Kavissos, Thrace, Greece, showing interlaced fiber-shaped crystals of mordenite (micrograph taken by B. de Vito); (c) scanning electron micrograph of a tuff sample from Chiaiano, near Naples, Italy, showing rod-shaped crystals of phillipsite (micrograph taken by M. Colella); (d) scanning electron micrograph of a tuff sample from S. Nicola La Strada, Caserta, Italy, showing pseudo-cubic crystals of chabazite (micrograph taken by M. Colella); (e) scanning electron micrograph of a tuff sample from Agua Prieta, Sonora, Mexico, showing bundles of needle-shaped erionite crystals (micrograph taken by M. Colella).

4.2.10.3
Applications of Sedimentary Zeolites in Adsorption Processes and Catalysis

Natural zeolite-bearing materials are generally less suitable than the synthetic zeolites for use in adsorption and catalysis. The reasons for this are several. From a technical point of view, the natural zeolites are disfavored mainly because of their small-port structure and lower adsorption capacity compared to the most "open" synthetic zeolites. Also, the presence of impurities in the parent rocks is often unacceptable, especially in catalysis. From the economic side, the lower cost of the natural zeolites is a profitable circumstance only in applications in which *ceteris paribus*, a massive use of zeolites is expected, such as the applications based on ion exchange and in the building industry, which will be dealt with in the next sections.

Nevertheless, there are some actual or perspective uses of sedimentary zeolites that are worthwhile to be mentioned and will be summarized in the following.

4.2.10.3.1 Selective Adsorption

Given the remarkable selectivity of zeolites for water vapor, there is no doubt that gas drying is the main application of natural zeolite-bearing rocks. It is apparent from the data of Table 2 that chabazite and mordenite, depending on the values of their void volume or window size, are the zeolites that present on the average the most useful structural features for such a type of application. In fact, rocks containing each of the two former zeolites are reported to be used as drying media of sour natural gas and other artificial gas streams [53]. But clinoptilolite-rich materials have also been proposed for adsorption processes, for instance they are widely used in Bulgaria in pressure-swing adsorption systems for drying air [54]. In addition, sedimentary zeolite-rich rocks are often employed simply as desiccants in place of silica gel, because of their higher adsorbing capacity at very low water vapor partial pressures (70 Pa or less) (see, e.g., ref. [55]). Examples of these applications are freon drying in the on-line water vapor traps of refrigeration circuits or air drying in the frames of the double window-panes.

Water desorption-adsorption cycles are involved also in the processes of storage in zeolitic tuffs of thermal energy coming from discontinuously running sources, such as the Sun [56]. Heat is stored during water desorption and released by driving moist air through the thermally activated adsorbent bed. The result is a warm, relatively dry, air supply useful for drying food-stuffs or fruit [57]. Adsorption-desorption cycles in sedimentary zeolites, such as chabazite- or clinoptilolite-rich tuffs, provide also heat pump systems, which have been proposed in air-conditioning and in refrigeration. As an example, effective solar refrigerators and other heating/cooling devices, based on a zeolite solar-cooling system (Fig. 4), have been produced and sold for some time by the Zeopower Co., Natick, Mass., USA [58].

Sedimentary zeolites are occasionally used (or have been proposed for use) also for removing gaseous pollutants from various gas emissions, for instance chabazite-rich materials are marketed for removing HCl from reformed H_2 streams [59], CO_2 from stack gas of oil- and coal-burning power plants emissions [59], H_2O, H_2S and CO_2 from low-grade natural gas or biogas in order to improve their caloric value and to make possible their injection in pipelines [60]. Several natural zeolite-bearing materials, namely almost all those referred to in Table 2, have been demonstrated to be highly selective for common pollutants such as SO_2 and NH_3 [61–63] and to be presumably very effective in the removal of these gases from industrial or agricultural gaseous mixtures.

4.2.10.3.2 Gas Separation

Mordenite-rich materials have been used in Japan since the end of the 1960s for the production of high-grade O_2 from air in generators operating by the pressure-

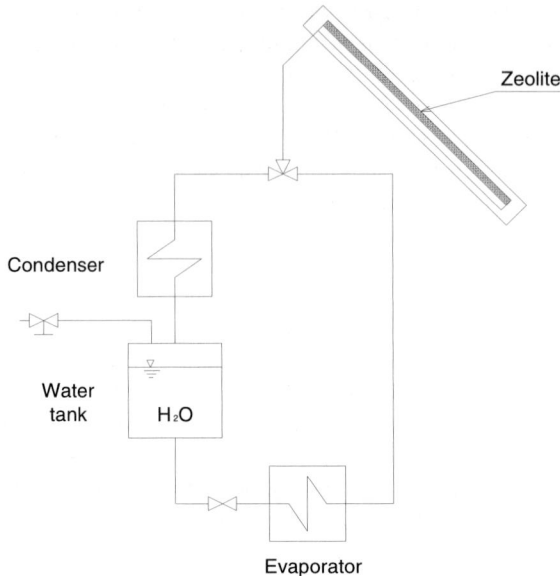

Fig. 4. Schematic diagram of a zeolite solar-cooling system. During the daytime part of the cycle (left side of the figure) the zeolite is heated by solar heat and water vapor is desorbed. Increasing vapor pressure in the container results in condensation of water, which is collected in the storage tank. During the night-time part of the cycle (bottom and right side of the figure) zeolite is cool enough to adsorb water, forcing liquid water to evaporate. The evaporation heat is subtracted from the neighboring space, which is cooled, whereas the adsorption heat is released into the atmosphere.

swing adsorption method [64]. This separation, which allows O_2-enriched air to be obtained for use in hospitals, in fish breeding and transportation, is based on the selectivity demonstrated by the dehydrated mordenite for N_2 over O_2, due to the nitrogen quadrupole moment. Also chabazite and clinoptilolite, pre-exchanged in various cationic forms, have been recognized as effective adsorbents to perform the oxygen-nitrogen separation [65, 66] and clinoptilolite-rich tuffs have been successfully tested in a pilot plant in Bulgaria [54]. Another interesting separation process, which is, however, still at the level of laboratory scale, is the water-ethanol vapor separation by phillipsite-rich volcanic tuffs [67]. This would be an easy and inexpensive procedure to enrich in ethanol the diluted solutions obtained through fermentation of glucose syrups.

4.2.10.3.3 Catalysis

No important uses of natural zeolites in catalysis are reported, apart from some minor applications, such as that developed by Mobil Oil Corporation, which is said to use a minor tonnage of high-grade erionite-clinoptilolite material from their deposit in Jersey Valley, Nevada, as a selective-forming catalyst [68]. Nevertheless, dozens of papers have been published on this subject over years. Hungarian scientists in particular have a long tradition of activity in this field, using either clinoptilolite or mordenite for several different reactions (see, e.g., ref. [69]). Recently, a very effective Cd-clinoptilolite-based catalyst has been proposed for the hydration of acetylene and homologues to acetaldehyde and derived products [70].

To reduce air pollution, various reactions catalyzed by natural zeolites have successfully been tested, for instance CO oxidation [71], NO disproportionation into N_2O and N_2O_3 [72] and H_2S oxidation to S [73] on chabazite-rich materials and NO selective reduction to N_2 with NH_3 on Cu-exchanged mordenite-rich tuff [74, 75].

4.2.10.4
Applications of Sedimentary Zeolites Based on Ion Exchange

Most present and potential applications of natural zeolites are based on their cation exchange properties [76, 77]. This is due to the fact that these applications often need less sophisticated technologies (or no technology at all) and can employ materials with a substantial presence of impurities and variable zeolite contents. It is to be noted, however, that in some cases the use of natural zeolites is connected to specific features, as an example, cation selectivity in wastewater purification, which are absent in the marketed synthetic species.

4.2.10.4.1 Wastewater Purification
The removal of cationic pollutants from wastes of industrial or municipal origin is one of the oldest and still one of the most promising fields of application of zeolitic tuffs. Specific favorable features of ion exchange compared with the usual chemical and biological procedures are: (a) it does not modify substantially water chemistry, (b) it allows the water purity requirements to be met regularly, because it works satisfactorily in any condition, also when the concentration of the contaminant is low, and (c) if performed in column operations it enables, upon regeneration of the zeolite bed, to concentrate the pollutant in a volume reduced by a factor of 10^3–10^4.

Following the extended work performed in the 1960s by Ames (a part of the relevant bibliography can be found in refs [76–78]) applications of clinoptilolite-rich rocks in ammonium removal from municipal wastewaters or Cs and Sr radioactive nuclides removal from nuclear power plants wastewaters were devised [78].

Removal of ammonium is commonly carried out in fixed-bed plants, in which the packed tuff grains are passed through by the sewage, allowing the purified solution to be collected. When exhausted, the beds are regenerated by elution with a concentrated NaCl solution or mixed $NaCl + CaCl_2$ solution. This process is being used in several plants, mostly in the United States, for example in Tahoe-Truckee Sewage District, California (amount of treated sewage: 22,500 m^3 per day) and in the Rosemont, Minnesota, Experimental Wastewater Treatment Plant (2250 m^3 per day) [79, 80]. Similar plants, but smaller as regards the volume treated per day (from some tens of m^3 per day to several hundred m^3 per day) have been reported

References see page 1183

to operate in Japan, for instance in Toba, Mie Prefecture (80 m^3 per day) [81]. Some other plants, still smaller, are working also in Europe, for instance in Vae, near Budapest (Hungary) (50 m^3 per day) [82]. The spent regenerant, after suitable alkalinization, is stripped to recover ammonia, which is absorbed in a sulfuric acid solution to give a concentrated $(NH_4)_2SO_4$ solution usable as fertilizer. An improved procedure to recover in the solution after regeneration both nutrients present in sewage, namely ammonium and phosphate, has been introduced with the so-called RIM-NUT process, which has been favorably tested in a 240 m^3 per day demonstration plant, operated in 1982 at the West Bari (Italy) Wastewater Treatment Plant and in 1987 in South Lyon, Michigan, U.S.A, under the auspices and with the support of the U.S. Environmental Protection Agency. In this process, after removing ammonium with a clinoptilolite-rich tuff bed and phosphate with a bed of an anionic resin, both pollutants are precipitated and recovered as $MgNH_4PO_4 \cdot 6H_2O$ [79] (Fig. 5).

Similar processes are used in countless small aquacultural plants to remove ammonium produced by fish metabolism [83]. Clinoptilolite-rich rocks have also been proposed for removing ammonia from drinking water, as demonstrated in a long-lasted experiment with a pilot-plant (50 m^3 per day) [84]. An integrated two-stage process for treatment of municipal wastewaters, based on clinoptilolite-rich tuffs, designed to allow both sedimentation of suspended solids and removal

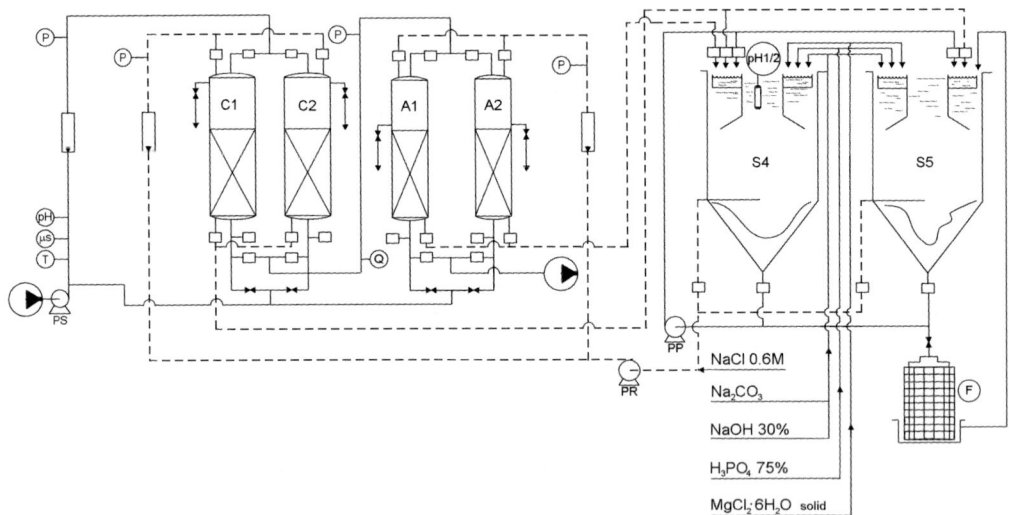

Fig. 5. Flow-sheet of the RIM-NUT demonstrative plant (redesigned from ref. [79]), composed of two clinoptilolite columns (C1, C2) and two Kastel A/510 anionic resin columns (A1, A2). Municipal wastewater, containing ammonium and phosphate ions, is passed through one of the two pairs of columns (C1-A1 or C2-A2), while the other pair is regenerated by the NaCl solution. Precipitation of $MgNH_4PO_4 \cdot 6H_2O$ is accomplished in the settler-thickeners (S4, S5) adding Na_2CO_3, NaOH, H_3PO_4 and $MgCl_2 \cdot 6H_2O$ to the spent regenerant. F = filtration apparatus; continuous line = service; dashed line = regeneration.

of ammonium is at pilot plant stage in Hungary. Ammonium, recovered as $NH_4H_2PO_4$, together with the sludges collected in the first stage, are planned to be used as fertilizers [85]. A further improved process, using an additive, composed of clinoptilolite-rich tuff, ferrous sulfate and limestone, to remove phosphate from sewage, has recently been designed [86]. Further, the use of Ca-exchanged clinoptilolite-rich tuff for ammonium removal from NASA's advanced life-support wastewater system has been proposed [87].

Phillipsite and chabazite exhibit remarkable selectivity for ammonium [88, 89] and may be utilized in place of clinoptilolite for municipal wastewater deammoniation, as also demonstrated by laboratory fixed-bed tests [90]. Possible use of phillipsite- (and/or chabazite-) rich tuffs for removing ammonium from industrial wastewaters is dependent on the presence of K^+, which is exchanged into both zeolites comparably well as or even better than ammonium [91].

Most natural zeolites exhibit a good selectivity for cesium and a fair selectivity for strontium and other radioactive cations [76]. This prompted a number of applications of zeolite-rich tuffs for the decontamination of radioactive wastes. Tuffs, charged by radioactive nuclides, are usually stored directly or after incorporation into concrete for long-term burial. A few examples of applications in this sector are reported in the following; major information and details may be found in refs [80, 92].

In the frame of the Hanford Atomic Energy Project, Richland, Washington, in order to avoid, after solidification of the waste, the excessive heat generation in the underground storage tanks, due to ^{137}Cs fission, a bed of chabazite-rich material was used to remove this cation from a liquid waste having a very concentrated Na^+ interfering matrix. The exhausted bed was regenerated using a concentrated $(NH_4)_2SO_4$ solution, from which either NH_4^+ was recovered by distillation or Cs^+ under the form of nitrate [78].

A low-level radioactive wastewater from an irradiated fuel-storage basin, containing ^{90}Sr and ^{137}Cs, was treated for several years with a local clinoptilolite-rich tuff at the Idaho National Engineering Laboratory, Idaho Falls, Idaho [78]. Interesting results have also been obtained in a bench-scale test, which employed several natural zeolites (clinoptilolite, mordenite, chabazite, erionite, ferrierite) for the treatment of process wastewaters from Oak Ridge National Laboratory, Oak Ridge, Tennessee. Chabazite, for instance, was able to remove amounts of Cs and Sr at the level of ppb from waters containing interfering cations at a concentration of 10^{-3} M [93]. On the basis of equilibrium data, application of phillipsite for the removal of cesium and strontium from nuclear power plants wastewaters must be considered feasible [94].

Natural zeolites have also helped in decontaminating waters and soils in the occasion of nuclear accidents. Mixed zeolite beds of a chabazite-bearing tuff and synthetic zeolite A were used for treating waters at Three Mile Island, Pennsylvania, in 1979 [95], whereas for similar treatments a local clinoptilolite-rich material was preferred in the occasion of the Chernobyl, Ukraine, accident in 1986 [96].

References see page 1183

Mordenite- and clinoptilolite-rich tuffs have been proposed also as buffer materials for stopping or retarding the migration and dissemination of radioactive nuclides in soil [97]. For analogous reasons, tuff deposits at Yucca Mountain, Nevada, containing clinoptilolite and mordenite, are considered potential repositories for high-level radioactive wastes [98].

The removal of heavy metals from wastewaters by natural zeolites, although potentially advantageous, given the good cation-exchange selectivity and kinetics of most sedimentary zeolites for some of them, for instance lead, [76, 80], is still at level of laboratory or pre-pilot plant scale. The failure in passing from basic science to technology is mainly due to the compositional complexity of many industrial liquid wastes, which results in a series of undesired interfering reactions and therefore in low process performance.

Prospects of application in fixed bed plants are therefore restricted to "ideal" wastes, containing a limited number of hazardous cations, for instance, to lead removal from wastewaters of some ceramic factories, having a rather simple interfering cation matrix, very similar to tap water. Laboratory-scale fixed-bed tests showed excellent dynamic selectivity and overall efficiency for lead removal by phillipsite-rich tuff (Fig. 6) and moderate to good values of dynamic parameters in the case of chabazite-rich tuff [99, 100]. This opens the way to the possible use of Neapolitan yellow tuff (see Sect. 4.2.10.2.2) for such a type of process.

Ion exchange may be devised also as a discontinuous process (addition of zeolite to liquid waste). This is particularly suitable when the exchanger is inexpensive,

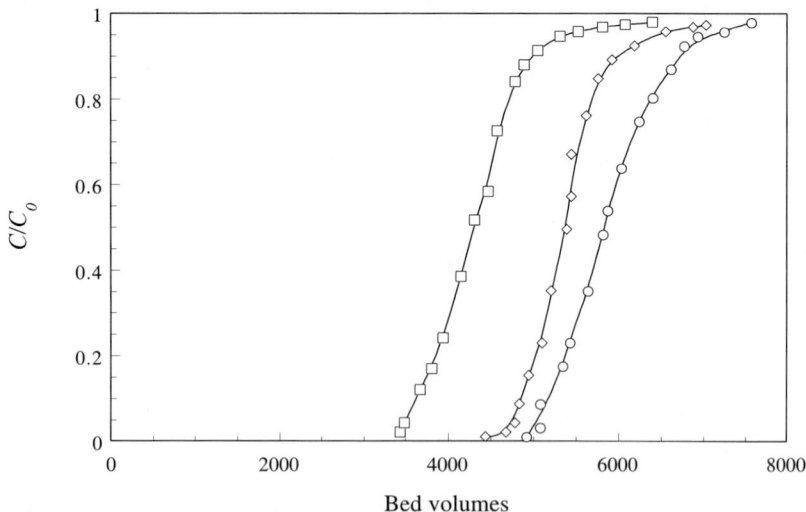

Fig. 6. Lead breakthrough curves obtained eluting a Na-exchanged tuff bed, containing roughly 61 % phillipsite, from Chiaiano (Naples, Italy) with 92 mg l^{-1} Pb^{2+} (C_0) solutions with no interfering cations (circles), and with 500 mg l^{-1} (rhombs) or 1000 mg l^{-1} (squares) Na^+ as interfering cation. Feed flowrate: 0.42 l h^{-1}. Calculated dynamic selectivity was 1, 0.86, and 0.71, respectively. Overall efficiency was 0.85, 0.74, and 0.58, respectively (data from Ref. [99]).

namely in the case of zeolite-rich rocks. Adding the powdered zeolite to wastewater gives the possibility to modulate the solid-to-liquid ratio to the specific process requirement: as an example, increasing such ratio enables removal from water of polluting cations for which the specific zeolite does not exhibit a high selectivity. Obviously, this practice does not eliminate pollution, but it simply transfers the problem from a disperse phase (liquid) to a dense phase (solid). To avoid the disposal in an expensive sanitary landfill, the exhausted material, namely, the pollutant-loaded zeolitic sludge, may be solidified-stabilized safely in a cement matrix, taking advantage of the excellent pozzolanic activity displayed by zeolitic tuffs (see Sect. 4.2.10.5.3). This method appears particularly worthy for radioactive nuclides (see above), but successful results have been obtained also in disposing of other polluting cations, such as Cd^{2+}, using phillipsite- and chabazite-bearing materials [101]. Given the low selectivity of the latter zeolites for cadmium [102], performance during the ion-exchange stage is very low, but this is counterbalanced by the low price of the material and the advantage of a possible reuse. For instance, to bring Cd concentration below the limit allowed by Italian law (0.02 mg l^{-1}), starting from a model tap water containing 10 mg l^{-1} Cd required not less than 50 g of powdered tuff, containing roughly 50 % zeolite, per liter of solution, which means that not more than a few tenths of 1 % of the total exchange capability was utilized (Fig. 7). But the cement-based products, obtained by hardening of the Cd-containing sludges/Portland clinker blends, showed, apart from satisfying leaching performances, values of compressive strengths comparable with those shown by commercial blended cements [101].

More information on the use of natural zeolites as cation exchangers in environmentally friendly applications can be found in two reviews having appeared recently in the literature [49, 103].

4.2.10.4.2 Soil Amendment and Soilless Substrate Preparation

The unconscious utilization of natural zeolitic materials as soil amendants dates back some millennia. This is proved by the practice, we have inherited from our ancestors, to grow specific crops in zeolite-rich soils, such as the cultivation of some types of vines in phillipsite- and/or chabazite-rich soils in the Campania Region (southern Italy).

The employment of sedimentary zeolites as additive in agriculture is, on the contrary, a rather recent achievement. Experiments in this fields began in the 1960s in Japan and then spread rapidly to many areas of the world, so much so that the use of natural zeolites in agronomy and horticulture must be considered today one of the most common uses, although mostly empirical and of local significance. The state-of-the-art of this type of application of natural zeolites was made in 1982, in Rochester, New York, where a specific conference was held on the use of natural zeolites in agriculture and aquaculture [104]. The updating of the research in this field may be found in the proceedings of the subsequent international conferences on natural zeolites [105–108].

References see page 1183

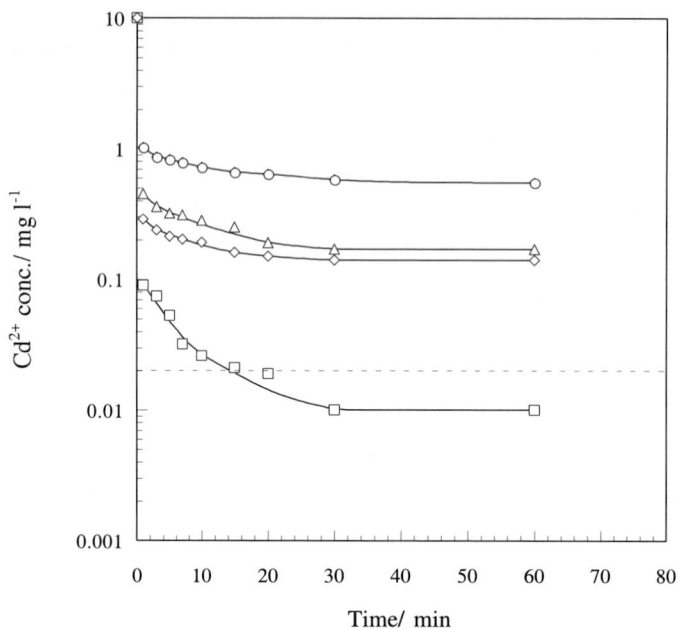

Fig. 7. Cadmium uptake curves from a model tap water containing 10 mg l^{-1} Cd^{2+} by addition of 10 g (circles), 20 g (triangles), 30 g (rhombs) and 50 g (squares) of phillipsite- and chabazite-rich tuff, containing roughly 50 % zeolite. Ordinates in logarithmic scale; dashed line = limiting value of Cd^{2+} concentration at the discharge (data from Ref. [101]).

The overall research carried out to date, focused mainly on clinoptilolite-rich materials, showed that the use of natural zeolites as a soil amendant in agriculture is beneficial for the growth of a large variety of plants, including spinach, radish, sugar beet, tomato, maize, rice, potato and cucumber (see references in ref. [103]). The reasons for this positive effect are not completely explained. Zeolites, when rich in potassium and/or ammonium (introduced by ion exchange), may be considered as slow-release fertilizers, but only if associated to organic amendment. More generally, it may be assumed that zeolites enhance the plant conditions in the soil environment by increasing the global soil fertility, namely water retention and cation exchange properties, particularly in the neighborhood of the rizosphere. Zeolites, in fact, as rigid porous media, can absorb and retain large amounts of water, becoming a sort of reservoir of available water. In addition, they increase the global cation exchange capacity of the soil, which normally does not exceed 0.3 meq g^{-1}, with the exception of some organic soils, which may exhibit CEC values > 1 meq g^{-1} [109]. This results in a series of advantages, such as (a) capture by ion exchange and retention of high amounts of nutrients, for instance ammonium produced by biochemical reactions, which are then made available slowly; (b) withdrawal from the vital cycle of plants of toxic elements, such as Al, Mn, Pb, Cd and others; (c) buffering action against the trend of soils to become acid by depletion and/or leaching of alkali. Possible negative side-effects, due to indiscriminate use of natural zeolites, may be: (a') excessive sodium release to soil, which may result in a series of troubles [103]; (b') withdrawal by ion exchange of a series of beneficial micro- or macro-elements, which become no longer available for plant uptake.

Natural zeolites, in connection with their absorption/exchange properties, can be used also in the remediation of soils, either in the case of incidental environmental pollution (for instance, after the Chernobyl nuclear accident [96]), or to control contaminants added to soil along with sewage sludges [110, 111]. Moreover, they could also be utilized as agrochemical carriers for various herbicides, fungicides, insecticides and growth regulators, enhancing their effectiveness and reducing their mobility, therefore reducing the environmental hazard [112].

Of remarkable interest is also the use of natural zeolites for the preparation of soilless substrates (zeoponic plant-growth systems) [110, 111]. This practice, introduced in Bulgaria in the early 1980s and then developed especially in Cuba and in the United States, allows the cultivation of plants in any artificial substrate as a growth medium, in which zeolite minerals constitute an important component. Contrary to the early systems, in which periodic treatment with fertilizers or nutrient solutions were devised, recent research has focused on developing synthetic substrates, such as NH_4- and K-exchanged clinoptilolite and either natural or synthetic apatite, which supply a balanced diet of plant nutrients for many production cycles without fertilizer addition. Promising results have been obtained with wheat; preliminary but interesting results in space applications of this technique (growth of plants during space missions) have been collected, using various crops, such as dwarf wheat, turnip, radish and cabbage [111].

4.2.10.5
Applications of Sedimentary Zeolites as Building Materials

As mentioned in the historical part (Sect. 4.2.10.1.1), the use of zeolitic tuff in construction has been one of the early applications of natural zeolites. This is also the most widespread use today, considering that a prudent estimate of the present worldwide tuff consumption in the building industry amounts to some ten million tons per year [1].

In the following, a brief account of the applications of the sedimentary zeolites as dimension stone, in manufacturing of lightweight aggregates and foamed materials, and as pozzolanic material is given, referring to a recent publication for a more extended treatment of the subject [1].

4.2.10.5.1 Dimension Stones
The useful properties of the zeolitized tuffs as construction material have been appreciated since ancient times. These rocks are, in fact, reasonably resistant to

References see page 1183

Tab. 3. Physical and mechanical properties of zeolitic tuffs, compared with ordinary bricks and other dimension stones (data from [1]).

Material	Zeolites	γ/kN m^{-3}	P	σ/MPa
Italian tuffs[1]	Phi and/or Cha	10–16	37–61	1–13
Rhenish tuff *(trass)*, Germany	Cha/Phi or Cha	13–17	31–43	5–20
Oya–ishi tuff, Japan[2]	Cli and Mor	16	n.d.	8
Transylvanian tuff, Romania	Cli	16–23	26–31	2–4
Thracian tuff, Greece[2]	Cli	16	36	26
Clay brick	–	15–24	8–42	10–45
Basalt	–	27–30	1–3	196–392
Limestone	–	23–26	5–15	49–147
Sandstone	–	18–26	4–20	39–127

[1] Phillipsite- and/or chabazite-rich tuffs from various Italian deposits.
[2] Data refer only to one sample. Legend: P = % porosity; γ = unit weight; σ = uniaxial compressive strength, Phi = phillipsite; Cha = chabazite; Cli = clinoptilolite; Mor = mordenite; n.d. = not determined.

erosion, but also sufficiently soft to be cut for preparing "natural" bricks. In addition, they show physical and mechanical properties suitable for use as dimension stones, as can be deduced from the data of Table 3.

Compared with other building materials, zeolitized tuffs exhibit on the average lower unit weight (γ), higher porosity (P) and poorer mechanical properties (σ), which accounts for their inferior durability.

Zeolitic tuffs are also excellent materials for controlling temperature and humidity ranges. Depending on their ability to adsorb and desorb water molecules reversibly as a function of temperature and/or humidity, zeolites, during diurnal heating/cooling cycles, can store/release both sensible heat and heat of adsorption. Masonry made with zeolitic tuffs keeps the environment cool during the day, when the zeolite stores heat by desorbing water molecules, and warm at night, when the zeolite releases heat by readsorbing water vapor. The action of thermal regulation is therefore coupled with that of regulation of the environmental humidity. This favorable thermal behavior is one of the reasons for the historic practice to excavate caves, burial chambers, refuges, stables, cellars in tuff or to use it as building stone.

Table 4 reports some specific thermal properties of the zeolitic tuffs, compared with other building materials. Note that, probably due to the high porosity and the presence of zeolitic water, zeolitic tuff is a better insulating material than other common building materials. In addition, due to water adsorption phenomena and the associated heat of adsorption, it presents a greater thermal storage capacity than the other nonadsorbent building materials.

The attractive physical and thermal properties of zeolitic tuffs are the main reason why these materials are still utilized today in Italy for masonry and the external walls of houses, even though new building technologies (structures in concrete)

Tab. 4. Thermal properties of some zeolited tuffs at room temperature, compared with other building materials (data from [1]).

Material	Zeolites	$k/W\ m^{-1}\ K^{-1}$	$c_p/kJ\ kg^{-1}\ K^{-1}$
Campanian ignimbrite, Italy	Cha + Phi	0.32	1.09
Neapolitan yellow tuff, Italy	Phi + Cha	0.32	–
La Rioja tuff, Argentina	Cli	0.38	–
Clay brick	–	0.69	0.84
Concrete	–	0.93	0.65
Granite	–	1.70–3.98	0.84
Limestone	–	0.69	0.92
Marble	–	2.08–2.94	0.88

Legend: k = thermal conductivity; c_p = specific heat; Phi = phillipsite; Cha = chabazite; Cli = clinoptilolite.

and new building materials (brick made with cement and lightweight aggregates) are progressively reducing this use. As regards other countries where deposits of zeolitic tuff are present, the use of this material as dimension stone is not as common as in Italy, but a remarkable tradition of use is present especially in Germany, where more than 200 monumental buildings were built in the past centuries using bare blocks of tuff [113], and to a lesser extent, in Bulgaria, Cuba, Greece, Hungary, Japan, Mexico, Romania, and Turkey.

4.2.10.5.2 Lightweight Aggregates and Foamed Materials

Sedimentary zeolites can also be used to manufacture lightweight materials for the building industry.

One possibility is to obtain low-density pellets having unit weights γ only one third that of the parent tuff stone and, in addition, depending on their porosity, exhibiting good sound-proofing and heat insulation. In fact, similarly to other hydrated materials, such as perlite, zeolite tuff grains expand on heating. The temperature required for obtaining expanded zeolitic tuffs is usually higher than that needed for other expandable materials (1500 K or above compared with about 1100 K, respectively), but the expanded zeolite products are reported to be stronger and more resistant to abrasion [92].

Most expanded materials have been prepared from clinoptilolite- [114–116] or mordenite-rich rocks [81], but recent promising results have been collected also with chabazite- and phillipsite-rich tuffs [117]. Reported unit weights range between 7–10 kN m^{-3}, porosities are well over 50 %. All these materials have been used as lightweight aggregates in mortars and concretes for manufacturing precompacted building blocks.

Zeolite tuffs can be used also as foaming agents for manufacturing cellular materials. Zeolites, in fact, can store large amounts of thermal energy upon dehydra-

References see page 1183

tion at temperatures of about 700 K or greater. When mixed with water and cement or similar materials (lime, gypsum, dolomite plaster), dehydrated zeolite powders generate large amounts of heated air through water adsorption, with resultant foaming and volume expansion. A lightweight material with elevated porosity is this way obtained upon hardening.

Clinoptilolite-rich rocks have usually been utilized in these processes. The obtained foamed materials presented unit weights ranging between 4 and 10 kN m^{-3} and compressive strengths in the 2–10 MPa range [118–120].

Recently, a thermal-insulating material named *Sibeerfoam* has been developed in Siberia by heating a powdered clinoptilolite-rich or heulandite-rich rock in the range 1400–1700 K in the presence of a gas-producing additive (silicon carbide). Sibeerfoam, which is prepared in the form of blocks or granules, is characterized by uniform closed porosity (0.5–5 mm), low unit weight (2–9 kN m^{-3}), high compressive strength (up to 18 MPa) and good thermal insulation (thermal conductivity ranging between 0.06 and 0.21 W m^{-1} K^{-1}) [121].

4.2.10.5.3 Pozzolana-like Materials

Pozzolana is an unconsolidated, mostly glassy, rock, originated from the rapid quenching of fluid lava shreds projected in air during a volcanic eruption. This product has been demonstrated to be the starting material, from which, especially in central-southern Italy, extended deposits of natural zeolites were generated through diagenesis [40].

Pozzolana and some other X-ray amorphous silica- and alumina-rich materials, such as fly ash (pulverized fuel ash), microsilica (condensed silica fume), heat-treated clay (also crushed bricks or tiles), zeolitic tuffs, diatomaceous earth and some vitreous rhyolites, can react with lime forming hydraulic compounds, namely products such as hydrated calcium silicates (*tobermorites*) and aluminates, which are able to harden in both aerial and aqueous environments [122]. This is why all these materials are called "pozzolanic materials" and the property to react with lime is called "pozzolanic activity" [122, 123].

Pozzolana and, in minor extent, crushed bricks have been used since ancient times (especially by Romans) in mixture with lime to prepare hydraulic mortars, which have demonstrated to be so durable that many Roman civil works are still preserved nowadays [7, 8].

The Roman cements are no longer prepared today, but in their place, since the early decades of the 20th century, a class of composite cements called "pozzolanic cements" are manufactured, in which the basic Portland clinker is partly replaced by a pozzolanic material. The use of pozzolanic cements involves some advantages, such as improvement of resistance to lime leaching by flowing waters and to expansion due to sulfate attack [124]. Further advantages are the decreased overall energy required in manufacturing the cement and the slower and decreased heat evolution during its hydration [125], that is beneficial in the setting-up of massive structures [124].

Shortage or lack of pozzolanas results in the possible use of zeolitic tuffs [126]. Natural zeolitized materials have in fact been proposed (and are frequently used)

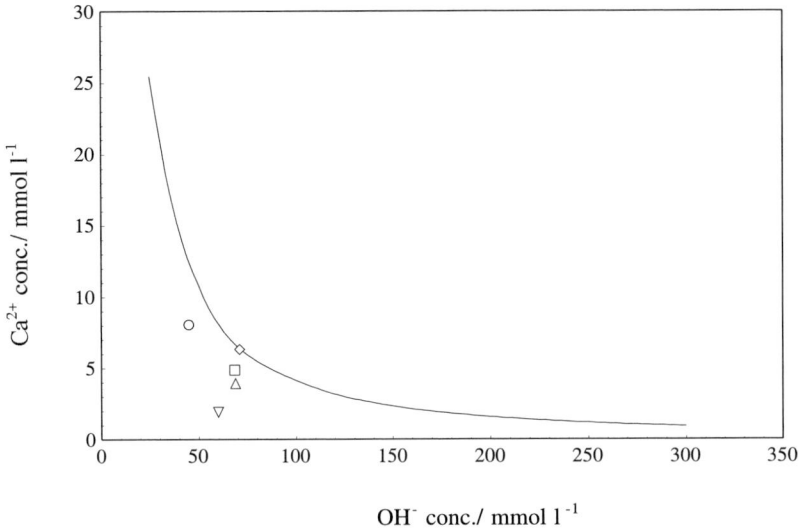

Fig. 8. Fratini's pozzolanicity test. Point on the Ca(OH)$_2$ solubility curve is representative of a saturated solution and is proof of the lack of pozzolanic activity. Points under the curve are representative of undersaturated solutions and therefore of a successful pozzolanic action. Rhomb = Portland cement; circle = blend of Portland cement/clinoptilolite-rich tuff from Pentalofos (Greece); square = blend of Portland cement/pozzolana from Baia (Naples, Italy); triangle = blend of Portland cement/Neapolitan yellow tuff from Chiaiano (Naples, Italy); upside-down triangle = blend of Portland cement/Campanian Ignimbrite from S. Nicola La Strada (Caserta, Italy).

as pozzolanic material in many countries, such as Bulgaria, China, Cuba, Germany, Greece, Italy, Jordan, Russia, Turkey, United States, and Yugoslavia [1].

One of the methods to evaluate the pozzolanic activity of the blended cements is Fratini's test, which is recognized in the European Standards, although it was devised some fifty years ago [127]. This test consists in the estimation of the amount of Ca(OH)$_2$ leached from a blended cement put into contact with deionized water. Experimental results are projected in a plot reporting the solubility curve of Ca(OH)$_2$, as a function of Ca^{2+} and OH$^-$ concentrations in solution. Points above the curve or on the curve are representative of oversaturated or saturated solutions, respectively, and are the proof of the lack of pozzolanic activity. On the contrary, points below the curve are representative of undersaturated solutions and therefore of the successful pozzolanic action played by the specific mineral addition to the ordinary Portland cement (Fig. 8).

In recent years much research has been carried out on the utilization of zeolite-bearing tuffs, in place of true pozzolanas, for manufacturing blended cements (see, e.g., [128–131]). Zeolite types that have been tested so far are those more frequently found in the sedimentary zeolite (tuff) deposits, namely, clinoptilolite, mordenite, phillipsite and chabazite [103]. It has been ascertained that:

References see page 1183

- zeolites are excellent pozzolanic materials, which often behave better than pozzolana itself [124] or a glass of identical composition [132] in reducing the lime coming from the clinker hydrolysis; zeolite reactivity is connected to its large external specific surface and metastability, which favors its dissolution into the saturated lime solution and the successive precipitation of hydrated calcium silicate and hydrated calcium aluminate phases [133];
- the hardened cement mortars containing zeolitized tuffs present, particularly at later ages, compressive strengths that are comparable with or greater than those of the mortars prepared using Portland cement or blended cements in which pozzolana or other pozzolanic materials have been used [126, 134, 135];
- replacement of Portland clinker for zeolitic tuff reduces workability [126] and increases water demand [136], but this effect is of little significance because it may be overcome by using a superplasticizer [137];
- replacement of Portland clinker for zeolitic tuff reduces the risk of alkali-silica reaction, which results in undesired expansion and cracking of concrete [125].

4.2.10.6
Applications of Sedimentary Zeolites in Life Sciences

The last frontier in the application of natural zeolites is the field of life sciences. Zeolitic tuffs, in fact, have been used for some thirty five years in animal science as dietary supplements in animal husbandry; more recently their usage has been extended to biology and medicine. The function played by zeolite in all these applications is in most cases not completely explained, although it is likely that both ion-exchange and adsorption properties are involved in the action mechanisms.

4.2.10.6.1 **Animal Science and Nutrition**
The use of sedimentary zeolites (namely, clinoptilolite- and/or mordenite-rich tuffs) as active additions to animal diets started in Japan in the middle of the 1960s [138], in imitation of the successful use of montmorillonite in slowing down the passage of nutrients in the digestive system of chicken with resultant improvement of the caloric efficiency [139]. Since then the applications of natural zeolites in this field have multiplied greatly, so much so that this use is nowadays consolidated everywhere.

The first ten years of research in animal husbandry have been reviewed by Mumpton and Fishman [140]. Substantial updating appeared in the proceedings of the subsequent international conferences on natural zeolites [104–108].

Many animal species, especially swine, ruminants and poultry, have been tested. Performances were in general positive in that two goals were constantly achieved: (1) improvement of the general health conditions with reduced incidence of diseases (e.g., diarrhea) and (2) moderate to good weight performance in terms of feed efficiency, that is weight gain/feed intake. Depending on the animal species, there are many possible physiological explanations for these effects, only some of which have been completely understood:

- the slowing down of the nutrients transport in the gastrointestinal tract allows a better assimilation and improves feed efficiency;
- ammonium withdrawal from the circulation, due to the good cation exchange selectivity of the natural zeolites for this cation (see Sect. 4.2.10.4.1), allows a better utilization of nitrogen by the animal, which results in a better conversion of feed-stuff nitrogen to animal protein;
- ammonium withdrawal is beneficial also for the control of the ammonia levels in blood;
- zeolite exerts a protection from tissue accumulation of toxic elements;
- zeolite particles might stimulate the lining of the stomach and intestinal tract, causing the animal to produce more antibodies and thereby to inhibit diseases such as scours.

4.2.10.6.2 Pharmaceutics

The first experiences on the use of natural zeolites as drugs date back some thirty years when tests were carried out in several laboratories to evaluate these materials as antacids, on the basis of their well-known alkaline nature and the supposed ability to act as neutralizing agent in hyperacidity and its associated gastric disturbances. At present clinoptilolite-rich materials are used is several countries (e.g., Cuba, Bulgaria) for preparing drugs having neutralizing properties, also in association with other gastric alkalinizants, such as Na_2CO_3 [141], or associated to aspirin to reduce the well-known gastrointestinal side effects that this drug causes in many patients [142].

Another recognized effect of natural zeolites is the antidiarrheic action, already referred to in the section dealing with animal nutrition (Sect. 4.2.10.6.1). It has been proved that clinoptilolite-rich materials are very effective in preventing diarrhea in animals and humans, because of their adsorption capability for many endogenous and exogenous substances, responsible for this and other diseases, such as bile acids [143], mycotoxins [144, 145], glucose [146]. Antidiarrheic drugs based on the above zeolitic material have already passed the clinical tests and are now on the market, for instance in Cuba.

Clinoptilolite-rich materials, pre-exchanged in various cationic forms, are claimed to be effective also as antihyperglycemic agents, because of the above-mentioned good adsorption selectivity for glucose, and as hypocholesterolemic agents, because of their effective action towards the bilious acids [147].

Natural zeolites in various cationic forms have been proposed also as effective bactericides for a wide spectrum of microorganisms. In this case zeolite does not appear to take part directly in the biocidal activity; it works as a carrier of bactericidal cations, for instance Ag^+, Cu^{2+} and Zn^{2+}, which are gradually released in the right amount that is necessary to play their action, avoiding useless or poisonous excesses [147, 148].

Considering the excellent results collected with synthetic NH_4^+-selective zeolites

References see page 1183

[149], natural zeolites, such as clinoptilolite, phillipsite and/or chabazite, are candidates to be used for the preparation of microcapsules of urease-zeolite useful to control the hematic urea levels. Urea is decomposed by the enzyme in ammonium and carbonate ions and the former ions are removed by zeolite through ion exchange.

4.2.10.6.3 Medical Applications

Sedimentary zeolites are occasionally used in filter tips of cigarettes (for instance in South Korea) to partly sorb some components of the tobacco smoke, likely nicotine and various carcinogenic agents, reducing the amount of toxic substances inhaled by smokers.

External applications of natural zeolite powders have been found to be effective to decrease the healing time of wounds and surgical incisions [150].

Patents have been granted on the possible use of NH_4^+-selective natural zeolites as filter media to remove ammonium from the dyalisate of kidney patients, allowing a recycle of the cleansed solution and the use of a portable dyalisis system [151].

4.2.10.7
Miscellaneous Uses of Sedimentary Zeolites

Several thousand tons per month of white clinoptilolite-rich tuff are marketed by Japanese companies as paper filler in place of kaolin [81]. The obtained paper is said to be bulkier, more opaque, easier to cut and less susceptible to ink blotting than that filled with clay [152]. It is supposed that the adsorptive zeolite, apart from acting as a bulking agent, controls pitch. A similar application of sedimentary zeolites is also reported in Hungary [153].

A series of consumer products are manufactured with natural zeolites and sold directly in the supermarkets in many American and European countries and in Japan. They are based on adsorption and/or ion-exchange properties of zeolites and are offered as deodorizing agents for house environments and stables or as pet litters to absorb water and especially to reduce the unpleasant odor due to ammoniac compounds reeking from animal urine. For similar reasons natural zeolites are sold, especially in Japan, for the treatment of animal-wastes. Among the consumer products it is also worth mentioning the sale of small bags of natural zeolites for soil amendment of indoor plants.

4.2.10.8
Summary and Prospects

The present bird's-eye survey points out that sedimentary zeolites are a natural resource of great interest. The large diffusion and the low cost of these materials (especially those containing clinoptilolite, mordenite, phillipsite and chabazite), the moderate-to-good grade, the interesting chemical and physical features make their application attractive in many fields that are only to a limited extent overlapped to those of the synthetic zeolites. Adsorption – with some exceptions, for instance those based to dehydration – and catalysis are, in general, precluded to zeolite-bearing tuffs, essentially for lack of purity. On the contrary, inconstancy of composition and the presence of impurities do not prevent utilizing the natural materials in fields of environmental and industrial interest, such as water pollution control, manufacture of cement and lightweight aggregates, and agriculture, including soil amendment, animal husbandry and aquaculture.

Future directions of research should focus especially on applications in which massive utilization is expected. Selected fields are:

- Removal of hazardous cations from wastewaters. Interesting prospects are foreseen for the direct addition procedure, followed by safe stabilization-solidification of the cation-loaded sludge (e.g., by using the sludge as a pozzolanic material).
- Applications in soil science and technology. Investigation is needed for better understanding the dynamics of macro- and microelements in soil amended with zeolites and their agronomic effects. Interesting developments are also awaited in the use of natural zeolites for the reconstruction of depleted soils and for the preparation of soilless substrates.
- Manufacture of blended cements and artificial hydraulic lime. Research is needed to better understand and to optimize the $Ca(OH)_2$-zeolite interaction during hydration and hardening of hydraulic mortars made with pozzolanic cements or lime-zeolite mixtures.
- Applications in life sciences and medicine. This is an emerging sector, which, apart from the consolidated use of natural zeolites as dietary supplement in animal husbandry, is still at the infancy. But results collected in recent years demonstrate that this is destined to become the new frontier in the applications of natural zeolites.

References

1 C. COLELLA, M. DE' GENNARO, R. AIELLO, in: Natural Zeolites: Mineralogy, Occurrence, Properties, Applications. D. L. BISH, D. W. MING (Eds), MSA Reviews in Mineralogy & Geochemistry, Vol. 45, Mineralogical Society of America, Washington, D.C., 2001, p. 551–587.

2 I. MÂRZA, N. MÉSZÁROS, in: Tufuríle vulcanice din bazinul Transilvaniei. (The volcanic tuffs of the Transylvanian basin), Editura Dacia, Cluj-Napoca, Romania, 1991, 11–22 (in French).

3 F. A. MUMPTON, in: Occurrence, Properties and Utilization of Natural Zeolites. D. KALLÓ, H. S. SHERRY (Eds), Akadémiai Kiadó, Budapest, 1988, p. 333–366.

4 M. DE' GENNARO, C. COLELLA, R. AIELLO, E. FRANCO, Industrial Minerals 1984, No. 204, 97–109.

5 V. CARDONE, Il tufo nudo nell'architettura napoletana. CUEN (Cooperativa Universitaria Editrice Napoletana), Napoli, Italy, 1990, p. 63–72 (in Italian).

6 Vitruvius [M. Vitruvius Pollio], On Architecture. [De Architectura], translated from Latin by F. Granger, Vol. II, Loeb Classical Library, Harvard University Press, Harvard, U.S.A., 1934, Book VI, Section 23.

7 R. SERSALE, *Atti Accademia Pontaniana (Naples)* **1991**, *40*, 257–275.

8 R. SERSALE, R. AIELLO, *L'Industria Italiana del Cemento* **1964**, *34*, 747–760 (in Italian).

9 A. LANGELLA, M. ADABBO, Field guide to the tuff deposits of northern Latium. 10th Int. Zeolite Conf., Supplement to Bollettino AIZ (Italian Zeolite Association) No. 3, De Frede, Napoli, 1994, 45 pp.

10 A. F. CRONSTEDT, *Kongl. Wetenskaps Acad. Handl. Stockh.*, **1756**, *17*, 120–123 (in old Swedish); A. F. CRONSTEDT, *Forsok til Mineralogie eller Mineral-Rikets Upstallning* (An Attempt to Mineralogy or Arrangement of the Realm of Minerals), Wildiska Tryckeriet, Stockholm, **1758**, Sections 108–112, p. 99–103 (in old Swedish).

11 G. GOTTARDI, *Eur. J. Mineral.* **1989**, *1*, 479–487.

12 C. F. NAUMANN, Elemente der Mineralogie. 12th edn, Wilhelm Engelmann, Leipzig, Germany, 1885, p. 705–726 (in German).

13 D. S. COOMBS, A. ALBERTI, T. ARMBRUSTER, G. ARTIOLI, C. COLELLA, E. GALLI, J. D. GRICE, F. LIBAU, H. MINATO, E. H. NICKEL, E. PASSAGLIA, D. R. PEACOR, S. QUARTIERI, R. RINALDI, M. ROSS, R. A. SHEPPARD, E. TILLMANNS, G. VEZZALINI, *The Canadian Mineralogist* **1997**, *35*, 1571–1606.

14 A. DAMOUR, *Ann. Chim. Phys.* **1858**, [3], *53*, 438–459 (in French).

15 G. FRIEDEL, *Bull. Soc. Franç. Minéral.* **1896**, *19*, 94–118 (in French).

16 F. GRANDJEAN, *C. R. Hebd. Séances Acad. Sci.* **1909**, *149*, 866–868 (in French).

17 R. SEELIGER, K. LAPKAMP, *Physik. Z.* **1921**, *22*, 563–568 (in German).

18 O. WEIGEL, E. STEINHOFF, *Z. Kristallogr.* **1924**, *61*, 125–154 (in German).

19 J. W. MCBAIN, *Kolloid-Zeitschrift* **1926**, *40*, 1–9 (in German).

20 O. SCHMIDT, *Z. Physik. Chem.* **1928**, *133*, 263–303 (in German).

21 J. SAMESHIMA, *Bull. Chem. Soc. Japan* **1929**, *4*, 96–103.

22 T. BABA, *Bull. Chem. Soc. Japan* **1930**, *5*, 190–202.

23 J. W. MCBAIN, The Sorption of Gases and Vapours by Solids. G. Routledge & Sons, Ltd., London, 1932, p. 167–176.

24 C. K. HERSH, Molecular Sieves. Reinhold Publishing Co., New York, 1961, p. 29–52.

25 R. M. BARRER, Zeolites and Clay Minerals as Sorbents and Molecular Sieves. Academic Press, London, 1978, 497 pp.

26 H. S. THOMPSON, *J. Royal Agric. Soc. Engl.* **1850**, *11*, 68–74.

27 J. T. WAY, *J. Royal Agric. Soc. Engl.*, **1850**, *11*, 313–379.

28 H. EICHHORN, *[Poggendorff's] Ann. Phys. Chem. (Leipzig)* **1858**, [4], *15*, 126–133 (in German).

29 R. GANS, *Jahrb. Königl. Preuss. Geol. Landesanst. und Bergakad. (Berlin)* **1905**, *26*, 179–211 (in German).

30 B. A. ADAMS, E. L. HOLMES, *J. Soc. Chem. Ind. (London)* **1935**, *54*, 1–6T.

31 I. ZOCH, *Chemie der Erde* **1914**, *1*, 219–269 (in German).

32 R. M. BARRER, *J. Chem. Soc.* **1950**, 2342–2350.

33 R. M. BARRER, in: Natural Zeolites: Occurrence, Properties, Use. L. B. SAND, F. A. MUMPTON (Eds), Pergamon Press, Elmsford, New York, 1978, p. 385–395.

34 R. M. MILTON, in: Molecular Sieves, Soc. Chem. Ind., London, 1968, p. 199–203.

35 F. A. MUMPTON, in: Proc. Sixth Int. Zeolite Conf. D. OLSON, A. BISIO (Eds), Butterworths, Guildford, U.K., 1984, p. 68–86.

36 F. A. MUMPTON, in: Mineralogy and Geology of Natural Zeolites. F. A. MUMPTON (Ed.), Short Course Notes Vol. 4, Mineralogical Society of America, Washington, 1977, p. 1–17.

37 R. A. SHEPPARD, A. J. GUDE, III, G. M. EDSON, in: Natural Zeolites: Occurrence, Properties, Use. L. B. SAND, F. A. MUMPTON (Eds), Pergamon Press, Elmsford, New York, 1978, p. 319–328.

38 T. SUDO, *J. Geol. Soc. Japan* **1950**, *56*, 13–16.

39 D. S. Coombs, *Trans. Royal Soc. New Zealand* **1954**, *82*, 65–109.

40 R. Sersale, *Rend. Accad. Sci. Fis. Mat. (Naples)* **1958**, *25*, 181–207 (in Italian).

41 R. L. Hay, in: Natural Zeolites: Occurrence, Properties, Use. L. B. Sand, F. A. Mumpton (Eds), Pergamon Press, Elmsford, New York, 1978, p. 135–143.

42 W. M. Meier, D. H. Olson, Ch. Baerlocher, Atlas of Zeolite Structure Types, 4th edn, *Zeolites* **1996**, *17*, p. 1–229.

43 D. B. Hawkins, in: Zeoagriculture. Use of Natural Zeolites in Agriculture and Aquaculture. W. G. Pond, F. A. Mumpton (Eds), Westview Press, Boulder, Colorado, 1984, p. 69–78.

44 G. Gottardi, E. Galli, Natural Zeolites. Springer-Verlag, Berlin, 1985, 409 pp.

45 M. de' Gennaro, M. Adabbo, A. Langella, in: Natural Zeolites '93: Occurrence, Properties, Use. D. W. Ming, F. A. Mumpton (Eds), International Committee on Natural Zeolites, Brockport, New York, 1995, p. 51–67.

46 R. A. Sheppard, *U. S. Geol. Survey Open-File Report* **1996**, No. 96–18, 24 pp.

47 A. Temel, M. N. Gündogdu, *Mineral. Deposita* **1996**, *31*, 539–547.

48 R. M. Barrer, Zeolites and Clay Minerals as Sorbents and Molecular Sieves. Academic Press, London, 1978, p. 24–27.

49 C. Colella, in: Natural Microporous Materials in the Environmental Technology. P. Misaelides, F. Macášek, T. Pinnavaia and C. Colella (Eds), Kluwer A.P., Dordrecht, The Netherlands, 1999, p. 207–224.

50 G. V. Tsitsishvili, in: Occurrences, Properties and Utilization of Natural Zeolites. D. Kalló, H. S. Sherry (Eds), Akadémiai Kiadó, Budapest, 1988, p. 367–393.

51 A. Langella, M. Pansini, P. Cappelletti, B. de Gennaro, M. de' Gennaro, C. Colella, *Microporous Mesoporous Mater.* **2000**, *37*, 337–343.

52 M. de' Gennaro, A. Langella, *Mineral. Deposita* **1996**, *31*, 452–472.

53 F. A. Mumpton, *Proc. Nat. Acad. Sci. USA* **1999**, *96*, 3463–3470.

54 I. M. Galabova, in: Proc. 3° Convegno Naz. Scienza e Tecnologia delle Zeoliti. R. Aiello (Ed.), Associazione Italiana Zeoliti, Napoli, Italy, 1995, p. 253–268.

55 R. Aiello, C. Colella, *Chim. Ind. (Milan)* **1972**, *54*, 303–307.

56 R. Aiello, A. Nastro, G. Giordano, C. Colella, in: Occurrence, Properties and Utilization of Natural Zeolites. D. Kalló, H. S. Sherry (Eds), Akadémiai Kiadó, Budapest, 1988, p. 763–772

57 G. Giordano, A. Nastro, R. Aiello, in: Proc. Int. Conf. Energy Options and the Rural Sector in Developing Countries. E. Lorenzini, D. Gennari, P. Vestrucci (Eds), T. P. Editrice, Bologna, Italy, 1988, p. 418–427.

58 D. Tchernev, in: Natural Zeolites '93: Occurrence, Properties, Use. D. W. Ming, F. A. Mumpton (Eds), International Committee on Natural Zeolites, Brockport, New York, 1995, p. 613–622.

59 Linde Molecular Sieve Bulletin F-1617, Union Carbide, Tarrytown, New York, 1962, 4 pp.

60 Brocure, NRG NuFuel Company, NRG, Los Angeles, California, 1975, 6 pp.

61 D. Kalló, J. Papp and J. Valyon, *Zeolites* **1982**, *2*, 13–16.

62 D. Caputo, B. de Gennaro, M. Pansini, C. Colella, in: Natural Microporous Materials in the Environmental Technology. P. Misaelides, F. Macášek, T. Pinnavaia, C. Colella (Eds), Kluwer A.P., Dordrecht, The Netherlands, 1999, p. 225–236.

63 D. T. Hayhurst, in: Natural Zeolites: Occurrence, Properties, Use. L. B. Sand, F. A. Mumpton (Eds), Pergamon Press, Elmsford, New York, 1978, p. 503–507.

64 H. Minato, T. Tamura, in: Natural Zeolites: Occurrence, Properties, Use. L. B. Sand, F. A. Mumpton (Eds),

Pergamon Press, Elmsford, New York, 1978, p. 509–516.

65 I. M. GALABOVA, G. A. HARALAMPIEV, B. ALEXIEV, in: Natural Zeolites: Occurrence, Properties, Use. L. B. SAND, F. A. MUMPTON (Eds), Pergamon Press, Elmsford, New York, 1978, p. 431–437.

66 P. CIAMBELLI, V. DE SIMONE, C. MARINO, *Ann. Chim. (Rome)*, **1984**, *74*, 435–446.

67 C. COLELLA, M. PANSINI, F. ALFANI, M. CANTARELLA, A. GALLIFUOCO, *Microporous Mater.* **1994**, *3*, 219–226.

68 K. G. PAPKE, *Nevada Bureau of Mines and Geological Bulletin* **1972**, *79*, 1–32; N.Y. Chen, US Patent 3,630,066 (1971).

69 D. KALLÓ, in: Occurrence, Properties and Utilization of Natural Zeolites. D. KALLÓ, H. S. SHERRY (Eds), Akadémiai Kiadó, Budapest, 1988, p. 601–624.

70 G. ONYESTYÁK, D. KALLÓ, in: Natural Zeolites for the Third Millennium. C. COLELLA, F. A. MUMPTON (Eds), De Frede, Napoli, Italy, 2000, p. 395–401.

71 G. GRECO, JR., F. ALFANI, G. IORIO, *Chim. Ind. (Milan)* **1980**, *62*, 481–486.

72 G. BAGNASCO, P. CIAMBELLI, *React. Kinet. Catal. Lett.* **1979**, *12*, 417–422.

73 E. ALABISO, P. CIAMBELLI, V. DE SIMONE, C. PORCELLI, R. VALENTINO, *React. Kinet. Catal. Lett.* **1979**, *12*, 451–456.

74 I. S. NAM, U. C. HWANG, S. W. HAM, Y. G. KIM, in: Catal. Sci. Technol. *1*, S. YOSHIDA, N. TAKEZAWA, T. ONO (Eds), Kodansha, Tokyo, 1991, p. 165–170.

75 M. TURCO, G. BAGNASCO, L. LISI, G. RUSSO, D. SANNINO, P. CIAMBELLI, in: Natural Zeolites '93: Occurrence, Properties, Use. D. W. MING, F. A. MUMPTON (Eds), International Committee on Natural Zeolites, Brockport, New York, 1995, p. 429–435.

76 C. COLELLA, *Mineral. Deposita* **1996**, *31*, 554–562.

77 R. T. PABALAN, F. P. BERTETTI, in: Natural Zeolites: Mineralogy, Occurrence, Properties, Applications. D. L. BISH, D. W. MING (Eds), MSA Reviews in Mineralogy & Geochemistry, Vol. 45, Mineralogical

Society of America, Washington, D.C., 2001, p. 453–518.

78 B. W. MERCER, L. L. AMES, in: Natural Zeolites: Occurrence, Properties, Use. L. B. SAND, F. A. MUMPTON (Eds), Pergamon Press, Elmsford, New York, 1978, p. 451–462.

79 L. LIBERTI, A. LOPEZ, V. AMICARELLI, G. BOGHETICH, in: Natural Zeolites '93: Occurrence, Properties, Use. D. W. MING, F. A. MUMPTON (Eds), International Committee on Natural Zeolites, Brockport, New York, 1995, p. 351–362.

80 M. PANSINI, *Mineral. Deposita* **1996**, *31*, 563–575.

81 K. TORII, in: Natural Zeolites: Occurrence, Properties, Use. L. B. SAND, F. A. MUMPTON (Eds), Pergamon Press, Elmsford, New York, 1978, p. 441–450.

82 A. DYER, An Introduction to Zeolite Molecular Sieves. John Wiley & Sons, Chichester, UK, 1988, p. 83.

83 R. G. PIPER, C. E. SMITH, in: Zeo-Agriculture: Use of Natural Zeolites in Agriculture and Aquaculture. W. G. POND AND F. A. MUMPTON (Eds), Westview Press, Boulder, Colorado, 1984, p. 223–228.

84 J. HLAVAY, J. INCZÉDY, K. FÖLDI-POLYÁK, M. ZIMONYI, in: Occurrence, Properties and Utilization of Natural Zeolites. D. KALLÓ, H. S. SHERRY (Eds), Akadémiai Kiadó, Budapest, 1988, p. 483–490.

85 D. KALLÓ, in: Natural Zeolites '93: Occurrence, Properties, Use. D. W. MING, F. A. MUMPTON (Eds), International Committee on Natural Zeolites, Brockport, New York, 1995, p. 341–350.

86 D. KALLÓ, J. PAPP, in: Porous Materials in Environmentally Friendly Processes. I. KIRICSI, G. PÁL-BORBÉLY, J. B. NAGY, H. G. KARGE (Eds), *Studies in Surface Science and Catalysis*, Vol. 125, Elsevier, Amsterdam, 1999, p. 699–706.

87 C. GALINDO, JR., D. W. MING, M. J. CARR, A. MORGAN, K. D. PICKERING, in: Natural Zeolites for the Third Millennium. C. COLELLA, F. A. MUMPTON (Eds), De Frede, Napoli, Italy, 2000, p. 363–371.

88 C. Colella, E. Torracca, A. Colella, B. de Gennaro, D. Caputo, M. de' Gennaro, in: Zeolites and Mesoporous Materials at the Dawn of the 21st Century. A. Galarneau, F. di Renzo, F. Fajula, J. Vedrine (Eds) *Studies in Surface Science and Catalysis*, vol. 135, Elsevier Amsterdam, 2001, p. 371, paper 01-O-05 (CD-ROM).

89 E. Torracca, P. Galli, M. Pansini, C. Colella, *Microporous Mesoporous Mater.* **1998**, *20*, 119–127.

90 V. Amicarelli, G. Baldassarre, G. Boghetich, L. Liberti, N. Limoni, in: Proc. 2nd Int. Conf. on Environmental Protection. CUEN, Napoli (Italy), 1988, p. [2.A] 75–82.

91 C. Colella, R. Aiello, in: Occurrences, Properties and Utilization of Natural Zeolites. D. Kalló, H. S. Sherry (Eds), Akadémiai Kiadó, Budapest, 1988, p. 491–500.

92 F. A. Mumpton, in: Natural Zeolites: Occurrence, Properties, Use. L. B. Sand, F. A. Mumpton (Eds), Pergamon Press, Elmsford, New York, 1978, p. 3–27.

93 S. M. Robinson, T. E. Kent, W. D. Arnold, in: Natural Zeolites '93. Occurrence, Properties, Use. D. W. Ming, F. A. Mumpton (Eds), International Committee on Natural Zeolites, Brockport, New York, 1995, p. 579–586.

94 M. Adabbo, D. Caputo, B. de Gennaro, M. Pansini, C. Colella, *Microporous Mesoporous Mater.* **1999**, *28*, 315–324.

95 E. D. Collins, D. O. Campbell, L. J. King, J. B. Knauer, *AIChE Symp. Ser.* **1982**, *213*, 9–15.

96 N. F. Chelishchev, in: Natural Zeolites '93: Occurrence, Properties, Use. D. W. Ming, F. A. Mumpton (Eds), International Committee on Natural Zeolites, Brockport, New York, 1995, p. 525–532.

97 G. V. Tsitsishvili, T. G. Andronikashvili, G. N. Kirov, L. D. Filizova, Natural Zeolites. Ellis Horwood, Chichester, UK, 1992, p. 235–262.

98 D. T. Vaniman, D. L. Bish, in: Natural Zeolites '93: Occurrence,

Properties, Use. D. W. Ming, F. A. Mumpton (Eds), International Committee on Natural Zeolites, Brockport, New York, 1995, p. 533–546.

99 D. Caputo, Ph. D. Thesis, Department of Materials and Production Engineering, University of Naples Federico II, Napoli, Italy, 1997, 135 pp.

100 C. Colella, M. Pansini, in: Perspectives in Molecular Sieve Science. W. H. Flank, T. E. Whyte (Eds), ACS Symposium Series No. 368, Washington, DC, 1988, p. 500–510.

101 R. Cioffi, M. Pansini, D. Caputo, C. Colella, *Environmental Technology* **1996**, *17*, 1215–1224.

102 C. Colella, M. de' Gennaro, A. Langella, M. Pansini, in: Natural Zeolites '93: Occurrence, Properties, Use. D. W. Ming, F. A. Mumpton (Eds), International Committee on Natural Zeolites, Brockport, New York, 1995, p. 377–384.

103 C. Colella, in: Porous Materials in Environmentally Friendly Processes. I. Kiricsi, G. Pál-Borbély, J. B. Nagy, H. G. Karge (Eds), *Studies in Surface Science and Catalysis*, Vol. 125, Elsevier, Amsterdam, 1999, p. 641–655.

104 W. G. Pond, F. A. Mumpton (Eds), Zeoagriculture: Use of Natural Zeolites in Agriculture and Aquaculture. Westview Press, Boulder, Colorado, 1984, 296 pp.

105 D. Kalló, H. S. Sherry (Eds), Occurrence, Properties and Utilization of Natural Zeolites. Akadémiai Kiadó, Budapest, 1988, 857 pp.

106 G. Rodriguez-Fuentes, J. A. Gonzales (Eds), Zeolites '91: Memoirs of the 3rd Int. Conf. on the Occurrence, Properties and Utilization of Natural Zeolites. Parts I&II, International Conference Center, Havana, Cuba, 1991, 310+322 pp (mostly in Spanish).

107 D. W. Ming, F. A. Mumpton (Eds), Natural Zeolites '93: Occurrence, Properties, Use. International Committee on Natural Zeolites, Brockport, New York, 1995, 622 pp.

108 C. Colella, F. A. Mumpton (Eds), Natural Zeolites for the Third

Millennium. De Frede, Napoli, Italy, 2000, 484 pp.

109 G. Sposito, in: Handbook of Soil Science. M. E. Sumner (editor in chief), P. M. Huang (Ed.), CRC Press LLC, Boca Raton, Florida, 2000, p. B241–B263.

110 E. R. Allen, D. W. Ming, in: Natural Zeolites '93: Occurrence, Properties, Use. D. W. Ming, F. A. Mumpton (Eds), International Committee on Natural Zeolites, Brockport, New York, 1995, p. 477–490.

111 D. W. Ming, E. R. Allen, in: Natural Zeolites for the Third Millennium. C. Colella, F. A. Mumpton (Eds), De Frede, Napoli, Italy, 2000, p. 417–426.

112 F. A. Mumpton, in: Zeoagriculture: Use of Natural Zeolites in Agriculture and Aquaculture. W. G. Pond, F. A. Mumpton (Eds), Westview Press, Boulder, Colorado, 1984, p. 3–27.

113 B. Fitzner, L. Lehners, in: Proc. 6th Int. Congr. Int. Assoc. Eng. Geol. D. G. Price, (Ed.), Balkema, Rotterdam, The Netherlands, 1990, p. 3181–3188.

114 D. Stojanovic, in: Proc. Serb. Geol. Soc. for 1968–1970. Belgrade, Yugoslavia, 1972, p. 9–20.

115 A. L. Bush, in: Minutes 25th Ann. Meet. Perlite Inst. Colorado Springs, Colorado, Apr. 18–23, 1974.

116 R. A. Gaioso Blanco, E. De Jongh Caula, C. Gil Izquierdo, in: Zeolites '91: Memoirs of the 3rd Int. Conf. on the Occurrence, Properties and Utilization of Natural Zeolites. G. Rodriguez-Fuentes, J. A. Gonzales (Eds), Part II, International Conference Center, Havana, Cuba, 1991, p. 203–207 (in Spanish).

117 R. de Gennaro, M. Dondi, P. Cappelletti, A. Colella, A. Langella, in: Proc. EUROMAT 2001. (CD-ROM), AIM (Associazione Italiana di Metallurgia), Milano, Italy, 2001.

118 J. Kasai, T. Urano, T. Kuji, W. Takeda, O. Machinaga, Japan. Kokai 73,066,123 (1973).

119 I. Escalona Ledea, in: Zeolites '91: Memoirs of the 3rd Int. Conf. on the Occurrence, Properties and Utilization of Natural Zeolites. G. Rodriguez-Fuentes, J. A. Gonzales (Eds), Part

II, International Conference Center, Havana, Cuba, 1991, p. 214–215 (in Spanish).

120 Y. Fu, J. Ding, J. J. Beaudoin, US Patent 5,494,513 (1996), assigned to National Research Council of Canada.

121 B. A. Fursenko, L. K. Kazantseva, I. A. Belitsky, in: Natural Zeolites for the Third Millennium. C. Colella, F. A. Mumpton (Eds), De Frede Editore, Napoli, Italy, 2000, p. 337–349.

122 H. F. W. Taylor, Cement Chemistry. Academic Press, London, 1990, p. 276–315.

123 F. Massazza, *Cement & Concrete Composites* **1993**, *15*, 185–214.

124 G. Malquori, in: Proc. 4th Int. Symp. on Chemistry of Cement. National Bureau of Standards, Monograph 43-Vol. II, U.S. Department of Commerce, Washington, 1962, p. 983–1006.

125 M. I. Sanchez de Rojas, M. P. Luxan, M. Frias, N. Garcia, *Cem. Concr. Res.* **1993**, *23*, 46–54.

126 R. Sersale, in: Natural Zeolites '93: Occurrence, Properties, Use. D. W. Ming and F. A. Mumpton (Eds), Int. Comm. Natural Zeolites, Brockport, New York, 1995, p. 603–612.

127 N. Fratini, *Ann. Chim. (Rome)* **1949**, *39*, 41–49; **1950**, *40*, 461–469 (in Italian).

128 L. Huizhen, in: Proc. 9th Int. Congr. on Chemistry of Cement. Vol. 3, Nat. Council for Cement and Building Materials, New Delhi, India, 1992, p. 128–134.

129 F. Nai-Qian, Y. Hsia-Ming, Z. Li-Hong, *Cem. Concr. Res.* **1988**, *18*, 464–472.

130 K. P. Kitsopoulos, A. C. Dunham, *Mineral. Deposita* **1996**, *31*, 576–583.

131 E. Alcantara Ortega, C. Cheeseman, J. Knight, M. Loizidou, *Cem. Concr. Res.* **2000**, *30*, 1641–1646.

132 Z. Zhang, J. Guo, C. Liang, in: Program and Abstracts, Zeolite '93, 4th Int. Conf. Occurr., Prop., Utiliz. Nat. Zeol. Boise, Idaho, USA, 1993, p. 221–223.

133 B. Drzaj, S. Hocevar, M. Slokan, A. Zajc, *Cem. Concr. Res.* **1978**, *8*, 711–720.

134 F. Urrutia Rodriguez, M. Gener Rizo, in: Zeolites '91: Memoirs of the 3rd Int. Conf. on the Occurrence, Properties and Utilization of Natural Zeolites. G. Rodriguez-Fuentes, J. A. Gonzales (Eds), Part I, International Conference Center, Havana, Cuba, 1991, p. 198–202 (in Spanish).

135 Y. Kasai, K. Tobinai, E. Asakura, N. Feng, in: Proc. 4th Int. Conf. Fly Ash, Silica Fume, Slag, and Natural Pozzolans in Concrete. Vol. 1, American Concrete Institute, Detroit, Michigan, 1992, p. 615–634.

136 D. Fragoulis, E. Chaniotakis, M. G. Stamatakis, *Cem. Concr. Res.* **1997**, *27*, 889–905.

137 V. S. Ramachandran, Cement Admixtures Handbook. Properties, Science and Technology. Noyes Publ., Park Ridge, New Jersey, 1984, p. 116.

138 H. Minato, *Koatsugasu* **1968**, *5*, 536–547 (in Japanese).

139 J. H. Quisenberry, *Clays & Clay Minerals* **1968**, *16*, 267–270.

140 F. A. Mumpton, P. H. Fishman, *J. Anim. Sci.* **1977**, *45*, 1188–1203.

141 A. Rivera, G. Rodriguez-Fuentes, E. Altshuler, *Microporous Mesoporous Mater.* **1998**, *24*, 51–58.

142 A. Lam, L. R. Sierra, G. Rojas, A. Rivera, G. Rodriguez-Fuentes, L. A. Montero, *Microporous Mesoporous Mater.* **1998**, *23*, 247–252.

143 G. Rodriguez-Fuentes, M. A. Barrios, A. Iraizoz, I. Perdomo, B. Cedré, *Zeolites* **1997**, *19*, 441–448.

144 R. B. Harvey, L. F. Kubena, M. H. Elissalde, T. D. Phillips, *Avian Diseases* **1993**, *37*, 67–73.

145 M. Tomasevic-Canovic, M. Dumic, O. Vukicevic, Z. Masic, O. Zurovac-Kuzman, A. Dakovic, in: Natural Zeolites Sofia '95. G. Kirov, L. Filizova, O. Petrov (Eds), Pensoft, Sofia-Moscow, 1997, p. 127–132.

146 B. Concepción-Rosabal, J. Balmaceda-Era, G. Rodriguez-Fuentes, *Microporous Mesoporous Mater.* **2000**, *38*, 161–166.

147 G. Rodriguez-Fuentes, A. Iraizoz, M. A. Barrios, A. Rivera, B. Concepción, J. C. Torres, R. Simon, I. Perdomo, V. Zaldivar, A. R. Ruiz-Salvador, B. Cedré, M. Mir, A. Lam, M. Gener, in: Program and Abstracts, Zeolite '97, 5th Int. Conf. Occurrence, Properties, Utilizations of Natural Zeolites. De Frede, Napoli, Italy, 1997, p. 258–260.

148 G. N. Kirov, G. Terziiski, in: Natural Zeolites Sofia '95. G. Kirov, L. Filizova, O. Petrov (Eds), Pensoft, Sofia-Moscow, 1997, p. 133–141.

149 M. V. Cattaneo, T. M. S. Chang, *Trans. Am. Soc. Artif. Intern Organs* **1991**, *37*, 80–87.

150 K. Maeda, European Patent Appl. EP 298,726 (1989), assigned to Karo Maeda, Japan; K. Ikegami, M. Koide, Japan Kokai Tokkyo Koho JP 05,285,209 (1993) assigned to Terumo Corp., Japan.

151 S. Andersson, I. Grenthe, E. Jonsson, L. Naucler, German Offen. 2,512,212 (1975), assigned to Gambro A.-G., Sweden; S.R. Ash, US Patent 4,581,141 (1986), assigned to Purdue Research Foundation, USA.

152 Y. Kobayashi, Japan Kokai 45,041,044 (1970), assigned to Jikuraitu Kagaku Kogyo Co., Japan.

153 F. A. Mumpton, in: Atti 3° Congresso Nazionale AIMAT, Omaggio Scientifico a Riccardo Sersale. C. Colella (Ed.), De Frede, Napoli, Italy, 1996, p. XXXI–LXIV.

4.3
Porous Metal-Organic Frameworks

Stefan Kaskel

4.3.1
Introduction

The first report of a cyanide inclusion compound appeared in 1897, when the German chemist Hofmann discovered the compound $Ni(CN)_2 \cdot NH_3 \cdot C_6H_6$ [1]. He already noticed the size-selectivity of the clathrate formation, although the structure was solved much later [2–4]. Since then, a large number of coordination polymers with occluded host molecules have been structurally characterized and specific host–guest interactions were studied. Nowadays, to a large extent, metal-organic framework structures are predictable on the basis of ligand design and known coordination patterns [5, 6]. However, these materials were not always investigated with respect to porosity, ion exchange and thermal stability. The recent interest in new porous materials with advanced functionality and large, adaptable pore sizes [7, 8] has focused on the design of porous network materials with the aid of crystal-engineering techniques and led to remarkable achievements [9–14].

Ligands that are engineered to achieve desired network topologies have been used and proven to result in stable highly porous compounds, potentially useful in heterogeneous catalysis, for selective sorption or detection of small molecules.

In this chapter we will focus on specific compounds and describe the structural chemistry in relation to their properties. Certainly this rather fast-developing field is still in its infancy, and therefore a review can not be more than a snapshot concerning materials. Still, the concepts, which will be outlined, are expected to be quite general. In particular, requirements for porosity and thermal stability are discussed. It should be pointed out that the number of synthetic three-dimensional (3D) coordination polymers is much higher than the few compounds presented here, but in many cases only structural features have been reported and adsorption measurements have not been carried out. In some cases, the valuable porous structure of 3D networks might have been overlooked. On the other hand, it would be impossible to cover the whole area of coordination polymers. We have therefore concentrated on recent developments and focused mainly on compounds for which porosity or exchange properties have been experimentally measured. The compar-

Molecular Complexes   ## Extended Solids

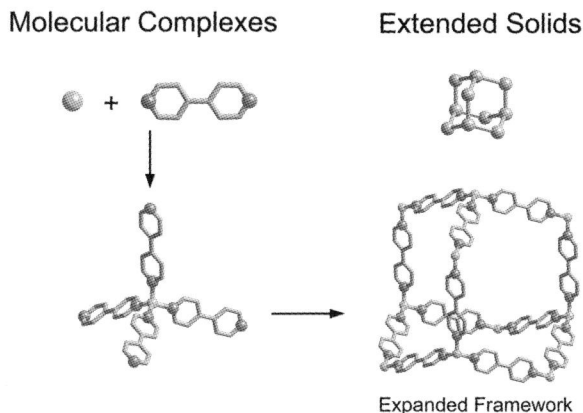

Expanded Framework

Fig. 1. Modular assembly of predesigned organic linkers (4,4'-bipyridine) and metal centers (Cu(I)) are used to achieve expanded network architectures. A diamond net is constructed by virtual replacement of C atoms by Cu and C–C bonds by two-connecting spacers (from Eddaoudi et al. [9]).

ison of specific compounds is not straightforward due to different techniques and algorithms used for specific surface area and pore-size assessment. BET theory does not apply to microporous solids and we will refer to specific pore volumes, whenever reported. However, in some cases other measures had to be used, since they appear in the literature and can not be converted, without access to primary data.

4.3.1.1
Modular Chemistry

The key in engineering metal-organic frameworks is the use of linkers that are designed to achieve desired network topologies by connecting transition-metal centers or polynuclear clusters that serve as the nodes of the network [9–11]. For example, the assembly of a tetrahedrally coordinated metal center and a linear organic linker like 4,4'-bipyridine results in structures with expanded diamond topology (Fig. 1). Each bond of the diamond network is replaced by a sequence of bonds that expands the networks and yields void space proportional to the length of the linker. Such rational syntheses of organic-inorganic frameworks allow the prediction of crystal structures within certain limits, hereby termed *crystal engineering*. The conception of the framework is facilitated by using predesigned modules, on the one hand organic linkers with rigid geometry and coordinating functional groups at the outer positions. On the other hand polynuclear units like dinuclear copper clusters prefer rigid and distinct arrangements of functional groups that predetermines the orientation of the linker. They represent secondary

References see page 1244

building units (SBUs). The modules can be combined in various ways to build frameworks. For two-connecting linkers metal centers serve as the nodes of a 2D or 3D net, a representation that simplifies the atomic structure and allows the comparison of net topologies. Molecules that connect more than two SBUs can no longer be represented by a line and become nodes of the net themselves.

4.3.1.2
Network Topologies

A convenient tool for the comparison of solid framework structures is simplification in terms of network topologies. A connecting ligand or a supramolecular building unit may be represented by a simple node indicating the connectivity. A 3D net may be derived from a stacking of 2D nets to demonstrate relations between structures with different stacking schemes. For regular nets, the term (n, p) introduced by Wells [15] is used, where p is the number of links meeting at each point and n is the number of points in the smallest circuit (Fig. 2). For example $(10,3)$ is a 3-connected net consisting of 10-gons. For less-regular nets the Schläfli

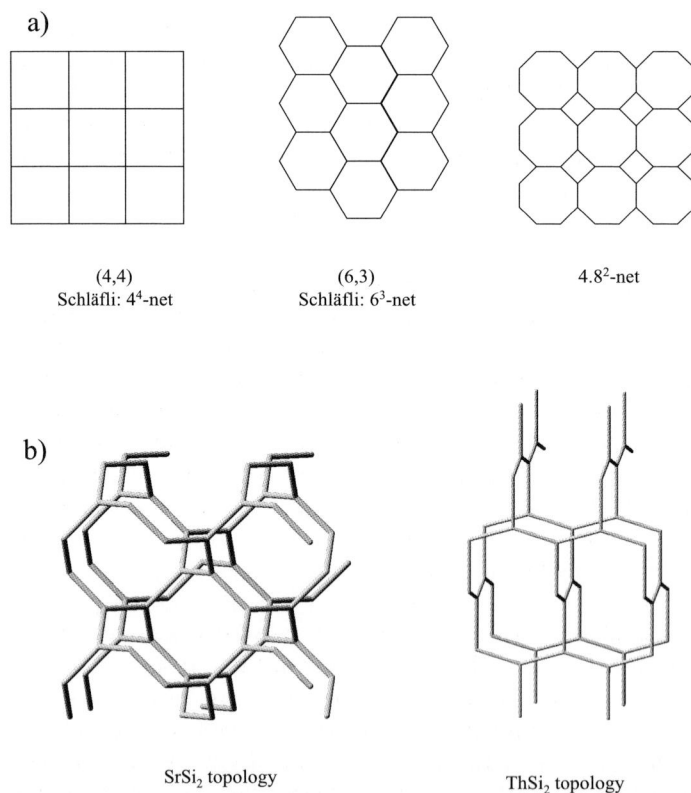

a)

| (4,4) | (6,3) | 4.8^2-net |
| Schläfli: 4^4-net | Schläfli: 6^3-net | |

b)

SrSi$_2$ topology   ThSi$_2$ topology

Fig. 2. Net topologies are characterized by a symbol (n, p) in which p is the number of links meeting at each point and n is the number of points in the smallest circuits. Less-regular topologies are indexed with Schläfli Symbols. Most of the coordination polymers can be simplified and assigned to highly symmetric network topologies; (a) 2D; (b) 3D (10,3) nets.

symbols $m^s.n^t$ are used in which s m-gons and t n-gons start out from one node and $s + t$ is the connectivity at this node [16]. Some simple 2D net topologies are shown in Fig. 2a. In comparison with four-connected zeolites metal-organic frameworks show a broader variety of net-topologies, such as 3-, 5-, 6-, 7- and 8-connected 3D nets, because of the variable coordination number of transition metals or poly-nuclear clusters. An interesting example is the (10,3)-a net (Fig. 2b), which is chiral and the only regular 3-connected net with a 3-fold axis at a vertice [10]. This topology is also termed $SrSi_2$, since the silicon sublattice in the Zintl-phase adopts the same topology. In the (10,3)-b net the symmetry of the node is reduced to $mm2$ ($ThSi_2$ topology, Fig. 2b). As compared with the $SrSi_2$ net one angle is topologically different from the other two. The $ThSi_2$ topology is common in structures with T- or Y-shaped building units.

4.3.1.3
Rigidity and Dynamics

Metal-organic frameworks are held together by metal–ligand interactions. The degree of robustness depends on the strength and dimensionality of the bonds. Robust networks that sustain guest-free cavities, in general, require coordinative bonds in all three dimensions. In many cases the dimension of the coordination-polymer is lower. However, 2D, 1D and even 0D topologies can also assemble through secondary interactions into porous networks. These interactions might be stronger (ionic) or weaker (hydrogen bond, van der Waals) than the metal–ligand bonds, but can lead to remarkably high thermal stability of the solid. For example substituted tetraphenylmethane derivatives can form purely hydrogen-bonded 3D nets. Up to 63 % of the guest molecules can be removed from these nets accompanied only by a small contraction in the cell parameters [17].

The X-ray powder pattern associated with the crystalline structure of a rigid framework serves, at best, to judge if the network remains intact after desorption of the guest molecules. The crystal morphology may be retained but small changes in the lattice constants and symmetry usually give rise to cracking and micro-crystallinity. In comparison with zeolites, coordination polymers are more flexible and less rigid. Desorption of the guest molecules often leads to a broadening of diffraction peaks or even X-ray amorphous samples, although the network is still intact. At least, when reimmersed in a guest liquid or in contact with guest vapor, crystallinity is restored. In theses cases it is difficult to assign porosity since re-crystallization can not be ruled out. Guest–exchange reactions may be used to prove the accessibility of less-rigid metal-organic frameworks. Recrystallization mechanisms can be ruled out if the host is insoluble in the media used and the morphology of the crystals is unaffected. Due to the flexibility of hydrogen-bonded networks, as an extreme, the guest-free unporous host may be transformed to the open structure only upon contact with a suitable guest molecule, whereas the network is not stable enough to sustain guest-free cavities. This so-called induced-fit adjustment may be a valuable feature of new flexible architectures.

References see page 1244

4.3.1.4
Interpenetration

The physical extension of known framework topologies by spacing metal centers apart with the aid of organic linkers does not necessarily result in a porous structure. Instead, the voids and channels are filled with a second identical network (or more) to give an interpenetrating structure. In some cases this can be avoided, but prediction is problematic. Interactions with the solvent and crystallization conditions seem to be crucial. However, even interpenetrating structures can leave residual voids suitable for the inclusion of molecules or ions. The best concept for designing interpenetrating porous systems is probably to start from a structure in which one net perfectly fills the pores of the second net and then, without affecting the topology, introduce a small extension in the organic spacer, long enough to create free voids but not too long, since this would allow for the insertion of yet another network. Interpenetration can also be beneficial for the creation of higher 3D architectures from low-dimensional polymers, such as grids or ladders that may lead to stable porous architectures.

4.3.1.5
Functionality

As compared with traditional molecular sieves, metal-organic frameworks suffer from low thermal stability, since most organic molecules decompose between 200 and 300 °C or become oxidized. Only some frameworks are stable up to 500 °C. On the other hand metal-organic frameworks are more variable in terms of pore size and shape. Various functional groups and residues can be used to tailor and adapt the inner surface of pores for different guest molecules. Hydrogen bonds or π–π-interactions may be used to direct the guest into preferred orientations. Transition metal centers provide active sites and functionality. In some cases polynuclear clusters resemble active sites of enzymes. Metal-organic frameworks provide flexibility, specific host–guest interaction and metal centers with a variety of redox, acid-base and even magnetic properties, and could therefore lead to new heterogeneous catalysts capable of mimicking enzymes. This is why porous coordination polymers should probably not be recognized as competitors to traditional molecular sieves, but rather new and highly functionalized materials for new developments in heterogeneous catalysis under mild conditions, for energy conversion, or opto- and magnetoelectronic devices like sensors.

4.3.2
Compounds

In terms of network topology, coordination geometry, and stability the design of organic modules is the key step in the development of metal-organic frameworks. Even though prediction is impossible, in general the ligands and their donor functions govern the degree of linking and the dimensionality of the net. The thermal stability and rigidity of the network is mostly associated with the structure and stability of the ligand. The same is true for the metal-to-ligand bonds. More importantly the formation of secondary building units like dinuclear clusters, for example in the copper and zinc carboxylates (Sect. 4.3.2.1), results from the reaction of bridging (carboxylate) functions from the linker and is therefore also determined by the choice of ligand. Additional hydrogen bonds between functional groups, which significantly enhance the network stability, are also specific for the organic part.

4.3.2.1
Carboxylates

Metal-organic frameworks (MOF) based on rigid carboxylate linkers have shown remarkable thermal stability. They represent the most promising candidates for heterogeneous catalysis and specific adsorption. This field has been mainly developed by the group of Yaghi [9–11, 18–21] who uses the term MOF-n ($n = 1$–14) as an identifier for the frameworks.

The intriguing feature of carboxylate chemistry is a variety of different secondary building units arising from clustering of transition metals. This is due to the various binding modes of the carboxylate group that can coordinate a metal center as bidentate, monodentate or di-monodentate. Bridging is achieved in the di-monodentate mode but also from single carboxylate oxygens. Typical mono- and polynuclear motifs observed in carboxylate frameworks are shown in Fig. 3.

The formation of polynuclear complexes is crucial to achieve stable frameworks. On the other hand in some cases prediction of the structure is complicated since the connectivity of the linker is not necessarily identical with the number of carboxylate groups and the formation of a new unexpected complex might obscure the initial net design.

4.3.2.1.1 Dicarboxylates
The most promising materials have been synthesized using p-BDC (para-benzene dicarboxylic acid, terephthalic acid) but various other linkers were also employed (Fig. 4). A typical microporous carboxylate containing p-BDC is Zn(p-BDC)(H$_2$O)·(DMF) [22] (MOF-2) (DMF = N,N′-dimethylformamide). Four carboxylate groups of the p-BDC building blocks are bonded to two clustered zinc atoms in a di-monodentate fashion (Fig. 5) to give extended 2D ML$_{4/2}$–layers (M = Zn$_2$-cluster, L = p-BDC). These layers are held together by hydrogen bonds of additional coordinated water ligands and carboxylate oxygens to give a microporous 3D framework with an extended 1D pore system. Single-crystal X-ray studies of the as-synthesized material reveal 5-Å-wide pores filled with DMF molecules. The guests and water ligands can be removed at 315 °C to give crystalline Zn(p-BDC) with permanent microporosity and a micropore volume of 0.094 cm^3 g^{-1}.

References see page 1244

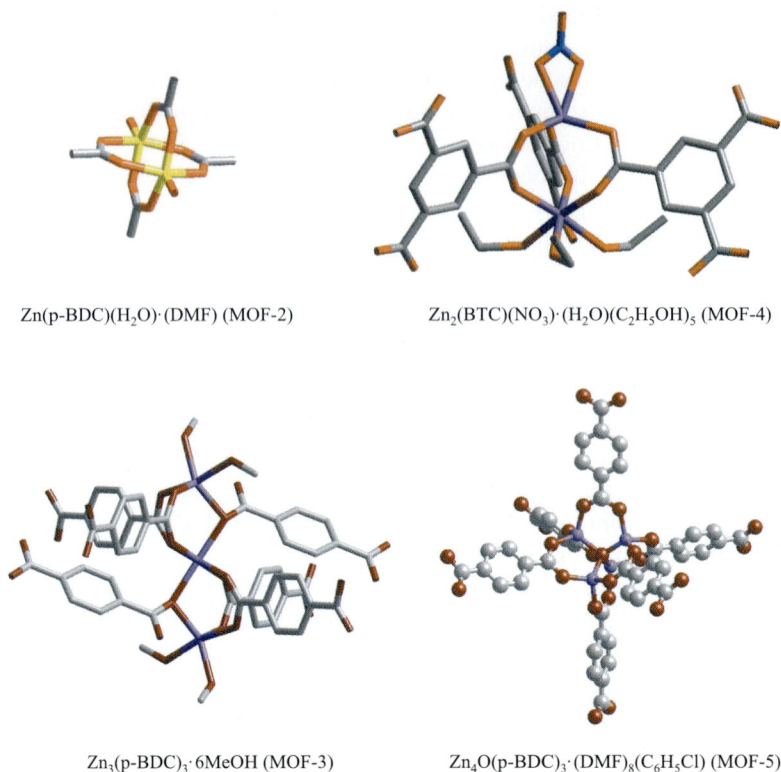

Zn(p-BDC)(H$_2$O)·(DMF) (MOF-2)

Zn$_2$(BTC)(NO$_3$)·(H$_2$O)(C$_2$H$_5$OH)$_5$ (MOF-4)

Zn$_3$(p-BDC)$_3$·6MeOH (MOF-3)

Zn$_4$O(p-BDC)$_3$·(DMF)$_8$(C$_6$H$_5$Cl) (MOF-5)

Fig. 3. Secondary building units (di-, tri- and tetranuclear clusters) in carboxylate chemistry (from Eddaoudi et al. [9]).

Cu(II)(p-BDC) is also a crystalline material with a high micropore volume of 0.22 cm^3 g^{-1} [23]. The architecture is probably related to MOF-2, but it was not possible to obtain single crystals suitable for structural analysis [24]. A wide variety of microporous dinuclear copper, molybdenum and ruthenium dicarboxylates was shown to have a high micropore volume by means of N$_2$-, Ar-, O$_2$-, and Xe-adsorption measurements [25].

The layers can be connected with the aid of N,N′-donor ligands like triethylenediamine (TED) and a 3D network is obtained. The structural model proposed for crystalline Cu(p-BDC)(TED)$_{0.5}$ (TED = triethylenediamine) [23] consists of Cu$_2$(CO$_2$)$_4$ paddle-wheels that are linked by the p-BDC ligands to form (4,4) nets. These layers are stacked directly upon each other and separated by a TED pillar to give a 3D structure that is stable up to 250 °C. A high micropore volume of 0.58 cm^3 g^{-1} was deduced from the type I argon-adsorption isotherm. The methane sorption capacity was 180 cm^3 g^{-1} at 35 bar.

The use of benzoate (BZ) instead of p-BDC and a spacer pyrazine (PYZ) results in a very similar but 1D structure [26]. Interestingly, this compound Cu(II)(BZ)$_2$(PYZ)$_{0.5}$·CH$_3$CN also becomes porous upon desolvation with a capac-

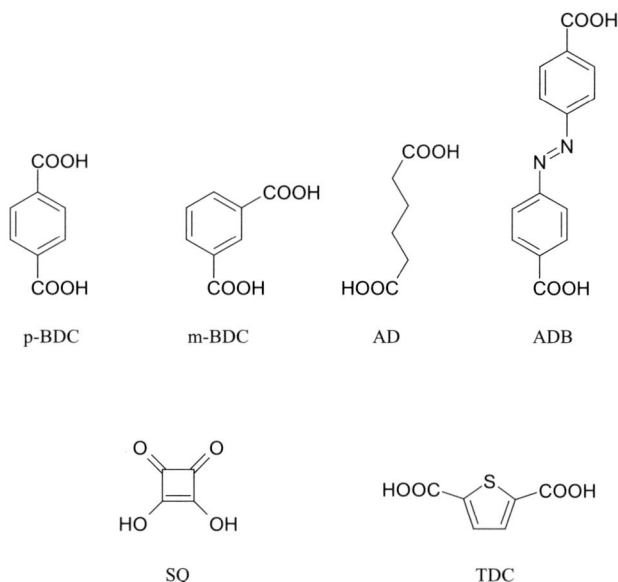

Fig. 4. Dicarboxylic acids. Abbreviations correspond to the completely deprotonated acid.

ity of three N_2 molecules per Cu_2-unit and a pore diameter close to 4.3 Å. The $Cu_2(BZ)_4$ paddle-wheels are only connected through the N-donor ligand and form infinite linear chains along the Cu–Cu vector but the stacking is such that micropores 4×5 Å are generated. Water-adsorption isotherms show practically no uptake due to the hydrophobicity of the micropores, which is a consequence of the structure since essentially only benzene rings from the benzoate groups constitute the inner surface.

The structure of $Zn_3(p\text{-BDC})_3 \cdot 6\text{MeOH}$ [20, 27] (MOF-3) is more complex (Fig. 6) since the Zn centers form trimeric $Zn_3C_6O_{12}$-clusters (Fig. 6a) in which six carboxylate groups hold the trimers together, four in a bidentate and two in a monodentate fashion to give an octahedral coordination of the central Zn atom (Fig. 6b). Four methanol molecules per formula unit are bonded to the external Zn centers and two are held in the channels through weak hydrogen bonds. Each of the six BDC ligands connects two trimeric units $(Zn_3(BDC)_{6/2})$, thus a 3D network with pores approximately 7 Å in diameter is obtained (Fig. 6c). Desolvation at 140 °C in vacuum allows removal of four methanol molecules but diffraction lines in the X-ray powder pattern after heat treatment indicate a slight deformation of the pore structure. However the initial pattern is restored instantaneously upon reintroduction of methanol into the pores. The material shows a clear preference for the inclusion of alcohols versus acetonitrile or benzene, indicating a potential for the separation of molecules from solution mixtures

References see page 1244

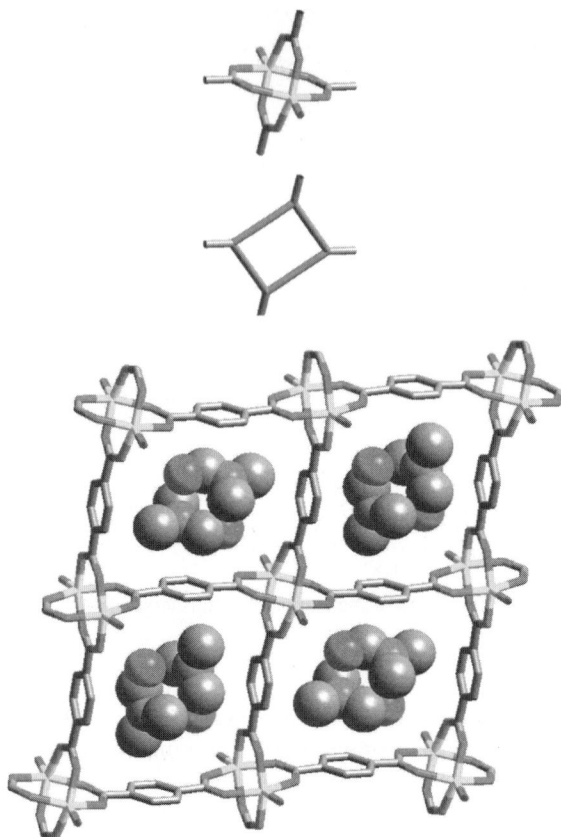

Fig. 5. 2D grids assembled from paddle-wheel clusters and p-BDC ligands in Zn(p-BDC)(H$_2$O)·(DMF), (MOF-2) (from Eddaoudi et al. [9]).

The synthesis of (Zn$_4$O)(p-BDC)$_3$·(DMF)$_8$(C$_6$H$_5$Cl) (MOF-5) [20, 28] turned out to be a breakthrough in metal-organic framework chemistry. Tetranuclear tetrahedral Zn$_4$O clusters are the key feature of the wide-open cubic structure (Fig. 7). The clusters form a primitive lattice. They are linked through capping carboxylate groups of six two-connecting p-BDC ligands to form an extended 3D six-connected net. The topology of the net is the same as in the ReO$_3$ structure with Zn$_4$O and p-BDC replacing Re and O, respectively. Approximately 55–61 % of the space is accessible to guest species and molecules up to 8.0 Å in diameter should be able to pass through the apertures of the intersecting channels. An extremely high thermal stability has been demonstrated by complete removal of intercalated molecules at 300 °C, without loss of crystallinity. A single-crystal X-ray diffraction study has been carried out on a desolvated form of this framework that showed mono-crystallinity with only little change in cell constants and atomic parameters. For a

Fig. 6. Trinuclear zinc clusters are linked via p-BDC ligands to give an open network in $Zn_3(p\text{-}BDC)_3\cdot6MeOH$ (MOF-3) (from Eddaoudi et al. [9]).

a)

b)

c)

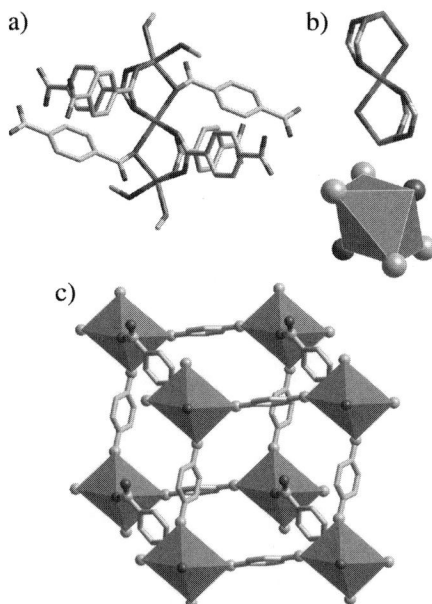

crystalline material the density of the desolvated material is extremely low (0.59 g cm^{-3}), which is a prerequisite for the exceptional high pore volume of 0.61–0.54 cm^3 cm^{-3} and the apparent Langmuir surface area of 2900 mg^2 g^{-1} of the material, a record unprecedented in zeolite chemistry.

In $Tb_2(p\text{-}BDC)_3(H_2O)_4$ (MOF-6) [29] Tb(III) is coordinated by 6 BDC ligands. Each carboxylate group is connected to two Tb ions in different unit cells and thus p-BDC serves as a four-connector. The compound has $M_2(CO_2)_3$-layers coplanar to the *ac*-plane of the triclinic cell which are stacked along *b* and separated by BDC linkers (Fig. 8a). Thus, a 3D condensed network with extended 1D channels along the crystallographic *b*-axis results. The coordinated water molecules point into the center of the channels and can be removed at only 115 °C, without collapse to give a thermally stable (up to 450 °C) porous material with a pore volume of 0.099 cm^3 g^{-1} estimated from water-sorption experiments. The pore size of the material is comparable to small-pore zeolites and allows for the sorption of water (2.60 Å) but not for N$_2$ (3.64 Å). Due to differences in the luminescence lifetime of the hydrated and the dehydrated compound a fluorescence detection of small molecules seems quite promising (see Sect. 4.3.3.3).

$Tb(p\text{-}BDC)NO_3\cdot2DMF$ [30] on the other hand has a completely different net topology related to the PtS-structure (Fig. 8b). Extended 1D Tb(p-BDC)$_2$-chains, in which each Tb ion is coordinated by four oxygen atoms of different BDC-ligands, are linked with each other perpendicular to the chain axis to give a porous 3D

References see page 1244

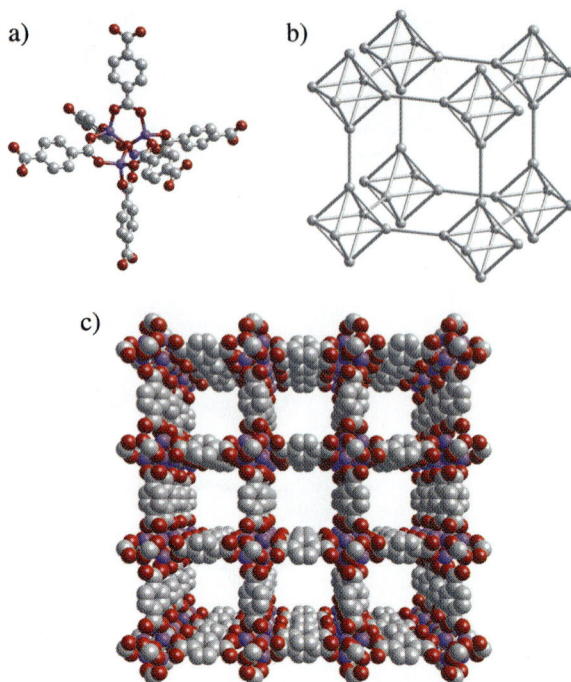

a)

b)

c)

Fig. 7. Evolution of the $(Zn_4O)(p\text{-}BDC)_3 \cdot (DMF)_8(C_6H_5Cl)$ (MOF-5) structure (from Eddaoudi et al. [9]): (a) The basic building unit is a Zn_4O-cluster surrounded by six p-BDC ligands; (b) the carboxylate C-atoms represent the connectivity pattern: two-connected octahedra form a 3D net (the same topology is found in the boron net of CaB_6); (c) space-filling representation.

structure. The DMF molecules are removed at 120 °C to give a microporous solid, stable up to 320 °C with a specific pore volume of 0.032 cm^3 g^{-1} and a pore size that allows for the adsorption of CH_2Cl_2 (4 Å), but not cyclohexane (6 Å). Contact with water converts the desolvated material into $Tb_2(p\text{-}BDC)_3(H_2O)_4$.

4,4′-azodibenzoate (ADB, Fig. 4) can be considered as an extended version of p-BDC. The ligand may be used to construct interpenetrating octahedral networks, such as $Tb_2(ADB)_3(DMSO)_4 \cdot 16DMSO$ (MOF-9) [31] in which $Tb_2C_6O_{12}$ clusters (Fig. 9a,b) can be considered topologically as octahedral SBUs. Four of the six carboxylate groups surrounding the cluster form bridges between the two metal centers in a di-monodentate fashion, whereas the remaining two are bidentate (Fig. 9a). The clusters are joined by long ADB-linkers to give two identical interpenetrating networks (Fig. 9c). The free volume is, at 71 % (calculated) extremely high for interpenetrating networks and the dimensions of the rectangular channels are 7.5 × 6.7 Å. All DMSO molecules (guest and coordinating) can be removed at 180 °C and reversible exchange of $CHCl_3$ and DMF has been demonstrated. A simple model shows that the size of linker and SBU is optimal to just miss the

a)

b)

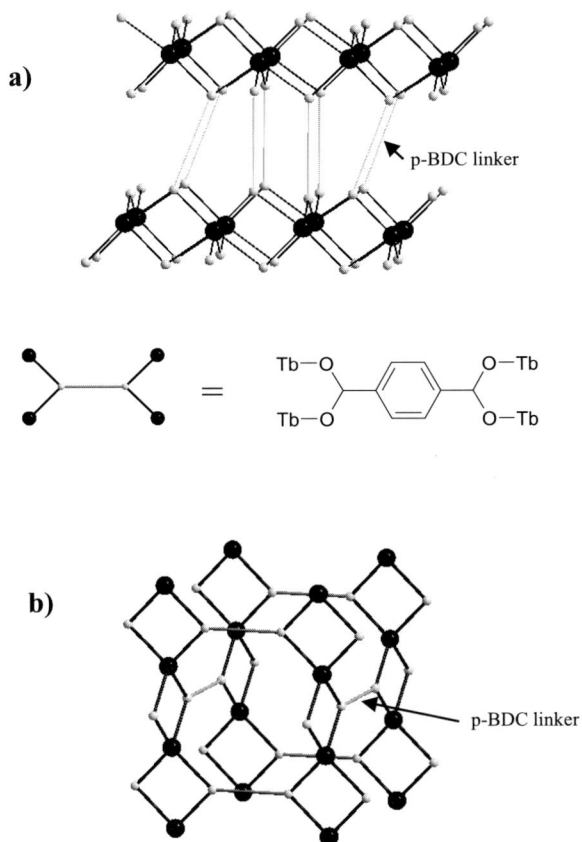

Fig. 8. Network topology in (a) $Tb_2(p\text{-}BDC)_3(H_2O)_4$; (b) $Tb(p\text{-}BDC)NO_3\cdot2DMF$.

formation of a third interpenetrating framework that would result in less free volume (40 %).

Finite metal-organic polyhedra (MOP) were obtained with m-BDC (Fig. 4) in $Cu_{24}(m\text{-}BDC)_{24}(DMF)_{14}(H_2O)_{10}\cdot(H_2O)_{50}(DMF)_6(C_2H_5OH)_6$ [32], obviously due to the 120° angle between the functional groups. Four $-CO_2-$ groups are linked to binuclear copper clusters (paddle-wheel) to form a square planar SBU. The resulting polyhedron is a truncated cuboctahedron 25 Å in diameter with 24 two-connectors (m-BDC) and 12 $Cu_2(CO_2)_4$-paddle-wheels of high thermal stability similar to extended porous structures, and might be well suited for further functionalization to connect the clusters in a secondary synthesis step. Preliminary experiments show that the polyhedral structure is stable in solution, as it is possible to dissolve the solid in refluxing DMF and to recrystallize it by cooling to room temperature.

References see page 1244

a)

b)

c)

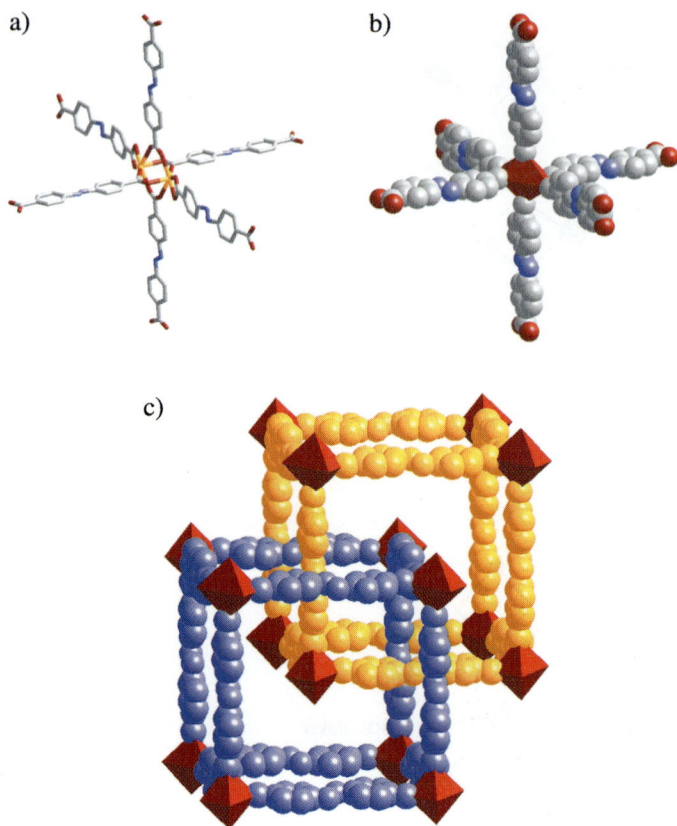

Fig. 9. Interpenetrating architecture of
$Tb_2(ADB)_3(DMSO)_4 \cdot 16DMSO$ (MOF-9): (a) Tb_2-SBU held
together from bidentate carboxylate groups; (b) space-filling
representation; (c) 3D network (from Eddaoudi et al. [9]).

A 3D porous coordination polymer $Mn(TDC)(H_2O)_2$ [33] has been synthesized
from $Mn(ClO_4)_2$ and thiophene-2,5-dicarboxylate (TDC) in which the ligand binds
in a di-monodentate fashion to four Mn(II) centers. The Mn centers thus form
layers of four-membered rings (Fig. 10a) bridged together by carboxylate groups
and separated from each other by the thiophene spacer perpendicular to the layers
(Fig. 10b). Two additional water molecules complete the distorted octahedral coordination of Mn(II). They can be removed at 185 °C and no further weight loss is
observed up to 350 °C, indicating the formation of a stable Mn(TDC) phase, but
physisorption experiments have not been carried out.

Aliphatic dicarboxylates like adipato ligands (AD, Fig. 4) are less rigid
and therefore disadvantageous in the synthesis of metal-organic frameworks.
$La_2(AD)_3(H_2O)_4 \cdot 6H_2O$ [34], for example, is porous and has large channels in
which water molecules form a sophisticated system of hydrogen bonds, but the

a)

b)

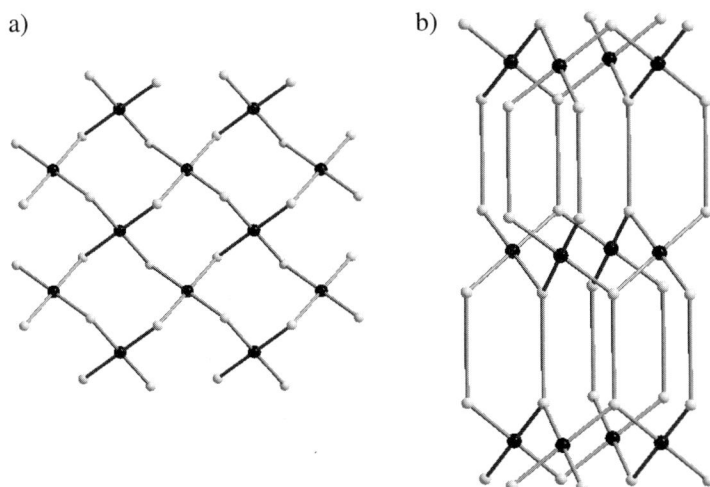

Fig. 10. 3D framework topology in Mn(TDC)(H$_2$O)$_2$, (a) view upon the layers; (b) stacking of the layers, only carboxaylate C and Mn atoms are shown.

compound dehydrates at 100 °C into crystalline nonporous La$_2$(AD)$_3$(H$_2$O)$_2$ with a surface area of only 8 m^2 g^{-1}. Interestingly the dehydration is reversible, which indicates that the framework is not irreversibly destroyed upon removal of the solvent molecules. Mn$_2$(AD)$_2$(H$_2$O) [35] has a layered structure composed of edge-sharing MnO$_6$ octahedra linked together by carboxylate groups. All adipate linkers are oriented along the a-axis and separate the layers from each other. The compound is antiferromagnetic below 15 K and stable up to 210 °C, but dehydrates irreversibly at this temperature.

The oxalate anion is the shortest dicarboxylate linker. Thus, a template is used for the synthesis of open-framework materials. With the aid of TMEDA, or guanidinium carbonate, two different structures were obtained [36] in which each oxalate anion connects two metal centers. [Me$_2$NH(CH$_2$)$_2$NHMe$_2$]Sn$_2$(C$_2$O$_4$)$_3$·H$_2$O (Fig. 11a) consists of puckered sheets in which SnO$_6$-units are connected through the oxalate groups and form six-metal-membered rings. The overall T-shaped arrangement of the ligands around the Sn atoms results in the formation of a brick-wall-like 2D pattern. The pores are filled with protonated template (Fig. 11b) and water molecules. The guanidinium compound [C(NH$_2$)$_3$]$_2$[Sn$_4$(C$_2$O$_4$)$_5$]·2H$_2$O contains even larger voids. Four- and six-coordinated Sn atoms are two-connected through oxalate anions to give ten-metal-membered rings that are condensed into a unique sawtooth layered network. However, the large apertures are stabilized by the occupation of guest molecules, which are necessary to provide the structural integrity of the frameworks. Removal of the guest molecules generally leads to the collapse of the framework structure.

References see page 1244

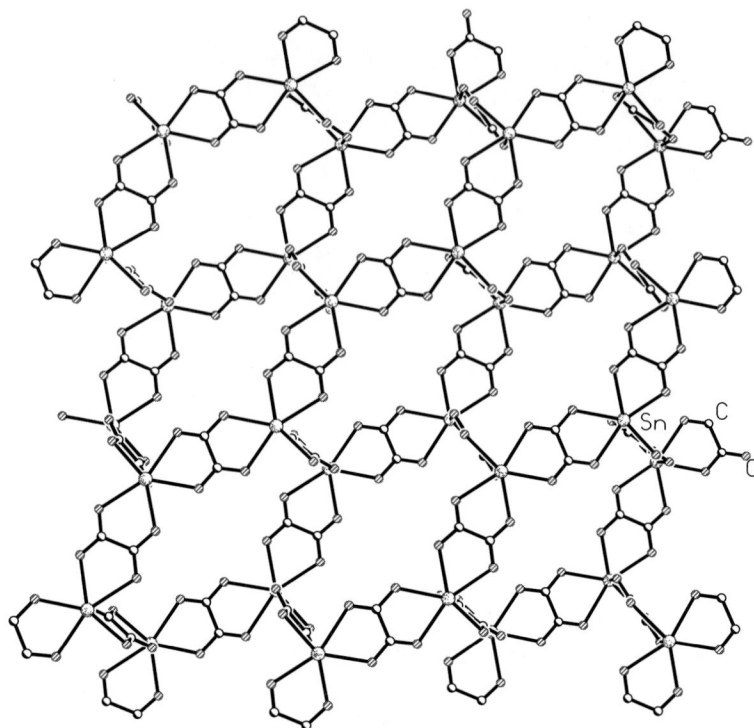

(a)

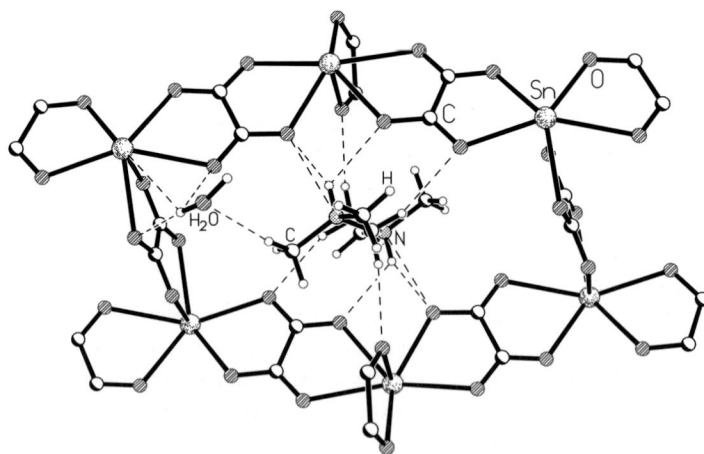

(b)

Fig. 11. (a) Tin-oxalate framework in
[Me$_2$NH(CH$_2$)$_2$NHMe$_2$]Sn$_2$(C$_2$O$_4$)$_3$·H$_2$O; (b) position of the
template (from Natarajan et al. [36]).

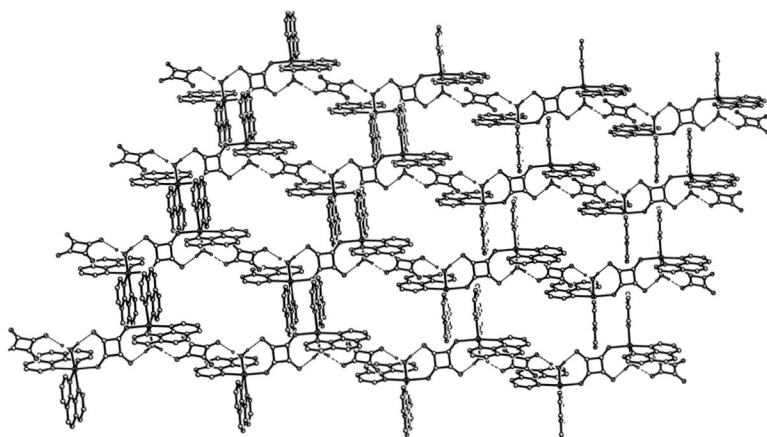

Fig. 12. Sheet-like network in $Co_2(phen)_4(\mu$-$C_4O_4)(H_2O)_2[C_4O_4]\cdot 8H_2O$ with strong interionic hydrogen bonds and π–π interactions (from Lai et al. [37]).

Architectures with thermal stability up to 200 °C were obtained using squaric acid. Strong interionic hydrogen bonds in chains composed of dinuclear $[Co_2(phen)_4(\mu$-$C_4O_4)(H_2O)_2]^{2+}$ (phen = 1,10-phenanthroline) cations and squarate anions as well as cooperative π–π interactions between adjacent chains (Fig. 12) are responsible for the rigidity of the planar sheets of $Co_2(phen)_4(\mu$-$C_4O_4)(H_2O)_2[C_4O_4]\cdot 8H_2O$ [37]. The sheets are stacked to give channels 12×7 Å (phen-to-phen distance) in size which are filled with water. Dehydrated samples show reversible N_2-adsorption and desorption. The presence of hydrophilic and hydrophobic void characteristic is an interesting feature of the supramolecular complex. The molecular $Mn(HC_4O_4)_2(H_2O)_4$ and chain-like $Zn(C_4O_4)(H_2O)_2(DMSO)_2$ can be converted into 3D extended cage systems $M(C_4O_4)(H_2O)_2$ [38] that contain cavities [39]. In $Cu(PYZ)(C_4O_4)(H_2O)_4$, linear $Cu(PYZ)$ chains are interlinked via hydrogen-bonded squarate anions to form a 3D-network [40].

4.3.2.1.2 Tricarboxylates

The free acids used for the synthesis of tricarboxylate frameworks are shown in Fig. 13. The best results were obtained with the BTC ligand (trimesic acid) but recently remarkably stable BTB networks were also obtained. In $Co(HBTC)(NC_5H_5)_2\cdot 2/3NC_5H_5$ [41] the transition metal is coordinated to three different 1,3,5-benzenetricarboxylates (BTC) as either mono- or bidentate ligands to form planar metal carboxylate sheets along the x–y plane of the hexagonal crystals. Two additional pyridine molecules complete the coordination and hold the layers together by π-bonds between the pyridine molecules to give a layer separation of 7 Å (Fig. 14).

References see page 1244

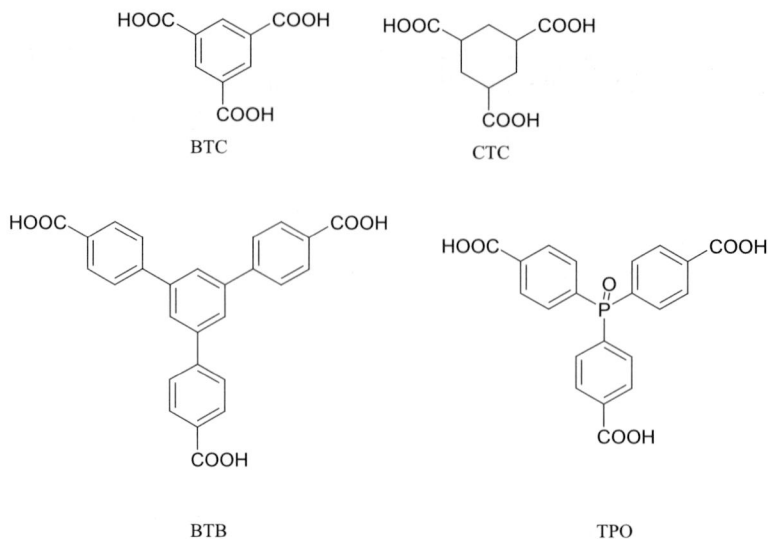

Fig. 13. Tricarboxylates. Shown are the free acids. The abbreviation corresponds to the respective anion.

Uncoordinated pyridine molecules fill the remaining space between the layers. Even though not covalent, the bonds in between adjacent sheets are surprisingly stable. Only pyridine molecules occupying the channels are lost upon heating, but no further weight loss appears up to 350 °C, which is remarkable, especially for a

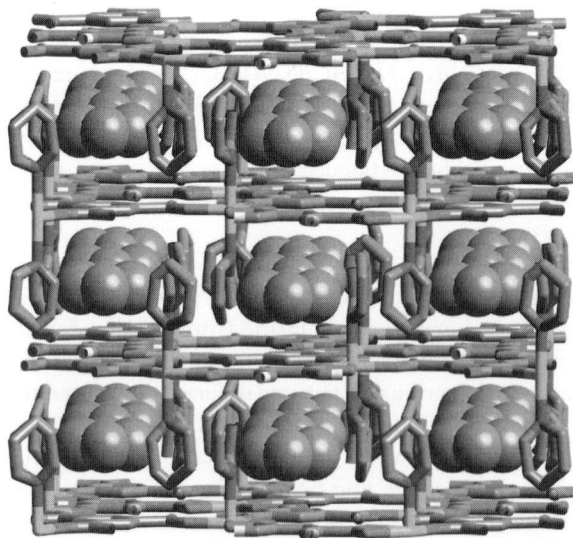

Fig. 14. The layers in $Co(HBTC)(NC_5H_5)_2 \cdot 2/3NC_5H_5$ are held together by $\pi-\pi$ interactions between coordinated pyridine molecules of adjacent layers (from Yaghi et al. [11]).

layered compound. The framework shows a high selectivity for the occlusion of aromatic solvents. From the conceptional point of view the use of metal–ligand bonds as well as π-bonds in different directions of a framework seems comparable to the situation in $Co_2(phen)_4(\mu\text{-}C_4O_4)(H_2O)_2[C_4O_4]\cdot 8H_2O$ [37] (Sect. 4.3.2.1.1). As in pillared clays further extension using condensed aromatic systems may be achieved.

The reaction of M(II) acetates with BTC in water results in the formation of the isostructural layered hydrates $M_3(BTC)_2\cdot 12H_2O$ (M = Co, Ni, Zn) [42]. BTC serves as a three-connector, but one of the carboxylate groups binds in a bidentate fashion to a metal center that is only coordinated to one BTC. Since the two residual carboxylate groups bind in a unidentate fashion to two axial positions of the second metal center, zigzag chains are formed and not a network. However, these are further connected via hydrogen bonds to yield a tightly held 3D framework with an extended 1D channel system filled with water ligands (4×5 Å without water molecules). Eleven water molecules per formula unit are reversibly removed upon heating and rehydration restores the crystalline structures. Especially for the zinc compound there is strong evidence that the poorly crystalline monohydrate maintains the structural arrangement of the porous framework using hydrogen bonds. A high affinity toward small guests with lone electron pairs, like ammonia, was observed but attempts to include other small molecules, such as CO, CO_2 or CH_3CN, were unsuccessful.

$Ca(HBTC)(H_2O)\cdot H_2O$ has a layered structure [43]. Infinite 2D layers of HBTC anions alternate with layers of calcium cations, which are coordinated by five distinct carboxylate anions. The HBTC ligand uses mono- and bidentate modes to bind the calcium. In addition, it forms a strong hydrogen bond, linking the HBTC anions in chains. Complete dehydration of the compound is achieved at 230 °C and rehydration restores crystallinity.

On the other hand the nickel compounds $Ni_3(BTC)_2(py)_9(H_2O)_3\cdot 3.3(C_5H_{12}O)\cdot 1.5(py)4H_2O$ and $Ni_3(BTC)_2(py)_6(BuOH)_6\cdot 2.2(BuOH)6.5H_2O$ [44] form 2D hexagonal (6,3) networks. The hydrogen bonding to auxiliary ligands connected to the metal controls the dimensionality of coordinated frameworks. Each hexagon contains six three-connecting BTC ligands linked via all three monodentate carboxylate groups by octahedral nickel centers. Hydrogen bonds between one coordinated water and carboxylate oxygen in the first compound induce a slight puckering of the hexagonal sheets. They have a separation distance of 7 Å and are stacked ABCA′B′C′. The voids in between layers are filled with guest molecules. In the second compound, the equatorial sites of Ni(II) are occupied by two pyridine and two 1-butanol molecules in a trans arrangement, which enforces the 2D motif and locks the BTC units in approximately the same plane [44]. Interlayer hydrogen bonding induces a more severe puckering and a new stacking, AAA, in which the hexagonal voids are aligned to form parallel channels with dimensions 13.7×13.4 Å. Heating to 150 °C results in loss of guests and ligands (approximately 41 % by weight). The desolvated frameworks are X-ray amorphous, but upon readsorption of pyridine crystallinity is restored.

References see page 1244

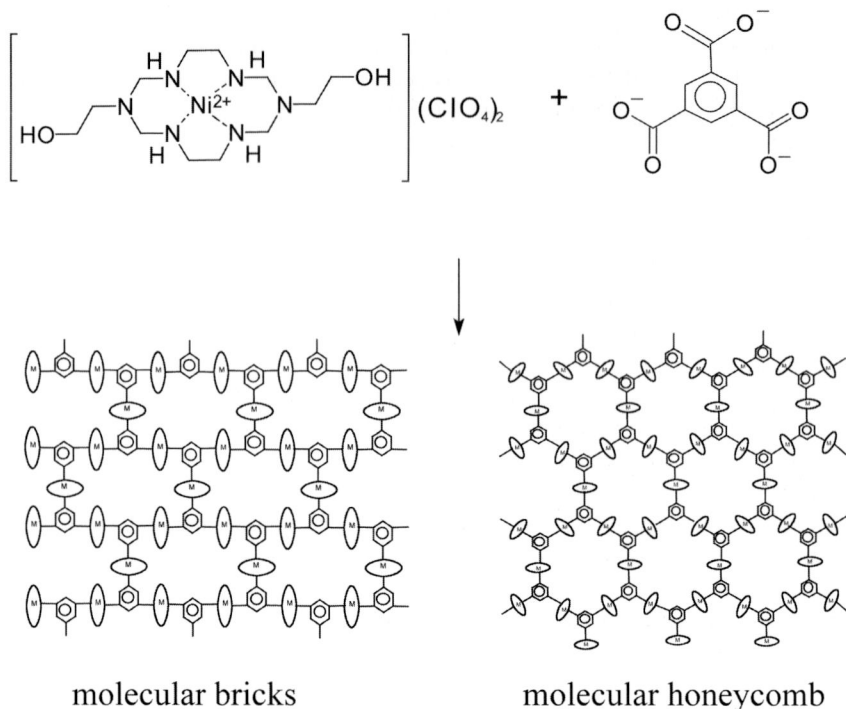

Fig. 15. Macrocyclic Ni-complexes may be used as secondary building blocks for the construction of functionalized networks [45].

Layered brick-wall and honeycomb patterns were also obtained using macrocyclic Ni(II) complexes and BTC [45]. This concept might be interesting because of the magnetic properties associated with the complex and additional functional groups attached to the macrocycle, which are valuable for recognition. Each of the BTC carboxylate groups binds in a monodentate fashion to one Ni(II) center to form double six-membered rings (Fig. 15). The layers are stacked at a distance of approximately 9 Å and held together by hydrogen-bonding interactions of different functional groups contained in the macrocycle. The resulting cavities are 6.7×13 Å and 11.4×11.4 Å and filled with H_2O and pyridine guest molecules. Magnetic measurements indicate a weak antiferromagnetic coupling of the Ni(II) macrocycles between the S = 1 paramagnetic centers.

$Cu_3(BTC)_2(H_2O)_3$ [46] has also a metal to BTC ratio of 3:2, but the formation of $Cu_2(CO_2)_4$ paddle-wheels leads to a 3D network in which the Cu_2-units function as four-connectors and the BTC ligands as three-connectors. The resulting double four- and six-membered rings constitute a wide open framework (Fig. 16) which is stable up to 240 °C. The intersecting 3D channel-system has large square-shaped pores 9×9 Å wide with a measured pore volume of 0.333 cm^3 g^{-1} which represents a void fraction of 40.7 %. The three water molecules per formula unit are

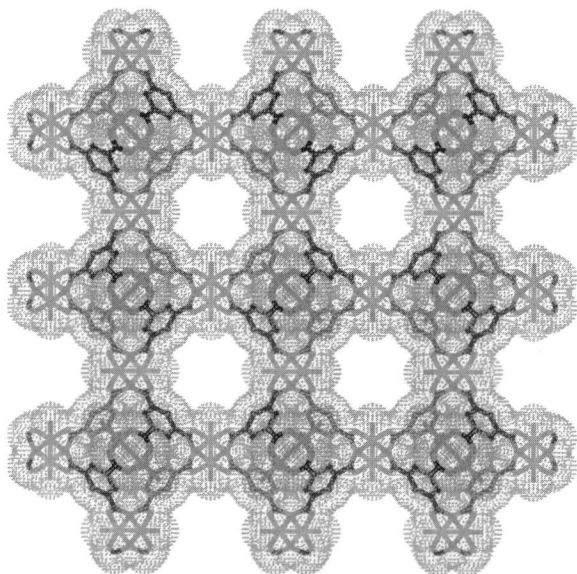

Fig. 16. Cu$_2$(COO)$_4$-paddle-wheels and three-connecting BTC ligands form a cubic wide-open framework in Cu$_3$(BTC)$_2$(H$_2$O)$_3$ (from Chui et al. [46]).

axial water ligands opposite to the Cu–Cu vector. They can be replaced by dehydration at 100 °C and subsequent addition of other donor ligands such as pyridine without loss of structural integrity.

Chiral networks are promising host matrices for enantio-selective binding and catalytic transformations of chiral or prochiral molecules. In particular the (10,3)-a net (SrSi$_2$-topology, Fig. 2b) attracts attention since the net topology is chiral *per se*. An impressive example of such a porous chiral host has been synthesized by Yaghi et al. [47]. Zn$_2$(BTC)(NO$_3$)·(H$_2$O)(C$_2$H$_5$OH)$_5$ (MOF-4) is composed of trigonal bipyramidal Zn$_2$(CO$_2$)$_3$-clusters, in which three carboxylate C atoms occupy the equatorial positions and each axial Zn is coordinated by three carboxylate O atoms in a monodentate fashion (Fig. 17a). The net is thus composed only of alternating Zn$_2$- and BTC-three-connecting units with their 3-fold axis almost perpendicular to each other (Fig. 17b). They form rings of 10 alternating three-connecting sub-units (Fig. 17c) that are fused together to form a 3D (10,3)-a framework with the SrSi$_2$ topology (Fig. 17c,d, see also Fig. 2). Water and two of the ethanol molecules are loosely intercalated in the 14-Å wide channels, and are mostly removed in vacuum at room temperature. The tightly bound nitrate ions can not be exchanged, but guest removal and selective binding of smaller alcohols and DMF has been demonstrated to be a size- and shape-selective inclusion process. Although gas adsorption was not observed, the kinetics of ethanol sorption indicate distinct coordi-

References see page 1244

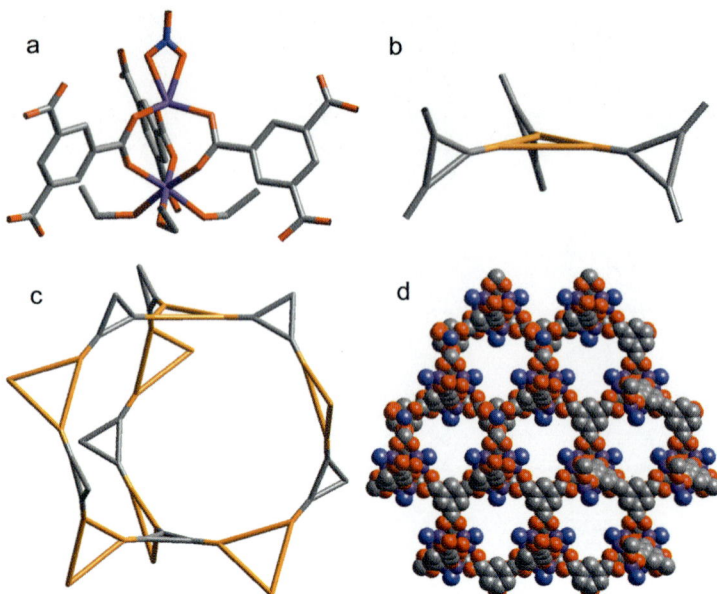

Fig. 17. Binuclear zinc clusters (a) in
$Zn_2(BTC)(NO_3)\cdot(H_2O)(C_2H_5OH)_5$ and BTC ligands, both
three-connecting subunits, are assembled into 10-membered
rings (b) and (c) to give a chiral porous (10,3)-a net (d) (from
Eddaoudi et al. [9]).

nation transitions [20], which demonstrate the principal accessibility of the Zn
centers for catalytic transformations.

In $M_3(BTC)_2(py)_6(eg)_6\cdot(eg)_x(H_2O)_y$ (M = Co(II), Ni(II), eg = ethylene glycol)
[48, 49] four of such chiral (10,3)-a networks are interpenetrating each other (Fig.
18b). This result is quite remarkable since the Flack parameter is 0.03(11) and thus
the crystal investigated represents a pure enantiomer. Although, without using
chiral auxiliaries the bulk samples are racemic mixtures, in principle the crystals of
one enantiomer might be useful as chiral heterogeneous catalysts. Ni is coordinated to only two BTC ligands (Fig. 18a). The benzene rings are arranged non-
coplanar and Ni(II) serves as a linear linker for the three-connectors. The same
(10,3)-a nets, somewhat distorted but now only as two interpenetrating networks
are obtained with 1,2-propanediol used instead of ethylene glycol in the synthesis.
The former coordinates the metal as bidentate and due to the directing character of
the hydrogen bonds (Fig. 18a) a small distortion from the ideal (10,3)-a net results.
In $Ni_3(BTC)_2(py)_6(eg)_6\cdot(eg)_x(H_2O)_y$ approximately 28.1 % of the crystal volume is
occupied by free solvent and each pore has a maximum diameter of 16 Å. The
compound may be desolvated completely at 115 °C accompanied by a color change
to leave a poorly crystalline material with empty chiral cavities [49] but without
irreversibly destroying the framework structure. Interestingly, the chiral bidentate
1,2-propanediol was refined as a single enantiomer, although a racemic mixture

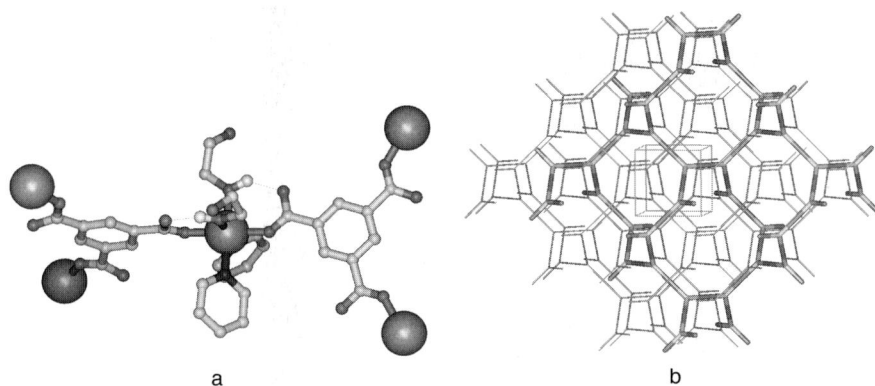

Fig. 18. The remarkable architecture of
$Ni_3(BTC)_2(py)_6(eg)_6 \cdot (eg)_x(H_2O)_y$: (a) coordinating ethylene
glycol ligands direct, via hydrogen bonds, the orientation of the
two benzene-rings with respect to each other; (b) the four
chiral interpenetrating (10,3)-a nets (from Kepert et al. [48]).

had been used in the synthesis. The enrichment was also proven by chiral GC experiments of dissolved enantiopure single crystals and demonstrates the specific templating of enantiopure chiral microporous phases. In fact, enantiopure bulk phases can be synthesized using enantiopure 1,2-propanediol [50]. The chiral host is a promising candidate for enantioselective sorption and catalysis.

Despite the structural similarity to BTC the saturated ring carboxylate CTC remains largely unexplored as a building block for coordination porous solids. $Zn_3[(CTC)(C_5H_5N)]_2 \cdot 2DMF$ [51] possesses channels having a pore diameter of nearly 6 Å. The principal building unit is composed of three Zn(II) centers bridged together by six carboxylate groups. They are arranged in layers nearly perpendicular to the crystallographic a-axis and linked together through the CTC ligands. The channels are filled with DMF molecules that are mobile and can be removed by heating to 140 °C.

A more recent example is the 3D lanthanide network of $Er(CTC)(H_2O)_2 \cdot 2.5H_2O$ [52] synthesized from a hydrothermal reaction. Each Er is coordinated by seven carboxylate oxygen atoms from two bidentate and three monodentate carboxylate groups, five different CTC ligands in total, and each CTC ligand also connects five Er ions. One of the additional coordinating water molecules also connects two Er centers to give a porous 3D network (Fig. 19). The 1D channels are filled with water and run along each of the crystallographic axes, of which the largest cavity size is 4.7 Å. Dehydration at room temperature in vacuum removes 2.5 H_2O per formula unit quantitatively accompanied by degradation of long-range crystallinity and retention of the framework, as judged from rehydration experiments. At 150 °C one of the coordinated water molecules can be removed without loss of the

References see page 1244

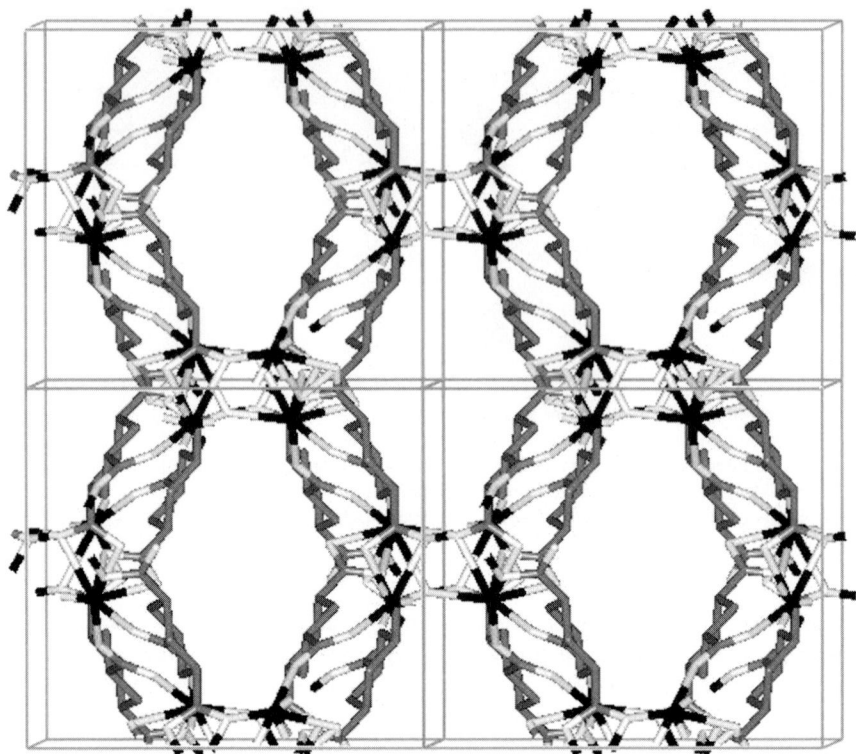

Fig. 19. 1D channel system in Er(CTC)(H$_2$O)$_2$·2.5H$_2$O (from Pan et al. [52]).

frameworks integrity, but above that temperature removal probably of the bridging H$_2$O irreversibly changes the Er-CTC framework structure.

Interwoven architectures with remarkably high accessible pore volume were synthesized recently from copper(II) nitrate and 4,4′,4″-benzene-1,3,5-triyl-tribenzoic acid (H$_3$BTB, Fig. 13), an expanded version of BTC. This results in the formation of Cu$_2$(CO$_2$)$_4$ paddle-wheels, which again serve as planar four-connecting SBUs. Cu$_3$(BTB)$_2$(H$_2$O)$_3$·(DMF)$_9$(H$_2$O)$_2$ (MOF-14) [53] has two discrete nets with the Pt$_3$O$_4$ topology, in which each four-connecting paddle-wheel replaces platinum and the three-connecting BTB replaces oxygen. The interweaving is responsible for the rigidity and thermal stability of the open framework up to 250 °C with two inter-penetrating labyrinths, each consisting of six orthogonal channels separated by a P-minimal surface (Fig. 20). The cavities are large enough to accommodate spheres up to 16.4 Å in diameter. They are filled with water and DMF guests and represent 67 % of the cell volume. Accessibility is guaranteed through the 7.66 × 14.00 Å wide apertures joining two cavities and removal of virtually all guests is accomplished in vacuum to give a porous framework with a pore volume of 0.53 cm^3 g^{-1} and type I isotherms for N$_2$, Ar, C$_6$H$_6$, and m-xylene. Even in organic solvents MOF-14 maintains its structural integrity.

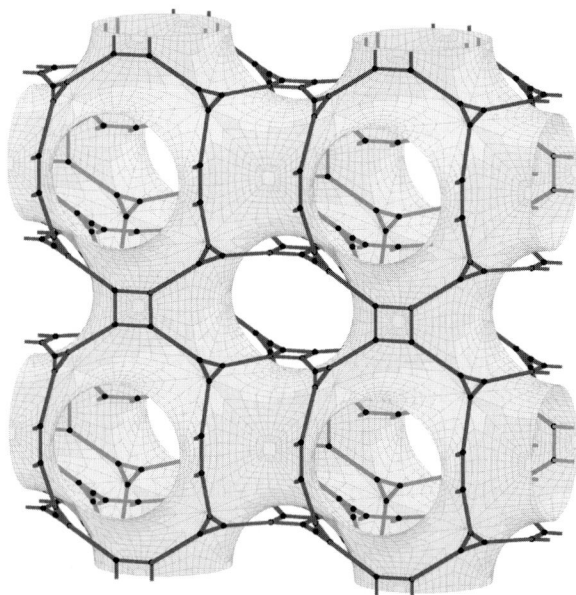

Fig. 20. Interpenetrating nets with the Pt$_3$O$_4$ topology in Cu$_3$(BTB)$_2$(H$_2$O)$_3$·(DMF)$_9$(H$_2$O)$_2$ (MOF-14); each square represents a Cu$_2$(CO$_2$)$_4$-paddle-wheel and each triangle a three-connecting BTB-ligand. The interwoven frameworks are separated by a P-minimal surface, a triply periodic minimal surface in Euclidean space (from Chen et al. [53]).

Tris(4-carboxylphenyl)phosphineoxide H$_3$TPO (Fig. 13) can also be seen as an extended version of BTC even though not planar. In combination with 2,2′-bipyridyl, which serves as a nonconnecting ligand, it forms ladders such as Co$_3$(TPO)$_2$(2,2′-BPY)$_2$(H$_2$O)$_6$·2.4H$_2$O [54, 55] with large channels. Here, the TPO ligand is the T-shaped three-connecting building block in the ladder and the cobalt centers are two-connecting expanders in the rungs and side chains. Residual coordination sites are filled with water in the rungs and 2,2′-BPY at the edges. The double chains associate via van der Waals forces between the 2,2′-bipyridyl rings. Noncoordinating water is loosely bound within channels and removable in vacuum without collapse of the framework. Porosity of the 1D coordination-polymer was also demonstrated on the basis of D$_2$O–H$_2$O exchange experiments, and the material showed superior performance as compared to Ca(HBTC)(H$_2$O)·H$_2$O [43] and Co$_3$(BTC)$_2$·12H$_2$O [42].

4.3.2.1.3 Tetracarboxylates

Less work has been carried out with the tetracarboxylates so far, obviously due to charge-matching and crystallization problems. The tetraphenylmethane derivatives are probably promising rigid candidates for tetrahedral networks. However, the

References see page 1244

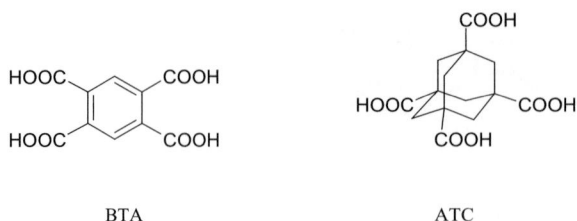

BTA   ATC

Fig. 21. Tetracarboxylates. Shown are the free acids. The abbreviation corresponds to the respective anion.

benzene tetracarboxylates obviously tend to form layered compounds. For example Fe(phen)(BTA)$_{0.5}$ (BTA = benzene-1,2,4,5-tetracarboxylate) [56] consists of 2D polymeric layers. All eight carboxylate oxygen atoms are used to link the ligand to six metal ions forming a grid with two kinds of rings, one consisting of two carboxyl groups and two metal centers and another with two BTA ligands and two Fe(II) ions. The layers are stacked at an average distance of 9.3 Å. Planar phenanthroline disks are oriented almost perpendicular to the layers and show weak π–π interactions which strengthen the network.

On the other hand 1,3,5,7-adamantane tetracarboxylate (Fig. 21) seems to be an ideal ligand to achieve four-connected nets because of the tetrahedral arrangement of the carboxylate groups in this molecule. In Cu$_2$(ATC)·6H$_2$O (MOF-11) [57] binuclear copper clusters are surrounded by four tetracarboxylate tetrahedra. Again, Cu$_2$(CO$_2$)$_4$ paddle-wheels can be considered as square planar secondary building units, which connect the adamantane tetrahedra to give a decorated PtS structure (Fig. 22). The coordination of copper is completed by two water molecules per Cu$_2$-cluster. They point into the channels that are 6.0 to 6.5 Å in diameter. The residual four water molecules are located in the pores. Interestingly the compound can be completely dehydrated at 260 °C without collapse to give a porous material with a free measured pore volume of 0.20 cm^3 g^{-1}, which can also take up organic molecules like C$_6$H$_{12}$ (6.2 Å).

4.3.2.1.4 Heterocyclic Carboxylates with N-Donor Functionality

The combination of ionic and nonionic functionality provides a higher variability in terms of connectivity patterns. However, the results obtained so far are less encouraging than those obtained from pure carboxylate or N-donor ligands. On the other hand, unbound functional groups might serve to further modify microporous host materials in postsynthesis treatments.

Heterocyclic ortho-dicarboxylates (Fig. 23) like pyrazine-2,3-dicarboxylate (PZDC) form layers, for example, with Cu(II) in which each metal center is connected to three carboxylate groups in a monodentate fashion to form neutral layers that can be pillared with neutral N-donors like PYZ, 4,4′-BPY or PIA (Fig. 24) to give porous stackings with 1D channels of 4 × 6, 9 × 6 and 10 × 6 Å, respectively [58]. The framework can be dehydrated at 100 °C without loss of crystallinity and is stable up to 260 °C. The methane-adsorption capacity increases with the size of the pillar up to 2.9 mmol g^{-1} for PIA.

Pyridine-2,5-dicarboxylic acid acts as a bidentate ligand using the N-donor atom

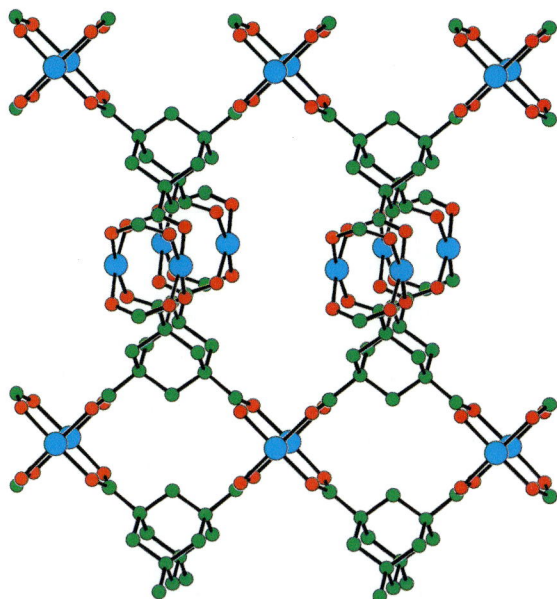

Fig. 22. Decorated PtS-framework in $Cu_2(ATC)\cdot 6H_2O$ (MOF-11) (from Eddaoudi et al. [9]).

and the carboxylate group close to the nitrogen. The second carboxylate group can form two additional monodentate links to give, for example, a layered structure such as $M(II)(2,5\text{-PDC})(H_2O)_2\cdot H_2O$ (M = Co, Ni) [59] in which the water ligands also participate in inter- and intralayer hydrogen bonds.

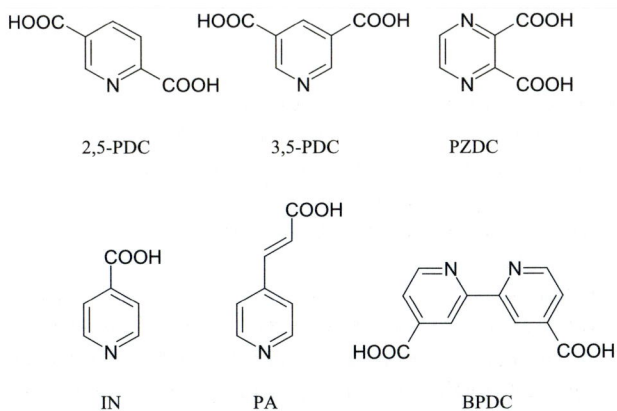

Fig. 23. Carboxylates derived from heterocycles. The heteroatoms provide additional donor functionality that may be used to achieve a higher connectivity.

References see page 1244

PYZ   4,4'-BPY   BPET   AZPY   DPDS   BPETHY   PIA   AZB

BPD

PYTZ   C₄-BPY   HMT   4,4'-BPZ   HPYMO

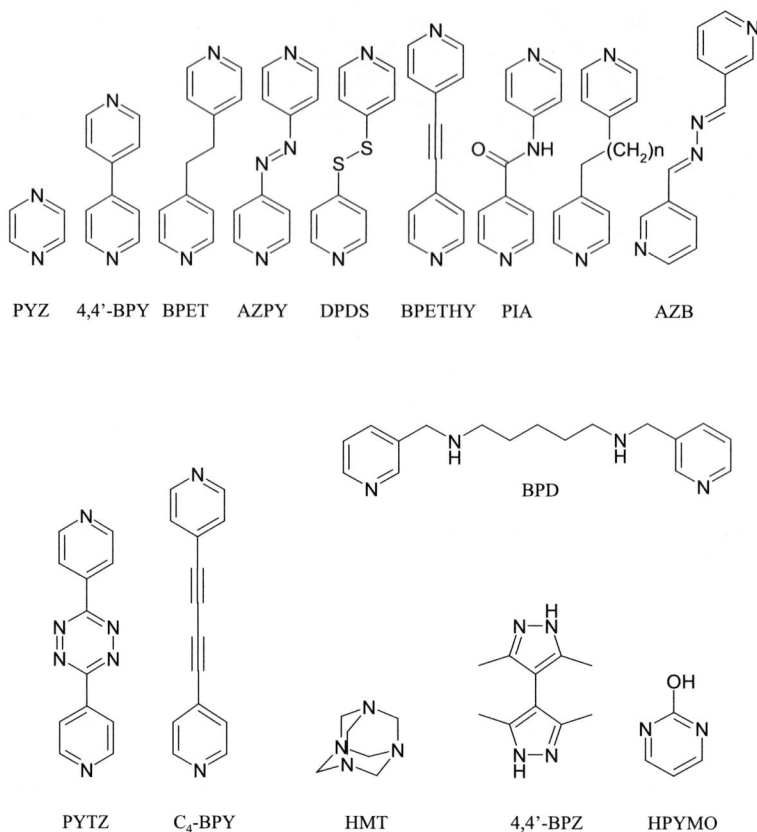

Fig. 24. N-donor ligands of the 4,4'-BPY type are rigid and variable in length and thus ideal building blocks for the assembly of porous coordination polymers.

Lamellar structures were also obtained with pyridine-2,5-dicarboxylic acid. $Gd_2Ag_2(2,5\text{-PDC})_4(H_2O)_4$ [60], for example, represents such a structure in which all PDC ligands connect three metal atoms. The layers are further linked by hydrogen-bond interactions forming rectangular cavities with a size of 6.3×7.9 Å.

Pyridine-3,5-dicarboxylic acid is more symmetric. In $Co(3,5\text{-PDC})(H_2O)_2$ [61] each metal center is coordinated by two bidentate carboxylate groups and one pyridine N atom to give a three-connected layered structure.

A more efficient linker is 2,2'-bipyridine-4,4'-dicarboxylate (BPDC). For example in $Mn(BPDC)\cdot 2H_2O$ [62] the carboxylate groups are bridging in a di-monodentate fashion that aligns the manganese centers in infinite chains. Further crosslinking is accomplished via N-coordination to give a 3D network with 6.87×5.86 Å (atom to atom) wide channels and high thermal stability up to 310 °C.

Chessboard tunnels have been found in $Cu(II)(IN)_2\cdot 2H_2O$ (IN = isonicotinate) [63] in which the square-pyramidal coordinated copper centers are linked by five IN ligands. The framework is readily understood as a square ML_2-grid in which L is a two-connector using the N-donor function and one of the carboxylate oxygens on the opposite side of the linker. The layers are stacked upon each other in a way that the second carboxylate oxygen forms the apex of the square pyramid in an adjacent layer. The resulting 3D network possesses chessboard-like tunnels. The guest water molecules can be removed at 200 °C. At 330 °C the framework starts to collapse. The key feature in the synthesis of the material is to start from 4-cyanopyridine that is hydrolyzed slowly during the hydrothermal treatment to give isonicotinic acid.

Trans-4-pyridylacrylic acid (HPA) is an extended version of isonicotinic acid and forms interpenetrating square grid architectures such as $M(PA)_2(H_2O)_2$ (M = Ni, Cu) or diamondoid networks in $Co(PA)_2$ [64]. In all three compounds it serves as a two-connecting linker, but the arrangement of the ligand around the metal produces square planar nodes in the case of Ni and Cu and a tetrahedral node for Co (two bidentate carboxylate groups and two cisoid N-donor groups surround the cobalt center). Despite interpenetration, the networks possess porosity potentially suitable for guest-molecule inclusion as judged from the X-ray structure analysis.

4.3.2.2
N-Donor Ligands

Organic linkers with N-donor functionality have been used widely to design porous framework materials. Important ligands are shown in Fig. 24. Most of them are two-connectors of the 4,4′-bipyridine type with more or less rigid extenders between the two pyridine rings. As compared to anionic carboxylate ligands the neutral bipyridines usually form charged networks. The counterion can either block coordination sites of the metal or intercalate in the channels. In some cases the anion serves as an additional linker and helps to achieve stable network architectures.

4.3.2.2.1 Four-connected Nets
Square-grid architectures are common for metals that prefer square planar coordination. With 4,4′-BPY the linking results in the formation of 2D polymers. In $Co(4,4'\text{-BPY})_2(NCS)_2 \cdot 2Et_2O$ [65] the two residual coordination sites are filled with NCS^- ligands (Fig. 25). Ether molecules fill the voids in between layers but they are lost immediately after removal of the solid from the mother liquor and further decomposition occurs as low as 140 °C. $Co(AZPY)_2(NCS)_2 \cdot 0.5EtOH$ has the same network structure but the four-rings adopt the shape of a rhombus [66]. The pores are smaller than expected from the size of the rhombus alone due to double-interpenetration (about 3×3 Å), and contain one ethanol molecule per two cobalt centers. Noninterpenetrating square grids are present in $Cd(AZPY)_2(NO_3)_2 \cdot (AZPY)$ [66], in which Cd is coordinated to two terminal NO_3^- anions and four AZPY ligands forming large channels of about 7×4 Å filled with AZPY guest molecules.

References see page 1244

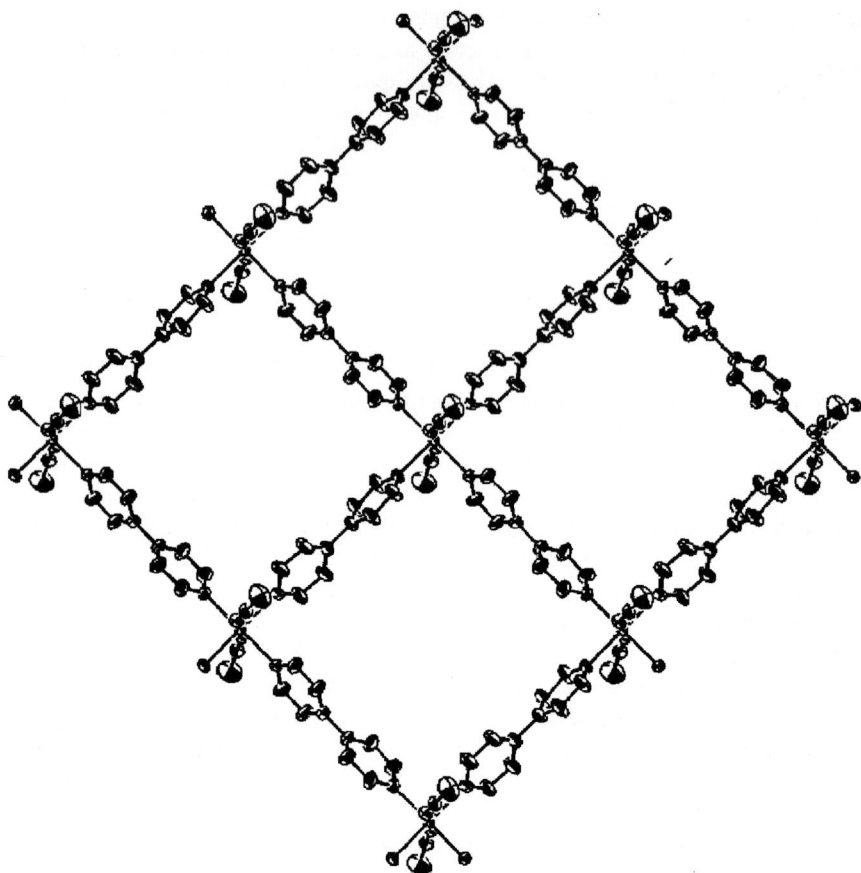

Fig. 25. Common square-grid architecture obtained with two-connecting ligands and square planar coordinated metal centers. Co(4,4′-BPY)$_2$(NCS)$_2$·2Et$_2$O has two terminal ligands completing the coordination sphere but the (4,4) nets may be further crosslinked (see Fig. 26) (from Lu et al. [65]).

One way to avoid complete pore blocking in interpenetrating systems is to lengthen the connecting units. For BPY alkane, alkene or alkyne linkers can be used and especially conjugated diynes serve as rigid linear units (Fig. 24). When copper or cobalt nitrate solutions are allowed to diffuse slowly into a methylene chloride solution of 1,4-bis-(4-pyridyl)butadiyne (C$_4$-BPY) isostructural polymeric grid architectures M(C$_4$-BPY)$_2$(NO$_3$)$_2$·CH$_2$Cl$_2$ (M = Cu, Co) result [67], in which four crystallographically equivalent two-connectors surround the metal center in a square planar fashion to form a porous net of interpenetrating flat grids. Even though interpenetration reduces porosity, the rhombic channels are large enough to accommodate solvent molecules, which can be removed at 180 °C without loss of crystallinity or collapse of the framework due to the interpenetration of the planar square nets at an angle of 60°, which enhances the stability.

Fig. 26. Schematic representation of the Zn(4,4'-BPY)$_2$SiF$_6$ structure; planar Zn(4,4'-BPY)-layers are crosslinked via μ-SiF$_6^{2-}$-anions along the tetragonal axis.

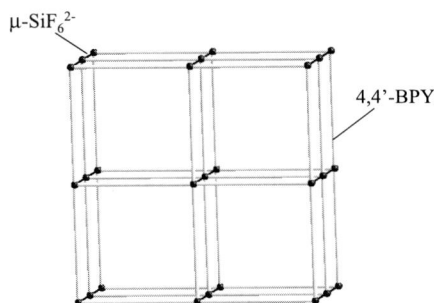

A rare example of a neutral framework with the counterions as part of the 3D net has been synthesized from ZnSiF$_6$ and 4,4'-BPY in DMF. Zn(4,4'-BPY)$_2$(SiF$_6$)·xDMF [68] has a layered structure in which Zn is square planar surrounded by four bridging BPY ligands to form an extended 2D tetragonal net. These planar nets are stacked along the crystallographic c-axis of the tetragonal structure in a primitive manner ($P4/mmm$) and separated only by bridging μ-SiF$_6^{2-}$ anions (Fig. 26). The 1D channels are 8 × 8 Å wide and constitute 50 % of the total volume of the structure. Unfortunately, neither physisorption measurements nor thermal or catalytic properties were reported, but recently Cu(4,4'-BPY)$_2$(SiF$_6$)·8H$_2$O has been described as a Jahn–Teller distorted structural analog [69] with square pores of 8 × 8 Å and a pore volume of 0.56 cm^3 g^{-1} as derived form argon-adsorption measurements. Smaller guest molecules, such as water, are beneficial since they are more easily removed as compared to DMF in the Zn compound and therefore essential to obtain microporous coordination polymers. Methane-adsorption experiments show that Cu(4,4'-BPY)$_2$(SiF$_6$) has a higher adsorption capacity than zeolite 5A (6.5 mmol g^{-1} at 36 atm). The porous structures are remarkable achievements since the dihydrates M(4,4'-BPY)$_2$(H$_2$O)$_2$(SiF$_6$) (M = Zn, Cd, Cu) [5, 70] are known to crystallize with two interpenetrating (4,4) nets perpendicular to each other, and thus a nonporous structure is the result. Obviously the equilibria in these systems strongly depend on the solvent and salt concentration.

Hydrogen bonds between coordinated water ligands and a spacer ligand can be strong enough to afford a noninterpenetrating stable framework. Cu(4,4'-BPY)(H$_2$O)$_2$(FBF$_3$)$_2$·4,4'-BPY [71] has a rectangular net similar to that of Co(NCS)$_2$(4,4'-BPY)$_2$·2Et$_2$O but one BPY is coordinated to copper to form infinite chains whereas the other is bound via hydrogen bonds to the two trans water ligands to give a 2D architecture (Fig. 27). The BF$_4^-$ anions complete the octahedral coordination. The effective cross-sectional area of the Cu$_4$ rings is 7.7 × 11.6 Å but because each sheet is displaced in a way that the BF$_4^-$ anions fit into these pores, no micropore volume, accessible to small molecules, is found in the as-synthesized structure. However, this compound has a remarkably high micropore volume of 0.51 cm^3 g^{-1}, as measured from N$_2$-adsorption isotherms [72]. Interestingly, the

References see page 1244

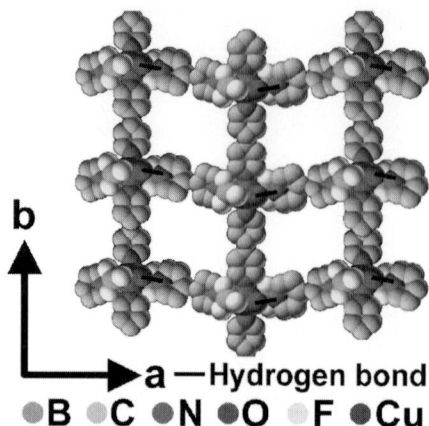

Fig. 27. The layered structure of Cu(4,4'-BPY)(H₂O)₂(FBF₃)₂·4,4'-BPY (from Li and Kaneko [72]).

adsorption begins at a definite relative pressure ("gate pressure"), which has been attributed to structural changes associated with the pretreatment and an additional structural change at the gate pressure. The unique adsorption behavior is obviously due to a rearrangement of the sheets at the gate pressure at which the four-rings are stacked in line. This produces a 1D channel that can adsorb molecules.

The 4,4'-BPZ ligand (3,3',5,5'-tetramethyl-4,4'-bipyrazolyl, Fig. 24) is a two-connector and has been used to engineer 1D and 2D coordination polymers with silver, cadmium and copper [73]. Moreover, the two additional NH-groups may form hydrogen bonds with counterions such as in $Cu(4,4'-BPZ)_2(H_2O)(BF_4)_2$·$0.5C_6H_5Br$ (Fig. 28) and $Cu(4,4'-BPZ)_2(H_2O)(HCOO)_2·2.5HCONH_2$ [73] with 2D four-connected $Cu(4,4'-BPZ)_2$ nets similar to those in $Zn(4,4'-BPY)_2(SiF_6)·xDMF$ but somewhat less regular due to the nonlinear BPZ ligand. In the fluoroborate compound the layers pack exactly one on top of another and BF_4^- anions are involved in weak hydrogen bonding with coordinated water molecules and NH groups (Fig. 28b). They separate the layers at a Cu–Cu distance of 7.75 Å to give large rectangular cages, populated with hydrophobic bromobenzene molecules (Fig. 28b). Interestingly the analogous formiate compound contains hydrophilic formamide guest molecules.

Remarkable 3D networks were obtained from CuX_2 salts (X = NO_3, Cl, $(SO_4)_{1/2}$) and 2-hydroxypyrimidine (HPYMO, Fig. 24) [74]. The resulting open-framework $Cu(PYMO)_2$ is poorly crystalline, but microporous, with a BET surface area of approximately 200 m^2 g^{-1} and is stable up to 250 °C. In the presence of salts, such as NH_4ClO_4, it forms highly crystalline clathrates of the type $Cu(PYMO)_2$·$(NH_4ClO_4)_{1/3}$ that can be heated to form $Cu(PYMO)_2·(HClO_4)_{1/3}$, an acidic material. The behavior is closely related to that of zeolites that, upon ammonium ion exchange and subsequent heat treatment, form acidic sites on the inner surface. The structure of $Cu(PYMO)_2·(NH_4ClO_4)_{1/3}$ consists of Cu^{2+} ions, which are square planar coordinated by four different PYMO ligands, each one connecting two copper centers to form four- and six-metal-membered rings. The pores, 8.1 Å in diameter, are defined by the $Cu_6(PYMO)_6$ rings and form a diamondoid network.

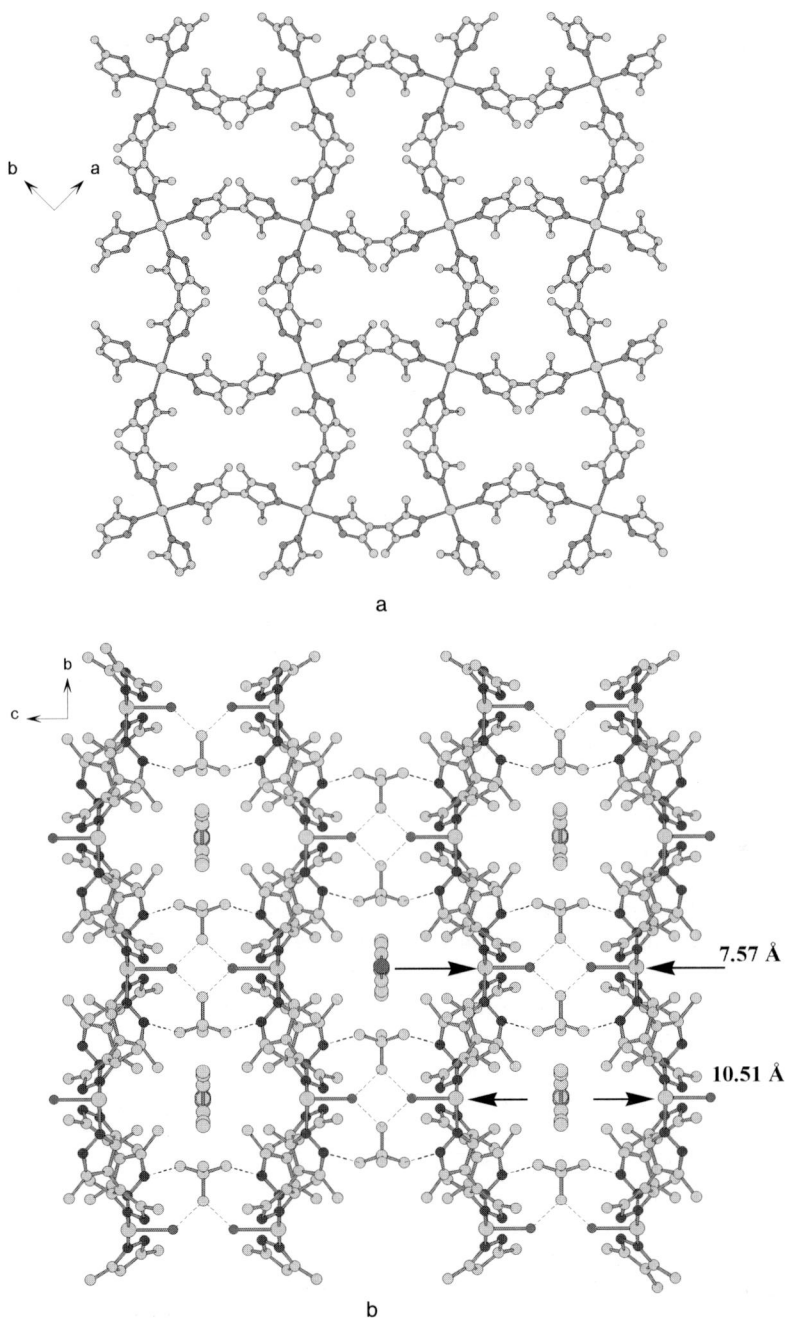

Fig. 28. 2D coordination layers in Cu(4,4′-BPZ)$_2$ compounds; (a) view from the top on the layers; (b) stacking scheme and hydrogen bonding with BF$_4^-$ anions and hydrophobic guest molecules (from Boldog et al. [73]).

2D layers of square meshes, in which the ligand acts as the four-connector, and silver, as a two-connecting bent spacer, have been observed in $Ag_2(\mu_4\text{-HMT})(Tos)_2$ (HMT = hexamethylenetetramine, Tos = p-toluenesulfonate) [75]. Ag(I) is coordinated by two N atoms of two HMT ligands, one oxygen, and a η^1-bonded aromatic ring and adopts, in total, a highly distorted tetrahedral coordination. The layers stack along the crystallographic c-axis with a separation of 14.23 Å and are held together through $\pi-\pi$ interactions between aromatic rings in adjacent layers to give a polymeric network stable up to 230 °C. Obviously, the self-assembly of the potentially tetradentate ligand HMT does not lead to the anticipated super-tetrahedral networks, probably due to steric limitations. For example, $Ag_2(\mu_3\text{-HMT})_2(CF_3SO_3)(H_2O)\cdot(CF_3SO_3)(H_2O)$ [75] consist of hexagonal meshes in which each hexagon contains three three-connecting Ag(I) centers and HMT ligands. The undulated layers are held together by hydrogen bonds between coordinated water molecules and coordinated anions to give a 3D network with additional anions and water intercalated in the channels, but the stability of the network is low. Hydrogen bonds are also responsible for the connection of hexagonal layers in $Ag_3(\mu_3\text{-HMT})_2(H_2O)_4\cdot(PF_6)_3$. Here, each hexagon contains six three-connected HMT ligands joined by six two-connecting spacer-like silver ions.

$Cu(I)(4,4'\text{-BPY})_2(PF_6)$ [76] forms a diamondoid framework (Fig. 1) with Cu–Cu separations of 11.16 Å in which Cu replaces carbon and the BPY ligand the C–C bonds, but four independent nets interpenetrate each other, which reduces the pore volume. Substituting half the BPY spacers by cyanide ions gives $Cu(CN)(4,4'\text{-BPY})\cdot2(4,4'\text{-BPY})$, a 3D network with 12.8×13.1 Å-wide channels filled with non-coordinated 4,4'-BPY guest molecules [77].

4.3.2.2.2 T-Shaped Building Blocks and other Three-connected Nets

The reaction of divalent metal nitrates $M(NO_3)_2$ (M = Co, Ni) with rigid 4,4'-BPY type linkers commonly results in the formation of $ML_{3/2}(NO_3)_2$-building blocks, in which the ligands are arranged in a T shape and nitrate ions complete in a bidentate fashion the metals coordination sphere (Fig. 29). T-shaped building blocks

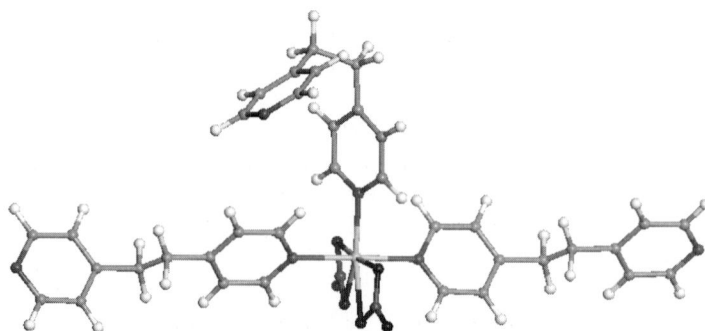

Fig. 29. Typical structure of the T-shaped $ML_{3/2}(NO_3)_2$ building block (L = BPET) (from Atencio et al. [86]).

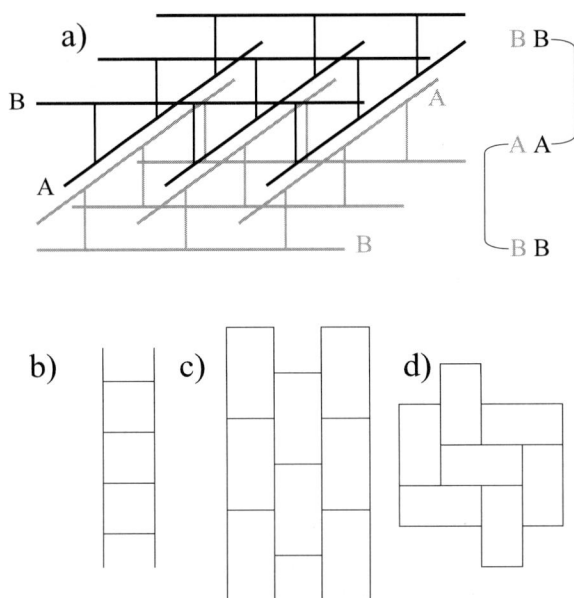

Fig. 30. Topologies from T-shaped building blocks: (a) tongue-and-groove in $M_2(4,4'\text{-BPY})_3(NO_3)_4 \cdot xH_2O$ (M = Co, Ni, Zn); (b) ladder; (c) brick-wall; (d) herringbone-type pattern.

can be connected in several ways to achieve stable 3D architectures or interwoven nets of lower dimensionality.

In $M_2(4,4'\text{-BPY})_3(NO_3)_4 \cdot xH_2O$ (M = Co, Ni, Zn) [78] T-shaped building blocks are linked to form double-layers (Fig. 30a). Linear $ML_{2/2}$ chains with different orientation A and B 11.3 Å apart from each other can be distinguished that are linked perpendicular to the chain axis and form double layers (Fig. 30a). The double layers are stacked along the c-axis and fit tightly into each other (*tongue-and-groove*) so that parallel chains have a distance of only 2.6 Å. The resulting channel-like voids along the a- and b-axes are 3 × 6 and 3 × 3 Å wide and filled with water molecules. Dehydration is achieved in vacuum at 25 °C without loss of crystallinity to give a microporous material with a type I isotherm and a methane sorption capacity of 2.3 mmol g^{-1} at 30 atm. The possibility of tailoring the specific interactions of gas molecules by adjusting the network composition and functionality could be beneficial compared to traditional molecular sieves [78].

T-shaped $ML_{3/2}(NO_3)_2$-building blocks may also be used to achieve topologies other than double-layers. Ladders (Fig. 30b) were found in $M_2(BPETHY)_3(NO_3)_4$ (M = Zn, Co) [79]. They interpenetrate each other to give an overall 3D architecture. In $Cd_2(PYTZ)_3(\mu\text{-}NO_3)(NO_3)_3(MeOH)$ (PYTZ = 3,6-bis-(pyridin-4-yl)-1,2,4,5-tetrazine) the interpenetrating ladders are linked via one bridging NO_3^- anion to give a unique knotted motif [80]. However, the corresponding

References see page 1244

$Cd_2(BPETHY)_3(NO_3)_4 \cdot CH_2Cl_2$ compound as well as $Cd_2(AZPY)_3(NO_3)_4$ [79] and $Cd_2(AZPY)_3(NO_3)_4 \cdot 2Me_2CO$ [66] have a layered herringbone structure composed of rectangular 2D (6,3)-nets, as shown in Fig. 30d, which are also interpenetrating. The methane-adsorption ability of $Cd_2(AZPY)_3(NO_3)_4 \cdot 2Me_2CO$ [66] is comparable to that of zeolite 13 X (about 1.8 mmol at 36 atm).

$Co_2(AZPY)_3(NO_3)_4 \cdot Me_2CO \cdot 3H_2O$ forms quadruply interpenetrating brick-wall networks (Fig. 30c) [66], but the acetone and water filled channels are small (4×3 Å). The methane-adsorption isotherm is not the Langmuir-type but the Henry-type, probably due to the small cross section of the channel.

Noninterpenetrating structures with T-shaped $ML_{3/2}(NO_3)_2$ building blocks have been obtained with 1,4-bis(3-pyridyl)-2,3-diaza-1,3-butadiene (AZB) with cobalt and cadmium nitrates [81] consisting of "all-equatorially" condensed cyclohexane-like $M_6(AZB)_6$ units. The framework, which is thermally stable up to 210 °C, is an undulated variant of the brick-wall pattern and resembles the structure of elemental arsenic in its gray form. Large cavities and channels with a metal-to-metal distance of 15 Å are filled with CH_2Cl_2 guests. Methylene chloride is reversibly desorbed at 160 °C without cavity volume collapse or symmetry change and can be re-adsorbed at room temperature.

A lengthened version of BPY has also been used for the synthesis of 2D polyrotaxanes [82]. Cucurbituril (CB[6]) is a macrocyclic cavitand with a hollow core of 5.5 Å and two identical portals surrounded by carbonyl groups (Fig. 31). The cavitand alone is known to form columnar 1D coordination polymers in the solid state [83]. The "bead" CB[6] can be thread on an appropriate "string" like N,N'-bis(3-pyridylmethyl)-1,5-diaminopentane (BPD) to give a pseudorotaxane (PR, Fig. 31).

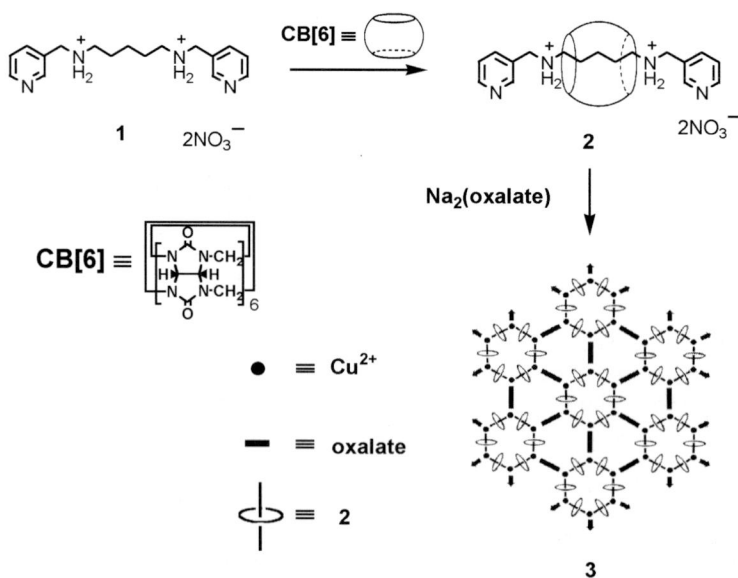

Fig. 31. Construction of 2D polyrotaxanes with supramolecular building blocks (from Lee et al. [82]).

This supramolecular building unit serves as a two-connector to form $Cu(\text{II})_6PR_6$ hexagonal necklaces that are further linked via oxalate ions into honeycomb like layers. The 2D nets are efficiently held together by van der Waals interactions with a distance of 12.9 Å from each other and stacked in such a way that the cavities are aligned to form 1D channels, with the largest cavity being 14 Å in diameter and a void volume of 48 %. However, removal of the guest molecules causes loss of crystallinity, which can be restored upon exposure to water vapor. Size-selective ion exchange and substitution of water by ammonia has been demonstrated. On the other hand, the polycatenated 2D polyrotaxane net, composed of edge-sharing, chair-shaped Ag_6PR_6 rings (PR = N,N'-bis(4-pyridylmethyl)-1,4-diaminobutane·CB[6]), has fully interlocked hexagons due to interpenetration [84].

The same concept has been used for the synthesis of a 3D Tb(III) polyrotaxane [85] network. In fact, the ligand N,N'-bis(3-cyanobenzyl)-1,4-diaminobutane again forms a similar pseudorotaxane with CB[6], but the cyano groups are hydrolyzed in the reaction with $Tb(NO_3)_3$ under hydrothermal conditions to give the dicarboxylate. Four di-monodentate and two bidentate carboxylate groups coordinate to binuclear terbium centers to give a six-connected 3D framework (compare also Fig. 9), which resembles the α-polonium topology, but each two-connecting carboxylate linker carries a CB[6] bead and the void space is filled with nitrate ions and a pseudorotaxane counterion. Unfortunately, sorption or ion-exchange experiments are still lacking.

Flexible architectures from T-shaped building blocks were obtained using 1,2-bis(4-pyridyl)ethane, BPET [86], a more flexible ligand as compared to 4,4'-BPY. In the anticonformation it functions as a linear two-connector to form $ML_{2/2}$-chains. The third chain-connecting ligand can adopt the gauche- (Fig. 32a) or anticonformation (Fig. 32b). The latter corresponds to the double layers in $M_2(4,4'-BPY)_3(NO_3)_4 \cdot xH_2O$ (M = Co, Ni, Zn) [78] and the former gives rise to new flexible $M_2(BPET)_3(NO_3)_4$-bilayer architectures (M = Co, Ni) [86], which contain channels

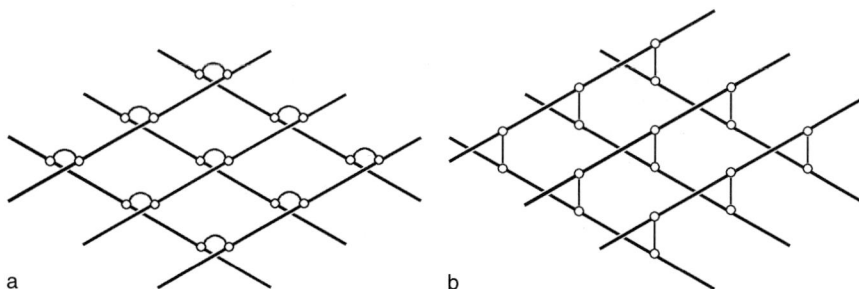

a   b

Fig. 32. Flexiblility of the 1,2-bis(4-pyridyl)ethane (BPET) ligand in architectures based on T-shaped building units allows for variable layer separations depending on the conformation of the third ligand; (a) gauche; (b) anticonformation (from Atencio et al. [86]).

References see page 1244

Fig. 33. Interpenetrating ThSi$_2$-networks from T-shaped
building units in Ag(4,4'-BPY)·NO$_3$. The rectangular pores are
occupied by nitrate ions (space-filling representation) (from
Yaghi et al. [11]).

suitable for the enclathration of small organic molecules like acetonitrile, CS$_2$ and
dioxane. Without guest molecules the effective volume of these channels is zero
but the conformational flexibility around the C$_2$-bond appears to allow for variable
size and shape of the channels. Thus, enclathration increases the layer separation
from 2.95 Å to 6.13 Å in the case of dioxane. Although the thermal stability is
probably lower for flexible systems they might show greater ability for molecular
recognition.

The slightly distorted trigonal planar coordination of three 4,4'-BPY ligands in
Cu(4,4'-BPY)$_{3/2}$·NO$_3$(H$_2$O)$_{1.25}$ [87] and the rotation of the two 4,4'-BPY rings with
respect to each other induces the formation of a stable 3D framework with large
rectangular channels (8 × 6 and 4 × 5 Å). The six interpenetrating (10,3)-b nets
with ThSi$_2$ topology do not fill all available voids completely but leave channels
occupied by nitrate ions that can be exchanged with SO$_4^{2-}$ and BF$_4^-$.

So far the T-shaped building blocks were derived from ML$_{3/2}$ fragments. Yaghi
and Li [88] demonstrated that a direct connection of metal centers via metal–metal
interactions can also result in a 3D network of T-shaped units. Ag(4,4'-BPY)·NO$_3$
has linear extended Ag(4,4'-BPY)$_{2/2}$ chains that are linked with each other via
Ag–Ag bonds (2.977(1) Å, attributed to d^{10}–d^{10} interactions) to form three interpenetrating 3D frameworks with 23 × 6 Å channels running along one of the
crystallographic axes (Fig. 33). The topology of a single framework in this compound is analogous to that of the Si net in ThSi$_2$ (Fig. 2). The vacant coordination
sites are occupied by nitrate anions in the channels, an interesting feature since
they are exchangeable with PF$_6^-$ without loss of crystallinity. The compound is
surprisingly insoluble in polar and nonpolar solvents and thermally stable up to
240 °C. The vacant silver sites are a unique feature of this coordination polymer
and should be useful for the catalytic conversion of small molecules.

[Cu(4,4'-BPY)$_{3/2}$(PPh$_3$)]BF$_4$·(THF)$_{1.33}$(CHCl$_3$)$_{0.33}$ [89] is a 2D coordination polymer with pentagonal cavities (Fig. 34). Each copper ion has a distorted tetrahedral

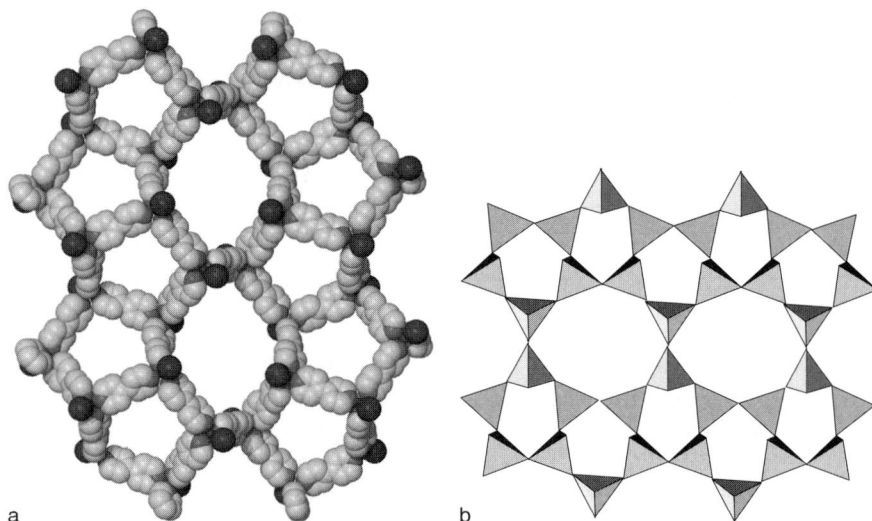

Fig. 34. Cu(4,4'-BPY)$_{3/2}$(PPh$_3$)$^+$ has pentagonal cavities (a), and resembles the network of nekoite (b), a hydrated sheet silicate (from Keller and Lopez [89]).

coordination and is linked to three copper ions by BPY ligands with a Cu–Cu distance of 11.2 Å. The Cu$_5$BPY$_5$ pentagons are fused into a net of five- and eight-membered rings. The net topology represents a metal-organic analog of the mineral nekoite, Ca$_3$Si$_6$O$_{15}$·7H$_2$O. Noncoordinating anions and solvent molecules fill the pentagonal cavities. Due to the bulky character of the PPh$_3$ the N–Cu–N angles are compressed from the tetrahedral angle. The PPh$_3$ group is vital to the formation of the structure as compared to other (T-shaped) coordination geometries obtained with stoichiometrically equivalent M(BPY)$_{3/2}$ frameworks.

4.3.2.2.3 Polynuclear Building Blocks
Clustering of metal centers gives rise to new topologies. A dinuclear Cu$_2$Cl$_2$ cluster in Cu(4,4'-BPY)Cl [90] should give a four-connected net if all BPY ligands would bridge different clusters, but two of the two-connectors form a double-bridge and thus a three-connected net of 16 × 26 Å wide six-rings results. Interpenetration of such 2D (6,3)-nets at an angle of 90° reduces the free space to channels that are approximately 2 × 4 Å in diameter. Even though this is too small for the sorption of small solvent molecules like acetonitrile the compound might be useful for gas separation. Irreversible decomposition occurs at temperatures as low as 175 °C.

Dinuclear Cu$_2$I$_2$-clusters in the 2D coordination polymer (CuI)$_2$(Mepyz)$_2$·Mepyz (Mepyz = 2-methylpyrazine) [91] are connected to four different two-connecting Mepyz-ligands to afford a (4,4)-net-topology with channels 8.75 × 8.54 Å in cross section, and additional noncoordinated 2-methylpyrazine molecules in the chan-

References see page 1244

nels. The guest molecules are readily lost in vacuum and exchange with benzonitrile is possible apparently without loss of the structures integrity. The framework is chiral which could be of potential interest in asymmetric catalysis. Obviously the asymmetrically substituted ligand can induce chirality. The use of chiral R-2-methylpiperazine has led to the formation of CuI(Mepip) [91], a 3D network with narrow channels composed of helical CuI chains, linked by bridging R-2-methylpiperazine ligands. In contrast, the corresponding chloride and bromide compounds CuX(Mepyz) (X = Cl, Br) are not chiral. They contain infinite CuX chains linked by the two-connecting ligand to form layered polymers with smaller 14-membered pores.

Di- and trinuclear zinc clusters were also found when saturated extended versions of 4,4'-BPY were used [92]. For example L = 1,3-bis(4-pyridyl)-propane affords $Zn_3(OH)_3(L_3)(NO_3)_3 \cdot 8.67H_2O$ with hydroxyl bridged trinuclear $Zn_3(OH)_3$ 6-membered rings and a 2D interwoven network. Using 1,3-bis(4-pyridyl)butane and 1,7-Bis(4-pyridyl)heptane dinuclear Zn_2OH subunits form, which are connected to give layered structures or 1D chains. The formation of polynuclear hydroxyl-bridged zinc clusters, which are also active sites in certain enzymes [93], is induced by the basicity of the ligand. The higher flexibility of the ligand, as compared to rigid BPY linkers, allows these clustered subunits to fit into 3D polymeric arrays.

4.3.2.2.4 Reduced Dimensionality and Porosity: Chains

1D coordination polymers with microchannels, 3×2 and $5 \times 4 \text{Å}$ wide, are present in $Mn(BPET)_2(NCS)_2$ (BPET = 4,4'-bipyridylethane) and $Cd(DPDS)_2(H_2O)_2 \cdot 2NO_3 \cdot$ $2EtOH \cdot 2H_2O$ (DPDS = 4,4'-bipyridyldisulfide) [94]. The flexibility of the two-connector allows for the formation of double-bridges between the octahedral metal centers. Two additional axial ligands complete the octahedral coordination. The BPET compound has very small channels and contains no guest molecules but the DPDS polymer contains ethanol that can be removed without decomposition of the framework.

$Mn(AZPY)(NO_3)_2(H_2O)_2 \cdot 2EtOH$ is also a 1D chain compound with the two-connecting AZPY ligand in axial positions of the octahedral manganese center, but the chains form arrays assembled via hydrogen bonding to give a log-cabin-type structure with large hydrophobic channels $8 \times 8 \text{Å}$ in cross section, filled with ethanol molecules [66]. However, the channels in these compounds lie in between the assembled chains and therefore porosity is usually lost upon drying. Subsequent gas adsorption is not effective. Open porous structures with 1D (and 2D) topologies were also obtained using tetrapyridylporphyrin and $HgBr_2$ [95].

Railroad-like topologies result when 4,4'-BPY serves as a two-connector and terminal ligand at the same time (Fig. 35). In $Ni(4,4'-BPY)_{2.5}(H_2O)_2(ClO_4)_2 \cdot 1.5(4,4'-BPY) \cdot 2H_2O$ [96] each Ni is coordinated by four BPY ligands in a square planar arrangement but only three of them serve as a two-connector to form a railroad-like motif. The extended strings are further connected through $\pi - \pi$ interactions of the terminal BPY ligands resulting in a porous sheet structure with an extended 1D channel network and $11 \times 11 \text{Å}$ pores. They are filled with perchlorate ions, water and guest 4,4'-BPY molecules. Ion exchange without loss of crystallinity has been

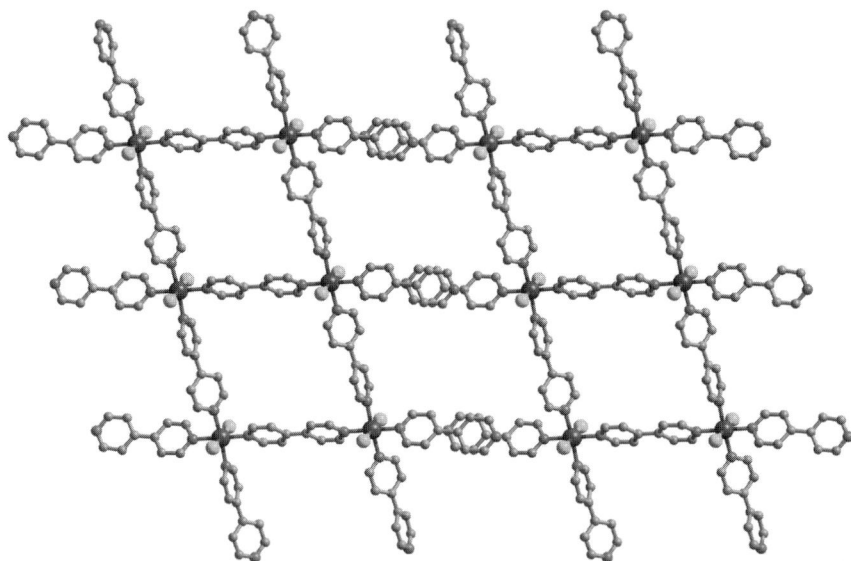

Fig. 35. Railroad-like structure of the $Ni_2(4,4'-BPY)_5$ network (from Yaghi et al. [96]).

demonstrated but the coordinated ligands are lost between 220 and 325 °C, indicating that the thermal stability of the polymer is low.

4.3.2.2.5 Porous Zero-Dimensional Frameworks?

An interesting switching behavior between polymorphs depending on the sorbent has been observed for molecular complexes. For example α-bis(1,1,1-trifluoro-5,5-dimethyl-5-methoxy-acetylacetonato)copper(II) is readily transformed into a molecular host–guest arrangement (β-[CuL$_2$]·G) upon exposure to appropriate gaseous or liquid guest molecules like benzene [97]. Single-crystal X-ray studies of the inclusion-derivative reveal straight channels in the framework with larger and smaller cavities alternating. This would be nothing else than an inclusion compound but remarkably with the aid of a suitable transforming gas like methyl bromide it is possible to remove the guest and retain the microporous β-form [97]. The reversible switching between α- and porous β-polymorphs can be monitored *in situ* with hyperpolarized ^{129}Xe NMR spectroscopy, which shows creation and destruction of pore space as well as the amount of adsorbate in the pore system of the open framework [98].

4.3.2.3
Phosphonates

A large number of phosphonates with promising optical, sorption and catalytic properties have been synthesized in recent years with a variety of phosphonic acids

References see page 1244

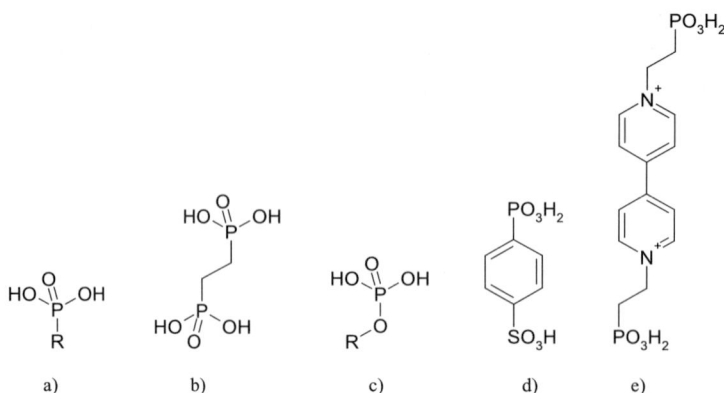

Fig. 36. (a) Alkyl phosphonic acid; (b) diphosphonic acid; (c) alkylphosphoric acid; (d) sulfonic phosphonic acid; (e) N,N'-bis(2-phosphonoethyl)-4,4'-bipyridine (PV).

(Fig. 36). A comprehensive description of the field is given by Clearfield [99, 100] who has made major contributions in this field over many years.

The zirconium phosphonates form layered structures in which the organic residue separates the layers from each other. The inorganic layered structure is related to the structure of α-zirconium phosphate, $Zr(HPO_4)_2 \cdot H_2O$ [101] in which the phosphate groups are arranged in a hexagonal pattern above and below the sheets of zirconium ions. Three oxygen atoms of each phosphate group are bonded to three different zirconium ions. Each zirconium ion is octahedrally coordinated from six different HPO_4^{2-} groups. The protonated oxygen of the phosphate points into the interlayer space. The layered phosphonates adopt the same bonding scheme, but the hydroxyl group is replaced by an organic residue R (Fig. 36a), pointing into the interlayer space. The organic group largely determines the separation distance of the layers. Biphosphonic acids (Fig. 36b) may be used to cross-link the inorganic octahedral layers and thus, by variation of the pillar's length, the interlayer spacing may vary from 7.8 Å in $Zr(O_3PCH_2CH_2PO_3)$ up to 18.5 Å in $Zr(O_3P(C_6H_4)_3PO_3)$. Usually, in such compounds the pillars are arranged too close to each other to create microporous materials. However, by mixing smaller and larger pillaring groups, microporous materials were obtained with surface areas in the range of 300–400 m^2 g^{-1}, but with significant contribution from mesopores probably due to structural defects [99, 102]. A clever tactic to avoid too-dense an arrangement of the pillars has been used by Alberti et al. [103]. The use of 3,3',5,5'-tetramethylbiphenylene bis(phosphonic acid) produces a layered microporous zirconium phosphonate with narrow pore-size distribution in which the methyl groups are too large to fit adjacent to each other. The compound has the idealized composition $Zr(HPO_3)_{1.33}(O_3P-R-PO_3)_{0.33}$ and has a BET-surface area of 375 m^2 g^{-1} with a maximum in the pore-size distribution at 5 Å as derived from a t-plot and ^{129}Xe NMR spectroscopy.

A second type of layered phosphonates is related to the inorganic layers in γzirconium phosphate, $Zr(PO_4)(H_2PO_4)\cdot 2H_2O$ [99]. Polyether derivatives were produced from the reaction of ethylene oxide with γ-zirconium phosphonate in the 1970s by Yamanaka [104]:

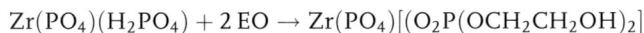

$$Zr(PO_4)(H_2PO_4) + 2\,EO \rightarrow Zr(PO_4)[(O_2P(OCH_2CH_2OH)_2]$$

Sulfophosphonates (Fig. 36d) have been shown to be proton conductors. The best conductivity was of the order of $10^{-3}\ \Omega^{-1}\ cm^{-1}$ in a wet nitrogen atmosphere. They are also valuable ion-exchange materials. Crown-ether-containing compounds were also described, leading to large interlayer distances [105, 106]. They could be useful as ion-selective materials.

Recent developments focus on new pillars, which are not only inert components, holding the inorganic layers apart, but rather functional and active parts of the porous material. For example in the lamellar $Zr_2(PO_4)(PV)X_3\cdot 3H_2O$ (X = halide) [107] N,N'-bis(2-phosphonoethyl)-4,4'-bipyridine (PV, "phosphonate viologen", Fig. 36e) functions as the spacer. According to structural information obtained from synchrotron X-ray powder data and Rietveld refinement, the inorganic lamellae are separated 13.5 Å from each other, in between having large pores filled with one halide ion and three water molecules per formula unit. The free halide ion can be exchanged with $PtCl_4^{2-}$ ions and, after reduction, colloidal metal particles are obtained inside the porous solid. The compound has been used for the photochemical splitting of water making use of the photoinduced charge separation in combination with the colloidal platinum, which acts as the catalyst for the reduction of water to hydrogen gas. Photoinduced charge separation in zirconium phosphonate viologen compounds causes photoreduction of the viologens, which could be useful as size and shape-selective reductants [108]. The reaction of zirconium viologen diphosphonates with phosphonates enhances the porosity but produces disordered phases [109].

The vanadyl phosphonates $VO(RPO_3)\cdot H_2O$ also form layered phases. VO_6-octahedra are corner-sharing to form parallel $-V{=}O{-}V{=}O$ chains and are linked via phosphonate oxygens above and below the layers, similar to the behavior observed for the zirconium compounds [99]. For a recent review on oxovanadium phosphonates see also [110]. They are prepared from V_2O_5 in hot alcohol that reduces V^{5+} to V^{4+}. The compounds usually intercalate solvent alcohol and exhibit molecular-recognition effects. For example, vanadyl hexylphosphonate, $VO(O_3PC_6H_{13})\cdot H_2O$, intercalates n-butanol and isobutyl alcohol but not sec-butanol [111]. The selectivity can be controlled by the type of phosphonic acid used in the synthesis.

The first microporous aluminumphosphonate with organic side-groups was synthesized by Maeda et al. [112] and has the composition $Al_2(CH_3PO_3)_3\cdot H_2O$ (AlMepO-β). Dehydration at 400 °C affords a microporous crystalline phosphonate with a pore volume of 0.117 $cm^3\ g^{-1}$. The 3D framework contains unidimensional 18-ring channels with an estimated diameter of 5.8 Å [113]. AlMepO-α is a polymorph with the same composition, which also contains tetrahedrally and octahedrally co-

References see page 1244

ordinated Al(III) [114, 115]. Each phosphonate anion binds to one octahedral and two tetrahedral Alcenters (Fig. 37). The phosphonate groups and Al centers alternate along the trigonal *c*-axis and form 1D channels, 7 Å wide, in which the methyl groups point to the center of the channel. The two polymorphs differ in the connectivity among the octahedral aluminum (Fig. 37) [115]. Steam treatment of AlMepO-β at 500 °C induces the topotactic transformation into AlMepO-α [116]. AlMepO-α is stable up to 500 °C without breakdown of the framework and re-adsorbs 2,2-dimethylpropane that has a kinetic diameter of 6.2 Å [117].

1D hexagonal tunnels covered with methyl groups on the inner surface are also present in β-Cu(CH$_3$PO$_3$) [118], in which copper has a distorted tetragonal pyramidal coordination. Two of the phosphonate oxygens bridge copper atoms resulting in infinite zigzag chains of copper parallel to the trigonal axis, which are linked together by O–P–O bridges in the *a–b* plane. The distance between two opposite methyl carbons is only 6 Å. This explains why derivatives with longer residues usually afford lamellar structures.

The first functionalized zeolite-like phosphonate was prepared in a hydrothermal reaction from zinc nitrate and (2-aminoethyl)phosphonic acid [119]. In Zn(O$_3$PC$_2$H$_4$NH$_2$) the zinc ions are coordinated by phosphonate oxygens and nitrogen. The framework consists of corner-sharing ZnO$_3$N and PO$_3$C tetrahedra that form 16-membered rings perfectly aligned along the *a*-axis, channels with windows 3.6×5.3 Å.

Another phosphonate with large 1D channels, whose dimensions are determined by the length of the organic spacer, is Zn$_3$(HO$_3$PC$_3$H$_6$PO$_3$)$_2$·2H$_2$O [120]. The structure is very similar to the sheet-like structures of other zinc phosphonates with a metal-to-phosphonate ratio of one, like Zn(HO$_3$PC$_3$H$_6$PO$_3$H) [120]. The important difference is an additional direct bonding of zinc atoms in one layer to O atoms in the neighboring layer, which produces large elliptical pores within the structure (Fig. 38).

4.3.2.4
Cyanide Frameworks

The zeolitic properties of cyanometallates have been known for many years [1]. A large number of inclusion compounds derived from the Hofmann-type clathrates Cd(NH$_3$)$_2$Ni(CN)$_4$ have been synthesized in which the NH$_3$ ligand is replaced by aliphatic amines, or pyridine. Bidentate amines, such as H$_2$N(CH$_2$)$_n$NH$_2$, allow bridging of the 2D motif of the Hofmann clathrate with much larger cavities that can accommodate substituted and fused-ring aromatic guest molecules [1].

Recent efforts have been devoted to the synthesis of extended versions of Prussian Blue Fe$_4$[Fe(CN)$_6$]$_3$ (Super Prussian Blues) and related cyanide networks. The basic motif of the Prussian Blue structure consists of a cubic arrangement of iron centers bridged by cyanide ions in which Fe(II) and Fe(III) alternate. The deviation from the ideal stoichiometry M(CN)$_3$ results from 25 % Fe(CN)$_6^{4-}$ vacancies in the structure. The degree of ordering varies with crystallization conditions of the compound.

(a)

(b)

AlMepO-α

AlMepO-β

c)

AlMepO-α

d)

AlMepO-β

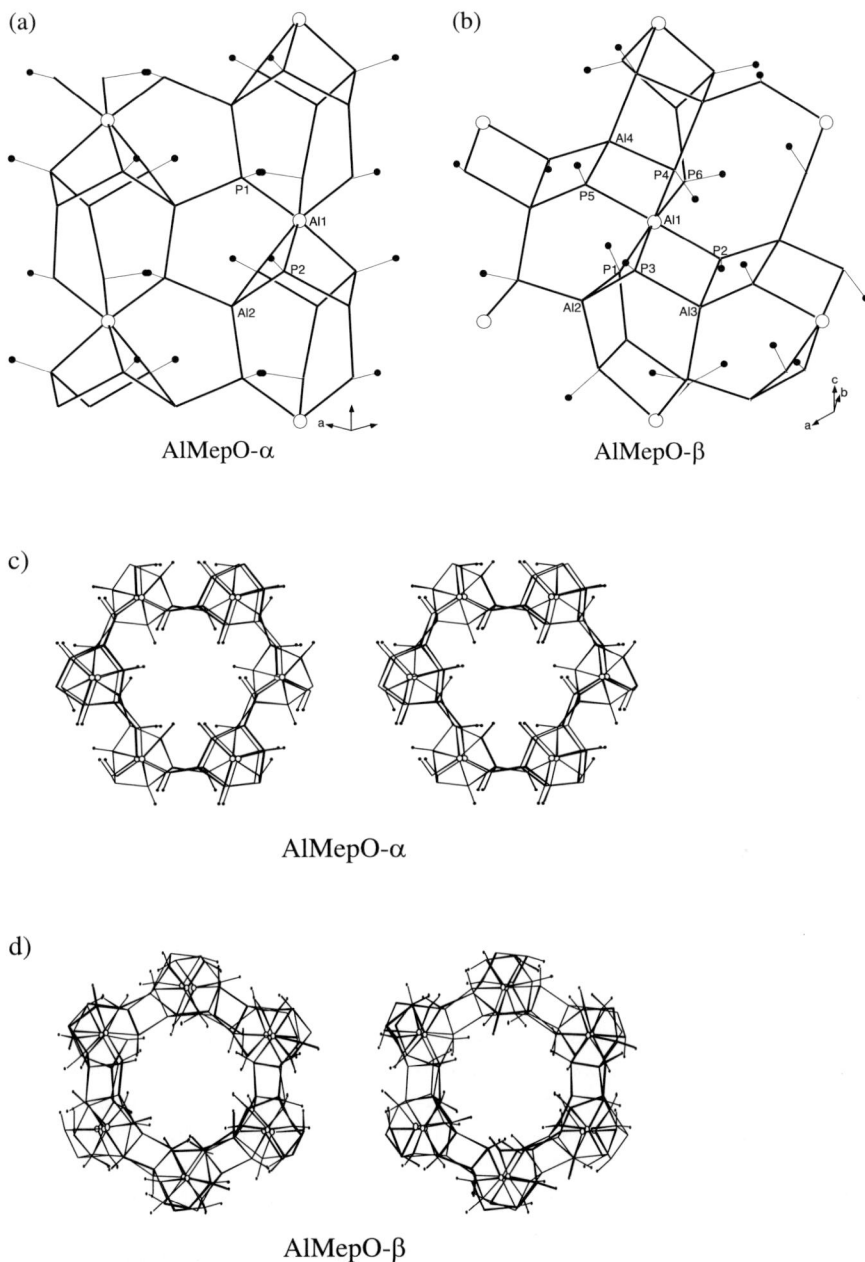

Fig. 37. Microporous aluminumphosphonates: (a) AlMepO-α; (b) AlMepO-β. Al in octahedral (open spheres) and tetrahedral (Al2–Al4) coordination. In both compounds the methyl groups point into the center of the hexagonal channels ((c) and (d)) (from Maeda et al. [117]).

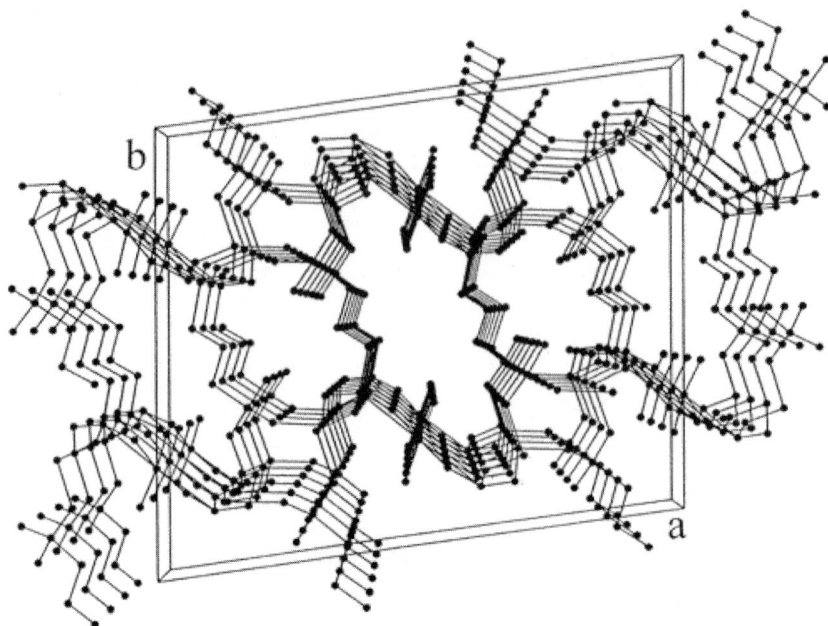

Fig. 38. Elliptical 1D tunnels in $Zn_3(HO_3PC_3H_6PO_3)_2 \cdot 2H_2O$. Coordinated water molecules are projecting towards the center of the pores to form a hydrophilic environment (from Poojary et al. [120]).

There are two remarkable approaches to extend the framework structure of Prussian Blue. On the one hand organometallic cations can be used to extend the CN^- bridges. The second method focuses on transition metal clusters, which are used as secondary building units and replace the metal centers in Prussian Blue and related networks.

4.3.2.4.1 Organometallic Extension

Most of the developments are due to the work of Fischer who has used Cp_3U^+ [121] and R_3Sn^+ cations [122, 123] as modules for linear extension of the Prussian Blue network. The basic extending building block is a trigonal bipyramide, the μ-$CNSnMe_3NC^-$ anion with CN^- in axial positions, which is generated from the reaction of Me_3SnCl and complex cyanides.

Supramolecular organolead and organotin compounds of the type $M(III)(\mu$-$CNSnMe_3NC)_3$ (M = Fe, Co) form 3D open networks in which the metal is octahedrally coordinated, and μ-$CNSnMe_3NC$ is a trigonal bipyrimidal two-connector [123–125]. The cobalt compound contains infinite interlinked $[-Co-C\equiv N-Sn-N\equiv C-]_\infty$ chains with straight channels 10×10 Å in cross section [124]. The wide-open 3D networks can accommodate metallocenium guests like $CoCp_2^+$ or Et_4N^+ [126]. Larger guest molecules, such as nBu_4N^+, result in a new stoichiometry $(nBu_4N)_{0.5}(Me_3Sn)_{3.5}Fe(CN)_6 \cdot H_2O$ [127], a chiral host material with even larger cavities that contains a few terminal $CNSnMe_3 \cdot H_2O$ groups. Suspension of the $M(III)(\mu\text{-}CNSnMe_3NC)_3$ compounds in aqueous solutions of R_4N^+ salts leads to cleavage of the Sn–N bonds and subsequent depolymerization of the network [128–130].

The reaction of Me_2SnCl_2 and $K_3[Co(CN)_6]$ produces $(Me_2Sn)_3[Co(CN)_6]_2 \cdot 6H_2O$ with hexacoordinated Sn and a planar ribbon-like structure, which might have interesting catalytic properties [131].

In $CuCN(Me_3SnCN)(4,4'\text{-}BPY)_{0.5}$ the 2D (4,4) nets are composed of $CuCN(Me_3SnCN)$ layers in which $Cu_2(\mu\text{-}CN)_2(CN)_2$ fragments are linked via trigonal planar Me_3Sn bridges to infinite quasi-planar sheets [132]. Here BPY functions as the pillar and links the sheets in the third dimension into two equivalent but interpenetrating 3D frameworks. In fact, each pillar of one framework passes through the center of a 24-atomic ring of the complementary framework. The interpenetration stabilizes the framework without loss of porosity to form straight channels 10×5 Å in cross section and a BET surface area of 1090 m^2 g^{-1}.

$(Bu_4N)[(Et_3Sn)_2Cu(CN)_4]$ has an anionic framework with cationic template molecules well ordered in the wide open channels [133]. The network topology is somewhat distorted from the initially envisioned super-diamondoid lattice. Interestingly Me_3SnCl affords $(Bu_4N)[(Me_3Sn)Cu_2(CN)_4]$, which has a layered structure, and trigonal $Cu(CN)_3$ building blocks [134].

4.3.2.4.2 Cluster Extension

M_6X_8 (M = Re, X = S, Se, Te) clusters can also be used to synthesize extended Prussian Blue compounds of high stability [135–139], which are useful for chemical sensing (see Sect. 4.3.3.3). The structural motif of the cluster is an M_6-octahedron with all eight triangular faces capped with a chalcogen ligand and six radial positions available for ligand exchange and connection to other network components.

Typical examples of cluster-expanded Prussian Blue analogs are $Fe_4[Re_6Te_8(CN)_6]_3 \cdot 27H_2O$ and $Ga_4[Re_6Se_8(CN)_6]_3 \cdot 38H_2O$ [137] in which the Fe(II) positions in Prussian Blue are substituted by Re_6X_8-clusters (Fig. 39). Only 75 % of the cluster sites are occupied. The void volumes are 1275 and 1580 $Å^3$ for the iron and gallium compound, which compares to 557 $Å^3$ in $Fe_4[Fe(CN)_6]_3$. These compounds are remarkably stable up to 300 °C. Dehydration is achieved around 125 °C and the porous frameworks can take up large amounts of methanol or ethanol. Bulkier alcohols are not absorbed due to the small openings (2.4×5.9 Å) of the structure.

Reaction of water soluble $[Re_6S_8(CN)_6]^{4-}$ with Fe^{2+} in the presence of Cs^+ yields the 2D phase $Cs_2[trans\text{-}Fe(H_2O)_2][Re_6S_8(CN)_6]$, a square lattice of alternating cluster cores and iron atoms linked by cyanide bridges [135]. However, the selenium analog affords 3D $Cs_2[trans\text{-}Fe(H_2O)_2]_3[Re_6Se_8(CN)_6]_2$ [135], which is characterized by cubic $[Re_6Se_8(CN)_3]_8Fe_6$ cages with Re_6-clusters at the corner and Fe atoms capping the faces of each cube. These open cages enclose a volume of 720 $Å^3$, filled with 15 water molecules. Reaction with Co^{2+} or Mn^{2+} pro-

References see page 1244

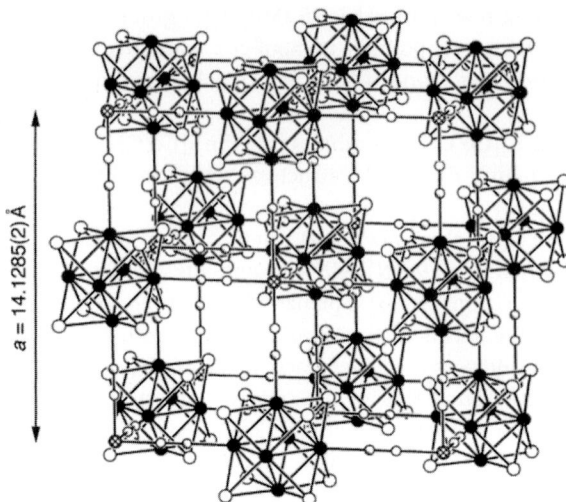

$a = 14.1285(2)\,\text{Å}$

Fig. 39. Prototype of the cluster-expanded Prussian Blue type frameworks; $Fe_4[Re_6Te_8(CN)_6]_3 \cdot 27H_2O$ (from Shores et al. [137]).

duces the analogous $Cs_2[\textit{trans}\text{-}Mn(H_2O)_2]_3[Re_6Se_8(CN)_6]_2 \cdot 9H_2O$ and $(H_3O)_2[\textit{trans}\text{-}Co(H_2O)_2]_3[Re_6Se_8(CN)_6]_2 \cdot 8.5H_2O$ [136]. The compounds can easily be dehydrated but irreversibly transform into new phases with yet unresolved structures.

$[Co(H_2O)_4]_3[W_4Te_4(CN)_{12}] \cdot 15.38H_2O$ [140] has a similar cubic arrangement of W_4-clusters bridged by pseudo-octahedrally coordinated Co^{2+} ions. The cubic ReO_3-type lattice has cuboctahedral voids $1520\,\text{Å}^3$ in volume, filled with highly disordered water molecules, but the network stability is very low and the compound becomes amorphous upon dehydration.

$[Mn(Pr^iOH)_2(H_2O)]_2[Re_6S_8(CN)_6] \cdot 2Pr^iOH$ [141] is an extended Prussian Blue version in which Mn_2- and Re_6S_8-units play the role of alternating Fe centers bridged by cyanide ions. The substitution results in an extension of the unit cell parameter from 10.2 in Prussian Blue to $16.8\,\text{Å}$ with a more hydrophobic cavity ($3120\,\text{Å}^3$).

The cluster extension approach was also successful for the expansion of $Na_2Zn_3[Fe(CN)_6]_2 \cdot 9H_2O$ [138], which itself has a porous framework suitable for ion exchange and can be dehydrated without loss of crystallinity. The extended framework $Na_2Zn_3[Re_6Se_8(CN)_6]_2 \cdot 24H_2O$ (Fig. 40) has an isotypic structure with large hexagonal bipyramidal cages and can be completely desolvated at 250 °C without pore collapse or loss of crystallinity. Due to the larger building blocks the cage size is increased from $421\,\text{Å}^3$ to $1340\,\text{Å}^3$. The hydrated sodium ions in the structure are exchangeable with a range of $[M(H_2O)_6]^{n+}$ species ($M = Mg^{2+}$, Cr^{3+}, Mn^{2+}, Ni^{2+}, Zn^{2+}), but larger species like $[Cr(en)_3]^{3+}$ also fit into the porous framework. However, dehydration of the corresponding porous $(H_3O)_2Zn_3[Re_6Se_8(CN)_6]_2 \cdot 20H_2O$ [142] transforms this compound below 100 °C into a new phase.

Heterometallic clusters in mixed sulfide-cyanide systems were studied by Zhang

Fig. 40. Large hexagonal bipyramidal cages in Na$_2$Zn$_3$[Re$_6$Se$_8$(CN)$_6$]$_2$·24H$_2$O (from Bennett et al. [138]).

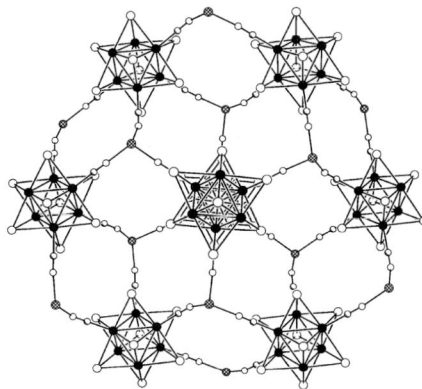

et al. [143–145]. Microporous [Et$_4$N]$_2$[MS$_4$Cu$_4$(CN)$_4$] (M = Mo, W) was obtained from CuCN and (Et$_4$N)$_2$MS$_4$ in pyridine [143, 144]. The anionic network has a diamondoid structure composed of MoS$_4$Cu$_4$-clusters that are linked via cyanide double-bridges. The compound shows strong third-order NLO absorption effects, but it is not clear if the cations can be removed from the lattice without collapse of the structure.

4.3.2.5
Other Ligands

Recently, the two-connector 4,4'-bipyridine-N,N'-dioxide (BPYD, Fig. 41) has been used in combination with La(III) centers to make unusual 3D network structures with eight-, seven-, six- and five-connected lanthanum nodes [146, 147]. La(BPYD)$_4$·(CF$_3$SO$_3$)$_3$·4.2MeOH forms a CsCl-like body centered cubic framework structure in which each La is connected to eight different neighboring metal centers via BPYD (Fig. 42a). However, in La(BPYD)$_4$·(BPh$_4$)·(ClO$_4$)$_2$·2.75MeOH each La is only connected to seven neighboring metal centers since a pair of ligands form a "double-bridge". The 3D net adopts the rare seven-connected topology consisting of (6,3)

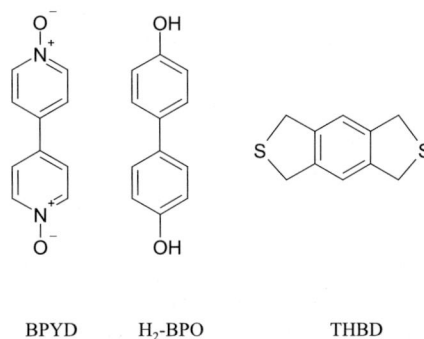

Fig. 41. 4,4'-bipyridine-N,N'-dioxide (BPYD), 4,4'-biphenoxide (BPO) and tetrahydro-benzodithiophene (THBD).

BPYD   H$_2$-BPO   THBD

References see page 1244

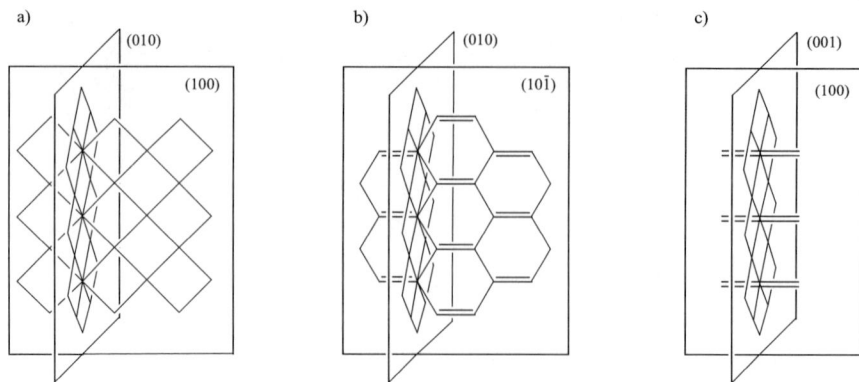

Fig. 42. Schematic representation of La(BPYD)$_4$-frameworks with (a) 0; (b) 1; (c) 2 "double-bridges" per metal center. The resulting seven-connected net (b) is a rare topology (from Long et al. [146]).

nets interlinked by (4,4) nets (Fig. 42b). In a similar manner the two "double-bridges" per metal center in La(BPYD)$_4$·([Co(C$_2$H$_{11}$B$_9$)$_2$])$_3$·0.5MeOH reduce the connectivity further to a six-connected NaCl-like 3D structure (Fig. 42c). In the five-connected network of [La$_4$(BPYD)$_{10}$(MeOH)$_{10}$Cl$_3$]Cl(BPh$_4$)$_8$·22MeOH and [La(BPYD)$_{2.5}$(MeOH)(Ph$_2$B(OMe)$_2$)](BPh$_4$)$_2$·4.5MeOH each lanthanum center is coordinated by five N-oxide ligands, which bridge the metals [147]. The structure can be described as a combination of (6,3) plane nets, which are linked through zigzag chains forming large "six-metal-membered" channels of effective cross section 8×18 Å. Even though porosity can be expected from these highly connected framework topologies, unfortunately, no sorption measurements have been carried out.

The first row transition metals Co, Ni, Cu, and Zn form isomorphous 3D coordination polymers with the composition [M(H$_2$O)$_4$(BPYD)][ClO$_4$]$_2$·2BPYD [148]. The covalent ML-chains are connected through hydrogen bonds of the coordinated water with additional ligand molecules to form interwoven 3D frameworks (Fig. 43). Chlorate anions are included in the channels.

Ti(BPO)$_{3/2}$(OPr)(HOPr) (BPO = 4,4′-biphenoxide) was obtained by treatment of Ti(OPr)$_4$ with dihydroxybiphenyl (H$_2$-BPO) in THF for 15 weeks [149]. The primary building blocks are bioctahedral dititanium units, which can be considered as the connecting points. They form a primitive 3D network with BPO ligands as the two-connectors.

Only in a few cases have thioethers been investigated for the construction of coordination polymers. The reaction of tetrahydrobenzodithiophene (THBD) with AgBF$_4$ in acetonitrile (AN) gives 2D cationic Ag(THBD)(AN)$_2$ nets [150], which are coplanar to each other and separated by BF$_4^-$ anions. The nitrile ligands are oriented perpendicular to the lamellae and can be substituted with more spacious nitriles. For example, exchange of acetonitrile with benzonitrile expands the layer separation from 10.085 to 13.890 Å. The calculated specific surface area of these

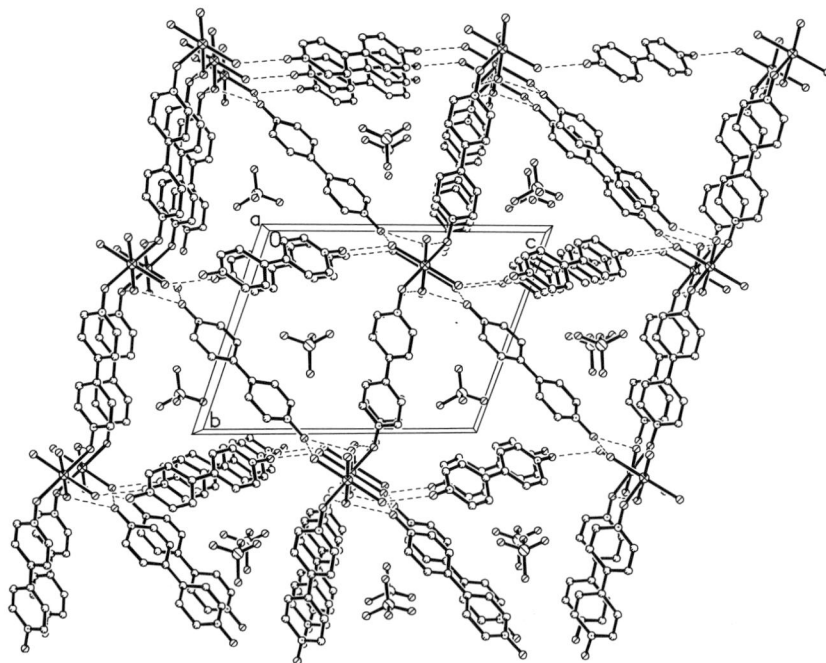

Fig. 43. 3D framework of [Co(H$_2$O)$_4$(BPYD)][ClO$_4$]$_2$·2BPYD with 1D covalent interactions (along *b*) and additional hydrogen-bonding interactions, responsible for the network formation (from Ma et al. [148]).

compounds is, at 1160 m^2 g^{-1}, quite high, but desorption of the nitrile ligands below 100 °C will probably preclude the development of applications in gas separation or heterogeneous catalysis.

The N-donor ligand DPDS (4,4′-dipyridyldisulfide) has been used in its protonated form for the construction of porous vanadates [151]. The pyridinium cations in (H$_2$DPDS)$_2$[V$_{10}$O$_{26}$(OH)$_2$]·10H$_2$O bind via hydrogen bonds to oxygen atoms in the decavanadate anion and to water molecules that are also connected to the anion. The resulting framework is a 1D coordination polymer with microchannels 2 × 4 Å wide and filled with water.

4.3.3
Properties and Applications

As compared to zeolite molecular sieves, metal-organic frameworks are thermally less stable but variable pore size and shape as well as complex functionality has led to wide interest in this rather new and promising class of materials. However, the

References see page 1244

potential of these recent developments is soon to become realized. Certainly, an interdisciplinary approach will be appropriate to evaluate the catalytic properties or construct sensor devices based on magnetic or optical response functions.

4.3.3.1
Sorption Properties and Host–Guest Inclusion

Since the sorption behavior is closely related to the specific structure and composition of the framework, the relevant properties were already discussed in Sect. 4.3.2. In the following, some general properties and peculiarities will be discussed. The porosity of metal-organic frameworks is compared in Table 1 on the basis of specific pore volumes.

Efficient guest removal without pore collapse is the key to obtaining accurate sorption measurements, but this has not always been achieved. In the case of the carboxylate compounds it was possible to obtain fully desolvated samples of some frameworks. Nitrogen physisorption measurements in general show type I isotherms and a narrow pore-size distribution [20, 21]. They indicate a high micropore volume for $Zn(p\text{-BDC})(H_2O)\cdot(DMF)$ (MOF-2), $Zn_3(p\text{-BDC})_3\cdot6MeOH$ (MOF-3) and $(Zn_4O)(p\text{-BDC})_3\cdot(DMF)_8(C_6H_5Cl)$ (MOF-5) (0.094, 0.038 and 1.04 cm^3 g^{-1}). The frameworks also show high sorption capacities for Ar, CO_2, CH_2Cl_2, $CHCl_3$, CCl_4, C_6H_6, and C_6H_{12}.

In the case of MOF-2 sorption measurements of the larger guest molecules like cyclohexane, slight irregularities are observed in the low-pressure region, indicating a distortion of the host lattice associated with guest inclusion. Thus, it is unclear whether the framework remains intact after cyclohexane removal. Interest-

Tab. 1. Sorption properties of metal-organic frameworks (increasing pore volume).

Compound	Specific pore volume/ (cm^3 g^{-1})	Pore diameter/ Å	Probe
Tb(p-BDC)NO$_3$·2DMF [30]	0.032	5	CO$_2$
Tb$_2$(ADB)$_3$(DMSO)$_4$·16DMSO (MOF-9) [31]	0.035	8	CO$_2$
Zn$_3$(p-BDC)$_3$·6MeOH (MOF-3) [20, 27]	0.038	8	N$_2$
Zn(p-BDC)(H$_2$O)·(DMF) (MOF-2) [20, 22]	0.094	7	N$_2$
Tb$_2$(p-BDC)$_3$(H$_2$O)$_4$ (MOF-6) [29]	0.099	4	H$_2$O
Al$_2$(CH$_3$PO$_3$)$_3$·H$_2$O [112–117]	0.117	6	N$_2$
Cu$_2$(ATC)·6H$_2$O (MOF-11) [57]	0.200	7	N$_2$
Cu(II)(p-BDC) [23]	0.220	6.0	Ar
Cu$_3$(BTC)$_2$(H$_2$O)$_3$ [46]	0.333	9	N$_2$
Cu(μ-4,4'-BPY)(H$_2$O)$_2$(FBF$_3$)$_2$·4,4'-BPY [71, 72]	0.510	–	N$_2$
Cu$_3$(BTB)$_2$(H$_2$O)$_3$·(DMF)$_9$(H$_2$O)$_2$ (MOF-14) [53]	0.530	16^a	N$_2$
Cu(4,4'-BPY)$_2$(SiF$_6$)·8H$_2$O [69]	0.560	8	Ar
Cu(p-BDC)(TED)$_{0.5}$ [23]	0.580	7.4	Ar
Zn$_2$(BTC)(NO$_3$)·(H$_2$O)(C$_2$H$_5$OH)$_5$ (MOF-4) [20, 47]	0.612	14	Ethanol
(Zn$_4$O)(p-BDC)$_3$·(DMF)$_8$(C$_6$H$_5$Cl) (MOF-5) [20, 28]	1.040	12	N$_2$

^{a} from the crystal structure.

ingly, the pore volume deduced from the nitrogen-adsorption isotherm is lower compared to the solvent vapor isotherms, which is attributed to the inability of the smaller molecules to diffuse freely into the pores at low temperature [20].

MOF-3 shows a similar behavior. Interesting selectivities are observed for partially desolvated samples. They show a high selectivity for alcohols driven by hydrogen bonding interactions between incoming alcohols and methanol molecules present in the channels.

For $Zn_2(BTC)NO_3 \cdot (EtOH)_5 H_2O$ (MOF-4) nitrogen-adsorption measurements were unsuccessful, only ethanol-sorption isotherms were obtained. Measurements at room temperature show a type V isotherm with large hysteresis. The structure is a truly 3D framework, but three of the ethanol groups are bound to zinc. The desorption and sorption processes are therefore associated with changes in the coordination of Zn. Obviously, the sorption involves an activation process, then coordination to zinc followed by pore filling. The desorption shows a large hysteresis due to the shielding of ethanol ligands by ethanol guests [20].

For MOF-5, an interesting method to obtain the fully desolvated sample was used [20, 28]. First, the guest molecules in the as-synthesized material are exchanged with chloroform. Due to the high mobility of the guests full exchange is achieved by immersing the sample for 18 h in $CHCl_3$. The chloroform is easily removed from the pores by evacuation at room temperature (24 h). The framework crystallinity remains intact. In fact, single-crystal X-ray studies revealed monocrystallinity for the desolvated samples. Nitrogen- and solvent-sorption isotherms show a type I isotherm. The pore volume varies between 1.04 cm³ g⁻¹ (N_2) and 0.92 cm³ g⁻¹ (C_6H_{12}), which is unprecedented and corresponds to an apparent surface area of 2900 m² g⁻¹. Metal-organic frameworks with such remarkable sorption behavior and crystallographically well-defined pore structure could also be valuable for the modeling and understanding of the adsorption process itself.

The copper(II) dicarboxylates reversibly adsorb N_2, Ar, O_2 and Xe [25]. The fumarate, terephthalate, naphthalenedicarboxylate and *trans*-1,4-cyclohexane-dicarboxylate are assumed to have a square grid architecture analogous to that of $Zn(p\text{-}BDC)(H_2O) \cdot (DMF)$ (MOF-2) with $M_2(CO_2)_4$ paddle-wheels. Their sorption capacity ranges from 0.8 to 1.8 mol N_2 per mol copper atoms and the pore sizes are typically 4.9 to 6.0 Å, as deduced from nitrogen physisorption using the Horvath–Kawazoe method. Comparable results are obtained for molybdenum(II) and Ru(II,III) carboxylates. $Mo_2(p\text{-}BDC)_2$ has the highest sorption capacity (2.0 mol N_2/mol of molybdenum) as compared to fumarate, pyridinedicarboxylate and trans-1,4-cyclohexane-dicarboxylate. ¹³C CP/MAS NMR investigations show a significant change of the local structure when *n*-butane is adsorbed [25]. The line width of *n*-butane is small, suggesting rapid motion of the adsorbed molecules but the chemical environment of the CH groups in the aromatic ring of p-BDC is changed significantly, indicating a reversible interaction with adsorbed molecules.

AlMepO-α and AlMepO-β differ remarkably in their sorption behavior [152]. AlMepO-β has a typical type I isotherm with a maximum adsorption capacity for nitrogen of 0.134 cm³ g⁻¹. In contrast, the α-polymorph shows one type I uptake

References see page 1244

in the low pressure region and a second in the relative pressure range of $0.02 < P/P_0 < 0.03$. The adsorption isotherm with two plateaus is attributed to a packing change of the adsorbed nitrogen molecules.

The two microporous aluminum methylphosphonates possess a highly hydrophobic inner surface due to the methyl groups that point into the center of the channels. Thus, AlMepO-α shows no water adsorption at 0 °C and AlMepO-β only shows an uptake at $P/P_0 > 0.6$.

4.3.3.2
Catalysis

Surprisingly, few applications in catalysis have been reported so far.

The use of organic host materials to enhance endoselectivity in Diels–Alder reactions has been reported by Aoyama [13, 153].

Cd(4,4'-BPY)$_2$(NO$_3$)$_2$, a square network material, is an efficient catalyst for the cyanosilylation of aldehydes [154] (Fig. 44). The framework is highly shape selec-

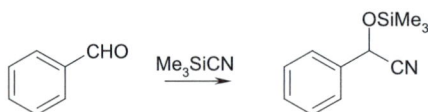

Fig. 44. The cyanosilylation of aldehydes is catalyzed by Cd(4,4'-BPY)$_2$(NO$_3$)$_2$ and proceeds shape specific.

Fig. 45. Cluster compounds as solid chemical sensors: The color changes of [Co$_2$(H$_2$O)$_4$][Re$_6$S$_8$(CN)$_6$]·10H$_2$O (top) and [Co(H$_2$O)$_3$]$_4$[Co$_2$(H$_2$O)$_4$][Re$_6$S$_8$(CN)$_6$]$_3$·44H$_2$O (bottom). From left to right: As-prepared material, THF and diethyl-ether-exposed samples (from Beauvais et al. [139]).

tive for the inclusion of ortho-dihalogenobenzenes (halogen = chlorine, bromine) whereas the meta- and para-isomers do not form clathrates. Shape specificity similar to that in clathration was observed in cyanosilylation. For example, the conversion of 2-tolualdehyde (40 %) is significantly higher as compared to 3-tolualdehyde (19 %).

Platinum or palladium colloids in the layered zirconium viologen phosphonates $Zr_2(PO_4)(PV)X_3$ were shown to be active catalysts for the production of H_2O_2 from H_2 and O_2 [155]. The photoinduced charge separation in viologen phosphonates and associated reduction reactions were already mentioned in Sect. 4.3.2.3.

4.3.3.3
Sensing

The high concentration of lanthanide or transition metals in porous frameworks allows the detection of adsorbed molecules based on the change of magnetic or optical properties upon adsorption.

The luminescence decay lifetime constants have been used in the case of $Tb_2(p\text{-}BDC)_3(H_2O)_4$ (MOF-6) [29] as a response to the intercalation of small molecules. This material has only small pores (<4 Å) and a pore volume of 0.099 cm^3 g^{-1} suitable for the sorption of H_2O and NH_3 but not N_2. The luminescence decays of the $^5D_4 \rightarrow {}^7F_5$ emission at 543 nm for the evacuated compound, the ammonia adduct and the hydrate are all single exponential. The decay constants are distinctly different and decrease from 1.13 ms^{-1} in the hydrate to 1.00 ms^{-1} in the ammonia adduct, whereas the desolvated material has a lower constant (0.74 ms^{-1}). Larger voids in Tb compounds could therefore lead to shape-selective chemical sensing host materials.

Cyano-bridged Re_6Q_8 clusters were proposed as chemical sensing devices [139] using vapochromic effects. The orange framework compounds $[Co_2(H_2O)_4]$ $[Re_6S_8(CN)_6]\cdot10H_2O$ and $[Co(H_2O)_3]_4[Co_2(H_2O)_4][Re_6S_8(CN)_6]_3\cdot44H_2O$ show distinct reversible color changes upon exposure to solvents like diethylether (blue-violet), THF (red-violet), ethanol (orange-red) or cyclohexane (orange) due to changes in the coordination geometry of Co^{2+} ions from octahedral to tetrahedral (Fig. 45 see p. 1242). The absorbance ratio at two different wavelengths allows to distinguish between various solvent molecules. The sensitivity is around 20 ppm and the response time is below 1 s. Even after 100 cycles no significant changes in the spectra were observed.

4.3.4
Syntheses

Metal-organic frameworks are usually synthesized in solution at room temperature or in refluxing solvents. Often, ligand design is more demanding and time con-

References see page 1244

suming than the synthesis of the coordination polymer. The framework is usually precipitated by combining a soluble metal salt with a solution of the ligand. Metal nitrate solutions are commonly used. In general, crystallization is the critical process since the solubility of the solid is low. Hydrothermal or solvothermal techniques are common to enhance the solubility of the framework.

A slow precipitation is beneficial. This can be achieved by slow diffusion of one reactant into a solution of the other. For example, crystallization in silica gels has been used for the synthesis of squarates in order to achieve slow diffusion rates of the reactants [40].

In carboxylate chemistry, the components are mixed and the precipitation of the solid is achieved by slow addition of a liquid base like an amine. This results in deprotonation of the carboxylic acids and precipitation of the metal-organic framework. A problem that seriously interferes with topology prediction is the subsequent deprotonation of di- and tricarboxylates that may lead to only partially deprotonated acids. These ligands may form insoluble salts with a lower degree of networking than desired.

An elegant homogeneous generation of carboxylate linkers is slow hydrothermal hydrolysis of nitriles that may lead to networks different from those obtained by direct addition methods [63, 85]. Since local supersaturations are avoided, the number of nuclei remains small and low concentrations of the ligand may be achieved over a long crystallization period without any sophisticated apparatus.

4.3.5
Outlook

Recent developments in metal-organic framework chemistry have shown the tremendous potential of new flexible architectures with high functionality and well-defined but variable pore structure. Pore sizes unattained in zeolite chemistry have been realized using predesigned organic building units, and the materials were shown to be highly stable at temperatures up to 300 °C. Chiral networks are promising candidates for enantioselective separations or catalytic transformations. In this first stage researchers have mainly focused on the design principles and structural elucidation of the coordination polymers using X-ray tools. Some researchers have already started to develop new applications in various fields. The time has now come to combine the expertise of organic, inorganic and organometallic chemists, crystallographers and industrial engineers and prove the usefulness of coordination polymers. A deep knowledge of the intrinsic properties will help to serve future needs in catalytic engineering, molecular sieving, sensing and fine-chemical industry.

References

1 K. R. Dunbar, R. A. Heintz, in: Progress in Inorganic Chemistry. K. D. Karlin (Ed.), Vol. 45, John Wiley, New York, 1997, p. 283–391.

2 F. V. Williams, *J. Am. Chem. Soc.* **1957**, *79*, 5876–5877.

3 H. M. Powell, J. H. Rayner, *Nature* **1949**, *163*, 566–567.

4 H. M. POWELL, *Nature* **1951**, *168*, 11–14.

5 S. R. BATTEN, R. ROBSON, *Angew. Chem. Int. Ed. Engl.* **1998**, *37*, 1461–1494.

6 S. R. BATTEN, *Curr. Opin. Solid State Mater. Sci.* **2001**, *5*, 107–114.

7 T. J. BARTON, L. M. BULL, W. G. KLEMPERER, D. A. LOY, B. MCENANEY, M. MISONO, P. A. MONSON, G. PEZ, G. W. SCHERER, J. C. VARTULI, O. M. YAGHI, *Chem. Mater.* **1999**, *11*, 2633–2656.

8 A. K. CHEETHAM, G. FEREY, T. LOISEAU, *Angew. Chem. Int. Ed. Engl.* **1999**, *38*, 3269–3292.

9 M. EDDAOUDI, D. B. MOLER, H. LI, B. CHEN, T. M. REINEKE, M. O'KEEFFE, O. M. YAGHI, *Acc. Chem. Res.* **2001**, *34*, 319–330.

10 M. O'KEEFFE, M. EDDAOUDI, H. LI, T. REINEKE, O. M. YAGHI, *J. Solid State Chem.* **2000**, *152*, 3–20.

11 O. M. YAGHI, H. LI, C. DAVIS, D. RICHARDSON, T. L. GROY, *Acc. Chem. Res.* **1998**, *31*, 474–484.

12 S. KITAGAWA, M. KONDO, *Bull. Chem. Soc. Jpn.* **1998**, *71*, 1739–1753.

13 Y. AOYAMA, *Top. Curr. Chem.* **1998**, *198*, 131–161.

14 A. NANGIA, *Curr. Opin. Solid State Mater. Sci.* **2001**, *5*, 115–122.

15 A. F. WELLS, 3D Nets and Polyhedra. Wiley-Interscience, New York, 1977, 268 pp.

16 H. VAN KONINGSVELD, in: Introduction to Zeolite Science and Practice. H. VAN BEKKUM, E. M. FLANIGEN AND J. C. JANSEN (Ed.), Vol. 58, Elsevier, Amsterdam, 1991, p. 35–76.

17 P. BRUNET, M. SIMARD, J. D. WUEST, *J. Am. Chem. Soc.* **1997**, *119*, 2737–2738.

18 O. M. YAGHI, *Chem. Innovation* **2000**, *30*, 3–4.

19 O. M. YAGHI, M. O'KEEFFE, M. KANATZIDIS, *J. Solid State Chem.* **2000**, *152*, 1–2.

20 M. EDDAOUDI, H. LI, O. M. YAGHI, *J. Am. Chem. Soc.* **2000**, *122*, 1391–1397.

21 M. EDDAOUDI, H. L. LI, T. REINEKE, M. FEHR, D. KELLEY, T. L. GROY, O. M. YAGHI, *Top. Catal.* **1999**, *9*, 105–111.

22 H. LI, M. EDDAOUDI, T. L. GROY, O. M. YAGHI, *J. Am. Chem. Soc.* **1998**, *120*, 8571–8572.

23 K. SEKI, S. TAKAMIZAWA, W. MORI, *Chem. Lett.* **2001**, 332–333.

24 W. MORI, F. INOUE, K. YOSHIDA, H. NAKAYAMA, S. TAKAMIZAWA, M. KISHITA, *Chem. Lett.* **1997**, 1219–1220.

25 W. MORI, S. TAKAMIZAWA, *J. Solid State Chem.* **2000**, *152*, 120–129.

26 R. NUKADA, W. MORI, S. TAKAMIZAWA, M. MIKURIYA, M. HANDA, H. NAONO, *Chem. Lett.* **1999**, 367–368.

27 H. LI, C. E. DAVIS, T. L. GROY, D. G. KELLEY, O. M. YAGHI, *J. Am. Chem. Soc.* **1998**, *120*, 2186–2187.

28 H. LI, M. EDDAOUDI, M. O'KEEFFE, O. M. YAGHI, *Nature* **1999**, *402*, 276–279.

29 T. M. REINEKE, M. EDDAOUDI, M. FEHR, D. KELLEY, O. M. YAGHI, *J. Am. Chem. Soc.* **1999**, *121*, 1651–1657.

30 T. M. REINEKE, M. EDDAOUDI, M. O'KEEFFE, O. M. YAGHI, *Angew. Chem. Int. Ed. Engl.* **1999**, *38*, 2590–2594.

31 T. M. REINEKE, M. EDDAOUDI, D. MOLER, M. O'KEEFFE, O. M. YAGHI, *J. Am. Chem. Soc.* **2000**, *122*, 4843–4844.

32 M. EDDAOUDI, J. KIM, J. B. WACHTER, H. K. CHAE, M. O'KEEFFE, O. M. YAGHI, *J. Am. Chem. Soc.* **2001**, *123*, 4368–4369.

33 B. L. CHEN, K. F. MOK, S. C. NG, M. G. B. DREW, *New J. Chem.* **1999**, *23*, 877–883.

34 V. KIRITSIS, A. MICHAELIDES, S. SKOULIKA, S. GOLHEN, L. OUAHAB, *Inorg. Chem.* **1998**, *37*, 3407–3410.

35 Y. KIM, D. P. JUNG, *Inorg. Chem.* **2000**, *39*, 1470–1475.

36 S. NATARAJAN, R. VAIDHYANATHAN, C. N. R. RAO, S. AYYAPPAN, A. K. CHEETHAM, *Chem. Mater.* **1999**, *11*, 1633–1639.

37 S. F. LAI, C. Y. CHENG, K. J. LIN, *Chem. Commun.* **2001**, 1082–1083.

38 O. M. YAGHI, G. M. LI, T. L. GROY, *J. Chem. Soc., Dalton Trans.* **1995**, 727–732.

39 A. WEISS, E. RIEGLER, C. ROBL, *Z. Naturforsch., B: Anorg. Chem.* **1986**, *41*, 1333–1336.

40 O. M. YAGHI, G. M. LI, T. L. GROY,

J. Solid State Chem. **1995**, *117*, 256–260.

41 O. M. YAGHI, G. LI, H. LI, *Nature* **1995**, *378*, 703–706.

42 O. M. YAGHI, H. LI, T. L. GROY, *J. Am. Chem. Soc.* **1996**, *118*, 9096–9101.

43 M. J. PLATERS, R. A. HOWIE, A. J. ROBERTS, *Chem. Commun.* **1997**, 893–894.

44 C. J. KEPERT, T. J. PRIOR, M. J. ROSSEINSKY, *J. Solid State Chem.* **2000**, *152*, 261–270.

45 H. J. CHOI, M. P. SUH, *J. Am. Chem. Soc.* **1998**, *120*, 10622–10628.

46 S. S. Y. CHUI, S. M. F. LO, J. P. H. CHARMANT, A. G. ORPEN, I. D. WILLIAMS, *Science* **1999**, *283*, 1148–1150.

47 O. M. YAGHI, C. E. DAVIS, G. M. LI, H. L. LI, *J. Am. Chem. Soc.* **1997**, *119*, 2861–2868.

48 C. J. KEPERT, T. J. PRIOR, M. J. ROSSEINSKY, *J. Am. Chem. Soc.* **2000**, *122*, 5158–5168.

49 C. J. KEPERT, M. J. ROSSEINSKY, *Chem. Commun.* **1998**, 31–32.

50 M. ROSSEINSKY, personal communication.

51 O. M. YAGHI, R. JERNIGAN, H. LI, C. E. DAVIS, T. L. GROY, *J. Chem. Soc., Dalton Trans.* **1997**, 2383–2384.

52 L. PAN, E. B. WOODLOCK, X. WANG, C. ZHENG, *Inorg. Chem.* **2000**, *39*, 4174–4178.

53 B. L. CHEN, M. EDDAOUDI, S. T. HYDE, M. O'KEEFFE, O. M. YAGHI, *Science* **2001**, *291*, 1021–1023.

54 M. J. PLATER, M. R. S. FOREMAN, E. CORONADO, C. J. GOMEZ-GARCIA, A. M. Z. SLAWIN, *J. Chem. Soc. Dalton Trans.* **1999**, 4209–4216.

55 M. J. PLATER, M. R. S. FOREMAN, T. GELBRICH, M. B. HURSTHOUSE, *J. Chem. Crystallogr.* **2000**, *30*, 155–158.

56 D. Q. CHU, J. Q. XU, L. M. DUAN, T. G. WANG, A. Q. TANG, L. YE, *Eur. J. Inorg. Chem.* **2001**, 1135–1137.

57 B. CHEN, M. EDDAOUDI, T. M. REINEKE, J. W. KAMPF, M. O'KEEFFE, O. M. YAGHI, *J. Am. Chem. Soc.* **2000**, *122*, 11559–11560.

58 M. KONDO, T. OKUBO, A. ASAMI, S.

NORO, T. YOSHITOMI, S. KITAGAWA, T. ISHII, H. MATSUZAKA, K. SEKI, *Angew. Chem. Int. Ed. Engl.* **1999**, *38*, 140–143.

59 M. J. PLATER, M. R. S. FOREMAN, R. A. HOWIE, E. E. LACHOWSKI, *J. Chem. Res., Synop.* **1998**, 754–755.

60 Y. LIANG, R. CAO, W. SU, C. HONG, *Chem. Lett.* **2000**, 868–869.

61 M. J. PLATER, A. J. ROBERTS, R. A. HOWIE, *J. Chem. Res., Synop.* **1998**, 240–241.

62 T. SCHAREINA, C. SCHICK, R. KEMPE, *Z. Anorg. Allg. Chem.* **2001**, *627*, 131–133.

63 Y. H. LIU, Y. L. LU, H. L. TSAI, J. C. WANG, K. L. LU, *J. Solid State Chem.* **2001**, *158*, 315–319.

64 Y. H. LIU, C. S. LIN, S. Y. CHEN, H. L. TSAI, C. H. UENG, K. L. LU, *J. Solid State Chem.* **2001**, *157*, 166–172.

65 J. LU, T. PALIWALA, S. C. LIM, C. YU, T. NIU, A. J. JACOBSON, *Inorg. Chem.* **1997**, *36*, 923–929.

66 M. KONDO, M. SHIMAMURA, S. NORO, S. MINAKOSHI, A. ASAMI, K. SEKI, S. KITAGAWA, *Chem. Mater.* **2000**, *12*, 1288–1299.

67 M. B. ZAMAN, M. D. SMITH, H.-C. zur LOYE, *Chem. Mater.* **2001**, *13*, 3534–3541.

68 S. SUBRAMANIAN, M. J. ZAWOROTKO, *Angew. Chem. Int. Ed. Engl.* **1995**, *34*, 2561–2561.

69 S. NORO, S. KITAGAWA, M. KONDO, K. SEKI, *Angew. Chem. Int. Ed. Engl.* **2000**, *39*, 2082–2084.

70 R. W. GABLE, B. F. HOSKINS, R. ROBSON, *J. Chem. Soc., Chem. Commun.* **1990**, 1677–1678.

71 A. J. BLAKE, S. J. HILL, P. HUBBERSTEY, W. S. LI, *J. Chem. Soc., Dalton Trans.* **1997**, 913–914.

72 D. LI, K. KANEKO, *Chem. Phys. Lett.* **2001**, *335*, 50–56.

73 I. BOLDOG, E. B. RUSANOV, A. N. CHERNEGA, J. SIELER, K. V. DOMASEVITCH, *Polyhedron* **2001**, *20*, 887–897.

74 L. C. TABARES, J. A. R. NAVARRO, J. M. SALAS, *J. Am. Chem. Soc.* **2001**, *123*, 383–387.

75 L. CARLUCCI, G. CIANI, D. M.

PROSERPIO, S. RIZZATO, *J. Solid State Chem.* **2000**, *152*, 211–220.

76 L. R. MacGILLIVRAY, S. SUBRAMANIAN, M. J. ZAWOROTKO, *J. Chem. Soc., Chem. Commun.* **1994**, 1325–1326.

77 O. TEICHERT, W. S. SHELDRICK, *Z. Anorg. Allg. Chem.* **2000**, *626*, 1509–1513.

78 M. KONDO, T. YOSHITOMI, K. SEKI, H. MATSUZAKA, S. KITAGAWA, *Angew. Chem. Int. Ed. Engl.* **1997**, *36*, 1725–1727.

79 L. CARLUCCI, G. CIANI, D. M. PROSERPIO, *J. Chem. Soc., Dalton Trans.* **1999**, 1799–1804.

80 M. A. WITHERSBY, A. J. BLAKE, N. R. CHAMPNESS, P. A. COOKE, P. HUBBERSTEY, M. SCHRODER, *J. Am. Chem. Soc.* **2000**, *122*, 4044–4046.

81 Y. B. DONG, M. D. SMITH, R. C. LAYLAND, H. C. ZUR LOYE, *Chem. Mater.* **2000**, *12*, 1156–1161.

82 E. LEE, J. KIM, J. HEO, D. WHANG, K. KIM, *Angew. Chem. Int. Ed. Engl.* **2001**, *40*, 399–402.

83 J. HEO, J. KIM, D. WHANG, K. KIM, *Inorg. Chim. Acta* **2000**, *297*, 307–312.

84 D. WHANG, K. KIM, *J. Am. Chem. Soc.* **1997**, *119*, 451–452.

85 E. S. LEE, J. S. HEO, K. KIM, *Angew. Chem. Int. Ed. Engl.* **2000**, *39*, 2699–2701.

86 R. ATENCIO, K. BIRADHA, T. L. HENNIGAR, K. M. POIRIER, K. N. POWER, C. M. SEWARD, N. S. WHITE, M. J. ZAWOROTKO, *Mater. Res. Bull.* **1998**, *33*, 203–212.

87 O. M. YAGHI, H. LI, *J. Am. Chem. Soc.* **1995**, *117*, 10401–10402.

88 O. M. YAGHI, H. LI, *J. Am. Chem. Soc.* **1996**, *118*, 295–296.

89 S. W. KELLER, S. LOPEZ, *J. Am. Chem. Soc.* **1999**, *121*, 6306–6307.

90 O. M. YAGHI, G. LI, *Angew. Chem. Int. Ed. Engl.* **1995**, *34*, 207–209.

91 B. ROSSENBECK, W. S. SHELDRICK, *Z. Naturforsch., B: Anorg. Chem.* **2000**, *55*, 467–472.

92 M. J. PLATER, M. R. S. FOREMAN, T. GELBRICH, M. B. HURSTHOUSE, *J. Chem. Soc., Dalton Trans.* **2000**, 1995–2000.

93 E. HOUGH, L. K. HANSEN, B. BIRKNES, K. JYNGE, S. HANSEN, A. HORDVIK, C.

LITTLE, E. DODSON, Z. DEREWENDA, *Nature* **1989**, *338*, 357–360.

94 M. KONDO, M. SHIMAMURA, S. NORO, Y. KIMURA, K. UEMURA, S. KITAGAWA, *J. Solid State Chem.* **2000**, *152*, 113–119.

95 C. V. K. SHARMA, G. A. BROKER, R. D. ROGERS, *J. Solid State Chem.* **2000**, *152*, 253–260.

96 O. M. YAGHI, H. LI, T. L. GROY, *Inorg. Chem.* **1997**, *36*, 4292–4293.

97 D. V. SOLDATOV, J. A. RIPMEESTER, S. I. SHERGINA, I. E. SOKOLOV, A. S. ZANINA, S. A. GROMILOV, Y. A. DYADIN, *J. Am. Chem. Soc.* **1999**, *121*, 4179–4188.

98 A. V. NOSSOV, D. V. SOLDATOV, J. A. RIPMEESTER, *J. Am. Chem. Soc.* **2001**, *123*, 3563–3568.

99 A. CLEARFIELD, in: Progress in Inorganic Chemistry. K. D. KARLIN (Ed.), Vol. 47, John Wiley, New York, 1998, p. 371–510.

100 A. CLEARFIELD, *Curr. Opin. Solid State Mater. Sci.* **1996**, *1*, 268–278.

101 J. M. TROUP, A. CLEARFIELD, *Inorg. Chem.* **1977**, *16*, 3311–3314.

102 M. E. THOMPSON, *Chem. Mater.* **1994**, *6*, 1168–1175.

103 G. ALBERTI, U. COSTANTINO, F. MARMOTTINI, R. VIVANI, P. ZAPPELLI, *Angew. Chem. Int. Ed. Engl.* **1993**, *32*, 1357–1359.

104 S. YAMANAKA, *Inorg. Chem.* **1976**, *15*, 2811.

105 A. CLEARFIELD, D. M. POOJARY, B. ZHANG, B. ZHAO, A. DERECSKEI-KOVACS, *Chem. Mater.* **2000**, *12*, 2745–2752.

106 B. ZHANG, A. CLEARFIELD, *J. Am. Chem. Soc.* **1997**, *119*, 2751–2752.

107 H. BYRD, A. CLEARFIELD, D. POOJARY, K. P. REIS, M. E. THOMPSON, *Chem. Mater.* **1996**, *8*, 2239–2246.

108 L. A. VERMEULEN, M. E. THOMPSON, *Chem. Mater.* **1994**, *6*, 77–81.

109 L. A. VERMEULEN, S. J. BURGMEYER, *J. Solid State Chem.* **1999**, *147*, 520–526.

110 M. I. KHAN, J. ZUBIETA, in: Progress in Inorganic Chemistry. K. D. KARLIN (Ed.), Vol. 43, John Wiley, New York, 1995, p. 1–149.

111 J. W. JOHNSON, A. J. JACOBSON, W. M.

BUTLER, S. E. ROSENTHAL, J. F. BRODY, J. T. LEWANDOWSKI, *J. Am. Chem. Soc.* **1989**, *111*, 381–383.

112 K. MAEDA, Y. KIYOZUMI, F. MIZUKAMI, *Angew. Chem. Int. Ed. Engl.* **1994**, *33*, 2335–2337.

113 K. MAEDA, J. AKIMOTO, Y. KIYOZUMI, F. MIZUKAMI, *J. Chem. Soc., Chem. Commun.* **1995**, 1033–1034.

114 K. MAEDA, J. AKIMOTO, Y. KIYOZUMI, F. MIZUKAMI, *Angew. Chem. Int. Ed. Engl.* **1995**, *34*, 1199–1201.

115 K. MAEDA, J. AKIMOTO, Y. KIYOZUMI, F. MIZUKAMI, in: Studies in Surface Science and Catalysis. H. CHON, S.-K. IHM, Y.S. UH (Ed.), Vol. 105, 1997, p. 197–204.

116 K. MAEDA, A. SASAKI, K. WATANABE, Y. KIYOZUMI, F. MIZUKAMI, *Chem. Lett.* **1997**, 879–880.

117 K. MAEDA, Y. KIYOZUMI, F. MIZUKAMI, *J. Phys. Chem. B* **1997**, *101*, 4402–4412.

118 J. LE BIDEAU, C. PAYEN, P. PALVADEAU, B. BUJOLI, *Inorg. Chem.* **1994**, *33*, 4885–4890.

119 S. DRUMEL, P. JANVIER, D. DENIAUD, B. BUJOLI, *J. Chem. Soc., Chem. Commun.* **1995**, 1051–1052.

120 D. M. POOJARY, B. ZHANG, A. CLEARFIELD, *Chem. Mater.* **1999**, *11*, 421–426.

121 R. D. FISCHER, G. R. SIENEL, *J. Organomet. Chem.* **1978**, *156*, 383–388.

122 P. BRANDT, A. K. BRIMAH, R. D. FISCHER, *Angew. Chem. Int. Ed. Engl.* **1988**, *27*, 1521–1522.

123 K. YÜNLÜ, N. HÖCK, R. D. FISCHER, *Angew. Chem. Int. Ed. Engl.* **1985**, *24*, 879–881.

124 U. BEHRENS, A. K. BRIMAH, T. M. SOLIMAN, R. D. FISCHER, D. C. APPERLEY, N. A. DAVIES, R. K. HARRIS, *Organometallics* **1992**, *11*, 1718–1726.

125 A. BONARDI, C. CARINI, C. PELIZZI, G. PELIZZI, G. PREDIERI, P. TARASCONI, M. A. ZORODDU, K. C. MOLLOY, *J. Organomet. Chem.* **1991**, *401*, 283–294.

126 P. SCHWARZ, E. SIEBEL, R. D. FISCHER, D. C. APPERLEY, N. A. DAVIES, R. K. HARRIS, *Angew. Chem. Int. Ed. Engl.* **1995**, *34*, 1197–1199.

127 P. SCHWARZ, S. ELLER, E. SIEBEL, T. M. SOLIMAN, R. D. FISCHER, D. C.

APPERLEY, N. A. DAVIES, R. K. HARRIS, *Angew. Chem. Int. Ed. Engl.* **1996**, *35*, 1525–1527.

128 P. SCHWARZ, E. SIEBEL, R. D. FISCHER, N. A. DAVIES, D. C. APPERLEY, R. K. HARRIS, *Chem.-Eur. J.* **1998**, *4*, 919–926.

129 E. M. POLL, S. SAMBA, R. D. FISCHER, F. OLBRICH, N. A. DAVIES, P. AVALLE, D. C. APPERLEY, R. K. HARRIS, *J. Solid State Chem.* **2000**, *152*, 286–301.

130 E. M. POLL, F. OLBRICH, S. SAMBA, R. D. FISCHER, P. AVALLE, D. C. APPERLEY, R. K. HARRIS, *J. Solid State Chem.* **2001**, *157*, 324–338.

131 E. SIEBEL, R. D. FISCHER, N. A. DAVIES, D. C. APPERLEY, R. K. HARRIS, *J. Organomet. Chem.* **2000**, *604*, 34–42.

132 A. M. A. IBRAHIM, E. SIEBEL, R. D. FISCHER, *Inorg. Chem.* **1998**, *37*, 3521–3525.

133 A. K. BRIMAH, E. SIEBEL, R. D. FISCHER, N. A. DAVIES, D. C. APPERLEY, R. K. HARRIS, *J. Organomet. Chem.* **1994**, *475*, 85–94.

134 E. M. POLL, J. U. SCHÜTZE, R. D. FISCHER, N. A. DAVIES, D. C. APPERLEY, R. K. HARRIS, *J. Organomet. Chem.* **2001**, *621*, 254–260.

135 L. G. BEAUVAIS, M. P. SHORES, J. R. LONG, *Chem. Mater.* **1998**, *10*, 3783–3786.

136 N. G. NAUMOV, A. V. VIROVETS, M. N. SOKOLOV, S. B. ARTEMKINA, V. E. FEDOROV, *Angew. Chem. Int. Ed. Engl.* **1998**, *37*, 1943–1945.

137 M. P. SHORES, L. G. BEAUVAIS, J. R. LONG, *J. Am. Chem. Soc.* **1999**, *121*, 775–779.

138 M. V. BENNETT, M. P. SHORES, L. G. BEAUVAIS, J. R. LONG, *J. Am. Chem. Soc.* **2000**, *122*, 6664–6668.

139 L. G. BEAUVAIS, M. P. SHORES, J. R. LONG, *J. Am. Chem. Soc.* **2000**, *122*, 2763–2772.

140 V. P. FEDIN, A. V. VIROVETS, I. V. KALININA, V. N. IKORSKII, M. R. J. ELSEGOOD, W. CLEGG, *Eur. J. Inorg. Chem.* **2000**, 2341–2343.

141 N. G. NAUMOV, D. V. SOLDATOV, J. A. RIPMEESTER, S. B. ARTEMKINA, V. E. FEDOROV, *Chem. Commun.* **2001**, 571–572.

142 N. G. NAUMOV, A. V. VIROVETS, V. E.

Fedorov, *Inorg. Chem. Commun.* **2000**, *3*, 71–72.

143 C. Zhang, Y. Song, Y. Xu, H. Fun, G. Fang, Y. Wang, X. Xin, *J. Chem. Soc., Dalton Trans.* **2000**, 2823–2829.

144 C. Zhang, G. Jin, Y. Xu, H. K. Fun, X. Xin, *Chem. Lett.* **2000**, 502–503.

145 Q. F. Zhang, W. H. Leung, X. Q. Xin, H. K. Fun, *Inorg. Chem.* **2000**, *39*, 417–426.

146 D. L. Long, A. J. Blake, N. R. Champness, C. Wilson, M. Schröder, *Angew. Chem. Int. Ed. Engl.* **2001**, *40*, 2444–2447.

147 D. L. Long, A. J. Blake, N. R. Champness, C. Wilson, M. Schröder, *J. Am. Chem. Soc.* **2001**, *123*, 3401–3402.

148 B. Q. Ma, S. Gao, H. L. Sun, G. X. Xu, *J. Chem. Soc., Dalton Trans.* **2001**, 130–133.

149 J. M. Tanski, E. B. Lobkovsky, P. Wolczanski, *J. Solid State Chem.* **2000**, *152*, 130–140.

150 G. K. H. Shimizu, G. D. Enright, C. I. Ratcliffe, J. A. Ripmeester, D. D. M. Wayner, *Angew. Chem. Int. Ed. Engl.* **1998**, *37*, 1407–1409.

151 M. Kondo, K. Fujimoto, T. Okubo, A. Asami, S. Noro, S. Kitagawa, T. Ishii, H. Matsuzaka, *Chem. Lett.* **1999**, 291–292.

152 K. Maeda, F. Mizukami, *Catal. Surv. Jpn.* **1999**, *3*, 119–126.

153 K. Endo, T. Koike, T. Sawaki, O. Hayashida, H. Masuda, Y. Aoyama, *J. Am. Chem. Soc.* **1997**, *119*, 4117–4122.

154 M. Fujita, Y. J. Kwon, S. Washizu, K. Ogura, *J. Am. Chem. Soc.* **1994**, *116*, 1151–1152.

155 K. P. Reis, V. K. Joshi, M. E. Thompson, *J. Catal.* **1996**, *161*, 62–67.

4.4
Layered Structures and Pillared Layered Structures

Pegie Cool, Etienne F. Vansant, Georges Poncelet, and Robert A. Schoonheydt

4.4.1
Introduction

Layered crystalline solids (LCS) are natural and synthetic inorganic compounds with sheet structures. Each elementary sheet or layer is about 1 nm thick and a few tens to hundreds of nanometers wide and long. The sheets are usually aggregated. Table 1 gives a survey of the materials of interest to this chapter. A complete list has been given by Alberti and Costantino [1]. The layers can be electrically neutral, positively or negatively charged. In the latter two cases, the layer charge is compensated by exchangeable anions and cations, respectively. These are located between the layers in the so-called interlamellar space, as schematically drawn in Fig. 1.

In aggregates of neutral sheets, the interaction between the sheets is of the van der Waals type. Because of the large number of atoms and the additive nature of the van der Waals forces, the sheet–sheet interaction is very strong. It is extremely difficult to break up the aggregates into individual sheets. This is the case for clay minerals such as kaolinite, serpentine, pyrophyllite, and talc. For charged layers with exchangeable cations or anions in the interlamellar space, the van der Waals interaction is weaker, but this is compensated by an electrostatic interaction. In addition, exchangeable cations can attract water molecules by cation-dipole interaction. If the latter is more important than the cation–layer interaction, water is taken up in the interlamellar space and the LCS swells. This is the case for smectite-type clay minerals with low layer charge. Examples are montmorillonite, hectorite, saponite, and beiddelite. If the layer charge is too high, the cation-sheet interaction is dominant and swelling in water does not occur. This is the case for all the other charged layers of Table 1. Swelling can be induced in these materials too, but this needs special techniques such as charge reduction or partial pre-exchange with short-chain alkylammonium cations [2].

LCSs occur as powders and have a surface area and pore volume. The surface area, as measured from N_2 adsorption-desorption isotherms, is due to the external

Tab. 1. Layered crystalline solids.

1. *Neutral layers*
   clay minerals: kaolinite
                    serpentine
                    pyrophyllite
                    talc
   graphite
   transition metal dichalcogenides, MX_2
   metal oxyphospates, $MOPO_4$
   transition metal oxyhalides, MOX
   vanadium pentoxide, V_2O_5
   molybdenum oxide, MoO_3

2. *Negatively charged layers*
   clay minerals:
   α- and γ-metal phosphates: α-$M(HPO_4)_2 \cdot H_2O$; γ-$M(PO_4)(H_2PO_4) \cdot 2H_2O$
   silicic acids, $H_2Si_2O_5$
   β-alumina, $NaAl_{11}O_{17}$
   alkali-transition metal oxides: $M(I)XO_2$

3. *Positively charged layers*
   hydrotalcite $[[M(II)]_6[Al(III)]_2(OH)_6]CO_3 \cdot 4H_2O$

surface of the aggregated particles. The size of the layers, the number of layers per aggregate, the ordering of the layers in the aggregates and the history of treatments all determine the measured surface areas. The pore volume is determined by the same parameters and by the organization of the aggregates into a microfabric. The result is a powder with micro- and mesoporosity. This is schematically shown in Fig. 2. Extensive studies have been performed on clay minerals [3]. Typical numbers are given in Table 2 [4].

Typical catalysts and adsorbents have much larger surface areas and pore volumes. In addition, these quantities should be more or less independent of the conditions to which the materials are submitted. This is not the case for LCS

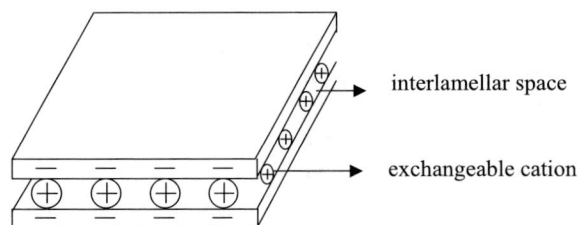

interlamellar space

exchangeable cation

Fig. 1. Schematic drawing of a layered crystalline solid with negatively charged layers and exchangeable cations in the interlamellar region.

References see page 1305

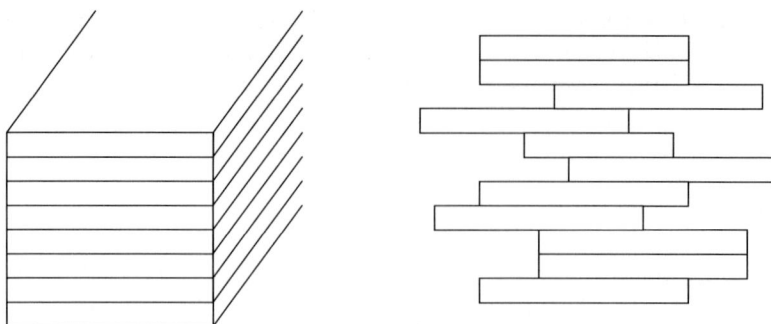

Fig. 2. Idealized drawings of aggregates.

materials. However, if the interlamellar surface is made available, a material is obtained with surface areas and pore volumes that are comparable with those of typical heterogeneous catalysts, such as zeolites, silica and alumina. This is done by introduction of thermally stable pillars between the elementary sheets of the aggregates. The process is called pillaring and consists of the following steps: (1) swelling of the LCS in a polar solvent, usually water; (2) ion exchange of a bulky, highly charged cation; (3) washing of the product and (4) calcination.

The LCS must swell in the solvent (usually water) and this is only possible for charged LCS with small to moderate layer charge, more specifically for smectite-type clay minerals. In other cases, pillaring can be achieved by preswelling with an organic cation or by ion exchange of a cation, which hydrolyzes and polymerizes into a bulky pillar in the interlamellar space. As a consequence, pillaring has only been reported for negatively charged LCS such as clay minerals or metal(IV) phosphates, and layered double hydroxides (LDH).

A clear distinction must be made between intercalation and pillaring [5]. Both processes result in the reversible insertion of guest species in the interlayer region

Tab. 2. Surface areas and pore volumes of montmorillonites.

Smectite	Cation drying	$S/m^2\ g^{-1}$	$V_t/\mu L\ g^{-1}$	V_{micro}	V_{meso}
Moosburg	Na$^+$ lyo	124	116	40	76
	sed	115	120	43	77
	Cs$^+$ lyo	157	120	49	71
	sed	151	118	59	59
	Ca^{2+} lyo	104	81	26	55
	sed	107	86	39	47
Otay	Na$^+$ lyo	144	142	54	88

Lyo = freeze-drying; sed = dried by sedimentation; V_t = total volume; V_{micro} = microporous volume; V_{meso} = mesoporous volume.

of the host LCS with retention of the structure of the latter. In addition, the pillared product has thermally stable micro- and mesoporosity. Thus, while all pillared compounds are intercalation compounds, the reverse is not true.

4.4.2
Layered Crystalline Solids

Layered crystalline solids are a class of compounds with a wide variety of chemical compositions (Table 1). In this chapter, only the three groups that have been converted to pillared products will be discussed.

4.4.2.1
Clay Minerals [6]

2:1 clay minerals are phyllosilicates. The elementary sheet contains a central layer of edge-shared Al- or Mg-octahedra, sandwiched between two layers of corner-shared Si tetrahedra. In the case of Al^{3+}, two out of three octahedral positions are occupied and the material is called dioctahedral. In the case of Mg^{2+}, all octahedral positions are occupied and the material is trioctahedral. Negative layer charge is introduced by isomorphous substitution in the octahedral layer or in the tetrahedral layer. In the latter case, Si^{4+} is substituted by Al^{3+} (some traces of Fe^{3+} might also occur). Octahedral substitution involves the replacement of Al^{3+} by Mg^{2+} (or Fe^{2+}) for dioctahedral clay minerals, or the replacement of Mg^{2+} by Li^+ for trioctahedral clay minerals.

Clay minerals have a pseudo-hexagonal structure. The unit cell is represented with eight Si atoms. Four groups of 2:1 clay minerals are distinguished, based on the layer charge (Table 3). If the layer charge is zero, the material has no cation-exchange capacity and does not swell. It cannot be pillared. Smectites have a low layer charge and swell easily in water. As a consequence, the interlamellar surface is easily available for ion exchange and pillaring is possible. Vermiculites and

Tab. 3. Classification of 2:1 clay minerals.

Charge/formula	Group	Subgroup	Name
0	pyrophillite talc	dioctahedral trioctahedral	pyrophyllite talc
0.25–0.60	smectite	dioctahedral trioctahedral	montmorillonite saponite/hectorite
0.6–0.9	vermiculite	dioctahedral trioctahedral	vermiculite vermiculite
1	mica	dioctahedral trioctahedral	muscovite talc

References see page 1305

micas have a high layer charge. They do not swell in water. Pillaring is only possible with special treatments, which aim at a reduction of the layer charge.

The cation-exchange capacity (CEC) is the amount of positive charge needed to compensate the negative layer charge. It is composed of a pH-independent part and a pH-dependent part. The former is the amount of positive charge needed to compensate the negative lattice charge, created by isomorphous substitution. The latter is the amount of positive charge needed to compensate for the negative charge at the edges of the layers. This is also the pH-dependent part of the CEC and amounts to 0.1–10 % of the experimentally measurable CEC, depending on the size of the layer, the site of isomorphous substitution (octahedral, tetrahedral), crystal size, and shape. Typical CEC values are 0.64–1.50 meq g^{-1} for smectites, 1.50–2.25 meq g^{-1} for vermiculites, and >2.25 meq g^{-1} for micas. The layer charge σ is then obtained from:

$$\sigma(C\ m^{-2}) = CEC(meq\ g^{-1})eFW/(1000\ 2ab) \tag{1}$$

with e the charge of the electron (1.6×10^{-19} C), FW the formula weight of the unit cell in g, and $2ab$ the surface area of the unit cell in m^2. The factor 1000 is needed because of the meq. The surface area S can be obtained from the dimensions of the unit cell:

$$S\ (m^2\ g^{-1}) = N_a 2ab/FW \tag{2}$$

with N_a, Avogadro's number. Alternatively, it can be obtained from the density ρ (g cm^{-3}) and the d_{001} spacing in nm from:

$$S = 2 \times 10^3/\rho d_{001} \tag{3}$$

With these formulae the edge surface is not taken into account. Experimentally, the surface area is obtained from N$_2$ adsorption-desorption isotherms. One then measures the external surface area, because the N$_2$ molecules cannot penetrate into the interlamellar region under the conditions of the experiments. These experimental surface areas are dependent on the size and shape of the elementary sheets, their stacking in the aggregates and the stacking of the aggregates into microfabrics. The same parameters determine the pore volumes, which are determined from the same isotherms.

Clay minerals are acidic. There are two sources of acidity: the edge sites and residual water molecules polarized by the exchangeable cations. In the latter case, the acidity is situated in the interlamellar region and is not easily available for incoming molecules. Another source of acidity is seen with NH$_4^+$-exchanged clay minerals. Upon heating, ammonium decomposes into ammonia, which is volatile, and protons that react with structural oxygens to form acidic OH groups, similar to the bridging OH groups in zeolites. However, as upon heating the expanded clay layers collapse, these OH groups are not accessible to incoming molecules.

Fig. 3. Structural representation of a LDH-like material.

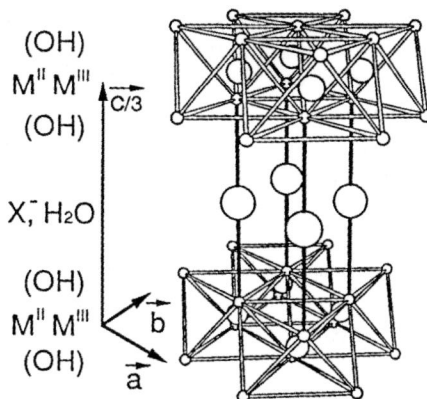

$$(OH)$$
$$M^{II} M^{III}$$
$$(OH)$$

$$X^-, H_2O$$

$$(OH)$$
$$M^{II} M^{III}$$
$$(OH)$$

4.4.2.2
Layered Double Hydroxides

Layered double hydroxides (LDHs) can be considered as a complementary class of the already extensively studied cationic clays, because the charge on the layers and the intercalated ions is the reverse of that found for smectite clays ("mirror-images").

They have a structure similar to that of brucite $(Mg(OH)_2)$, where each Mg^{2+} ion is octahedrally surrounded by six OH^- ions (Fig. 3). Different octahedra share edges to form infinite sheets, which are stacked on top of each other and held together by weak hydrogen bonding [7, 8]. Because of the isomorphous substitution of Mg^{2+} cations for others of higher charge, a positive charge is generated on the LDH layers. To maintain the electrical neutrality, anions are located in the inter-layer region.

The following general formula reflects the atomic contents of the LDH structures:

$$[M(II)_{1-x}M(III)_x(OH)_2]^{x+}[A^{n-}_{x/n}] \cdot mH_2O.$$

It is possible to synthesize a large number of compounds with different compositions and stoichiometries, however, only a few of these variations appear in nature: x is generally equal to 0.25, and carbonate is invariably the preferred anion [9, 10]. Table 4 gives an overview of the most commonly known LDH minerals that occur in nature.

Theoretically, all $M(II)$ and $M(III)$ ions that can be accommodated into the holes of the close-packed configurations of OH-groups in the brucite-like layers, can form LDH structures. Examples are:

References see page 1305

Tab. 4. Mineral composition of naturally occuring LDHs.

Name	Chemical composition	Symmetry
Hydrotalcite	$Mg_6Al_2(OH)_{16}CO_3 \cdot 4H_2O$	3R
Manasseite	$Mg_6Al_2(OH)_{16}CO_3 \cdot 4H_2O$	2H
Pyroaurite	$Mg_6Fe_2(OH)_{16}CO_3 \cdot 4.5H_2O$	3R
Sjögrenite	$Mg_6Fe_2(OH)_{16}CO_3 \cdot 4.5H_2O$	2H
Stichtite	$Mg_6Cr_2(OH)_{16}CO_3 \cdot 4H_2O$	3R
Barbertonite	$Mg_6Cr_2(OH)_{16}CO_3 \cdot 4H_2O$	2H
Takovite	$Ni_6Al_2(OH)_{16}CO_3,OH..4H_2O$	3R
Reevesite	$Ni_6Fe_2(OH)_{16}CO_3 \cdot 4H_2O$	3R
Desautelsite	$Mg_6Mn_2(OH)_{16}CO_3 \cdot 4H_2O$	3R

$M(II)$: Be, Mg, Cu, Ni, Co, Zn, Fe, Mn, Cd, Ca.
$M(III)$: Al, Ga, Ni, Co, Fe, Mn, Cr, V, Ti, In.

For the degree of isomorphous substitution (value of x), different ranges have been reported in the literature, depending on the nature of the structural cations present. Although it has been reported that LDH structures can exist for values of x between 0.1 and 0.5 ($M(II)/M(III)$ between 9/1 and 1/1), the purity of these LDHs with extreme $M(II)/M(III)$ ratios can be doubted. These materials mainly consist of pure hydroxides (one cation), or compounds with different structures [11, 12].

The possible interlayer anions are numerous [8]:

- Inorganic halides and oxoanions: F^-, Cl^-, Br^-, ClO_4^-, NO_3^-, OH^-, CO_3^{2-}, SO_4^{2-}, WO_4^{2-}, $S_2O_3^{2-}$, ...
- Isopolyanions: $[V_4O_{12}]^{4-}$, $[V_{10}O_{28}]^{6-}$, $[Mo_7O_{24}]^{6-}$, etc.
- Heteropolyanions: $[PMo_{12}O_{40}]^{3-}$, $[PW_{12}O_{40}]^{3-}$, $[SiV_3W_9O_{40}]^{7-}$, $[H_2W_{12}O_{40}]^{6-}$, etc.
- Complex anions: $Fe(CN)_6^{4-}$, $Fe(CN)_6^{3-}$, etc., and metal-organic complexes.
- Organic acids: oxalic, succinic, malonic, carboxylates and dicarboxylates, alkylsulfates, etc.
- Layered compounds, as in the mineral chlorite, $(Mg,Fe,Al)_6[(Si,Al)_4O_{10}](OH)_8$.

Because of the presence of large numbers of exchangeable anions between the layers, LDHs are considered to be one of the few inorganic anion exchangers with a high capacity (around 3 meq g^{-1}). For pillaring purposes, LDHs containing monovalent anions, such as nitrate or chloride, are considered to be the best precursors.

Layered double hydroxides can be synthesized by various techniques: coprecipitation (constant pH), precipitation at variable pH, deposition/precipitation reactions, hydrothermal synthesis, anion exchange on existing LDH structures, structure reconstruction based on the memory effect, electrochemical methods, and hydrolysis reactions [8, 9, 13–17].

When applying the method of precipitation, a pH higher than or equal to the precipitation pH of the most soluble hydroxide is chosen for the LDH preparation

[9, 18, 19]. The hydroxides of most metals form LDHs in a pH range between 7 and 10. Higher values initiate the dissolution of first Al, then other metals.

Hydrothermal synthesis involves the treatment of freshly precipitated layered hydroxides or a mixture of the metal oxides with water. The conditions for this treatment are high pressure (autoclave), at high (>373 K) or low (<373 K) temperatures. The objective is to improve the crystallinity of the LDH material, which is composed of small crystallites or an amorphous precipitate [20, 21].

Another very interesting preparation method is based on the memory effect of calcined LDH structures, which allows a reconstruction of the LDH by contacting the mixed oxide with a solution containing the interlayer anion of interest. Although this procedure does not follow the classic ion-exchange methods, it can be classified indirectly as a multistep exchange method. It is not only restricted to the creation of LDHs containing simple inorganic anions, but also relatively large organic ions and anionic complexes can be introduced into the LDH. By exposing the calcined LDH to the appropriate solution, LDHs with a range of different interlayer anions can be prepared, from bulky organic ions, such as sebacic acid and p-toluene sulfonate, to large inorganic polyoxometalate complexes [22].

The basic LDH solids possess only a poor porosity, due to the large amount of charge-compensating interlayer anions and the small free interlayer spacing between the sheets. The N_2-adsorption isotherms at 77 K, measured by Nijs [23] on (NO_3)-MgAl-LDHs and (NO_3)-ZnAl-LDHs, indeed reveal very low BET surface areas between 6 and 21 m^2 g^{-1} and no microporosity. Even by lowering the charge density on the layers, the interlayer free spacing between hydroxide sheets is still too small to create microporosity. However, this influences the size of the mesopores that are created by an irregular stacking of the different aggregates of parallelly assembled layers (also called a "stack" of layers). The size of those stacks is dependent on the electrostatic interactions between the interlayer anions and the layers. The higher the charge density on the LDH layers, the stronger the bonding with the interlayer NO_3^- ions, resulting in larger stacks and larger mesopores between them. The average radii of these mesopores as a function of the Mg/Al ratios range between 2.9–6.4 nm, and between 3.9–7.5 nm as a function of the Zn/Al ratio.

4.4.2.3
Zirconium Phosphates

Metal oxide-phosphate materials constitute another class of layered zeolite-like materials that can be used for pillaring in order to create a porous solid. The most studied member of the group of metal(IV) phosphates is alpha-zirconium phosphate (α-ZrP) [24–27]. The discovery of its synthesis and the characterization of its structure have initiated research for other metal phosphates. α-ZrP is built up of a layered structure, in which each layer can be described as a central Zr-atom plane placed between two phosphate layers. The Zr-atoms are octahedrally coordinated

References see page 1305

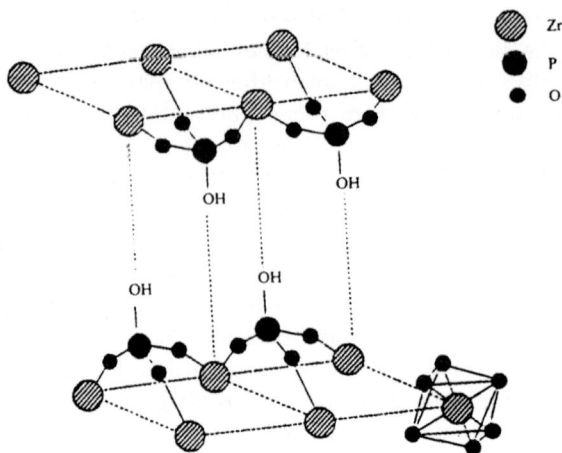

Fig. 4. Schematic representation of the framework of zirconium phosphate (ZrP).

by six oxygen atoms belonging to six different phosphate groups. The P-atoms are tetrahedrally surrounded by three oxygens bonded to three different Zr-atoms, while the fourth oxygen coordination is made up of a hydroxyl group, pointing into the interlayer space. These POH groups are the active sites of α-ZrP. Alternating Zr and phosphate groups form 12-membered rings in a crown conformation. The final arrangement results in a zeolite-like 6-sided cavity (Fig. 4). The layers are held together by weak van der Waals forces with an interlayer distance of 0.76 nm. The structural formula of α-ZrP is $Zr(HPO_4)_2 \cdot H_2O$.

Amorphous ZrP gel can be easily prepared by adding diluted phosphoric acid to a Zr salt (e.g. $Zr(NO_3)_4$, etc.). In order to crystallize the obtained gel, two methods have been described. In the method established by Clearfield [28], the gel is refluxed in concentrated H_3PO_4, which greatly improves the crystallinity. The second synthesis procedure, as introduced by Alberti and Torracca [29], is referred to as the precipitation method.

The most general interest of Zr-phosphate is its high cation-exchange capacity (CEC = 6.64 meq g^{-1}), resulting from the exchange of protons of the phosphate groups.

4.4.3
Pillaring

4.4.3.1
Definition and Criteria

Pillaring is the process by which an inorganic layered compound is transformed into a thermally stable micro- and mesoporous material with retention of the layer structure. The pillared material is a pillared layered solid or PLS. In the case of clay minerals, one obtains a pillared clay mineral; LDHs are transformed into pillared LDH and the same holds for zirconium phosphates. A PLS is a subgroup of the group of intercalation compounds, which meet the following criteria [5]:

1. the minimum increase of basal spacing is equal to the diameter of the N_2 molecule, commonly used to measure surface areas and pore volumes (0.315–0.353 nm);
2. the layers are separated vertically and do not collapse upon removal of the solvent;
3. the pillaring agent has molecular dimensions and is laterally spaced on a molecular length scale;
4. the interlamellar space is porous. The minimum size of the pore opening is the diameter of the N_2 molecules; there is no maximum limit to the size of the pores.

These four criteria mean that a PLS has chemical and thermal stability; that there is a molecular distribution of pillars in the interlamellar region, but ordering of the pillars is not required. The elemental platelets or lamellae of the crystalline layered solid must be ordered so as to give an XRD pattern, which allows the determination of the d_{001} spacing. Finally, the interlamellar region of the PLS must be accessible to molecules at least as large as N_2.

4.4.3.2
Al-Pillared Smectite Clays

The preparation of pillared clay minerals involves the following steps: (1) swelling of the clay mineral in water; (2) ion exchange of the pillaring agent or its precursor; (3) washing of the product; (4) calcination. The chemistry of each step has been very difficult to study insofar that even for the most commonly used pillaring cation, Al^{3+}, the chemistry is not fully understood, nor is there much known about the distribution of the pillars in the interlamellar region and their shapes. Butruille and Pinnavaia [30] have reviewed the alumina pillared clays up to 1996, while Gil and Gandia [31] have discussed the pillaring with solutions containing two or more cations, one being Al^{3+}. Both review papers stress the catalytic applications of the pillared materials. Here, we discuss the chemistry of each step in the pillaring process with Al^{3+} as an example.

4.4.3.2.1 Aqueous Chemistry of Al^{3+} and the Ion-Exchange Reaction
Al^{3+} readily hydrolyzes in water to form a variety of products [32]. One of them is the Keggin ion $[Al_{13}O_4(OH)_{24}(H_2O)_{12}]^{7+}$, Al_{13}^{7+}, which is stable under a broad range of conditions: $[Al^{3+}] < 0.01$ M, $[OH^-]/[Al^{3+}] < 2.4$ and pH < 5. It consists of a central AlO_4 tetrahedron with every O connected to three $Al(OH)_4(H_2O)_2$ oc-

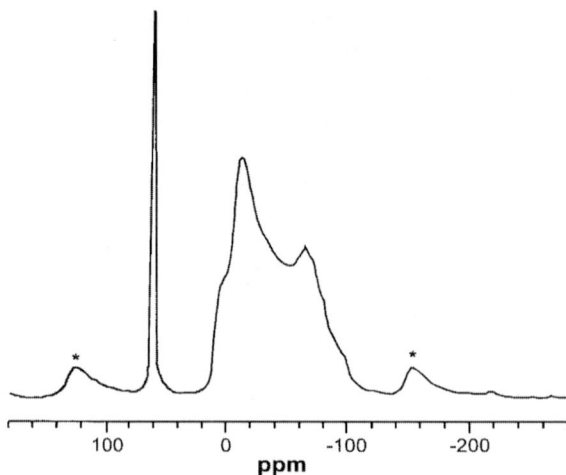

Fig. 5. The ^{27}Al MAS/NMR spectrum of the Al_{13}^{7+}-sulphate salt. Spinning side bands are indicated by an asterisk.

tahedra. In solution, the ^{27}Al NMR spectrum consists of a single sharp line at 60 ppm. The Keggin ion can be precipitated as a sulfate salt and, in the solid state, the ^{27}Al NMR spectrum contains the sharp tetrahedral line at 60 ppm and a complex line system of octahedral Al between 20 and -100 ppm (Fig. 5).

Further hydrolysis of Al_{13}^{7+} gives dimers, trimers, polymers, and finally pseudoboehmite, which precipitates out of the solution. All these species can be distinguished on the basis of their Al–OH vibrations, as given in Table 5 [4].

Most of the solutions used for pillaring contain a variety of Al^{3+} species with Al_{13}^{7+} as the majority species. These solutions are slightly acidic. Thus, several species of Al^{3+} and H^+ might be exchanged simultaneously. The relative amount of each species on the clay surface will depend on their relative amounts in solution, the selectivity of each species in the ion-exchange reaction and its chemical stability at the clay surface. To understand fundamentally the process, one species has to be selected and studied in detail. This is what has been done with the Al_{13}^{7+} Keggin ion [33]. The isotherm of saponite is shown in Fig. 6.

It confirms that the reaction is an ion-exchange reaction:

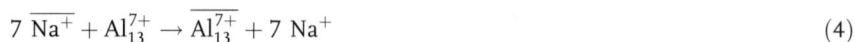

$$7 \overline{Na^+} + Al_{13}^{7+} \rightarrow \overline{Al_{13}^{7+}} + 7 Na^+ \tag{4}$$

Tab. 5. AlOH vibrations of oligomeric Al.

Species	$\delta(OH)/cm^{-1}$	$\gamma(OH)/cm^{-1}$
Al_{13}^{7+}	980	780
Associated Al_{13}	960–980	720–730
Polymeric Al_{13}	880–960	–
Pseudoboehmite	1070–1080	740–760

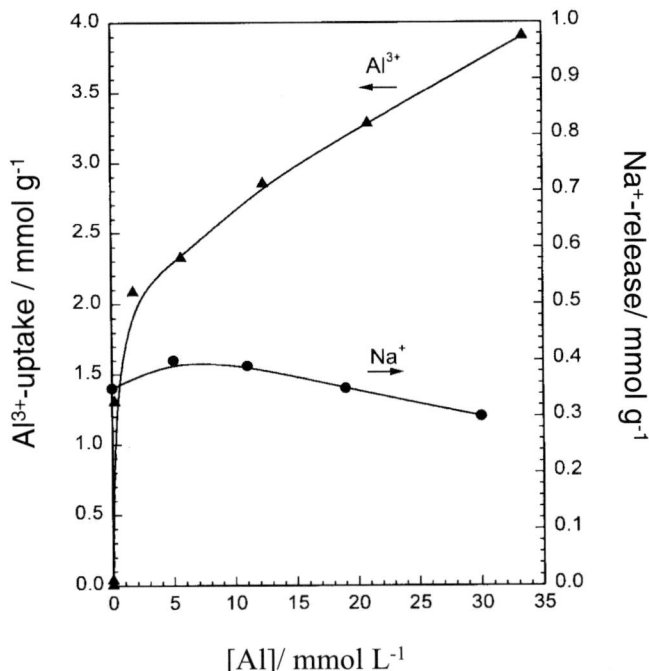

Fig. 6. Uptake of Al and release of Na upon ion exchange of Al_{13}^{7+} with saponite.

There is a quantitative uptake of Al^{3+} up to almost 2 mmol g^{-1} and a release of 1.5 mol g^{-1} of Na^+. The numbers indicate that the expected ratio $[Al]/[Na] = 13/7$ is not attained. More Na^+ is released in solution than required for a stoichiometric ion-exchange reaction. This can be ascribed either to exchange of protons from the slightly acidic solution or hydrolysis of the Keggin ion in contact with the clay surface:

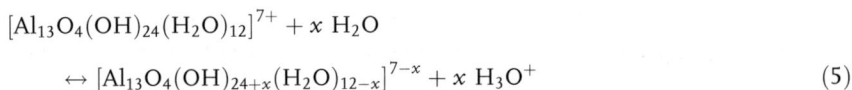

$$[Al_{13}O_4(OH)_{24}(H_2O)_{12}]^{7+} + x \; H_2O$$

$$\leftrightarrow [Al_{13}O_4(OH)_{24+x}(H_2O)_{12-x}]^{7-x} + x \; H_3O^+ \tag{5}$$

In the second stage of the isotherm, the amount of Al taken up increases proportionally with the amount offered in solution, but the amount of Na^+ released decreases. This indicates that the ion exchange is partial and that part of the Al_{13}^{7+} is precipitated on the clay. The relative amount of precipitated Al_{13}^{7+} increases with the amount offered in the exchange solution.

4.4.3.2.2 The Pillaring Process

After ion exchange, washing or dialysis is a necessary step to obtain a pillared clay mineral. Three regimes can be distinguished: (1) if the loading of Al is less than

References see page 1305

1.4 mmol g^{-1}, the d_{001} spacing is 1.25 nm and does not jump to 1.80–1.90 nm upon washing; (2) at Al loadings above 2.5 mmol g^{-1}, the d_{001} spacing is 1.80–1.90 nm before washing and does not change upon washing; (3) at the intermediate Al loadings, the d_{001} spacing is 1.25 nm before washing and increases to 1.80–1.90 nm upon washing. Most of the preparations of pillared clay minerals have these intermediate loadings and washing is required to obtain a pillared product.

The washing process releases Cl$^-$, Al^{3+}, and Na$^+$ into solution especially with the first washing. At the same time the pH of the clay suspensions in the dialysis bags increases from about 4.2 to 5–5.5 for beiddelite, montmorillonite, and saponite, and to 7–8 for hectorite and saponite [34]. These data suggest a complex chemistry accompanying washing or dialysis. The presence of Cl$^-$ indicates that Cl$^-$ either substitutes for OH$^-$ in the Keggin ion or that Al-chloro complexes are adsorbed, which then hydrolyze upon washing:

$$Al–Cl + 2H_2O \rightarrow Al–OH + H_3O^+ + Cl^- \tag{6}$$

and the proton exchanges for Na$^+$. The increase of the pH of the suspension in the dialysis tube and the concomitant (small or negligible) increase of the pH in the surrounding solution also indicate release of protons due to hydrolysis. Aceman et al. [34] suggest that the Keggin ion is adsorbed on the external surface of the clay particles, hydrolyzes and breaks up into smaller oligomers. These Al-oligomers diffuse in the interlamellar space together with Cl anions and hydrolyze to form the pillaring Al species. Whatever the exact chemistry, the experimental data indicate that pillaring is a complex process involving Al Keggin ions and other Cl-containing Al species in solution. Once adsorbed, these species undergo hydrolysis reactions and washing or dialysis is necessary to obtain the pillared product with the characteristic 1.80–1.90 nm spacing. At the same time, the d_{001} line sharpens, which is indicative of an increase in the degree of ordering in the pillared product.

The final step in the pillaring process is calcination. This converts the pillaring species into an Al-oxide pillar with retention of the d_{001} spacing. The thermal stability of this spacing depends on the crystallinity of the product obtained after washing: the sharper the d_{001} line, the more thermally stable is the pillared clay mineral. In any case, the structure of the oxidic cluster is unknown. In the starting Keggin ion, the octahedral Al: tetrahedral Al ratio is 12:1 and it is unlikely that this ratio exists in the cluster. A detailed NMR study of the pillaring of an Al-free smectite, such as hectorite, would shed some light on this problem. The formation of a chemical bond between the cluster and the Si-tetrahedra of the clay mineral has been discussed on the basis of NMR studies. Several authors have invoked such a Si–O–Al bond with inversion of the Si-tetrahedron for smectites with isomorphous substitution in the tetrahedral layer [35].

For industrial production of pillared clays, the pillaring process has to be optimized and upscaled. Sonication of the clay suspension, acid and base treatments, and competitive exchange, are some of the techniques found in the literature [36]. In laboratory preparations large amounts of water are needed. Thus pillaring of concentrated aqueous smectite suspensions has been tried in order to find better conditions for upscaling [37].

In conclusion, pillaring of smectites is a well-established process in laboratory conditions. There is room for further investigation, mainly in two areas: determination of the structure of the oxidic Al cluster and optimization of the pillaring conditions for upscaling. An essential condition for pillaring is the swelling of the smectites in aqueous suspension. Swelling of clay minerals with a much higher charge density than that of smectites does not occur spontaneously in aqueous suspensions. Very recently, one of the authors has established procedures for pillaring of highly charged clay minerals. This is discussed in the next section.

4.4.3.3
Al-Pillared Vermiculite and Phlogopite Clays

Phlogopites are trioctahedral 2:1 sheet silicates with all the possible octahedral positions occupied mostly by divalent cations. The numerous octahedral and tetrahedral layer substitutions are responsible for the high net negative layer charge of micas. The end-member phlogopite has the following general formula: $K_2Mg_6(Si_6Al_2)O_{20}(OH,F)_4$. Natural phlogopites generally contain Mg^{2+}, Fe^{2+}, Al^{3+}, and Fe^{3+} in the octahedral layer, and may also have substantial amounts of structural fluorine (replacing hydroxyls), conferring higher resistance to weathering, hardness, and thermal stability. The potassium ions, the dominant charge-compensating interlayer cations, are located between unit layers, with adjacent layers being stacked in such a way that the potassium ion is equidistant from 12 oxygens, 6 of each tetrahedral layer [6]. In their original state, natural micas do not swell in the presence of water because the hydration energy of the interlayer potassium ions is insufficient to overcome the cooperative structural forces at the coherent edges of a cleavage surface [6] and, hence, ion exchange does not occur. Two obstacles have to be overcome in order to pillar natural micas: (1) the high selectivity of mica for the potassium ions makes it very difficult to exchange other cations, and (2) even a Na-exchanged mica, when brought into contact with an Al-pillaring solution, selectively retains Al^{3+} instead of the bulkier Al_{13}^{7+} species from the pillaring solution [38].

Vermiculites may be considered as swelling trioctahedral micas with Al for Si substitutions in the tetrahedral layers, and Fe and Al for Mg substitutions in the octahedral layers. They constitute intermediate minerals in the natural weathering sequence of micas to smectites [39], with a negative layer charge density between that of micas and smectites, and where hydrated cations, most often Mg^{2+} and Ca^{2+}, have replaced the charge-neutralizing K^+ ions of the parent mica, providing partial swelling properties. Straight pillaring of vermiculite suspensions with Al_{13}-containing solutions gives materials with interlayer spacings of about 1.4 nm at room temperature [40–42], namely, half the gallery height (or interlayer free spacing) commonly achieved in 1.8-nm Al-pillared smectites (0.4 nm versus about 0.8 nm [43, 44], a failure attributed to the location of the negative charge on the basal oxygens of the tetrahedral layers, preventing the selective intercalation of the bulkier Keggin-type Al_{13}^{7+} cations [42]. A preliminary treatment of vermiculite with

References see page 1305

an ornithine solution prior to pillaring treatment results in a mixture of mostly nonpillared material and a small pillared fraction [45].

Successful pillaring of vermiculites and phlogopites has been recently achieved after a preliminary treatment aiming at a reduction of the layer charge density of the minerals. This treatment is comparable with a controlled "accelerated weathering" process, which mimics the natural alteration sequence: micas → vermiculites → smectites. Bringing the charge-reduced minerals in contact with the pillaring solution results in 1.8-nm pillared materials [46, 47]. The higher structural stability of vermiculites and micas to temperature, compared with smectites, is of considerable interest in obtaining pillared materials with high resistance to thermal treatments, a weakness shared by all the smectite-based pillared materials. From the catalytic point of view, the higher number of Al for Si tetrahedral substitutions compared with, for example, saponites and beidellites makes those minerals very attractive because it may be anticipated that more Brønsted acid sites can be generated upon pillaring.

4.4.3.3.1 Pillaring Method

Pillaring of vermiculites and phlogopites was only possible after a treatment aiming at the charge reduction of the minerals. In setting up the pillaring method, three requirements were considered by the authors: (1) there should be no need for an additional grinding treatment of the mineral; (2) the mineral dispersions should not be too diluted, and (3): the method should be applicable to vermiculites and phlogopites from different deposits.

The conditioning treatment consisted of the following four successive steps:

(1) acid leaching with diluted nitric acid in controlled conditions; (2) calcination in air at 873 K for 4 h; (3) acid leaching with a diluted solution of a complexing acid; and (4): saturation of the residual exchange positions with sodium (or calcium) ions. At the end of this "conditioning" treatment, the overall negative charge of the minerals is reduced by about one third. Charge reduction mainly occurs in steps (1) and (2), whereas step (3) is carried out in order to eliminate extra-framework elements produced in the preceding steps, which block the exchange sites in the interlamellar space. This treatment is not indispensable but, when carried out, it significantly improves the characteristics of the final product. Thorough Na- (or Ca-) exchange is essential for the quality of the pillared materials. The charge-reduced vermiculites and phlogopites are then brought into contact with the Al-pillaring solution, as in the case of the pillaring of smectites. Details on the operating variables of the conditioning treatment and pillaring process of these minerals can be found in refs [46, 47].

4.4.3.3.2 Interlayer Spacings and Thermal Stability

Adequately Al-pillared smectites are characterized by an interlayer distance of about 1.8 nm (or gallery height of about 0.8 nm), which does not significantly decrease after a calcination at temperatures of 673–773 K. Typical XRD patterns illustrative of Al-pillared vermiculite and Al-pillared phlogopite are shown in Fig. 7.

Both materials exhibit narrow peaks with an interlayer distance of 1.88–1.9 nm

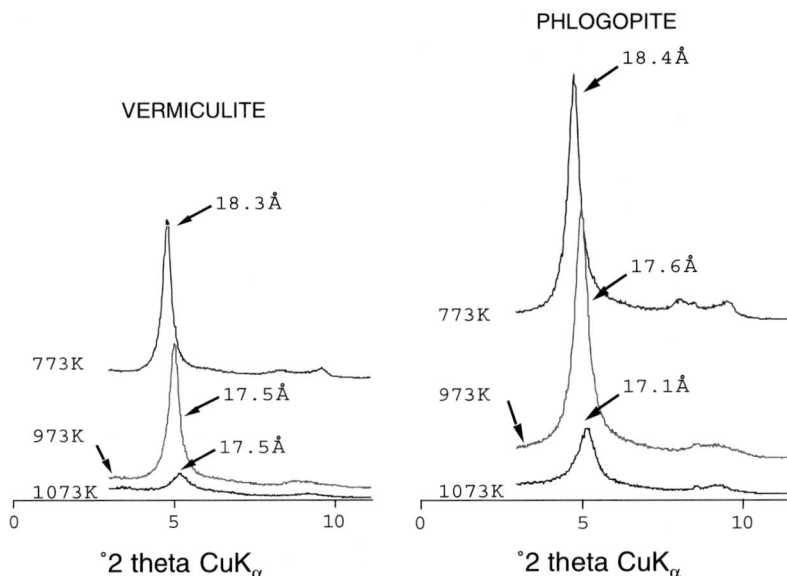

Fig. 7. X-ray diffraction patterns of Al-pillared vermiculite and phlogopite after calcination at 773, 973, and 1073 K.

after heating at 473 K. Spacings of 1.83–1.84 nm are found after heating at 773 K, with a slight decrease after a calcination at 973 K and 1073 K (1.7–1.71 nm). The maintenance of the spacings at these high temperatures is in contrast to equivalent materials obtained with smectites, which generally collapse at lower temperatures owing to the lower intrinsic stability of the clay itself. Framework dehydroxylation of smectites generally occurs at around 873–923 K compared with 1073 K and for vermiculite and 1373 K for phlogopite, a difference that is reflected in the thermal stability of the corresponding pillared forms. It has been shown that the structural stability is lowered upon pillaring, as a result of the protolysis of some bonds in the octahedral layers and, consequently, a weakening of the thermal framework stability.

4.4.3.4
Pillared Layered Double Hydroxides

In order to create porosity in the interlayer region, the concept of pillaring first introduced for the cationic clays has also been applied to the LDHs in recent years. The only difference is that the pillaring species used here are anions instead of cations. Creating porosity in LDH materials is not straightforward as is the case for the cationic clays, for different reasons: (1) due to the high charge density on the sheets, pillaring causes stuffing of the interlayer space. As a consequence, the

References see page 1305

anionic pillars must possess a sufficiently large charge; (2) decomposition of the pillars and/or the LDH itself can occur when the stability ranges (pH) of the LDH and the anionic complexes do not overlap, resulting in decomposition processes during the pillaring reaction. Most LDH solids are not stable in strong acidic conditions. While most of the polyoxometalate (POM) pillaring species are formed in the low-pH region, pillaring often results in multiphased materials.

Two main categories of pillars have been used for the pillaring reaction:

1. *hetero- and isopolyoxometalates or polyoxoanions* [48]: these robust complexes can be considered as ideal pillaring agents. Their structure consists of multiple layers of space-filling oxygens and they are characterized by a wide range of possible compositions and charge densities. A typical example of a heteropolyoxometalate is the anion α-$(XM_{12}O_{40})^{n-}$ of the Keggin-type, giving rise to gallery heights of approximately 1.0 nm. In general, isopolyoxometalates are smaller. Examples are $[V_{10}O_{28}]^{6-}$, $[V_4O_{12}]^{4-}$, $[W_7O_{24}]^{6-}$, etc., resulting in gallery heights of 0.4–0.8 nm.
2. *hexacyanometal complexes* [49–51]: for example $Fe(CN)_6^{3-}$, $Fe(CN)_6^{4-}$, etc. $[Fe(CN)_6]$-LDHs have a *d*-spacing between 0.6–0.65 nm.

4.4.3.4.1 Pillaring with POMs

Woltermann [52] was one of the first to report on polyoxometalate pillared LDHs ([POM]-LDHs) prepared via direct anion exchange. Since then, various studies on the intercalation of vanadate complexes in LDHs have been reported [48]. A study by Twu and Dutta [53] on the exchange of vanadates in (Cl)-LiAl-LDHs emphasizes the effect of pH on the nature of the exchangeable pillaring species. A gallery height of 0.6 nm was obtained for the pillared solids. The exact nature of the intragallery complexes could be identified by Raman spectroscopy. For LDHs prepared at pH 8–11, mainly containing $[V_2O_7]^{4-}$ species, the anions can be arranged in two different positions, corresponding to different gallery heights (0.50 nm and 0.78 nm) (Fig. 8). Based on spatial and thermodynamical considerations, arrangement [II] appears to be the most favorable. This value is in agreement with the experimental *d*-spacing when the interlayer zone is filled with one layer of water molecules.

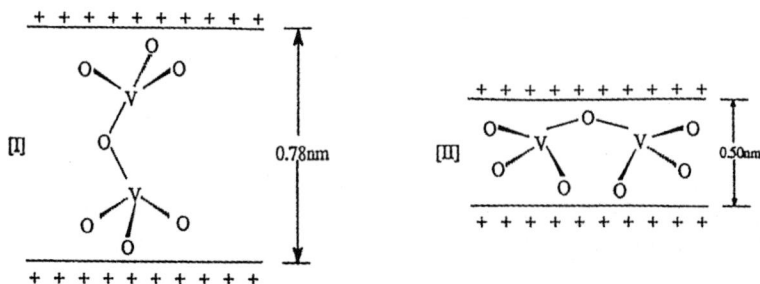

Fig. 8. Possible arrangements of $[V_2O_7]^{4-}$ in an LDH interlayer region [23].

Over the last decade, much attention has been devoted to the pillaring complexes of the Keggin type α-$(XM_{12}O_{40})^{n-}$ [48]. Their advantages over the V(v)- and Mo(vi)-isopolyoxometalates is the possibility to create larger gallery heights. By incorporating them in LDHs, stable materials can be developed combining both the catalytic properties of the POMs and the porous character of pillared LDHs. Pinnavaia and coworkers [54, 55] reported the pillaring of (NO$_3$)-ZnAl-LDHs with α-$[H_2W_{12}O_{40}]^{6-}$ and α-$[SiV_3W_9O_{40}]^{7-}$. They claimed that the complete exchange for the POMs can only proceed when the negative charge of the Keggin ion equals or exceeds 6. Contrary to these results, Clearfield and coworkers [56] performed a complete exchange of $[PW_{12}O_{40}]^{3-}$ in (NO$_3$)-MgAl- and (NO$_3$)-NiAl-LDHs, despite the low charge of the Keggin complex. However, these results need to be interpreted with care since the used acidic pillaring conditions will cause a partial dissolution of the LDH structure.

In a recent paper by Yun and Pinnavaia [57], different pillaring methods and reaction conditions were compared for the intercalation of POM complexes with various structures. So far, mainly the intercalation of nearly spherical POM pillars has been reported. For the symmetrical Keggin-type α-$[H_2W_{12}O_{40}]^{6-}$, only one gallery height (1.0 nm) was observed, irrespective of the reaction conditions, whereas two different gallery heights were found for the cylindrical Dawson anion α-$[P_2W_{18}O_{62}]^{6-}$ (1.45 nm and 1.28 nm) and the Finke anion $Co_4(H_2O)_2(PW_9O_{34})_2^{10-}$ (1.33 and 1.26 nm), depending on the reaction temperature. The difference in POM orientation was rationalized in terms of different electrostatic and hydrogen-bonding interactions between the pillars and the LDH layers. After a thermal treatment exceeding 373 K, the intercalated Dawson and Finke POM ions exhibited only one gallery orientation. In Fig. 9 a schematic presentation of the formation of these pillared LDHs is given.

In general, based on reported literature, it is clear that pillaring with POMs via a direct anion exchange results in rather poorly pillared LDHs because of pH-related

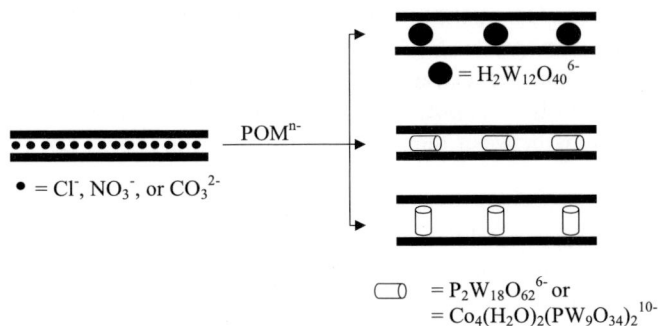

Fig. 9. Schematic representation of the formation of LDH derivatives pillared by Keggin POM^{n-} ions with a nearly spherical symmetry, and by Dawson and Finke POM^{n-} ions with a cylindrical symmetry [23].

References see page 1305

structural deformations during the pillaring reaction. Surface areas and micropore volumes in the ranges of 38–70 m^2 g^{-1} and 0.002–0.013 cm^3 g^{-1} have been reported. However, when co-precipitation is used to directly synthesize POM pillared LDHs, the reaction of host and guest does not cause any decomposition if the LDH itself is acidic, since most POM systems are also acidic [58].

4.4.3.4.2 Organic Preswelling

Since LDHs do not exhibit the same good swelling capacities as the smectite clays, the exchange for very large pillaring species can be promoted by separating the sheets with large organic anions, prior to the actual pillaring step. By this organic preswelling step, the hydrolysis of the POMs and the decomposition of the LDH structure is minimized. Terephtalate- and p-toluenesulfonate-containing LDHs were considered to be the best precursors. Until now mainly the pillaring with iso-polyanions, [V$_{10}$O$_{28}$]-, [W$_7$O$_{24}$]- and [Mo$_7$O$_{24}$]-MgAl-LDHs has been reported [59, 60]. Despite the improved pillaring method, these pillared [iso-POM]-LDHs exhibited rather poor porosity characteristics, with specific surface areas not exceeding 120 m^2 g^{-1}.

Nijs et al. [61] studied the pillaring of pre-exchanged terephtalate (T)-MgAl- and (T)-ZnAl-LDHs with a series of POM pillars. The reaction results in the formation of single-phased, pillared LDH solids. Moreover, reducing the charge density on the LDH layers (by changing the ratio M^{II}/M^{III}) influences the porosity of pillared LDHs in a positive way. This is a consequence of the reduced amount of large complexes in the interlayer space, creating porosity. Maximal surface areas and micropore volumes of 166 m^2 g^{-1} and 0.047 cm^3 g^{-1} have been reported for [PV$_2$W$_{10}$O$_{40}$]$^{5-}$-ZnAl-LDH with a Zn^{2+}/Al^{3+} ratio of 4.22.

Combining structure reconstruction with the intercalation of long-chain organic acids produced single, well-crystallized pillared LDHs, according to Dimotakis and Pinnavaia [62]. Structure reconstruction via (OH)-MgAl-LDHs was carried out in the presence of an extra swelling agent, glycerol. The obtained toluenesulfonate-LDHs were exchanged afterwards for polyoxometalate anions, such as [SiW$_{11}$O$_{39}$]$^{8-}$ and [H$_2$W$_{12}$O$_{40}$]$^{6-}$, leaving microporous MgAl-LDHs with a free interlayer spacing of 1.00 nm and BET surface areas of 155 and 107 m^2 g^{-1}, respectively. This combined method was claimed to be superior to the direct ion-exchange techniques, since unwanted side reactions between the LDH and the pillaring agents can be avoided.

4.4.3.4.3 Pillaring with Hexacyanometal Complexes

Focusing on the group of hexacyanometal complexes, higher porosity values are obtained despite the relatively low charge and limited size of the complexes. Pillaring is easy, since no precautions concerning the pH need to be taken while the intercalation of the Fe(CN)$_6^{3-}$ species takes place with a total conservation of LDH and complex structure.

Highly porous pillared [Fe(CN)$_6$]-MgAl-LDHs with a uniform pore-size distribution have been synthesized by Nijs et al. [61, 63]. Moreover, a good correlation between the Mg/Al ratio in the LDH sheets and the resulting microporosity is obtained. With increasing Mg/Al ratio, creating lower charge densities on the layers, the micropore volume is enhanced. An optimum is found at a ratio of 3.33, resulting in a surface area of 499 m^2 g^{-1} (Langmuir) and a micropore volume of 0.177 cm^3 g^{-1}. The maximum in the Horvath–Kawazoe micropore-size distribution is located around 0.6 nm diameter.

4.4.3.5
Pillared Zirconium Phosphates

Four methods are being used to pillar layered phosphates [64]:

1. Direct intercalation of large, highly charged inorganic complexes.
2. Intercalation of organic base coordinating ligands, followed by an *in situ* complex pillar formation.
3. Derivatization of the phosphate groups.
4. Intercalation of inorganic polyoxocations, followed by calcination.

The first method consists of the exchange of the protons of the POH groups for highly charged complexes. Although a large number of such complexes exists, only a few authors have followed this route. The pillaring of α-ZrP with Ru(bpy)$_3^{2+}$ (bpy:bipyridine) has been attempted, but required severe conditions [65].

The second procedure involves the insertion of an organic base, the dimensions of which ensure an overlap with neighboring phosphates, followed by an ion-exchange reaction [66]. Subsequently, the inserted cations can complex with the organic bases (*in situ*) with the formation of pillars. Complexation will only take place if the basic groups of the organic ligands prefer to coordinate to the Lewis acidic transition metal ions instead of the Brønsted acidic POH groups. Typical examples of the organic bases are: 2,2'-bipyridine and 1,10-phenantroline. The moderate Brønsted basic strength and the large dimensions of these ligands do not allow a straightforward interaction with the POH groups of α-ZrP, but require a preswelling of the structure. The third method describes the formation of organically pillared phosphates by the substitution of the POH groups for PR groups with R an inorganic (H) or organic group (CH$_3$, C$_6$H$_5$, etc.) [67, 68]. This method allows the synthesis of highly uniform, microporous materials. Four different materials can be distinguished, the Zr-phosphonates [Zr(RPO$_3$)$_2$·nH$_2$O], the mixed Zr-phosphate-phosphonates [Zr(HPO$_4$)$_{2-x}$ (RPO$_3$)$_x$·nH$_2$O], the Zr-diphosphonates [Zr(O$_3$P-R-O$_3$P)·nH$_2$O], and the mixed Zr-diphosphonate-phosphonates [Zr(O$_3$P-R-PO$_3$)$_{1-x}$(O$_3$P-R')$_{2x}$·nH$_2$O]. It must be pointed out that only the Zr-diphosphonates and their mixed compounds are really organic pillared materials, since their adjacent layers are interconnected through the organic backbone (carbon chain).

The last procedure is analogous to the pillaring of clays and will be discussed in more detail. Though, as layered phosphates distinguish themselves from the

References see page 1305

smectite type clays by their high CEC, they need a preswelling step, which can be induced by the insertion of amines. Afterwards, the protonated amines can be exchanged for the large, inorganic polyoxycations. However, this "simple" ion-exchange reaction is restricted by two important effects:

1. Interfering ions, present in the pillaring solution, may be preferred.
2. Due to the high CEC, the pillaring and/or interfering cations may occupy the complete interlayer region resulting in a low porosity.

The latter problem has been observed with Al-pillaring of butylamine ZrP-intercalates [69]. The Al-solution, prepared by a controlled hydrolysis of $AlCl_3$ with NaOH, contains Al-species and Na^+ ions. Both ions can, therefore, be inserted into α-ZrP. To avoid the presence of interfering ions, Jones et al. [70] synthesized an Al-solution from the hydrolysis of $AlCl_3$ with an amine, which was previously used as the swelling agent for α-ZrP. Clearfield and Roberts [71] used a commercial Al-solution. These authors suggested that the Al-Keggin ion, which is believed to be the Al-pillaring precursor, was intercalated, but a nonporous substrate ($SA_{BET} = 30$–35 m^2 g^{-1}) was formed.

Peeters et al. [72] investigated the partial intercalation of butylamines into the parent α-ZrP-substrate before Al-pillaring. A partial intercalation leaves residual free POH groups in the interlayer space. The mobility of the alkylammonium ions and the protons is enhanced and this is thought to facilitate the exchange of the Al-Keggin ions. However, this improved mobility is only achieved if the creation of free POH groups does not give rise to a phase segregation after pillaring, and this is indeed what occurs. The same problem of phase segregation was encountered when pre-exchanged α-ZrP was used as the pillaring precursor. Because protons of the POH groups are difficult to exchange, the proposed method involves the exchange of the POH groups for PO^-M^+ (M = monovalent cation, e.g. Li). M^+ ions can be more easily exchanged for the Al_{13} pillaring species. The parent α-ZrP was therefore converted into [LiHZrP; $ZrLiH(PO_4)_2 \cdot 4H_2O$] before adding the Al-pillaring solution. However, no evidence for the exchange of Li for Al_{13} was found, due to the small interlayer distance of the LiHZrP phase. In an attempt to increase the interlayer distance, small amounts of butylamine (10–20 % of the CEC) were added, leading to a phase segregation as pictured in Fig. 10.

This segregation process is promoted by the low affinity of the Li^+ ions for the PO^- groups, which causes the Li^+ ions to jump from one exchange site to another, when butylamine is added. As a result, all the Li^+ ions migrate to the center of the ZrP-crystals, while the butylamine molecules are positioned at the edges. The flexibility of the ZrP-layers is thought to be the main reason for the segregation process [73].

Because of the problems pointed out so far, the Al-pillaring of zirconium phosphate has resulted in only limited success with moderate microporosities.

Besides Al, the elements Zr, Ti, and Fe are also common pillar species for clays, though for the metal (IV) phosphates these pillars cannot be applied, because the pillaring solutions are too acidic (pH = 1–2). At this pH, the amines are more

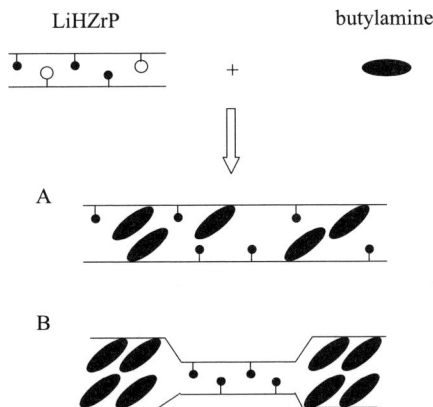

Fig. 10. A: mixed LiBuAZrP phase; B: phase segregation (LiHZrP and BuAZrP). Open circles represent the POH groups, black dots the Li$^+$ ions.

easily exchanged for protons than for the pillaring ions, giving rise to the formation of the parent metal(IV) phosphate [71].

Much effort has been devoted to the reduction of the CEC of α-ZrP, because this parameter was addressed as the major reason for the lack of microporosity in its pillared analogs. Some elegant methods are known, allowing the reduction of the CEC to a value comparable to the CEC of smectite clays.

The intercalation of aminosilanes in ZrP, followed by subsequent calcination results in porous Si-pillared ZrP materials [74, 75]. This procedure has several advantages: (1) aminosilanes are easily intercalated, because of the high affinity of the amine groups for the structural POH groups, and (2) the bulky organic groups effectively space the Si-pillars. This method does not really decrease the number of POH groups, but inactivates some of them. When the aminosilane interacts with a POH group, its large organic group causes a shielding of the neighboring POH groups. The most successful pillaring was obtained by refluxing the hydrolyzed aminosilane together with a propylamine-ZrP suspension, resulting in solids with a SA$_{BET}$ = 232 m^2 g^{-1} with the highest contribution of the porosity in the microporous region [76].

Another way to reduce the CEC is by partial hydrolysis of the POH groups. This hydrolysis is, however, difficult to control, though Alberti and Marmottini [76] developed a method to hydrolyze these POH groups in a controlled manner by means of the addition of methylamine. Another inventive technique to reduce the CEC is based on a partial substitution of the P-OH for P-OR groups (R = organic radical), which then remain inactive towards further pillaring [77, 78]. The PR/POH ratio can be controlled by the molar ratio of the corresponding acids (RPO$_3$/H$_3$PO$_4$) in the synthesis. The potential towards pillaring of these materials is still not fully explored.

Recently, syntheses using surfactants, originally utilized for the synthesis of mesoporous silicates, have also been applied to produce porous metalphosphate

Phosphate + Surfactant

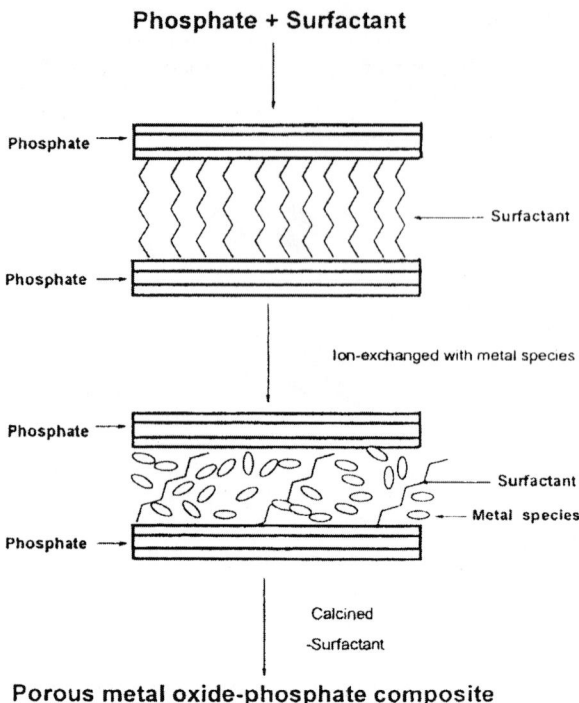

Porous metal oxide-phosphate composite

Fig. 11. Scheme for synthesis of porous metal-oxide-phosphate compounds [79].

materials. This type of synthesis usually involves three steps, as shown schematically in Fig. 11. The steps are as follows: (1) long-chain surfactants are intercalated between the phosphate layers; (2) intercalated phosphates are ion exchanged with metal species; (3) the pillared phosphates are calcined to remove the remaining surfactants and to obtain porous metal oxide-phosphate composites.

Based on this method, highly porous Ga(III)-ZrP composites were prepared with a combined micro- and mesoporosity [80]. The products had surface areas in the range from 295 to 366 $m^2 g^{-1}$, pore size in the range 1.5–1.6 nm, and a micropore volume between 0.117 and 0.143 $cm^3 g^{-1}$ (20–40 % of the total pore volume). Cassagneau et al. [81] synthesized porous SiO_2-ZrP with surface areas from 175 to 390 $m^2 g^{-1}$ and a total pore volume of 0.139–0.289 $cm^3 g^{-1}$ with most of the pores having radii in the range 2.2–3.6 nm.

4.4.3.6
From Micro- to Mesoporous Pillared Materials

Different strategies are reported in the literature to obtain mesoporous materials out of a layered silicate: (1) hydrothermal pillaring with rare-earth-Al pillar moieties [82]; (2) adsorption of colloidal sols [83]; (3) gallery templated polymerization

Tab. 6. Mesoporous pillared layered silicates.

Material	d_{001}/nm	Surface area BET/m^2 g^{-1}	Pore volume/cm^3 g^{-1}	
			Meso	Micro
La, Al-PM	2.6	493	0.13	0.15
Ce, Al-PB	2.57	431	0.23	–
La, Al-PB-AD	2.54	426	0.24	–
Ti, Si-PB-AD	–	403	0.50	–
Ti, Si-PB-SCD	–	501	4.8	–
Zr, Si-PB-AD	–	354	0.54	–
Zr, Si-PB-SCD	–	521	5.0	–
Ti, Si-PB-AD	–	125	0.03	–
Ti, Si-PB-SCD	–	467	1.5	–
Imogolite-PM	–	480	0.20	–
Natural-PCH	1.5–3.0	450–900	0.20–0.70	0.10–0.40
Synthetic-PCH	1.5–2.0	650–1100	0.10–0.70	0.30–0.60
FSM-16	3.5–4.5	900–1200	0.7–1.3	–

PM: pillared montmorillonite, PB: pillared beidellite, AD: air dried,
SCD: supercritically dried, PCH: porous clay heterostructure, FSM:
folded sheet material.

of, for example, Si(OEt)$_4$ [84, 85]; (4) topochemical transformation of layered silicates in the presence of surfactants [86, 87].

In the first method commercial, partially neutralized Al^{3+} solutions are modified by addition of La^{3+} or Ce^{3+}, and the pillaring process of a smectite clay mineral is performed under reflux or under hydrothermal conditions. A material is obtained that is, at least partially, expanded to a d_{001} spacing of 2.6 nm. The structure and composition of the pillaring oligomer is unknown nor is it known to what extent the La^{3+} or Ce^{3+} cations are incorporated into the pillar. In view of the pH conditions of the pillaring process, only traces are expected to be incorporated in the pillar; some may be ion exchanged or adsorbed on the pillar. Despite this, the data of Table 6 clearly show the increase in surface area and pore volume with respect to the regularly pillared materials.

When sols of TiO$_2$–SiO$_2$, SiO$_2$–Al$_2$O$_3$ or ZrO$_2$–SiO$_2$ are used in combination with supercritical drying, pillared smectites are obtained that are almost X-ray amorphous. The materials can be seen as expanded clay structures with a 5–10-fold increase of pore volume, mainly due to the effect of supercritical drying (Table 6). When imogolite, a tubular aluminosilicate, is used as pillaring agent, two types of micropores are formed: intratubular and intertubular (Fig. 12) with a very small mesoporosity [88].

True mesoporous clay-derivatives can be obtained via two routes, resulting in different structures depending on the clay dimensions. The first one, proposed by Pinnavaia's group describes the synthesis of porous clay heterostructures or briefly

References see page 1305

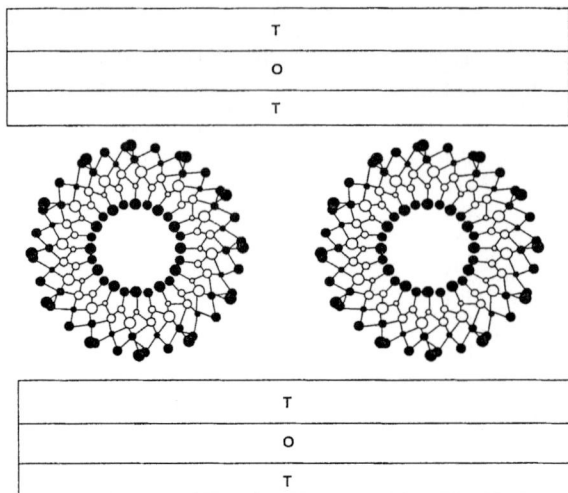

Fig. 12. Cross-sectional view of Na^+ montmorillonite intercalated with a monolayer of imogolite ($SiAl_2O_3(OH)_4$).

PCHs [84]. It consists of an expansion of the individual clay mineral sheets in aqueous suspension with a cationic surfactant in conjunction with a neutral co-surfactant, followed by interlamellar hydrolysis and condensation polymerization of neutral inorganic precursors, e.g. $Si(OEt)_4$. The surfactants form intragallery micelles around which the controlled hydrolysis and condensation of $Si(OEt)_4$ takes place (Fig. 13). The mechanism of micelle formation and silicate polymerization is analogous to the liquid crystal templating (LCT) mechanism for the preparation of the well-known class of mesoporous MCM-materials, except that here the reaction takes place in the interlayer region of a layered host. By a final

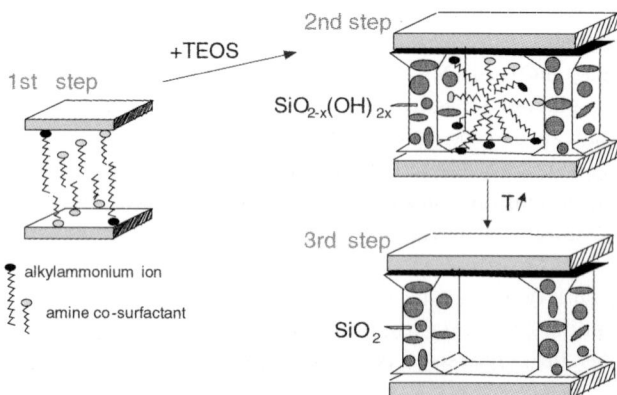

Fig. 13. Formation mechanism of mesoporous natural PCH material.

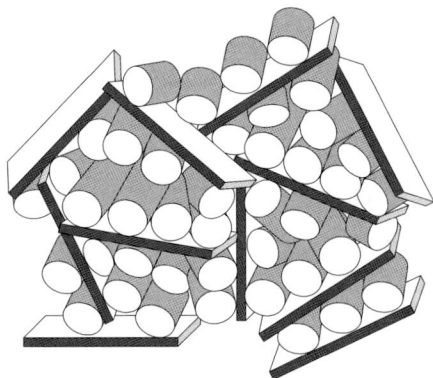

Fig. 14. The structure of mesoporous synthetic PCH material.

calcination, the surfactants are removed and SiO_2 pillars are formed. The structure proposed above is only valid for large-sized natural clays (~ 1 μm), like montmorillonite, for which a parallel or face-to-face orientation of the clay sheets is favored. In the case of the smaller-sized synthetic clays (~ 0.05 μm), like laponite or saponite, the edge-to-face and edge-to-edge interactions between plates are dominant, resulting in a different heterostructure (Fig. 14). As a result of these types of stackings, individual laponite plates play an active role in the micelle formation, and, in combination with surfactant and co-surfactant, determine the final pore structure of the PCH. Upon calcination, a 3D synthetic PCH structure is obtained, in which clay sheets are linked by the SiO_2-network. Micropores exist here as void spaces between individual micelles and between micelles and clay sheets as a result of the nonparallel stacking of clay plates. Both PCH materials exhibit supermicropores or mesopores between 1.5–3.0 nm and surface areas exceeding that of typical zeolites. Cool and coworkers [85, 89] reported the existence of a unique combined micro- and mesoporosity for PCHs derived from natural montmorillonite and synthetic saponite. By changing the synthesis conditions, a controlled tuning of the ratio micro-/mesoporosity of these materials becomes possible (ratio varies between 0.05–3 for natural PCH and between 2–7 for synthetic PCH). Moreover, the mesopore dimensions of these materials can be easily tailored by changing the chain lengths of the used surfactant and co-surfactant, since they determine the size of the formed micelles and the created pore size of the PCH after calcination. When using $C_{12}H_{25}NH_2$ as co-surfactant instead of $C_6H_{13}NH_2$, in conjunction with the surfactant $C_{16}H_{33}N(CH_3)_3^+$ for the reaction on Li^+-fluorohectorite host, the PCH pore size can be increased from 1.50 nm to 2.34 nm.

An alternative synthesis route has been developed by Kuroda and coworkers [87], who modified monolayered silicates such as kanemite with cationic surfactants. Under controlled pH conditions, the intercalated surfactants form micelles and the thin kanemite sheets undergo a topotactic transformation by folding around the

References see page 1305

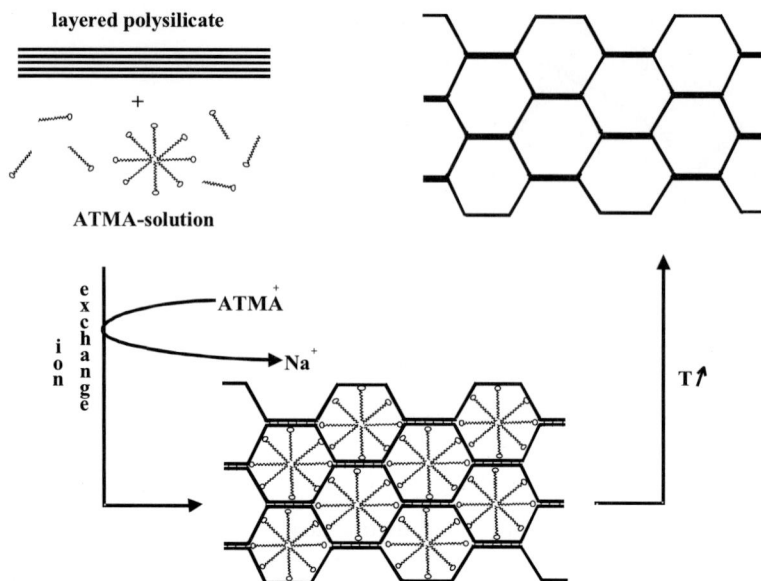

layered polysilicate

ATMA-solution

Fig. 15. Formation mechanism of mesoporous FSM materials.

micelles (Fig. 15). After calcination, the organics are removed, resulting in a highly porous mesoporous material. They are termed folded sheet mesoporous (FSM) solids, after their mechanism of formation. Similarly with the PCH materials, the mesopore size can also be tuned by a variation of the chain-length of the used surfactant. For an increase in chain-length of the surfactant from $C_{10}H_{21}N(CH_3)_3^+$ (giving rise to FSM-10) to $C_{16}H_{33}N(CH_3)_3^+$ (FSM-16), the mesopore size can be enhanced from 3.2 nm to 5.4 nm.

4.4.4
Porosity

Introduction of inorganic pillars between the clay sheets transforms nonmicroporous layered silicates to microporous ones. As the pillared edifices resist temperature, so does the microporosity. The development of micropores brings about an increase of the total specific surface area. Surface area and porosity are important characteristics of porous pillared layered solids for applications in the fields of adsorption and catalysis. Total surface area and pore volume are crucial criteria in heterogeneous catalysis since they determine the accessibility of the active sites, and are therefore related to the catalytic activity. The pore architecture of a porous solid controls transport phenomena and governs shape selectivity. Properties such as pore volume and pore-size distribution are therefore essential parameters to be analyzed.

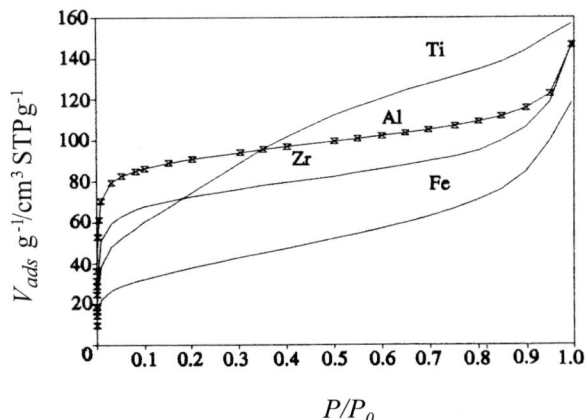

Fig. 16. N$_2$-adsorption isotherms at 77 K obtained for montmorillonite pillared with different pillaring species (Al, Zr, Fe, and Ti).

4.4.4.1
Isotherms

The surface area of a solid is commonly derived from its nitrogen adsorption/desorption isotherm at 77 K. Brunauer et al. [90] have defined five different types of isotherms. Type I isotherms are characteristic of microporous adsorbents, while type IV isotherms are typical for mesoporous solids. A significant increase of volume adsorbed in the low P/P_0 region in type IV isotherms indicates the presence of micropores associated with mesopores. In the case of pillared clays, the uptake of N$_2$ in the low-pressure region is large due to the large amount of micropores that have been generated by the pillaring process. The shape of the isotherms of pillared clays is dependent on the pillar species introduced [91] (see Fig. 16).

The isotherms of Al-PILC and Zr-PILC are of type I, indicating the small dimensions of the pillars. The shape of the adsorption isotherm of Fe-PILC resembles a type II isotherm (nonporous). The isotherm of Ti-PILC is close to type IV, but with a distortion in the relative pressure range 0.05–0.5 followed by capillary condensation. The distortion is due to the presence of pores on the border region between micro- and mesopores (1.5–3.5 nm). Isotherms of intermediate shape can also be observed, as for example alumina PILC doped with lanthania (LaAl-PILC) [92]. For some PILC materials, the presence of a large hysteresis loop in the isotherm might be observed, indicating the importance of mesopores in the structure. From the adsorption isotherm data, several parameters describing the porosity can be quantitatively derived.

References see page 1305

4.4.4.2
Surface Area

In the case of PILCs, three main methods have been applied to calculate the surface area based on the isotherm data: the Langmuir method, the BET method, and the t-plot or α_s-plot (t-plot method: see Sect. 4.4.4.3.).

Physical adsorption in microporous solids shows type I isotherms because the micropores limit the adsorption to a few molecular layers. Using a kinetic approach, Langmuir [93] described the type I isotherm considering that adsorption was limited to a monolayer. This approach assumes that the adsorption energy is constant and independent of the fraction of the surface occupied by the adsorbed molecules.

Brunauer, Emmett and Teller (BET) [94] extended Langmuir's theory to multi-layer adsorption. The BET theory assumes that the uppermost molecules in the adsorbed stacks are in dynamic equilibrium with the vapor, and that the adsorbed molecules, not in direct interaction with the surface, are all equivalent to the liquid state. Discussions on the validity of the assumptions, as well as a complete description of the BET theory may be found in the literature and in Sect. 2.5.2 [95–98]. For PILCs, the Langmuir surface area is always higher than the BET surface area, by about 30 to 40 % (Table 7). It is difficult to determine which one is closest to the real situation, though the reliability of specific surface areas can be checked by setting an upper limit for the monolayer capacity of porous solids [99]. Since clay sheets have a silica surface, a monolayer of adsorbed N_2 on a nonporous silica surface is completed at a P/P_0 of 0.09, according to Gregg and Sing [96]. Since the adsorption force in pores should be stronger than on an open surface, the monolayer capacity of a PILC should be lower than its uptake at $P/P_0 = 0.09$. The BET and Langmuir monolayer capacities of PILCs can be compared now to their adsorption at $P/P_0 = 0.09$. Zhu and Vansant [99] performed these calculations on a series of different Al-PILC samples (Table 7) and concluded that the Langmuir surface areas are too high and the BET data are lower than in the real situation. For these Al-PILCs, after the filling of pores of about 0.8 nm, the adsorption changes slightly with the increase of P/P_0. In this case, the slope of the BET plot will in-

Tab. 7. BET and Langmuir surface areas and monolayer capacities of PILCs, in comparison with their adsorption at P/P_0 of 0.09 ($V_{0.09}$) [99].

Sample	$SA_{BET}/$ $m^2\ g^{-1}$	$SA_L/$ $m^2\ g^{-1}$	$V_{BET\text{-}monolayer}/$ $cm^3\ STP\ g^{-1}$	$V_{L\text{-}monolayer}/$ $cm^3\ STP\ g^{-1}$	$V_{0.09}/cm^3$ $STP\ g^{-1}$
Al-PILCs					
– Al-PILC1	223.9	295.7	55.1	68.0	63.1
– Al-PILC2	284.0	377.1	71.0	86.6	80.2
– Al-PILC3	337.9	447.9	77.6	102.9	93.5
Fe-PILC	129.4	178.0	29.7	40.9	29.5
Ti-PILC	251.5	348.0	57.8	91.2	54.4

crease steeply,

$$s = 1/[V(1 - P/P_0)] \tag{7}$$

resulting in an underestimation of the BET surface area. Therefore, the deviation is caused by the space restriction of the very fine pores.

4.4.4.3
Microporosity

De Boer et al. [100] have developed the *t-method* to obtain the micropore volume and the external surface area of microporous solids. The method is based on the comparison of adsorption isotherm data of a porous sample and of a nonporous sample of identical chemical composition and surface character (reference isotherm). It is assumed that at relative pressures above a certain value, the micropores are completely filled, and only the adsorption on the external surface influences the isotherm. When the adsorbed N_2 volume is plotted against the statistical thickness t (in nm) of the adsorbed N_2 layer (V_{ads} versus t), a linear relation can be obtained. If both reference and sample isotherm are identical, as is the case for nonporous solids, a straight line passing through or close to the origin is obtained (Fig. 17). Horizontal departures from the straight line indicate the presence of micropores.

The microporous volume is obtained from a straight line extrapolated to a positive intercept on the ordinate. The estimation of t should be based on the reference isotherm, which yields the same BET constant (C_{BET}) as that of the material tested. However, C_{BET} of the external surface of a microporous solid is rather difficult to determine [101]. The choice of the standard t function and the part of the $V - t$ curve used for linear fitting drastically affect the values of the external surface and micropore volume. The strong influence of the chemical composition of the surface on the evolution of t as a function of P/P_0 was evidenced [101]. To avoid the above complications, Sing [102] introduced and developed the α_s method in which the normalized adsorption, α_s ($= n/n_{0.4}$), is derived from the isotherm of a reference material by using the amount adsorbed at a relative pressure of 0.4 ($n_{0.4}$) as the normalization factor. Lecloux and Pirard [103] have shown that the α_s- and t-methods, when based on the same standard isotherm, only differ by a constant factor and must give the same estimation of the micropore volume and external surface.

In Table 8 an overview of the micropore parameters derived from t-plots of PILCs is given.

4.4.4.3.1 Textural Characteristics of Al-Pillared Clays
For Al-pillared smectites, micropore volumes and specific surface areas in the ranges of 0.06–0.130 cm^3 g^{-1} and 250–350 m^2 g^{-1} are obtained, respectively cal-

References see page 1305

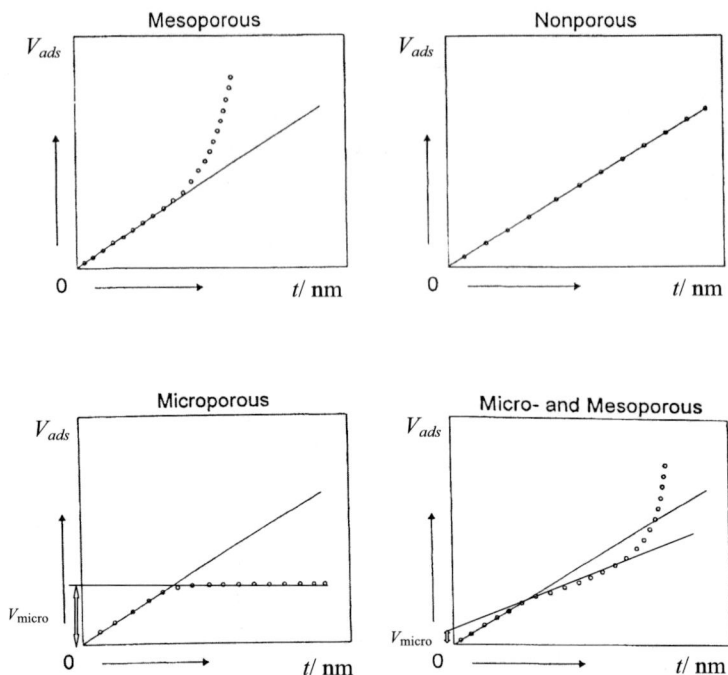

Fig. 17. *t*-plots of mesoporous, nonporous, microporous and combined micro-/mesoporous solids.

culated with the *t*-plot and BET method [104]. Figure 18 clearly illustrates the development of the microporosity resulting from the Al-pillaring of vermiculite and phlogopite. Nitrogen adsorption-desorption isotherms of the starting minerals (1), after the conditioning treatment (prior to Al-pillaring) (2), and of pillared vermiculite and phlogopite calcined at 773 K (3) and 973 K (4) are shown.

Typical values of the specific surface areas and micropore volumes of Al-pillared phlogopite and vermiculite calcined at 773 K are 322 m^2 g^{-1} and 0.099 cm^3 g^{-1} for a pillared vermiculite [46], and 315 m^2 g^{-1} and 0.102 cm^3 g^{-1} for pillared phlogopite [47].

Tab. 8. Micropore parameters of pillared clays derived from *t*-plots.

Sample	SA_{ext}/m^2 g^{-1}	$SA_{micropore}/m^2$ g^{-1}	$V_{micropore}/cm^3$ g^{-1}	$d_{micropore}/nm$
Al-PILC	57.8	223	0.092	0.82
Ti-PILC	70.0	193	0.133	1.38
Zr-PILC	40	177	0.078	0.88
LaAl-PILC	37	185	0.082	0.89

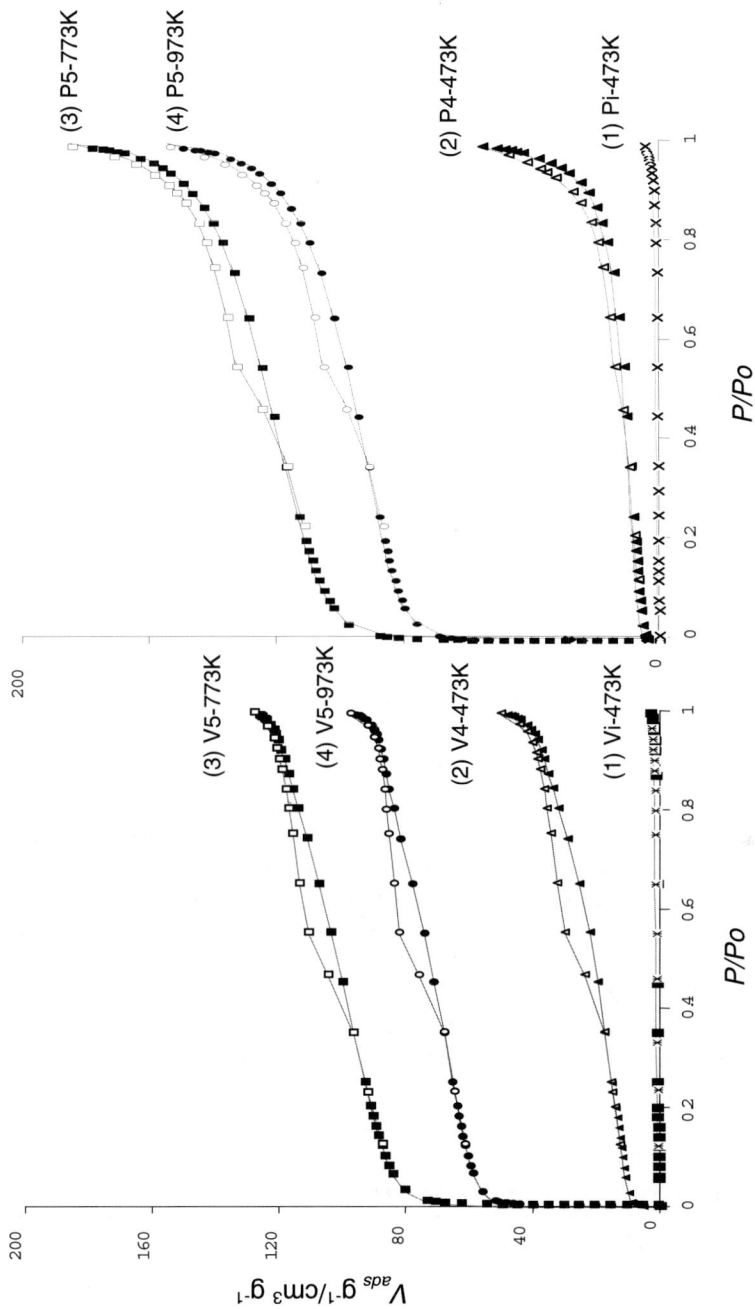

Fig. 18. Nitrogen adsorption-desorption isotherms at 77 K of (1) initial vermiculite (left) and phlogopite (right); after the conditioning treatment (2), and of the Al-pillared forms after calcination at 773 K (3) and 973 K (4).

4.4.4.4
Mean Micropore Width and Mesopore Size

The ratio of volume and surface area of the pores has the dimension of a pore size. When the surface area and volume of the micropores as well as the pore geometry are known, the micropore size can be obtained by a simple calculation. The micropores in the PILC have a slit shape. Therefore, the mean micropore width is calculated with the following formula:

$$d_{mi} = 2r_{mi} = 2V_{mi}/SA_{mi} = 2V_{mi}/(SA_{total} - SA_{ext} - SA_{me}) \tag{8}$$

For the calculation of the mesopore size, the same routine can be followed based on the mesopore SA (SA_{me}) (difference between the total SA and the micropore SA and the external SA: $SA_{total} - SA_{mi} - SA_{ext}$) and the mesopore volume (difference between the total volume and the micropore volume and the external volume: $V_{total} - V_{mi} - V_{ext}$). However, it is not certain that the pore geometry of the mesopores in PILCs is slit-shaped. The mesopores are mainly interparticle voids rather than free spacings between two parallel clay sheets.

4.4.4.5
Micropore-Size Distributions

Micropore-size distributions are derived from physisorption isotherm data and provide valuable microstructural information. Methods to determine micropore-size distributions are briefly introduced here. A more detailed discussion of the methods and limitations can be found in Sect. 2.5.2.

The adsorption data can be interpreted using the thermodynamic *Dubinin–Radushkevich (DR)* equation that holds for a variety of adsorption systems [105]. This equation is based on the assumption that adsorbed molecules experience an adsorption potential, A, that governs the fractional pore filling. The DR equation is expressed by:

$$V_a/V_0 = \exp[-(A/\beta E_0)^2] \tag{9}$$

with V_a, V_0, E_0, and β being the amount of adsorption at equilibrium pressure, P, the micropore volume, the characteristic adsorption energy, and the affinity coefficient characterizing the adsorbate, respectively.

The term "adsorption potential" (A) was introduced by Polanyi, and it corresponds to the change of a molar free energy caused by the change of a vapor pressure [106]:

$$A = RT \ln(P_0/P) = -\Delta G \tag{10}$$

with ΔG and R being the differential free energy of adsorption and the universal gas constant, respectively.

The total heterogeneity, i.e. the sum of the structural and surface heterogeneities of an adsorbent, is usually characterized by a distribution function (X) of the adsorption potential and may be evaluated by means of the condensation approximation method [107] as:

$$X(A) = -d\Theta(A)/dA \tag{11}$$

with Θ being the relative adsorption.

The intercept of the linear part with the lower slope of the plot $\ln V$ versus $\ln^2(P_0/P)$ yields the micropore volume. The micropore filling of a system with a distribution of pore sizes can be described by the sum of the contributions from individual pore groups that are characterized by their own pore volume, V_i, and adsorption energy E_i. Adsorption on microporous solids having two kinds of micropores can be described by a two-term DR equation [106, 108], expressed as:

$$V = V_1 \exp[-(R^2T^2 \ln^2(P_0/P)/E_1^2)] + V_2 \exp[-(R^2T^2 \ln^2(P_0/P)/E_2^2)] \tag{12}$$

The *Dubinin–Stoeckli (DS)* [109] relation describes an empirical relationship between the characteristic adsorption energy (E_0) and the average half-width of the slit-shaped micropores (χ_0).

$$E_0 = k/\chi_0, \tag{13}$$

where a constant value of 10 or 12 is used for χ_0 [106].

This relation is used for the transformation of the characteristic adsorption energy to the half-width of the slit-shaped micropores. A Gaussian distribution of these pore widths is given by

$$dV_a/d\chi_0 = [V_0/\delta(2\pi)^{1/2}] \exp[-(\chi_0 - \chi)^2/2\delta^2], \tag{14}$$

with δ and χ being the dispersion of the Gaussian distribution and the value of the pore half-width [106], respectively. The micropore-size distribution can be derived from the relationship $dV_a/d\chi$ versus χ. The DR method usually works well for solids in which the pore-size distribution is Gaussian. If the distribution deviates strongly from Gaussian, as in many pillared clay solids, physically unacceptable data are obtained.

Horvath and Kawazoe [110] described a method for calculating the effective micropore volume distribution of slit-shaped pores. The interaction between adsorbent and adsorbate using the interaction parameter, depending on their nature, was taken into account. This model relates the relative pressure (P/P_0) to a micropore radius described by a Lennard–Jones potential. In the case of samples containing small pores (<1 nm), the results obtained by the HK method are in good agreement with the data obtained by other approaches [92, 111]. For zeolites hav-

References see page 1305

ing small pores, as well as for Al-pillared clays with a free spacing of 0.7–0.8 nm, the HK approach gives reasonable results for the pore-size distributions [92]. Some PILCs have a wider pore-size distribution with the main part of the pores situated in the large micropore region, i.e. the supermicropores with a pore size > 1 nm. In this case, the results calculated from both the DR and HK method are unreasonable. According to the authors, applying the HK model to solids with pores > 1.2 nm is unsuccessful.

The MP method is an extension of the t-method of De Boer. The slope of the t-plot at a given point can be related to the surface areas that are still available for adsorption [112] by

$$S_r = k_s/k_{ref} \cdot S_{ref} = k_s \cdot C_{ref} \tag{15}$$

with S_r = surface area still available for adsorption; k_s = tangent slope; k_{ref} and S_{ref} the slope of the t-curve and the specific surface area (both constants). The surface area of a group of pores of similar pore size can be obtained from the difference in slopes of the tangents drawn at two adjacent points:

$$S_{group} = C_{ref}(k_{s,i} - k_{s,i+1}) \tag{16}$$

The MP method calculates mean group pore sizes directly from the average of the t-values for two points. The volume of this pore group is then calculated from the surface area and the size of the pore group with an assumption on the pore shape. When this analysis is continued until there is no further decrease in the slope of the t-plot, the pore volume and pore surface-area distributions over the micropore range are obtained. However, using the t value read directly from the t-plot as the pore size is questionable. As pointed out by Gregg and Sing [96], t is a statistical thickness rather than a real value in any specific pore. At a low relative pressure, the adsorbed layer in narrow micropores can be thicker than that on an open surface of the same sample because the adsorption energy in the pores is very strong [113]. As a consequence, the real pore size of the micropores must be greater than that obtained from the t-thickness of the adsorbed film over the whole solid surface.

Instead of using the t-values as in the original MP method, Zhu et al. [114] recently developed a *new improved MP method* by applying the well-known principle, throughout the literature, for the calculation of micropore volume, V_{micro}, using the t-plot [100] by an extrapolation to the adsorption axis ($t = 0$). The method allows the determination of pore-size distributions of PILCs from multistep t-plots. If the overall surface area of the micropores, S_{micro}, is also obtained, an average pore size of the micropores can be derived from the data of S_{micro} and V_{micro}. At this stage, it is necessary to assume the pore type so that the hydraulic radius [115], defined as the ratio of pore volume to pore surface area $r_h = 2(V_{pore}/S_{pore})$, can be converted to pore size. For instance, the micropores of pillared clays are generally visualized to be slit-shaped. The width of the slit-shaped pores, w, is equal to r_h, while for cylindrical pores, the diameter of a pore is $d_c = 2r_h$. This procedure involves the following calculations. First, dV_a/dt is evaluated as the slope, k_s, at various t points. The intercept at each point, I_i, is then obtained by

$$I_i = V_{a,i} - k_{s,i}.t = V_{a,i} - (dV_{a,i}/dt_i).t_i. \tag{17}$$

The volume and surface area per pore group are derived from slope differences in tangents at two adjacent points using Eq. (16), which can now be written:

$$S_{group,i} = C_{ref}(k_{s,i} - k_{s,i+1}) = C_{ref}(dV_{a,i}/dt_i - dV_{a,i+1}/dt_{i+1}) \tag{18}$$

and

$$V_{a_{group},i} = C_{LN_2}(I_{i+1} - I_i) \times const \tag{19}$$

where C_{LN2} is a volume-conversion factor and point $(i + 1)$ is at higher P/P_0 than point i. The mean hydraulic diameter, $d_{h,i}$ for each pore group is the volume/surface area ratio, and micropore-size distributions are plots of differential adsorbed volume, $dV_{a,i}/d(d_{h,i})$ against $d_{h,i}$. Examples of pore-size distributions obtained following this improved MP method are given in Fig. 19 for Al-PILC, Ti-PILC and Fe-Cr-PILC. An advantage of this method is that it allows pore size estimates for ultra-micropores. However, it still suffers from the problem of other methods, i.e. poorly estimated pore sizes for irregular pore shapes.

Recently, another new method was proposed by Zhu and coworkers [92, 99, 116] using the low-pressure part of the adsorption isotherm for the determination of the micropore-size distribution of PILCs. The N_2-adsorption isotherm has been re-plotted on a logarithmic P/P_0 scale to obtain a curve that represents the adsorption potential, $\Delta G_{ads} = RT \ln P/P_0$, which is highly dependent on the pore size and shape, and the presence of active sites. The curve thus obtained contains steps that are thermodynamically related to the filling of pores of a certain width. By locating and further investigating these steps, more detailed information on the pore size can be obtained. This is done by taking the derivative of the logarithmic curve, $dV/d \log(P/P_0)$. A large step in the logarithmic curve reflects the filling of the main pore type, and results in a clear maximum in the derivative plot (Fig. 20).

The pressure range of these maxima can now be correlated to the interlayer free spacings of the pillared clays, by assuming that the relative pressure at which the micropores are filled depends on the number of N_2 layers that can be accommodated in these pores. Consequently, the micropore range is divided into 5 pore groups, in which N_2 can be adsorbed in 1 to 5 layers with the thickness of 1 layer being 0.354 nm. Investigation of PILCs with different known slit widths (from XRD analysis) situated in different pore-groups, makes it possible to correlate the pressure ranges of the maxima in the derivative curves to a certain pore-size range (see Table 9).

In this way, a straightforward micropore-size distribution is obtained for pillared

References see page 1305

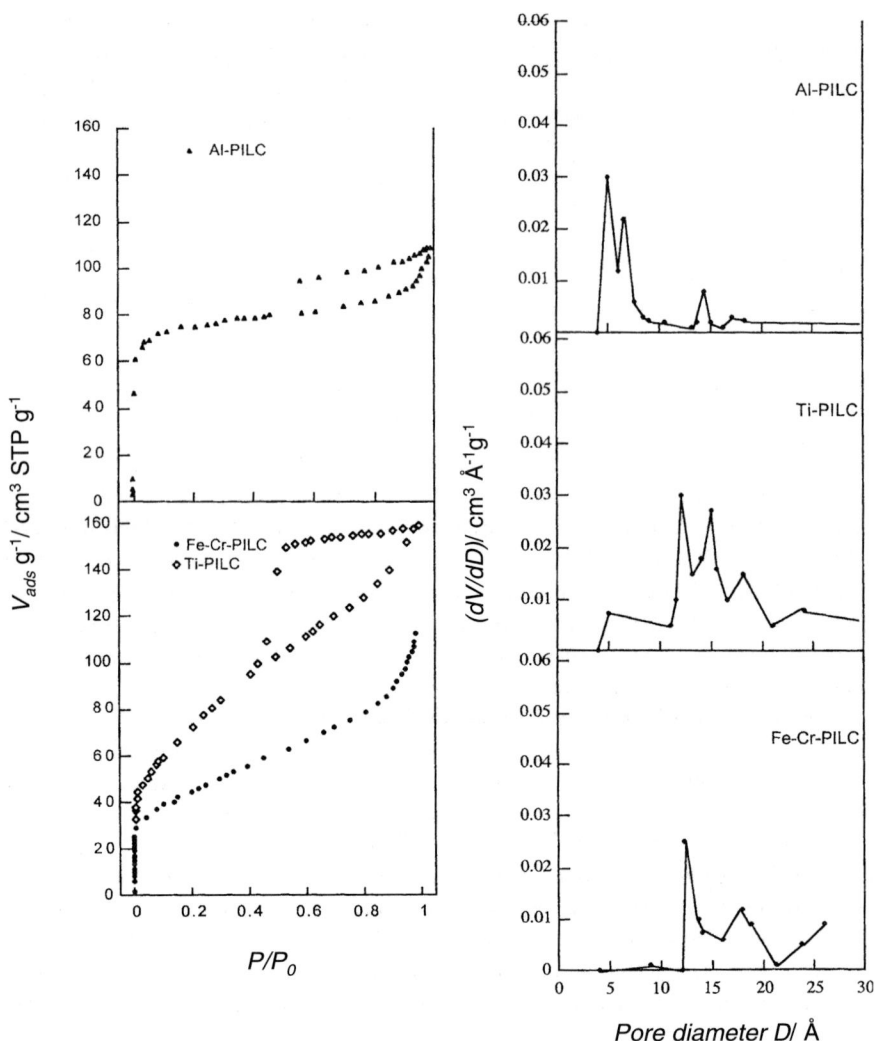

Fig. 19. N$_2$ adsorption-desorption isotherms of different PILCs
(left); corresponding pore-size distributions obtained following
the improved MP method (right) [114].

materials, which can be used in the entire micropore range. Compared to other
methods for PSD determination, this model gives better results in the super-
micropore region. Another advantage of this approach is that an estimation of the
surface area can be made independently of the BET and Langmuir methods by
summing the surface area of each type of pore group (1 to 5).

In Figs. 21 and 22, the calculated micropore-size distributions of Al- and Ti-PILC
are presented and compared to the PSD obtained with the HK method.

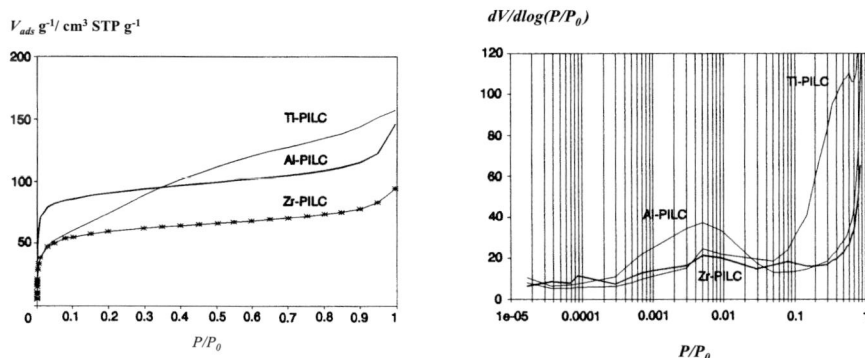

Fig. 20. Normal N$_2$ adsorption isotherms of Al-, Ti-, and Zr-PILC (left); Derivative curves of the logarithmic adsorption isotherms of Al-, Ti- and Zr-PILC (right) [116].

In the case of Al-PILC, there is a good correlation between the two methods. Both Zhu's method and HK give reasonable results. For Ti-PILC with larger pores no reliable HK-PSD can be derived, only the small peak in the distribution makes it clear that smaller pores are also present.

4.4.4.6
Evidence of Porosity by Modern Microscopic Techniques

Transmission and scanning electron microscopy (TEM/SEM) can be used to characterize the micro- and mesoporosity of pillared clays, though they are not routine techniques. Basal spacings observed using TEM/SEM agree well with X-ray diffraction (XRD) results. Electron microscopy can provide information on the stacking and ordering of clay sheets and, as a consequence, also on the porosity since the contribution of micro- and mesoporosity in the PILC structure is largely determined by the orientation of the clay sheets. For pillared montmorillonite, for instance, there is a characteristic long-range face-to-face stacking of silicate layers,

Tab. 9. Correlation between the relative pressure ranges and the corresponding pore-size ranges [116].

P/P_0 range	Pore-size range/nm	Maximum amount of adsorbed N$_2$-layers
<0.0003	<0.71	1
0.0003–0.003	0.71–1.06	2
0.003–0.03	1.06–1.42	3
0.03–0.2	1.42–1.77	4
0.2–0.5	1.77–2.12	5

References see page 1305

Pore Volume / cm³ g⁻¹

Pore-size range / Å

(dV/dr)/ cm³ Å⁻¹ g⁻¹

Pore radius / Å

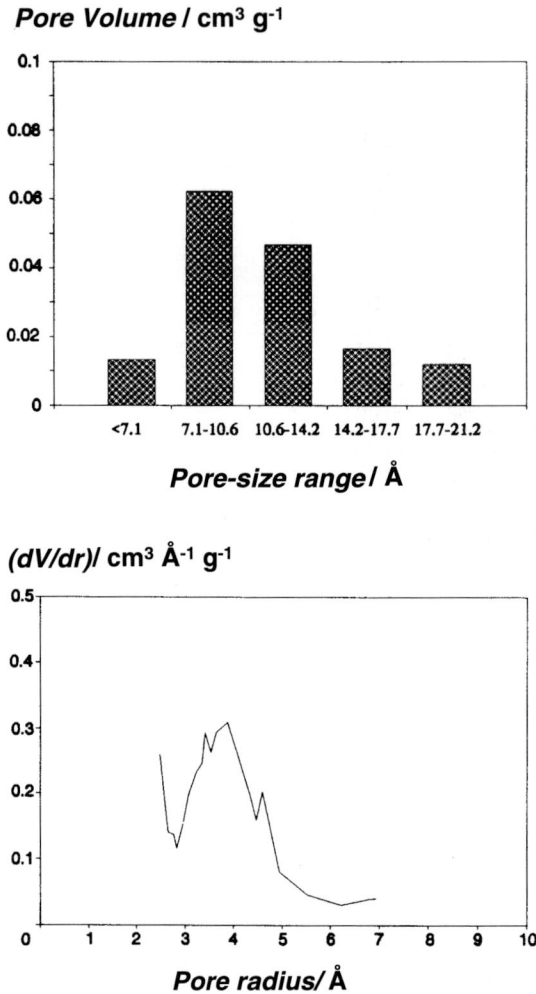

Fig. 21. Micropore-size distributions for Al-PILC (top: Zhu's model; bottom: HK method) [116].

resulting in a large microporosity [117]. Delaminated laponite clays on the other hand, are characterized by short stackings of 10–15 layers. Face-to-edge and edge-to-edge orientation of these smaller aggregates generate the meso/macropores present in delaminated structures [117]. Not only the clay ordering but also the existence of a possible external phase after pillaring or of nonintercalated clay plates can be elucidated using the TEM technique. In the case of Ti-pillared montmorillonite, black dots in the TEM picture clearly indicate the presence of an external TiO_2 phase as particles of approximately 3 nm [118]. For Al-pillared montmorillonite, a more homogeneous structure is depicted in the image without evidence of an external phase [118]. Ahenach et al. [119] studied the influence of the water concentration on the pillaring of montmorillonite with an organosilane

Pore Volume / cm³ g⁻¹

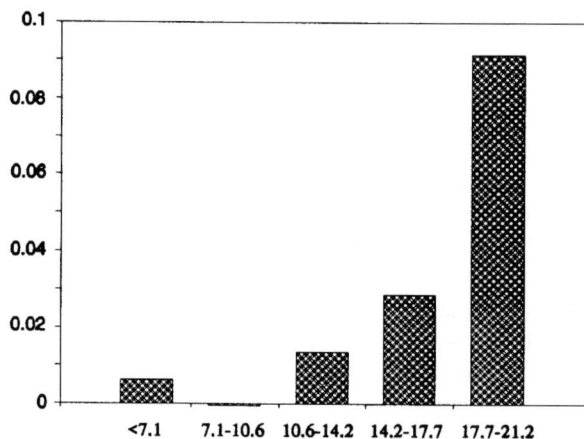

Pore-size range/ Å

(dV/dr)/ cm³ Å⁻¹ g⁻¹

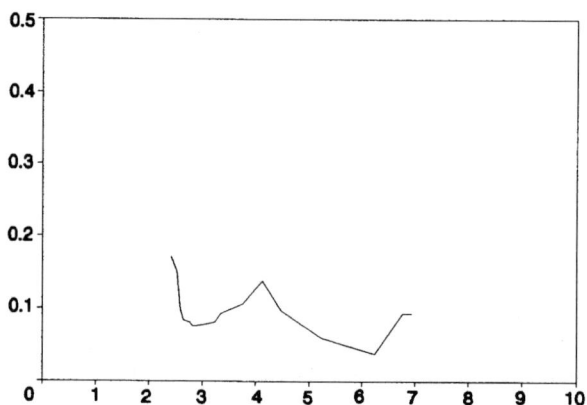

Pore radius/ Å

Fig. 22. Micropore-size distributions for Ti-PILC (top: Zhu's model; bottom: HK method) [116].

as the pillaring precursor. The TEM images proved that with an increasing H_2O concentration to enhance the hydrolysis of the silane, particles of a certain size (appearing as black spots in the image) are formed at the external surface of the clay. The TEM result also indicates the contribution of nonintercalated clay platelets. In the field of porous pillared clay membranes, Vercauteren et al. [120]

References see page 1305

showed that SEM can be very useful to obtain information on the Al-pillared clay top layer deposited on an Al_2O_3 support membrane. Smooth and uniform top layers, which are very thin, can be seen in the SEM pictures. No cracks are present. From the images, it was possible to calculate the top layer thicknesses, which were between 0.1 and 0.5 μm, depending on the concentration of the clay suspension used in the dipping procedure.

Since the invention of the atomic force microscope (AFM) by Binning et al. [121], which can generate atomic-scale images of materials, this new tool is often used in combination with electron microscopy to examine the surface and porosity of pillared clays. The principle of this technique is based on scanning the surface of a sample with a very sharp tip, brought within close proximity of the sample, to map the contours of the surface. Hartman et al. [122] demonstrated the ability of the AFM to image molecular-scale features of montmorillonite and illite. In 1993, Occelli et al. [123] conducted a profound characterization of Al-pillared montmorillonite with AFM. The AFM imaged the clay surface structure and provided detail of the sorption of polyoxycations in pillared clay catalysts. Atomic-scale-resolution images of the clay surfaces before and after pillaring consist of hexagonal arrays of bright spots (representing the 3 basal oxygens of a SiO_4 unit). The average nearest-neighbor distance (the spot-to-spot distance a) before and after pillaring with Al_{13} ions could be calculated, and it was consistently observed that pillared montmorillonites possess larger nearest-neighbor distances than the parent materials. The pillaring reactions seem to increase the center-to-center distance between bright spots (from 0.515 ± 0.05 nm to 0.545 ± 0.05 nm), so it is assumed that by replacing the parent-clay charge-compensating cations with Al_{13} ions in the interlamellar space, stretching of the SiO_4 layers occurs. With AFM, direct observation of Al_2O_3 clusters (the pillars) between the layers is not possible, though the expansion of the clay layers as a result of the pillaring is visualized. The mean basal spacing, as calculated from the distribution of expanded layers, is 0.90 ± 0.01 nm. This result is in agreement with the basal spacing obtained by XRD (0.85 nm). In a very recent study by the same authors [124] on the characterization of the microstructure of Al-pillared montmorillonite by the AFM technique, the images clearly indicate that the PILC surface is free from adsorbed Al-species and that all the Al-ions are located in the interlamellar space. From the cross-sectional analysis of the surface of the atomic scale image, the next-neighbor distance a has an average value of 0.53 nm and the average lateral distance l is 0.93 nm. The AFM images thus allow a direct determination of the unit cell surface area of Al-PILC: $al = 0.4929$ nm^2. The total interlamellar surface area per gram of clay calculates to: $[2 \times 0.4929$ nm$^2 \times 6.02 \times 10^{23}]/[747$ g clay$] = 794$ m^2 g^{-1} clay. For pillared clay sheets with an interlayer height h of 0.84 nm, the interlamellar void volume per unit cell is given by hal, or 0.414 nm^3 corresponding to 0.33 cm^3 g^{-1} clay (with the unit cell weight determined by elemental analysis). Knowing there are 0.88 g clay g^{-1} PILC, the ideal interlamellar volume can be expressed as $0.33 \times 0.88 = 0.29$ cm^3 g^{-1} PILC and the ideal interlamellar surface area is 699 m^2 g^{-1} PILC. The authors used atomic force microscopy in combination with high-resolution electron microscopy (HREM) to investigate the hydrothermal stability of Al-pillared montmorillonite. A treatment under 100 % steam for 5 h at 1033 K decomposes the pillared clay structure. AFM shows the absence of the repeated distances characteristic for the siloxane layers, thus steam aging causes O–Si–O bond breakage. HREM images illustrate the structural damage of the Al-PILC by the appearance of wavy layers, curled edges, discontinued (broken) layers and the formation of amorphous patches between discontinued layers.

4.4.5
Adsorption

4.4.5.1
Principles of Adsorption

The typical microporous volume and pore sizes make pillared clay minerals good candidates for adsorption and molecular sieving. When the size of the adsorbed molecules is comparable to the pore size of the adsorbent, the porous materials can be used for separation of gas mixtures into its components. Adsorptive separation can be classified into three groups: the first is *equilibrium separation*, in which separation is achieved based on the differences in affinity of the gas molecules for the porous solid (also thermodynamical separation); the second is *steric separation*, which is realized based on the differences in shapes and sizes of the molecules in the mixture (molecular sieving effect); the third is *kinetic separation*, which is achieved by virtue of the difference in the diffusion rate of adsorbate molecules into the adsorbent materials. Equilibrium separation is the most commonly used adsorptive separation method in industrial applications, for which understanding of adsorption equilibria of gas mixtures in microporous materials is of crucial importance.

The temperature of adsorption is a very important parameter. At low adsorption temperatures (77 K), mainly a nonlocalized adsorption occurs by a condensation of the gases in the pores of the substrate. At higher temperatures, localized adsorption effects gain in importance. The adsorption between the gas and adsorption sites on the surface or in the pores of the substrate can be described as electrostatic interactions [125]. These interaction forces are divided into specific and nonspecific forces.

Nonspecific interaction forces are present in any gas/substrate system, independent of the properties of the adsorbate or adsorbent. They include attractive dispersion forces (ϕ_D, also called London forces, which arise from the rapid fluctuations in electron density within one atom which induce an electrical moment in a neighboring atom), and the short-range repulsion forces (ϕ_R).

ϕ_D is always dominant and consists of the sum of instantaneous dipole forces, induced dipole–induced quadrupole and induced quadrupole–induced quadrupole interactions. They can be represented by the following equation:

References see page 1305

$$\phi_D = -\frac{A_1}{r_{12}^6} - \frac{A_2}{r_{12}^8} - \frac{A_3}{r_{12}^{10}} \tag{20}$$

A_i are dispersion constants and r_{12} is the distance between the centers of two interacting molecules. The latter two terms of the equation are usually omitted because they are relatively small compared to the dipole-dipole term. The coefficient A_1 might be expressed by the formulae of Kirkwood–Muller, London or Slater–Kirkwood [126].

The repulsion force ϕ_R equals $B/(r_{12})^{12}$, in which B is another empirical constant.

The combination of both nonspecific forces leads to a Lennard–Jones potential function:

$$\phi = -\frac{A}{r^6} + \frac{B}{r^{12}} \tag{21}$$

or generally denoted as:

$$\phi = 4\varepsilon \left[\left(\frac{\sigma}{r}\right)^{12} - \left(\frac{\sigma}{r}\right)^6 \right] \tag{22}$$

in which σ is the kinetic diameter (the intermolecular distance of closest approach for two molecules colliding with zero initial kinetic energy). ε is the potential energy at the van der Waals diameter, σ_{min}, and represents the maximum energy of attraction between two molecules. Moreover, $\sigma_{min} = 2^{1/6}\sigma$.

The specific adsorption forces on the other hand arise from electrostatic interactions between permanent electrical moments in a gas molecule and the adsorption sites (hydroxyl groups, ions, etc.) on a surface. The most important specific forces are:

ϕ_P = arising from the polarization of the gas molecule
ϕ_μ = field–dipole interaction
ϕ_Q = field–quadrupole interaction

They are expressed by:

$$\phi_P = -\frac{1}{2}\alpha^2 E \tag{23}$$

$$\phi_\mu = -\mu E \tag{24}$$

$$\phi_Q = \frac{1}{2}Q\frac{\partial E}{\partial r} \tag{25}$$

For physisorption processes, the total interaction energy (ϕ) between a gas and the porous structure is then expressed as a sum:

$$\phi = \phi_D + \phi_R + \phi_\mu + \phi_Q + \phi_P \tag{26}$$

Depending on whether a molecule has a high polarizability, dipole moment or large quadrupole moment, the total interaction force between the molecule and the polar surface of the solid increases. In many cases, these properties are used to separate gases, especially if the porous substrate is modified in order to increase this interaction effect.

4.4.5.2
Adsorption in Pillared Clays

In recent years, a number of investigations has been carried out to study adsorption in pillared clays, almost all of which are experimental in nature. Despite their many potential applications and the existence of valuable information that can be obtained through extensive experimental efforts, very few fundamental theoretical studies have been undertaken so far to investigate transport and adsorption processes in PILCs. A few computer simulation studies have been carried out [127–131]. Fundamental molecular simulations of adsorption and diffusion in model pillared clays have been performed by the group of Yi et al. [132–134]. Using grand-canonical-ensemble Monte Carlo and molecular dynamics (MD) simulation, the adsorption of single-component and binary gas mixtures (CH_4/N_2 mixture) in a porous model PILC with interconnected pores at room temperature has been studied [134]. The authors investigated in detail the effect of various factors, such as the morphology of the pore space and the adsorbent–adsorbate interactions. A new mean-field statistical mechanical theory of adsorption has been developed, providing very accurate predictions for the simulation results over wide ranges of pressure, temperature and porosity of the system.

Some adsorption studies deal with liquid systems, and clearly show the potential use of PILCs as adsorbents for environmental applications.

One example is the uptake of toxicants such as chlorophenols on pillared and delaminated clay structures [135]. Al-delaminated laponite is more effective in adsorbing pentachlorophenol (PCP) than the Al-pillared montmorillonite. At an equilibrium time = 24 h, an equilibrium pH = 4.7 and an initial PCP concentration of 38 µmol L^{-1}, the maximal adsorption of PCP was 27 µmol g^{-1} on Al-delaminated laponite, and 12 µmol g^{-1} on Al-pillared montmorillonite. The binding of PCP onto the substrate is attributed to interactions between the PCP molecule and the immobilized Al_2O_3 species. The greater adsorption capacity for the delaminated structure arises from the greater dispersion and availability of the Al-oxide aggregates in the clay.

In order to adsorb organics selectively from aqueous solution, it is very important that the adsorbents are hydrophobic. Therefore, Shu et al. [136] investigated the adsorption of the same organic toxicants on a hydrophobic surfactant (Tergitol)-modified Zr-pillared montmorillonite. The surfactant-modified external sur-

References see page 1305

faces have a high affinity for the organics and this affinity is related to the surfactant loading. An adsorption capacity of phenol equal to 0.8 mmol g^{-1} has been reached at an equilibrium concentration of 7 mg mL^{-1}.

Porous pillared clays can also be used for the removal of color-generating compounds such as "humic" and "fulvic" acids from water [137]. The adsorption of "humic acid" from both distilled and tap water was studied. A maximum adsorption capacity of 23.4 mg g^{-1} "humic acid" can be obtained out of a tap water solution with an initial concentration of 160 mg L^{-1} on Al-PILC thermally treated at 453 K. The efficiency of Al-PILCs is pH dependent. Thermal treatment of the PILC also has an influence on the adsorption properties. At 453 K the Al-PILCs are better adsorbents than those treated at 673 K, due to the larger expansion of the clay layers at a lower calcination temperature.

For the adsorption of inorganic gases on pillared clays, most often, nitrogen adsorption at 77 K is described and the data are used for the calculation of the surface area and pore volume of PILCs. Studies of the affinity, capacity and selectivity of PILCs towards inorganic gases were first described by Barrer and coworkers [138, 139]. In these studies, Co(en)$_3$-pillared fluorhectorites (en = ethylenediamine) were used to adsorb H$_2$, D$_2$, Ne, N$_2$, O$_2$, Ar, Kr, and CO$_2$. Low-temperature uptake rates showed the potential molecular-sieving effects of the microporous materials. Gameson et al. [140] investigated low-temperature adsorption of Xe and Kr on Al-PILC, which resulted in a model for the coverage mechanism in the pores.

Most work on the adsorption of gases at ambient temperature on various PILCs was performed by Baksh and Yang [141, 142], who proposed that the pores in pillared clays consist of narrow and tall paths limited by the interpillar spacing rather than the interlayer free spacing. For these slit-shaped micropores, the electrostatic fields of top and bottom clay layers overlap in the middle of the pore and result in a very high interaction field. In a first study, Zr-PILCs were used for testing their adsorption properties (N$_2$, O$_2$, hexane, benzene) [141]. The equilibrium isotherms for N$_2$ and O$_2$ at 298 K on Zr-PILC are shown in Fig. 23, together with comparison data of 5A zeolite, which is widely used for air separation. The N$_2$/O$_2$ equilibrium selectivity on Zr-PILC is high but still smaller than that of zeolite 5A (approximately 3), as a result of the limiting interpillar spacing in Zr-PILC, not allowing multilayer buildup. In an attempt to increase the interpillar spacing and to tailor the pore-size distribution of the Zr-PILCs for specific separations, Ca^{2+} ions were introduced during the pillaring process since the cations are competitive with the pillaring precursors during the synthesis. This seemed to have a positive influence on the benzene adsorption of the Zr-pillared clays.

The same authors tested the adsorption of O$_2$, N$_2$, CH$_4$, CO$_2$, SO$_2$, and NO on five different pillared clays (Zr-, Al-, Cr-, Fe-, and Ti-PILCs) [142]. The adsorption isotherms at 298 K showed that the PILC materials exhibit a high and selective adsorption towards the gases. The adsorption data were used to determine diffusion constants and potential energy profiles in the slit-shaped pores of the PILCs. The equilibrium selectivity of CH$_4$/N$_2$ on Al-PILC is greater than 5.0 (at 1013 hPa), exceeding all known sorbents by a large margin (Fig. 24). In addition, high SO$_2$/CO$_2$ equilibrium selectivities are observed on the pillared clays.

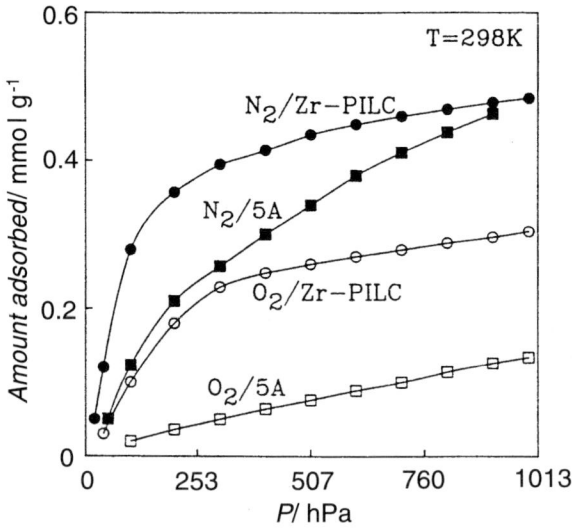

Fig. 23. Equilibrium isotherms of N_2 and O_2 on Zr-PILC compared with 5A zeolite [141].

A systematic investigation has recently been undertaken to tailor the micropore dimensions of pillared clays for enhanced gas adsorption [143]. Two clays with different CECs are used for Al-pillaring, namely Arizona montmorillonite (CEC = 1.4 meq g^{-1}) and Wyoming montmorillonite (CEC = 0.76 meq g^{-1}), resulting in a different pillar density and interpillar distance. It is shown that the CH_4 adsorp-

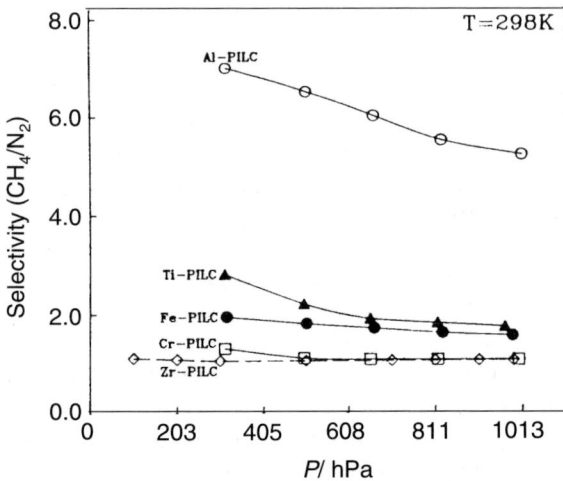

Fig. 24. CH_4/N_2 equilibrium selectivities at 298 K on Zr, Al, Cr, Fe, and Ti-PILCs [142].

References see page 1305

Tab. 10. Adsorption properties of Al-pillared clay minerals.

Type of PILC (calcined at 873 K)	Amount adsorbed/mmol g^{-1}	
	CH$_4$	N$_2$
Arizona montmorillonite	0.087	0.037
Wyoming montmorillonite	0.048	0.032

tion on the Arizona pillared clay can be nearly doubled by the smaller interpillar spacing, due to the back-to-back overlapping potential in the micropores (Table 10). The N$_2$ adsorption is not significantly influenced because of the low polarizability of N$_2$. The CH$_4$/N$_2$ selectivity ratio for the Al-pillared Arizona montmorillonite calcined at 873 K is 2.35, which is adequate for separation by the pressure-swing adsorption process [144].

In order to obtain a porous pillared material with improved adsorption properties, some additional modification techniques on PILCs, both during or after the synthesis, have been developed.

The incorporation of metals in the pillars, when performed during the synthesis, results in the formation of mixed oxide-pillars. Heylen et al. [145] observed an enhanced adsorption at 273 K of cyclohexane, CCl$_4$ and CO$_2$ on Fe/Cr-pillared montmorillonite compared to the pure Fe-pillared montmorillonite. By the synthesis of mixed-oxide-pillars, specific adsorption sites are created in the PILC, exerting a positive influence on the adsorption capacity and selectivity towards gases.

A way to increase the porosity of pillared clays and to modify their adsorption behavior is the preadsorption of organic templates prior to the ion exchange with the pillaring precursors. n-Alkylammonium ions have therefore been pre-exchanged on the clay in an amount equal to a fraction of the CEC. As a result, the pillar density decreases, since part of the interlayer space is occupied by the templates, which are only removed in the final calcination step. It is shown that this indirect pillaring procedure induces an increase in the adsorption capacity towards inorganic gases. Heylen et al. [146, 147] investigated the influence of the preadsorption of butylammonium ions on Fe- and Al-pillared montmorillonite (in an amount $= \frac{1}{2}$ of the CEC). The surface area and micropore volume of Fe-PILC was 2.5-times higher than the unmodified Fe-PILC (Table 11). Cool et al. [148, 149] performed a templated synthesis of Zr-pillared laponite using ethylenediamine in an amount exceeding the CEC of laponite. Here, the amines not only have a positive influence on the pillar distribution, but also favor the parallel orientation of the clay sheets, resulting in a more homogeneously pillared structure with increased microporosity. Thus, on laponite, the main function of the template is not to block exchange sites, but to influence the stacking of clay layers. The porosity and adsorption characteristics of the different pillared clays are summarized in Table 11.

The unmodified and modified pillared clays, synthesized using amines or with mixed pillars, have an intermediate hydrophilic-hydrophobic character and thus some potential for the removal of chlorinated hydrocarbons. In Fig. 25 the adsorp-

Tab. 11. Porosity and adsorption characteristics (N_2 and O_2; at 273 K on Al- and Zr-PILC and at 194 K on Fe-PILC) of pillared clays, prepared without and with templates.

Type of PILC	$SA_{BET}/m^2\ g^{-1}$	$MicroPV/cm^3\ g^{-1}$	Amount adsorbed/mmol g^{-1}	
			N_2	O_2
Al-montmorillonite	340	0.112	0.058	0.057
BuA-Al-montmorillonite	361	0.131	0.110	0.105
Fe-montmorillonite	134	0.037	0.000	0.030
BuA-Fe-montmorillonite	233	0.121	0.230	0.171
Zr-laponite	425	0.227	0.203	0.175
Etdiam-Zr-laponite	482	0.340	0.286	0.219

BuA = butylammonium; Etdiam = ethylenediamine.

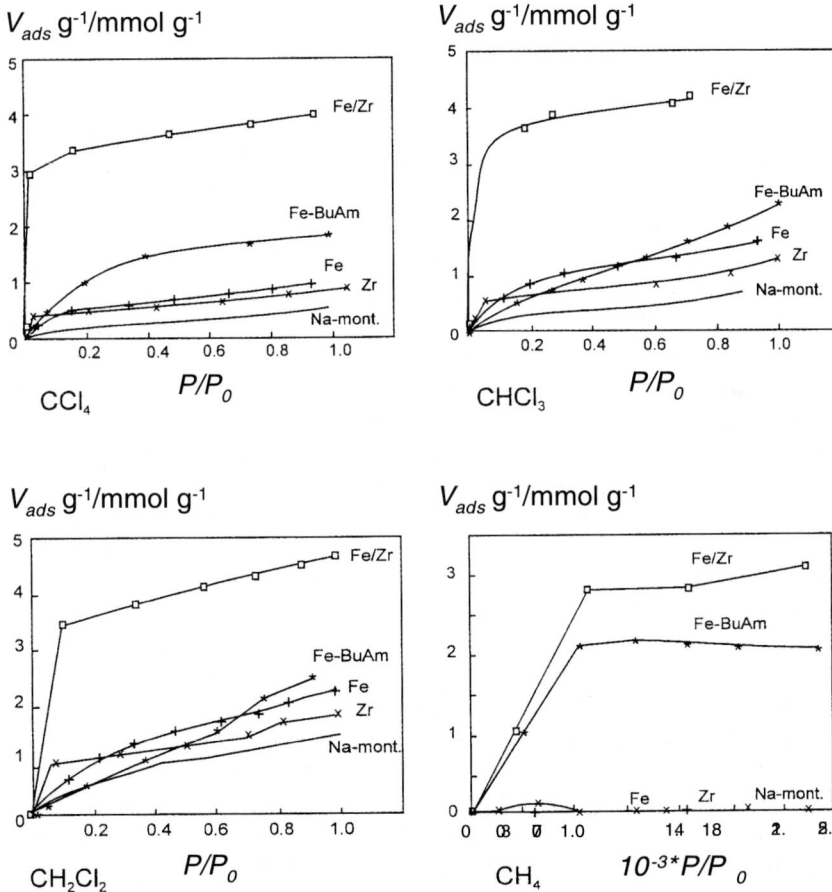

Fig. 25. Adsorption isotherms at 273 K of chlorinated hydrocarbons (CCl_4, $CHCl_3$, CH_2Cl_2, CH_4) on different pillared clays and on the parent Na-montmorillonite [147].

tion isotherms of some hydrocarbons on different pillared clays are compared to the adsorption isotherm on Na-montmorillonite [147]. In all the cases, the adsorption capacity of the mixed Fe/Zr PILC is a factor 4 higher than that of the pure Fe-PILC. Besides the adsorption capacity, the isotherm type also changes.

Another modification technique of PILCs, found to be useful to improve the adsorption properties, is the incorporation of specific cations in the porous structure of pillared clays, serving as specific adsorption sites in certain applications. It is known that the cation-exchange capacity (CEC) of montmorillonite greatly decreases after pillaring with alumina pillars, due to the nonexchangeable H^+ cations in the clay structure that are formed during the calcination. It is, however, possible to restore the CEC of a PILC in two ways [150, 151].

The direct ion exchange procedure involves the exchange of a desired cation or anion from an alkaline or acidified salt solution, respectively. This is possible because of the amphoteric character of the hydroxyl groups, present on the pillars and on the clay layer edges [152, 153]. At low pH, the $-OH$ groups protonate and act as anion exchangers, while at high pH they deprotonate and become cation exchangers.

The indirect ion-exchange method consists of a first modification of the PILC with ammonia. The reaction of the gas with the H^+ ions present in the clay structure after calcination, results in the formation of NH_4^+ ions which can subsequently be exchanged for any other desired cation from a salt solution. These ion modifications result in the introduction of cations in the PILC pores, and they alter the adsorption properties of the substrate.

Molinard and Vansant [154, 155] prepared a series of cation (Mg^{2+}, Ca^{2+}, Sr^{2+}, Ba^{2+}) modified Al-PILCs and tested the adsorption capacity/selectivity for the air components N_2 and O_2 (Fig. 26). The introduction of cations has a clear positive influence on the adsorption behavior of the pillared clays. Compared to oxygen, nitrogen has a higher quadrupole moment, resulting in additional interaction with the adsorption sites on the pillared clay surface. The difference in affinity between O_2 and N_2 is most pronounced on Sr^{2+}-Al-PILC.

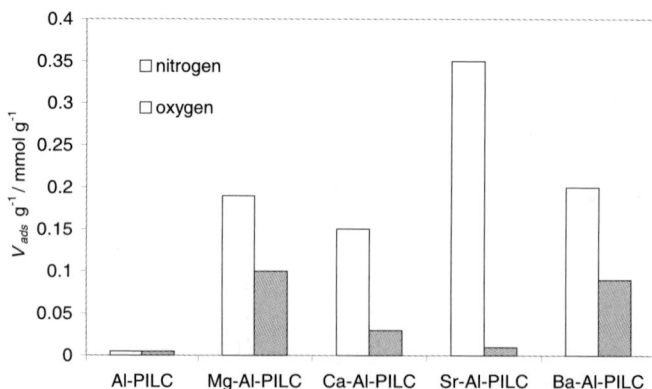

Fig. 26. Comparison of the adsorption capacity for N_2 and O_2 on cation-modified Al-PILCs ($T = 194$ K; $p = 51332.67$ Pa).

V_{ads} g^{-1}/ mmol g^{-1}

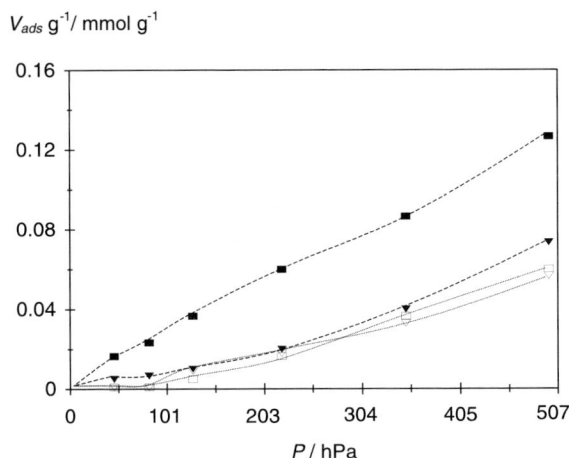

Fig. 27. Adsorption isotherms at 273 K on Al-PILC ($\square$N$_2$; $\triangledown$O$_2$) and Sr-Al-PILC ($\blacksquare$N$_2$; $\blacktriangledown$O$_2$) [156].

The adsorption isotherms of nitrogen and oxygen on Al-PILC and Sr^{2+}-Al-PILC above 273 K are presented in Fig. 27. For the original Al-PILC, there is an equal uptake of both gases over the entire pressure range, while after the introduction of cationic adsorption sites (Sr^{2+}) in the PILC structure, the N$_2$ capacity at 507 hPa is doubled. As a result, the selectivity ratio N$_2$/O$_2$ can be significantly increased by ion modification.

In order to study the influence of the amount of exchanged cations on the gas-adsorption properties of the porous substrate, two different PILCs have been prepared with a different Ca^{2+}-loading. The N$_2$/O$_2$ ratios of the Ca^{2+}-Al-PILCs at 273 K are determined as a function of the pressure (Fig. 28). Both PILCs have a higher preference for nitrogen than the parent Al-PILC (N$_2$/O$_2 \approx 1$) and the PILC with the highest cation amount (2.23 wt %) exhibits a better N$_2$/O$_2$ selectivity. Compared to the Sr^{2+}-exchanged PILCs, the capacities obtained on the Ca^{2+}-Al-PILCs are higher.

In a recent review by Zhu and Lu [157], the many different cation doping techniques on PILCs, described throughout the literature in order to tailor the pore structure and properties of pillared clays, have been discussed. The adsorption of water, air components and organic vapors on cation-doped pillared clays are studied. It is shown that by the presence of cations on the PILC surface, their hydrophilicity can be enhanced. By comparing water adsorption on Al-PILC before and after loading with Ca^{2+} ions, it becomes clear that introducing cations enhances the water sorption, particularly at low vapor pressures [158]. Moreover, there is the possibility to adjust the shape of the water isotherm by loading various amounts of cations. This technique will be very useful in developing PILCs-based dessicants and adsorbents for dehumidification and cooling applications.

References see page 1305

ratio N$_2$/O$_2$ adsorption

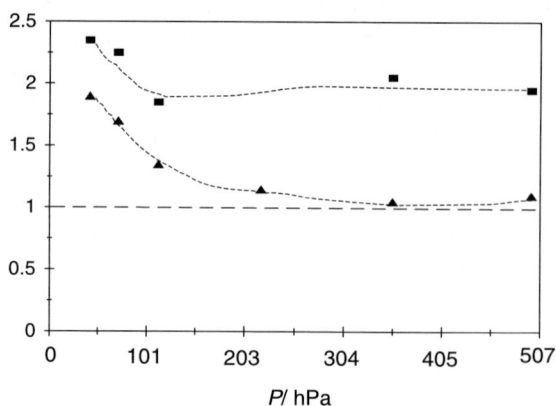

Fig. 28. N$_2$/O$_2$ adsorption ratio of two Ca-Al-PILCs, containing 1.26 wt % Ca^{2+} (▲) and 2.23 wt % Ca^{2+} (■) [156].

Adsorption of organic molecules on pillared clays was also investigated by the authors [157]. Usually, the adsorption of organic molecules of various dimensions on pillared clays is measured to provide details of the PILC pore structure, since the pores are in the range of the molecular sizes of many organic vapors. Zhu modified an Al-PILC with slit-shaped pores of ~0.8 nm with various amounts of Na$^+$ ions, in order to fine tune the size of the interlayer micropores. The adsorption results of p- and m-xylene on the Na-doped samples at 298 K are presented in Fig. 29. When the Na$^+$ content is low, the PILC shows almost no selectivity to-

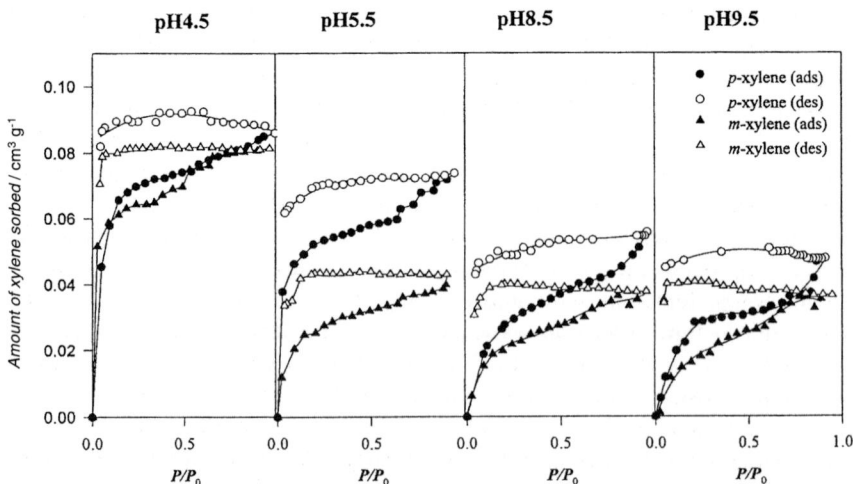

Fig. 29. Adsorption-desorption isotherms of p- and m-xylene on Al-PILC doped with various amounts of Na$^+$ ions. The pH value of the dispersion for loading the Na$^+$ into the PILC structure is given on top of the plots [157].

wards the isomers, indicating that it cannot distinguish between p- and m-isomers. As the amount of Na^+ increases, the uptake of both p- and m-isomer decreases. A sudden drop is observed for the m-isomer at a Na^+ loading of only 0.066 mmol g^{-1}. The uptake ratio of p- over m-xylene reaches a maximum of approximately two here. Since they are slightly larger molecular entities, m-isomers are more subjected to sterical hindrance than p-isomers and at this cation loading, quite large portions of the micropores are no longer accessible to m-isomers. The uptake ratio further decreases with increasing cation loading, since both isomers are excluded from the smaller micropores then. The results indicate that the modification of micropores with cations is a simple and powerful strategy to tailor the adsorption properties of the pillared clay structure. Adjusting the amount of doped ions is of crucial importance to fine tune the opening size of the narrow micropores.

4.4.6
Surface Properties

4.4.6.1
Acidity

Pillaring also confers acidic properties to the resulting materials. In Al-pillared smectites, Lewis acidity has generally been associated with the aluminum pillars, whereas the origin of the Brønsted acidity (protons) is more diverse. Protons can be formed at the pillaring step, as a result of polymerization processes of the pillaring species occurring in the interlayer space [44]; protons are also produced at the calcination step as a result of the transformation of the hydrated pillar precursors to oxidic pillars [159]. In the case of Al-pillared clays obtained from smectites with Al for Si substitutions (saponites, beidellites), Brønsted acidity has also been associated with Si–OH–Al groups resulting from the proton attack of Si–O–Al linkages of the tetrahedral layers. Such acid groups give rise to a new OH stretching band at 3440 cm^{-1} in Al-pillared beidellite and at 3595 cm^{-1} in Al-pillared saponite. These bands disappear upon pyridine adsorption and are restored after desorption of the base [160–162]. This feature differs from pillared clays without tetrahedral substitutions where such acid groups are absent.

Two main methods have been employed to characterize the acidity of solids: temperature-programmed desorption of ammonia and infrared spectroscopy of adsorbed bases. Temperature-programmed desorption (TPD) over acid solids previously exposed to ammonia allows quantification of the acid content and provides information on the strength of the acid sites. However, the distinction between Lewis and Brønsted acidities is not straightforward and is often hampered by physisorbed ammonia. The validity of this method has recently been questioned [163]. Infrared spectroscopy of adsorbed pyridine and ammonia have been most extensively used to investigate the acid properties of solids including the pillared clays. This approach allows us to distinguish and quantify the species that interact

References see page 1305

with Lewis and Brønsted acid sites, owing to the fact that both types of interaction give rise to distinct IR bands. The quantification of the Lewis and Brønsted acid contents ($q_{B,L}$) commonly is based on the integrated intensities of the IR bands at 1448 cm^{-1} characteristic of pyridine in interaction with Lewis sites (Py-L), and at 1545 cm^{-1} for pyridine adsorbed on Brønsted sites (Py-B), by means of the following equation:

$$q_{B,L} = (A_I \pi D^2)(4w\varepsilon_{B,L})^{-1} \tag{27}$$

where A_I, D, w, and $\varepsilon_{B,L}$ represent the integrated band area of Py-B and Py-L, the diameter of the wafer (cm), the sample weight, and the extinction coefficient of Py-B (1.67 ± 0.12 cm µmol^{-1}), and Py-L (2.22 ± 0.21 cm µmol^{-1}) [164], respectively. After outgassing under vacuum at 373 K, values of 51 (Py-B) and 254 (Py-L) µmol g^{-1} were found for Al-pillared phlogopite, and 39 (Py-B) and 183 (Py-L) µmol g^{-1} for Al-pillared vermiculite. After outgassing at 473 K, these values decreased to 29 (Py-B) and 155 (Py-L) µmol g^{-1} for the pillared mica, and 22 (Py-B) and 112 (Py-L) µmol g^{-1} for the pillared vermiculite. Using a similar method, an Al-pillared saponite had 15 (Py-B) and 122 (Py-L) µmol g^{-1} after outgassing at 473 K [165]. Both Al-pillared vermiculite and phlogopite have more Brønsted acid sites than Al-pillared saponite, in line with prediction. The pillared mica contains more B and L acid sites than the pillared vermiculite, with the latter one and Al-pillared saponite having similar L acid content. Figure 30 compares the variation of the normalized Py-B and Py-L band areas versus outgassing temperature relative to the values found after an outgassing in vacuum at 373 K for Al-pillared vermiculite and phlogopite.

Both materials have B and L sites with similar acid strength. NH$_3$-TPD runs over Al-pillared phlogopite gave a B + L acid content of 290 µmol g^{-1}, which is in fair agreement with the values obtained from pyridine adsorption (305 µmol g^{-1}). For comparison, the acid contents (B and L) established from NH$_3$-TPD measurements over different Al-pillared smectites were in the range 110–150 µmol g^{-1} for Al-pillared montmorillonites, and between 180 and 280 µmol g^{-1} for Al-pillared materials prepared with saponites from different deposits [104].

4.4.6.2
Catalytic Activity

Pillared clays with pillars based on Al and also different elements (e.g. Cr, Fe, etc.), or with mixed pillars, have been extensively investigated in numerous catalytic reactions of hydrocarbons, principally in proton-catalyzed reactions. Catalytic aspects have been reviewed in several articles [31, 166, 167]. Comparison of the catalytic performances of acid solids with potential catalytic applications preferably with a reference catalyst is a straightforward way to obtain preliminary information on their efficiency. With this approach, the activity of Al-pillared materials prepared with different smectites and pillaring methods have been compared, using the hydroisomerization of linear paraffins over Pt-impregnated pillared samples. In these bifunctional catalysts, the metal function is necessary to dehydrogenate the paraf-

NB-P5
NL-P5
NB-V5
NL-V5

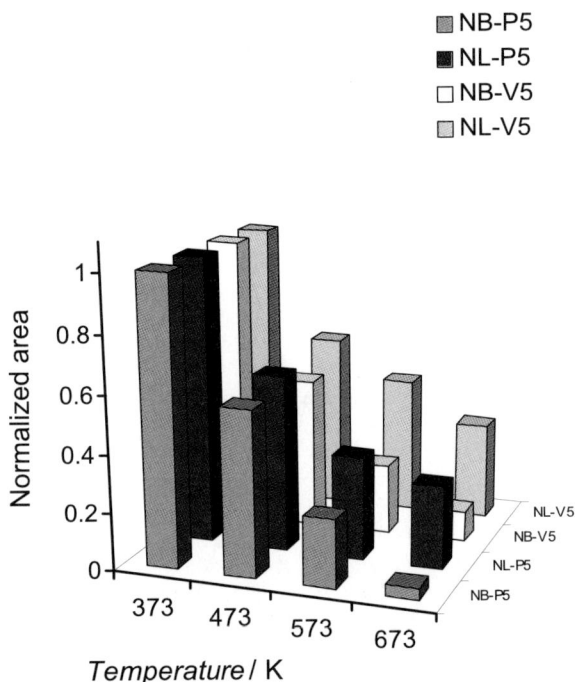

Fig. 30. Relative variation of the normalized band areas of the Py-B (at 1545 cm^{-1}) and Py-L (1448 cm^{-1}) bands versus outgassing temperature for Al-pillared vermiculite (V5) and Al-pillared phlogopite (P5).

fin and re-hydrogenate the branched olefins, while the protons operate the formation of carbenium ions which isomerize via protonated cyclopropane structures [168]. In typical reactions carried out at increasing temperatures, the isomers are first formed at yields that increase with temperature to reach a maximum, and then decrease at still higher temperatures. At a given temperature, mainly depending on the paraffin chain length and catalyst acidity, the isomers start to undergo β-scission giving lower alkanes. At still higher temperatures, cyclization occurs. In such a reaction, the best catalysts, of course, are those exhibiting high selectivities to the isomers at high conversions. Comparative results of heptane, octane, and decane hydroisomerization obtained over Al-pillared forms of montmorillonite, hectorite, beidellite and saponite have been reported in several studies [104, 169–174]. These studies have shown, without exception, that the Al-pillared forms of smectites with Al for Si substitutions in the tetrahedral layers (beidellite, saponites) were significantly more efficient catalysts compared with clays exempt of such substitutions (montmorillonites, hectorite), independent of the paraffin used. The higher activity of Al-pillared beidellites and saponites has been attributed to the higher strength of the Si–OH–Al acid sites of the tetrahedral layers compared with Al-pillared montmorillonite and hectorite where such sites are absent.

References see page 1305

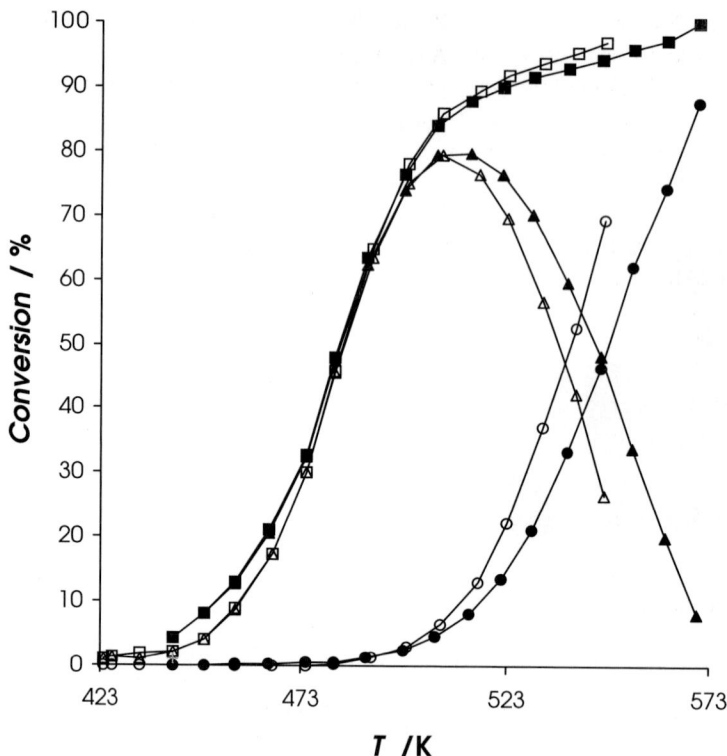

Fig. 31. Variation of octane conversion (squares), yields of C8 isomers (triangles) and cracking products (circles) versus reaction temperature for Al-pillared vermiculite (full symbols) and Al-pillared phlogopite (open symbols).

Illustrative conversion–time curves of octane hydroisomerization obtained over Al-pillared vermiculite and phlogopite are shown in Fig. 31.

As seen, total conversion is achieved at about 573 K. The yields of C8 isomers reach a maximum of 80 % at 503 K for the pillared phlogopite, and at 513 K for Al-pillared vermiculite, for overall conversions of 85 and 88 % and selectivities to C8 isomers of 94 and 91 %, respectively. Under similar conditions, Al-pillared mont-morillonite and saponite exhibited a maximum yield of isomers of 61.5 % at 574 K, and 70.4 % at 524 K, respectively [175]. The results obtained over the pillared vermiculite and phlogopite are the best ever obtained with Al-pillared materials and most of the commercial zeolites.

4.4.7
Conclusions

Research on pillared layered structures has been abundant during the last ten years. Materials are obtained that are fine tuned towards adsorption, separation or catalysis. Further development can be expected, but it is based on an extension of our fundamental knowledge of the pillaring process. We lack information on the size, shape, and distribution of the pillars in the interlamellar region. In the case of Al-pillars, solid-state NMR of the ^{27}Al nucleus will be very helpful, because modern NMR techniques allow quantitative distinction of different Al-species. If this chemistry were better known, one could start the optimization of the pillaring process towards uniformly distributed pillars with the same size and shape. Pillaring could then be adjusted in order to obtain predesigned pore diameters in the pillared layered solids. We are still far from achieving this ultimate goal.

Acknowledgment

Pegie Cool acknowledges the FWO Vlaanderen (Fund for Scientific Research-Flanders-Belgium) for financial support.

References

1 G. ALBERTI, U. COSTANTINO, 'Layered Solids and their Intercalation Chemistry' in: Comprehensive Supramolecular Chemistry, J. L. ATWOOD, D. D. MACNICOL, F. VOGTLE, J.-M. LEHN (Eds), Vol. 7; *Solid-State Supramolecular Chemistry: Two- and Three-Dimensional Inorganic Networks*, G. ALBERTI, T. BEIN (Eds), Pergamon-Elsevier, Oxford, 1996, p. 1–23.

2 D. P. VLIERS, D. COLLIN, R. A. SCHOONHEYDT, F. C. DE SCHRYVER, *Langmuir*, 1986, 2, 165.

3 M. S. STUL, L. VAN LEEMPUT, *Surface Technology*, 1982, 16, 89; ibid., 101.

4 K. Y. JACOBS, R. A. SCHOONHEYDT, in: 'Clays: from two to three Dimensions' in Introduction to Zeolite Science and Practice, H. VAN BEKKUM, P. A. JACOBS, E. M. FLANIGEN, J. C. JANSEN (Eds), *Studies in Surface Science and Catalysis*, Vol. 137, Elsevier, Amsterdam, (in press).

5 R. A. SCHOONHEYDT, T. PINNAVAIA, G. LAGALY, N. GANGAS, *Pure Appl. Chem.* 1999, 71, 2367.

6 R. E. GRIM, Clay Mineralogy 1968, 596 pp.; E. Nemecz, Clay Mineralogy. 1981, 547 pp.; A.C.D. Newman, Chemistry of Clays and Clay Minerals, 1987, 480 pp.

7 W. T. REICHLE, *Chemtech.* 1986, 16, 58.

8 A. DE ROY, C. FORANO, K. EL MALKI, J. P. BESSE, in: Expanded Clays and other Microporous Solids, M. L. OCCELLI, H. E. ROBSON (Eds), Vol. 2, Reinhold, New York, 1992, p. 108.

9 F. CAVANI, F. TRIFIRO, A. VACCARI, *Catal. Today*, 1991, 11, 173.

10 V. A. DRITS, T. N. SOKOLOVA, G. V. SOKOLOVA, V. I. CHERKASHIN, *Clays Clay Miner.* 1987, 35, 401.

11 M. C. GASTUCHE, G. BROWN, M. MORTLAND, *Clay Miner.* 1967, 7, 177.

12 G. MASCOLO, O. MARINO, *Miner. Mag.* 1980, 43, 619.

13 H. SCHAPER, E. B. M. DOESBURG, J. M. C. QUARTEL, L. L. VAN REIJEN, in: Preparation of Catalysts III. G. PONCELET, P. GRANGE, P. A. JACOBS (Eds), *Studies in Surface Science and Catalysis*, Vol. 16, Elsevier, Amsterdam, 1983, p. 301.

14 C. J. SERNA, J. L. WHITE, S. L. HEM, *Clays Clay Miner.* 1977, 25, 384.

15 P. COURTHY, C. MARCILLY, in: Preparation of Catalysts III. G. PONCELET, P. GRANGE, P. A. JACOBS (Eds), *Studies in Surface Science and Catalysis*, Vol. 16, Elsevier, Amsterdam, 1983, p. 485.

16 R. M. TAYLOR, *Clays Clay Miner.* 1984, 19, 591.

17 H. C. B. HANSEN, R. M. TAYLOR, *Clays Clay Miner.* 1991, 26, 507.

18 S. Miyata, *Clays Clay Miner.* **1975**, *23*, 369.

19 F. Trifiro, A. Vaccari, G. Del Pierro, in: Characterization of Porous Solids, Proc. of the IUPAC Symposium (COPS I), Bad Soden a. TS., April 26–29, 1987, K. K. Unger, J. Rouquerol, K. S. W. Sing, H. Kral (Eds), *Studies in Surface Science and Catalysis*, Vol. 39, Elsevier, Amsterdam, 1988, p. 571.

20 S. Miyata, *Clays Clay Miner.* **1980**, *28*, 50.

21 E. C. Kruissink, L. L. van Reijen, J. R. H. Ross, *J. Chem. Soc., Faraday Trans. 1*, **1981**, *77*, 649.

22 S. Carlino, *Solid State Ionics*, **1997**, *98*, 73.

23 H. Nijs, A Contribution to the Development of Microporous Pillared Layered Double Hydroxides, PhD thesis, University of Antwerp, Belgium, 1999.

24 A. Clearfield, G. D. Smith, *Inorg. Chem.* **1969**, *8*, 431.

25 J. M. Troup, A. Clearfield, *Inorg. Chem.* **1977**, *16*, 3311.

26 N. J. Clayden, *J. Chem. Soc., Dalton. Trans.* **1987**, 1877.

27 A. Clearfield, in: Inorganic Ion Exchange Materials. A. Clearfield (Ed.), CRC Press, Florida, 1982, p. 2.

28 A. Clearfield, J. A. Stynes, *J. Inorg. Nucl. Chem.* **1964**, *26*, 117.

29 G. Alberti, E. Torracca, *J. Inorg. Nucl. Chem.* **1968**, *30*, 317.

30 J. R. Butruille, T. J. Pinnavaia, 'Alumina pillared Clays' in: Comprehensive Supramolecular Chemistry, J. L. Atwood, J. E. D. Davies, D. D. Macnicol, F. Vogtle, J.-M. Lehn (Eds), Vol. 7; *Solid-State Supramolecular Chemistry: Two- and Three-Dimensional Solid-State Networks*, G. Alberti, T. Bein (Eds), Pergamon-Elsevier, Oxford, 1996, p. 219–250.

31 A. Gil, L. M. Gandia, *Catal. Rev., Sci. Eng.* **2000**, *42*, 145; J. T. Kloprogge, *J. Porous Mater.* **1998**, *5*, 5.

32 V. Seefeld, R. Bertram, P. Starke, W. Gessner, *Silikattechnik*, **1988**, *39*, 239; V. Seefeld, R. Bertram, H. Gorz, W. Gessner, S. Schonherr, *Z. Anor. Allg. Chem.* **1991**, *603*, 129; V.

Seefeld, R. Bertram, D. Muller, W. Gessner, *Silikattechnik*, **1991**, *42*, 305; J. T. Kloprogge, D. Seykens, J. B. H. Jansen, J. W. Geus, *J. Non-Cryst. Solids*, **1992**, *142*, 94; J. T. Kloprogge, D. Seykens, J. W. Geus, J. B. H. Jansen, *J. Non-Cryst. Solids*, **1992**, *142*, 87; J. T. Kloprogge, D. Seykens, J. B. H. Jansen, J. W. Geus, *J. Non-Cryst. Solids*, **1993**, *152*, 207; J. T. Kloprogge, D. Seykens, J. W. Geus, J. B. H. Jansen, *J. Non-Cryst. Solids*, **1993**, *160*, 144; G. Furrer, C. Ludwig, P. W. Schindler, *J. Coll. Interf. Sci.*, **1992**, *149*, 56.

33 R. A. Schoonheydt, H. Leeman, A. Scorpion, I. Lenotte, P. G. Grobet, *Clays Clay Miner.* **1997**, *45*, 518.

34 S. Aceman, N. Lahav, S. Yariv, *Appl. Clay Sci.* **2000**, *17*, 99.

35 D. Plee, F. Borg, L. Gatineau, J. J. Fripiat, *J. Am. Chem. Soc.* **1985**, *107*, 2362.

36 M. De Bock, H. Nijs, P. Cool, E. F. Vansant, *J. Porous Mater.* **1999**, *6*, 323; S. P. Katdare, V. Ramaswany, A. V. Ramaswany, *J. Mater. Chem.* **1997**, *7*, 2197; J. Ahenach, P. Cool, E. F. Vansant, *Microporous Mesoporous Mater.* **1998**, *26*, 185.

37 L. Storaro, M. Lenarda, M. Perissinotto, V. Luchini, R. Ganzerla, *Microporous Mesoporous Mater.* **1998**, *20*, 317.

38 E. Maes, L. Vielvoye, W. Stone, B. Delvaux, *Eur. J. Soil Sci.* **1999**, *50*, 107.

39 L. A. Douglas, in: Minerals in Soil Environments, J. B. Dixon, S. B. Weed (Eds), Soil Sci. Soc. of America, Madison, Wisconsin, 1977, p. 259.

40 P. H. Hsu, T. F. Bates, *Soil Sci. Soc. Am. Proc.* **1964**, *28*, 763.

41 P. H. Hsu, *Clays Clay Miner.* **1992**, *40*, 300.

42 B. d'Espinose de la Caillerie, J. J. Fripiat, *Clays Clay Miner.* **1991**, *39*, 270.

43 N. Lahav, U. Shani, J. Shabtai, *Clays Clay Miner.* **1978**, *26*, 107.

44 A. Schutz, W. E. E. Stone, G. Poncelet, J. J. Fripiat, *Clays Clay Miner.* **1987**, *35*, 251.

45 L. J. Michot, D. Tracas, B. S. Lartiges, F. Lhote, C. H. Pons, *Clay Miner.* **1994**, *29*, 133.

46 F. J. del Rey-Perez-Caballero, G. Poncelet, *Microporous Mesoporous Mater.* **2000**, *37*, 313.

47 F. J. del Rey-Perez-Caballero, G. Poncelet, *Microporous Mesoporous Mater.* **2000**, *41*, 169.

48 V. Jeannin, M. Fournier, M. T. Pope, Heteropoly and Isopoly Oxometalates, Springer-Verlag, Berlin, 1983, 180 pp.

49 S. Miyata, T. Hirose, *Clays Clay Miner.* **1978**, *26*(6), 441.

50 I. Crespo, C. Barriga, V. Rives, M. A. Ulibarra, *Solid State Ionics*, **1997**, *729*, 101.

51 H. C. Bruun Hansen, C. B. Koch, *Clays Clay Miner.* **1994**, *42*(2), 170.

52 G. M. Woltermann, US Patent 4,454,244, assigned to Ashland Oil Inc. (Ashland, KY), 1984; *Chem. Abstr.* **1984**, *101*, 60992.

53 J. Twu, P. K. Dutta, *J. Phys. Chem.* **1989**, *93*, 7863.

54 T. Kwon, T. J. Pinnavaia, *Chem. Mater.* **1989**, *1*, 381.

55 T. Kwon, T. J. Pinnavaia, *J. Mol. Catal.* **1992**, *74*, 23.

56 J. Wang, Y. Tian, R. C. Wang, A. Clearfield, *Chem. Mater.* **1992**, *4*, 1276.

57 S. K. Yun, T. J. Pinnavaia, *Inorg. Chem.* **1996**, *35*, 6853.

58 E. Gardner, T. J. Pinnavaia, *Appl. Catal. A*, **1998**, *167*, 65.

59 M. A. Drezdzon, *Inorg. Chem.* **1988**, *27*, 4628.

60 J. Twu, P. K. Dutta, *J. Catal.* **1990**, *124*, 503.

61 H. Nijs, P. Cool, E. F. Vansant, *Interface Sci.* **1997**, *5*, 83.

62 E. D. Dimotakis, T. J. Pinnavaia, *Inorg. Chem.* **1990**, *29*, 2393.

63 H. Nijs, M. De Bock, E. F. Vansant, *Microporous Mesoporous Mater.* **1999**, *30*, 243.

64 A. A. G. Tomlinson, in: Pillared Layered Structures: Current Trends and Applications, I. V. Mitchell (Ed.), Elsevier, London, 1990, p. 91.

65 D. P. Vliers, R. A. Schoonheydt, F. C. De Schrijver, *J. Chem. Soc., Faraday Trans.* **1985**, *81*, 2009.

66 M. A. Massucci, A. La Ginestra, A. A. G. Tomlinson, C. Ferragina, P. Patrono, P. Cafarelli, in: Pillared Layered Structures: Current Trends and Applications, I. V. Mitchell (Ed.), Elsevier, London, 1990, p. 127.

67 G. Alberti, U. Constantino, F. Marmottini, R. Vivani, P. Zappelli, in: Pillared Layered Structures: Current Trends and Applications, I. V. Mitchell (Ed.), Elsevier, London, 1990, p. 119.

68 G. Alberti, in: Multifunctional Mesoporous Inorganic Solids, C. A. C. Sequeira, M. J. Hudson (Eds), Kluwer Academic Publishers, Netherlands, 1993, p. 179.

69 K. Peeters, P. Grobet, E. F. Vansant, *J. Mater. Chem.* **1996**, *6*(2), 239.

70 D. J. Jones, J. M. Leloup, Y. Ding, J. Rozière, *Solid State Ionics*, **1993**, *61*, 117.

71 A. Clearfield, D. B. Roberts, *Inorg. Chem.* **1988**, *27*, 3237.

72 K. Peeters, R. Carleer, J. Mullens, E. F. Vansant, *Microporous Mater.* **1995**, *4*, 475.

73 G. Alberti, *Tetravalent Metal Salts*, **1978**, *11*, 163.

74 L. Li, X. Liu, J. Klinowski, *J. Phys. Chem.* **1991**, *95*, 5910.

75 P. Sylvester, R. Cahill, A. Clearfield, *Chem. Mater.* **1994**, *6*, 1890.

76 G. Alberti, F. Marmottini, *J. Coll. Interf. Sci.* **1993**, *157*, 513.

77 G. Alberti, U. Constantino, J. Kornyei, M. L. Giovagnotti, *React. Polym.* **1985**, *4*, 1.

78 J. D. Wang, A. Clearfield, G. Z. Peng, *Mater. Chem. Phys.* **1993**, *35*, 208.

79 C. T. Kresge, M. E. Leonowicz, W. J. Roth, J. C. Vartuli, J. S. Beck, *Nature*, **1992**, *359*, 710.

80 J. Jimenez, P. Maireles-Torres, P. Olivera-Pastor, A. Jimenez-Lopez, *Langmuir*, **1997**, *13*, 2857.

81 T. Cassagneau, D. J. Maireles-Torres, J. Roziere, M. L. Occelli, H. Kessler, Synthesis of Porous Materials, Vol. 59, Marcel Dekker, New York, 1977, p. 509.

82 E. Booij, J. T. Kloprogge, J. A. R. van Veen, *Appl. Clay Sci.* **1996**, *11*, 115.

83 M. L. Occelli, P. A. Peaden, G. P. Ritz, P. S. Iyer, M. Yokoyama, *Microporous Mater.* **1993**, *1*, 99.

84 A. Galarneau, A. Barodawalla, T. J. Pinnavaia, *Nature*, **1995**, *374*, 529.

85 P. Cool, J. Ahenach, E. F. Vansant, in: Adsorption Science and Technology, Proc. 2nd Pacific Basin Conference on Adsorption Science and Technology, Brisbane, Australia, 14–18 May 2000, D. D. Do (Ed.), World Scientific Publishing Co., Singapore, 2000, p. 638.

86 S. Inagaki, Y. Fukushima, K. Kuroda, in: Zeolites and Related Microporous Materials: State of the Art 1994, J. Weitkamp, H. G. Karge, H. Pfeifer, W. Hölderich (Eds), *Studies in Surface Science and Catalysis*, Vol. 84, Elsevier, Amsterdam, 1994, p. 125.

87 T. Yanagisawa, T. Shimizu, K. Kuroda, C. Kato, *Bull. Chem. Soc. Jpn.* **1990**, *63*, 988.

88 T. J. Pinnavaia, T. Kwon, S. K. Kun, in: Zeolite Microporous Solids: Synthesis, Structure and Reactivity, E. G. Derouane, F. Lemos, C. Naccache, F. Ramao-Ribeiro (Eds), NATO ASI Series C, Vol. 352, Kluwer, Dordrecht, 1992, p. 91.

89 J. Ahenach, P. Cool, E. F. Vansant, *Phys. Chem. Chem. Phys.* **2000**, *2*, 5750.

90 S. Brunauer, L. S. Deming, E. Deming, E. Teller, *J. Am. Chem. Soc.* **1940**, *62*, 1723.

91 N. Maes, I. Heylen, P. Cool, E. F. Vansant, *Appl. Clay Sci.* **1997**, *12*, 43.

92 H-Y Zhu, N. Maes, E. F. Vansant, in: Characterization of Porous Solids III, J. Rouquerol, F. Rodriguez-Reinoso, K. S. W. Sing, K. K. Unger (Eds), *Studies in Surface Science and Catalysis*, Vol. 87, Elsevier, Amsterdam, 1994, p. 457–466.

93 I. Langmuir, *J. Am. Chem. Soc.* **1915**, *40*, 1361.

94 S. Brunauer, P. H. Emmet, E. Teller, *J. Am. Chem. Soc.* **1938**, *60*, 309.

95 S. Lowell, Introduction to Powder Surface Area. Wiley Interscience, Chichester, 1979, 199 pp.

96 S. J. Gregg, K. S. W. Sing, Adsorption, Surface Area and Porosity, Academic Press, London, 1982.

97 R. D. Vold, M. J. Vold, Colloid and Interface Chemistry, Addison-Wesley, Reading, MA 1983, chapter 3.

98 K. S. W. Sing, D. H. Everett, R. A. W. Haul, L. Moscow, R. A. Pierotti, J. Rouquérol, T. Siemieniewska, *Pure Appl. Chem.* **1985**, *57*(4), 603.

99 H-Y Zhu, E. F. Vansant, *J. Porous Mater.* **1995**, *2*, 107.

100 J. H. de Boer, B. G. Linsen, T. J. Osinga, *J. Catal.* **1964**, *4*, 643; J. C. P. Broekhoff, J. H. de Boer, *J. Catal.* **1968**, *10*, 391.

101 M. J. Remy, A. C. Vieira Coelho, G. Poncelet, *Microporous Mater.* **1996**, *7*, 287.

102 K. S. W. Sing, *Chem. Ind.* **1967**, *20*, 829.

103 A. Lecloux, J. P. Pirard, *J. Coll. Interf. Sci.* **1980**, *18*, 355.

104 S. Moreno, R. Sun Kou, G. Poncelet, *J. Phys. Chem.* **1997**, *101*, 1569.

105 M. M. Dubinin, *Carbon*, **1985**, *23*, 373.

106 K. Kaneko, *J. Membr. Sci.* **1994**, *96*, 59.

107 G. F. Cerofilini, *Surf. Sci.* **1971**, *24*, 391.

108 A. Gil, M. Montes, *Langmuir*, **1994**, *10*, 291.

109 M. M. Dubinin, H. F. Stoeckli, *J. Coll. Interf. Sci.* **1980**, *75*, 34.

110 G. Horvath, K. Kawazoe, *J. Chem. Eng. Jpn.* **1983**, *16*, 470.

111 M. S. A. Baksh, E. S. Kikkinides, R. T. Yang, *Ind. Eng. Chem. Res.* **1992**, *39*, 303.

112 B. C. Lippens, J. H. De Boer, *J. Catal.* **1965**, *4*, 319.

113 D. H. Everett, J. C. Powl, *J. Chem. Soc., Faraday Trans. 1*, **1976**, *72*, 619.

114 H. Y. Zhu, G. Q. Lu, N. Maes, E. F. Vansant, *J. Chem. Soc., Faraday Trans.* **1997**, *93*(7), 1417.

115 S. Brunauer, R. S. H. Mikhail, E. E. Bodor, *J. Coll. Interf. Sci.* **1967**, *24*, 451.

116 N. Maes, H. Y. Zhu, E. F. Vansant, *J. Porous Mater.* **1995**, *2*, 97.

117 M. L. Occelli, D. J. Lynch, J. V. Sanders, *J. Catal.* **1987**, *197*, 557.

118 N. Maes, I. Heylen, P. Cool, M. De Bock, C. Vanhoof, E. F. Vansant, *J. Porous Mater.* **1996**, *3*, 47.

119 J. Ahenach, P. Cool, E. Vansant, O. Lebedev, J. Van Landuyt, *Phys. Chem. Chem. Phys.* **1999**, *1*, 3703.

120 S. Vercauteren, K. Keizer, E. F. Vansant, J. Luyten, R. Leysen, *J. Porous Mater.* **1998**, *5*, 241.

121 G. Binning, C. F. Quate, C. Gerber, *Phys. Rev. Lett.* **1986**, *56*, 430.

122 H. Hartman, G. Sposito, A. Yang, S. Manne, S. A. C. Gould, P. K. Hansma, *Clays Clay Miner.* **1990**, *38*(4), 337.

123 M. L. Occelli, B. Drake, S. A. C. Gould, *J. Catal.* **1993**, *142*, 337.

124 M. L. Occelli, J. A. Bertrand, S. A. C. Gould, J. M. Dominguez, *Microporous Mesoporous Mater.* **2000**, *34*, 195.

125 F. Rouquerol, J. Rouquerol, K. Sing, Adsorption by Powders & Porous Solids, Academic Press, London UK, 1999.

126 D. M. Ruthven, Principles of Adsorption and Adsorption Processes, J. Wiley & Sons, New York, USA, 1984.

127 P. A. Politowicz, J. J. Kozak, *J. Phys. Chem.* **1988**, *92*, 6078.

128 P. A. Politowicz, L. B. S. Leung, J. J. Kozak, *J. Phys. Chem.* **1989**, *93*, 923.

129 M. Sahimi, *J. Chem. Phys.* **1990**, *92*, 5107.

130 Z-X Cai, S. D. Mahanti, S. A. Solin, T. J. Pinnavaia, *Phys. Rev. B.* **1990**, *42*, 6636.

131 B. Y. Chen, H. Kim, S. D. Mahanti, T. J. Pinnavaia, Z-X Cai, *J. Chem. Phys.* **1994**, *100*, 3872.

132 X. Yi, S. Shing, M. Sahimi, *AIChE. J.* **1995**, *41*, 456.

133 X. Yi, S. Shing, M. Sahimi, *Chem. Eng. Sci.* **1996**, *51*, 3409.

134 X. Yi, J. Ghassemzadeh, S. Shing, M. Sahimi, *J. Chem. Phys.* **1998**, *108*, 5.

135 R. C. Zielke, T. J. Pinnavaia, *Clays Clay Miner.* **1988**, *36*(5), 403.

136 H. T. Shu, D. Li, A. A. Scala, Y. H. Ma, *Separation and Purification Technology,* **1997**, *11*, 27.

137 R. Wibulswas, D. A. White, R. Rautiu, *Environm. Techn.* **1998**, *19*, 627.

138 R. M. Barrer, Zeolites and Clay Minerals as Sorbents and Molecular Sieves, Academic Press, London, 1978.

139 R. M. Barrer, R. J. B. Craven, *J. Chem. Soc., Faraday Trans.* **1992**, *88*(4), 645.

140 I. Gameson, W. J. Stead, T. Rayment, *J. Phys. Chem.* **1991**, *95*, 1727.

141 R. T. Yang, M. S. A. Baksh, *AIChE J.* **1991**, *37*(5), 679.

142 M. S. A. Baksh, R. T. Yang, *AIChE J.* **1992**, *38*(9), 1357.

143 L. S. Cheng, R. T. Yang, *Microporous Mater.* **1997**, *8*, 177.

144 R. T. Yang, Gas Separation by Adsorption Processes, Butterworth, Boston, 1987, 352 pp.

145 I. Heylen, N. Maes, A. Molinard, E. F. Vansant, in: Separation Technology, Process Technology Proceedings, vol. 11, E. F. Vansant (Ed.), Elsevier, Amsterdam, 1994, p. 355.

146 I. Heylen, C. Vanhoof, E. F. Vansant, *Microporous Mater.* **1995**, *5*, 53.

147 I. Heylen, E. F. Vansant, *Microporous Mater.* **1997**, *10*, 41.

148 P. Cool, E. F. Vansant, *Microporous Mater.* **1996**, *6*, 27.

149 P. Cool, E. F. Vansant, in: Adsorption Science and Technology. Proc. 2nd Pacific Basin Conference on Adsorption Science and Technology, Brisbane, Australia, 14–18 May 2000, D. D. Do (Ed.), World Scientific Publishing Co., Singapore, 2000, p. 633.

150 D. E. W. Vaughan, J. S. Magee, R. J. Lussier, US Patent, 4,271,043, assigned to W.R. Graces Co. (New York, NY), **1981**.

151 A. Molinard, K. K. Peeters, N. Maes, E. F. Vansant, in: Separation Technology, Process Technology Proceedings, Vol. 11. E. F. Vansant (Ed.), Elsevier, Amsterdam, 1994, p. 445.

152 S. Yamanaka, P. B. Malla, S. Komarneni, *J. Colloid Interface Sci.* **1990**, *134*, 51.

153 P. B. MALLA, S. YAMANAKA, S. KOMARNENI, *Solid State Ionics*, **1989**, *32*(33), 354.

154 A. MOLINARD, E. F. VANSANT, in: Separation Technology, Process Technology Proceedings, vol. 11, E. F. VANSANT (Ed.), Elsevier, Amsterdam, 1994, p. 423.

155 A. MOLINARD, E. F. VANSANT, *Adsorption*, **1995**, *1*, 49.

156 P. COOL, A. CLEARFIELD, R. M. CROOKS, E. F. VANSANT, *Advances in Environmental Research*, **1999**, *3*(2), 139.

157 H. Y. ZHU, G. Q. LU, *J. Porous Mater.* **1998**, *5*, 227.

158 H. Y. ZHU, W. H. GAO, E. F. VANSANT, *J. Colloid Interface Sci.* **1995**, *171*, 377.

159 D. E. W. VAUGHAN, R. J. LUSSIER, in Proceedings of the 5th International Conference of Zeolites, L. V. REES (Ed.), Heyden and Sons, London, 1980, p. 94.

160 G. PONCELET, A. SCHUTZ, in: Chemical Reactions in Organic and Inorganic Constrained Systems, R. SETTON (Ed.), Reidel, Dordrecht, 1986, p. 165.

161 S. CHEVALIER, R. FRANCK, H. SUQUET, J.-F. LAMBERT, D. BARTHOMEUF, *J. Chem. Soc. Faraday Trans.* **1994**, *90*, 667.

162 S. MORENO, R. SUN KOU, G. PONCELET, *J. Catal.* **1996**, *162*, 198.

163 R. J. GORTE, *Catal. Lett.* **1999**, *62*, 1.

164 C. A. EMEIS, *J. Catal.* **1993**, *141*, 347.

165 F. KOOLI, W. JONES, *J. Mater. Chem.* **1998**, *8*, 2119.

166 F. FIGUERAS, *Catal. Rev. Sci. Eng.* **1988**, *30*, 457.

167 J.-F. LAMBERT, G. PONCELET, *Topics Catal.* **1997**, *4*, 43.

168 J. A. MARTENS, P. A. JACOBS, in: Theoretical Aspects of Heterogeneous Catalysis, J. B. MOFFAT (Ed.), Van Nostrand-Reinhold, New York, 1990, p. 52.

169 A. SCHUTZ, D. PLÉE, F. BORG, P. JACOBS, G. PONCELET, J. J. FRIPIAT, in: Proceedings Intern. Clay Conf. Denver, 1985, L. G. SCHULTZ, H. VAN OLPHEN, F. A. MUMPTON (Eds), The Clay Mineral Soc., Bloomington, Indiana, 1987, p. 305.

170 A. VIEIRA COELHO, G. PONCELET, in: Pillared Layered Structures: Current Trends and Applications, I. V. MITCHELL (Ed.), Elsevier, London, 1990, p. 185.

171 S. MORENO, E. GUTIERREZ, A. ALVAREZ, N. G. PAPAYANNAKOS, G. PONCELET, *Appl. Catal. A: General*, **1997**, *165*, 103.

172 S. MORENO, R. SUN KOU, R. MOLINA, G. PONCELET, *J. Catal.* **1999**, *182*, 174.

173 R. MOLINA, S. MORENO, A. VIEIRA COELHO, J. A. MARTENS, P. A. JACOBS, G. PONCELET, *J. Catal.* **1994**, *148*, 304.

174 R. MOLINA, S. MORENO, G. PONCELET, in: 12th International Congress on Catalysis. Proc. of the 12th ICC, Granada, Spain, July 9–14, 2000, A. CORMA, F. V. MELO, S. MENDIOROZ, J. L. G. FIERRO (Eds), *Studies in Surface Science and Catalysis*, Vol. 130, Elsevier, Amsterdam, 2000, p. 983.

175 F. J. DEL REY-PEREZ-CABALLERO, M. L. SANCHEZ-HENAO, G. PONCELET, in: 12th International Congress on Catalysis. Proc. of the 12th ICC, Granada, Spain, July 9–14, 2000, A. CORMA, F. V. MELO, S. MENDIOROZ, J. L. G. FIERRO (Eds), *Studies in Surface Science and Catalysis*, Vol. 130, Elsevier, Amsterdam, 2000, p. 2417.